OXFORD IB DIPLOMA PROGRAMME

IN COOPERATION WITH ib

GEOGRAPHY

COURSE COMPANION

Garrett Nagle
Briony Cooke

OXFORD
UNIVERSITY PRESS

Great Clarendon Street, Oxford OX2 6DP

Oxford University Press is a department of the University of Oxford.
It furthers the University's objective of excellence in research,
scholarship, and education by publishing worldwide in

Oxford New York

Auckland Cape Town Dar es Salaam Hong Kong Karachi
Kuala Lumpur Madrid Melbourne Mexico City Nairobi
New Delhi Shanghai Taipei Toronto

With offices in

Argentina Austria Brazil Chile Czech Republic France Greece
Guatemala Hungary Italy Japan Poland Portugal Singapore
South Korea Switzerland Thailand Turkey Ukraine Vietnam

First published 2011

British Library Cataloguing in Publication Data

Data available

ISBN: 978-0-19-838917-0
10 9 8 7 6 5 4 3

Printed in Malaysia by Vivar Printing Sdn. Bhd.

Acknowledgments

Cover image: Konstantin Kalishko/Alamy; P184: Tony Waltham/Robert Harding World Imagery/Getty Images; P188: Blake Little/The Image Bank/Getty Images; P189: Gail Johnson/Fotolia; P191: Sipa Press/Rex Features; P196: Colin Monteath/Hedgehog House/Minden Pictures/Getty Images; P198: Francesco Ruggeri/Photographer's Choice/ Getty Images; P207: Ajay Verma/Reuters; P241: Enrique Aguirre/Oxford Scientific/Photolibrary; P243: AA World Travel Library; P245: Photogerson/Shutterstock; P246: Ho New/Reuters; P248: Stefano Rellandini/Reuters; P258: Claudia Dewald/Istockphoto; P264: Ho New/Reuters; P277: CDC/James Gathany; P279: Images of Africa Photobank/ Alamy; P282: © WHO 2011; P283: Claro Cortes/Reuters; P292l: Ho New/Reuters; P293: Ng Han Guan/AP Photo; P294: Ming Ming/Reuters; P299l: Daniel Berehulak/Getty Images; P299r: Punit Paranjpe/Reuters; P302: Danny Lehman/Corbis; P305: Chris Sharp/Oxford Scientific/Photolibrary; P310: © Julius Mwelu/IRIN; P348: Doha Development Agenda; P355b: Waitrose Foundation; P362b: Greenpeace; P363: © WWF; P364br: ©2010-2011 McDonald's; P369t: Picture by Mona Kensik; P369b: David Pearson/Alamy; P373: © Olivier Blaise; P386: Avaaz; P391: Robert Convery/Alamy; P394: Charistoone-stock/Alamy.

All other photos provided by Garrett Nagle & Briony Cooke.

The author and publisher are grateful to the following for permission to reprint the copyright material listed:

Cengage Learning EMEA Ltd, Tayor & Francis Group for Table 8.6 from Keith Smith and David N Petley: Environmental Hazards (5e, Routledge 2009) and Table 18.1 from Warnick E Murray: Geographies of Globalization (Routledge, 2006).

Hodder Education Ltd for Table 3.4 from Nick Middleton: The Global Casino (4e, Hodder Education, 2008).

World Health Organization for table from The Global Burden of Disease 2009, WHO

We have made every effort to trace and contact all copyright holders before publication but this has not been possible in all cases. If notified, the publisher will rectify any errors or omissions at the earliest opportunity.

Course Companion definition

The IB Diploma Programme Course Companions are resource materials designed to provide students with support through their two-year course of study. These books will help students gain an understanding of what is expected from the study of an IB Diploma Programme subject.

The Course Companions reflect the philosophy and approach of the IB Diploma Programme and present content in a way that illustrates the purpose and aims of the IB. They encourage a deep understanding of each subject by making connections to wider issues and providing opportunities for critical thinking.

The books mirror the IB philosophy of viewing the curriculum in terms of a whole-course approach; the use of a wide range of resources; international-mindedness; the IB learner profile and the IB Diploma Programme core requirements; theory of knowledge, the extended essay, and creativity, action, service (CAS).

Each book can be used in conjunction with other materials and indeed, students of the IB are required and encouraged to draw conclusions from a variety of resources. Suggestions for additional and further reading are given in each book and suggestions for how to extend research are provided.

In addition, the Course Companions provide advice and guidance on the specific course assessment requirements and also on academic honesty protocol.

IB mission statement

The International Baccalaureate aims to develop inquiring, knowledgable and caring young people who help to create a better and more peaceful world through intercultural understanding and respect.

To this end the IB works with schools, governments and international organizations to develop challenging programmes of international education and rigorous assessment.

These programmes encourage students across the world to become active, compassionate, and lifelong learners who understand that other people, with their differences, can also be right.

IB learner profile

The aim of all IB programmes is to develop internationally minded people who, recognizing their common humanity and shared guardianship of the planet, help to create a better and more peaceful world. IB learners strive to be:

Inquirers They develop their natural curiosity. They acquire the skills necessary to conduct inquiry and research and show independence in learning. They actively enjoy learning and this love of learning will be sustained throughout their lives.

Knowledgable They explore concepts, ideas, and issues that have local and global significance. In so doing, they acquire in-depth knowledge and develop understanding across a broad and balanced range of disciplines.

Thinkers They exercise initiative in applying thinking skills critically and creatively to recognize and approach complex problems, and make reasoned, ethical decisions.

Communicators They understand and express ideas and information confidently and creatively in more than one language and in a variety of modes of communication. They work effectively and willingly in collaboration with others.

Principled They act with integrity and honesty, with a strong sense of fairness, justice, and respect for the dignity of the individual, groups, and communities. They take responsibility for their own actions and the consequences that accompany them.

Open-minded They understand and appreciate their own cultures and personal histories, and are open to the perspectives, values, and traditions of other individuals and communities. They are accustomed to seeking and evaluating a range of points of view, and are willing to grow from the experience.

Caring They show empathy, compassion, and respect towards the needs and feelings of others. They have a personal commitment to service, and act to make a positive difference to the lives of others and to the environment.

Risk-takers They approach unfamiliar situations and uncertainty with courage and forethought, and have the independence of spirit to explore new roles, ideas, and strategies. They are brave and articulate in defending their beliefs.

Balanced They understand the importance of intellectual, physical, and emotional balance to achieve personal well-being for themselves and others.

Reflective They give thoughtful consideration to their own learning and experience. They are able to assess and understand their strengths and limitations in order to support their learning and personal development.

A note on academic honesty

It is of vital importance to acknowledge and appropriately credit the owners of information when that information is used in your work. After all, owners of ideas (intellectual property) have property rights. To have an authentic piece of work, it must be based on your individual and original ideas with the work of others fully acknowledged. Therefore, all assignments, written or oral, completed for assessment must use your own language and expression. Where sources are used or referred to, whether in the form of direct quotation or paraphrase, such sources must be appropriately acknowledged.

How do I acknowledge the work of others?

The way that you acknowledge that you have used the ideas of other people is through the use of footnotes and bibliographies.

Footnotes (placed at the bottom of a page) or endnotes (placed at the end of a document) are to be provided when you quote or paraphrase from another document, or closely summarize the information provided in another document. You do not need to provide a footnote for information that is part of a "body of knowledge". That is, definitions do not need to be footnoted as they are part of the assumed knowledge.

Bibliographies should include a formal list of the resources that you used in your work. "Formal" means that you should use one of the several accepted forms of presentation. This usually involves separating the resources that you use into different categories (e.g. books, magazines, newspaper articles, Internet-based resources, CDs and works of art) and providing full information as to how a reader or viewer of your work can find the same information. A bibliography is compulsory in the extended essay.

What constitutes malpractice?

Malpractice is behaviour that results in, or may result in, you or any student gaining an unfair advantage in one or more assessment component. Malpractice includes plagiarism and collusion.

Plagiarism is defined as the representation of the ideas or work of another person as your own. The following are some of the ways to avoid plagiarism:

- Words and ideas of another person used to support one's arguments must be acknowledged.
- Passages that are quoted verbatim must be enclosed within quotation marks and acknowledged.
- CD-ROMs, email messages, web sites on the Internet, and any other electronic media must

be treated in the same way as books and journals.

- The sources of all photographs, maps, illustrations, computer programs, data, graphs, audio-visual, and similar material must be acknowledged if they are not your own work.
- Works of art, whether music, film, dance, theatre arts, or visual arts, and where the creative use of a part of a work takes place, must be acknowledged.

Collusion is defined as supporting malpractice by another student. This includes:

- allowing your work to be copied or submitted for assessment by another student
- duplicating work for different assessment components and/or diploma requirements.

Other forms of malpractice include any action that gives you an unfair advantage or affects the results of another student. Examples include, taking unauthorized material into an examination room, misconduct during an examination, and falsifying a CAS record.

Contents

Introduction

Geography is all around us. Every day, we are affected by physical geography and by human geography. Climate change, air quality, natural and technological hazards, impacts of globalization, the food we eat, and the clothes we wear are all geographical products. As the world's population continues to grow, there will be increasing difficulties in managing the world's resources for the benefit of all people. Geography is constantly changing and asking us to make decisions about the world in which we live.

During your IB Diploma Programme studies you will notice how geography overlaps with other subjects. It has close links with biology (ecology and climate change), mathematics (statistical testing), chemistry (environmental chemistry), environmental systems and society (ecosystems, human populations, pollution, and climate change) and economics (development, inequalities). Sociocultural exchanges in geography relate to languages and culture, and the decision making aspect is firmly rooted in TOK. You will learn about geographical issues in many subjects and in many different ways.

You are encouraged to make links between geography and the other subjects on your Diploma Programme 'hexagon'. This book follows the IB Diploma Programme subject guide for geography and also contains information on exam preparation, writing skills, and skills for fieldwork. No geography book should ever claim to be comprehensive – this one certainly is not. You are encouraged to read a variety of resources including newspapers, search the internet (some references are given throughout this book) to keep up to date and to use examples that relate to your own local area.

We hope that this book will help stimulate a real desire to find out more about geography. We have learned a great deal in the writing of this book. We would like to thank the many students and colleagues who have helped with comments on early drafts.

Dedication:
To Angela, Rosie, Patrick and Bethany for their help, good humour and patience.

To Henry for his advice and encouragement.

Garrett Nagle and Briony Cooke

Part 1 Core theme – patterns and change

The core theme provides an overview of some of the key global issues of today. The aim is to provide a broad factual and conceptual introduction to each topic and to the United Nations' Millennium Development Goals (MDGs), in particular those regarding poverty reduction, gender equality, improvements in health and education and environmental sustainability.

The core examines the likely impact of global climate change, a major current issue of international significance and population issues such as global and regional inequities, and sustainable development, for example. The core emphasizes the concepts underlying global issues, and the consequent regional and global patterns. Attention is given to positive as well as negative aspects of change, and the need to accept responsibility for seeking solutions. The emphasis in the core theme is on national, regional and global trends and patterns.

1 Populations in transition

By the end of this chapter you should be able to:

- explain population trends, including birth rates, fertility rates, mortality rates and life expectancy
- analyse population pyramids and the impacts of youthful and ageing populations
- evaluate pro-natalist and anti-natalist policies
- discuss the causes and effects of forced and voluntary migration
- explain the impacts of international migration
- examine gender inequalities in culture, education, employment and life expectancy.

Exponential growth – an increasing or accelerating rate of growth.

World population growth

The world's population has been growing increasingly rapidly, or exponentially, and most of this growth is quite recent i.e. since the mid nineteenth century. The world's population doubled between 1650 and 1850, 1850 and 1920, and 1920 and 1970. It is thus taking less time for the population to double. Up to 95% of population growth is taking place in developing countries. However, the world's population overall is expected to stabilize by around 2050–80. In all but two regions – North and Latin America – the population is predicted to be falling by the end of the 21st century.

Although in 2002 the world population increased by 74 million people – the population of Egypt – this is well below the peak of 87 million added in 1989–90. At 1.2% a year, the increase is also well below the 2.2% annual growth seen 40 years ago. The slowdown in global population growth is linked primarily to declines in fertility (see Figure 1.5). The price to be paid is an increase in the number of elderly people in the world.

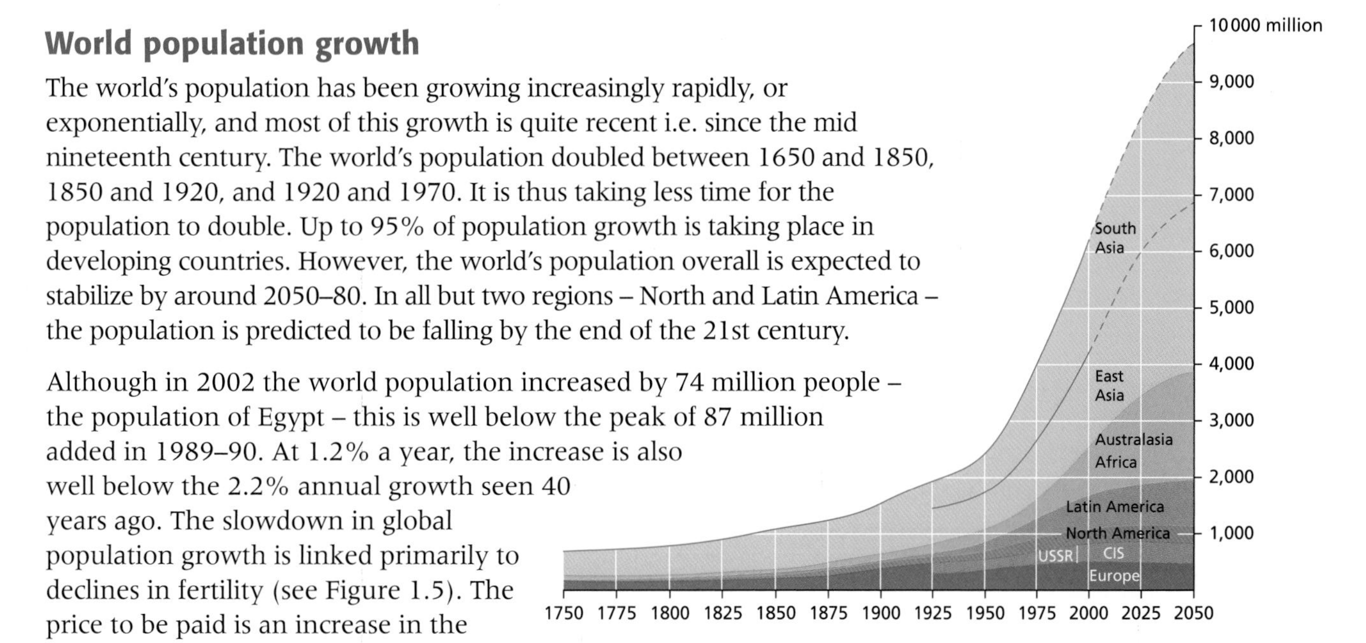

Figure 1.1 Growth of the world's population, 1750–2050

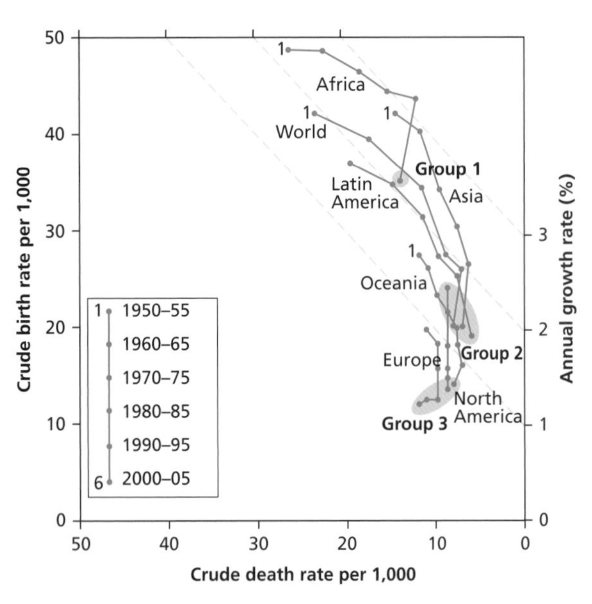

Figure 1.2 Demographic paths of the world's major regions

Figure 1.2 is a complex graph. Notice that the scale for crude birth rates is on the vertical axis while the scale for crude death rates is in reverse on the horizontal axis. Natural increase is the difference between birth rates and death rates. These are shown by the lines at 45°. Birth and death rates are shown in rates per 1,000 (‰), whereas natural increase is shown in per cent (%). A growth rate of 1% is the same as 10‰.

To do:

Study Figure 1.2, which shows changes in birth rates and death rates in different regions of the world.

a Describe the demographic path of Africa between 1950 and 2005.

b Explain how changes in crude death rate have affected the trends in population growth in one region.

In 1990 the world's women, on average, were giving birth to 3.3 children each. By 2010 this had dropped to 2.56 children, slightly above the level needed to replace the population. The level of fertility for the world as a whole is projected to fall below replacement level before 2050. The projections also suggest that AIDS, which has killed more than 20 million people in the past 20 years, is already lowering average life expectancy at birth to around 40 years in some countries. AIDS continues to have its greatest impact in the developing countries of Asia, Latin America and especially sub-Saharan Africa. Botswana and South Africa are among countries that may see population decline because of AIDS related deaths.

- The world's population is likely to peak at between 10 billion and 12 billion in 2070. By 2100, it could be down to 9 billion.
- North America (the US and Canada) will be one of only two regions in the world with a population still growing in 2100. It will have increased from 314 million today to 454 million, partly because first-generation immigrant families tend to have more children. The other expanding region is Latin America, forecast to increase from 515 to 934 million.
- Despite disease, war and hunger in some areas the population of Africa will grow from a billion today to 1.6 billion in 2050. By 2100 it will be 1.8 billion, although it will have begun to decline, and more than a fifth of Africans will be over 60, more than in western Europe today.
- The China region (China and Hong Kong together with five smaller neighbouring nations) will see its population shrink significantly by 2100, from 1.4 billion to 1.25 billion. Because of its education programme, by 2020, when China is reaching its population peak of about 1.6 billion, it will have more well-educated people than Europe and North America combined.
- India will overtake China as the world's most populous nation by 2020.
- Europe – including Turkey and the former Soviet Union west of the Urals – will see its population fall from 813 million now to 607 million in 2100: from 13% of the world's population to just 7%. Eastern European countries such as Russia have already seen their populations fall; western Europe's is likely to peak in the next few decades.

Annual growth rate – found by subtracting the crude death rate (‰ – per 1,000) from the crude birth rate (‰) and then expressed as a percentage (%).
Crude birth rate (CBR) – the number of births per 1,000 people in a population.
General fertility rate (GFR) – the number of births per 1,000 women aged 15–49 years.
Age-specific birth rate (ASBR) – the number of births per 1,000 women of any specified year groups.
Standardized birth rate (SBR) – a birth rate for a region on the basis that its age composition is the same as for the whole country.
Total fertility rate (TFR) – the average number of births per 1,000 women of childbearing age.

• About 10% of the world's population is over 60. By 2100, that proportion will have risen to one-third.
• In 1950 there were thought to be twice as many Europeans as Africans. By 2050 the proportions will be reversed.

Table 1.1 Key predictions from the global population forecast

Birth rates and fertility rates

The crude birth rate (CBR) is easy to calculate and the data are readily available. However, it does not take into account the age and sex structure of the population. Total Fertility rate is the average number of children a woman would have if she expenend the current age-specific fertility rates through her life time, and she were to servive through child birth.

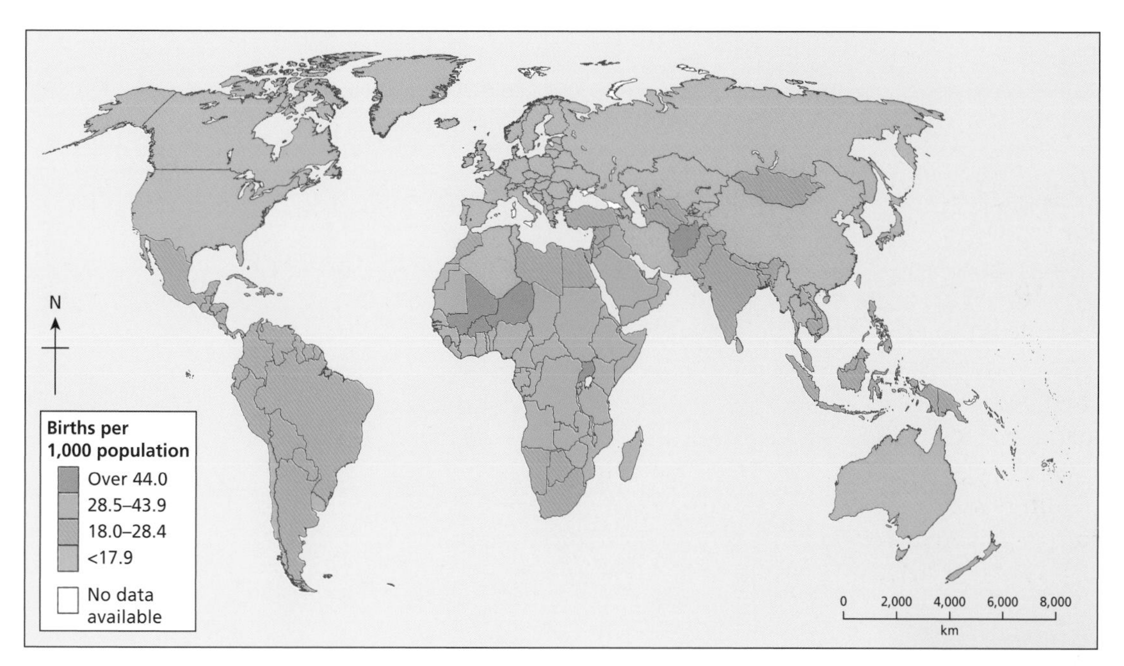

Figure 1.3 World birth rates, 2009

In general, the poorest countries have the highest fertility rates, and few of them have made the transition from high to low birth rates. In most developed countries, birth rates, and fertility rates have fallen. This means that their decline in population growth is not due to changing population structure.

Why do you get high birth rates?

Parents want children

- for labour
- to look after them in old age
- to continue the family name
- for prestige
- to replace other children who have died (a high infant mortality rate)
- children are net contributors to family income.

Why do birth rates come down?

- Children are costly.
- The government looks after people through pensions and health services.

Figure 1.4 Africa's youthful population has implications for growth

- More women want their own career and have higher status.
- There is widespread use of family planning.
- As the infant mortality rate comes down there is less need for replacement children.
- Urbanization and industrialization are associated with social changes and a decline in traditional beliefs and customs.

Infant mortality rate (IMR) – the number of deaths of children less than one year old per 1,000 live births.
Child mortality rate – the number of deaths in children under the age of 5 per 1000 children.

Birth rates are generally higher in developing countries, and there is an implicit suggestion that circumstances are different in developed countries: Many of the key data are shown in the table below.

Factor	Brazil	China	India	Japan	UK	USA
Population (millions)	198.3	1,338	1,156	127	61	307
Area (thousand km^2)	8,515	9,596	3,287	378	244	9,827
Population density (people/km^2)	22	139	317	337	254	32
Population growth (%/year)	1.2	0.66	1.4	-0.2	0.3	1.0
Birth rate (per 1,000)	18	14	22	7	11	14
Death rate (per 1,000)	6	7	8	9	10	8
Infant mortality rate (per 1,000)	23	20	51	3	5	6
Life expectancy	72	73	66	82	79	78
Urban population (%)	86	43	29	66	90	82
Fertility rate (no. of children per woman)	2.2	1.79	2.6	1.2	1.7	2.0
Use of contraception (%)	77	84	48	56	84	76
Age structure						
0–14 years	27	20	31	14	17	20
15–59 years	67	72	64	64	67	67
>60 years	6	8	5	22	16	13
Employment structure						
Agriculture	20	11	52	4	1	1
Industry	14	49	14	28	18	20
Services	66	40	34	68	81	79
Average income ($US)	10200	660	31,00	32600	35200	46400
Literacy (%)	88	91	61	99	99	99
Spending on education (as % of GNP)	4	1.9	3.2	3.5	5.6	5.3
Spending on military (as % of GNP)	2.6	4.3	2.5	0.8	2.4	4.0

Table 1.2 Selected statistics to help explain variations in the birth rate

More material available:

www.oxfordsecondary.co.uk/ibgeography

Skills: Using evidence to support an answer

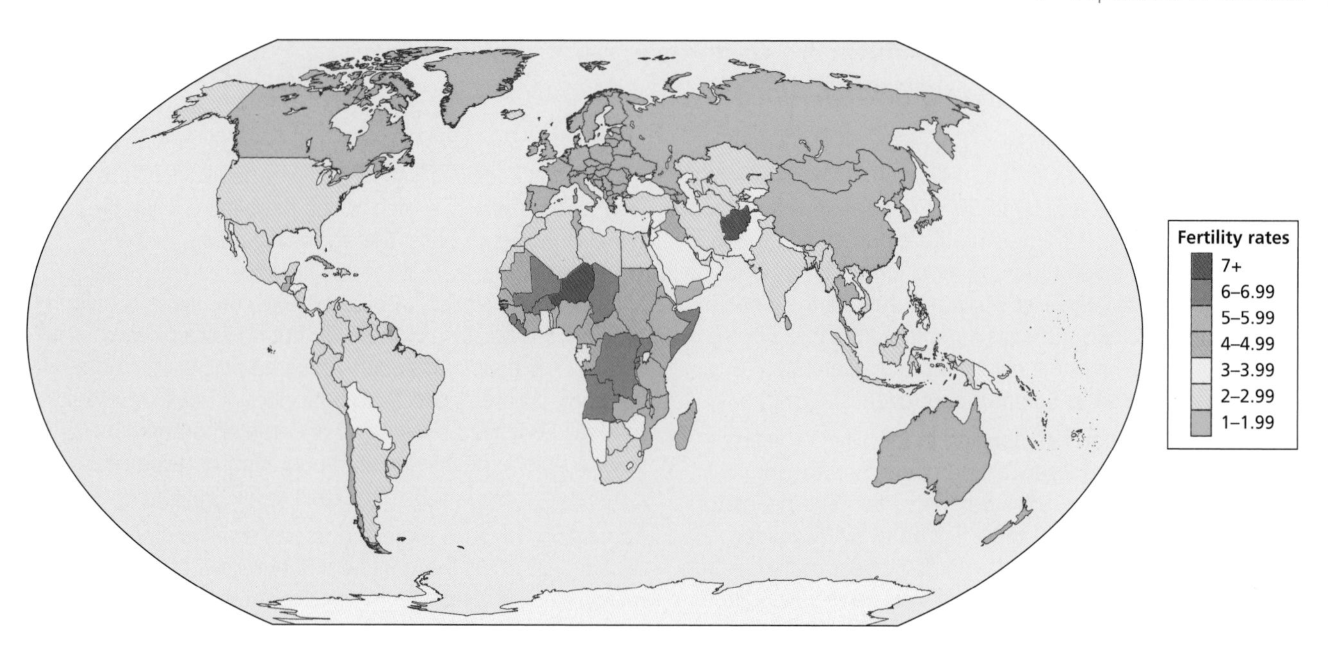

Figure 1.5 Fertility rates

TOK Link

Why use rates? Why are some rates "crude"?

In China the total number of births in any one year is going to be much greater than in, say, the Netherlands, partly because China is a much larger country with a larger population. In order to compare the number of births in different countries, we have to standardize the number of births by giving an average figure such as the number of births that every 1,000 people would produce in a given year. This allows us to compare two or more countries. However, the figure does not take into account the age structure of the population. A country with a younger population would be expected to have a higher birth rate than a country with an elderly population. Consequently, the birth and death rate are referred to as crude rates.

Be a critical thinker

Whose statistics to use?

The statistics that different organizations use can produce different results. For example, the world map of fertility rates varies depending on whether you use the World Bank's (using UN data) figures or the CIA's. Go to the wikipedia website at http://en.wikipedia.org and search for "List of countries and territories by fertility rate" to view two lists and maps of world fertility rates as defined by the UN and by the CIA. Alternatively, go to www.nationmaster.com and search for "total fertility rate" to see a bar graph (CIA statistics) or www.worldbank.org/indicator for their statistics.

The World Bank total fertility rate (TFR) ranking is a list of countries by TFR: the expected number of children born per woman in her childbearing years. Figures are from the 2008 revision of the UN World Population Prospects Report. Only countries/territories with a population of 100 000 or more in 2007 are included.

The CIA TFR ranking is also done using countries' TFR, based on 2008 age-specific fertility rate data. Sovereign states and countries are ranked. Some countries might not be listed because they are not recognized as countries at the time of this census. Figures are from the CIA World Factbook.

- Whose statistics produce the higher fertility rates?
- How do the rates for Uganda vary between the two?
- Suggest reasons why they differ.
- Is this important and if so, why?

Case study: *Population growth in Africa*

Africa has the fastest-growing population in the world: in 1850 its population was 110 million, in 2009 its billionth* person was born, and by 2050 its population is projected to be 1.6 billion. Pessimists predict a human tide that will put an unbearable burden on food, jobs, schools, housing and healthcare. However, Africa has 20% of the world's landmass but only 13% of its population. It has a bulge of young people, and that brings to the marketplace a huge workforce (whereas Europe's population is ageing).

Africa's population has doubled in the past 27 years, with Nigeria's and Uganda's numbers climbing the fastest. Whereas in 1950 there were two Europeans for every African, by 2050 there will be two Africans for every European. Women in Africa still bear more children than in other regions. The world's highest fertility rate is in Niger, where women have on average 7.4 children.

Africa's population continues to rise because of low life expectancy. Traditionally in all societies, fertility tends to be high when mortality is high. When people are dying the population tries to offset that by having more children. Whereas globally 62% of married women of childbearing age use contraception, in parts of Africa the figure is only 28%. This is because women do not necessarily have control over their own bodies (in some places men and laws do). Women's access to reproductive health services may still be limited because of underdevelopment, poverty and sometimes limited education or resources.

Sub-Saharan Africa has the world's most youthful population, which is projected to stay that way for decades. In 2050 the continent is expected to have 349 million people aged 15–24, or 29% of the world's total, compared with 9% in 1950.

*one billion = one thousand million (US definition)

Case study: *Falling populations in the Balkans*

Population is shrinking in the Balkans. Serbia has 7.2 million people but the population is declining by 30 000 a year, and could shrink by mid-century to only 6 million. This is not because Serbs are becoming rich and want smaller families, but because the war years and economic hardship have led to fewer children, mass emigration and high abortion rates. Croatia, now with 4.4 million, is also shrinking, if not so drastically.

Kosovar fertility is dropping, too. In 1950 it was 7.8; as late as 1990 it was still 3.6, but now it is about 2.2. Fertility in Albania has fallen from 2.0 in 2000 to 1.33 in 2007. In 1999 Albania's population was 3.3 million; by 2008 it had fallen to 3.1 million. This is not because Albanians are having fewer children, but because so many women of childbearing age have emigrated.

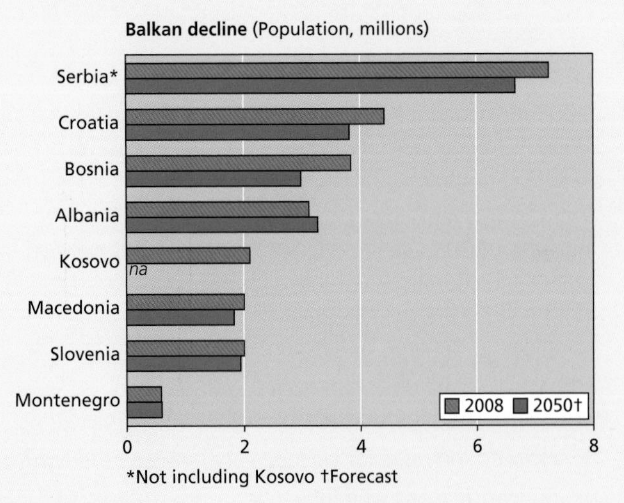

Figure 1.6 Demography of the Balkans

To do:

Study Figure 1.6.

a Which country is projected to lose most population by 2050?

b Which country is projected to gain population?

c Suggest reasons why the Balkan countries are projected to lose populations.

To research

Visit www.imf.org

This is a useful site for a number of features in global demographic (population) trends, such as population growth, changes in birth and death rates, migration ratios, and contrasts between rich and poor countries. Choose two to three countries at different levels of economic and social development and research data relevant to their fertility rates, for example, and track changes over time.

To do:

Describe the pattern of world death rates as shown in Figure 1.7.

Death rates

Death rates are generally high when there is a lack of clean water and food, poor hygiene and sanitation, overcrowding, contagious diseases such as diarrhoea and vomiting, and respiratory infections – in short, the conditions associated with poverty. This is why death rates are highest in poor rural areas, shanty towns, refugee camps and areas of relative and absolute poverty.

Death rates decline when there is clean water, a reliable food supply, good hygiene and sanitation, lower population densities, better vaccination and health care – in short, rising standards of living. That is why, despite other hazards, authorities can reduce death rates by providing access to clean water, food, shelter and sanitation.

The crude death rate is a poor indicator of mortality trends. Populations with a large number of aged people, as in most developed countries, will have a higher CDR than countries with more youthful populations. For 2005–10 the world CDR was 8.7%. (data.un.org).

Crude death rate (CDR) – the number of deaths per 1,000 people in a population.

To research

Visit www.indexmundi.com for a map of variations in the infant mortality rate around the world.

a Describe the variations in the infant mortality rate around the world.

b Suggest reasons for the differences you have noted. (Use the drop-down menus to look at other maps.)

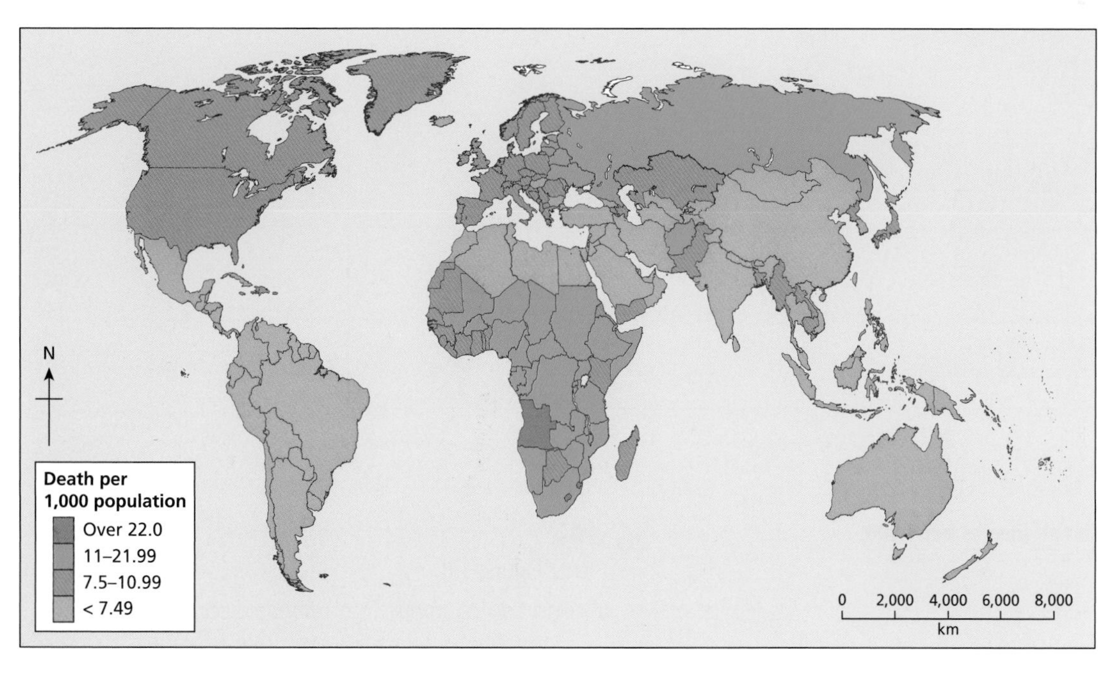

Figure 1.7 World death rates, 2009

Case study: Risk of early death

Researchers looked at early deaths (deaths under 60 years) in 187 countries in 1970, 1990 and 2010. During that time the overall risk fell by 34% in women and 19% in men, reflecting the progress in medicine and rising affluence. South Asia saw the most rapid decline for women and Australasia for men.

Sub-Saharan Africa currently has some of the highest early death rates, with half of people dying early compared to one in 20 in some developed countries. Some parts of the continent even saw rates get worse, reflecting the spread of HIV in recent years.

In the UK, 58 deaths per 1,000 among women are before the age of 60, while for men the figure stands at 93. In western Europe only Danish and Belgian women have a higher risk than those in the UK. Factors that influence the death rate include age structure, poverty, occupation, natural hazards, war and diet.

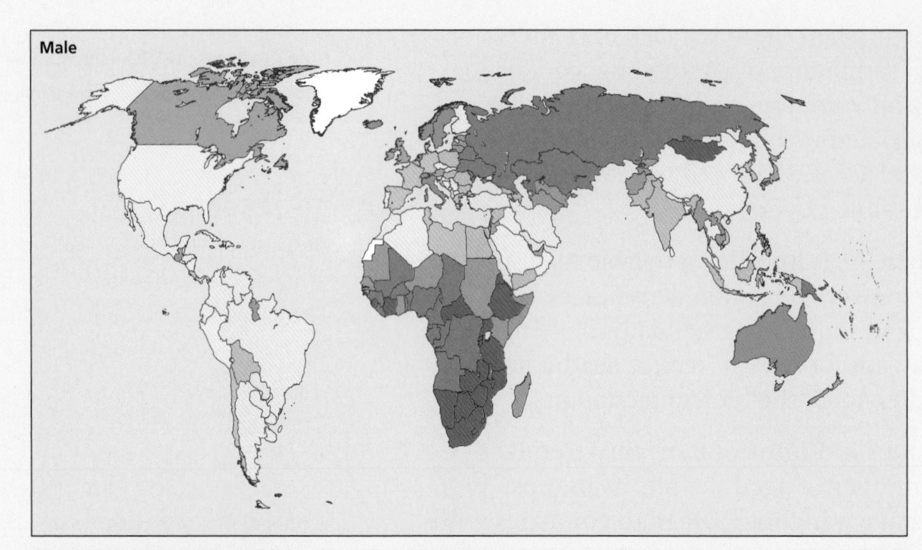

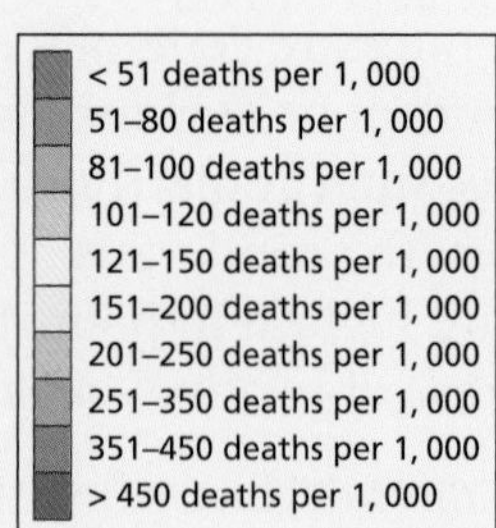

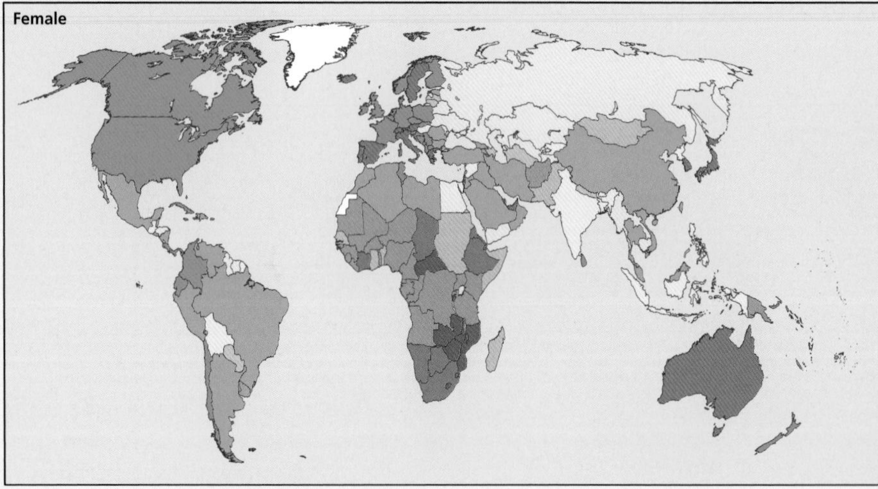

Figure 1.8 Worldwide mortality, 2010

Lowest early death risk (deaths per 1,000)

Women

Cyprus: 38
Japan, Italy and Greece: 41
South Korea: 40

Men

Iceland: 65
Sweden: 71
Malta and the Netherlands: 73

To do:

Study Figure 1.8.

a Find two countries from different parts of the world where the male death rate is higher than the female death rate.

b Find two countries from different continents where the male and female death rate are the same.

Life expectancy

Between 1960 and the late 1980s almost every country in the world showed continual increases in life expectancy. In South Asia it increased from 42 years to 60, and in North Africa from 47 years to 65. The exception was in sub-Saharan Africa. Until the late 1980s life expectancy increased slowly in western, central and eastern Africa and slightly faster in southern Africa, where it rose from 46 years to about 60. Since then, however, the HIV/AIDS epidemic has caused a large increase in mortality, bringing life expectancy in southern Africa below its level in 1960. Nine of the 10 countries showing the worst trends are in sub-Saharan Africa, most of them in southern or southeastern Africa.

Life expectancy (E_0) – average number of years that a person can be expected to live, usually from birth, if demographic factors remain unchanged.

Life expectancy in rich countries

In rich countries life expectancy is increasing so fast that half the babies born in 2007 will live to be at least 104, and half the Japanese babies born that year will reach the age of 107. If current trends continue, most babies born in the past few years in the UK will live to be over 100. People could be living not only longer but also better. People are surviving chronic illnesses, such as cancers and heart conditions, because they are diagnosed earlier and get better treatment. Shorter working weeks might further increase health and life expectancy.

Over the 20th century there were huge increases in life expectancy – more than 30 years – in most developed countries. Until the 1920s breakthroughs in saving babies from infectious diseases and mothers from the complications of childbirth were responsible for the increases in life expectancy. Then people started to live to greater ages. Since the 1950s, and especially since the 1970s, mortality at ages 80 years and older has continued to fall, in some countries even at an accelerating pace.

To do:

a Which country has seen the greatest (i) decrease and (ii) increase in life expectancy between 1970 and 2005?

b Compare the changes in life expectancy in Vietnam with those of South Africa.

c Suggest why life expectancy in so many African countries has decreased, whereas in Middle Eastern countries it has increased.

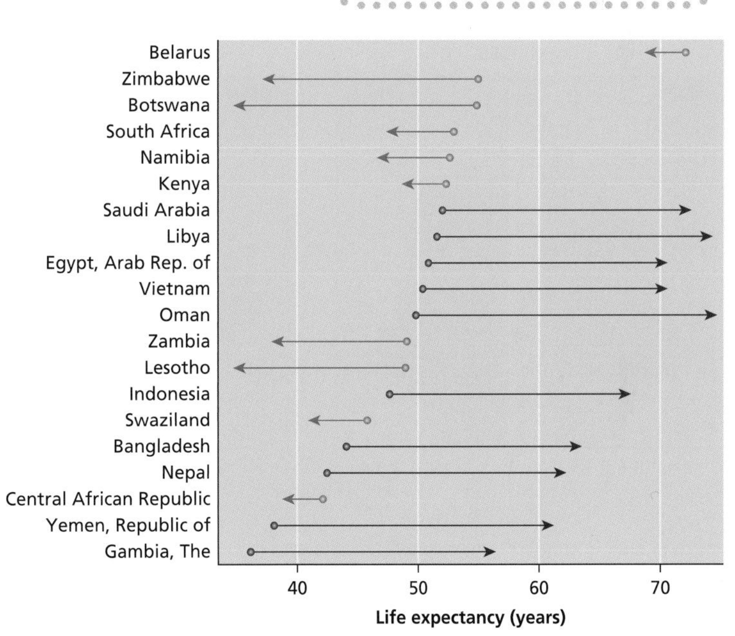

Figure 1.9 Changes in life expectancy in Africa, 1970–2005

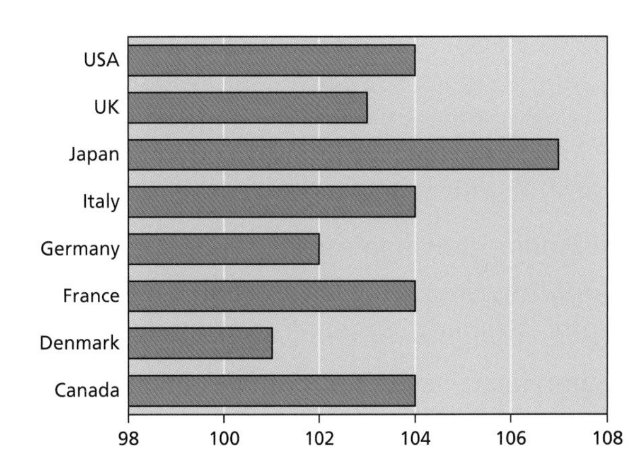

Figure 1.10 Life expectancy of 50% of babies born in 2007

Population pyramids

Population pyramids tell us a great deal about the age and sex structure of a population. They also tell us about population growth. They help planners find out how many services and facilities, such as schools and hospitals, will be needed in the future.

- A wide base suggests a high birth rate.
- A narrowing base indicates a falling birth rate.
- Straight or near-vertical sides show a low death rate.
- A concave slope suggests a high death rate.
- Bulges in the slope indicate high rates of in-migration. For instance, excess males of 20–35 years could be economic migrants looking for work. An excess of both male and female cohorts

(age groups) could be due to a baby boom. Excess elderly, usually female, might indicate a community of retired people.
- Deficits in the slope show out-migration or age-specific or sex-specific deaths (such as epidemics or war).

Population pyramids can also be used to show the racial composition of a population or the employed population group.

Population structure or population composition – any measurable characteristic of the population. This includes the age, sex, ethnicity, language, religion and occupation of the population.

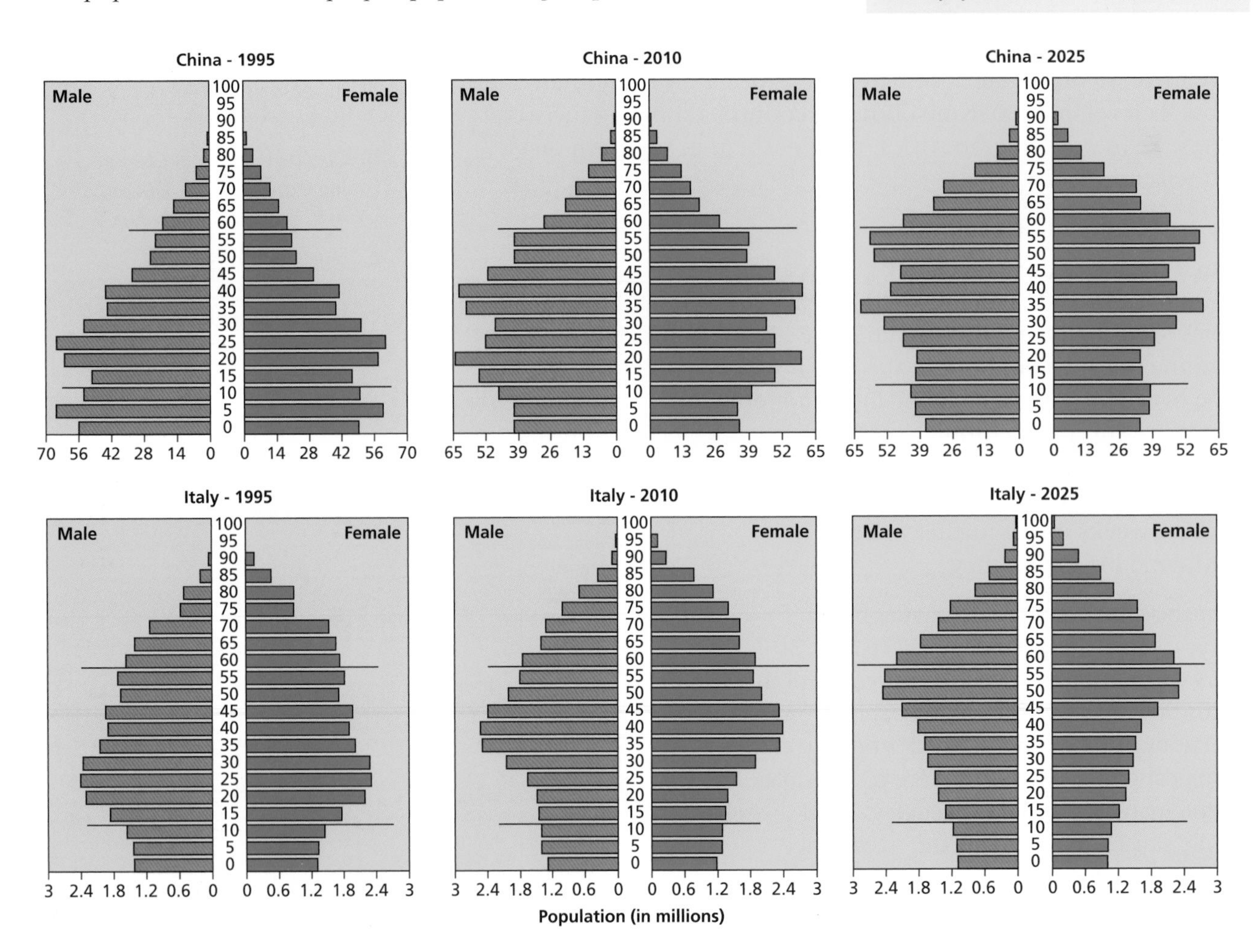

Figure 1.11 Population pyramids for China and Italy

Growth rates

The growth rate is the average annual percentage change in the population, resulting from a surplus (or deficit) of births over deaths and the balance of migrants entering and leaving a country. The rate may be positive or negative. The growth rate is a factor in determining how great a burden would be imposed on a country by the changing needs of its people for infrastructure (e.g. schools, hospitals, housing, roads), resources (e.g. food, water, electricity) and jobs.

Measuring growth rates

Doubling time is the length of time it takes for a population to double in size, assuming its natural growth rate remains constant.

Approximate values for it can be obtained by calculating the formula:

Doubling time (years) = 70/growth rate in percentage.

To do:

a Label the population pyramids for China and Italy for 2010 and 2025. (Remember to use terms such as elderly, youth, adult, birth rate, death rate, falling, in-migration, out-migration, impact of one-child policy, and gender differences. Use the scale, but be careful: the scale varies from 1995 to 2010 to 2025.)

b How does China's population change over time?

c Define and illustrate the following terms: population momentum, doubling time, life expectancy.

Population momentum is the tendency for population to grow despite a fall in the birth rate or fertility levels. It is also the tendency for a population to continue to fall despite a rise in the birth rate. It occurs because of a relatively high concentration of people in the pre-childbearing and childbearing years. As these young people grow older and move through their reproductive years, the greater number of births will exceed the number of deaths in the older populations, and so the population will continue to grow.

Population projections are predictions about future population based on trends in fertility, mortality and migration.

To research

Visit the US Census Bureau International database at www.census.gov and submit a query for the population pyramids of countries that interest you. Try to annotate the pyramids to describe and explain the changes in them.

Country	Growth rate (%)	Doubling time
Burundi	3.69	19 years
China	0.66	106 years
South Korea	0.27	259 years
Austria	0.05	1,400 years

Table 1.3 Doubling times for selected countries, 2010

Responses to high and low fertility

Dependency ratios

The dependency ratio measures the working population and the dependent population. It is worked out by a formula:

$$\frac{\text{Population aged} <15 + \text{population aged} >64 \text{ (the dependents)}}{\text{Population aged 16–64 (the economically active)}}$$

It is very crude. For example, many people stay on at school after the age of 15 and many people work after the age of 60. However, it is a useful measure to compare countries or to track changes over time.

- In the developed world there is a high proportion of elderly.
- In the developing world there is a high proportion of youth.

These can be shown on a triangular graph.

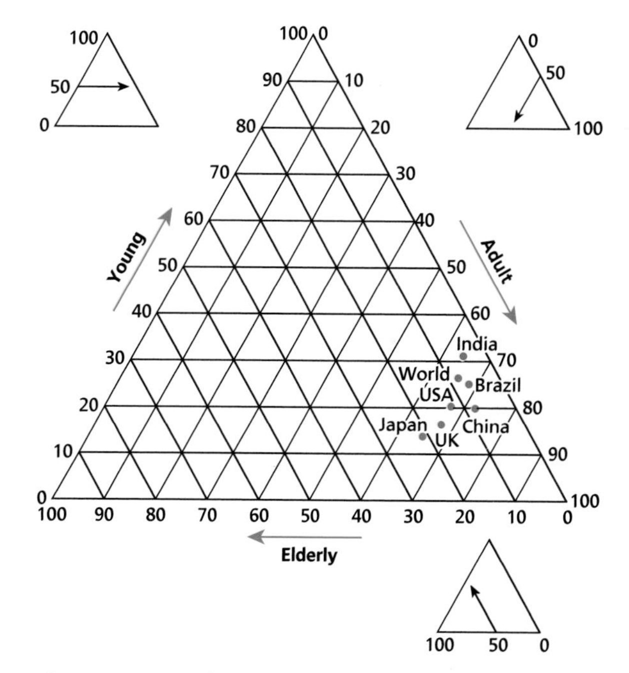

Figure 1.12 A triangular graph

Triangular graphs

Triangular graphs are used to show data that can be divided into three parts. As well as showing the age structure of population, triangular graphs are commonly used to show variations in soil structure (sand, clay, silt) and employment (primary, secondary and tertiary sectors). They are useful for several reasons.

- They can show a large amount of data.
- Groupings can be recognized easily.
- Dominant characteristics can be shown.
- Classifications can be made.

Sometimes only part of the graph is shown, but at a larger scale.

To do:

From Figure 1.12, work out the population structure for China, USA and the world.

On a copy of Figure 1.12, add a dot to locate the population age structure for Ethiopia, Vietnam and Zimbabwe.

	Youthful	Adult	Elderly
Ethiopia	46	51	3
Vietnam	26	68	6
Zimbabwe	44	52	4

Potential advantages	Potential disadvantages
Large potential workforce	Cost of supporting schools and clinics
Lower medical costs	Need to provide sufficient food, housing and water to a growing population, e.g. Kibera, Nairobi
Attractive to new investment	High rates of unemployment
Source of new innovation and ideas	Large numbers living in poor-quality housing, e.g. in shanty towns
Large potential market for selected goods	High rates of population growth
Development of services such as schools, crèches	High crime rates

Table 1.4 Some potential advantages and disadvantages of a youthful population

Figure 1.13 Youthful population in South Africa

Ageing populations

An ageing population has certain advantages. The elderly may have skills (including social skills) and training, and some employers, especially supermarkets and home improvement/furniture stores, prefer them to younger workers. The elderly may look after their grandchildren and therefore allow both parents to work. This is important in Japan, and also in South Africa (with its 'youthful' population structure) where a "granny culture" occurs in many areas. In rich countries the elderly are often viewed as an important market – the "grey economy" – and many firms, ranging from holiday companies to healthcare providers, have developed to target this market.

The shift to an ageing population is due to the time-delayed impact of high fertility levels after the Second World War and to more recent improvements in health that are bringing down death rates at older

Older dependency ratio (ODR) – the number of people aged 65 and over for every 100 people aged 20 to 64.

More material available:

www.oxfordsecondary.co.uk/ibgeography

Read about the implications of Africa's youthful population.

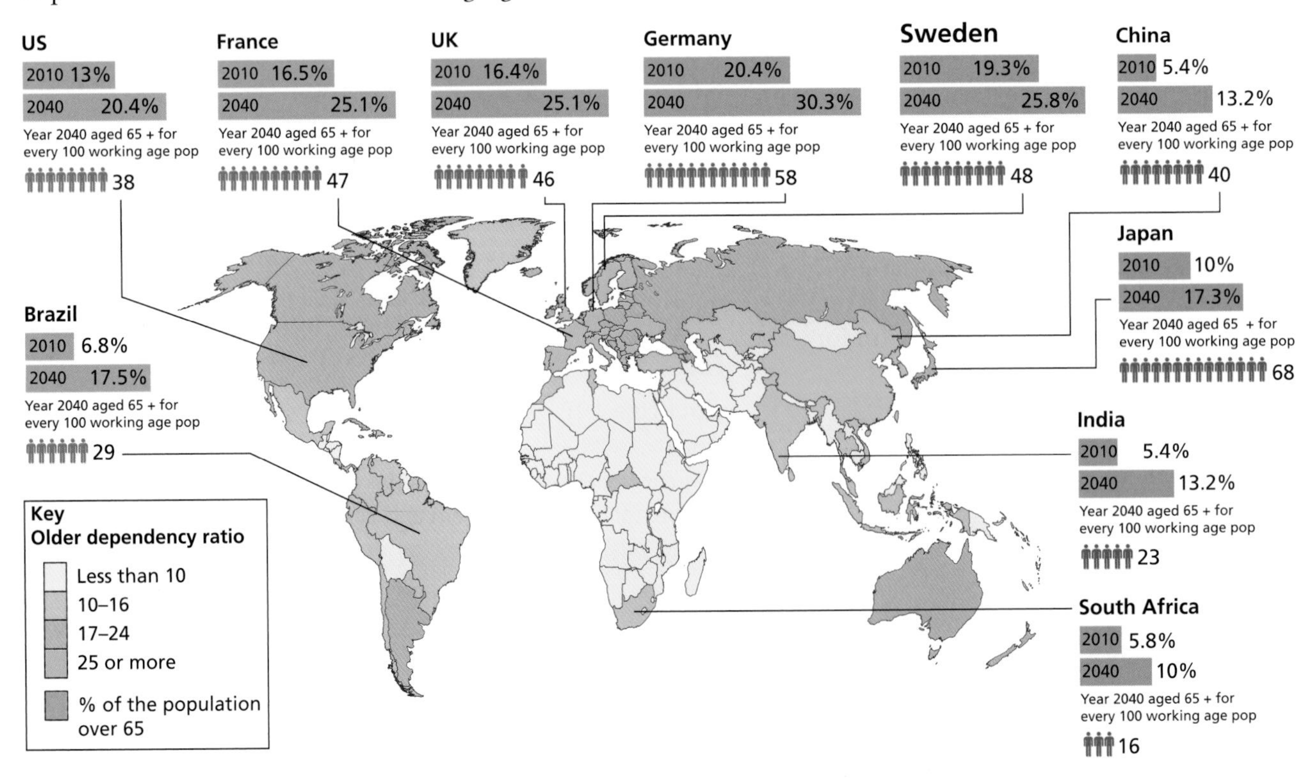

Figure 1.14 The global pattern of the ODR

ages. The change will bring widespread challenges at every level of human organization, starting with the structure of the family, which will be transformed as people live longer. There will in turn be new burdens on carers and social services providers, while patterns of work and retirement will similarly have huge implications for health services and pensions systems.

Europe is the greyest continent, with 23 of the world's 25 oldest countries. Such dominance of the regional league table will continue. By 2040 more than one in four Europeans is expected to be at least 65, and one in seven at least 75. Japan's life expectancy at birth – 82 years – is matched only by Singapore's, though in western Europe, France, Switzerland, Sweden and Italy all have life expectancies of more than 80 years.

The contrast in life expectancy between rich and poor nations remains glaring. The report shows that a person born in a developed country can expect to outlive his or her counterpart in the developing world by 14 years. Zimbabwe has the lowest life expectancy, which has been cut to 40 through a combination of AIDS, famine and dictatorship. Nevertheless, more than 80% of the increase in older people in the year up to July 2008 was seen in developing countries. By 2040 the poor world is projected to be home to more than a billion people aged 65 and over – 76% of the world total.

The **older dependency ratio**, or ODR, acts as an indicator of the balance between working-age people and the older population that they must support. It varies widely, from just six per 100 workers in Kenya and seven in Bangladesh, to 33 in Italy and also Japan. Countries with a high ODR are already suffering under the burden of funding prolonged retirement for their older population. In France, life expectancy after retirement has already reached 21 years for men and 26 years for women. With women living on average seven years longer than men, more older women are living alone. Around half of all women aged 65 and over in Germany, Denmark and Slovakia are on their own, with all the consequent issues of loneliness and access to care.

Japan's ageing population

Since 1945 the age structure of Japan's population has greatly changed, largely due to a decrease in both birth and death rates. The population is ageing much more rapidly than that of other countries

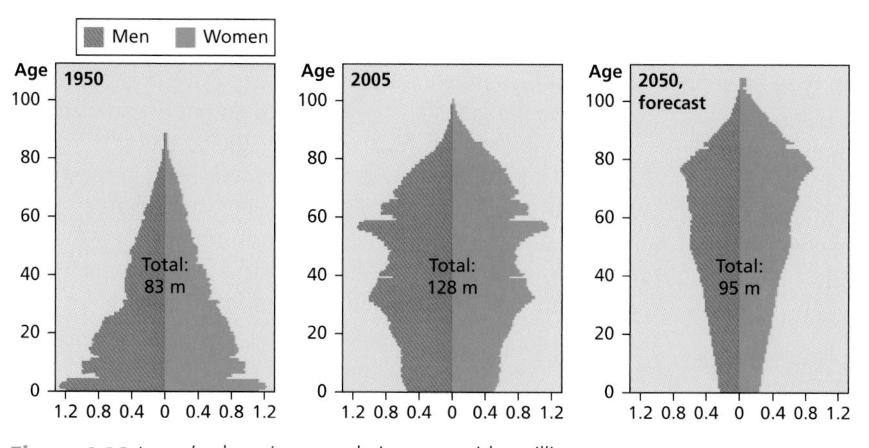

Figure 1.16 Japan's changing population pyramids, millions

Figure 1.15 Japan's ageing population

Be a critical thinker

An ageing world: advantage or disadvantage?

Over the next 30 years the number of over-65s is expected almost to double, from 506 million in 2008 to 1.3 billion – a leap from 7% of the world's population to 14%, and already the number is increasing at an average of 870 000 each month. The proportion of the global population aged 65 years and over is set to outnumber that of children under five for the first time.

A report by the US Census Bureau, An Ageing World 2008, highlights a huge shift towards not just an ageing but also an old population, with formidable consequences for rich and poor countries. New challenges for families and policymakers range from how to care for older people living alone to how to pay for unprecedented numbers of pensioners.

(although a number of European countries, such as Italy and Greece, are not far behind). The number of elderly people living alone in Japan increased from 0.8 million in 1975 to over 2.5 million in 2000. Since 1975 the percentage of young people has gradually declined, and by 1995 they accounted for only about 16% of the population.

At present under 15% of the Japanese population are aged over 65, whereas by 2020 the figure will be over 25%. This rapid ageing of the population is creating a huge burden on pension funds and social welfare programmes, especially health care. Problems include:

- inadequate nursing facilities
- depletion of the labour force
- deterioration of the economy
- a trade deficit
- migration of Japanese industry to overseas
- the cost of funding pensions and healthcare
- falling demand for schools and teachers
- new jobs needed for the elderly
- new leisure facilities needed for the elderly
- an increase in the burden on the working population to serve the dependent population
- reduced demand for goods from the smaller working population although some pensioners are spenders not savers, and the 'grey market' is an important and growing market
- a need for in-migration to fuel any increase in the workforce.

To research

Visit www.oeaw.ac.at for the European Demographic Research Report 2006 No. 1.

To do:

a Describe the global pattern of the ODR as shown in Figure 1.14.

b Comment on the change in the ODR between 1990 and 2040, as shown in Figure 1.14.

c Outline the advantages of an ageing population.

d Describe the changes in Japan's population pyramids, as shown in Figure 1.16.

e Briefly explain the disadvantages of ageing populations.

Managing population change

There are a number of ways in which governments attempt to control population numbers. Their strategies will depend on whether the country wishes to increase its population size (**pro-natalist**) or limit it (**anti-natalist**). Family planning methods include contraceptives such as the pill and condoms, as well as drastic methods such as forced sterilization, abortion and infanticide.

Family planning – attempts to limit family size.

Be a critical thinker

Should governments decide on family size? China's one-child policy

China operates the world's most severe and controversial family planning programme with its one-child policy. The policy, imposed in 1979, had a drastic impact: the birth rate fell to 17 per 1,000 from its 1970 rate of 33. The rule is estimated to have reduced population growth in a country of 1.3 billion by as much as 300 million people over its first 20 years alone, and so far the policy has prevented as many as 400 million births. The Chinese government has predicted that the population will peak at 1.5 billion in 2033.

Such Draconian family planning has resulted in a disparate ratio of 118 male to every 100 female births, above the global norm of between 103 and 107 boys to every 100 girls. This reflects the fact that many people in China value female infants less highly than males, and millions of females have been aborted, or died as a result of neglect, abandonment, or even infanticide.

The one-child policy is not an all-encompassing rule because it is restricted to ethnic Han Chinese living in urban areas, who comprise only about 36% of the population. In urban areas most families have only one child, and the growing middle classes do not discriminate against daughters as much. However, the countryside remains traditionally focused on male heirs, and in most provincial rural areas couples may have two children if the first one is a girl. In other provinces, parents may have two children regardless of the sex of the first child, and in a few areas the rules are even more relaxed.

Special more recent provisions also allow millions of other couples to have two children legally. If a couple is composed of two people without siblings, then they may have two children of their own, thus preventing too

dramatic a population decrease. Families who lost a child in the Sichuan earthquake of 2008 were permitted to have another child.

Several experts have now called for a move to a uniform two-child policy. One reason for this is the prospect of an ageing society in which one worker is left to support two parents and four grandparents. By 2050 China will have nearly 450 million people aged over 60, and 100 million over 80; there will be just 1.6 working adults to support every person aged over 60 compared with more than seven in the 1970s. At its peak in the 1960s the fertility rate was 5.8 babies for each woman of childbearing age and it is now just 1.8, well below the replacement rate of 2.1.

Another reason why China is now considering axing its one-child policy is that it has become divisive. In May 2007, in a riot sparked by the state's policy, thousands of villagers in southwest China attacked family planning officials, overturned cars and set fire to government buildings. It was seen as unfair that complex exemptions, enforcement inconsistencies and financial penalties allowed some people to have larger families than others. The policy has also become a symbol of the wealth gap in China: government officials have now admitted that many rich families violate the rules because they can afford the fines.

To research

Visit http://wiki.idebate.org for opposing views on China's one-child policy.

To do:

Study Figure 1.17, which shows different population scenarios for various levels of fertility in China.

- **a** By how much larger would China's population be in 2080 if it had had a two-child policy rather than a one-child policy?
- **b** Outline the main advantages of the one-child policy.
- **c** What are the main disadvantages of the one-child policy?
- **d** What are your views on the one-child policy? Justify your answer.

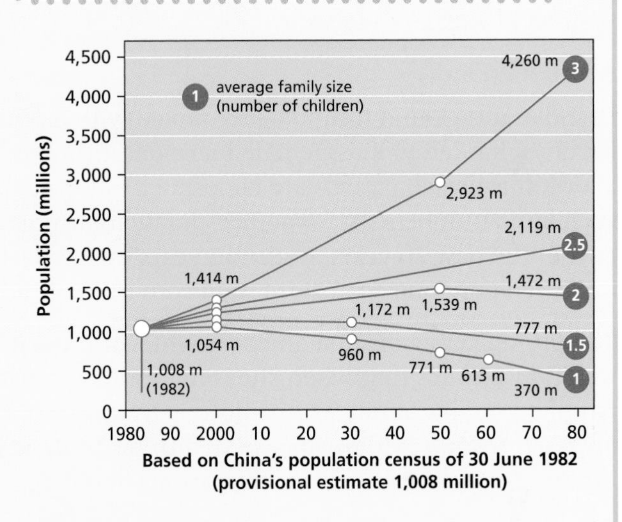

Figure 1.17 Five possible options for China's future population

Pro-natalist policies in Romania

In the 1960s Romania was approaching zero population growth, which had implications for its future labour supply and for industrialization. In 1966 the government decided to ban abortion on demand and introduced other pro-natalist policies to increase birth rates. Abortion became a criminal offence with penalties for those who sought or performed such operations. People who remained childless after the age of 25 were liable for a special tax amounting to 10–20% of their income. The government also made divorce much more difficult.

There were some positive pro-natalist policies. Family allowances were raised, and monetary awards were given to mothers on the birth of their third child. Moreover, the income-tax rate for parents of three or more children was reduced by 30%.

The pro-natalist policies had an immediate impact. The number of live births increased by 92.8%, from 273 687 in 1966 to 527 764 in 1967. As contraceptives were not manufactured in Romania, and importation of them had stopped, the sudden unavailability of

To research

Visit www.country-studies.com for a study of Singapore's population, including its changing population policies. Contrast these with those of Romania, which adopted a pro-natalist policy, at: http://countrystudies.us

abortion made birth control extremely difficult. Legal abortions fell from a million in 1965 to 52 000 in 1967. However, rising maternal and infant mortality rates associated with the restrictions on abortion marred the policy's initial success.

The increase in live births was short-lived. After the police returned to more normal duties, the number of abortions increased. The incentives provided by the state were not enough to sustain an increase in birth rate, which again began to decline. By 1983 the birth rate had fallen to 14.3 per 1,000, the rate of annual increase in population had dipped to 3.7 per 1,000, and the number of abortions again exceeded the number of live births.

Romanian demographic policies continued to be unsuccessful largely because they ignored the relationship of socio-economic development and demographics. Currently, Romania's birth rate is 10.3 per 1,000 (185th in the world) and its total fertility rate is 1.39 children per woman (198th in the world).

Abortion in India

In India, ultrasound technology, coupled with a traditional preference for boys, has led to mass female foeticide. Although gender-based abortion is illegal, parents are choosing to abort female foetuses in such large numbers that experts estimate that India has lost 10 million girls in the past 20 years. In the 12 years since selective abortion was outlawed only one doctor has been convicted of the crime.

The prejudices against having a daughter run deep in India, where tradition dictates that when she is married a woman's family must pay the groom's family a large dowry. By contrast a son is considered an asset. Even leaving aside the wealth his bride will bring, a boy will inherit property and care for his parents in old age.

Selective abortion has been accelerating in a globalizing India. Wealthier and better-educated Indians still want sons: a survey revealed that female foeticide was highest among women with university degrees. The urban middle classes can also afford the ultrasound tests to determine the sex of the foetus. For the two years up to 2005 India had just 880 female babies born to each 1,000 males.

The shortage of women has had some negative social effects: unmarried young men are turning to crime, and violence against women has increased. Some men in the rich northern state of Haryana have taken to buying brides from other parts of India. Many of these wives become slaves and their children are shunned.

Migration

People migrate for a number of different reasons. Most voluntary migrants are people moving either for work (this is especially true for

More material available:

www.oxfordsecondary.co.uk/ibgeography

Read how unsafe abortions kills 70 000 women a year

To do:

Study Figure 1.18, which shows safe and unsafe abortions per 1,000 women aged 15–44, in 1995 and 2003.

- **a** Define the terms safe and unsafe abortion.
- **b** In which regions are the most unsafe abortions found?
- **c** In which regions are most safe abortions found?
- **d** Suggest reasons for the varied pattern of abortions in Asia.
- **e** Comment on the changes in safe and unsafe abortions between 1995 and 2003.
- **f** "Female foeticide is a social necessity." Discuss.

To research

Visit http://news.bbc.co.uk for the article "Europe's terms for terminations" and see how access to abortion varies across Europe. Comment on the differences in access to abortion within the EU.

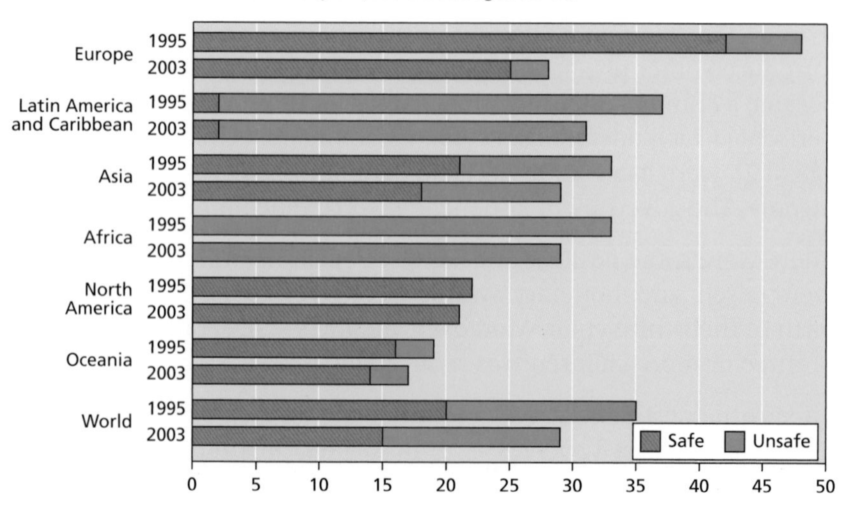

Figure 1.18 Global abortions, 1995 and 2003

young people), to retire to a small town or coastal area (this is especially true in some rich countries) or to live in a smaller urban area for a better quality of life than they had in a large urban area. Others may migrate for educational or health reasons. In contrast, forced migrations may be due to civil conflict, environmental damage or some form of persecution.

Migrations are commonly divided into a number of types:

- forced or voluntary
- long distance or short distance
- international or internal.

Migration – the movement of people, involving a change of residence. It can be internal or external (international) and voluntary or forced. It is usually for an extended period (more than a year) and does not include temporary circulations such as commuting or tourism.

Remittances – transfer of money or goods by foreign workers to their home countries.

Theories of migration

A number of laws and theories relate to patterns of migration. One of the earliest was that of Ernest Ravenstein, who investigated migration in the northwest of Britain up to and during the 1880s. He found that:

1. Most migrants proceed over a short distance. Due to limited technology/transport and poor communications, people know more about local opportunities.
2. Migration occurs in a series of steps or stages, typically from rural to small town, to large town to city, i.e. once in an urban area, migrants become "locked in" to the urban hierarchy.
3. As well as movement to large cities, there is movement away from them (dispersal). The rich move away from the urban areas and commute from nearby villages and small towns (an early form of suburbanization and counter-urbanization).
4. Long-distance migrants are more likely to go to large cities. People will know only about the opportunities in large cities of distant places.
5. Urban dwellers migrate less than rural dwellers, since there are fewer opportunities in rural areas.
6. Women are more migratory than men over short distances, especially for marriage and in societies where the status of women is low.
7. Migration increases with advances in technology such as transport, communications and the spread of information.

Figure 1.19 The Roman Catholic population in Brunei contains many Philippinos.

Another model of migration is Zipf's **inverse distance law**, which states that the volume of migration is inversely proportional to the distance travelled, i.e. $N_{ij} \propto \frac{1}{D_{ij}}$, where N_{ij} is the number of migrants between two towns, i and j, and D_{ij} is the distance between them. Stouffer refined this model in the 1940s using the idea of **intervening opportunities**. He stated that the number of migrants going to a place was proportional to the number of opportunities at that location, but inversely proportional to the number of opportunities that existed between the two places. Thus $N_{ij} \propto \frac{O_j}{O_{ij}}$, where O_j are the number of opportunities at j, and O_{ij} the number of opportunities that existed between i and j.

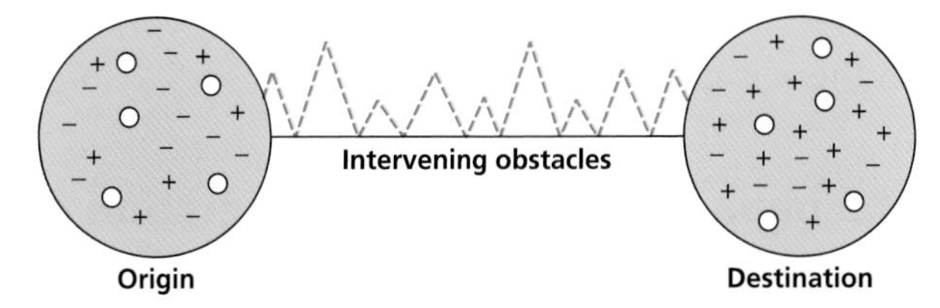

Figure 1.20 Lee's push–pull model of migration

One of the most widely used models is that of Everett Lee (1966), who describes migration in terms of

push and pull factors (Figure 1.20). Push factors are the negative features that cause a person to move away from a place. These include unemployment, low wages and natural hazards. Pull factors are the attractions (real or imagined) that exist at another place. Better wages, more jobs and good schools are pull factors. The term "perceived" means what the migrant imagines exists, rather than what actually exists. This may be quite close to or very different from the reality.

All these models are simplifications, and contain hidden assumptions that may be unrealistic. For example:

- Are all people free to migrate?
- Do all people have the skills, education and qualifications that allow them to move?
- Are there barriers to migration, such as race, class, income, language or gender?
- Is distance a barrier to migration?

To do:

Study Figure 1.21, which shows some of the world's most important current migration routes.

Using an atlas, if needed, identify

i the countries sending migrants to South Africa

ii the continent destinations of migrants from China

iii the sources of migrants to Russia

iv the sources of migrants to the Middle East.

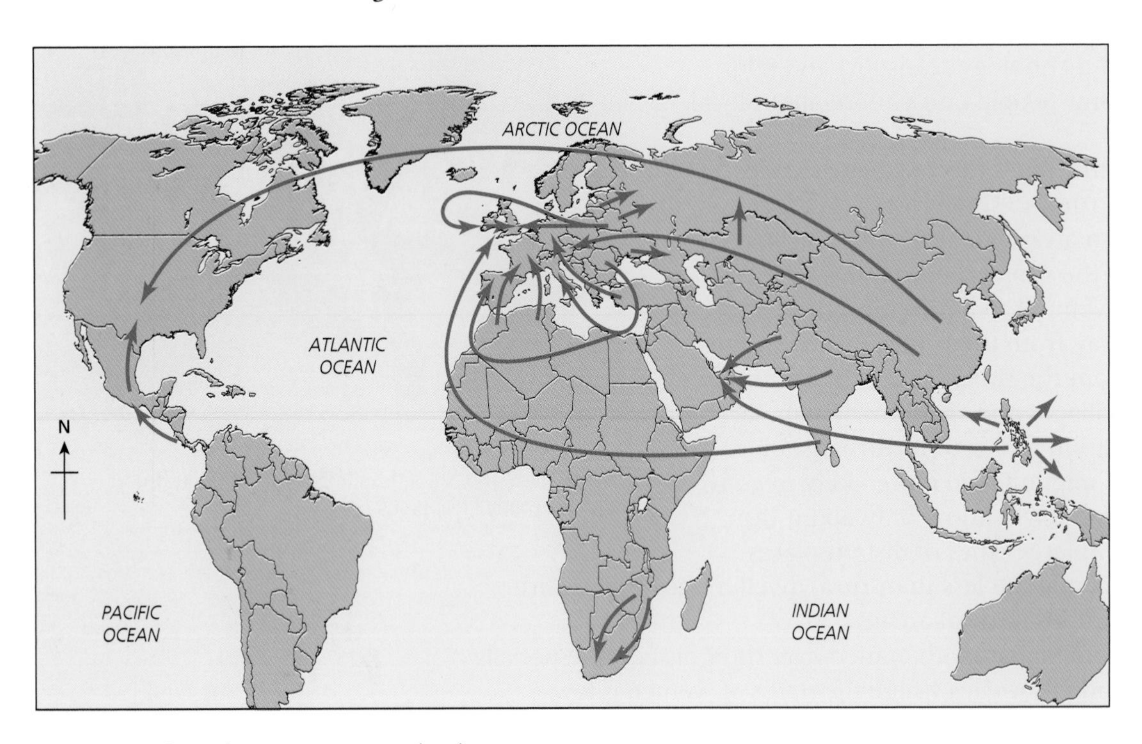

Figure 1.21 Some important current migration routes

The number of international migrants in the world today, both legal and illegal, is thought to total perhaps 200 million. That adds up to only 3% of the world's population, so there is great potential for growth.

The impact of international migration

International migrations can have a range of positive and negative impacts on both the source area and the destination. Some economies could not function without foreign workers. In the United Arab Emirates, for instance, they make up an astonishing 85% of the population. For the moment few other countries rely so heavily on outsiders, but in a number of developed countries, including the UK and the USA,

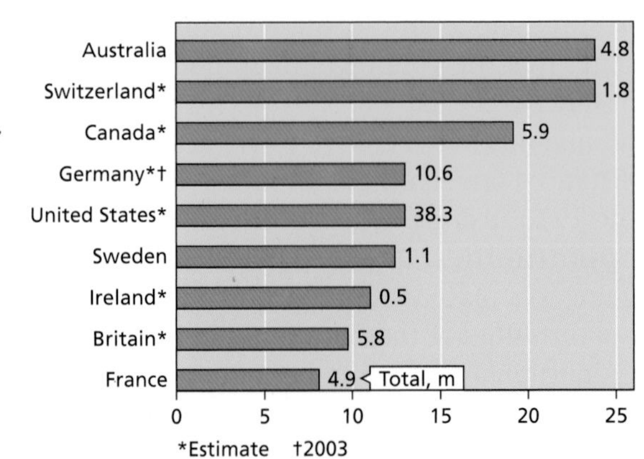

Figure 1.22 Foreign-born population in selected OECD countries, % of total population, 2005

foreigners typically make up 10–15% of the labour force and their share is rising. Migrants fill around half the new jobs created in Britain today, often because they have skills that locals lack (from plumbing to banking) or because natives scorn the work (from picking fruit to caring for the elderly).

Migrants help to create jobs, because a good supply of labour encourages those with capital to invest more. In contrast, countries where migrants have been kept at arm's length, such as Germany, complain about a chronic shortage of skilled workers such as engineers, scientists and programmers.

OECD – Organization of Economic Co-operation and Development, largely composed of developed countries.

The view from poor countries

According to the United Nations Conference on Trade and Development (UNCTAD), many of the world's least developed countries are losing large parts of their already shallow pool of skilled professionals, hindering their ability to pull themselves out of poverty. UNCTAD showed that in 2004 a million educated people emigrated from poor countries out of a total skilled pool of 6.6 million – a loss of 15%. Haiti, Samoa, Gambia and Somalia are among those that have lost more than half their university-educated professionals in recent years. The health sector, in particular, has suffered. In Bangladesh, 65% of all newly graduated doctors seek jobs abroad. However, remittances make up a significant part of those countries foreign earnings.

The problem is heightened by many developed countries such as the US and the UK actively gearing their employment policies to welcome more migrant workers in an attempt to make up for labour shortages. In 2005 between a quarter to a third of all practising doctors in countries such as the UK, the US, Canada and Australia were trained in another country. Whereas sub-Saharan Africa on average has only 13 doctors for 100 000 people, the US level is close to 300. Africa, in particular, suffers from large outflows of labour due to political conflict, unstable economic conditions and low wages.

The "brain drain"

Migrants to rich countries are often better educated than the native population. According to the OECD in 2008 fewer than 20% of locals in OECD countries are university educated compared with almost 25% of foreign-born workers. However, immigrants find it harder to match their skills to a job than locals do. The more educated migrants are, the more likely they are to be overqualified for their work. In Greece, for example, migrants are three times as likely to be overqualified for their job as locals. Immigrants to Spain, Sweden, Italy and Denmark are twice as likely as locals to be overqualified. New Zealand's immigrants tend to be better matched to their jobs than the native population is.

China suffers the worst brain drain in the world. Despite the booming economy and government incentives to return, an increasing number of the country's brightest minds are relocating to wealthier

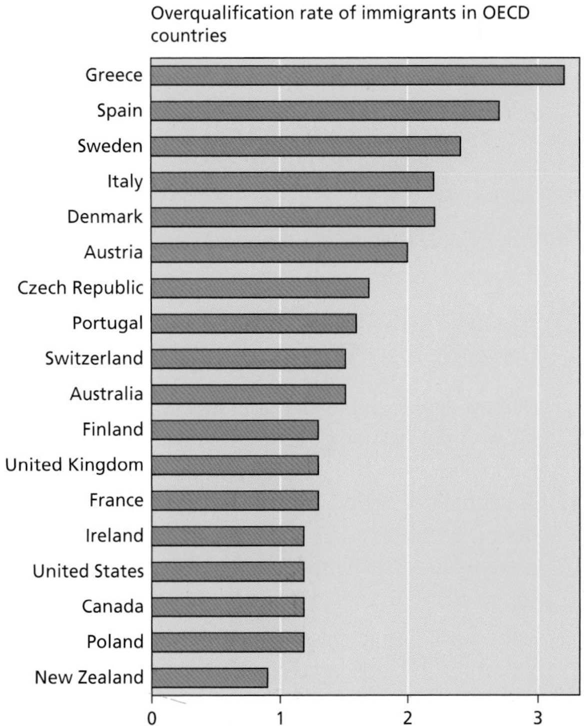

Figure 1.23 The brain drain

nations, where they can usually benefit from higher living standards, better career opportunities and the freedom to have as many children as they wish. Over a million Chinese went to study overseas between 1978 and 2007, but only 275 000 returned. The rest took postgraduate courses, found work, got married or changed citizenship. Unlike illegal migrants from the countryside – many of whom are poorly schooled – the students are usually welcomed by western institutions.

	Benefits		Costs	
	For the individual	*For the country*	*For the individual*	*For the country*
Emigrant countries	1 Increased earning and employment opportunities	1* Increased human capital with return migrants	1 Transport costs	1 Loss of social investment in education
	2* Training (human capital)	2 Foreign exchange for investment via migrant remittances	2 Adjustment costs abroad	2 Loss of "cream" of domestic labour force
	3* Exposure to new culture	3 Increased output per head due to flow of unemployed and underemployed labour	2 Separation from relatives and friends	3* Social tensions due to raised expectations of returning migrants
		4 Reduced pressure on public capital stock		4* Remittances generate inflation by easing pressure on financing public sector deficits
Immigrant countries	1* Cultural exposure	1 Permits growth with lower inflation	1 Greater labour market competition in certain sectors	1* Dependence on foreign labour in particular occupations
		2 Increased labour force mobility and lower unit labour costs		2 Increased demands on the public capital stock
		3 Rise in output per head for indigenous workers		3* Social tension with concentration of migrants in urban areas

*Indicates uncertain effects. (Source: *The Economist*, 15 November 1988)

Table 1.5 Benefits and costs of international migration for individuals and countries

Case study: *Singapore and immigration*

With an ageing population and low fertility, Singapore's government has long courted foreigners to plug gaps in its workforce. In 1990 citizens made up 86% of Singapore's 3 million people. Today, the share is 64% of around 5 million. More than one in three people are foreigners (either permanent residents, known as PRs, or non-residents).

In the past, immigrants were concentrated at the top or bottom of the jobs ladder, performing work that Singaporeans could not or did not want to do. Today, foreigners compete on almost every rung. Some, like geneticists, bring in useful skills. Others – it is feared – displace local skills and depress wages at the bottom. Such fears are especially sharp during a recession. High immigration has coincided with a widening income gap – higher than in China and America. The contrast between the glitzy downtown and the "heartlands" is glaring, and more damaging in tiny, dense Singapore than it would be in a big country.

Migrant workers

The movement of migrant workers – those who migrate to find work – can be:

- permanent or temporary
- long or short distance
- internal or across an international boundary.

Migrant labour has been vital for economic development in many countries, and it remains important today for many developed countries such as the USA, Australia and the UK.

Case study: *The impact of migration on the US economy*

America's economic boom of the 1990s coincided with a large influx of immigrants – more than 13.5 million – accounting for 40% of the USA's population growth. The new migrants accounted for more than 50% of the growth of the labour market. This was, in part, due to the ageing of America's population. The migrants helped fill the gap left by the 4.5 million decline in the number of workers aged between 25 and 34. Some 2.8 million foreign workers in that age group joined the economy during the 1990s. Without them, the labour force in that age group would have declined by 21%. In contrast, relatively few of the new immigrants were elderly, and male workers outnumbered females.

- Although fewer newcomers arrived in the northeast than the west, they accounted for the bulk of the population growth in New York, New Jersey, Connecticut, Massachusetts and Rhode Island.
- Without the new arrivals the northeastern states would have had a much smaller share of the economic boom.
- Up to 9 million arrivals were undocumented workers, living illegally in the USA.

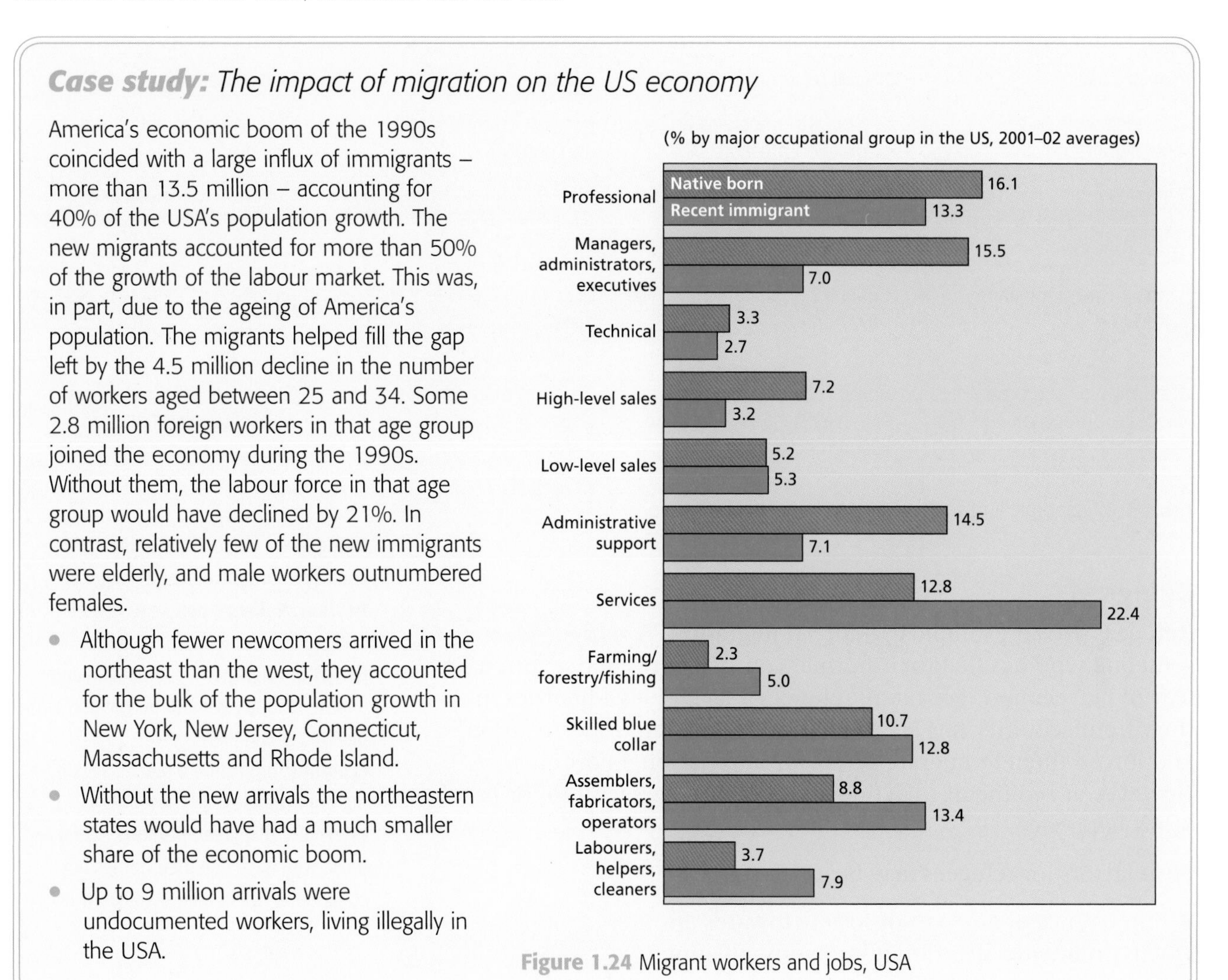

Figure 1.24 Migrant workers and jobs, USA

Trends

The main trends with migrant labour are:

- the globalization of migration
- the acceleration of migration
- the differentiation of migration into different types
- the feminization of migration, e.g. the migration of nurses from the Philippines to the UK.

Source		Destination	
Economic costs	*Economic benefits*	*Economic costs*	*Economic benefits*
Loss of young labour	Reduced un-/under-employment	Costs of educating children	Often fill undesirable posts
Loss of skilled labour slows development	Returning migrants bring back new skills	Displaced local labour	Skills gained at little cost (e.g. doctors to USA)
Out-migration leads to vicious circle of decline	Money sent home (remittances)	Money sent to country of origin; pension outflow	Some retirement costs transferred to source country
Loss of skilled labour deters investment	Less pressure on resources such as land	Increased pressure on resources	Dependence on guest workers
Social costs	*Social benefits*	*Social costs*	*Social benefits*
Creates a culture of out-migration	Lower birth rates and reduced population pressure	Racism, discrimination and conflict	Creation of multicultural societies
Females left as head of household, mother and main provider	Remittances may improve welfare and education	Male-dominated states, e.g. oil-rich economies	Cultural awareness and acceptance
Unbalanced population pyramid	Retiring population may build new homes	Loss of cultural identity, especially among second generation	Providers of local services
Returning on retirement places a burden on services	Some may develop new activities such as recreation, leisure and tourism	Creation of ghettos and ghettoized schools	Growth of ethnic retailing and restaurants

Table 1.6 Economic and social costs and benefits of migration

Refugees

Refugees are an example of a forced migration. A refugee is someone who has fled their normal country of residence, often for fear of persecution. For many refugees a lack of civil liberties in their home country has led to political repression or persecution, and forced them to migrate across an international border. However, it is difficult to define and measure persecution, or prove it has happened.

Push factors for refugees may be:

- intolerance of one part of society towards another
- environmental deterioration (overpopulation)
- state persecution
- natural disasters
- wars.

There are between 15 and 30 million refugees, depending upon the definition. According to the UN High Commission on Refugees (UNHCR) this definition is vital for the distribution of aid. There are over 3 million refugees in Africa. In addition, Africa contains over half of the world's internally displaced persons (IDPs): Sudan alone has over 4 million, while Congo has around 2 million and Somalia at least 1.3 million. Overall, there are about 12 million IDPs in Africa.

Refugee – a person fleeing their home country in order to escape danger.
Asylum seekers – people who seek refugee status in another country.
Illegal immigrants – people who enter another country without permission and plan to remain there.
Economic migrant – a person seeking job opportunities.
Internally displaced persons (IDPs) – those who have fled their homes but continue to live in their own countries.

To research

Visit www.unhcr.org, the home page of the United Nations High Commission for Refugees, and look for its photo gallery for refugees from natural disasters. The site also provides data on global trends for refugees and asylum levels.

Case study: Crisis in Darfur

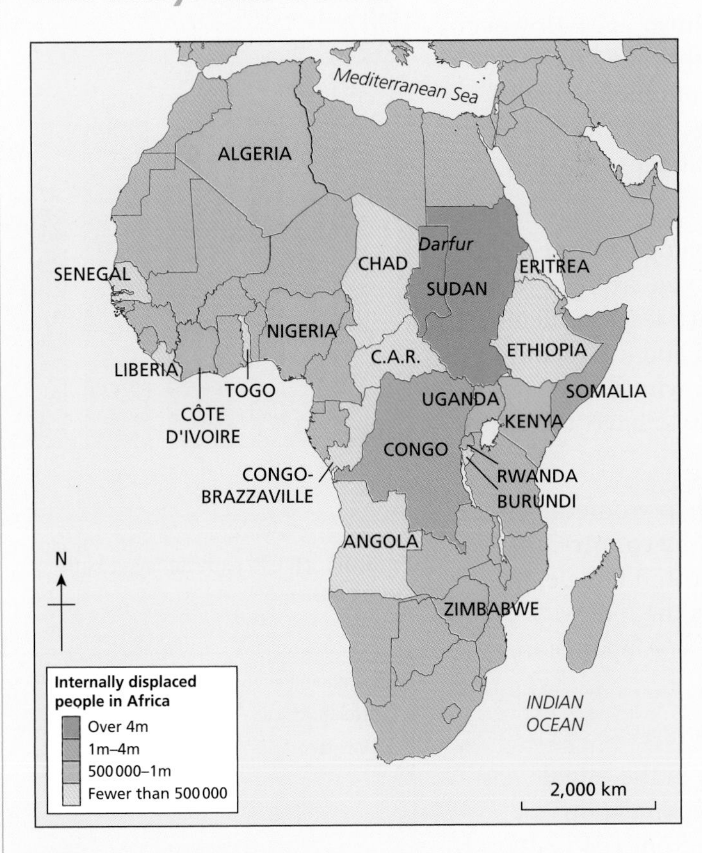

Figure 1.25 Internally displaced people in Africa (location of Darfur)

In August 2009 the UN's outgoing military commander in Sudan said that the six-year conflict in Darfur was effectively over. The conflict started in early 2003 after the Sudan Liberation Army (SLA) and Justice and Equality Movement (Jem) began attacking government targets, accusing Sudan's government of favouring Arabs and discriminating against black Africans. The government responded to the rebellion by mobilizing "self-defence militias" following rebel attacks. However, it denied links to the Arab Janjaweed militia, who were accused of trying to "cleanse" black Africans from large parts of the region.

More than 2.7 million of Darfur's population have fled their homes and now live in camps near Darfur's main towns. However, they are not necessarily safe there, claiming that the Janjaweed patrol outside the camps. Around 200 000 people have sought safety in neighbouring Chad, but many are camped along a 600-kilometre stretch of the border and remain vulnerable to attacks from the Sudan side. Chad's eastern areas have a similar ethnic make-up to Darfur and the violence has spilled over the border, with the neighbours accusing one another of supporting each other's rebel groups. The situation is complex.

Many aid agencies working in Darfur are unable to get access to vast areas because of the insecurity. Several agencies were banned after the International Criminal Court issued an arrest warrant for President Bashir for alleged war crimes. While the UN claims that up to 300 000 have died from the combined effects of war, famine and disease, President Bashir puts the death toll at 10 000. The international community lays much of the blame on President Bashir, accusing him of supporting the pro-government militias.

Gender and change

According to the Population Reference Bureau (PRB), numerous studies have demonstrated the positive impact of girls' education on child and maternal mortality, health, fertility rates, poverty and economic growth. Yet less than two cents of every dollar spent on international development is directed specifically toward adolescent girls, and they remain at the margins of international development programmes.

Early marriage compromises girls' development and often results in early pregnancy and social isolation. Child marriage also reinforces the vicious cycle of low education, high fertility and poverty. Most countries in the Middle East and North Africa have laws on the minimum legal age for marriage, but some families take advantage of religious laws that permit an earlier age.

Gender and life expectancy

Worldwide, men have higher mortality and greater disability than women. In nearly every country they die at younger ages. However, just because women live longer does not mean that they have better health. The female health disadvantage stems from the biological differences between the sexes (women are subject to risks related to pregnancy and childbearing) and gender (the distinct social and cultural roles played by men and women).

Gender bias is most apparent and acute in developing countries where there is a strong cultural preference for sons. Women are disadvantaged from birth in many countries, which leads to lifelong health problems. Girls receive less nutritious food and less medical care, perpetuating a cycle of poor health. Women who are undernourished during pregnancy are more likely to have low birth-weight babies and undernourished children.

While in most countries men live shorter lives than women, gender differences in mortality and life expectancy vary by country. In Russia, for instance, the difference between male and female life expectancy is large, whereas in the United States the male disadvantage is smaller. In some countries, such as Afghanistan, there is little or no male disadvantage.

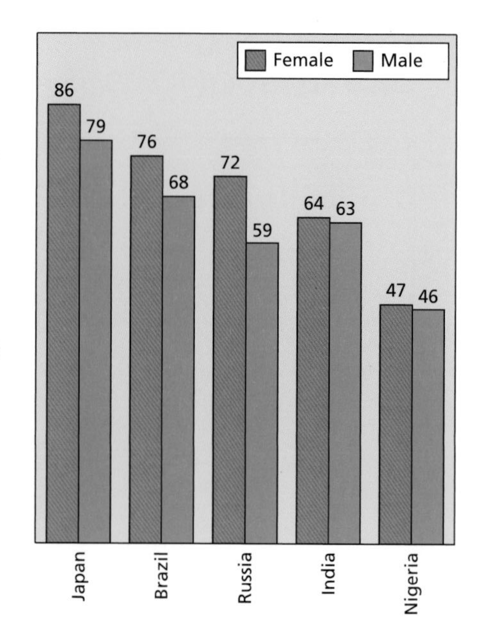

Figure 1.26 Life expectancy at birth by sex, 2007

Changing fertility in the Middle East and North Africa

Some of the world's fastest-growing countries are in the Middle East and North Africa (MENA), and the region saw its population quadruple in the second part of the 20th century, from 104 to 400 million. It added another 32 million by 2007 (see Figure 1.27). However, a revolution in marriage and childbearing in recent decades has slowed growth of the mainly Arab countries of the Middle East and North Africa. While a young population structure ensures momentum for future growth, the pace has slackened thanks to fertility declines in some of the region's largest countries. MENA's total fertility rate (TFR), or average number of children born per woman, declined from about seven children in 1960 to three in 2006.

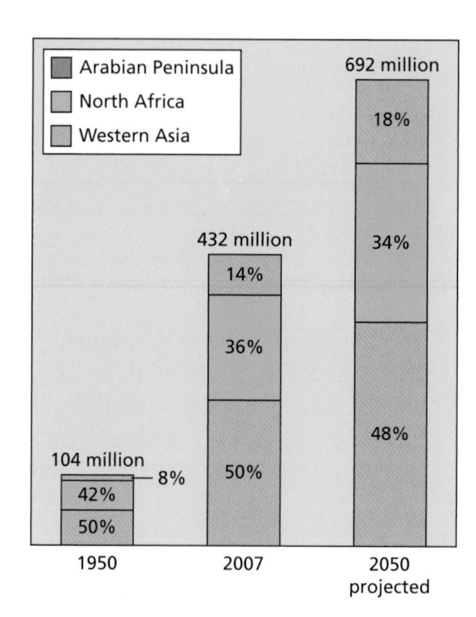

Figure 1.27 Population growth in the MENA regions: 1950, 2007 and 2050

The Arab performance in improving women's health is also unmatched. Female life expectancy is up from 52 years in 1970 to more than 70 in 2004. The number of children borne by the average Arab woman has fallen by half in the past 20 years, to a level scarcely higher than world norms. In Oman, fertility has fallen from 10 births per woman to fewer than four. A main reason for this is a dramatic rise in the age at which girls marry. A generation ago, three-quarters of Arab women were married by the time they were 20. That proportion has dropped by half. In large Arab cities, the high cost of housing, added to the need for women to pursue degrees or start careers, is prompting many to delay marriage until their 30s.

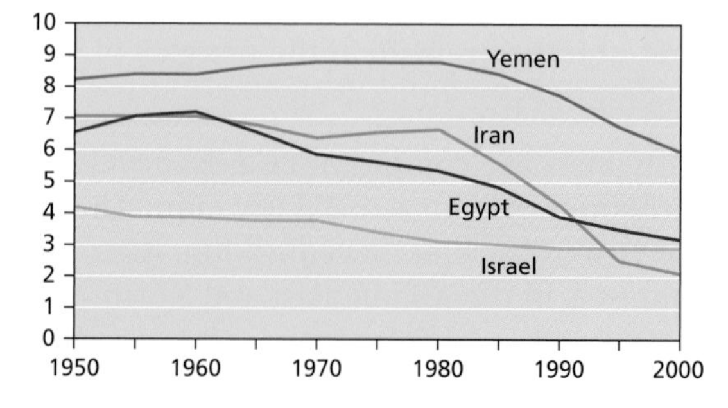

Figure 1.28 Patterns of fertility decline in selected MENA countries: children per woman, 1950–2000

The feminization of migration

There are over 200 million people around the world living in a country other than the one in which they were born. Data collected by governments indicate that women now account for almost half of immigrants around the world, but the proportion of international migrants who are male or female varies (Table 1.7) by country and region.

Although conventional wisdom had it that the vast majority of international migrants were male, recent assessments show that this does not describe global migration during the last half of the 20th century. In some countries more than half of migrants are female. The female proportion is higher in countries that have long been open to immigration, including the USA, Canada and Australia. For example, 54% of recent legal immigrants to the US are women, in part because they are more likely to be migrating spouses, and 70% of all Filipino labour migrants are women. In Mexico, on the other hand, 69% of emigrants are men.

Geographic area	1960	1980	2000
World	46.6	47.4	48.8
More developed countries	47.9	49.4	50.9
Less developed countries	45.7	45.5	45.7
Europe	48.5	48.5	52.4
North America	49.8	52.6	51.0
Oceania	44.4	47.9	50.5
North Africa	49.5	45.8	42.8
Sub-Saharan Africa	40.6	43.8	47.2
Southern Asia	46.3	45.9	44.4
Eastern/ southeastern Asia	46.1	47.0	50.1
Western Asia	45.2	47.2	48.3
Caribbean	45.3	46.5	48.9
Latin America	44.7	48.4	50.5

Table 1.7 Percentage of female migrants among total number of migrants

The exploitation of female migrant workers

There are more than 86 million economically active migrants around the world, two-thirds of whom have moved from developing to developed countries. Migrants from developing countries tend to concentrate at the bottom and the top of the employment ladder.

The International Labour Organization provides significant evidence of the feminization of labour migration, a trend most evident in Asia, where hundreds of thousands of women emigrate each year. The main sending countries are Indonesia, the Philippines, Sri Lanka and Thailand, while the main destinations are Hong Kong (China), Malaysia, Singapore and the Middle East. Many Asian migrants are teachers and nurses, but even more are employed as domestic workers or recruited to work in "sweatshops". Sweatshops have increased because of the globalization of international brands of garments, shoes, toys and sports equipment.

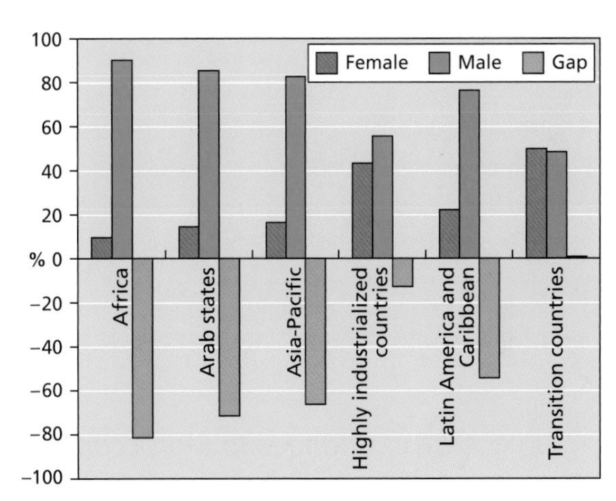

Figure 1.29 Female–male gaps in formal participation in cities

The Populations Reference Bureau's analysis of migrant women and occupation found that in 2007 the most skilled jobs were done by men, for example:

- men accounted for 74% of the science and engineering labour force
- over 90% of electrical, geological, mechanical, and naval engineers were men.

Women and unions

Working women are increasingly becoming unionized. In India, for example, SEWA (the Self-Employed Women's Association) operates as a trade union and as an economic empowerment group. Labour unions have historically been a male preserve, but women are now making up an increasing share of membership in these unions. The involvement of women in paid employment has also led to the politicization of women.

Women and tenure

Comparatively few African countries have legislation in place to assure women's access to land and property. Those that do include Burkina Faso, Malawi, Mozambique, Niger, Rwanda, South Africa,

Tenure – the way in which the rights, restrictions and responsibilities that people have with respect to land (and property) are held.

To do:

a What is Shariah law? In which countries does it apply?

b Identify the impacts that Shariah law has on women's rights.

Tanzania, Uganda and Zimbabwe. Eritrea and Ethiopia lack specific laws, but there are no impediments for women to own land. In Kenya, women have limited access to land. Formal law, traditional legal systems and societal norms, including customary and religious laws, often deny women the right to acquire and inherit property, particularly in countries where Shariah law applies.

Women and family law in the MENA (Middle East – North Africa) region

Despite their impressive gains in education and health, women in the MENA region still face gender discrimination that prevents them from reaching their potential. To varying degrees, discrimination against women is built into the culture, government policies and legal frameworks. In particular, the region's family laws codify discrimination against women and girls, placing them in a subordinate position to men within the family. This position is then replicated in the economy and society.

Region	Discrimination against women owning land	Discrimination against women inheriting land and housing	Discrimination against women taking mortgages in their own name
Africa	41%	70%	31%
Arab states	29%	29%	21%
Asia–Pacific	24%	24%	24%
Highly industrialized countries	11%	0%	0%
Latin America & Caribbean	24%	11%	34%
Transition countries	9%	6%	8%
World	24%	26%	24%

Table 1.8 Impediments to women's tenure rights 5%

Case study: *Changes to Moroccan family law*

The Moroccan *mudawana*, or family code, was drafted in 1957, based mainly on the Maliki School of Islamic jurisprudence. Changes introduced in 1993 did improve the position of women, but they still remained subordinate to men. As Moroccan civil society became increasingly organized and more women's associations were formed, a movement began to raise awareness about women's rights.

The appointment in 1998 of a progressive prime minister and a minister for women and family affairs who was committed to women's rights led to the formulation of the National Action Plan for the Integration of Women in Development, which called for reforming the *mudawana*. In the 2002 elections, 35 women entered the Moroccan parliament, assisted by a new quota system adopted by the political parties. The campaign to reform the *mudawana* was the work of more than a decade, and it was finally approved in 2004.

Some of the main features of the new Moroccan family law are:

- Husband and wife share joint responsibility for the family.
- The wife is no longer legally obliged to obey her husband.
- The right to divorce is a prerogative of both men and women, exercised under judicial supervision.
- The woman has the right to impose a condition in the marriage contract requiring that her husband refrain from taking other wives.
- In the case of divorce, the woman is given the possibility of retaining custody of her child even upon remarrying or moving out of the area where her ex-husband lives.
- For both men and women, the minimum legal age of marriage is 18 years.

Country	Non-discrimination and access to justice	Autonomy, security and freedom of the person	Economic rights and equal opportunities	Political rights and civic voice	Social and cultural rights
Algeria	3.0	2.4	2.8	3.0	2.9
Bahrain	2.2	2.3	2.9	2.1	2.8
Egypt	3.0	2.8	2.8	2.7	2.4
Iraq	2.7	2.6	2.8	2.2	2.1
Jordan	2.4	2.4	2.8	2.8	2.5
Kuwait	1.9	2.2	2.9	1.4	2.8
Lebanon	2.8	2.9	2.8	2.9	2.9
Libya	2.3	2.1	2.3	1.2	1.8
Morocco*	3.2	3.2	3.1	3.0	3.0
Oman	2,0	2.1	2.7	1.2	2.1
Palestine	2.6	2.7	2.8	2.6	2.9
Qatar	2.0	2.1	2.8	1.7	2.5
Saudi Arabia	1.2	1.1	1.4	1.0	1.6
Syria	2.7	2.2	2.8	2.2	2.3
Tunisia	3.6	3.4	3.1	2.8	3.3
UAE	1.7	2.1	2.8	1.2	2.3
Yemen	2.4	2.3	2.3	2.6	2.1

* This study covers development up to the end of 2003. It does not take into account the new developments in Moroccan family law reforms.

Table 1.9 Women's rights in MENA (5 is the highest rank, 1 the lowest)

To research

Visit the Population Reference Bureau www.prb.org for the 2009 World Population Data sheet, and see how life expectancy varies for the countries of your choice. The site also gives the world population highlights from the data sheet. Choose two contrasting countries, such as one in Africa or Asia and one in Europe or North America. Research the issue of gender inequality in these countries and comment on the findings you have made. Click on the link to the presentation to find out about global population issues and the problems for young women in Peru, Zambia and Mali.

The following website refers to data for the Netherlands. See section K for links to gender inequalities: http://hdrstats.undp.org

To do:

a What is the difference in life expectancy between men and women in (i) Russia, (ii) Japan, (iii) Brazil and (iv) India?

b In which region is the proportion of women migrants (i) the highest and (ii) the lowest?

c How has the proportion of women migrants changed since 1960 and 2000?

d Which region of MENA provides most rights for women? Which provides fewest rights for women? Comment on the information in Table 1.9 (Women's rights in MENA).

2 Disparities in wealth and development

By the end of this chapter you should be able to:

- understand the different ways of measuring regional and global inequalities
- explain the origin of disparities and how these are being addressed
- describe the purpose and achievements of the millennium development goals
- explain trends in life expectancy, education and income
- examine the impact of aid and debt relief to poorer countries.

Measurement of regional and global inequalities

The Human Development Index

The United Nations (UN) has encouraged the use of the Human Development Index (HDI) as a measure of development, as it is more reliable than single indicators such as gross national income (GNI) per head. (GNI was previously known as gross national product – GNP.) For 2009 Norway had the highest HDI score at 0.971, followed by Australia (0.970), Iceland (0.969), Canada and Ireland (both above 0.965). The lowest values were for Niger (0.340), Afghanistan (0.352) and Sierra Leone (0.365).

The Human Development Index (HDI) – a composite measure of development. It includes three basic components of human development:

1. longevity (life expectancy)
2. adult literacy and average number of years' schooling
3. standard of living – income adjusted to local cost of living, i.e. purchasing power.

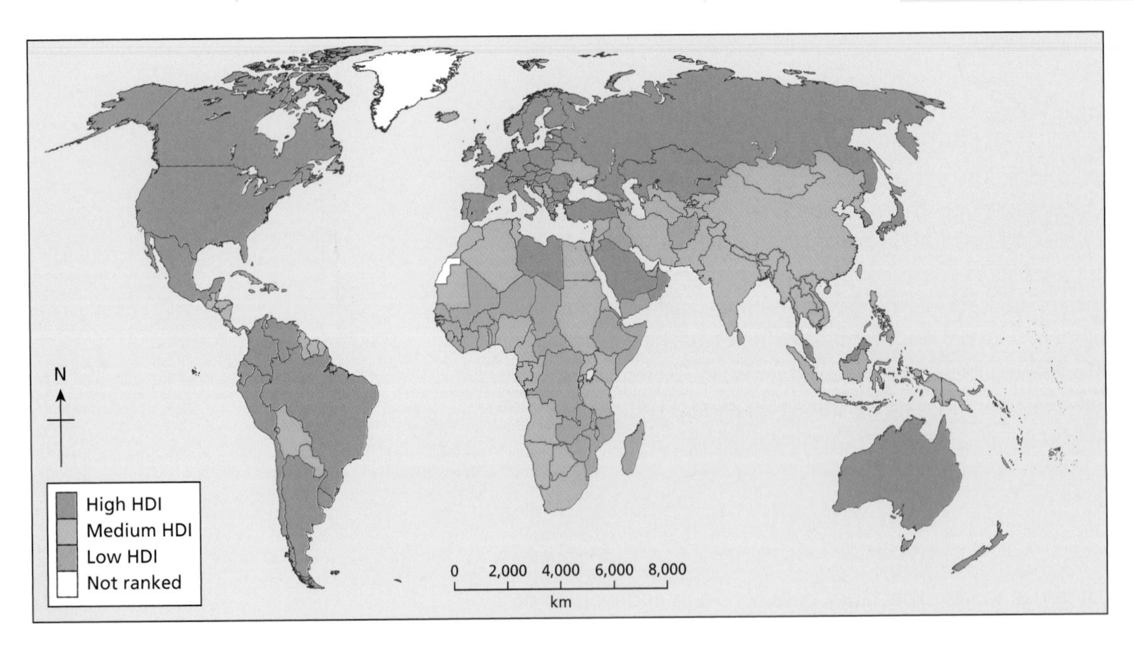

Figure 2.1 Global HDIs for 2009 (based on 2007 data)

The HDI can highlight the successes of some countries. For example, in 1975 Venezuela had a higher HDI score than Brazil, but Brazil has made much faster progress, and Finland had a lower HDI score than Switzerland in 1975, but is similar today.

High levels of human development can be achieved without high incomes, and high incomes do not guarantee high levels of human development. Pakistan and Vietnam have similar incomes, but Vietnam has a higher HDI score. Similarly, Jamaica has an income comparable to Morocco's but a much better HDI score.

Most regions have seen steady improvement in HDI over the past 15 years, with East Asia and the Pacific performing particularly well. Arab states have also seen substantial growth, exceeding the average increase for developing countries. Sub-Saharan Africa, by contrast, has been almost stagnant. Two groups of countries have suffered setbacks:

- CIS countries (the former Soviet republics) going through a long, painful transition to market economies
- poor African countries whose development has been hindered or reversed for a variety of reasons, including HIV/AIDS and internal and external conflicts.

Comparing the ranking of developing counties by their HDI and gross national income shows some interesting patterns. Many oil-producing countries, for example, have much lower HDI rankings than their GNI rank, while some poor countries rank relatively high in the HDI because they have deliberately devoted scarce resources to human development. Cuba, Costa Rica, Vietnam, and Sri Lanka fall into this category.

To research

Visit http://hdr.undp.org to see how the Human Development Index has changed over time. The site gives HDI data for each country, organized into high, medium and low HDI.

See also Gapminder at www.gapminder.org for national and international variations in levels of development. (Figure 2.5, showing global variations in the proportion of populations undernourished, is from this site.)

CIS – Commonwealth of Independent States, made up of the former Soviet republics.

Case study: Racial inequality in the USA

Over the past 25 years the existing inequalities between black and white people in the US have become even greater. A typical white family is now five times richer than its African-American counterpart of the same class. White families typically have assets worth $100 000 up from $22 000 in the mid-1980s. African-American families' assets stand at just $5,000, up from around $2,000. A quarter of black families have no assets. The richest 1% of the population owns 40% of its wealth.

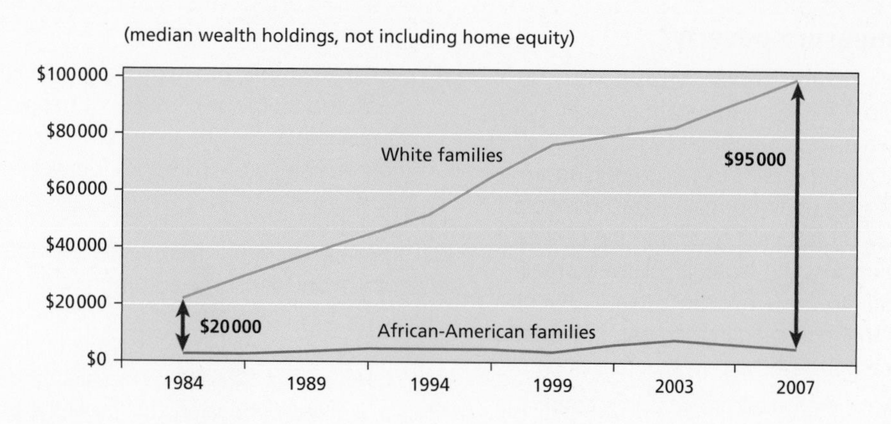

Figure 2.2 Inequalities in the USA

Other measures of development

Infant mortality rate (IMR)

The IMR is widely used as an indicator of development for a number of reasons:

- High IMRs are found in the poorest countries.
- Most of the causes of death in areas with a high IMR are often preventable.
- Where water supply, sanitation, housing, nutrition and basic (primary) healthcare are adequate, IMRs are low.

Infant mortality rate (IMR) – the number of children in a country who die before their first birthday per 1,000 live births.

Gender-related development index (GDI)

The GDI measures achievements in the same dimensions and using the same indicators as the HDI, but examines inequalities between women and men. It is simply the HDI adjusted for gender inequality.

Gender empowerment measure (GEM)

The GEM reveals whether women can take an active part in economic and political life. It exposes inequality in opportunities in selected areas. It focuses on participation, measuring gender inequality in key areas of economic and political decision-making. It tracks the percentages of women in parliament, among legislators, and professional and technical workers, and the gender disparity in earned income, reflecting economic independence. Norway is ranked first in the GEM (0.837) and Yemen 70th (0.127).

The Education Index

The Human Development Report 2009 provides an index of education. This varies from a minimum of 0.0 to a maximum of 1.0. The Education Index is based on the adult literacy rate and the combined enrolment for primary, secondary and tertiary schools. Adult literacy is defined as the proportion of the adult population aged 15 and over that can read and write a short, simple statement on their everyday life.

The Education Index varies from a high of 0.993 for countries such as Finland, Australia, New Zealand and Cuba to 0.354 for Afghanistan, 0.334 for Chad, 0.331 for Mali, 0.301 for Burkina Faso and 0.282 for

To research

Visit www.cia.gov and go to the World Factbook, where the guide to country comparisons lists IMRs ranked from highest to lowest. How does this compare with the list of countries arranged by HDI?

Figure 2.3 Child graves in South Africa

TOK Link

How can we measure poverty?

Many of the world's poorest people live in poverty so acute that it is difficult for those fortunate enough to live in developed countries to comprehend what it means to be poor. The UN Development Programme's **Human Poverty Index** is based on three main measurements (Table 2.1). The UNDP has calculated that the cost of eradicating poverty across the world is relatively small compared to global income – not more than 0.3% of world GDP – and that political commitment, not financial resources, is the real obstacle to poverty eradication.

While the HDI measures overall progress in a country in achieving human development, the Human Poverty Index (HPI) reflects the distribution of progress and measures the backlog of deprivation that still exists. The HPI measures deprivation in the same dimensions of basic human development as the HDI. It is divided into two indices, HPI-1 and HPI-2.

HPI-1

The HPI-1 measures poverty in developing countries. It focuses on deprivations in three dimensions:

- the probability at birth of not surviving to age 40
- knowledge, as measured by the adult illiteracy rate
- overall economic provisioning, public and private, as measured by the percentage of people without sustainable access to an improved water source and the percentage of children underweight for their age.

HPI-2

Because human deprivation varies with the social and economic conditions of a community, a separate index, the HPI-2, has been devised to measure human poverty in selected OECD countries, drawing on the greater availability of data. The HPI-2 focuses on deprivation in the same three dimensions as the HPI-1 and one additional one, social exclusion. The indicators are:

- the probability at birth of not surviving to age 60
- the adult functional illiteracy rate
- the percentage of people living below the income poverty line
- the long-term unemployment rate (12 months or more).

Index	Longevity	Knowledge	Decent standard of living	Participation or exclusion
HDI	Life expectancy at birth	1 Adult literacy rate 2 Combined enrolment ratio	GDP per capita (PPP US$)	-
HPI-1	Probability at birth of not surviving to age 40	Adult illiteracy rate	Deprivation in economic provisioning, measured by: 1 Percentage of people without sustainable access to an improved water source 2 Percentage of children under five underweight for age	-
HPI-2	Probability at birth of not surviving to age 60	Percentage of adults lacking functional literacy skills	Percentage of people living below the income poverty line (50% of median-adjusted disposable household income)	Long-term unemployment rate
GDI	Female and male life expectancy at birth	1 Female and male adult literacy rates 2 Female and male combined primary, secondary and tertiary enrolment ratios	Estimated female and male earned income, reflecting women's and men's command over resources	-

Table 2.1 HDI, HPI-1, HPI-2, GDI: same components, different measurements

Niger. The prevalence of sub-Saharan African countries in having a low index is noticeable. Likewise, the prevalence of rich countries in having a high index is clear. Interestingly, Cuba scores highly. This shows that a country that invests in its education system may have a high index without necessarily being rich.

Figure 2.4 Relative poverty: selling the *Big Issue*

Nutrition: global variations in hunger

Overall, there are 815 million hungry people living in developing countries. Up to 10 million people die every year of hunger and hunger-related diseases. Of the total number of undernourished people:

- 221.1 million live in India
- 203.5 million live in sub-Saharan Africa
- 142.1 million live in China.

Three-quarters of all hungry people live in rural areas. Overwhelmingly dependent on agriculture for their food, these populations have limited alternative sources of income or employment and, as a result, are particularly vulnerable to crises.

The Food and Agriculture Organization (FAO) calculates that of the 815 million hungry:

- Half are farming families, surviving off marginal lands prone to natural disasters like drought or flood.
- One in five belong to landless families dependent on farming.
- About 10% live in communities whose livelihoods depend on herding, fishing or forest resources.
- The remaining 25% live in shanty towns on the periphery of the biggest cities in developing countries.

To research

Visit http://hdr.undp.org/en and go to the statistics section. Click on interactive features and "Human development world maps" to see a map of the world Human Poverty Index.

Statistics here include the world GDI for 2009 (based on 2007 figures). This site also shows global variations in the GEM, and data for the 2009 HDR Education Index.

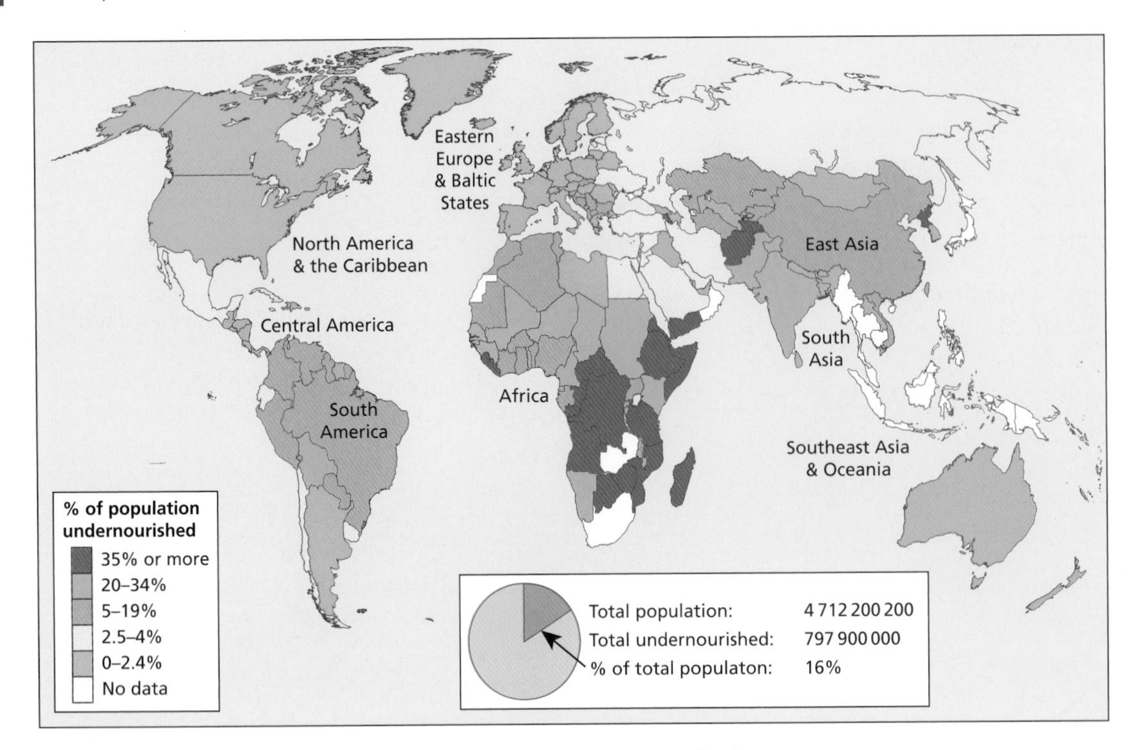

Figure 2.5 Global variations in proportion of population undernourished

The numbers of poor and hungry city dwellers are rising rapidly along with the world's total urban population.

Child hunger

Hunger still claims lives and scars the lives of those who survive it, especially young children (Figure 2.7). Malnutrition contributes to 53% of the 10.6 million deaths of children under five each year in developing countries. This amounts to one child dying every five seconds. An estimated 167 million children under five years of age in the world are underweight. This means that one in five of all hungry people are children aged less than five.

Figure 2.6 Absolute poverty in Peelton, South Africa

All too often, child hunger is inherited: up to 17 million children are born underweight annually, the result of inadequate nutrition before and during pregnancy. Undernourished infants lose their curiosity, motivation and even the will to play. Millions leave school prematurely. Chronic hunger also delays or stops the physical and mental growth of children. According to the FAO, every year that hunger continues at present levels costs five million children their lives. Economists estimate that hunger is responsible for reducing the GNP of some developing countries by 2–4%.

Women and hunger

Women are the world's primary food producers, yet cultural traditions and social structures often mean that women are much more affected by hunger and poverty than men. Seven out of 10 of the world's hungry are women and girls. Maternal stunting and underweight are also among the most prevalent causes of giving birth to a low birth-weight child.

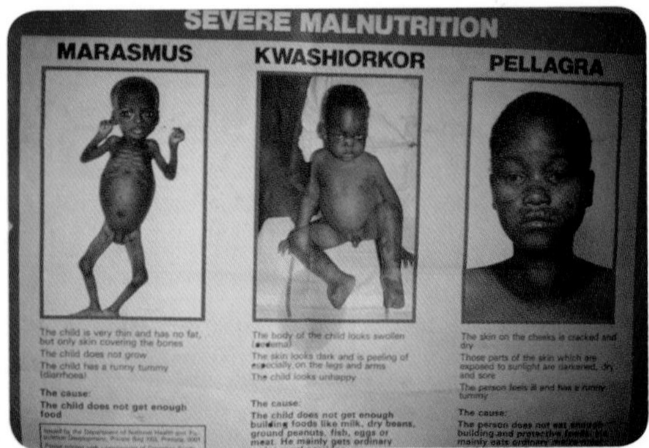

Figure 2.7 Types of malnutrition

The origin of disparities

Inequalities in development

Although some parts of the world have experienced rapid growth and improvement in living standards in recent years, poverty remains entrenched in other areas. Despite considerable economic growth in many regions, the world is more unequal than it was a decade ago.

Some formerly poor countries are now relatively well off. NICs such as South Korea and Taiwan have quite high levels of GNP per capita. The development of the original Asian "tiger economies" is the result of a combination of state-led industrialization, spontaneous industrialization and industrialization led by transnational corporations (TNCs). However, many countries remain locked in poverty, especially those in sub-Saharan Africa.

LEDCs – less economically developed countries
NIC – newly industrializing country.

Living conditions

More than two-thirds of the world's urban population is now in Africa, Asia, Latin America and the Caribbean. The population in urban areas in developing countries is expected to grow from 1.9 billion in 2000 to 3.9 billion in 2030. In the developed world, however, the urban population is expected to increase very slowly, from 0.9 billion in 2000 to 1 billion in 2030.

The global population growth rate for that period is 1%, while the growth rate for urban areas is nearly twice as fast, at 1.8%. At that rate, the world's urban population will double in 38 years. Growth will be even more rapid in the urban areas of the developing world averaging 2.3% per year, with a doubling time of 30 years.

Efforts to improve the living conditions of slum dwellers (especially within poor countries) peaked during the 1980s. However, renewed concern about poverty has recently led governments to adopt a specific target on slums in the United Nations Millennium Declaration, which aims to improve significantly the lives of at least 100 million slum dwellers by the year 2020.

What is meant by malnutrition?

Starvation: limited or non-existent food intake.

Deficiency diseases: lack of specific vitamins or minerals

Kwashiorkor: a lack of protein

Marasmus: a lack of calories/energy food

Obesity: too much energy and/or protein foods

Malnutrition means poor (bad) nutrition. It can be over- or undernutrition.

In 2001 the total number of slum dwellers in the world stood at about 924 million people. This represents about 32% of the world's total urban population. At that time, 78.2% of the urban population in LEDCs were slum dwellers. In some LEDC cities slums are so pervasive that it is the rich who have to segregate themselves behind small gated enclaves.

There are negative and positive aspects of slums. On the negative side, slums have high concentrations of poverty and social and economic deprivation, which may include broken families, unemployment and economic, physical and social exclusion. Slum dwellers have limited access to credit and formal job markets due to stigmatization, discrimination and geographic isolation. Slums are often recipients of the city's nuisances, including industrial effluent and noxious waste. People in slum areas suffer inordinately from waterborne diseases such as typhoid and cholera, as well as more

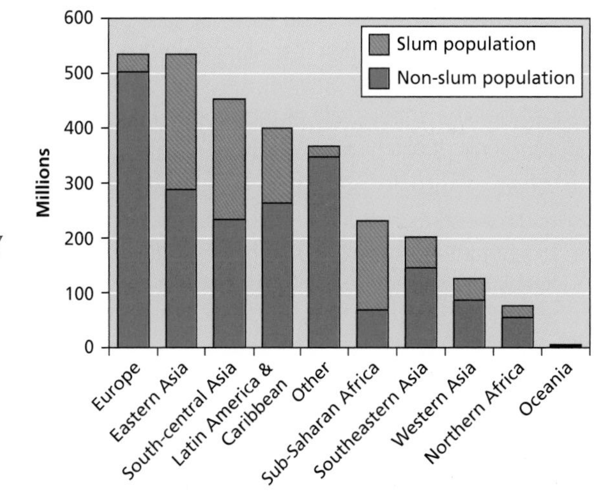

Figure 2.8 Proportion of slum dwellers in urban population by region

opportunistic ones that accompany HIV/AIDS. Slums have the most intolerable of urban housing conditions (Figure 2.9), which frequently include:

- insecurity of tenure
- lack of basic services, especially water and sanitation
- inadequate and sometimes unsafe buildings
- overcrowding
- location on fragile, dangerous or polluted land – land that no one else wants.

Slum women – and the children they support – are the greatest victims of all. Slum areas are also commonly believed to be places with a high incidence of crime, although this is not universally true since slums with strong social control systems will often have low crime rates.

Figure 2.9 Poor-quality housing, Grahamstown, South Africa

On the positive side, slums are the first stopping point for immigrants – they provide the low-cost and only affordable housing that will enable the immigrants to save for their eventual absorption into urban society. They are also the place of residence for low-income employees, and keep the wheels of the city turning in many different ways. Many informal entrepreneurs operating from slums have clienteles extending to the rest of the city. Most slum dwellers are people struggling to make an honest living, within the context of extensive urban poverty and formal unemployment. Slums are also places in which the vibrant mixing of different cultures frequently results in new forms of artistic expression.

Case study: *Race and ethnicity: the case of Australia's aborigines*

Belonging to a racial or ethnic minority can disadvantage a person. For example, Ninga Mia is an Aboriginal shanty town in the shadow of one of the world's biggest gold mines. A third of the houses lack bathrooms and toilets, and even those with these basic facilities are overcrowded and often insanitary. A house the size of a large bedsit typically provides a home for seven or more people. The sheer scale of the problem often seems to confound public health experts.

Whereas Australia as a whole ranks between Sweden and Iceland as ninth in the world for life expectancy, indigenous Australians come between the Cambodians and Sudanese at 178th. Those born Aboriginal can expect to:

- die 20 years earlier than their non-indigenous compatriots
- have an IMR nearly four times higher than that of the general population
- be seven times more likely to catch meningitis
- be 10 times more likely to catch tuberculosis.

An Aboriginal child is twice as likely to be born with a low birth weight and five times more likely to have a mother younger than 17. Poor housing exacerbates other problems: dusty and unhygienic surroundings mean that deaths from respiratory disease are four times more common among indigenous Australians; and deaths from heart disease and strokes are three times more common. Aborigines are 45 times more likely to become victims of domestic violence and 40% more likely to commit suicide.

Poverty, racism and dispossession are at the heart of the problem. The tragedy of Aboriginal ill health is that its causes are well understood and could be prevented without the need for dramatic technological breakthroughs.

(Adapted from *The Guardian*, 29 July 2004)

Land ownership (tenure) in South Africa and Cambodia

The decline of black subsistence agriculture has traditionally been put down to the shortage of land relative to the growing population and the increasing poverty of that population. Shortage of land led to overcrowding, overgrazing, use of poor land, soil erosion, denudation, and, ultimately, declining yields. In some places such as South Africa, the availability of land per person is falling (Table 2.2), partly as a result of population growth and partly as a result of land seizures in the 19th century (Figure 2.10).

	1970	1980	1990	2000	2020
Cultivated land per person (ha)	0.6	0.5	0.4	0.3	0.2*
Other land, e.g. pastoral, per person (ha)	5.5	4.2	3.2	2.4	1.5*

*Estimated

Table 2.2 Changes in the amount of land per person in South Africa, 1970–2020

To research

Visit COHRE at www.cohre.org to find out about housing and evictions in Cambodia.

Prior to c.1820 | c.1820–70 | c.1870–90 | c.1890 to present

Pre-colonial pre-capitalist period

Subsistence production

Small commercial farmers

Subsistence production plus a surplus for sale

Subsistence production

Subsistence-level farmers forced out as migrant labourers

Landless labour force formed through, for example, land expropriations and taxation

Figure 2.10 The causes of rural underdevelopment in rural communities in South Africa

Another example comes from Cambodia, where the Centre on Housing Rights and Evictions (COHRE) reports that many vulnerable (poor) communities face land tenure insecurity and forced displacement. In contrast, the rich have little difficulty in acquiring land titles in high-value areas in which poor communities live. These are often acquired through "unofficial" fees.

Land tenure insecurity is a major problem in shanty towns. The definition of a shanty town – an illegal settlement on land not owned by the householder – makes their inhabitants vulnerable to resettlement. Some residents of Dharavi, Mumbai, for example, have seen their homes knocked down to make way for new office developments.

Parental education and inequality

The link between investment in education and poverty is one of the most important dimensions of policies towards poverty. Education may affect poverty in two ways: it may raise the incomes of those with education, and also, by promoting growth in the economy, raise the incomes of those with given levels of education. Those with higher education tend to have fewer children.

Employment inequalities

Though some parts of the world have experienced unprecedented growth and improvement in living standards in recent years, poverty remains entrenched, and much of the world is trapped in an inequality predicament.

Formal sector	Informal sector
Large scale	Small scale
Modern	Traditional
Corporate ownership	Family/individual ownership
Capital intensive	Labour-intensive
Profit-oriented	Subsistence-oriented
Imported technology/inputs	Indigenous technology/inputs
Protected market (e.g. tariffs, quotas)	Unregulated/competitive markets
Difficult entry	Ease of entry
Formally acquired skills (e.g. school/college education)	Informally acquired skills (e.g. in home or craft apprenticeship)
Majority of workers protected by labour legislation and covered by social security	Minority of workers protected by labour legislation and covered by social security

Table 2.3 Common characteristics of the formal and informal employment sectors

Be a critical thinker

Working with rubbish: a hazardous occupation

Many people in developing countries make a living collecting, sorting, recycling and selling materials recovered from waste dumps. In some of the world's larger cities, thousands of people live and work in municipal dumps – 20 000 in Calcutta, 12 000 in Manila and 15 000 in Mexico City.

According to the World Bank, 1% of the global urban population, many of them women and children, earn a living from waste collection and/or recycling (Figure 2.11). In the least developed countries, up to 2% of the urban population make their living in this way.

Waste pickers tend to have low social status and face public scorn, harassment and sometimes violence. Waste pickers are also vulnerable to exploitation by the middlemen who buy their recovered material. In some cities of Colombia, India and Mexico, waste pickers can receive as little as 5% of the prices that industry pays for the recyclables, with the rest going to middlemen.

On account of their low earnings, waste pickers tend to live in deplorable conditions, lacking water, sanitation and other basic infrastructure. Their poor working and living conditions also make them vulnerable to health and safety risks, including exposure to dangerous waste, and various illnesses and disease. Not surprisingly, life expectancy rates are low in waste-picking communities. In Mexico City, for example, dumpsite waste collectors live an average of 39 years, compared with an average of 69 years for the general population.

(Source of data: www.wiego.org)

According to a survey of OECD countries, the employment disadvantage faced by the less qualified part of the labour force is mainly related to educational qualifications. The survey found that differences in employment rates for the most- and least-educated quartiles varied substantially within Europe, but are not on average higher than those in the USA. In developing countries there is a marked difference between those who work in the formal sector (government officials, teachers, health workers) and those who work in the informal sector; the latter have no job security and generally receive small wages (Table 2.3). Many people work in both sectors.

Figure 2.11 Woman picking garbage in Zwelitsha, South Africa

Analysing disparities and change

Two theories stress how development issues in poor countries are closely linked to what has happened in rich countries.

Dependency theory

According to the **dependency theory**, countries become more dependent upon more powerful, frequently colonial powers, as a result of interaction and "development". As the more powerful country exploits the resources of its weaker colony, the colony becomes dependent upon the stronger power. Goods flow from the colony to support consumers in the overseas country.

André Frank (1971) described the effect of capitalist development on many countries as "the development of underdevelopment". The problem of poor countries is not that they lack the resources, technical know-how, modern institutions or cultural developments that lead to development, but that they are being exploited by capitalist countries.

The dependency theory is a very different approach from most models of development:

- It incorporates politics and economics in its explanation.
- It takes into account the historical processes of how underdevelopment came about, that is, how capitalist

To do:

- **a** Comment on the information about slum dwellers as shown in Figure 2.8.
- **b** Describe the main characteristics of the housing shown in Figure 2.9.
- **c** Describe the changes in land availability per person in South Africa, 1970–2020.
- **d** Briefly explain how rural areas in South Africa became underdeveloped over time.
- **e** Outline the main differences between the formal and informal sectors.

development began in one part of the world and then expanded into other areas (imperial expansion).
- It sees development as a revolutionary break, a clash of interests between ruling classes (bourgeoisie) and the working classes (proletariat).
- It stresses that to be developed is to be self-reliant and to control national resources.
- It believes that modernization does not necessarily mean westernization, and that underdeveloped countries must set goals of their own, appropriate to their own resources, needs and values.

World systems analysis

World systems analysis is identified with Immanuel Wallerstein (1974), and is a way of looking at economic, social and political development. It treats the whole world as a single unit. Any analysis of development must be seen as part of the overall capitalist world economy, and not approached country by country. Wallerstein argued that looking at individual countries in isolation was too simplistic and suffered from **developmentalism**. The developmentalism school assumes that:

- each country is economically and politically free (autonomous)
- all countries follow the same route to development.

As such, developmentalism was **ethnocentric**, believing that what happened in North America and Europe was best and would automatically happen elsewhere.

According to Wallerstein, the capitalist world system has three main characteristics:

- a global market
- many countries, which allow political and economic competition
- three tiers of countries.

The tiers are defined as **core**, largely rich countries (MEDCs,) the **periphery**, which can be identified with poor countries (LEDCs), and the **semi-periphery** (Figure 2.12). The semi-periphery is a political label, referring to countries undergoing class struggles and social change, such as Latin America in the 1980s and eastern Europe in the late 1980s and early 1990s.

Wallerstein argued that capitalist development led to cycles of growth and stagnation. One of these cycles is a long-term economic cycle known as a Kondratieff cycle. This identifies cycles of depression at roughly 50- to 60-year intervals. The last two were during the 1920s and 1930s and during the late 1980s. Stagnation is important for the restructuring of the world system and allows the semi-periphery to become involved in the development process.

Core and periphery – the concept of a developed core surrounded by an undeveloped periphery. The concept can be applied at various scales.
Gross national income (GNI) – the total value of goods and services produced within a country, together with the balance of income and payments from or to other countries (now used in preference to gross national product (GNP)).
MEDCs – more economically developed countries

Trends in global inequalities

Income inequalities

According to the International Labour Organization (ILO), despite strong economic growth that has produced millions of new jobs since

Figure 2.12 World systems: global core and periphery

the early 1990s, income inequality grew dramatically in most regions of the world and is expected to increase because of the global financial crisis of 2008 onwards.

According to *World of Work Report 2008: Income inequalities in the age of financial globalization,* a major share of the cost of the financial and economic crisis will be borne by hundreds of millions of people who have not shared in the benefits of recent growth. It shows that the gap between richer and poorer households has widened since the 1990s. This reflects the impact of financial globalization and the weaker ability of domestic policies to enhance the income position of the middle class and low-income groups.

The report notes that while a certain degree of income inequality is useful in rewarding effort, talent and innovation, huge differences can be counter-productive and damaging for most economies.

The *World of Work Report* is the most comprehensive study to date of global income inequalities, and examines wages and growth in more than 70 developed and developing countries. Among its other conclusions, the report says:

- Employment growth has occurred alongside a redistribution of income away from labour. The largest decline in the share of wages in GDP took place in Latin America and the Caribbean (-13 percentage points), followed by Asia and the Pacific (-10 percentage points) and the advanced economies (-9 percentage points).
- Between 1990 and 2005 approximately two-thirds of the countries experienced an increase in income inequality. The income gap between the top and bottom 10% of wage earners increased in 70% of the countries for which data are available.

To research

Look up the *World of Work Report 2009, The Global Job Crisis and Beyond*. Research the global job crisis and ways forward.

- The gap between high and low incomes is also widening at an increasing pace. For example, in the United States in 2007 the chief executive officers (CEOs) of the 15 largest companies each earned 520 times more than the average worker. This is up from 360 times more in 2003. Similar patterns have been registered in Australia, Germany, Hong Kong (China), the Netherlands and developing countries in Africa.

The report also added that excessive income inequalities could be associated with higher crime rates, lower life expectancy and, in the case of poor countries, malnutrition and an increased likelihood of children being taken out of school in order to work.

Gini coefficients

Income inequality is usually measured by a country's Gini coefficient, in which 0 is perfect equality and 1 is perfect inequality. China's Gini coefficient rose from 0.41 in 1993 to 0.47 in 2004, the highest in Asia after Nepal (Figure 2.13). Gini coefficients in Latin America are based on income; those in Asia are mainly based on expenditure, because reliable income data are often not available.

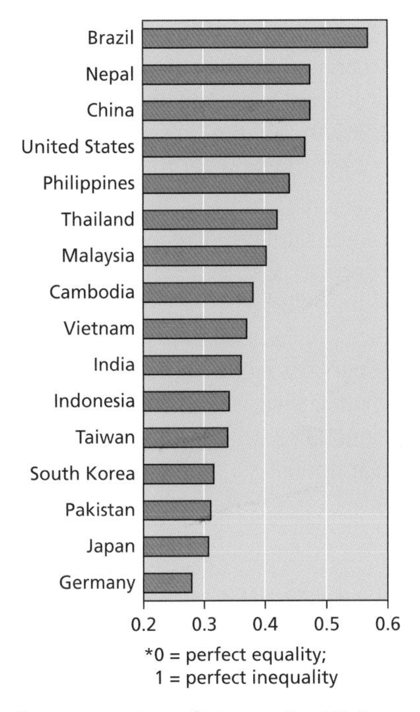

Figure 2.13 Income inequality (Gini co-efficient) for selected countries

Many Asian countries could now be nudging Latin American levels if measured on a comparable basis. India's Gini coefficient is in the lower half of the chart, yet health and education measures suggest that the country suffers from wide disparities. In the richest 20% of households, only 5% of children are severely underweight, compared with 28% in the poorest 20%. In India's richest state 99.8% of the population has access to clean water, but only 2% does in the poorest. The comparable figures for China, where income inequality is officially much greater, are 100% and 75%.

The main cause of increased inequality, especially in China, is the differing fortunes of rural and urban households. Productivity, and hence income, is growing much more slowly in agriculture. A second factor is the widening gap between those with and without skills. The shift from socialism to a market economy in China and India has increased the financial benefits of an education.

Does rising inequality matter so long as poverty is falling? Even where inequality has increased sharply, the poorest 20% of households are still better off in real terms than they were 10 years ago everywhere except in Pakistan. The number of people living on less than $1 a day has fallen everywhere except in Pakistan and Bangladesh. One reason to worry about widening inequality is that it can threaten growth if it results in social unrest.

Populist measures to "soak the rich" are not the answer: they could restrict growth. The Asian Development Bank (ADB) instead recommends that governments focus on policies that lift the incomes of the poor, such as improving rural access to health, education and social protection. More investment in rural infrastructure could boost productivity in farming and increase job opportunities for the poor.

Case study: Income inequality in emerging Asia

"Growth with equity" was the aim of the Asian tigers during the three decades to the 1990s. Unlike Latin America, most of them combined speedy economic growth with relatively low and sometimes even falling income inequality, thereby spreading the economic gains widely. More recently, Asian economies have continued to enjoy the world's fastest growth, but the rich are now growing richer much faster than the poor.

According to the ADB, income inequality has increased over the past decade or so in 15 of the 21 countries it has studied. The two main exceptions are Thailand and Indonesia, the countries worst hit by the 1997 financial crisis. The biggest increases in inequality were in China, Nepal and Cambodia.

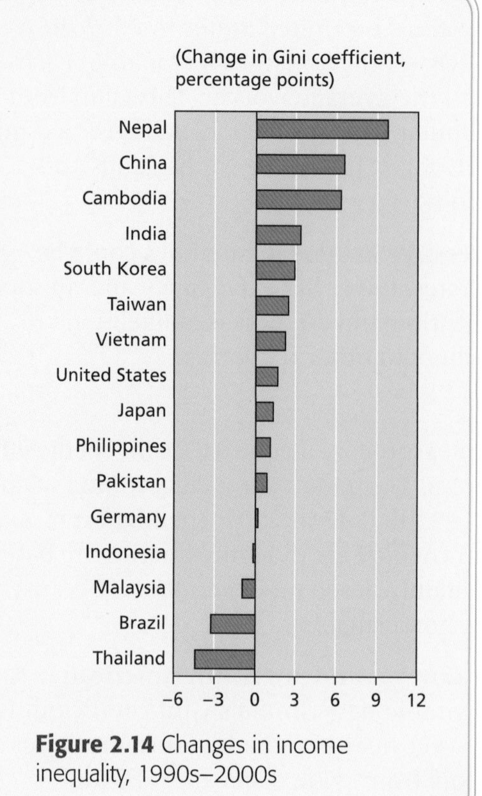

Figure 2.14 Changes in income inequality, 1990s–2000s

Trends in life expectancy

By the year 2025, according to the World Health Organization, 26 countries will have a life expectancy at birth of above 80 years. It is expected to be highest in Iceland, Italy, Japan and Sweden (82 years) followed by Australia, Canada, France, Greece, the Netherlands, Singapore, Spain and Switzerland (81 years). The countries with the lowest life expectancies in 2025 will be Angola, Burkina Faso, Burundi, Chad, Mozambique, Niger and Somalia (60 years); Mali and Uganda (59 years); Gambia and Guinea (58 years); and Afghanistan, Sierra Leone and Swaziland (50 years).

Asian tigers – four economies (Hong Kong, Singapore, South Korea and Taiwan) that were the first NICs, and were associated with very high growth rates and industrialization between the 1960s and 1990s.

There are certain interesting trends in life expectancy:

- In developed countries, not only do more people survive to old age, but those who do can expect to live longer than their predecessors. In the 1950s female life expectancy continued to rise but gains in male life expectancy slowed or levelled off. Women outlive men by an average of 5–9 years.
- The oldest old (aged 80+) are the fastest growing segment of many nations' populations. For the Scandinavian countries, France and Switzerland, the 80+ are approximately 4% of the total population. Increases in life expectancy are not uniform for all people living within a country. Indigenous populations living in developed countries have population pyramids that are more typical of developing countries. For example, American-Indian, Inuit and Aleut populations have an age structure like Morocco rather than the US.

More people than ever before now have access to at least minimum health care, to safe water supplies and sanitation facilities. Most of the world's children are now immunized against the six major diseases of childhood. The late 20th century has seen overall socioeconomic progress accompanied by steady and sometimes dramatic advances in the control and prevention of diseases, the development of vaccines and medicines, and countless medical and scientific innovations.

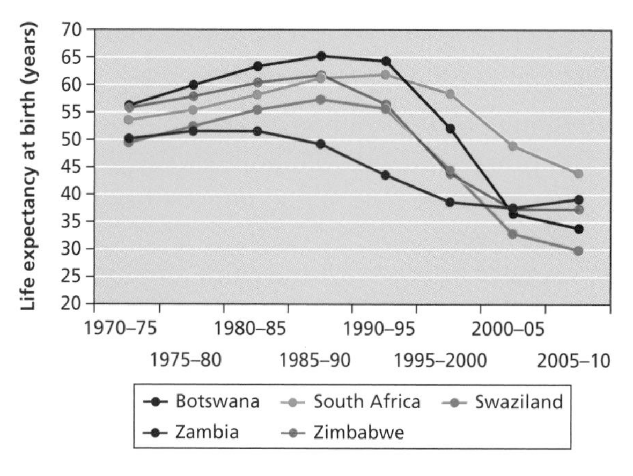

Figure 2.15 The impact of AIDS on life expectancy in five African countries

- Spectacular progress in reducing deaths among children under five in the last few decades is projected to continue. There were 21 million such deaths in 1955, about 10 million in 1997, and that figure should decline to 5 million by 2025.
- Gaps between the richest and poorest countries remain huge, but are gradually closing, at least in terms of premature deaths. For example, in 1995, 76% of people who died in the World Health Organization's (WHO) African region were under 50. By 2025 the proportion will fall to 57%.

However, progress has been far from universal. Some gaps in health between rich and poor are at least as wide as they were half a century ago, and are becoming wider still. While people in most countries are living longer, life expectancy is actually decreasing in some others. Between 1975 and 1995, 16 countries with a combined population of 300 million experienced such a decrease. Many of them were African countries in the grip of an HIV/AIDS epidemic.

Trends in education

The 2010 Education for All (EFA) Global Monitoring Report, *Reaching the marginalized,* shows some spectacular advances in education in many of the world's poorest countries over the past decade, a striking contrast with the "lost decade" of the 1990s. Since 1999 the number of children not attending school has fallen by 33 million, and more children are completing a full cycle of primary education.

Sub-Saharan Africa has increased enrolment at five times the rate achieved in the 1990s, with countries such as Benin and Mozambique registering rapid advances. In developing countries and western Asia the number of children out of school has more than halved, partly through policies aimed at getting more girls into school. In India, the number of children not in school fell by almost 15 million in just two years, from 2001 to 2003. The gender gap has also narrowed. In the space of one primary school generation, Senegal has moved from 85 girls for every 100 boys to an equal number of girls and boys.

Figure 2.16 Primary education in Japan

Education systems in many of the world's poorest countries are now experiencing the aftermath of the global economic downturn. The 2010 EFA Report warns that some 72 million children are still out of school, and a combination of slower economic growth, rising poverty and budget pressures could erode the gains of the past decade. Whereas rich countries nurture their economic recovery, many poor countries face the imminent prospect of education reversals.

The Report's data set reveals stark differences in education opportunity within countries:

- Being born into a poor household significantly raises the risk of deprivation. In the Philippines there is a four-year education gap between the richest and poorest households. The gap in India is seven years.
- Gender interacts with wealth and location. In Nigeria, the average youth aged 17–22 has received seven years in education. For poor rural Hausa females, that figure drops to less than six months.
- Disparities within countries are often bigger than disparities between countries. In Mexico a quarter of the young adults in the southern state of Chiapas have fewer than four years of education, a figure that falls to 3% for the Federal District.
- Some groups face acute disadvantage. In Kenya 51% of male Somali pastoralists and 92% of females aged 17–22 have less than two years in school.
- Language and ethnicity often reinforce marginalization. Turkey has made rapid progress in education, but Kurdish-speaking females from poor households average around three years in school, on a par with the national average for Chad.

More material available:

www.oxfordsecondary.co.uk/ibgeography

Read more materials on the MDGs

UN millennium development goals

The eight UN millennium development goals (MDGs) were agreed at the UN Millennium Development Summit in September 2000. Nearly 190 countries have signed up to them.

Case study: Progress towards MDGs in Mali and India

It is still possible to get things done, even if not at the pace that the MDGs demand. Mali, for instance, a landlocked country straddling the Sahel region and the Sahara desert, should be one of the least promising countries for development on earth. Mali's government, led by Amadou Toumani Touré, devotes most of its limited resources to what it calls the "struggle against poverty". As a result Mali is one of only five African countries to have fully qualified for America's Millennium Challenge Account (MCA), with its stringent criteria for good governance; that alone will bring in $460 million over five years, a huge amount in a country with a government budget of only about $1.5 billion.

Mali's government has made agriculture and infrastructure a priority. Rather than depending on aid, the government wants to raise growth to lift people out of poverty. With greater mechanization and irrigation, the country's 3.5 million farmers could become self-supporting, growing much more than the traditional crop of cotton. In a desert country irrigation is probably the most important anti-poverty tool of all. Around Timbuktu, for example, villagers have been shown how to build irrigation canals to capture the floodwaters of the huge Niger river. The village of Adina Koira has 5 kilometres of irrigation canals. On the irrigated lands they have been able to grow traditional crops such as cotton and rice, as well as new ones such as tomatoes and onions. So successful has some of this irrigation work been that the villagers have even reversed the usual patterns of immigration. Increasingly, Mali farmers are looking to have better access to European and American markets.

This MDG-approach is attractive, with its mix of inspiration (saving lives and educating minds) and accusation (why, then, is the rich world not doing enough?). But this thinking may be simplistic. Efforts to tackle the plight of the poor do not always win their favour. In 1980, for example, the UN proclaimed the "sanitation decade". The Indian government set about improving sanitation in villages where people still defecated by rivers and under trees. The solution seemed obvious: a toilet with a brick cubicle, squatting slab and two pits. The government set its budget and began building. However, as the only concrete structure in many homes, the toilets were often used for storing grain, keeping hens or even displaying deities. In the late 1980s, UNICEF realized they could not tackle the sanitation need until they first drummed up demand. Tackling some of the MDGs may need a fresh approach.

Be a critical thinker

How to make development goals work

In 1990 over 25% of people lacked access to safe water, according to the United Nations. By 2015 it will be less than half that number, if the MDGs succeed. However, the UN is better at making goals than meeting them. In 1977 in Mar del Plata Argentina, the world urged itself to provide safe water and sanitation for all by the end of the 1980s. In 1990 the UN extended the deadline to 2000. There is still a long way to go. In 1978 at Alma Ata (now Almaty, Kazakhstan), governments promised "health for all" by 2000. In 1990 in Jomtien, Thailand, they called for universal primary schooling by 2000, a goal pushed back to 2015 a decade later. In 2005, at the height of a "make poverty history" campaign, only 3% of Britons thought the world would meet the 2015 goal of halving poverty, defined as the proportion of people who live on less than the equivalent of a dollar a day.

However, the world is making unprecedented progress against poverty. As a result mainly of economic growth in China and India, the first MDG target should be met. Almost 32% of people in the developing world lived on less than a dollar a day in 1990. In 2004 that figure was 19.2%, and it should fall below 16% by 2015 (Figure 2.17), although only 57 out of 163 developing countries have counted the poor more than once since 1990. Ninety-two have not counted them at all.

Although the extreme-poverty rate in Africa has fallen from about 46% in 1999 to 41% in 2004, that is still way off the 2015 target of 22%. Hunger and malnutrition remain widespread: the proportion of under-fives who are underweight has declined only marginally, from 33% in 1990 to 29% in 2005. Death rates, though falling, are still highest in sub-Saharan Africa (Figure 2.18). Despite dramatic gains, Africa will not meet the goal of universal primary enrolment either; the rate is up from 57% in 1999 to 70% in 2005. Africa lags behind partly because its population is growing so rapidly. In 1990 there were 237 million Africans under 14; today that figure is 348 million, and by 2015 it is expected to top 400 million.

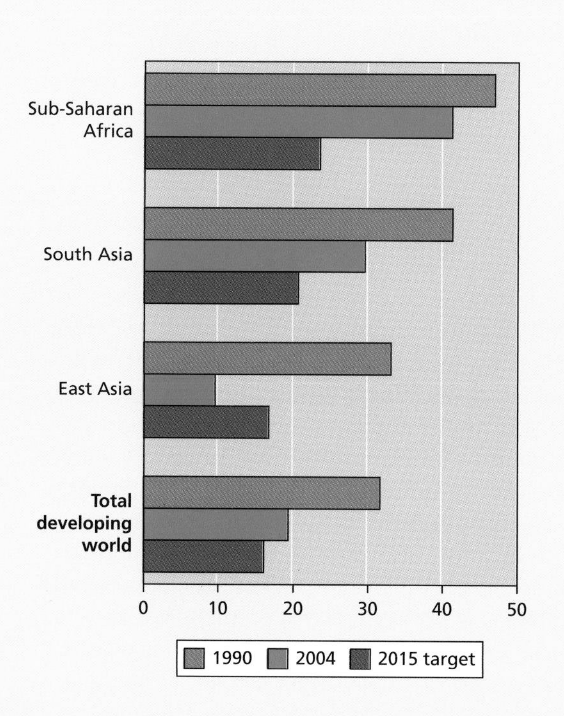

Figure 2.17 Changes in the percentage of people living on less than $1 a day

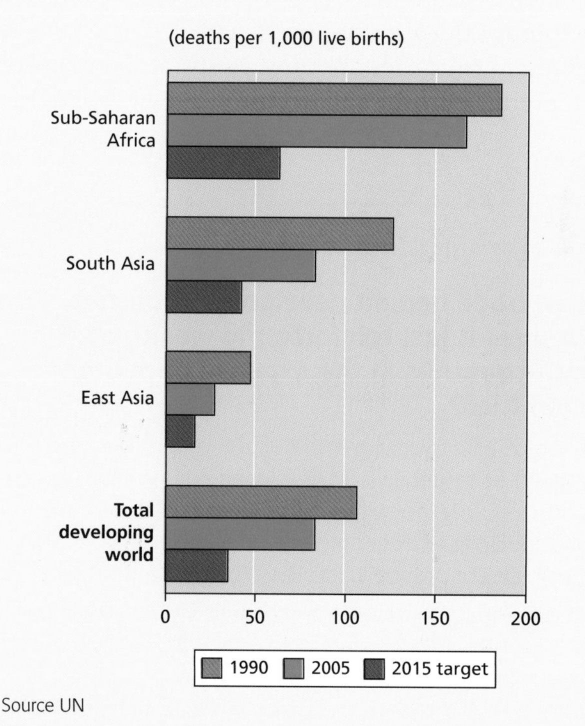

Source UN

Figure 2.18 Changes in the under-five mortality rates

Reducing disparities

Expanding trade

Expansion of trade links the labour markets of developed countries more closely with those of developing countries. This greater economic integration has had large benefits, raising average living standards in developed countries and accelerating development in developing countries. However, it has hit unskilled workers in the developed countries, reducing their wages and pushing them out of jobs. Governments in developed countries must take action to solve this problem, or these countries will continue to suffer from rising inequality and mass unemployment, and the developing countries will suffer from barriers to trade.

Over the past few decades changes have ocurred in the pattern of trade between developed and developing countries. Developing countries have ceased to be merely exporters of primary products, and their exports of manufactured goods to developed countries have increased massively. The old pattern of trade, in which manufactured goods from developed countries were exchanged for primary products from developing countries, has thus largely been replaced by a new pattern, in which both developed and developing countries each specialize in different sorts of manufactured goods. This needs to continue if developing countries are to improve their standards of living. The old one-way flow from developed to developing countries has been replaced by a two-way flow. The developing countries are now a substantial exporter of services such as shipping, tourism and even routine key-punching.

More material available:

www.oxfordsecondary.co.uk/ibgeography

See progress in achieving Millennium Development Goals

To research

Visit www.mdgmonitor.org to track progress towards the eight Millennium Development Goals, and for interactive maps of the MDGs. Look at the country profiles on this site for the facts about the country of your choice.

To do:

- Find out about the Millennium Challenge Account (run by the Millennium Challenge Corporation).
  - **a** Who are they?
  - **b** What do they do?
  - **c** What countries do they work with?
  - **d** What programmes and activities do they have?

Be a critical thinker

Can trade benefit developing countries, or does it just reinforce the wealth of rich countries at the expense of poorer countries?

There is controversy over the effects of the changing pattern of trade between developed and developing countries. In general, the consequences of such trade may be seen as either predominantly good (by optimists) or predominantly bad (by pessimists).

In developing countries, the optimistic view – based on the remarkable success of Korea and Taiwan – is that expansion of labour-intensive manufactured exports offers an ideal path to prosperity. It is ideal in that it promises not only rapid growth but also, by creating many jobs and increasing wages, less inequality. The pessimists, by contrast, regard export-oriented manufacturing as exploitation by foreigners, with very low wages and miserable working conditions.

In developed countries the optimists emphasize the efficiency gains from more trade with developing countries. Increased imports of labour-intensive manufactures release workers from low-productivity sectors. These imports also provide developing countries with more foreign exchange to spend on sophisticated exports from developed countries, which raises employment in high-productivity sectors. The optimists acknowledge that changes in the structure of employment will be painful for workers in contracting sectors, but this problem is seen as localized and temporary. The pessimistic view is that there is a vast global labour surplus. More trade with developing countries exposes workers in the developed countries to the consequences of this surplus, driving down their real wages and degrading their conditions of work. Leading eventually to job losses and increasing unemployment.

Trading blocs

A trading bloc is an arrangement among a group of nations to allow free trade between member countries but to impose tariffs (charges) on other countries wishing to trade with them. The European Union (EU) is an excellent example of a trading bloc. Many trading blocs were established after the Second World War, as countries used political ties to further their economic development. There are a number of regionally based trading blocs.

Within a trading bloc, member countries have free access to each other's markets. Thus, within the EU, the UK has access to the other countries of the EU, and they in turn have access to Britain's market. Being a member of a trading bloc is important, as it allows access to a much bigger market – in the case of the EU this amounts to over 490 million wealthy consumers.

Some critics believe that trading blocs are unfair as they deny access to non-members. Thus, for example, countries from the developing world have more limited access to the rich markets of Europe. This makes it harder for them to trade and to develop. In order to limit the amount of protectionism, the World Trade Organization has tried to promote free trade to and from all markets.

To do:

a Name the two trading blocs in South America.

b Name a trading bloc that consists of rich countries (MEDCs) and a relatively poor country (NIC).

c Outline the main advantages and disadvantages of trading blocs.

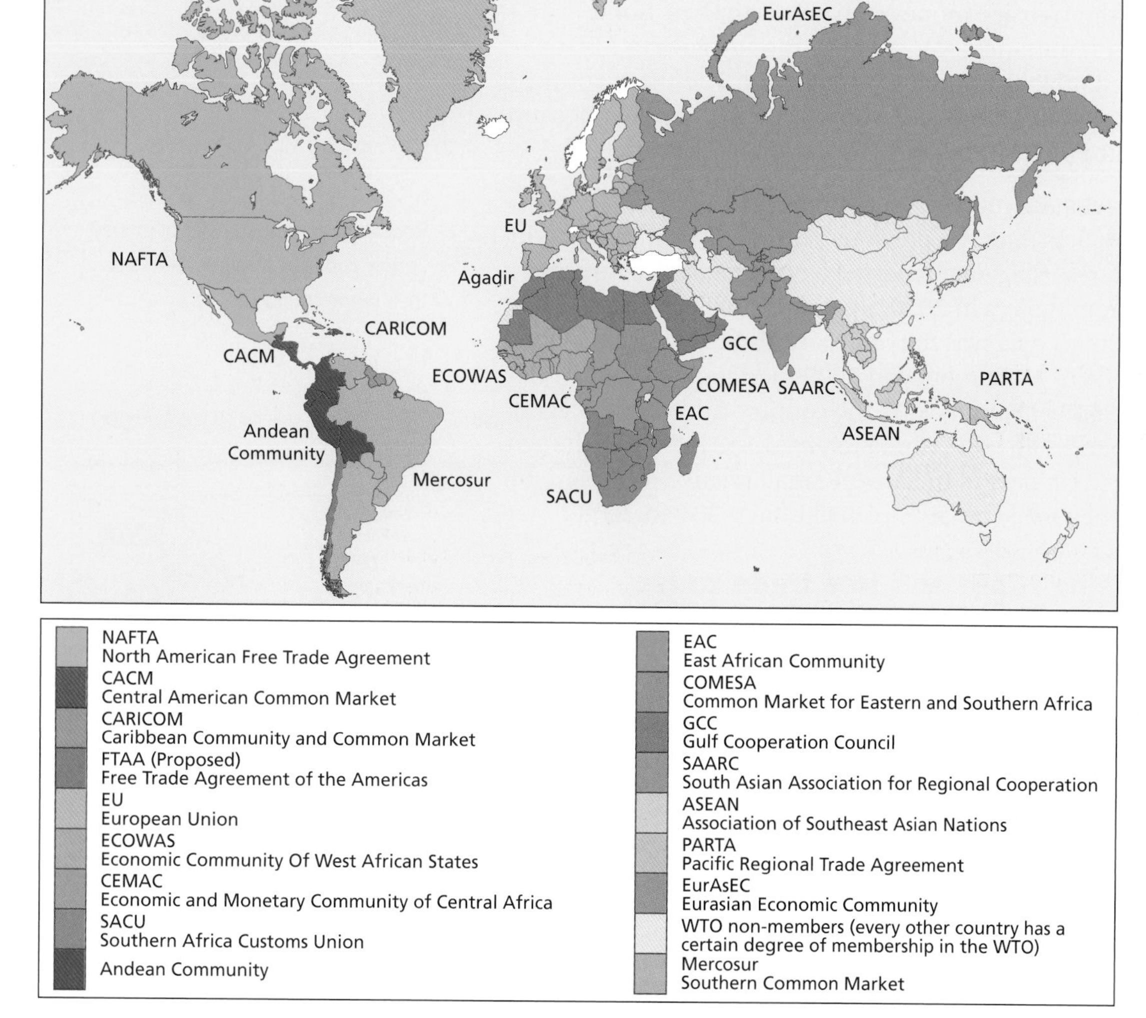

Figure 2.19 Major regional trading blocs

Fair or ethical trade

People Tree is an example of a company that follows the principles of fair trade as set out by the WFTO. Working closely with 50 fair trade groups in 15 countries, People Tree is a textiles company aiming to bring benefits to people at virtually every step of the production process – from growing cotton to weaving, dyeing, embroidery and stitching – helping to alleviate poverty in some of the world's most marginalized communities.

People Tree use ecologically sound methods of production, designing garments to be produced by hand wherever possible to minimize environmental impact. Most of their cotton is certified organic as well as fair trade, and dyed using safe and natural dyes. They source as many products as they can locally, choosing natural and recycled products over toxic, synthetic and non-biodegradable materials.

The People Tree Eco Policy is to:

- promote natural and organic farming
- avoid polluting substances
- protect water supplies
- use biodegradable substances where possible
- recycle materials where possible.

Fair, or ethical, trade – trade that attempts to be socially, economically and environmentally responsible. It is trade in which companies take responsibility for the wider impact of their business. Ethical trading is an attempt to address failings of the global trading system.

Figure 2.20 A fair trade rugby ball

To research

Visit the World Fair Trade Organization at www.wfto.com and http://www.peopletree.co.uk for People Tree.

Remittances

Foreign workers can transfer money and goods to their home countries through remittances. Most of the money goes to LEDCs, and is more than double the value of foreign aid. The three countries receiving most are India, China and Mexico, which together account for nearly one-third of remittances to the developing world.

Migrants from developing countries sent home $316 billion during 2009, or 6% less than the previous year. The World Bank, which compiles these figures using data reported by central banks, forecasts a 6.2% increase in remittances in 2010, up to $335 billion, just below their 2008 peak. South Asia was the only part of the developing world to which migrants sent more money in 2009 than in 2008. Indians abroad sent home $49 billion, making the country the world's biggest recipient of remittances in 2009. China received $48 billion. Although these remittances to China and India were small relative to the size of their economies, they made up 50% of Tajikistan's GDP in 2008.

Remittances – the transfer of money and/or goods by foreign workers to their home country.

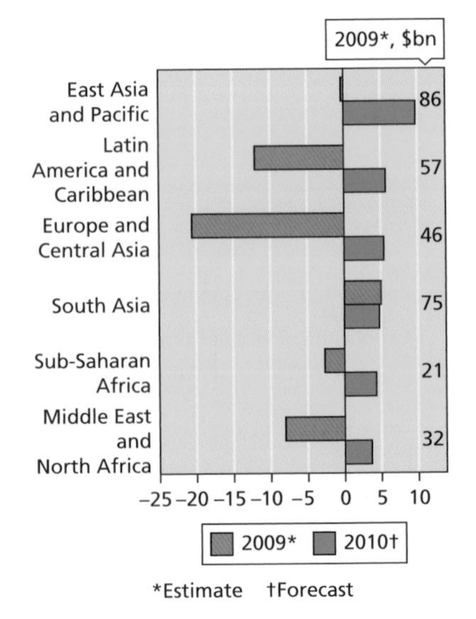

Figure 2.21 Changes in remittances sent, 2008–9 and 2009–10

Export processing zones and free trade zones

Export processing zones (EPZs) and free trade zones (FTZs) are important parts of the so-called new international division of labour, and represent what are seen as relatively easy paths to industrialization. By the end of the 20th century more than 90 countries had established EPZs as part of their economic strategies.

The popularity of EPZs is due to three groups of factors that link the economies of LEDCs with those of the world economy in general and the advanced economies in particular. These are:

1. problems of indebtedness and serious foreign exchange shortfalls in LEDCs since the 1980s
2. the spread of new liberal ideas in the 1990s that encouraged open economies, foreign investment and non-traditional exports

Case study: Incheon free economic zone, South Korea

South Korea's national government started the northeast Asian countries' business hub project in April 2002. Free economic zones were created to attract more direct foreign investment. Incheon is one of these special economic zones with minimized regulation and maximized business incentives and opportunities for foreign investment.

The free economic zone (FEZ) is the special area where, compared to the rest of Korea, exceptional measures such as tax breaks and other incentives exist to attract foreign-invested firms and expand foreign exchange circulation.

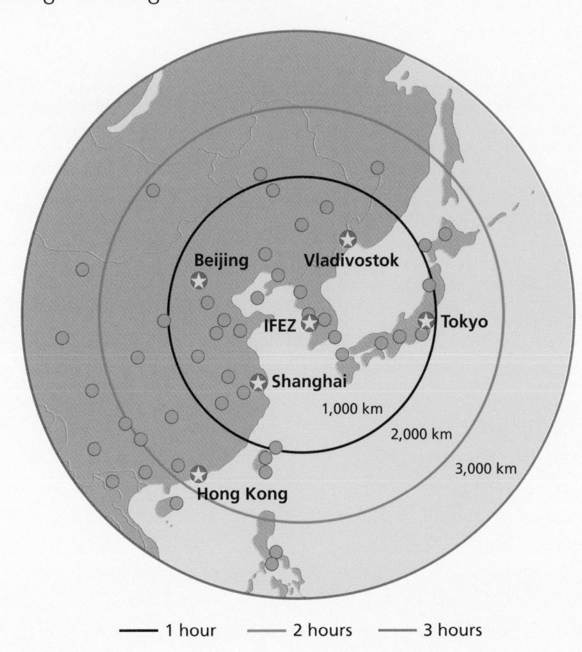

Figure 2.22 The accessibility of Incheon: areas within three hours' flight time

Overview of IFEZ

The northeast Asian region is one of the fastest-growing economic blocs in the world. Three countries of the region, Korea, China and Japan, are the driving force for this growth and development. Korea is strategically located in the centre of the vast markets between China and Japan. The best strategy for Korea's survival is therefore to make the best use of this economic and geopolitical location, and to fulfil the role of "facilitator" or "intermediary" between China and Japan as an "economic hub in northeast Asia".

Incheon is located to the west of Seoul and is partly built on land reclaimed from the sea. The Incheon FEZ covers an area of over 200 square kilometres, and will be completed by 2020. The IFEZ project comprises three separate districts within Incheon Metropolitan City – the Cheongna reclaimed area, Songdo and Yeongjong, which surrounds Incheon international airport. The total cost of the scheme is over $21 billion and over 510 000 people are expected to live there. IFEZ's goal is "to create the most desirable conditions for both living and business".

To do:

a What is an export processing zone?

b What is a free trade zone?

c Study Figure 2.22.

- Approximately how far are Tokyo and Beijing from IFEZ?
- Name a location that is within two hours' flight time from Korea.
- What does this map suggest about Korea and IFEZ?

SWOT (strengths, weaknesses, opportunities and threats) analysis of IFEZ

Incheon has many advantages ideal for establishing a free economic zone. It is a coastal city with an internationally competitive seaport and an airport that is the main entry point to Korea. Two billion people, or 32% of the world's population, and 61 cities each with over a million people live within 3.5 hours of flight time from Incheon. Sixty-nine airlines serve 160 cities in 49 countries. In addition, Incheon is adjacent to Korea's capital, Seoul, so it has a well-qualified workforce.

However, some of the incentives offered in the free economic zones may lead to expensive foreign schools in the IFEZ, which would widen the education gap between the rich and the poor. North Korea is a detriment to the development of IFEZ because foreign investors perceive a possible threat of a second Korean war: North Korea's nuclear experiments in 2006 and 2009 surprised the world.

Figure 2.23 The IFEZ at Incheon

3 the search by multinational corporations (MNCs) for cost-saving locations, particularly in terms of wage costs, in order to shift manufacturing, assembly and component production from locations in the advanced economies.

To do:

a How do remittances help poor countries?

b Where was the largest percentage increase between 2009 and 2010?

c Where was the largest percentage increase between 2008 and 2009?

d Describe the trend in remittances as shown in Figure 2.21.

The impact of aid and debt relief

A number of types of development (and development aid) can help countries grow. Two of the most well known are top-down and bottom-up development. Emergency relief is considered top down.

Top-down development	Bottom-up development
Usually large in scale	Small in scale
Carried out by governments, international organizations and "experts"	Labour intensive: common projects include building earthen dams, creating cottage industries
Done by people from outside the area	Involves local communities and local areas
Imposed upon the area or people by outside organizations	Run by locals for locals
Often well funded and quickly responsive to disasters	Limited funding available
Does not involve local people in the decision-making process	Involves local people in the decision-making process

Table 2.4 Types of development

Export processing zones (EPZs) – labour-intensive manufacturing centres that involve the import of raw materials and the export of factory products.
Free trade zones – zones in which manufacturing does not have to take place in order to gain trading privileges; such zones have become more characterized by retailing.

Most aid organizations working in the field would say that bottom-up development is best. However, it is hampered by limited funding and political ties. There are occasions when top-down aid can be effective, too.

When aid is effective	When aid is ineffective
It provides humanitarian relief.	It might allow countries to postpone improving economic management and mobilization of domestic resources.
It provides external resources for investment and finances projects that could not be undertaken with commercial capital.	It replaces domestic saving, direct foreign investment and commercial capital as the main sources of investment and technology development.
Project assistance helps expand much-needed infrastructure.	The provision of aid might promote dependency rather than self-reliance.
It contributes to personnel training and builds technical expertise.	Some countries have allowed food aid to depress agricultural prices, resulting in greater poverty in rural areas and a dependency on food imports. It has also increased the risk of famine in the future.
It can support better economic and social policies.	Aid is sometimes turned on and off in response to the political and strategic agenda of the donor country, making funds unpredictable, which can result in interruptions in development programmes.
	The provision of aid might result in the transfer of inappropriate technologies or the funding of environmentally unsound projects.
	Emergency aid does not solve the long-term economic development problems of a country.
	Too much aid is tied to the purchase of goods and services from the donor country, which might not be the best or the most economical.
	A lot of aid does not reach those who need it, that is, the poorest people in the poorest countries.

Table 2.5 The effectiveness of different types of aid provision

Poor countries' debt

Sub-Saharan Africa (SSA) includes most of the 42 countries classified as heavily indebted, and 25 of the 32 countries rated as severely indebted. In 1962 SSA owed $3 billion (£1.8 billion). Twenty years later it owed $142 billion. Today its debt is about $235 billion. The most heavily indebted countries are Nigeria ($35 billion), Côte d'Ivoire ($19 billion) and Sudan $18 billion).

Many developing countries borrowed heavily in the 1970s and early 1980s, encouraged to do so by western lenders, including export credit agencies. They soon ran into problems:

- low growth in industrialized economies
- high interest rates between 1975 and 1985
- oil price rises
- falling commodity prices.

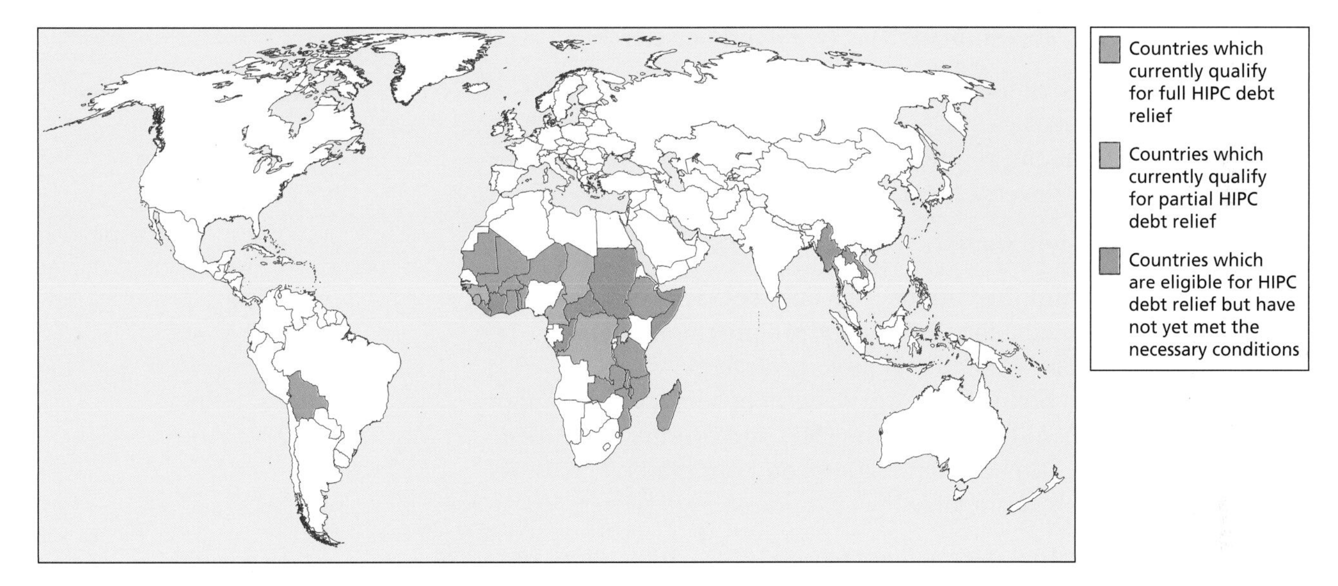

Figure 2.24 Heavily indebted poor countries (HIPC)

What has been done to deal with debt?

Since 1988 the Paris Club of government creditors has approved a series of debt relief initiatives. In addition, the World Bank has lent more through its concessional lending arm, and the International Development Agency has given loans for up to 50 years without interest but with a 3–4% service charge. Lending has risen from $424 million in 1980 to $2.9 billion, plus a further $928 million through the African Development Bank. The IMF has also introduced a **soft loan facility**, conditional on wide-ranging economic reforms.

Structural adjustment programmes

Structural adjustment programmes (SAPs) were designed to cut government expenditure, reduce the amount of state intervention in the economy, and promote liberalization and international trade. SAPs were explicit about the need for international trade. They consist of four main elements:

1. greater use of a country's resource base
2. policy reforms to increase economic efficiency

3 generation of foreign income through diversification of the economy and increased trade
4 reducing the active role of the state.

However, some people argue that these measures have made the situation worse.

The measures were sometimes divided into two main groups: the **stabilization measures**, short-term steps to limit any further deterioration of the economy (e.g. wage freeze, reduced subsidies on food, health and education); and adjustment measures, longer-term policies to boost economic competitiveness (tax reductions, export promotion, downsizing of the civil service, privatization and economic liberalization).

The Heavily Indebted Poor Countries (HIPC) initiative

The Heavily Indebted Poor Countries (HIPC) initiative, launched in 1996 by the International Monetary Fund (IMF) and the World Bank and endorsed by 180 governments, has two main objectives:

- to relieve certain low-income countries of their unsustainable debt to donors
- to promote reform and sound policies for growth, human development and poverty reduction.

Debt relief occurs in two steps:

- At the decision point the country gets debt **service relief** after having demonstrated adherence to an IMP programme and progress in developing a national poverty strategy.
- At the completion point the country gets debt **service relief** upon approval by the World Bank and the IMF. The country is entitled to at least 90% debt relief from bilateral and multilateral creditors to make debt levels sustainable.

"Debt service" is required over a given period for the repayment of interest and principal on a debt – monthly mortgage payments are a good example. "Stock relief" is the cancelling of specific debts; this achieves a reduction in debt service over the life of a loan.

Of the 42 countries participating in the initiative, 34 are in sub-Saharan Africa. None had a **purchasing power parity** (PPP) above $1,500/head in 2001, and all rank low on the HDI. Between 1990 and 2001 HIPCs grew by an average of just 0.5% a year. HIPCs have been overindebted for at least 30 years: by poor-country standards their ratios of debt to exports were already high in the 1980s. At the same time, HIPCs have received considerable official development assistance. Net transfers of such aid averaged about 10% of their GNP in the 1990s, compared with about 2% for all poor countries. To date 16 HIPCs have reached the decision point and eight have reached the completion point (Benin, Bolivia, Burkina Faso, Mali, Mauritania, Mozambique, Tanzania and Uganda).

Purchasing power parity (PPP) – the measure of average earnings in relation to local prices, i.e. how much you can buy for your money.

Expanding market access

Expanding market access is essential to help countries diversify and expand trade. Trade policies in rich countries remain highly discriminatory against developing country exports. Average OECD tariffs on manufactured goods from developing countries are more than four times those on manufactured goods from other OECD (Organisation for Economic Cooperation & Development) countries. Moreover, agricultural subsidies in rich countries lead to unfair competition. Cotton farmers in Benin, Burkina Faso, Chad, Mali and Togo have improved productivity and achieved lower production costs than their richer country competitors. Still, they can barely compete. Rich country agricultural subsidies total more than $300 billion a year, nearly six times official development assistance.

MEDCs should set targets to:

- increase official development assistance to fill financing gaps (estimated to be at least $50 billion)
- remove tariffs and quotas on agricultural products, textiles and clothing exported by developing countries
- remove subsidies on agricultural exports from developing countries
- agree and finance, for HIPCs, a compensatory financing facility for external shocks, including collapses in commodity prices
- agree and finance deeper debt reduction for HIPCs that have reached their completion points, to ensure sustainability.

To research

Visit www.imf.org for a fact sheet on debt relief under the Heavily Indebted Poor Countries (HIPC) initiative, and for facts on Mozambique and debt service.

Be a critical thinker

Should the debt be cancelled?

There is debate about whether or not to cancel the debt of the poorest African countries. In 2000 a coalition of groups under the Jubilee 2000 banner called on the G7 summit meeting to cancel the debt. As a result, a number of countries received full or partial debt relief. These countries were then able to use the money saved to improve conditions for their people; for example Tanzania introduced free schooling, built more schools and employed more teachers.

In 2005 the Make Poverty History campaign was instrumental in persuading the G8 summit meeting to alleviate the burden of debt by further reducing repayment obligations. The campaign continues to call for extending debt cancellation to all countries that otherwise are unable to meet basic human needs. In 2006 a new agreement called the Multilateral Debt Reduction Initiative (MDRI) came into force, writing off the entire $40 billion debt owed by 18 HIPCs to the World Bank, the IMF and the ADF. The annual saving in debt payments amounts to just over $1 billion. However, countries outside Africa are not included in this agreement.

Rich countries have now promised to cancel the debt of 42 countries and give an extra £5 million of aid by 2050. Critics of debt relief state that it does not help the poor, that it does not help countries that do not get into debt, and that it encourages countries to overspend.

The achievements of developing countries

It can be easy to forget about the achievements of the developing world. For example:

- average real incomes in the poor world have more than doubled in the past 40 years, despite population growth
- under-five death rates have been cut by 50% or more in every region over the past 40 years
- average life expectancy has risen by more than a third in every region since 1950
- the percentage of people with access to safe water has risen from about 10% to 60% in rural areas of the developing world since 1975.

To do:

a Briefly explain the HIPC initiative.

b Study Figure 2.24 on page 49.

i In which continent are most of the HIPC countries found?

ii Name a South American country that qualifies for HIPC debt relief.

iii Name an Asian country that is eligible for debt relief but has not yet made the necessary conditions.

3 Patterns in environmental quality and sustainability

By the end of this chapter you should be able to:

- explain the atmospheric energy budget and the causes and implications of climate change
- understand the causes and impacts of soil degradation, and how this is being addressed
- describe the increasing demands on the global water supply and the management strategies used to conserve water
- understand the importance of biodiversity and the key role of rainforests and the effects of deforestation.

Global warming – the increase in temperatures around the world that have been noticed over the last 50 years or so, and in particular since the 1980s.

The greenhouse effect – the process by which certain gases (water vapour, carbon dioxide, methane and chlorofluorocarbons (CFCs)) allow short-wave radiation from the sun to pass through the atmosphere and heat up the earth, but trap an increasing proportion of long-wave radiation from the earth. This radiation leads to a warming of the atmosphere.

The enhanced greenhouse effect – the increasing amount of greenhouse gases in the atmosphere as a result of human activities, and their impact on atmospheric systems, including global warming.

Atmosphere and change

The atmosphere is an open energy system receiving energy from both sun and earth. Although the earth's energy is very small it has an important local effect, as in the case of urban climates. **In**coming **sol**ar radi**ation** is referred to as **insolation**.

The energy that drives all weather systems and climates comes from the sun. The earth absorbs most of this energy in the tropical areas, whereas there is a loss of energy from more polar areas. To compensate for this there is also a redistribution of energy to higher latitudes from lower latitudes, caused by the wind circulation and the ocean circulations.

The atmospheric energy budget

The earth's atmosphere constantly receives solar energy. Nevertheless, there are long-term and short-term variations in the earth's climates. However, in recent decades, the rise in global temperatures – known as global warming – has been linked with human activities, and is discussed in depth in chapter 14.

Be a critical thinker

Global warming: real or make believe?

Global warming challenges views of certainty within the sciences. In the popular perception, global warming is having a negative impact on the world. However, people disagree about who – if anyone – is causing it, who suffers most, and what should be done to solve it. There is also some confusion between global warming and the greenhouse effect. The greenhouse effect is a natural process, without which there would be no life on earth. There is, however, an enhanced or accelerated greenhouse effect, which is synonymous with global warming. The enhanced greenhouse effect is largely due to human (anthropogenic) forces, although feedback mechanisms may trigger some natural forces, too.

Some lobby groups and politicians may take views that suit their own economic and political ends, and it is possible to hide behind the uncertainties around global warming (causes, consequences and potential solutions). In the USA, the strength of the oil companies during the George W. Bush presidency was seen by many as an example of economic groups – and the politicians they supported – choosing a side that was not in the long-term environmental, social or economic interest of the world. However, in the short term it benefited the oil companies and the politicians they supported.

Under "natural" conditions the balance is achieved in three main ways:

- radiation – the emission of electromagnetic waves such as X-ray, short wave and long wave; as the sun is a very hot body, radiating at a temperature of about 5,700°C, most of its radiation is in the form of very short wavelengths such as ultraviolet and visible light
- convection – the transfer of heat by the movement of a gas or liquid
- conduction – the transfer of heat by contact.

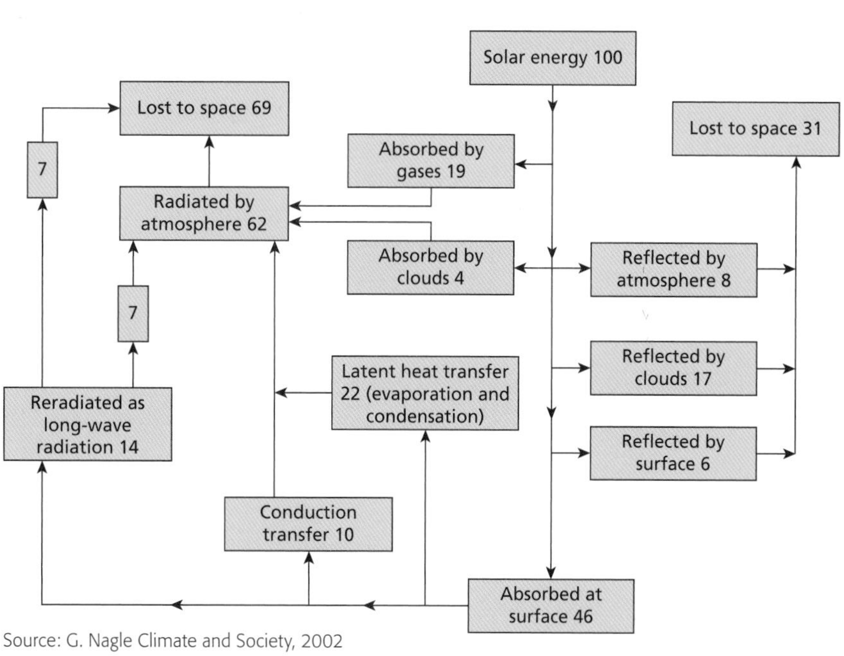

Source: G. Nagle Climate and Society, 2002

Figure 3.1 The earth's atmospheric energy budget

Only about 46% of the insolation at the top of the atmosphere actually gets through to the earth's surface. About 35% of insolation is reflected back to space and a further 19% is absorbed by atmospheric gases, especially oxygen and ozone at high altitudes, and CO_2 and water vapour at low altitudes. Scattering accounts for a net loss of 6%, and clouds and water droplets reflect 23%. In fact, clouds can reflect up to 80% of total insolation. Reflection from the earth's surface (known as the **planetary albedo**) is generally about 7%.

Energy received by the earth is reradiated at long wavelength. (Very hot bodies such as the sun emit short-wave radiation, whereas cold bodies such as the earth emit long-wave radiation.) Of this, 8% is lost to space. Clouds absorb some energy and reradiate it back to earth. Evaporation and condensation account for a loss of heat of 22%. In addition, a small amount of conduction occurs (carried up by turbulence). Thus, heat gained by atmosphere from the ground amounts to 39 units.

The atmosphere is largely heated from below. Most of the incoming short-wave radiation is let through, but CO_2 traps the outgoing long-wave radiation. This is known as the greenhouse principle.

The moon is an airless planet that is almost the same distance from the sun as the earth. However, daytime temperatures on the moon may reach as high as 100°C, whereas by night they may be −150°C. Average temperatures on the moon are about -18°C compared with about 15°C on earth. The earth's atmosphere therefore raises temperatures by about 33°C.

The earth's temperature changes for a number of reasons, one of the most obvious of which is a change in the output of energy from the sun. There is evidence of an 11-year solar cycle, and longer periods of changes such as the Milankovitch Cycle also occur. Slow variations in the earth's orbit affect the seasonal and latitudinal distribution of solar radiation, and are responsible for initiating the ice ages. On a shorter timescale, changes in atmospheric

Surface	Albedo (%)
Water (sun's angle above 40°)	2–4
Water (sun's angle less than 40°)	6–80
Fresh snow	75–90
Old snow	40–70
Dry sand	35–45
Dark, wet soil	5–15
Dry concrete	17–27
Black road surface	5–10
Grass	20–30
Deciduous forest	10–20
Coniferous forest	5–15
Crops	15–25
Tundra	15–20

Table 3.1 Albedo values

composition are linked to an increase in global temperature. The earth's atmosphere is vital for life, and changes to it disrupt the natural balance of the earth's energy budget, in terms of both the amount and type of radiation. Changes in reflectivity (**albedo**) also affect global climate change: for example, as ice melts and is replaced by darker-coloured vegetation, the amount of insolation absorbed increases, and temperatures rise.

To research

Visit the Guardian's Science Weekly podcast at www.guardian.co.uk/science/blog to find out more about the sun's effect on climate change.

Global climate change

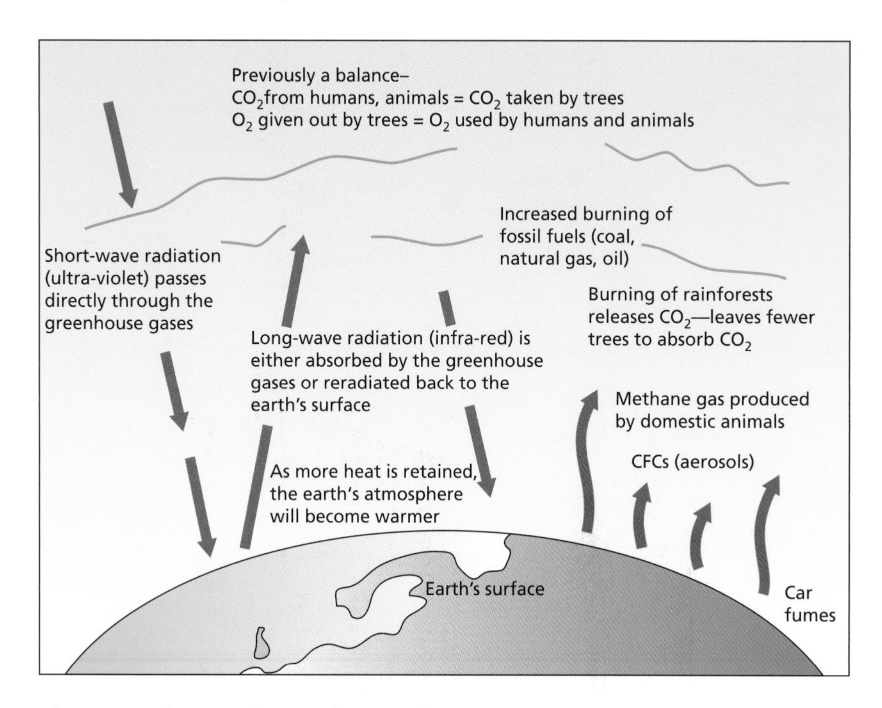

Figure 3.2 The greenhouse effect

To do:

a What type of radiation is emitted by (i) the sun and (ii) the earth?

b Briefly explain the meaning of the terms convection, conduction and radiation.

c Why has the earth a warmer climate than the moon?

d Describe the earth's energy budget.

One concern about global warming is the build-up of certain greenhouse gases. **Water vapour** is a greenhouse gas and accounts for about 95% of greenhouse gases by volume. However, the gases mainly implicated in global warming are carbon dioxide, methane and chlorofluorocarbons.

Global climate change – the changes in the global patterns of rainfall and temperature, sea level, habitats and the incidence of drought, floods and storms, resulting from changes in the earth's atmosphere, believed to be mainly caused by the enhanced greenhouse effect.

Carbon dioxide (CO_2) levels have risen from about 315 ppm in 1950 to 355 ppm today, and are expected to reach 600 ppm by 2050. The increase is due to human activities: burning fossil fuel (coal, oil and natural gas) and deforestation. Deforestation of the tropical rainforest is a double blow, since it not only increases atmospheric CO_2 levels, but it also removes the trees that convert CO_2 into oxygen.

Methane is the second-largest contributor to global warming, and is increasing at a rate of 1% per annum. It is estimated that cattle convert up to 10% of the food they eat into methane, and emit 100 million tonnes of methane into the atmosphere each year. Natural wetland and paddy fields are another important source: methane-rich paddy fields emit between 20 and 100 million tonnes of methane annually. As global warming increases, bogs trapped in permafrost will melt and release vast quantities of methane.

Chlorofluorocarbons (CFCs) are synthetic chemicals that destroy ozone, as well as absorbing long-wave radiation. CFCs, which are increasing at a rate of 6% per annum, are up to 10 000 times more efficient at trapping heat than CO_2.

The implications of climate change

Global warming is predicted to have various far-reaching effects on the natural, social and economic environment:

- Sea levels will rise, causing flooding in low-lying areas such as the Netherlands, Egypt and Bangladesh; up to 200 million people could be displaced.
- Storm activity will increase (owing to more atmospheric energy).
- Agricultural patterns will change, e.g. the USA's grain belt will decline, but Canada's growing season will increase.
- There will be less rainfall over the USA, southern Europe and the CIS.
- Up to 40% of species of wildlife will become extinct.

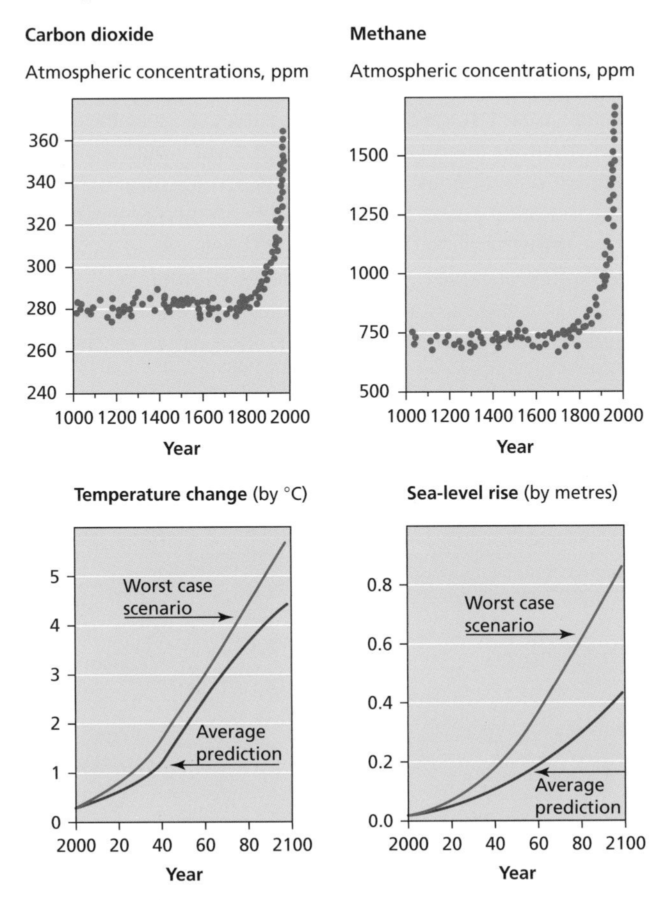

Figure 3.4 The causes and consequences of global warming

Nevertheless, there is a great amount of uncertainty in geography, and nobody knows exactly what the impact of climate change will be. The different scenarios are based on different possible temperature changes. Some people even suggest that certain areas might get colder, such as northwest Europe if the Gulf Stream shuts down. We do not know what will happen, and the results may be very different from the predictions.

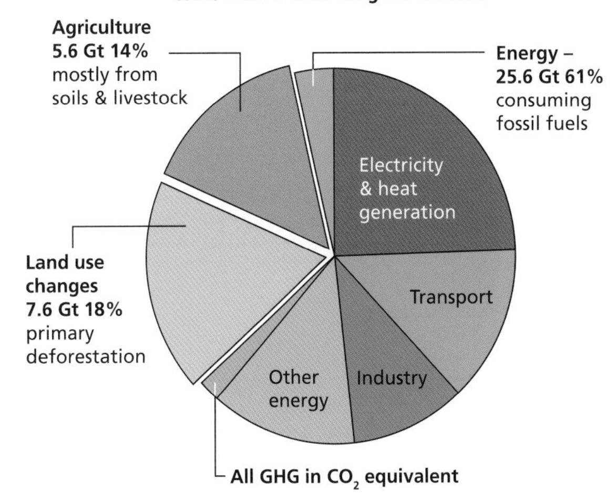

Figure 3.3 The main sources of CO_2 emissions

To do:

Study Table 3.2, which shows total carbon emissions for the top 20 countries, 1997–2007.

a Identify the country that generated the largest carbon emissions.

b Identify the country that had the highest relative increase in carbon emissions between 1997 and 2007.

c Comment on the geographic variations of the 10 largest carbon polluters.

Rank	Country	Million tonnes 97–07	Percent change 97–07
1	United States	64 166	7
2	China	45 301	102
3	Russia	17 360	15
4	Japan	13 342	9
5	India	11 870	60
6	Germany	9 487	–6
7	Canada	6 385	8
8	United Kingdom	6 281	–1
9	South Korea	5 059	21
10	Italy	4 997	9
11	South Africa	4 504	17
12	France	4 466	5
13	Mexico	4 302	30
14	Australia	4 203	37
15	Iran	4 128	69
16	Brazil	3 881	22
17	Spain	3 740	41
18	Ukraine	3 722	3
19	Saudi Arabia	3 663	70
20	Poland	3 308	–11

Source: The Guardian 5 December 2009

Table 3.2 Carbon emissions for selected countries, 1997–2007

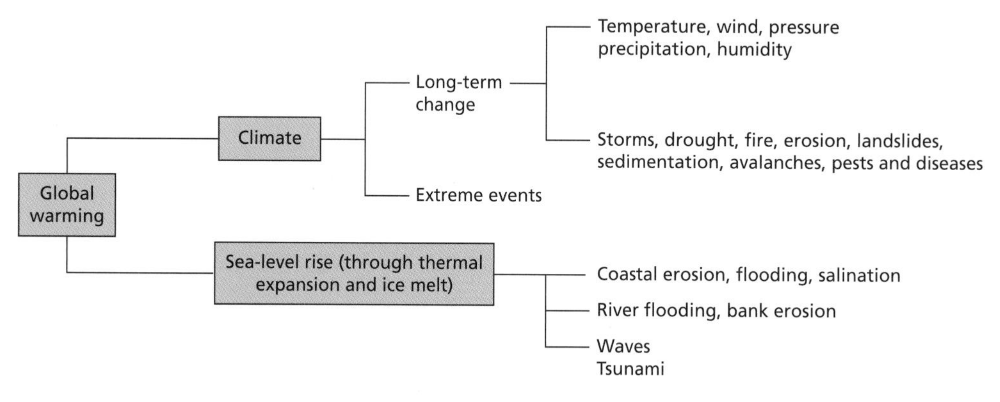

Figure 3.5 Some environmental implications of climate change

Case study: *The potential impacts of global temperature rise*

1°C – Vital for low-lying island states, but virtually impossible to achieve this limit

The Arctic sea ice is already disappearing and, after a 1°C global average temperature rise, it would disappear for good in the summer months. Heatwaves and forest fires would become more common in the subtropics, and worst-hit would be the Mediterranean region, southern Africa, Australia and the southwest United States. Most of the world's corals will die, including the Great Barrier Reef. Glaciers that provide fresh water for crops for 50 million people will begin to melt, and climate-related diseases such as malaria and diarrhoea will affect 300 000 people every year.

2°C – The temperature limit the scientists want

The heatwaves seen in Europe during 2003, which killed tens of thousands of people, will come back every year with a 2°C global average temperature rise. Southern England would regularly see summer temperatures of around 40°C. The Amazon would turn into desert and grasslands, while increasing CO_2 levels in the atmosphere would make the world's oceans too acidic for any remaining coral reefs and thousands of other marine life forms. More than 60 million people, mainly in Africa, would be exposed to higher rates of malaria. Agricultural yields around the world would drop, exposing half a billion people to a greater risk of starvation. The west Antarctic ice sheet would collapse, the Greenland ice sheet would melt and the world's sea level would begin to rise by 7 metres over the next few hundred years. Glaciers all over the world would recede, reducing the fresh water supply for major cities including Los Angeles. Coastal flooding would affect more than 10 million extra people. A third of the world's species will become extinct if the 2°C rise changes their habitats too quickly for them to adapt.

4°C – Possible with an extremely weak climate deal

At this level of increase the Arctic permafrost would enter the danger zone, and the methane and carbon dioxide currently locked in the soils would be released into the atmosphere. At the Arctic itself, the ice cover would disappear permanently, meaning extinction for polar bears and other native species that rely on the presence of ice. Further melting of Antarctic ice sheets would mean a further 5-metre rise in sea level, submerging many island nations. Italy, Spain, Greece and Turkey would become deserts and central Europe would reach desert temperatures of almost 50°C in summer. Southern England's summer climate could resemble that of southern Morocco today.

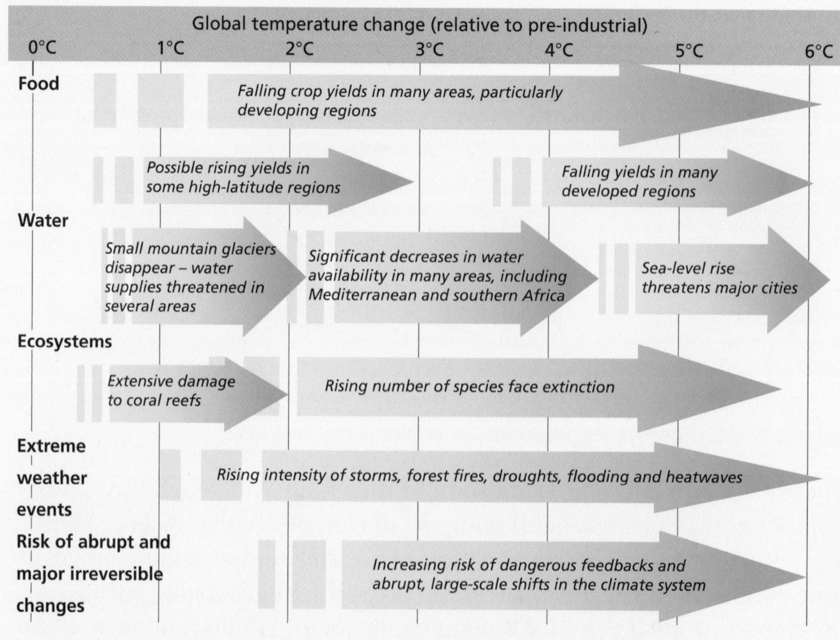

Figure 3.6 The impacts of climate change will vary with the scale of the temperature changes

Soil and change: degradation

Soil degradation is the decline in quantity and quality of soil. It includes:

- erosion by wind and water
- biological degradation (the loss of humus and plant/animal life)
- physical degradation (loss of structure, changes in permeability)
- chemical degradation (acidification, declining fertility, changes in pH, salinization and chemical toxicity).

Water and wind erosion account for more than 80% of the 20 million square kilometres degraded worldwide. **Water erosion** accounts for nearly 60% of soil degradation. There are many types of water erosion including surface, gully, rill and tunnel erosion.

Acidification is the change in the chemical composition of the soil, which may trigger the circulation of toxic metals. **Salt-affected soils** are typically found in marine-derived sediments, coastal locations and hot arid areas where capillary action brings salts to the upper part of the soil. Soil salinity has been a major problem in Australia following the removal of vegetation in dryland farming.

Predicting soil erosion

The universal soil loss equation **A = RKLSCP** is an attempt to predict the amount of erosion that will take place in an area, on the basis of certain factors that increase susceptibility to erosion.

Factor	Description
Ecological conditions	
Erosivity of soil **R**	Rainfall totals, intensity and seasonal distribution. Maximum erosivity occurs when the rainfall occurs as high-intensity storms. If such rain falls when the land has just been ploughed or full crop cover is not yet established, erosion will be greater than when falling on a full canopy. Minimal erosion occurs when rains are gentle, and fall on to frozen soil or land with natural vegetation or a full crop cover.
Erodibility **K**	The susceptibility of a soil to erosion. Depends upon infiltration capacity and the structural stability of soil. Soils with high infiltration capacity and high structural stability, which allow the soil to resist the impact of rain splash, have the lowest erodibility values.
Length-slope factor **LS**	Slope length and steepness influence the movement and speed of water down a slope, and thus its ability to transport particles. The greater the slope, the greater the erosivity; the longer the slope, the more water is received on the surface.
Land-use type	
Crop management **C**	Most control can be exerted over the cover and management of the soil, and this factor relates to the type of crop and to cultivation practices. Established grass and forest provide the best protection against erosion; of agricultural crops, those with the greatest foliage and thus greatest ground cover are optimal. Fallow land or crops that expose the soil for long periods after planting or harvesting offer little protection.
Soil conservation **P**	Soil conservation measures, such as contour ploughing, bunding, use of strips and terraces, can reduce erosion and slow runoff water.

Table 3.3 Factors relating to the universal soil loss equation (USLE).

To research

Visit the Guardian's Science Weekly podcast on climate change at www.guardian.co.uk/science/blog and listen to the talk about the climate change map (as a class). What are the potential impacts of a 4°C rise in temperature?

To do:

a Outline the causes of the greenhouse effect.

b Compare the impacts of a rise in temperature of 2°C with those of a 4°C rise.

c Suggest why low-lying island states are particularly at risk from climate change.

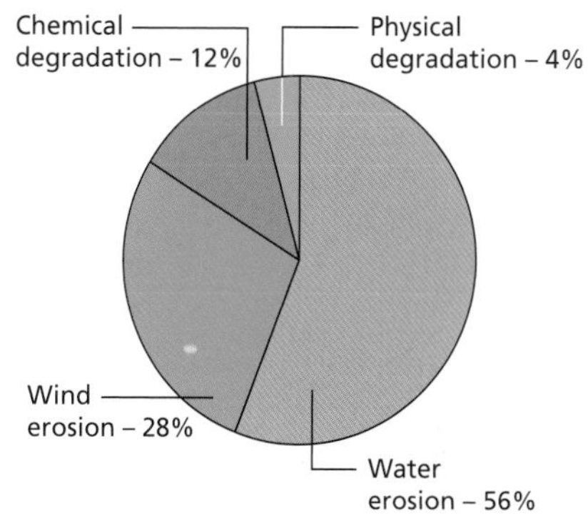

Figure 3.7 Types of soil degradation

Soil degradation – a severe reduction in the quality of soils, often through soil erosion, salinization or soil exhaustion (loss of soil fertility).

Figure 3.8 Soil degradation in Sicily

Causes of land degradation

The causes of soil or land degradation include:

- the reduction of the natural vegetative cover which renders the topsoil more susceptible to erosion, as when huge areas of forest are cleared for logging, fuelwood, farming or other human uses
- unsustainable land-use practices such as excessive irrigation, inappropriate use of fertilizers and pesticides, and overgrazing by livestock
- **groundwater over-abstraction,** leading to dry soils, leading to physical degradation
- **atmospheric deposition** of heavy metals and persistent organic pollutants, making soils less able to sustain the original land cover and land use.

Climate change will probably intensify the problem. Climate change is likely to affect hydrology and hence land use.

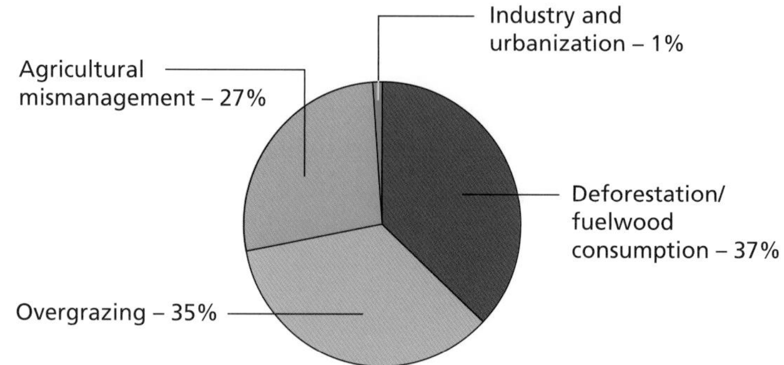

Figure 3.9 Causes of soil degradation

Underlying cause	leads to...
Natural disaster	Degradation due to biogeophysical causes, or "acts of God"
Population change	Degradation occurs when population growth exceeds environmental thresholds (neo-Malthusian), or when decline causes collapse of adequate management
Underdevelopment	Resources exploited to benefit world economy or developed countries, leaving little profit to manage or restore degraded environments
Internationalism	Taxation and other forces interfere with the market, triggering overexploitation
Colonial legacies	Trade links, communications, rural–urban linkages, cash crops and other "hangovers" from the past promote poor management of resource exploitation
Inappropriate technology and advice	Promotion of wrong strategies and techniques that result in land degradation
Ignorance	Linked to inappropriate technology and advice: a lack of knowledge of what leads to degradation
Attitudes	People's or institutions' attitudes blamed for degradation
War and civil unrest	Overuse of resources in national emergencies and concentrations of refugees, leading to high population pressures in safe locations

(Source Middleton, N. 1995. *The global casino.* London, UK. Arnold).

Table 3.4 Underlying causes of land degradation

Figure 3.10 Soil degradation in Brunei

The results of land degradation

There are more than 20 million square kilometres of degraded land worldwide. Overgrazing and agricultural mismanagement affect more than 12 million square kilometres of this land, and 20% of the world's pasture and rangelands have been damaged. The situation is most severe in Africa and Asia.

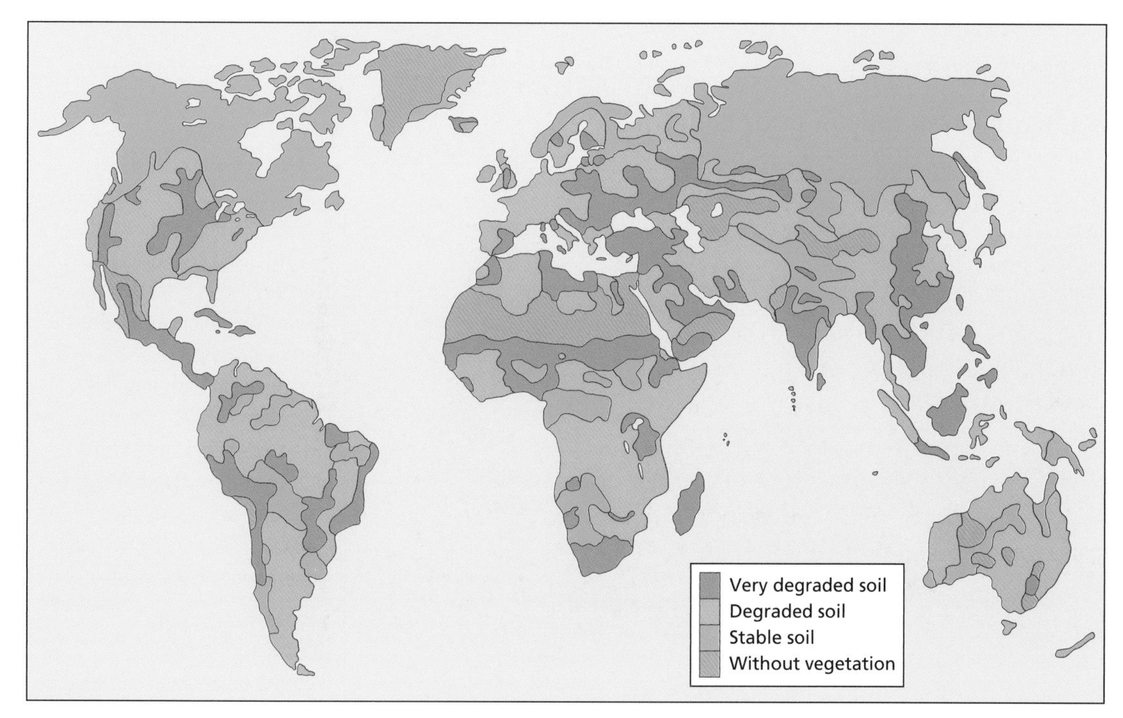

Source: qed. princeton edu

Figure 3.11 Global soil degradation

The removal of vegetation and topsoil frequently results in:

- increased surface runoff and stream discharge
- reduction of water infiltration and groundwater recharge
- development of erosional gullies and sand dunes
- change in the surface microclimate that enhances aridity
- drying up of wells and springs
- reduction of seed germination of native plants.

Socioeconomic impacts include a lack of farm productivity, rural unemployment, migration, silting of dams and reservoirs, hunger and malnutrition. The effects of soil degradation are not limited to poor countries.

Figure 3.12 Check dam to retain water and soil, Eastern Cape Province, South Africa

Be a critical thinker

Does soil degradation occur only in poor countries?

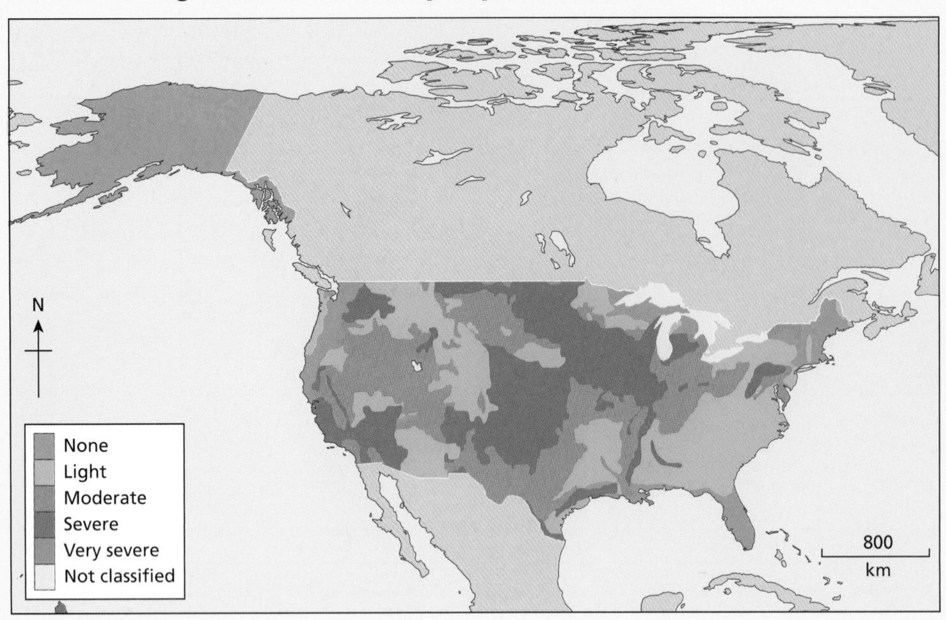

Figure 3.13 Soil degradation in the USA

The simple answer is no. There is considerable soil degradation in richer countries such as the USA and Australia. The USA, for example, experiences many of the features and processes associated with land degradation, such as:

- increased dust storm frequency
- reactivation of sand dunes
- erosion of steep gullies
- falling groundwater levels
- ground subsidence
- waterlogging and salinization
- overgrazing
- invasion of alien plants
- chemical contamination
- sedimentation in dams and reservoirs.

The most famous example is that of the Dust Bowl of the 1930s. This resulted from a combination of hot, dry years coinciding with a rapid colonization of the prairies by settler farmers. These farmers had been attracted into the region by a succession of wetter years during the 1920s, which had enabled higher levels of production.

Degradation is still a major problem in the USA. Centre-pivot irrigation has reduced groundwater levels in places such as the High Plains, while irrigation in the Central Valley of California has resulted in widespread salinization. High salt levels have resulted in reduced agricultural yields in about 20% of irrigated land. Shelterbelts and windbreaks have been removed to expand the cultivated areas; military training and urban growth have had more local impacts.

Alien plants such as the salt cedar have colonized parts of the Rio Grande valley, and use up almost 45% of the total water available. Ultimately, the rapid growth in population remains the fundamental cause of degradation in the USA. Degradation has followed human occupancy into dryland areas, and moved westwards over time.

However, it is also common in poorer countries such as Tunisia, where land degradation has increased rapidly since independence in 1958. In an attempt to control the rapidly increasing population and provide more food, the government embarked on a programme of construction, building villages, schools and clinics, and drilling new boreholes. The result was the settlement of many nomads in expanded villages and new villages.

Alongside the increase in population was an increase in the number of livestock. However, an important change was the decline of collective management of livestock, and an increase in privately managed flocks. In addition, settlers kept their stocks close to the village, resulting in overgrazing and degradation close to many villages.

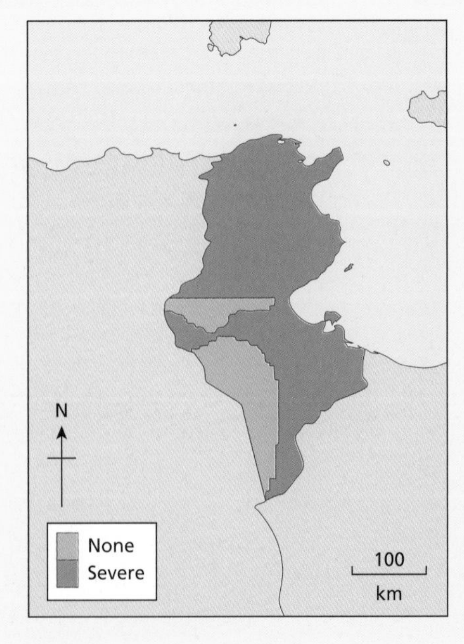

Figure 3.14 Soil degradation in Tunisia

Managing soil degradation

The European Union Soil Charter

In 1990 European Union ministers met to discuss the problems of soil pollution. Over 30% of European soils were considered to be moderately or severely damaged. As a result they drew up the Soil Charter. Its main points are:

- Soil is one of the world's most important resources.
- Soil is a non-renewable resource that is easily destroyed.
- Farmers must use methods to preserve soil quality.
- Soil needs to be protected against erosion and pollution.
- Urban development causes great damage to soil.
- Soil must be managed in a sustainable way.

Abatement strategies such as afforestation for combating accelerated soil erosion are lacking in many areas. To reduce the risk of soil erosion, farmers are encouraged towards more extensive management practices such as organic farming, afforestation, pasture extension and benign crop production. Nevertheless, there is a need for policymakers and the public to intensify efforts to combat the pressures and risks to the soil resource.

Tackling soil erosion

Methods to reduce or prevent erosion can be mechanical, e.g. physical barriers such as embankments and windbreaks, or they may focus on vegetation cover and soil husbandry. Overland flow can be reduced by increasing infiltration.

Mechanical methods include bunding, terracing and contour ploughing, and shelterbelts such as trees or hedges. The key is to prevent or slow the movement of rainwater downslope. Contour ploughing takes advantage of the ridges formed at right angles to the slope to act to prevent or slow the downward accretion of soil and water.

On steep slopes and in areas with heavy rainfall such as the monsoon in southeast Asia, contour ploughing is insufficient and terracing is undertaken. The slope is broken up into a series of flat steps, with bunds (raised levées) at the edge. The use of terracing allows areas to be cultivated that would not otherwise be suitable. In areas where wind erosion is a problem, shelterbelts of trees or hedgerows are used. The trees act as a barrier to the wind and disturb its flow. Wind speeds are reduced, and this reduces the wind's ability to disturb the topsoil and erode particles.

Preventing erosion by different cropping techniques largely focuses on maintaining a crop cover for as long as possible; keeping in place the stubble and root structure of the crop after harvesting; and planting a grass crop. Grass roots bind the soil, minimizing the action of the wind and rain on a bare soil surface. Increased organic content allows the soil to hold more water, thus preventing aerial erosion and

To research

Find out about the Dust Bowl of the 1930s. Whereabouts was it? What was the effect on the natural environment? What was the effect on population?

To do:

a Describe the distribution of severely degraded soils as shown in Figure 3.11.

b Comment on the causes of soil degradation as shown in Figures 3.7 and 3.9.

c In what ways do the causes of soil degradation in rich countries differ from those in poor countries?

d Briefly explain the environmental and socioeconomic impacts of soil degradation.

e In what ways is it possible to tackle soil degradation?

To research

Visit www.fao.org, the portal for the land and water division of the FAO. The website gives access to soil degradation assessments by country. Data on soil degradation have been linked with population numbers and densities in the mapping units. The extent of each severity class of soil degradation and the associated population numbers (LandScan 2000) in the area mapped are given for each country.

stabilizing the soil structure. In addition, care is taken over the use of heavy machinery on wet soils and when ploughing on soil sensitive to erosion, to prevent damage to the soil structure.

Tackling contaminated soils

There are three main approaches in the management of salt-affected soils:

- flushing the soil and leaching the salt away
- applying chemicals such as gypsum (calcium sulphate) to replace the sodium ions on the clay and colloids with calcium ones
- reducing evaporation losses to lessen the upward movement of water in the soil.

Equally specialist methods are needed to decontaminate land made toxic by chemical degradation.

Water usage and change

Over the past century the world's population has tripled, while our water use has increased sixfold. Some rivers that once reached the sea, such as the Colorado in the USA, no longer do so. Moreover:

- half the world's wetlands have disappeared in the same period
- 20% of freshwater species are endangered or extinct
- many important aquifers are being depleted
- water tables in many parts of the world are dropping at an alarming rate.

To do:

a Identify how water is used at a regional scale.

b Examine environmental and human factors affecting patterns and trends in physical and economic water scarcity.

c Examine the factors affecting access to safe drinking water.

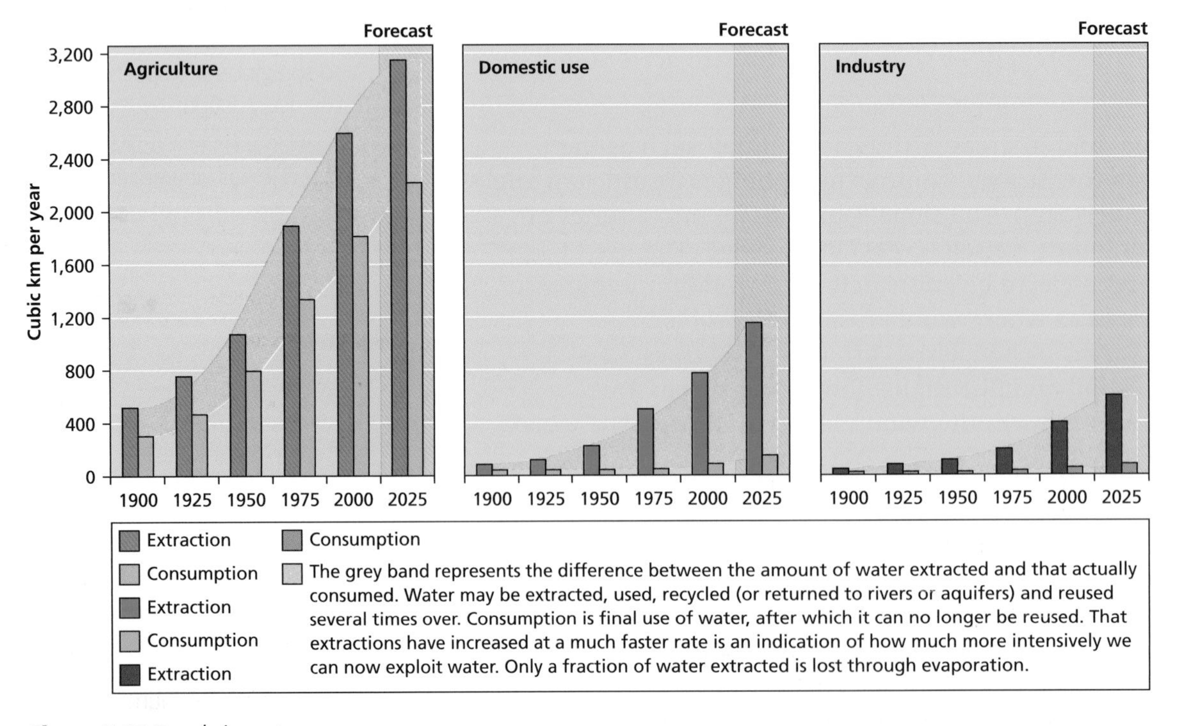

Figure 3.15 Trends in water use

World water use is projected to increase by about 50% in the next 30 years. By 2025, 4 billion people – half the world's population at that time – will live under conditions of severe water stress, specially in Africa, the Middle East and south Asia. Disputes over scarce water resources may lead to an increase in armed conflicts. Currently, about 1.1 billion people lack access to safe water, 2.6 billion are without adequate sanitation, and more than 4 billion do not have their wastewater treated to any degree. These numbers are likely to grow worse in the coming decades.

Water supply

Water supply depends on several factors in the water cycle, which include:

- rates of rainfall
- evaporation
- the use of water by plants (transpiration)
- river and groundwater flows.

Less than 1% of all water is available for people to use (figure 3.16). Most fresh water is stored as ice sheets and glaciers. Globally, around 12,500 cubic kilometres of water are considered available for human use on an annual basis. This amounts to about 6,600 cubic metres per person per year. If current trends continue, only 4,800 cubic metres will be available in 2025. The world's available freshwater supply is not distributed evenly around the globe, either seasonally or from year to year:

- About three-quarters of annual rainfall occur in areas containing less than a third of the world's population
- Two-thirds of the world's population live in the areas receiving only a quarter of the world's annual rainfall.

Water stress

When per capita water supply is less than 1,700 cubic metres per year, an area suffers from "water stress" and is subject to frequent water shortages. In many of these areas today, water supply is actually less than 1,000 cubic metres per capita, which causes serious problems for food production and economic development. Today, 2.3 billion people live in water-stressed areas. If current trends continue, water stress will affect 3.5 billion people – 48% of the world's projected population – in 2025.

Water use

The quantity of water used today for all purposes exceeds 3,700 cubic kilometres per year. Agriculture is the largest user, consuming almost two-thirds of all water drawn from rivers, lakes and groundwater. Since 1960, water use for crop irrigation has risen by 60–70%. Industry uses about 20% of available water, and the municipal sector uses about 10%.

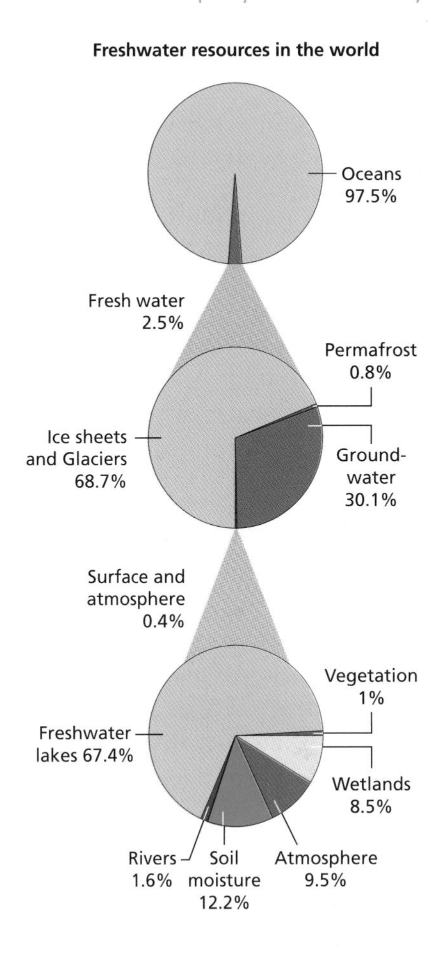

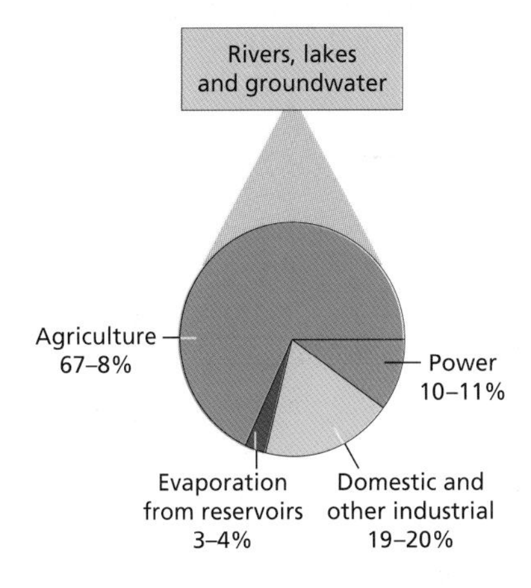

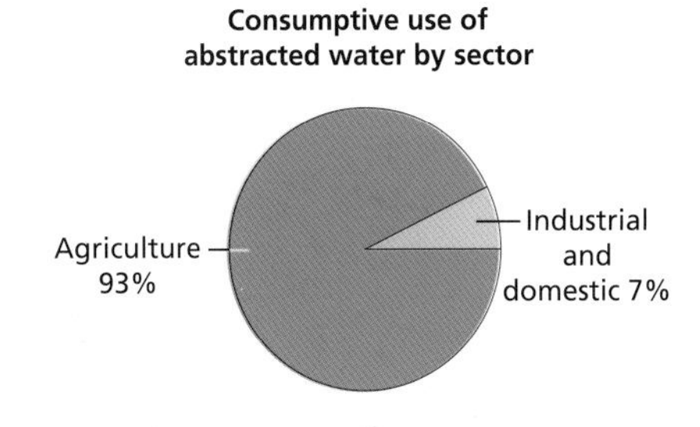

Figure 3.16 Availability of fresh water supplies

Population growth, urbanization and industrialization have increased the use of water in these sectors. As world population and industrial output have increased, water use has accelerated, and this is projected to continue. By 2025 global availability of fresh water may drop to an estimated 5,100 cubic metres per person per year, a decrease of 25% on the 2000 figure.

TOK Link

Most water collection is done by women and children. What are the health implications of this?

Water scarcity

The level of water scarcity in a country depends on precipitation and water availability, population growth, demand for water, affordability of supplies and infrastructure. Where water supplies are inadequate, two types of **water scarcity** exist:

- **physical water scarcity,** where water consumption exceeds 60% of the usable supply. To help meet water needs some countries such as Saudi Arabia and Kuwait have to import much of their food and invest in desalinization plants.
- **economic water scarcity,** where a country physically has sufficient water to meet its needs, but requires additional storage and transport facilities. This means having to embark on large and expensive water-development projects, as in many sub-Saharan countries.

In addition, the exhaustion of traditional water sources such as wells and seasonal rivers is what most affects access to adequate water supplies. In many poorer countries farmers use, on average, twice as much water per hectare as in industrialized countries, yet their yields can be three times lower.

Physical water scarcity – lack of available water where water resource development is approaching or has exceeded unsustainable levels; it relates availability to demand and implies that arid areas are not necessarily water scarce.
Economic water scarcity – lack of water where water is available locally, but not accessible for human, institutional or financial capital reasons.

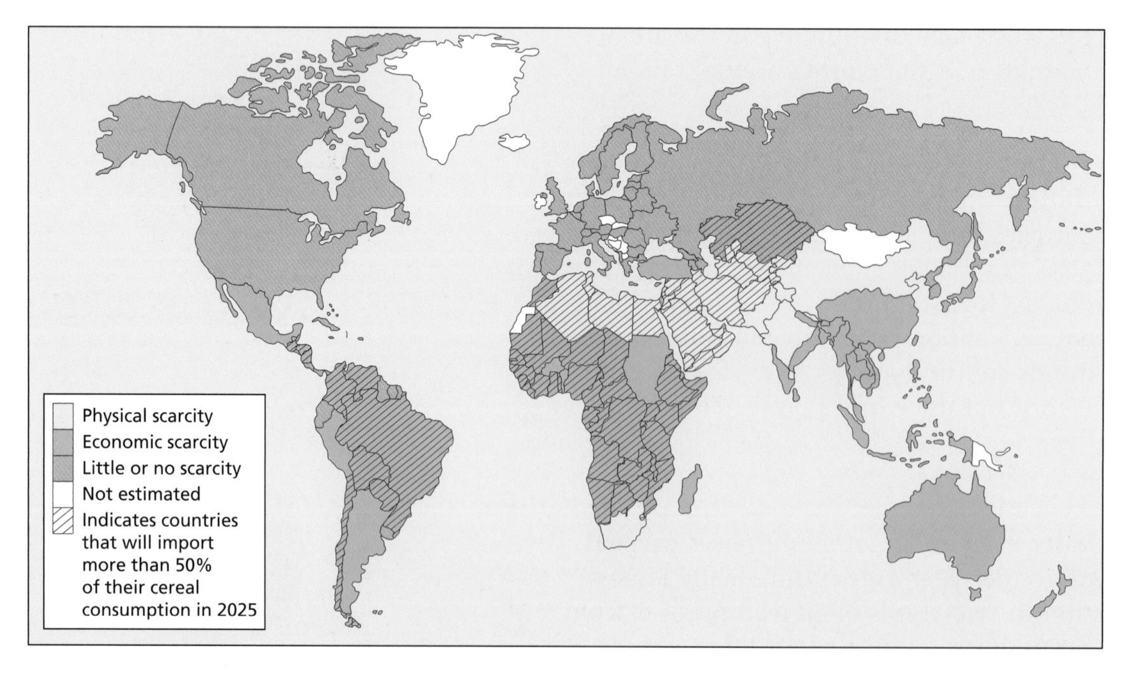

Figure 3.17 Estimated water scarcity, 2025

Water quality

Water also needs to be of an adequate quality for consumption. However, in developing countries too many people lack access to safe and affordable water supplies and sanitation. An estimated 4 million deaths each year are from water-related disease, particularly cholera, hepatitis and malaria and other parasitic diseases. Water quality may be affected by sewage, by fertilizers and pesticides from farming, and by heavy metals and acids from industrial processes and transport. Factors affecting access to safe water include:

- water availability
- water infrastructure
- cost of water.

	World population (billions)			Proportion of world population	
	total (billions)	*with access (billions)*	*without access (billions)*	*with access (%)*	*without access (%)*
Urban water supply	2.8	2.7	0.1	94	6
Rural water supply	3.2	2.3	0.9	71	29
Total water supply	6.1	5.0	1.1	82	18
Urban sanitation	2.8	2.4	0.4	86	14
Rural sanitation	3.2	1.2	2.0	38	62
Total sanitation	6.1	3.6	2.5	60	40

NB Figures have been rounded to the nearest 100 million.

Table 3.5 Global water supply and sanitation coverage

To research

Visit the official UN site for the MDG indicators, http://mdgs.un.org, and look up "Proportion of the population using improved sanitation facilities" to see the proportions of the urban and rural populations with access to improved sanitation facilities.

For data on water and sanitation in countries in which Unicef is working, visit www.unicef.org. Click on the icon for the countries of your choice, and investigate their access to water and sanitation. The site explores some thought-provoking issues regarding access to water:

a Why are women and girls more vulnerable to water shortages?

b Why are there rural and urban differences in access to water?

c Why are the young and old most vulnerable?

d How are water and sanitation linked to disease?

Urban areas are better served than rural areas, and countries in Asia, Latin America and the Caribbean are better off than African countries. In the case of Asia, China and India alone comprised some 2,280 million people in the year 2000, or over 60% of the region's total population. Many piped water systems, however, do not meet water quality criteria, leading more people to rely on bottled water bought in markets for personal use (as in major cities in Colombia, India, Mexico, Thailand, Venezuela and Yemen).

In some cases the poor actually pay more for their water than the rich. For example, in Port-au-Prince, Haiti, surveys have shown that households connected to the water system typically

To do:

Choose two or more contrasting countries and describe:

a how access to sanitation has changed over time

b the difference in access between rural and urban dwellers.

paid around $1.00 per cubic metre, while unconnected customers forced to purchase water from mobile vendors paid from $5.50 to a staggering $16.50 per cubic metre.

Sanitation and population growth

Fewer people have adequate sanitation than safe water, and the global provision of sanitation is not keeping up with population growth. Progress towards the MDG sanitation goal has slowed down, and between 1990 and 2000 the number of people without adequate sanitation rose from 2.6 billion to 3.3 billion. Least access to sanitation occurs in Asia (48%), especially in rural areas.

The following technologies are considered to be "improved":

Water supply	Sanitation
Household connection	Connection to a public sewer
Public standpipe	Connection to a septic system
Borehole	Pour-flush latrine
Protecting dug well	Simple pit latrine
Protecting spring	Ventilated improved pit latrine
Rainwater collection	

The following technologies are considered "not improved":

Water supply	Sanitation
Unprotected well	Service or bucket latrines (where excreta are manually removed)
Unprotected spring	
Vendor-provided water	Public latrines
Bottled water	Open latrine
Tanker truck provision of water	

Table 3.6 Water supply and sanitation technologies considered to be "improved" and "not improved"

In areas of rapid population growth in many LEDCs, rivers are now little more than open sewers. In Latin America only 2% of sewage receives any treatment at all. With squatter settlements in many of the world's poorest cities expanding rapidly, and local authorities unable to or legally prevented from providing sanitation, the situation is likely to deteriorate. Some 160 000 people are moving to cities from the countryside every day. At least 600 million people in Africa, Asia and Latin America now live in squatter settlements without any sanitation whatever, and governments are unable to cope.

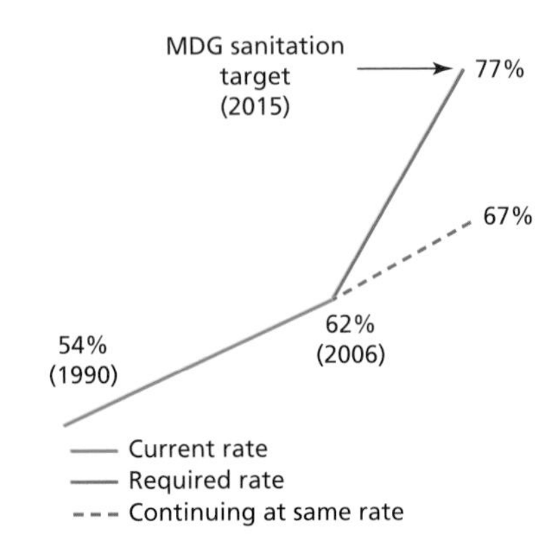

Figure 3.18 Progress towards better sanitation

Be a critical thinker

WDR 2010: development, climate change and water resources

According to the World Development Report 2010, climate change will alter the timing, availability and quality of water resources, making it harder to produce enough food for the world's growing population. Countries will need to cooperate to manage shared water resources to improve food trade.

While the world as a whole will get wetter as warming speeds up the hydrological cycle, increased evaporation will make drought conditions more prevalent (Figure 3.19). Most places will experience more unpredictable, intense and variable precipitation, with longer dry periods between them, so that traditional agricultural and water management practices are no longer useful.

To cope with climate change, countries will need to produce more from water and protect it better. Increasing knowledge about the world's water will improve management. To manage water well, it is crucial to know how much water is available in any basin and what it is used for.

Applying and enforcing sound water policies will be even more important, in order to allocate water efficiently and limit consumption to safe levels. Tradable water rights could improve water management in the long term but are not realistic short-term options in most developing countries. Climate change will require investing in new technologies and improving the application of existing

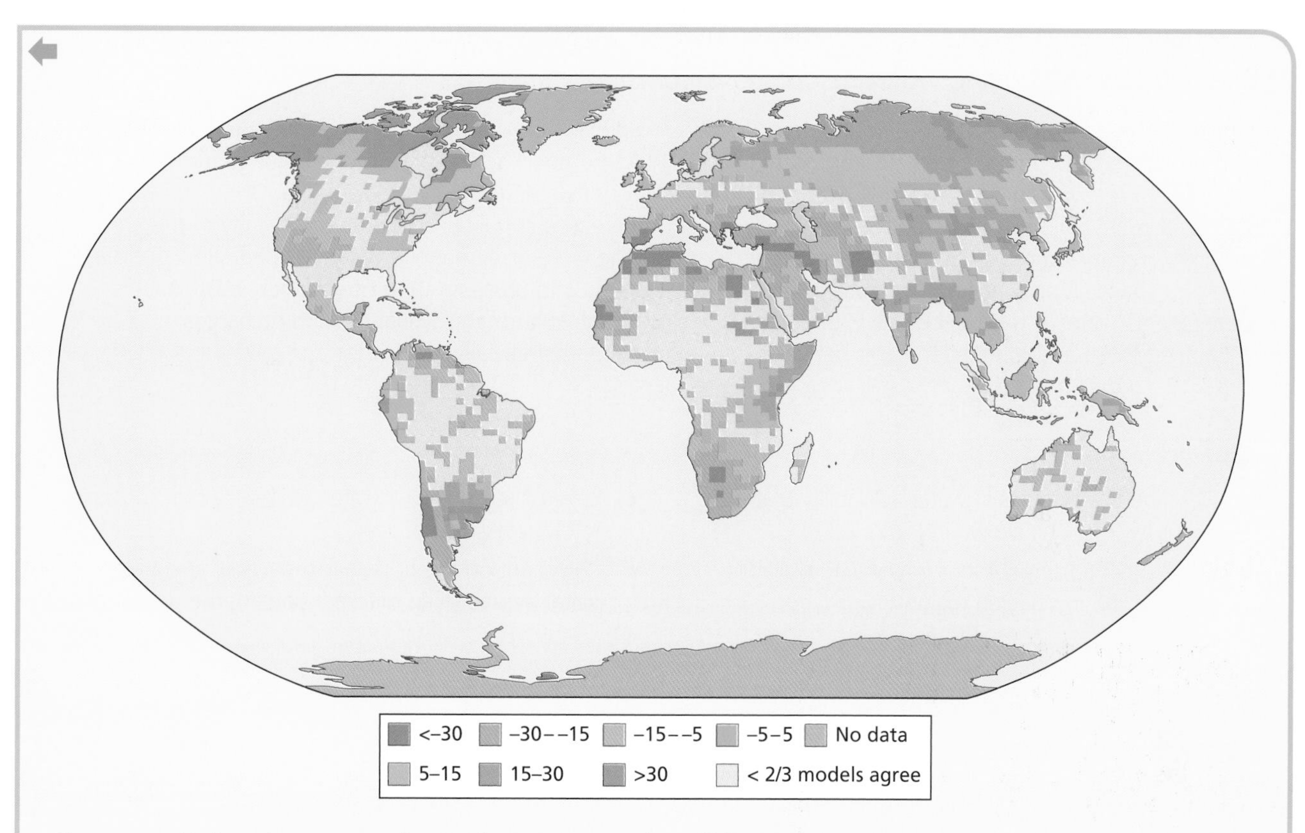

Figure 3.19 Projected percentage changes in average annual runoff, by 2050

technologies. Non-conventional technologies can increase water availability in some water-scarce regions. Water supplies can be enhanced by desalinating seawater or brackish water and reusing treated wastewater.

Water storage can help with increased variability. Groundwater is a "cushion" for coping with unreliable public supplies and rainfall. For example, in India it supplies 60% of irrigated agriculture and 85% of rural drinking water as well as half the drinking water received by households in Delhi. Well managed, groundwater can continue to act as a natural buffer. But it is far from well managed. In arid regions across the world, aquifers are overexploited. Up to a quarter of India's annual agricultural harvest is estimated to be at risk because of groundwater depletion.

There appears to be scope for increasing the productivity of water in rainfed agriculture, which provides livelihoods for the majority of the world's poor, generates more than half of the gross value of the world's crops, and accounts for 80% of the world's crop water use. Options include mulching, conservation tillage and similar techniques that retain water in the soil so that less is lost to evaporation and more is available to plants. Other options involve small-scale rainwater storage, or water harvesting (Figure 3.20).

Climate change will depress agricultural yields, and so we will need to increase productivity while protecting the environment. Efforts to mitigate climate change will put more pressure on land, and with growing populations, more carnivorous tastes and increased demand for dairy products, the world will require highly productive and diverse agricultural landscapes.

Figure 3.20 Water harvesting in Antigua

Case study: *Managing water resources in Tunisia*

Tunisia is a good example of the demands on water managers in countries that are approaching the limits of their resources. With only 400 cubic metres of renewable resources per capita, which are highly variable and distributed unevenly over time and space, Tunisia has a huge challenge in managing its water. It has built dams with conduits to connect them and to transfer water between different areas of the country. The stored water can be pumped across the mountain range into the country's principal river basin. The new water both increases supply and dilutes the salinity in the area where water demand is highest. In addition, Tunisia treats and reuses a third of its urban wastewater for agriculture and wetlands, and recharges aquifers artificially.

To do:

a Describe the changes in water availability that may occur by 2050, as shown in Figure 3.19.

b Suggest contrasting reasons why water availability may decline in selected regions.

c Suggest reasons why water availability may increase in other regions.

d Comment on the potential impacts of changing water availability in different parts of the world.

Case study: *Water diversion in China*

China's controversial South–North water diversion project channels water from the south of Hebei province to Beijing in the north. The water level at Wangkuai Reservoir, one of the biggest in the province, increased in 2008 in spite of drought, as the province sacrificed its water needs to Beijing's. The reservoir's water is sent to Beijing along a 307-kilometre-long waterway that cost more than $2 billion, along channels stretching for more than 1,000 kilometres.

Some complain that the scheme has exacerbated poverty in Hebei, forcing water-hungry and polluting industries to close and farmers to forsake rice growing for less water-intensive but also less profitable maize.

The extent of Beijing's predicament is not in doubt. In June 2008 Beijing's reservoirs were down to a 10th of their capacity. Beijing draws two-thirds of its water supply from underground and the water table is dropping by a metre a year, threatening geological disaster. Beijing has been trying to reduce demand by increasing water tariffs. Beijing's industries are now recycling 15% of their water, compared with 85% in developed countries.

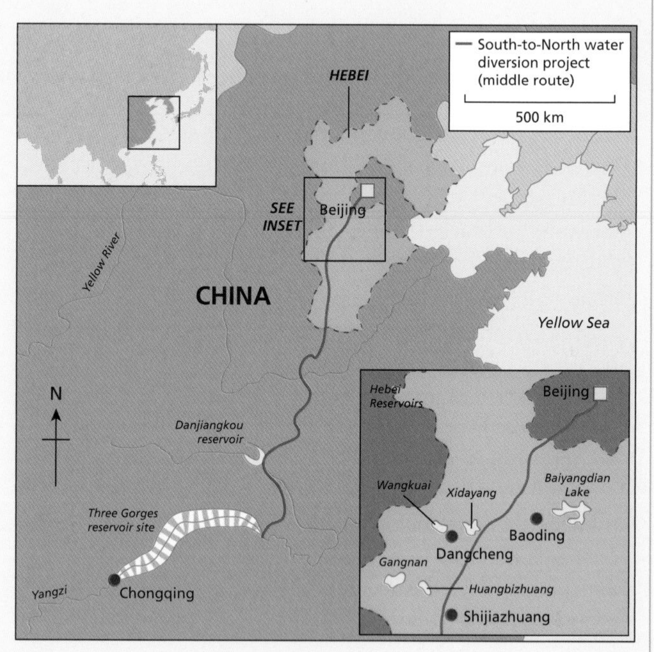

Figure 3.21 A water diversion scheme in China

Biodiversity and change

Biodiversity means biological diversity. It is the variety of all forms of life on earth – plants, animals, and microorganisms. It refers to species (species diversity), variations within species (genetic diversity), interdependence within species (ecosystem diversity), and habitat diversity. It is estimated that there are up to 30 million species on earth, of which only 1.4 million have yet been identified.

Figure 3.22 Natural tropical rainforest canopy, Temburong National Park, Brunei

Figure 3.23 Tropical rainforest, biodiversity and recreation, Temborung National Park, Brunei

The importance of the tropics

The tropics are the richest area for biodiversity (Figures 3.22 and 3.23), and tropical forests contain over 50% of the world's species in just 7% of the world's land. They account for 80% of the world's insects and 90% of primates. Tropical rainforests therefore provide immense support to life on earth. They not only have impacts on biodiversity, climate buffering, rainfall generation and soil stabilization, but also have health impacts, as well as industrial and recreational benefits. Increasingly, however, tropical rainforests are being reduced and fragmented. The Amazon rainforest is the main exception, although it is being destroyed rapidly.

Rainforests contain a vast store of water, and act as environmental air-conditioners by releasing vast amounts of water into the atmosphere. In Brazil alone, the Amazon releases about 20 billion tonnes each day. During the drought in Brazil in 2005, sea surface temperatures in the Caribbean rose by 3°C, possibly helping in the formation of Hurricane Katrina.

Be a critical thinker

Why is the destruction of the rainforest so detrimental?

A tree releases 8–10 times more moisture into the atmosphere than a square metre of ocean water. Remove the trees and the store of water disappears. As deforestation progresses, less water re-evaporates from the vegetation, and so the recycling of water diminishes. Evapotranspiration (EVT) rates from savanna grasslands are estimated to be only about a third of that of tropical rainforest. Thus mean annual rainfall is reduced, and the seasonality of rainfall increases.

Rainforests have a major impact on soil quality and quantity. Deforestation disrupts the closed system of nutrient cycling within tropical rainforests. Inorganic elements are released through burning and are quickly flushed out of the system by the high-intensity rains.

Soil erosion is also associated with deforestation. As a result of soil compaction, infiltration decreases, and overland runoff and surface erosion increases. Sandification is a process of selective erosion, when raindrop impact

washes away the finer particles of clay and humus, leaving behind the coarser and heavier sand. Evidence of sandification dates back to the 1890s in Santarem, Rondonia (Brazil).

As a result of the intense surface runoff and soil erosion, rivers have a higher flood peak and a shorter time lag. However, in the dry season river levels are lower, the rivers have greater turbidity (murkiness due to more sediment), an increased bed load, and carry more silt and clay in suspension.

Deforestation takes place for many reasons, some of which are summarized in Table 3.7.

Industrial uses	Ecological uses	Subsistence uses
Charcoal	Watershed protection	Fuelwood and charcoal
Saw logs	Flood and landslide protection	Fodder for agriculture
Gums, resins and oils		Building poles
Pulpwood	Soil erosion control	Pit sawing and saw milling
Plywood and veneer	Climate regulation, e.g. CO_2 and O_2 levels	Weaving materials and dyes
Industrial chemicals		Rearing silkworms
Medicines		Beekeeping
Genes for crops		Special woods and ashes
Tourism		Fruit and nuts

Table 3.7 The uses of tropical rainforest

Figure 3.24 Forest products, Sarawak

Case study: Deforestation of the tropical rainforest in southeast Asia

Forests in southeast Asia cover some 136 million hectares. It is predicted that by 2050 this natural forest will have been converted into agricultural land, forest plantation and other non-forested uses. Rates of annual forest loss in southeast Asia from 2000–2005 range from 0.5% for Papua New Guinea to 2% for Indonesia. In 2007, Indonesia was reclassified as the third largest contributor to global greenhouse gas emissions after the US and China.

Major threats to southeast Asian forests

Rapid population and economic growth is driving the degradation of these forests. Specific threats include:

- industrial logging concessions, worth about US $10.4 billion per annum
- illegal logging, especially in Indonesia, leading to an estimated US $4 billion in lost government revenues per annum
- agriculture, predominantly palm oil and rubber, worth about US $17.8 billion per annum and using some 7.6 million hectares of land
- burning and drainage of carbon-rich forested peatlands, particularly in Malaysia and Indonesia
- mining and petroleum, particularly in Papua New Guinea where it contributes 25% of GDP annually.

The effect of deforestation in southeast Asia is likely to have a major impact on rainfall generation. In addition to the carbon stored in its forests, at least 42 billion tonnes of soil carbon are stored in the forested tropical peatlands of southeast Asia. The region is also home to four of the world's 25 biodiversity hotspots. On account of the region's unique geological history, Indonesia and parts of Malaysia reach 60% endemism for plants and reptiles and 80% for amphibians; Papua New Guinea tops 80% endemism for mammals, reptiles and amphibians.

Figure 3.25 The natural rainforest replaced by a rubber plantation, a form of monoculture

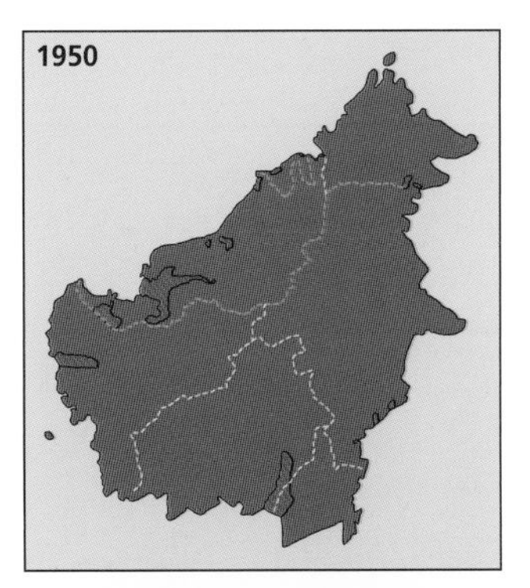

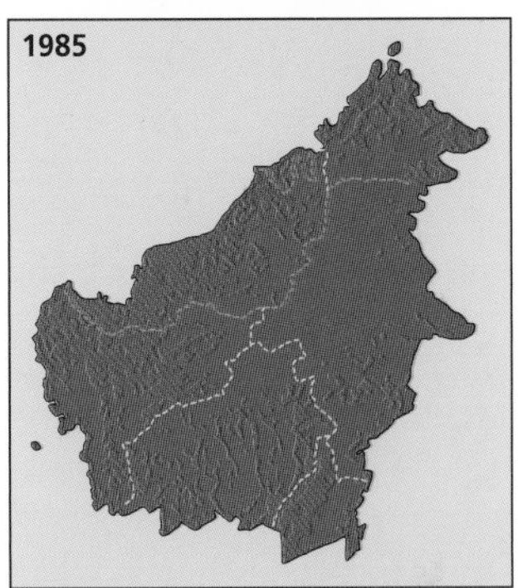

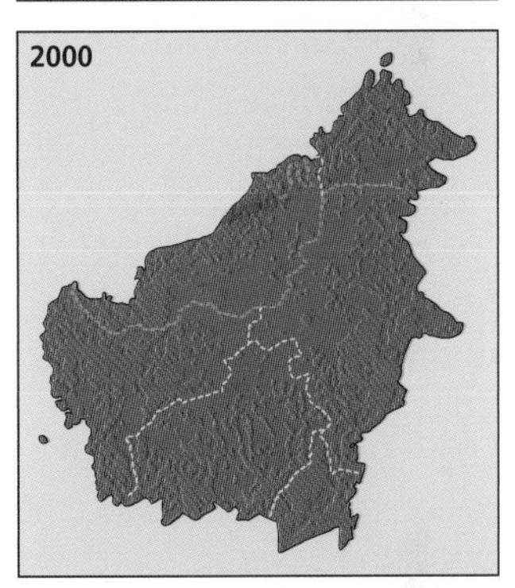

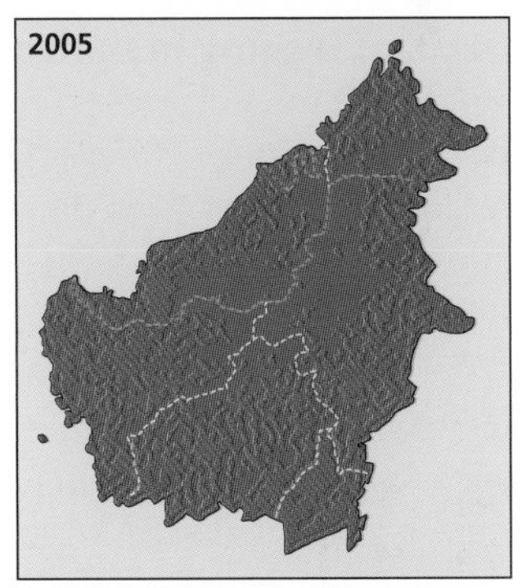

Figure 3.26 Deforestation in Borneo, 1950–2020

To do:

Describe the changes in forest cover in Borneo, as shown in Figure 3.26.

TOK Link

What do we mean by deforestation?

Deforestation can take many forms. Some are more severe than others. Some forms may lead to irreversible change, while other forms cause short-term changes.

- Is it complete removal of trees from an area?
- Is it a change in vegetation?
- What if natural vegetation succession (change) is allowed to occur?
- Does it include selective logging?
- Does it include the cutting of only parts of trees?
- Does it include cutting and then replanting?

Case study: *"Social" forestry in the Congo*

Lamoko on the Maringa river is on the edge of the virgin rainforest in the centre of the Democratic Republic of the Congo (DRC). In 2005 representatives of a major timber firm arrived to negotiate a contract with the traditional landowners. The company promised to build three simple village schools and pharmacies for Lamoko and other communities in the area. It was the kind of "social responsibility" agreement encouraged by the World Bank, but when the villagers found out that their forest had been "sold" so cheaply, they were furious. The company has driven logging roads deep into the forests near Lamoko (Figure 3.27), in order to extract and export teak and sapele trees, but the villages have yet to see their schools and pharmacies.

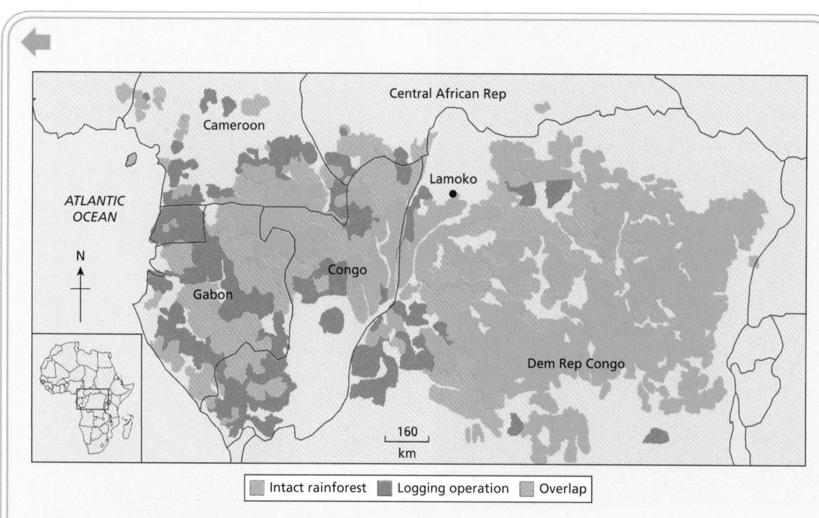

Figure 3.27 Deforestation in the Congo

The Lamoko agreement is just one of many contracts, or concessions, that European companies have signed with tribal chiefs in the DRC as the country begins to recover from a decade of civil wars and dictatorship. As many as 40 million of the poorest people in Africa depend on the Congolese forests. All the concessions handed out by the transition government are in inhabited areas. More than a third are home to pygmy communities.

It is believed that 20 foreign-owned forestry companies are active in the DRC, and that Chinese and other logging groups are also seeking to gain concessions. Most of the major logging companies, including Danzer, Trans-M, TB, NST, Olan, and Sicobois, have concession contracts signed after the World Bank moratorium. The companies, which export both logs and sawn timber, supply wood all over Europe, mostly as finished products such as flooring, windows, furniture and doors. African teak wood is protected by global agreement and cannot be exported from some tropical countries such as Cameroon, which have few trees left, but there are still no restrictions on its export from the DRC.

Effects of logging in the Congo

According to the Congolese Timber Industries Association, 80% of the logging in the Congo is illegal. Over 20 million hectares have been given over to logging firms. If the forests are felled, up to 34 billion tonnes of carbon could be released. The Congo rainforest stores up to 8% of the earth's carbon, and as much as 40% of the Congo's rainforest will be lost if illegal logging continues to take place.

Greenpeace and other international forestry groups say the fate of the Congo forests depends on the World Bank and other donors, including Britain, rejecting industrial logging, demanding a comprehensive land-use plan for a country that is effectively lawless. The bank accepts that logging could destroy the forests in a short time, leading to immense social problems.

To research

Visit www.greenpeace.org for the Greenpeace report, Carving up the Congo.

Case study: *Biodiversity and conservation at Batang Ai National Park, Sarawak*

Batang Ai National Park is part of Borneo's largest transnational protected area for tropical rainforest conservation. The 240-square-kilometre park adjoins the Lanjak-Entimau Wildlife Sanctuary in Malaysia and the Bentuang-Karimun National Park in Indonesia. Together these protected areas cover almost 10 000 square kilometres and form a sanctuary for one of the few viable orangutan populations in Borneo (estimated at at least 1,000 animals) as well as many other endangered species. For conservation reasons, Batang Ai National Park is the only part of this area open to visitors, but it has the highest orangutan population density in central Borneo (up to 1.7 animals per square kilometre).

The whole area has remarkable biodiversity, with over 1,000 tree species and almost 200 herbs, shrubs and climbers recorded in Batang Ai and the adjoining forests. As well as orangutans the park is also home to other primates, including Bornean gibbons, white-fronted langurs, maroon or red langurs, long-tailed macaques, pig-tailed macaques, and the nocturnal western tarsier and slow loris. Other mammals include the rare clouded leopard, two species of civet cat, sun bears, bearded pigs, barking deer, sambar deer, mouse deer, martens, weasels, otters, porcupines and giant squirrels, as well as a host of smaller squirrels, other rodents and tree shrews. With the exception of the macaques, most of these animals are shy and hard to spot.

Five of Sarawak's eight hornbill species are found in Batang Ai, including the rhinoceros hornbill. Other vertebrates found at Batang Ai and its adjoining forests include 13 snake species, 12 lizard species, two types of river turtle, spiny hill turtles, 52 species of frog, two burrowing legless amphibians or caecilians and over 80 species of fish. The incredible variety of insects and other invertebrates defies description.

Batang Ai has a unique mix of terrain. The park is primarily mixed lowland dipterocarp forest, with hill dipterocarp forest more than 500 metres above sea level, but in the southern edge of the park there is also old secondary forest which is on its way to becoming primary forest, and active areas of shifting cultivation dotted with ancient burial grounds. Batang Ai's biodiversity is very important for the local Iban population, who gather over 140 different kinds of medicinal plant from the forest, and eat 114 varieties of wild fruit and 36 varieties of jungle vegetable. Forest trees and plants are also important as a source of wood, fibres, rattan, bamboo and aromatic resins.

The Iban people have been settled in and around the park for over 400 years, and form an integral part of its ecosystem. They have historically played a major role in orangutan conservation as they have a strict taboo against harming these animals. Local communities were involved in the planning process before the park was established in 1991, and agreed to limit their activities in the park to farming of previously farmed areas and sustainable gathering of jungle produce. In return they benefit from employment in the park, and have formed their own community cooperative (Kooperasi Serbaguna Ulu Batang Ai) to provide transport, accommodation and guiding for visitors, and to market their handicrafts.

Figure 3.28 Shifting cultivation and the dam created at Batang Ai, Sarawak

Figure 3.29 Tropical rainforest along with shifting cultivation: rice growing in Sarawak

Environmental sustainability

The Environmental Sustainability Index (ESI) is produced by environmental experts from Yale and Columbia Universities. Using 21 indicators and 76 measurements, including natural resource endowments, past and present pollution levels and policy efforts, the index creates a "sustainability score" for each country, with higher scores indicating better environmental sustainability. In 2008 the term Environmental Performance Index was used in place of the ESI. The EPI focuses on two overarching environmental objectives:

- reducing environmental stresses to human health
- promoting ecosystem vitality and sound natural resource management.

The 10 most sustainable countries, as ranked by the ESI and the EPI, are dominated by wealthy, sparsely populated nations with an abundance of natural resources. Switzerland has been ranked first, with Norway, Sweden and Finland all figuring in the top five. The only developing nations in the top 10 are Costa Rica and Colombia, both of which have relatively low population densities and an abundance of natural resources. Conversely, the only densely populated countries that have received even above average rankings are Japan, Germany, the Netherlands and Italy, some of the richest countries on the list.

The effects of climate change on sustainability

Environmental sustainability is essential for helping poor people. They are highly dependent on the environment and its resources (fresh water, crops, fish, etc.), which provide roughly two-thirds of household income for the rural poor.

Climate change is dramatically reshaping the environment upon which poor people depend. The effects of climate change include increased rainfall variability (meaning more droughts and increased flooding), reduced food security, spread of disease, increased risk of accidents and damage to infrastructure. The poor are most vulnerable to these changes and have limited capability to respond to them.

The effects of climate change require a response at global, national and local levels. Most countries already fail to manage their environmental resources in a sustainable way. Climate change makes this an even more urgent priority.

Some facts and figures

- Overfishing has led to the collapse of many fisheries. A quarter of global marine fish stocks are currently overexploited or significantly depleted.
- About 60% of the ecosystem services resources evaluated by the UN's Millennium Ecosystem Assessment are being degraded or used unsustainably.
- Between 10% and 30% of mammal, bird, and amphibian species face extinction.
- Global timber production has increased by 60% in the past four decades. This means that roughly 40% of forest area has been lost, and deforestation continues at a rate of 13 million hectares per annum.

Challenges and solutions

Environmental concerns are fundamental to long-term sustainable development. We need to improve our understanding of the environmental impact of development strategies and recognize the link between environmental degradation and poverty.

To research

Visit www.globalcanopy.org for information about the Forests First campaign in the fight against climate change, the pressures on rainforests, the ABC (Asia, Brazil and Congo) of rainforests, and the call to action.

To do:

a Outline the value of tropical rainforests.

b Describe and explain why tropical rainforests are at risk from deforestation.

c Evaluate methods used to conserve and/or manage rainforests.

EPI Rank	Country	EPI score
1	Switzerland	95.5
3	Sweden	93.1
13	Germany	86.3
21	Japan	84.5
39	USA	81.0
47	Mexico	79.8
51	South Korea	79.4
95	Zimbabwe	69.3
105	China	65.1
120	India	60.3
123	Ethiopia	58.8
125	Bangladesh	58.0
134	Iraq	53.9
149	Niger	39.1

Table 3.8 EPIs for selected countries, 2008

To research

Visit http://sedac.ciesin.columbia.edu for the 2008 EPI rankings. Investigate the EPIs of contrasting countries. Comment on the results that you find.

The poor, who are most dependent on natural resources and most affected by environmental degradation, lack the information or the access to participate in decision-making and policy development. In contrast, those who are most influential in policy development have little understanding of the costs and benefits associated with environmental policy. Economic growth and the environment are often still viewed as competing objectives, even though investing in environmental management can be cost-effective, and contributes to improving livelihoods.

The cost of environmental inaction

The conventional constraint on government and private sector action in taking new environmental protection measures has been concern about the costs. This narrow preoccupation has overshadowed the equally important consideration of the mounting economic, social and ecological costs of not acting. A recent World Bank study provides a stark assessment of the risks and enormous costs if no remedial action is taken. For example, the annual losses to Nigeria of not acting on its growing environmental problems are estimated to be more than US$ 5,000 million annually (see Table 3.9).

Annual costs of inaction	(US$ million/year)
Soil degradation	3,000
Water contamination	1,000
Deforestation	750
Coastal erosion	150
Gully erosion	100
Fishery losses	50
Water hyacinth	50
Wildlife losses	10
Total	**5,110**

Table 3.9 Costs to Nigeria of environmental inaction

To research

Visit www.epa.gov/sustainability, the USA's Environmental Protection Agency's (EPA) website on sustainability.

When is the UN Decade of Education for Sustainable Development? What is planned for this decade? Investigate this site and find examples of local or national strategies to achieve sustainable development.

Case study: Managing the Korup National Park, Cameroon

In 1986 the government of the Cameroon created the Korup National Park with the support of the WWF. Management of Korup is important, since it contains over 400 species of trees, 425 species of birds, 120 species of fish and 100 mammal species. Over 60 species occur only in Korup, and 170 species are considered to be endangered or vulnerable. Under Cameroon law, human activity in the park is limited to tourism, research and recreation. The project is designed to "protect and manage the National Park and integrate it into the local economy and regional development plans".

One example of sustainable development in Korup is the community forests project. Villagers obtain and manage large areas of forest for a long period of time, with regular reviews by the government and the WWF to ensure that they are using the forest in a sustainable way. Other projects have included those for natural resources (more than 30), management committees, village infrastructural development (more than 40) and income-generating and credit activities, such as sustainable forms of agroforestry (more than 70).

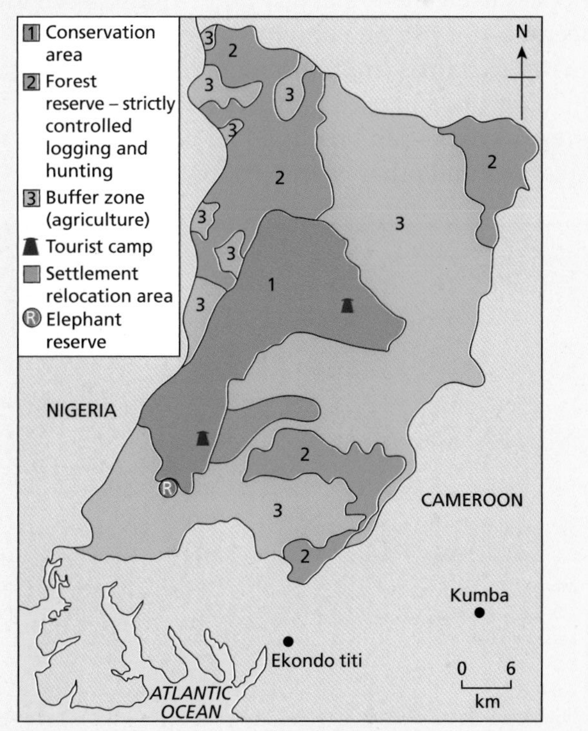

Figure 3.30 Location of Korup National Park

4 Patterns in resource consumption

By the end of this chapter you should be able to:

- understand the significance of ecological footprints, and differing theories about population and resources
- explain the changing patterns of energy consumption, with reference to fossil fuels and renewables
- describe the geopolitical and environmental impacts of oil, and trends in the production and consumption of oil
- describe conservation strategies and national and global initiatives to tackle climate change.

Ecological footprints

Everything we use for our daily needs and activities comes from natural resources. An ecological footprint is a measure of how much of those resources a population unit needs to support its consumption and take care of its wastes. The ecological footprint, measured in acres or hectares, calculates the amount of the earth's bioproductive space – ecologically productive land and water – a given population is consuming. The calculation takes into account the following:

Ecological footprint – the theoretical measurement of the amount of land and water a population requires to produce the resources it consumes and to absorb its waste, under prevailing technology.

- **arable land** – the amount of land required for growing crops
- **pasture land** – resources required for growing animals for meat, hides, milk, etc.
- **forests** – for fuel, furniture and buildings, and also for ecosystem services like climate stability and erosion prevention
- **oceans** – for fish and other marine products
- **infrastructure** – transportation, factories and housing, based on the built-up land used for these needs
- **energy costs** – land required for absorbing carbon dioxide emissions and other energy wastes.

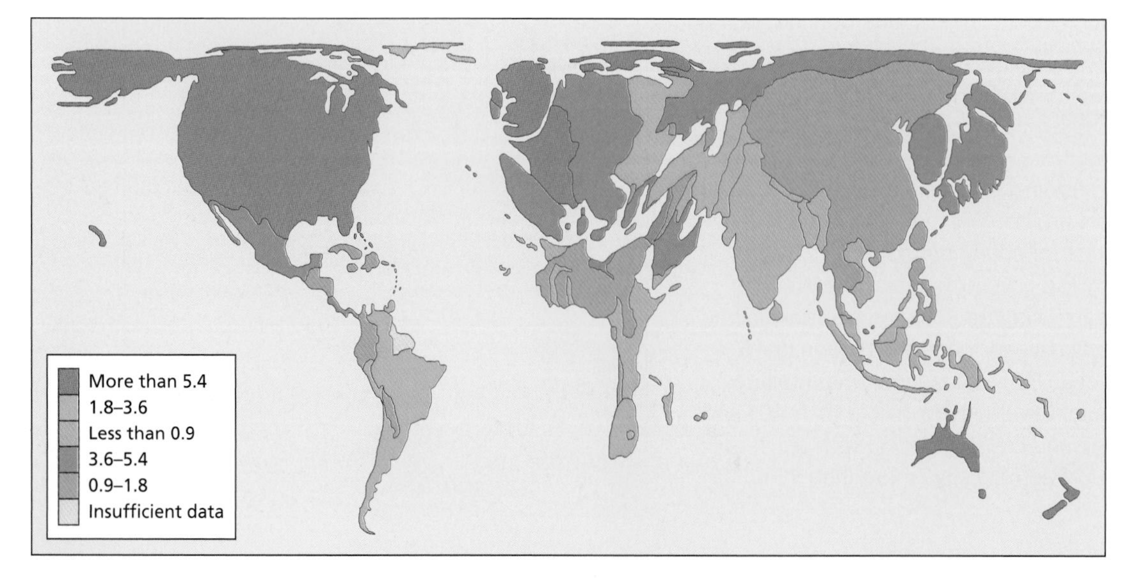

Figure 4.1 Global ecological footprints, global hectares per person

Species extinction and toxic pollution of the air, water and land are not yet taken into account in calculating ecological footprints.

Our global and national footprints

The planet's biological productive capacity (biocapacity) is estimated at 1.9 hectares or 4.7 acres per person. Currently, we are using up 2.2 hectares per person, so we are living beyond the planet's biocapacity to sustain us by 15%, or by a deficit of 0.4 hectares (one acre) per person. This deficit is showing up as failing natural ecosystems – forests, oceans, fisheries, coral reefs, rivers, soils and water – and global warming.

The planet's biocapacity is affected by the global population as well as the rate of consumption. Higher consumption depletes the planet's carrying, renewal and regeneration capacity. Estimates indicate that, if global population trends continue, the ecological footprint available to each person would be reduced to 1.5 hectares per person by 2050, and if the rest of the world adopted the same consumption rates as rich western countries, we would need four or five more planets to sustain ourselves.

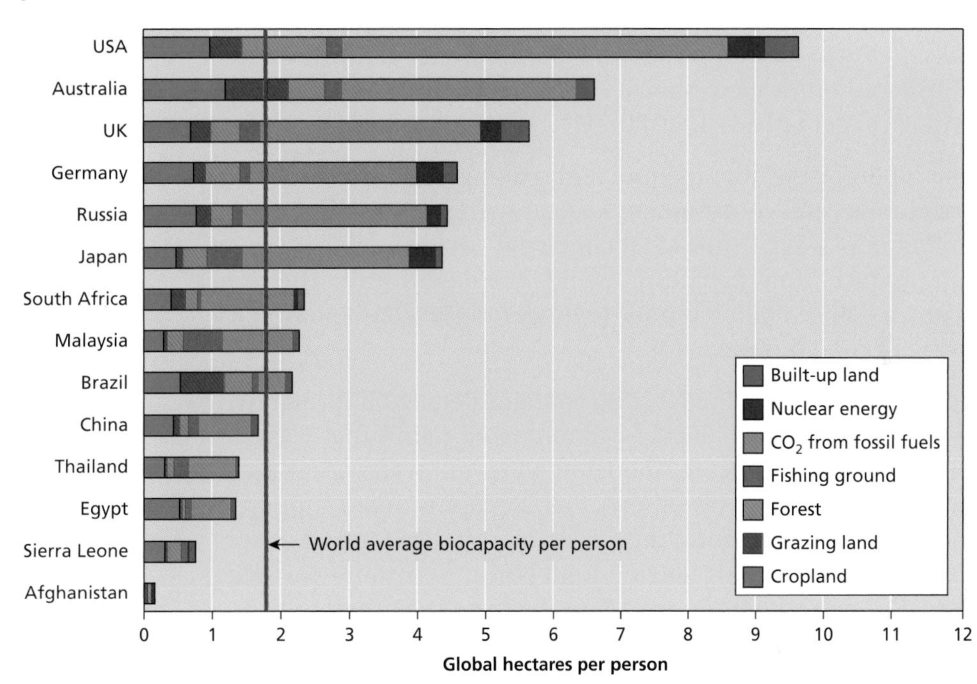

Figure 4.2 Ecological footprints, per person, by country, 2003

A person's footprint ranges vastly across the globe, from eight or more "global hectares" (20 acres or more) for the biggest consumers in the United Arab Emirates, the US, Kuwait and Denmark, to half a hectare in the Democratic Republic of Congo, Haiti, Afghanistan, Bangladesh and Malawi. The USA is the country with the largest per capita footprint in the world – a footprint of 9.57 hectares. If everyone on the planet were to live like an average American, our planet's biocapacity could only support about 1.2 billion people. On the other hand, if everyone lived like an average person in Bangladesh, the earth could support roughly 22 billion people.

The global ecological footprint has grown from about 70% of the planet's biological capacity in 1961 to about 120% of its biological capacity in 1999. Future projections show that humanity's footprint is likely to grow to about 180–220% of the earth's biological capacity by the year 2050.

To research

Visit www.worldmapper.org and find out the answers to the following:

1. Identify the seven countries with the highest ecological footprints.
2. Name one other Asian country that has a very high ecological footprint in relation to its area.
3. Name the European countries with the highest ecological footprints.

The Living Planet Index

The Living Planet Index measures trends in the earth's biological diversity. Based on figures from 2005, the index indicates that global biodiversity has declined by nearly a third since 1970. The US and China account for more than two-fifths of the planet's ecological footprint, with 21% each.

According to the Living Planet Report, 2008, the world is heading for an "ecological credit crunch" far worse than the financial crisis that began between 2008 and 2009. The report calculates that humans are using 30% more resources than the earth can replenish each year, which is leading to deforestation, degraded soils, polluted air and water, and dramatic declines in the numbers of fish and other species. As a result, people are running up an ecological debt of $4–4.5 trillion.

The report claims that the problem is also getting worse: populations and consumption keep growing faster than our abiliy to find new ways – through technology – of expanding what we can produce from the natural world. The downturn in the global economy since 2008 is a stark reminder of the consequences of living beyond our means.

For the first time the report also contains detailed information on the "water footprint" of every country, and claims that 50 countries are already experiencing "moderate to severe water stress on a year-round basis". It also shows that 27 countries are "importing" more than half the water they consume – in the form of water used to produce goods ranging from wheat to cotton – including the UK, Switzerland, Austria, Norway and the Netherlands.

To do:

a Study Figure 4.1, the map of global ecological footprints.

- **i** Describe how the map alters the size of countries to show their ecological footprint.
- **ii** Identify the countries with an ecological footprint of over 5.4 hectares per person.
- **iii** Identify three countries with an ecological footprint of less than 0.9 hectares per person.

b Study Figure 4.2, the divided bar chart, which shows the composition of the ecological footprint for countries in 2003.

- **i** Identify the main factor leading to the ecological footprint of the USA.
- **ii** Describe how the contribution from cropland to global ecological footprints varies by country.
- **iii** Describe how the contribution of grazing land to global ecological footprints varies by country.
- **iv** Compare the contributions of different sectors to the ecological footprints of South Africa and Malaysia.

Theories of population and resources

Thomas Malthus's theory of population

In 1798 the Reverend Thomas Malthus produced his *Essays on the Principle of Population Growth*. He believed that there was finite optimum population size in relation to food supply, and that any increase in population beyond this point would lead to a decline in the standard of living and to "war, famine and disease". He based his theory on two principles:

- in the absence of checks, population will grow at a geometric or exponential rate, i.e. 1, 2, 4, 8, 16...etc. and could double every 25 years
- food supply at best only increases at an arithmetic rate, i.e. 1, 2, 3, 4, 5... etc. (Figure 4.3).

If the time intervals were 25 years, in 100 years the ratio of population to food would be 16:5. Lack of food is therefore argued to be the ultimate check on population growth.

Malthus's principles used **potential** and not actual growth figures for population and food production. Because there is a limit to the amount of food that can be produced, it determines a "ceiling" to the population growth in a given country. Malthus suggested preventive and positive checks as two main ways by which population would be

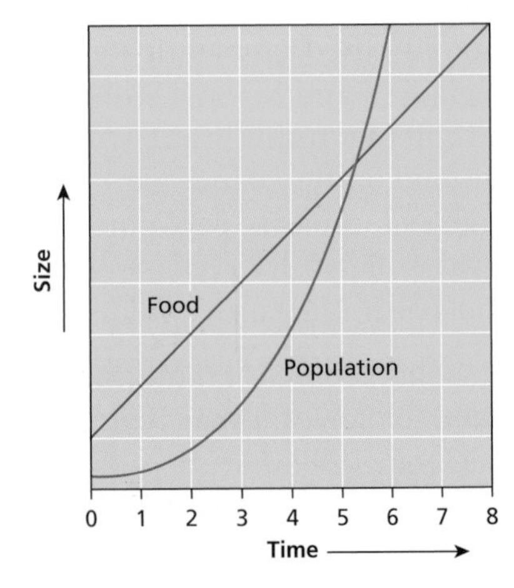

Figure 4.3 Malthus's views on the relationship between population growth and resources

curbed once this ceiling had been reached. Preventive checks include abstinence from marriage or a delay in the time of marriage and abstinence from sex within marriage, all of which would reduce the fertility rate. Malthus noticed strong negative correlations between wheat prices and marriage rate. Positive checks such as lack of food, disease and war directly affected death rates.

Figure 4.4 Overcrowding in India

Malthus suggested that the optimum population exists related to resources and the level of technology. This is now related to the concept of overpopulation and underpopulation rather than the optimum population. Optimum population is difficult to identify, and may vary as technology improves, resource availability changes, and attitudes change.

Even though the geometric rate of population increase is possible, it is rarely observed. The preventive and positive checks suggested by Malthus do not affect population growth, although contraception has since been introduced and should be considered a factor.

During the industrial revolution, agricultural production grew at a rapid rate (greater than arithmetic rate), exceeding the rate of population growth. Industrial development affected agricultural production through intensification (labour and capital) and extensification (more land); industry helped both. Malthus's ceiling limit on production was always ahead of and moving away from the population.

Since Malthus's time people have increased food production in many ways. These include draining marshlands, reclaiming land from the sea, cross-breeding of cattle, use of high-yielding varieties of crops, terracing steep slopes, growing crops in greenhouses, using more sophisticated irrigation techniques such as hydroponics, growing new foods such as soya, making artificial fertilizers, farming native crop and animal species, and fish farming.

Since the 1950s there have been two main phases of agricultural production, namely a mix of high yielding varieties (HYVs) improved yields in many countries and extensification in some temperate countries, and since the 1980s, intensification. These have created environmental issues such as loss of habitat, use of agrochemicals and high-energy farming.

Esther Boserup's theory of population

A different view – an anti-Malthusian one – is that of Esther Boserup (1965). She believed that people have the resources to increase food production. The greatest resource is knowledge and technology. When a need arises someone will find a solution (Figure 4.5). Although she was researching very small pre-technological villages, her views are widely held to be applicable to modern societies.

Figure 4.5 The Boserup approach: human ingenuity is the key (sign at the Posco Iron and Steel Works, Pohang, Korea)

Whereas Malthus thought that food supply limited population size, Boserup suggests that, on the contrary, in a preindustrial society an increase in population stimulated a change in agricultural techniques so that more food could be produced. Population growth has thus enabled agricultural development to occur.

Figure 4.6 Cultivation of steep slopes using terracing

She examined different land-use systems and their intensity of production. This was measured by the frequency of cropping. At one extreme was the forest fallow association with shifting cultivation; at its least intensive, any one piece of land would be used less than once every 100 years. At the other extreme was the multicropping system, with more than one harvest per year. She suggested that there was a close connection between the agricultural techniques used and the type of land-use system. The most primitive was shifting cultivation and fallow ploughing, with fallow reduction and increased cropping frequency occurring when higher yields were needed. She considered that any increase in the intensity of productivity by the adoption of new techniques would be unlikely unless population increased. Thus, population growth led to agricultural development and the growth of the food supply.

Boserup's theory was based on the idea that people knew the techniques required by a more intensive system, and adopted them when the population grew. If knowledge were not available then the agricultural system would regulate the population size in a given area.

Emile Durkheim's theory of population

Emile Durkheim was a French sociologist who thought that an increase in population density would lead to a greater division of labour, which would allow greater productivity to be attained (1893). He even suggested that population pressure was necessary to increase the division of labour. Borrowing from the Malthusian theory of population density and observing the changes in labour around him at a time of increasing industrialization, he noted that labour differentiation tended to increase in proportion to the social complexity and size of the population.

To do:

Study Figure 1.1 on page 1. Describe the growth of the world's population. To what extent is population growth occurring in (a) developed countries and (b) developing countries?

The Limits to Growth model

This study, also referred to as the Club of Rome model (1970), is a neo-Malthusian model, in that it has the same basic view as Malthus, although it suggests some potential solutions that Malthus would not have approved. The study examined the five basic factors that determine and therefore ultimately limit growth on the planet:

- population
- agricultural production
- natural resources
- industrial production
- pollution.

Many of these factors were observed to grow at an exponential rate, such as food production and population, until the diminishing resource base forces a slowdown in industrial growth. However, positive factors, such as the rate of technological innovation, only grow at a constant (arithmetic) rate.

The authors of the model illustrated exponential growth by considering the growth of lilies on a pond. The lily patch doubles in area every day. When the pond is half covered by lilies, it will only be another day until the pond is covered totally. This emphasized the apparent suddenness with which the exponential growth of a phenomenon approaches a fixed limit. It also shows that there is only a short period of time within which to take corrective action. If the predicted growth of world population is correct, there is an alarmingly short time available for preventive action.

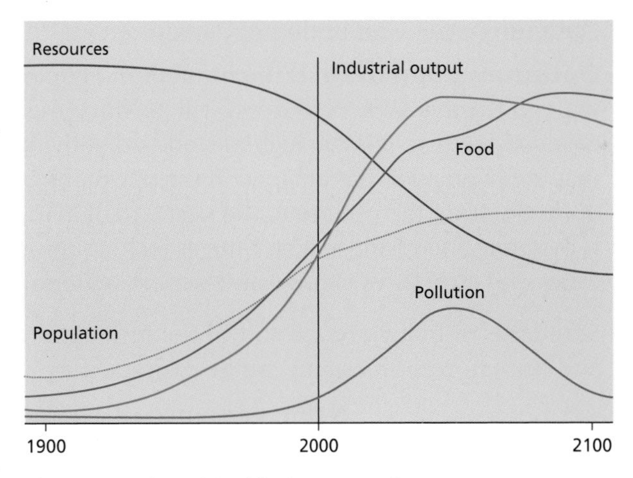

Figure 4.7 The original limits to growth

The team then examined the limits of growth:

- physical necessities that support all physiological and industrial activity, e.g. food, raw materials and fuels
- social necessities such as peace, stability and education.

The team assumed that there would be no great changes in human values in the future (Figure 4.7), and concluded that:

- If present trends continued, the limits to growth would be reached in the next 100 years. The result would probably be a sudden and uncontrollable decline in population and industrial capabilities.
- It is possible to alter these growth trends and establish a condition of ecological and economic stability that is sustainable into the future.

Much of this is remarkably reminiscent of Malthus's predictions.

There are a number of criticisms of the Limits to Growth model:

- It does not distinguish between different parts of the world.
- It ignores the spatial distribution of population and resources, of agricultural and industrial activity, and pollution. People and resources do not always coincide.
- The model emphasizes exponential growth and not the rate of discovery of new resources or of new users of resources.

To do:

Study Figure 4.7.

a Describe the curve for the change in resources over time.

b Suggest reasons to explain the nature of this curve.

c Compare and contrast the curves for population and pollution.

Carrying capacity

The concept of a population ceiling, first suggested by Malthus, is of a saturation level where population equals the carrying capacity of the local environment. There are three models of what might happen as a population growing exponentially approaches carrying capacity (Figure 4.8):

1. The rate of increase may be unchanged until the ceiling is reached, at which point the increase drops to zero. This highly unlikely situation is unsupported by evidence from either human or animal populations.
2. Here the population increase begins to taper off as the carrying capacity is approached, and then to level off when the ceiling is reached. It is claimed that populations that are large in size, with long lives and low fertility rates, conform to this S-curve pattern.
3. The rapid rise in population overshoots the carrying capacity, resulting in a sudden check, e.g. famine, birth control; after this the population recovers and fluctuates, eventually settling down at the carrying capacity. This 'j' shaped curve appears more applicable to small populations with short lives and high fertility rates.

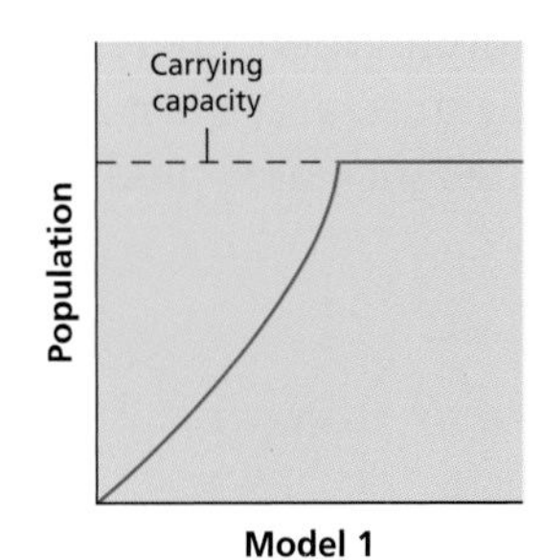

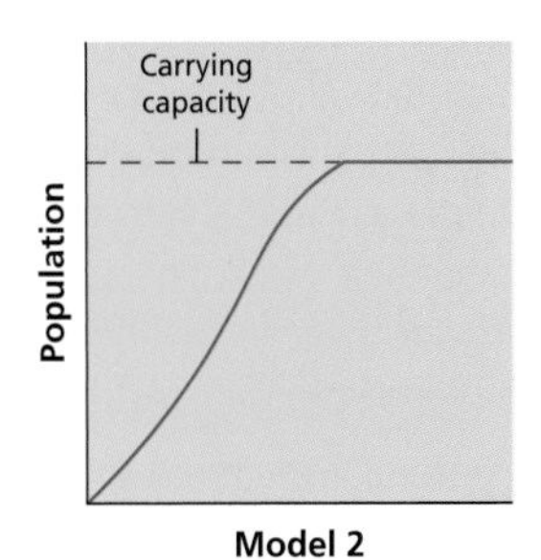

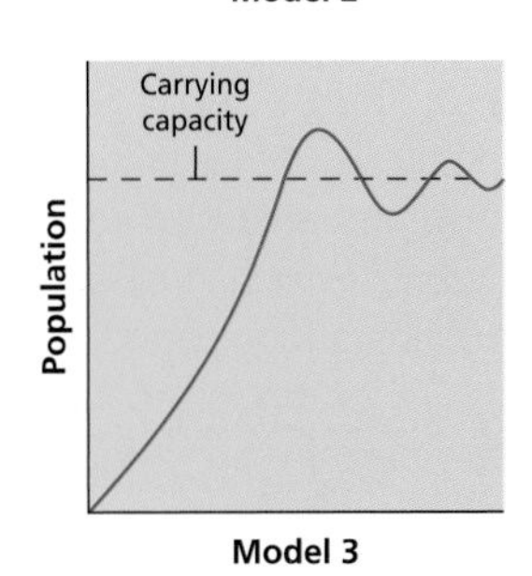

Figure 4.8 Carrying capacity

Optimum, over- and underpopulation

Optimum population is the number of people who, when working with all the available resources, will produce the highest per-capita economic return. It is the highest standard of living and quality of life that the population has attained. If the size of the population increases or decreases from the optimum, the standard of living will fall. This concept is dynamic and changes with time as techniques improve, as population totals and structures change and as new materials are discovered.

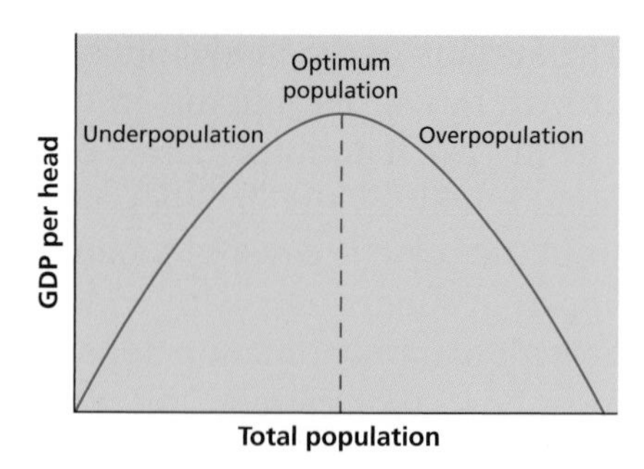

Figure 4.9 Optimum, over- and underpopulation

Standards of living are a result of the interaction between physical and human resources and can be expressed as:

$$\text{Standard of living} = \frac{\text{Natural resources} \times \text{technology}}{\text{Population}}$$

Overpopulation occurs when there are too many people, relative to the resources and technology locally available, to maintain an adequate standard of living. Bangladesh, Somalia and parts of Brazil and India are overpopulated. They suffer from natural disasters such as drought and famine and are characterized by low incomes, poverty, poor living conditions and a high level of emigration. Bangladesh is one of the most density populated countries, with over 1,000 people per square kilometre. Somalia with only 12 people/km^2 is also considered to be overpopulated.

Underpopulation occurs when there are far more resources in an area (such as food, energy and minerals) than can be used by the people living there. Canada could theoretically double its population and still maintain its standard of living. Countries like Canada and Australia can export their surplus food, energy and mineral resources, they have high levels of immigrants, and it is possible that their standard of living would increase through increased production if population were to increase.

Reserves – resources that are accessible and usable.

Resources – anything useful to humans such as soil, oil, water and minerals.

To do:

Study Figure 4.11, which shows changes in oil production by region.

a Estimate total world oil production in 2010 and 2015.

b Comment on the change in oil production in
(i) North America,
(ii) the Middle East and
(iii) the Asia–Pacific region.

Be a critical thinker

Oil and poor countries: friend or foe?

In his book *Crude World: The Violent Twilight of Oil* (Allen Lane, 2009), Peter Maass argues that the story of oil has many villains: greedy oil-company executives, rapacious dictators, shady middlemen and the like. And it has many victims: a warming atmosphere, sullied soils and water, and fragile societies. He suggests that oil weakens the bonds between governments and people by flooding public coffers with money, removing the need for wise spending. And it makes societies vulnerable to civil war.

Oil has the potential to create great wealth, but not what most poor countries need: jobs. Once wells or refineries are built, they take few men to run them. So the money just pours out of the ground, and into the hands either of foreign oil companies or of greedy and corrupt regimes. Oil has damaged the countries it comes from, with its capacity to generate large-scale environmental and social problems. It fills the streams in Ecuador where Chevron, an oil concessionaire, is accused of dumping its wastewater during a long period of drilling. Natural gas flares around the Niger Delta, with local populations such as the Ogoni fighting Shell, the biggest oil company in the region.

What do you think – is oil a curse or a blessing? How should it be managed?

Changing patterns of oil production and consumption

Oil production

In 2009 global oil production was around 84 million barrels per day. Eight producers – Saudi Arabia, the USA, Russia, Iran, China, Venezuela, Mexico and Norway – accounted for over 50% of global production. Oil production is marginal or non-existent in a large number of countries, including many in Western Europe and Africa.

Figure 4.10 A pump jack or "nodding donkey"

Oil consumption

In 2010 the global demand for oil was about 86 million barrels per day. Seven countries – the USA, Japan, China, Germany, Russia, Italy and France – accounted for over 50% of the global oil demand. A country's oil demand is roughly a function of population and level of development, and also the state of the world's economy, as shown by the 2008–09 banking crisis and economic recession.

Over 80% of oil refining now takes place in Europe, North America and Japan. However, the separation between production and refining causes problems.

Oil was considered a cheap fuel and many countries became dependent upon it. However, as a result of the oil price rise in 1973 many countries had to reassess their energy policy.

At present rates of production and consumption, reserves could last for another 40 years. Nearly two-thirds of the world's reserves are found in the Middle East, followed by Latin America (12.5%) and then equally in the developed world, centrally planned economies (CPEs) and developing countries.

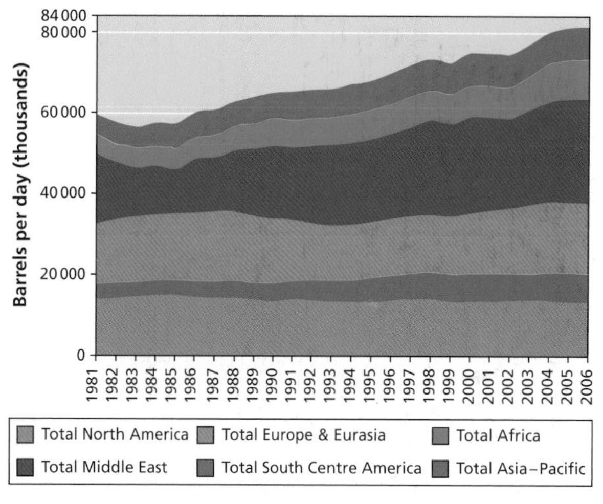

Figure 4.11 Global oil production by region

The geopolitical implications of Middle East oil

The Middle East is critically important as a supplier of oil. Involvement in the Gulf War (1991) is a case in point. Although the invasion of Iraq by US and UK forces was about freedom and democracy, a significant minority suggested that the invasion was about securing reliable sources of oil. The Organization of Petroleum Exporting Countries (OPEC) strongly influences the price of crude oil, and this has increased its economic and political power. It has also increased the dependency of all other regions on the Middle East. Countries that depend on the region for their oil need to:

- help ensure political stability in the Middle East
- maintain good political links with the Middle East
- involve the Middle East in economic cooperation.

This situation is also an incentive for developed countries to increase energy conservation or develop alternative forms of energy. There is a need to reassess other energy sources such as coal, nuclear power and renewables, and use energy less wastefully.

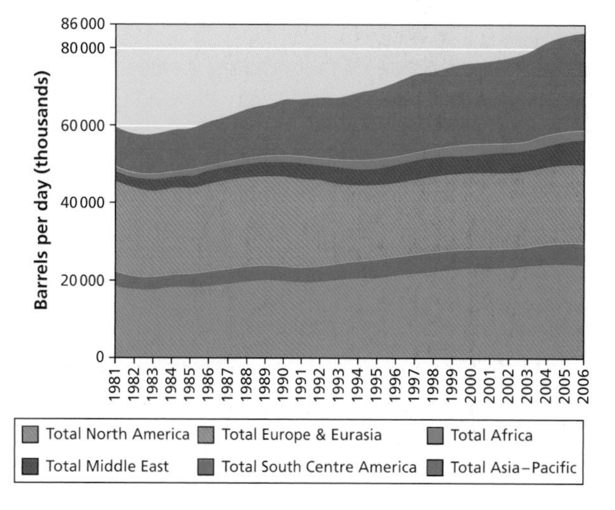

Figure 4.12 Global oil consumption by region

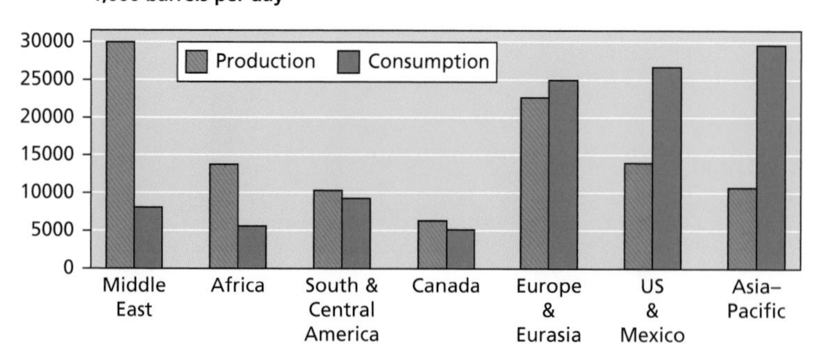

Figure 4.13 Contrasts in oil production and consumption, 2007

To do:

Study Figures 4.11 and 4.12.

a Identify the area that has shown the greatest increase in oil production between 1981 and 2006.

b Describe the global pattern and trend in the production and consumption of oil.

c What was the total consumption of oil in 2008?

d Comment on the trends in (i) overall consumption, (ii) the Asia–Pacific region and the Middle East, and (iii) North America and Europe–Eurasia.

Be a critical thinker

Peak oil

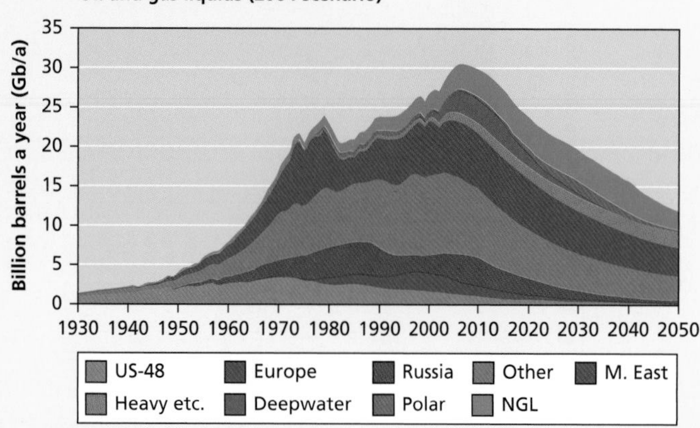

Figure 4.14 Peak oil by region

We depend on oil for many things: we use it for fuel, transport and heating, as a raw material in the plastics industry, and for fertilizer in food production. As oil production decreases after peak oil, so will all of these, unless we can find new materials and alternatives.

Peak oil varies country by country. The peak of oil discovery occurred in the 1960s, and by the 1980s the world was using more oil than was being discovered. Since then the gap between use and discovery has been increasing, and many countries have passed their peak oil production. However, reliable data is hard to come by, and some data is jealously guarded.

When will global peak oil occur?

The International Energy Agency (IEA) suggests that global peak oil will occur between 2013 and 2037. In contrast, the US Geological Survey suggests it will not occur until 2059. M. King Hubbert, who popularized the theory of peak oil, predicted that it would occur in 1995 "if there were no changes in contemporary trends". The Association for the Study of Peak Oil (ASPO) suggests it will be 2011. They claim that in 1950 the world consumed 4 billion barrels of oil per annum and the average discovery was 30 billion barrels per annum. Now, they say, the figures are reversed: new discoveries are around 4 billion barrels per year with consumption of 30 billion barrels.

To research

Visit http://bigpicture.typepad.com/comments/files/xtralargeposter2.gif for a poster on the oil industry.

1 Describe the projected trends in oil production between 2015 and 2050.

2 Describe the growing gap between oil production and oil discovery.

3 Comment on the distribution of oil reserves as shown on the map.

The future of oil production

According to the analyst Chris Ruppel (2006), the period from 1985 to 2003 was an era of energy security, and since 2004 there has been an era of energy insecurity. He claims that following the energy crisis of 1973 and the Iraq War (1990–91), there was a period of low oil prices and energy security. Insecurity has arisen for a number of reasons, including:

- increased demand, especially by NICs
- decreased reserves as supplies are used up
- geopolitical development: countries such as Venezuela, Iran and Russia have "flexed their economic muscle" in response to their oil resources and the decreasing resources in the Middle East and North Sea
- global warming and natural disasters such as Hurricane Katrina (see below), which have increased awareness about the misuse of energy resources
- terrorist activity, such as in Nigeria and Iraq.

Energy insecurity can cause and be the result of geopolitical tension. For most consumers a diversified energy mix is the best policy, rather than depending on a single source, especially from a single supplier.

According to IEA figures, in its report *World Energy Outlook*, future oil production is likely to reach 105 million barrels a day (mbd) by 2030. However, a 2009 study from Uppsala University in Sweden states that some of the IEA's assumptions drastically underplay the scale of future oil shortages. The authors of the Uppsala report, *The Peak of the Oil Age*, claim that oil production is more likely to be 75 mbd, and describe the IEA's report as a "political document" developed for countries with a vested interest in low prices.

Peak oil production – the year in which the world or an individual oil-producing country reaches its highest level of production, with production declining thereafter.

To do:

Study Figure 4.14.

a Identify the year when peak oil will occur.

b Describe the changes in production of oil products.

Energy security – a country's ability to secure all its energy needs.
Energy insecurity – a lack of security over energy sources.

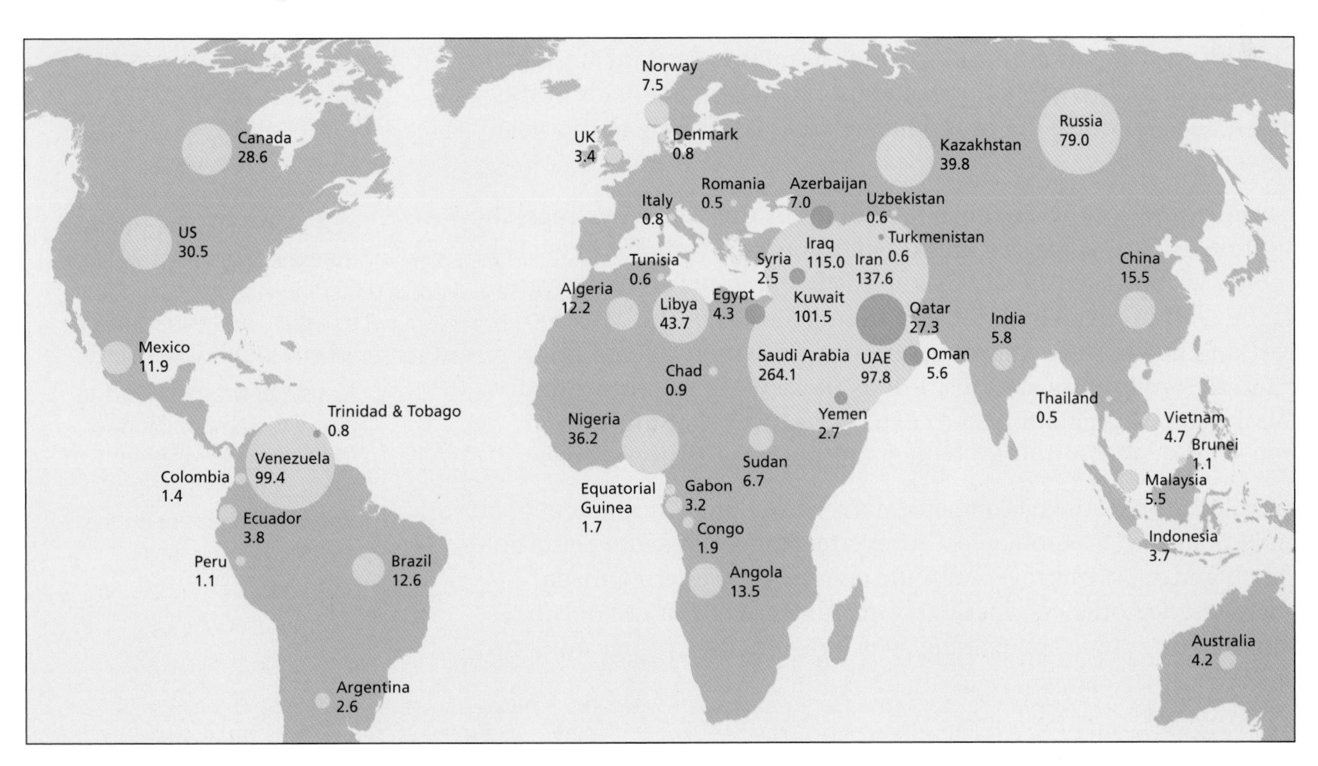

Figure 4.15 The world's proven oil reserves (billion barrels), 2009

Claims that the world is running out of oil sit uneasily with several massive new oil finds. Discoveries have been made off the southern states of USA and Brazil, as well as in completely new areas such as Ghana and Uganda. BP has made a "giant" find in the Gulf of Mexico (4 billion barrels in 1,219 metres of water) and BG, the former exploration arm of British Gas, talked of its "supergiant" off South America (2 billion recoverable of an estimated 150 billion barrels in very deep water).

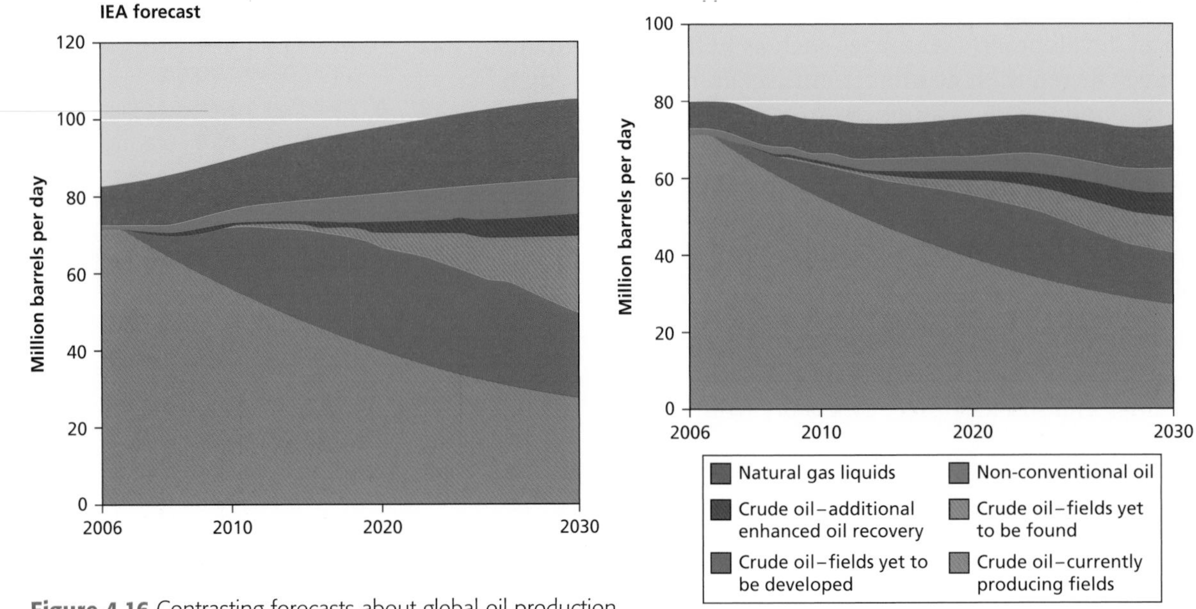

Figure 4.16 Contrasting forecasts about global oil production

Even so, "peak oil" believers say the high point of oil output could have already passed. The IEA admits that the world still needs to find the equivalent of four new Saudi Arabias to feed increasing demand, at a time when the depletion rate in the old fields of the North Sea and other major producing areas is running at 7% year on year.

As the output of the established oilfields of the world declines, the oil industry is well aware that the days of "easy" oil are over. The energy groups used to rely on the easily exploited shallow waters in the Gulf of Mexico, politically friendly areas of the Middle East and geologically simple reservoirs off Britain to feed their refineries and petrol stations. But as these wells begin to run dry, the industry is being forced to exploit reserves that are in ever more physically or politically demanding – and costly – areas.

In the meantime, the oil companies have moved into "unconventional" projects such as "gas-to-liquids" (converting natural gas into petrol and diesel) and, most controversially, the tar sands of western Canada. These reserves offer enormous new quantities of oil but can only be extracted by mining and refining, which require use up large amounts of energy and water.

To do:

a Describe the distribution of the world's proven oil reserves, as shown in Figure 4.15.

b Compare the trend in natural gas liquids in the IEA model and the Uppsala forecast (Figure 4.16). How do the contrasting models compare in their estimates of enhanced oil recovery?

c Construct two pie charts to show the proportion of oil production for different sources for each of the models. Comment on the results that you have produced.

Case study: *OPEC – an oil cartel*

The Organization of Petroleum Exporting Countries (OPEC) was established in 1960 to counter oil price cuts by American and European oil companies. Founder nations included Iran, Iraq, Kuwait, Saudi Arabia and Venezuela. Qatar, Indonesia and Libya joined in 1962 and the United Arab Emirates, Algeria, Nigeria, Ecuador and Gabon joined later. By agreeing to restrict supply, the members of OPEC were able to force up the world price of oil during the 1970s, thereby increasing revenue to their countries.

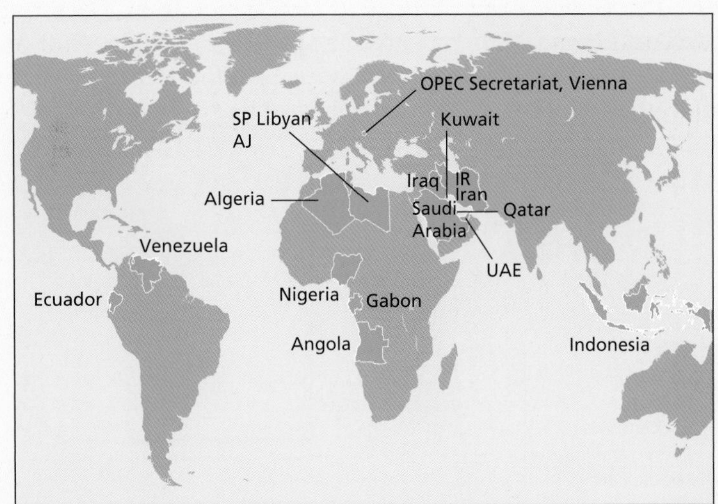

Figure 4.17 OPEC members

In 1979 the OPEC countries produced 65% of world petroleum, but only 36% by 2007. As early as 2003, concerns that OPEC members had little excess pumping capacity sparked speculation that their influence on crude oil prices would begin to slip. However, revenues from oil have allowed member countries to invest in and diversify their economies, and generate wealth over the last 40 years or so.

From a western perspective, the development of OPEC and the control of oil had important implications. As OPEC controlled the price of oil and much of the production in the 1970s and 1980s, Middle Eastern countries increased their economic and political power. It also increased the dependency of all other regions on the Middle East.

Arguably this might provide an incentive for old industrialized countries to increase energy conservation or develop alternative forms of energy.

The importance of oil meant that countries needed to maintain favourable relationships with OPEC countries and that the Middle East would be involved in economic cooperation and development with industrialized countries. It also means that there needs to be political stability in the Middle East.

Oil's environmental impact

On an average day, over 100 million tonnes of oil are transported around the world. Inevitably, some of this oil leaks or spills into the sea, causing pollution. Pollution hotspots (Figure 4.19) are most common where the greatest densities of oil routes are located (Figure 4.18).

The importance of oil as the world's leading fuel has thus had many negative effects on the natural environment. Examples include:

- oil slicks from tankers such as the *Exxon Valdez* (1987) and the Deepwater Horizon well in the Gulf of Mexico (2010)
- damage to coastlines, fish stocks and coastal communities
- water pollution caused by illegally washing out tankers
- the disposal of "retired" platforms.

Oil spills also occur inland, when oil wells and rigs are damaged or destroyed. During the Gulf War in 1991 oil stores and oil wells were targets for destruction, and the resulting fires and explosions caused immeasurable environmental damage.

Environmental disasters continue to affect the oil industry. Among the worst of these were the *Exxon Valdez* tanker disaster of 1989 and the Deepwater Horizon explosion of 2010 (see case studies on page 89).

Geopolitics – political relations among nations, particularly relating to claims and disputes pertaining to borders, territories and resources.

Cartel – an organization of people who supply the same good and join together to control the overall supply of the product. The members of a cartel can force up the price of their good either by restricting its supply on the world market or by agreeing on a particular supply price and refusing to sell the good for any less.

More material available:

www.oxfordsecondary.co.uk/ibgeography

Find out how Russia is increasing its political power through oil and gas

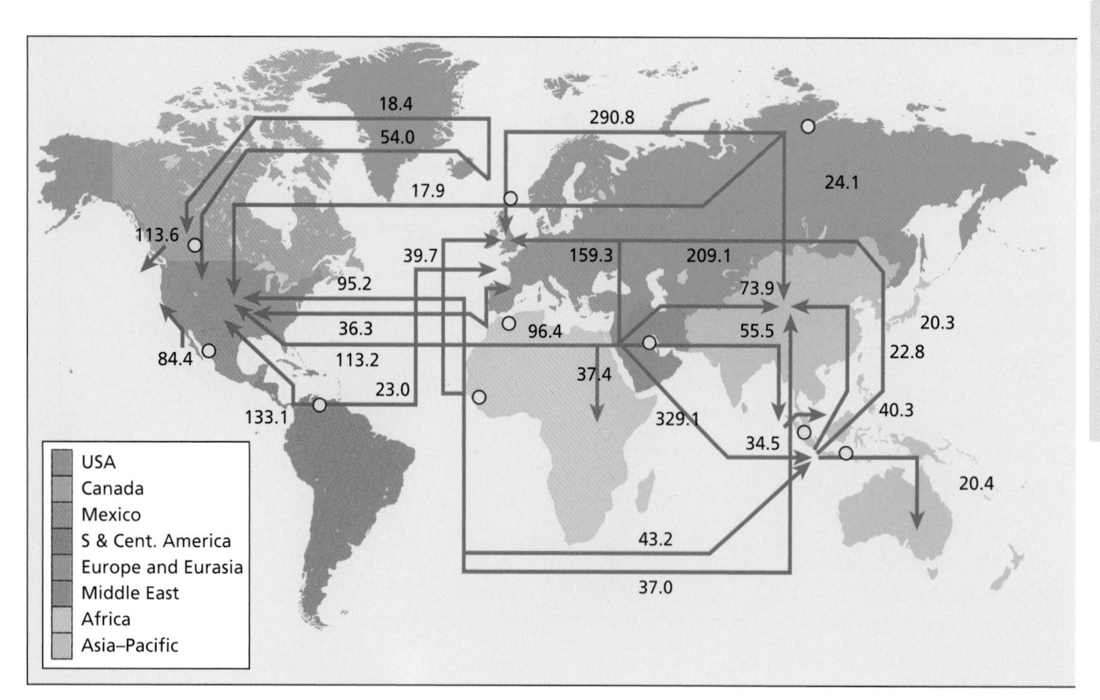

Figure 4.18 Oil trade routes (million tonnes)

To research

Visit the home page of the OPEC website at www.opec.org.

Find out about OPEC's mission and history (About us).

Visit the Data/Graphs section and find out about the OPEC share of oil reserves in 2009.

On 17 January 2001 the tanker *Jessica* ran aground on the Galapgos island of San Cristobal. However, the weather and ocean currents quickly dispersed most of the 3 million litres of diesel and bunker oil, suggesting a limited impact on the islands' famous wildlife. Only a few marine animals died immediately after the spill but, by December, 62% of the marine iguanas on nearby Santa Fe Island had died. In a normal year, only 2–7% mortality is expected. The Galapagos spill was not severe by oil-industry standards: diesel fuel evaporates quickly and is far less toxic than crude oil.

To research

Visit www.washingtonpost.com to see a map of the world's oil trade routes (26 July 2008).

To do:

Study Figures 4.18 and 4.19.

Suggest reasons for the distribution of oil pollution, as shown in Figure 4.19.

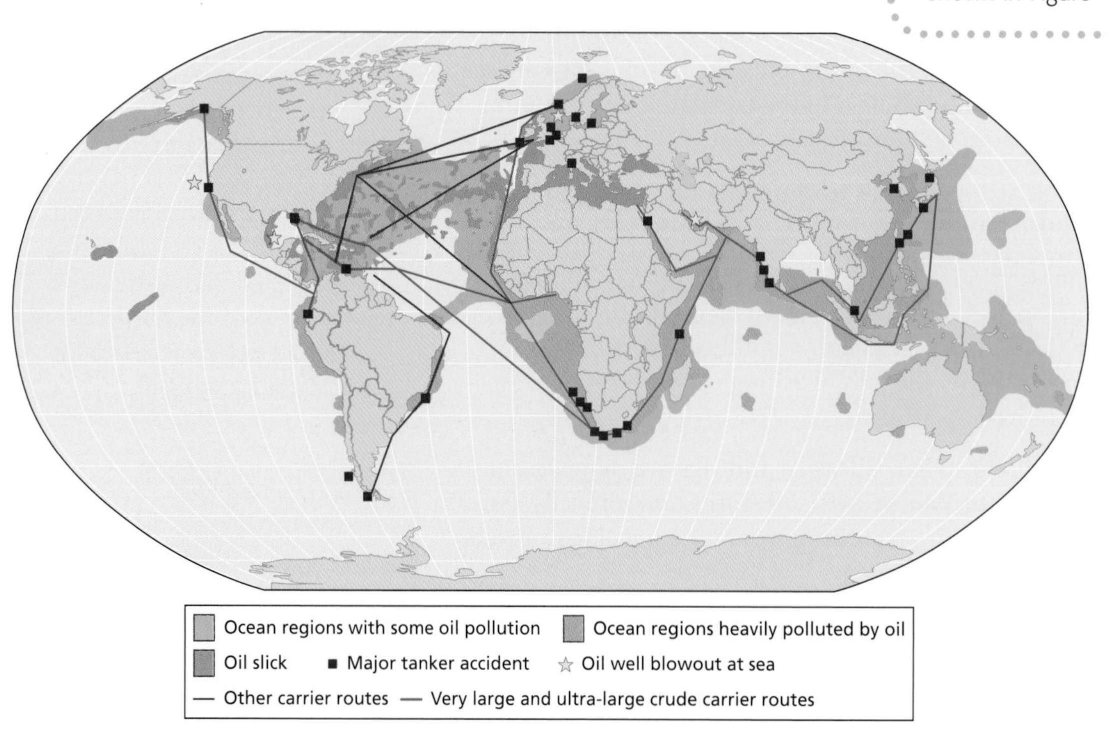

Figure 4.19 Oil pollution in the world's oceans

In 2002 the *Prestige* oil tanker shed 60 000 tonnes of oil off the coast of Spain. The slick spread as far north as England and as far south as the Canary Islands. Over 900 kilometres of coastline were closed to fishing and over 300 000 seabirds died. The clean-up costs and loss of revenue amounted to over $6 billion.

Case study: *The Exxon Valdez disaster*

In 1989 the *Exxon Valdez*, a supertanker laden with 1.2 million barrels of crude oil, ran aground in Prince William Sound, Alaska (USA). The ship, which was being piloted illegally, had just left Valdez, the terminal at the southern end of the Alaska oil pipeline. Over a quarter of a million barrels of oil leaked out during the 10 hours it took to set up the first oil containment booms and before oil-removing equipment reached the scene of the accident. The oil spread over some 25 000 square kilometres of offshore waters.

By the following year 35 000 dead seabirds had been found, a small proportion of the total killed. Some 10 000 otters, 16 whales and 147 bald eagles were killed by the spillage, and fish spawning grounds were also decimated. Wildlife continued to suffer effects into this century.

The massive clean-up cost is believed to have cost the Exxon company over $2 billion. In addition, in 1993, a US federal judge fined the company a record $5 billion in damages. This money was mainly used to compensate 34 000 fishermen affected by the spill.

To research

Visit www.faegre.com, and use the search button to find article 2881, an update on litigations in the *Exxon Valdez* case. This may be especially useful for anyone considering a career in environmental management or environmental law.

Case study: *The Deepwater Horizon disaster*

The Deepwater Horizon oil spill is the largest in US history. In April 2010 an explosion ripped through the Deepwater Horizon oil rig in the Gulf of Mexico, 80 kilometres off the Louisiana coast. Two days later the rig sank, with oil pouring out into the sea at a rate of up to 62 000 barrels a day at its peak. The oil threatened wildlife along the US coast as well as livelihoods dependent on tourism and fishing. Over 160 kilometres of coastline were affected, including oyster beds and shrimp farms.

The extent of the environmental impact is likely to be severe and last a long time. A state of emergency was declared in Louisiana. The cost to BP, who operated the rig, may reach $20 billion. BP's attempts to plug the oil leak were eventually successful. Dispersants were used to break up the oil slick but BP was ordered by the US government to limit their use, as they could cause even more damage to marine life in the Gulf of Mexico. By the time the well was capped (in July 2010), about 4.9 million barrels of crude oil had been released into the sea.

Be a critical thinker

Who is to blame?

Is BP, the oil producer, to blame for the Deepwater Horizon disaster, or Transocean, the owner of the rig? Who should be in charge of maintenance of the rig, the owner or the user?

To research

Visit http://www.guardian.co.uk/environment/interactive/2010/apr/30/deepwater-horizon-oil-spill-wildlife for an interactive map of the area affected by the Deepwater Horizon disaster, and see which species are at risk from the oil spill.

The changing importance of renewable energy sources

Renewable resources are those that can be used more than once. They include solar, wind, tidal and hydroelectric power (Figure 4.20). There is also considerable potential for geothermal power, although the areas where this can take place are limited.

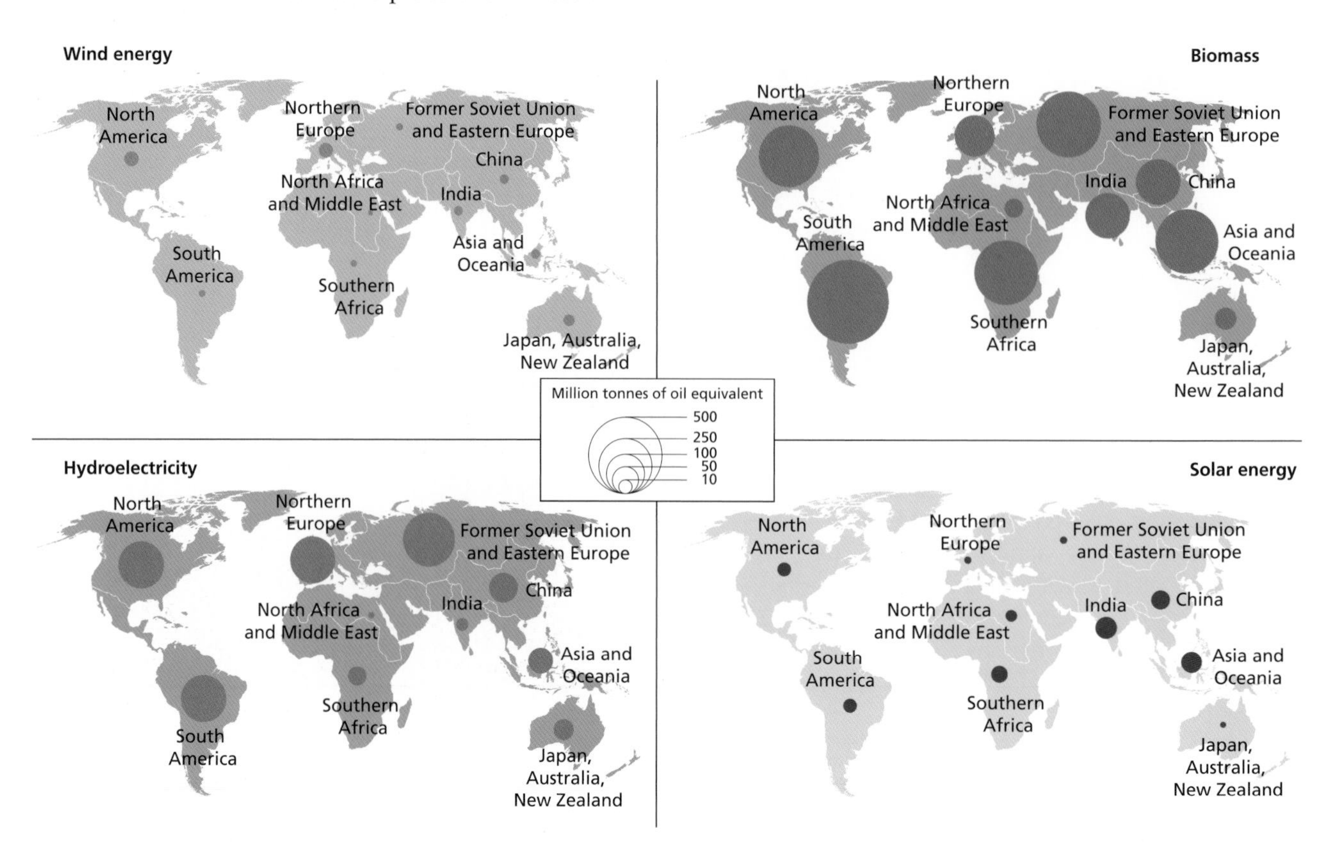

Figure 4.20 World potential renewable energy

Globally, renewable energy is growing fast. However, it is growing from a very low base. The rates of development for renewable energy sources are far exceeding those of fossil fuels such as oil, coal and natural gas. In 2006 wind and solar development grew by 20% and 40% respectively. Renewable energy will become increasingly important as the world attempts to reduce greenhouse gas emissions to the levels necessary for curbing global warming.

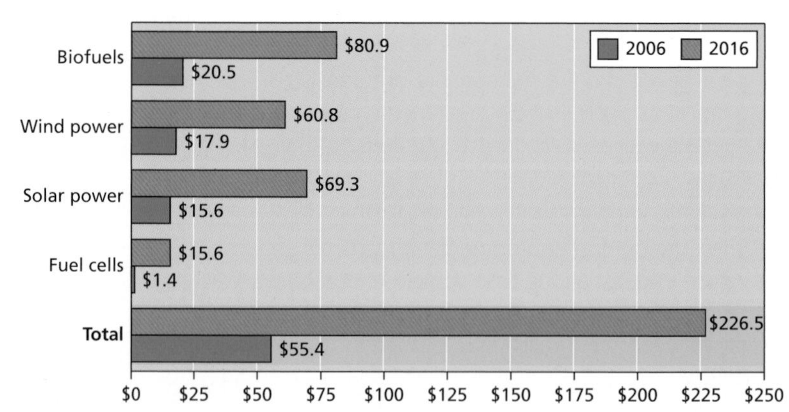

Figure 4.21 Trends in renewable energy, US $ billions (value of market)

The sixth annual Clean Energy Trends Report for 2007 found that the market for renewable energy sources was about $55 billion worldwide in 2006, and forecast growth to $226 billion by 2016.

Solar power

Energy from the sun is clean, renewable and so abundant that the amount of energy received by the earth in 30 minutes is the equivalent to all the power used by humans in one year.

	Levels				Growth
	Million tonnes of oil equivalent (mtoe)				*% p.a.*
	2007	2010	2020	2030	2007–30
Oil	4,045	3,967	4,457	4,902	0.8
Coal	3,129	3,225	3,871	4,438	1.5
Gas	2,479	2,551	3,124	3,808	1.9
Nuclear	736	759	873	1,065	1.6
Hydro	268	289	366	448	2.3
Biomass	394	446	618	840	3.4
Other renewables	59	73	151	303	7.4
Total	**11 109**	**11 310**	**13 461**	**15 804**	**1.5**
	Fuel share (%)				
	2007	2010	2020	2030	
Oil	36.4	35.1	33.1	31.0	
Coal	28.3	28.5	28.8	28.2	
Gas	22.3	22.6	23.2	24.1	
Nuclear	6.6	6.7	6.5	6.7	
Hydro	2.4	2.6	2.7	2.8	
Biomass	3.5	3.9	4.6	5.3	
Other renewables	0.5	0.6	1.1	1.9	
Total	**100**	**100**	**100**	**100**	

(Source OPEC, *World oil outlook*, 2009, p. 40)

Table 4.1 World supply of primary energy, 2007–30

To research

Visit http://earthtrends.wri.org, a free database focusing on sustainable development and the environment, including issues around renewable energy. Go to www.cleanedge.com for information on emerging clean energy trends.

To do:

Choose an appropriate technique to show percentage change in energy type over time. Study the data in the table (Table 4.1).

- **a** Identify the energy source that has shown the greatest percentage increase between 2007 and 2030.
- **b** Identify the energy source that has shown the greatest percentage decrease between 2007 and 2030.
- **c** Compare and contrast the trends shown.

However, the high costs of solar power make it difficult for the industry to achieve its full potential. Each unit of electricity generated by solar energy currently costs 4–10 times as much as that derived from fossil fuels. At present it does not make a significant contribution to energy efficiency. Although solar energy is increasing at a rate of 15–20% per year it is from a tiny starting base, and currently the annual production of photovoltaic (PV) cells is only enough to power one small city.

Advantages	Disadvantages
no finite resources involved – less environmental damage	affected by cloud cover, seasons, night time
no atmospheric pollution	not always possible when demand arises
suitable for small-scale production	high costs

Table 4.2 Advantages and disadvantages of solar power

Wind power

Wind power is good for small-scale production. It needs an exposed site, such as a hillside, flat land or land close to the coast or an off-shore location. It also requires strong, reliable winds. Such conditions are found at Altamont Pass, California, for example. Large-scale development is hampered by the high cost of development, the large number of wind pumps needed, and the high cost of new transmission grids. Suitable locations for wind farms are normally quite distant from centres of demand, as with hydroelectric power (HEP).

Advantages	Disadvantages
no pollution of air, ground or water	visual impact
no finite resources involved	noisy
reduces environmental damage elsewhere	winds may be unreliable
can be located offshore	impact on wildlife, e.g. migrating birds

Table 4.3 Advantages and disadvantages of wind power

Tidal power

Tidal power is a renewable, clean energy source. It requires a funnel-shaped estuary, free of other developments, with a large tidal range. The River Rance in Brittany has the necessary physical conditions.

Case study: Harnessing the Sahara for Europe

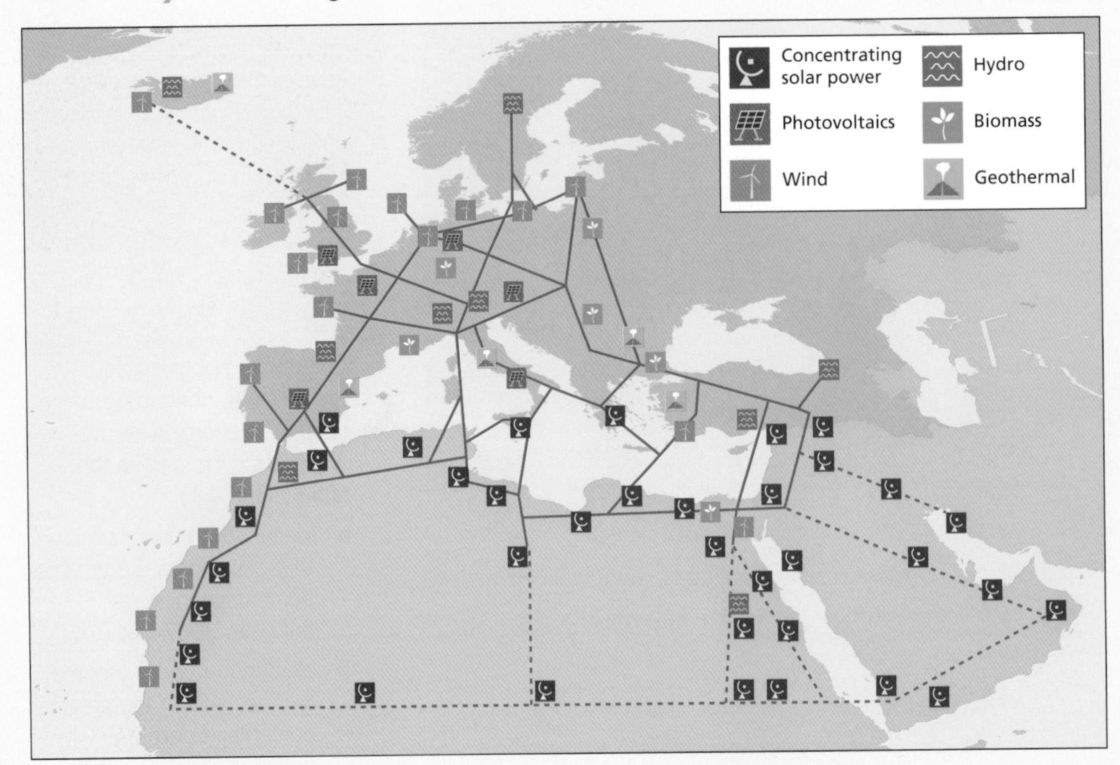

Figure 4.22 A supergrid for Europe

According to the European Commission's Institute for Energy, just 0.3% of the light falling on the Sahara and Middle East deserts would be enough to meet all of Europe's energy needs. The scientists are calling for the creation of a series of huge solar farms, producing electricity either through photovoltaic cells or by concentrating the sun's heat to boil water and drive turbines, as part of a plan to share Europe's renewable energy resources across the continent.

A new electricity supergrid would allow Europe to use the energy from the Sahara, and countries such as the UK and Denmark would also ultimately be able to export their wind energy at times of surplus supply to other parts of Europe, as well as import from other green sources such as geothermal energy from Iceland (Figure 4.22).

Harnessing the Sahara would be particularly effective because the sunlight in this area is more intense: solar photovoltaic (PV) panels in northern Africa could generate up to three times more electricity than similar panels in northern Europe. Southern Mediterranean countries including Portugal and Spain have already invested heavily in solar energy, and Algeria has begun work on a vast combined solar and natural gas plant. Algeria aims to export 6,000 megawatts of solar-generated power to Europe by 2020.

Scientists admit that it will take many years and huge investment to generate enough solar energy from north Africa to power Europe, but envisage that by 2050 it could produce 100 gigawatts, with an investment of around €450 billion.

Figure 4.23 Solar panels, Aero Island, Denmark

The energy would be transmitted along high-voltage direct-current (HVDC) transmission lines, the most efficient way to move electricity over long distances. Energy losses on DC power lines are far lower than on the traditional alternating current (AC) lines, which make transmission of energy over long distances uneconomic. HVDC can also be used to transfer electricity between different countries that might use AC at differing frequencies. In addition, HVDC cables can synchronize AC produced by renewable energy sources.

Large-scale production of tidal energy is limited for a number of reasons:

- high cost of development
- limited number of suitable sites
- environmental damage to estuarine sites
- long period of development
- possible effects on ports and industries upstream.

Many of these impacts are similar to the impacts of nuclear power.

Nuclear power

Although fast-breeder reactors can provide renewable energy, most nuclear power is not a renewable form of energy. However, it is often grouped with renewables since the amount of raw material (plutonium formed from uranium) needed to produce a large amount of energy is very small.

Nuclear power has a number of advantages. Those in favour of it claim that it is a cheap, reliable and abundant source of electricity. The production and running costs of nuclear power are low as long as there are no accidents. However, when the costs of construction and decommissioning, long-term waste disposal and costs of accidents are factored in, it is actually very expensive. Its main advantage is the security of supply. Unlike coal and oil, which have reserves estimated to last around 300 and 50 years respectively, there is enough uranium for it to be considered a renewable form of energy. Uranium fuel is available from countries such as the USA, Canada, South Africa, France and Australia, so the West would not have to rely on potentially unstable regions such as the Middle East, or former rivals such as Russia, for its energy needs. The European Union is in favour of nuclear power, and estimates that it will need to provide 40% of the EU's electricity (15% of total energy).

Figure 4.24 Wind turbines

To research

Find out about the nuclear disasters at Three Mile Island and Chernobyl. Compare their causes, effects and any management issues that arose.

However, uranium is a radioactive material and so the nuclear power industry is faced with the hazards of waste disposal and the problems of decommissioning old plants and reactors. Rising environmental fears concerning the safety of nuclear power and nuclear testing are based on experience: disasters have happened at Three Mile Island (USA) in 1979, and at Chernobyl (USSR) in 1986, although there have been no large-scale disasters since then.

Since the recession of the 1990s and 2000s the demand for energy has fallen, and less energy development is now required. The EU, for example, has a diverse range of energy suppliers, and so the threat of disruption to any one source is less worrying than it used to be. (Gas supplies from Russia may be an exception.)

Hydroelectric power (HEP)

HEP is a renewable form of energy that harnesses fast-flowing water with a sufficient **head (drop in height)**.

Factors affecting the location of a HEP scheme	Explanation
relief	a valley that can be dammed
geology	a stable, impermeable bedrock
river regime	a reliable supply of water
climate	a reliable supply of water
market demand	to be profitable
infrastructure	to transport the energy
The site for a high-head HEP power station depends upon	
local valley shape	narrow and deep
local geology	strong, impermeable rocks
lake potential	a large head of water
local land use	non-residential
local planning	lack of restrictions

Table 4.4 Factors affecting the location and site of a hydroelectric power scheme

A low-head dam is one with a drop of less than 13 metres and a generating capacity of less than 15 000 kilowatts. Low-head stations are located in the lower course of rivers where discharge or tidal flow is strong, e.g. on the Rhine. The main advantage of low-head dams is their ability to generate power near where it is needed, reducing loss in transmission.

However, HEP plants have some disadvantages:

- They are very costly to build.
- Only a small number of places have a sufficient head of water.
- The market is critical because the plant needs to run at full capacity to be economic. In some cases a market is created. For example, aluminium smelters are often located close to HEP plants in order to use up the available energy.

The Three Gorges Dam

The Three Gorges Dam on China's Yangtze river is the world's largest HEP dam, at over 2 kilometres long and 100 metres high. The lake behind the dam is over 600 kilometres long. The dam was built to help meet China's ever-increasing need for electricity and water storage, as the population moves from a sustainable existence to a more western-style urban culture. Over a million people were moved out of the valley to make way for the dam and the lake. The Yangtze river basin provides 66% of China's rice and contains 400 million people. The river drains 1.8 million square kilometres and discharges 700 cubic kilometres of water annually.

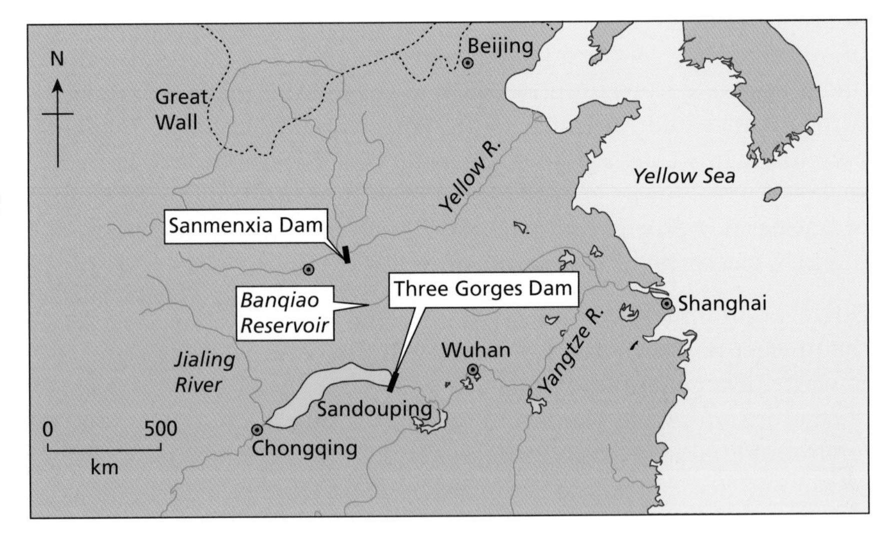

Figure 4.25 The location of the Three Gorges Dam

Chinese hydropower lobbyists are now calling for construction of an even bigger hydroelectric project on the upper reaches of the Brahmaputra river, as part of a huge expansion of renewable power in the Himalayas. A massive dam on the great bend of the Yarlung Tsangpo has been proposed, despite the likely concerns of downstream nations, India and Bangladesh, which access water and power from the river. Planners suggest a HEP scheme with a capacity of 38 gigawatts (more than half as big again as the Three Gorges Dam).

As a renewable form of energy to reduce China's dependency on coal, this source of hydroelectric power could save 200 million tonnes of carbon each year. That would be over a third of the UK's entire emissions. The mega-facility is among more than 28 dams on the river that are either planned, completed or under discussion by China. However, given the huge expense, technical difficulties and political sensitivities of the scheme, it is far from certain of final approval by the government.

Advantages	Disadvantages
The dam will generate up to 18 000 megawatts, eight times more than the Aswan Dam and 50% more than the world's next largest HEP dam, the Itaipu in Paraguay.	Up to 1.2 million people had to be moved to make way for the dam. Dozens of towns, for example Wanxian and Fuling with 140,000 and 80,000 people respectively, had to be flooded. Much of the land available for resettlement is over 800 metres above sea level, and is colder with infertile thin soils and on relatively steep slopes.
It will enable China to reduce its dependency on coal.	To reduce the silt load, afforestation is needed but resettlement of people will cause greater pressure on the slopes above the dam.
It will supply energy to Shanghai, population 13 million, one of the world's largest cities, and Chongqing, population 3 million, an area earmarked for economic development.	Up to 530 million tonnes of silt are carried through the Gorge annually: the first dam on the river lost its capacity within seven years and one on the Yellow River filled with silt within four years. The port at the head of the lake may become silted up as a result of increased deposition and the development of a delta at the head of the lake. The mouth of the river may be starved of silt, and so erosion of the coastline may result.
It will protect 10 million people from flooding (over 300 000 people in China have died as a result of flooding in the 20th century).	Most floods in recent years have come from rivers which join the Yangtze below the Three Gorges Dam.
It will allow shipping above the dam: the dams have raised water levels by 90 metres, and turned the rapids in the gorge into a lake.	The region is seismically active and landslides are frequent. The weight of the water behind the lake may contribute to seismic instability.
It has generated thousands of jobs, both in the construction and the industrial development associated with the availability of cheap energy.	Archaeological treasures were drowned, including the Zhang Fei temple. The dam will interfere with aquatic life: the Siberian Crane and the White Flag Dolphin are threatened with extinction
	It has cost as much as $70 billion.

Table 4.5 Advantages and disadvantages of the Three Gorges Dam

Figure 4.26 Aerial view of one of China's dams

Conservation, waste reduction, recycling and substitution

The recycling of paper, glass, and some metals and plastics saves scarce raw materials and helps reduce pollution. In Europe, there are high rates of recycling in Austria, Germany, the Netherlands and Switzerland (Figure 4.27). Reuse is usually more energy- and

Recycling – the processing of waste so that materials can be reused.
Reuse – the multiple use of a product.
Reduction – using less of a resource.
Substitution – using one resource rather than another.
Landfill – the burying of waste in big pits.

resource- efficient than recycling. It can also involve individual reuse of items such as containers, bags and pots. Reduction (or "reduce") includes using less energy such as turning off lights when not needed, or using only the amount of water needed when boiling a kettle. Substitution of a resource for another might mean using renewable rather than non-renewable resources, and would be of major benefit to the environment.

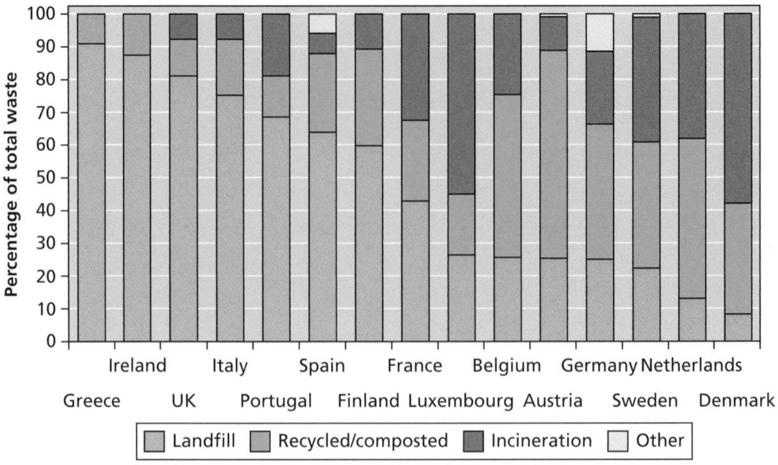

Percentage of total waste

100 90 80 70 60 50 40 30 20 10 0

Greece Ireland UK Italy Portugal Spain Finland France Luxembourg Belgium Austria Germany Sweden Netherlands Denmark

Landfill Recycled/composted Incineration Other

Figure 4.27 Waste management in the European Union

Landfill involves the burying of mainly domestic waste in the ground, and then covering over the filled pit with soil and other material. Landfill may be cheap but it is not always healthy, and will eventually run out. Much of the waste does not decompose easily, and some of it may be hazardous.

Dumping of waste and old equipment is an increasing problem. There are many reasons for the increase:

- increased standard of living and therefore more goods to dispose of
- increased cost of landfill
- more goods such as TVs, computers and refrigerators classified as hazardous and subject to restrictions on how they are disposed of
- the introduction of strict new EU regulations means that a high proportion of new products must be recycled; this can be costly to manufacturers and purchasers.

Figure 4.28 Recycling

Waste imports in China

A fairly new environmental problem is the dumping of old computer equipment. To make a new PC requires at least 10 times its weight in fossil fuels and chemicals. This can be as high as 240 kilograms of fossil fuels, 22 kilograms of chemicals and 1,500 kilograms of clean water. Old PCs are often shipped to China for recycling of small quantities of copper, gold and silver. PCs are placed in baths of acid to strip metals from the circuit boards, a process highly damaging to the environment and to the workers who carry it out.

China imports more than 3 million tonnes of waste plastic and 15 million tonnes of paper and cardboard each year. Containers arrive in the UK and other countries with goods exported from China, and load up with waste products for the journey back. A third of the UK's waste plastic and paper (200 000 tonnes of plastic rubbish and 500 000 tonnes of paper) is exported to China each year. Low wages and a large workforce mean that this waste can be sorted much more cheaply in China, despite the distance it has to be transported.

China is increasingly aware that this is not "responsible recycling" and that countries are exporting their pollution to them. They have begun to impose stricter laws on what types of waste can be imported.

To do:

Find out what materials can be:

(i) reused, (ii) recycled, and
(iii) reduced in the amount being used.

To research

For information on waste management in the European Union, visit the environment section of www.defra.gov.uk and click on the waste management icon.

Visit www.unescap.org for a consideration of issues relevant to resource consumption and management in the Asia–Pacific region (environment and sustainable development section of the Committee on Managing Globalization).

Figure 4.29 Waste management at the Casuarina Hotel, Barbados

Figure 4.30 More waste management at the Casuarina Hotel

Tackling climate change

Emissions of the main anthropogenic greenhouse gas, CO_2, are influenced by:

- the size of the human population
- the amount of energy used per person
- the level of emissions resulting from that use of energy.

A variety of technical options which could reduce emissions, especially from use of energy, are also available. Reducing CO_2 emissions can be achieved through:

- improved energy efficiency
- fuel switching
- use of renewable energy sources
- nuclear power
- capture and storage of CO_2.

These options are most easily applicable to stationary plant. Another measure involves increasing the rate at which natural sinks take up CO_2 from the atmosphere, for example by increasing forest cover.

International initiatives to protect climate

There have been a number of attempts at reducing the human impact on climate change. Although many of these attempts have been international in their focus, they have also included, in some cases, national targets for countries to follow. The evolution of climate change policy has not been an easy process.

In 1988 the United Nations Environment Programme (UNEP) and the World Meteorological Organization established the Intergovernmental Panel on Climate Change (IPCC). The IPCC has three working groups, each covering a different aspect of climate change, and their reports are intended to aid policymakers.

In 1992 the UN Conference on the Environment and Development (UNCED) was held in Rio de Janeiro. It covered a range of subjects and produced a number of statements, including the Framework Convention on Climate Change (FCCC), signed by more than 150 nations. This came into force in March 1994, and its ultimate objective was:

"...to achieve...stabilization of greenhouse gas concentrations in the atmosphere at a level that would prevent dangerous anthropogenic interference with the climate system."

To do:

Study Figures 4.29 and 4.30, which show types of waste management at the Casuarinas Hotel in Barbados.

a Identify each of the types of waste management shown in the photos.

b Outline the advantages of each type of waste management shown.

c Identify the types of waste management carried out in (i) your home and (ii) your school. In what ways would it be possible to increase waste management in both areas?

d Identify, and comment on, the obstacles to improving waste management in both environments.

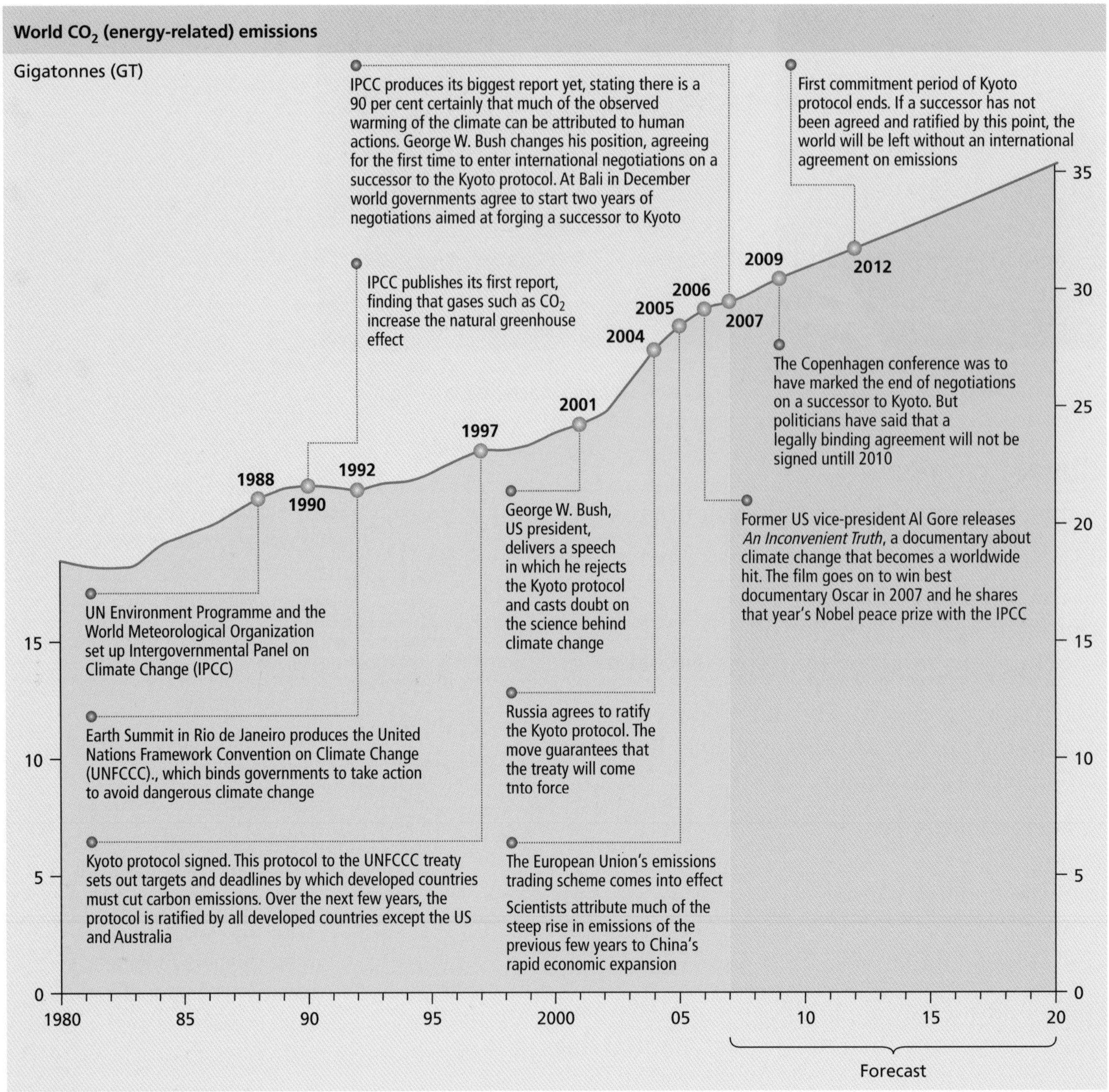

World CO_2 (energy-related) emissions

Gigatonnes (GT)

UN Environment Programme and the World Meteorological Organization set up Intergovernmental Panel on Climate Change (IPCC) (1988)

IPCC publishes its first report, finding that gases such as CO_2 increase the natural greenhouse effect (1990)

Earth Summit in Rio de Janeiro produces the United Nations Framework Convention on Climate Change (UNFCCC)., which binds governments to take action to avoid dangerous climate change (1992)

Kyoto protocol signed. This protocol to the UNFCCC treaty sets out targets and deadlines by which developed countries must cut carbon emissions. Over the next few years, the protocol is ratified by all developed countries except the US and Australia (1997)

George W. Bush, US president, delivers a speech in which he rejects the Kyoto protocol and casts doubt on the science behind climate change (2001)

Russia agrees to ratify the Kyoto protocol. The move guarantees that the treaty will come tnto force (2004)

The European Union's emissions trading scheme comes into effect (2005)

Scientists attribute much of the steep rise in emissions of the previous few years to China's rapid economic expansion

Former US vice-president Al Gore releases *An Inconvenient Truth*, a documentary about climate change that becomes a worldwide hit. The film goes on to win best documentary Oscar in 2007 and he shares that year's Nobel peace prize with the IPCC (2006)

IPCC produces its biggest report yet, stating there is a 90 per cent certainly that much of the observed warming of the climate can be attributed to human actions. George W. Bush changes his position, agreeing for the first time to enter international negotiations on a successor to the Kyoto protocol. At Bali in December world governments agree to start two years of negotiations aimed at forging a successor to Kyoto (2007)

The Copenhagen conference was to have marked the end of negotiations on a successor to Kyoto. But politicians have said that a legally binding agreement will not be signed untill 2010 (2009)

First commitment period of Kyoto protocol ends. If a successor has not been agreed and ratified by this point, the world will be left without an international agreement on emissions (2012)

Forecast: 2007–2020

Source: IEA

Figure 4.31 International policy to deal with climate change

The Kyoto agreement

The Kyoto Protocol (1997) was an addition to the Rio Convention. Unlike the agreement signed in Rio, which was voluntary, it gave all MEDCs legally binding targets for cuts in emissions to the 1990 level by 2008–12. The EU agreed to cut emissions by 8%, Japan by 7%, and the USA by 6%. Some countries found it easier to make cuts than others. The 1998 Toronto conference on climate change went further, calling for a 20% cut in CO2 emissions by 2005.

There are three main ways for countries to keep to the Kyoto target without cutting domestic emissions:

- plant forests to absorb carbon or change agricultural practices, e.g. keeping fewer cattle
- install clean technology in other countries and claim carbon credits for themselves
- buy carbon credits from countries such as Russia where traditional heavy industries have declined and the national carbon limits are underused.

Even if greenhouse gas production is cut by 60–80%, there could still be enough greenhouse gas in the atmosphere to raise temperatures by 4°C. The Kyoto agreement was meant to be only the beginning of a long-term process, not the end of one, and the guidelines for measuring and cutting greenhouses gases were not finished in Kyoto.

When George W. Bush was elected President of the USA, he rejected the Kyoto Protocol on the grounds that it would hurt the US economy and employment. Although the rest of the world could proceed without the USA, the USA emits about 25% of the world's greenhouse gases. So without the USA, and NICs such as China and India, the reduction of carbon emission would be seriously hampered. According to the Kyoto rules, 55 countries must ratify the agreement to make it legally binding worldwide, and 55% of the emissions must come from developed countries. If the EU, eastern Europe, Japan and Russia agree, they could just about make up 55% of the developed world's emissions. Without the USA (and Australia and Canada, who were also against cutting emissions), it is difficult to achieve this goal.

In November 2007 Australia signed the Kyoto agreement to limit CO_2 emissions, at once distancing itself from the US and ending a 10-year diplomatic exile on the issue. The decision took place on the first day of the UN conference in Bali. The USA still backed voluntary targets to fight climate change.

Copenhagen, 2009

Many people were hoping that the Copenhagen climate change talks of 2009 would result in a deal that would compel the nations of the world to address the climate crisis in a meaningful way. However, the talks failed to achieve a legally binding agreement. With no deal in sight, a number of countries (the USA, China, India, Brazil and South Africa) did manage to publish a summary document, which offered to enhance cooperative action against climate change, and recognized the need to help poor countries make adaptations.

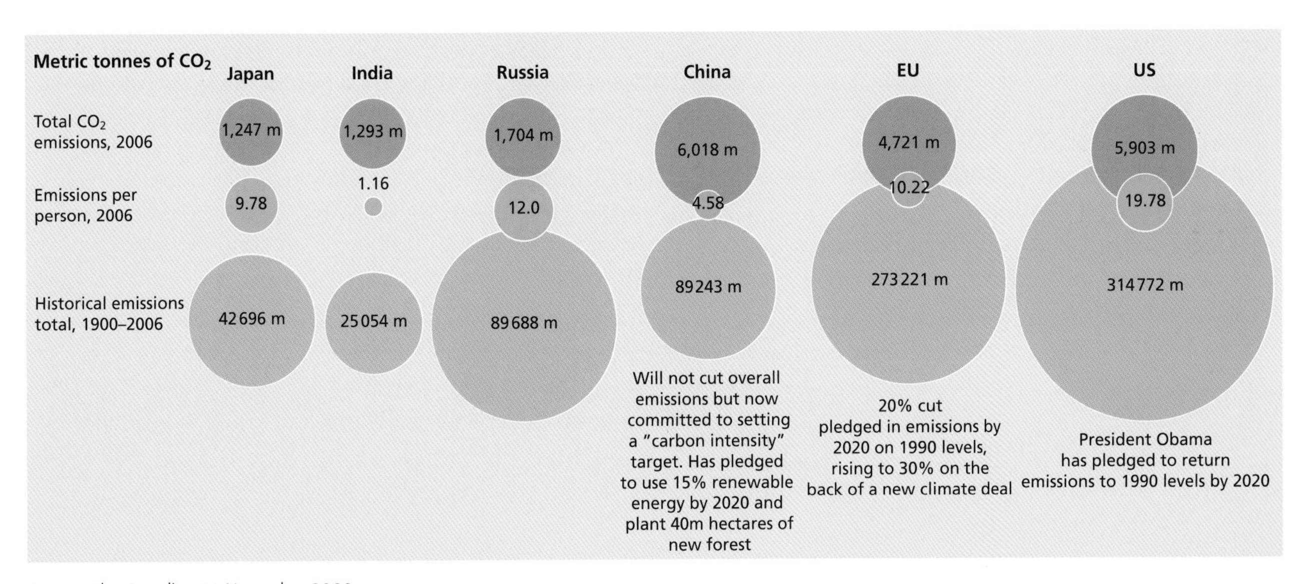

Source: The Guardian 30 November 2009

Figure 4.32 CO_2 emissions compared

The agreement, called the Copenhagen Accord, provides a way to bring together the offers of emission reductions made by various countries before the conference began, and gives a special status to the idea of holding global warming to no more than 2°C. It offers short-term funding for projects in developing countries of $30 billion, and aspires to a long-term system that would, in principle, provide $100 billion a year for mitigation and adaptation from 2020 onwards. And, perhaps the component of clearest value from outside the world of climate politics, it moves forward on the plan for reducing deforestation, known as "REDD".

To many environmentalists, the accord's great deficiency is that it sets no targets for emissions; earlier drafts had room for specific figures for developed-country reductions in 2020 and both developed-country and global reductions by 2050. It requires nothing from developing nations, even China, now (since 2007) the world's largest emitter of CO_2; and it requires nothing of the USA.

To research

Find out about the other main criticisms of the Copenhagen Accord.

To research

Find out about China's role in global warming, and China's role in tackling global warming.

Be a critical thinker

China and climate change talks: two different views

In December 2009 the former UK Deputy Prime Minister, John Prescott, who also helped negotiate the Kyoto protocol in 1997, defended China's role in the climate change summit, saying that the blame for its flawed outcome must lie with the USA. Mr Prescott feared that the Chinese would walk away from the talks if they continue to be singled out for blame. He criticized the US climate change special envoy, who claimed that China was projected to emit 60% more CO_2 than the US by 2030.

The Chinese news agency Xinhua avoids mention of how and why China blocked attempts to impose 2050 targets for reducing emissions. Beijing has consistently rejected such long-term goals, which it sees as a threat to its economic growth. It also failed to address claims that China vetoed the inclusion of a 1.5°C maximum global temperature rise, requested by small island states and African nations. Instead, it says, the Chinese prime minister Wen showed sincerity by accepting a rise of no more than 2°C by 2050.

In another report, Mark Lynas (author of *Six degrees – our future on a hotter planet*, 2007) claimed that China wrecked the Copenhagen summit, intentionally humiliated US president Barack Obama, and demanded a "deal" so poor that western leaders would carry the blame. He claims that it was China's representative who insisted that industrialized country targets, previously agreed as an 80% cut by 2050, be taken out of the deal. A 2020 peaking year in global emissions, essential to restrain temperatures to 2°C, was removed and replaced by woolly language suggesting simply that emissions should peak "as soon as possible". The target of global 50% cuts by 2050 was also excised.

How did China manage to pull off this coup? First, it was in an extremely strong negotiating position. China didn't need a deal. There was also a complete lack of civil society political pressure on China: campaign groups never blame developing countries for failure. China not only rejected targets for itself, but also refused to allow any other country to take on binding targets. China, with its polluting coal-based economy, may have wanted to weaken the climate regulation regime now in order to avoid the risk that it might be called on to be more ambitious in a few years' time.

Part 2 Optional themes

The seven optional themes explore the interaction between human and physical factors and processes, using contemporary case studies drawn from a variety of environments. Whichever of the themes you choose, each will raise your awareness of social injustice and uneven access to resources and how such problems may be overcome through sustainable management practices, from the global to the individual level.

5 Freshwater – issues and conflicts

By the end of this chapter you should be able to:

- understand the physical geography of freshwater, and why water on the land is a scarce resource
- consider the human impacts on water quality and the ways in which humans respond to the challenges of managing the quantity and quality of freshwater
- explain the consequences of water management, positive and negative.

Precipitation – the transfer of moisture (as dew, hail, rain, sleet or snow) to the earth's surface from the atmosphere.
Interception – the capture of raindrops by plant cover, which prevents direct contact with the soil.
Runoff – precipitation that does not soak into the ground but flows over it into surface waters.
Groundwater – water held underground in soil or porous rock, often feeding springs and wells.
Evapotranspiration (EVT) – the loss of water from vegetation and water surfaces to the atmosphere.

The global hydrological cycle

The global hydrological cycle is the transfer of water between sea, air and land. It comprises evaporation from oceans, water vapour, condensation, precipitation, runoff, groundwater and evapotranspiration.

- If 100 units represent global precipitation (on average 860 millimetres per annum), 77% falls over the oceans, 23% on to the land.
- About 84 units enter the atmosphere by evaporation via the oceans; thus there is a horizontal transfer of seven units from the land to the sea.
- Of precipitation over the land, 16 units are evaporated or transpired; seven units are runoff to the oceans.

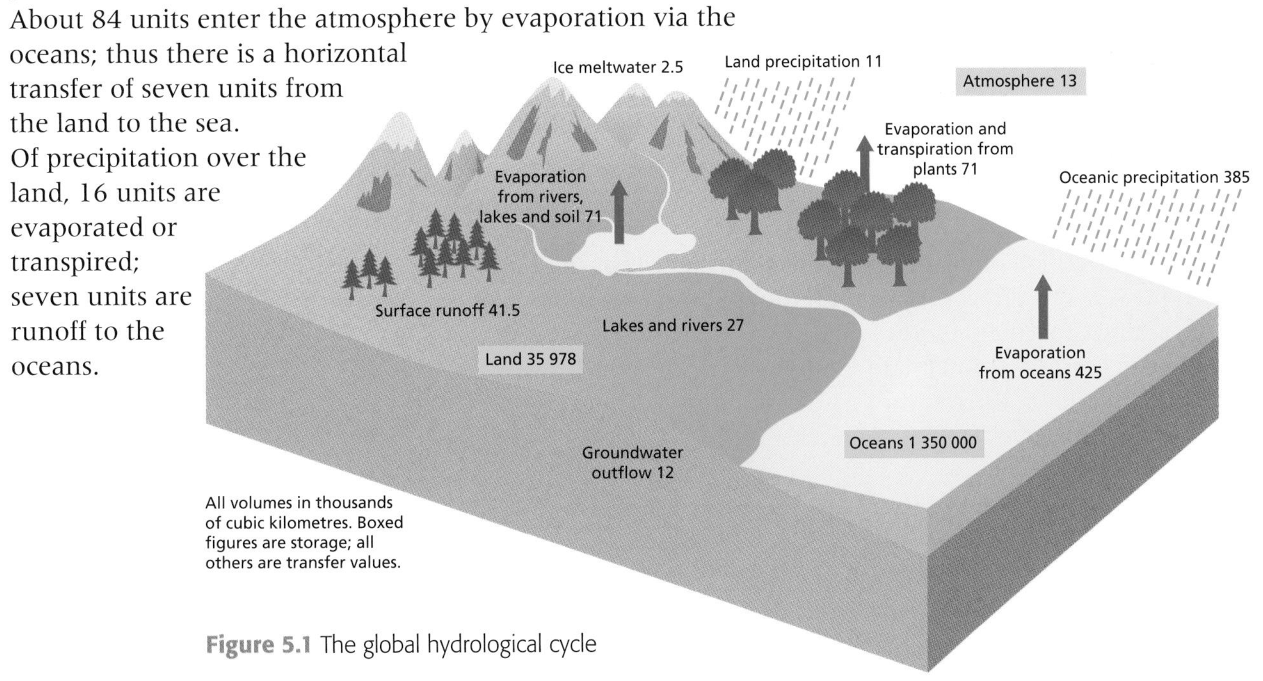

Figure 5.1 The global hydrological cycle

- There may be a time lag between precipitation and eventual runoff. About 98% of all free water on the globe is stored in the oceans.

Potential evapotranspiration (pEVT) – the rate of water loss from an area if there were no shortage of water.

Evaporation

This is the physical process by which a liquid becomes a gas. It is a function of:

- vapour pressure
- air temperature
- wind
- rock surface (e.g. bare soils and rocks have higher rates of evaporation than surfaces with a protective tilth, where rates are low).

High rates of evaporation are recorded in hot deserts. For example, in Atbara (Sudan) the **potential evapotranspiration** – the rate of water loss if there were no shortage of water – is 6,250 millimetres per annum, and in Helwan (Egypt) it is 2,390 millimetres. Rates are much lower in tropical rainforests because of the high humidity (500–750 millimetres) and in cold climates such as the UK (London's rate is 330 millimetres per year).

To research

Visit http://geography.about.com for articles about the hydrological cycle in the physical and cultural section (water and ice), and for links to some excellent sites on hydrology and rivers.

Try the matching quiz (for the hydrological cycle) at

http://highered.mcgraw-hill.com/sites/0072402466/student_view0/chapter10/matching_quiz.html

In parts of Egyypt, **evapotranspiration** rates – the rates at which water transfers from the earth into the atmosphere – are less than 250 millimetres per year, because Egypt's annual rainfall is less than 250 millimetres. However, if Egypt received, for example, 2,000 millimetres of rain each year, the evapotranspiration rate would increase because of the very high temperatures there. Thus, if there were no shortage of water in Egypt, potential evapotranspiration could be as high as 2,000 millimetres per year.

Condensation

Condensation is the process by which a gas (water vapour) becomes a liquid. It occurs when air cools to its dew point or becomes saturated by evaporation into it. Further cooling leads to condensation on nuclei, to form water droplets or frost. Precipitation occurs when so much water has condensed that the air can no longer hold it, so the water falls to earth as rain, hail, sleet or snow.

Figure 5.2 Condensation

Interception

When raindrops fall on plant cover, the plants intercept the rain, which prevents its direct contact with the soil. If rain is prolonged, the retaining capacity of leaves will be exceeded and water will drop to the ground (throughfall). Some will trickle along branches and down the stems or trunk (stemflow). The water retained on the leaves later evaporates.

Permanent snow	9,700 years
Oceans	2,500 years
Groundwater	1,400 years
Lakes	17 years
Swamp water	5 years
Soil moisture	1 year
Streams	16 days
Atmospheric moisture	8 days

Table 5.1 Average water renewal cycles for different water bodies

The world's changing water balance

The amount of water stored in the oceans and in ice varies with temperature change. This can be on a long-term scale – over millions of years – or on a short-term scale, such as with accelerated global warming. On a long-term scale, sea levels change in connection with the growth and decay of ice sheets.

Eustatic change refers to a global change in sea level. At the height of glacial advance, 18 000 years ago, sea level was 100–150 metres below current sea level.

The level of the land also varies in relation to the sea. Land may rise as a result of tectonic uplift or the removal of an ice sheet. The change in the level of the land relative to the level of the sea is known as **isostatic adjustment,** or **isostacy**. Parts of Scandinavia and Canada are continuing to rise at rates of up to 20 millimetres a year.

As the world's temperature rises, the amount of water stored in the world's glaciers and ice sheets decreases and global sea levels rise. In addition, there is also the Steric effect. This is the phenomeneon whereby seawater expands with higher temperatures. Thus, even if ice sheets and ice caps did not melt, sea levels would rise in a warmer world.

Some scientists have predicted that global worming may push the Greenland ice sheet over a threshold where the entire ice mass would melt within a few hundred years, causing sea level to rise by 7.2m. This would flood many of the world's coastal cities and many islands, such as the Maldives.

Reservoir	Value (km^3 × 10^3)	Percentage of total
Ocean	1 350 000.0	97.403
Atmosphere	13.0	0.00094
Land	35 977.8	2.596
Of which		
• Rivers	1.7	0.00012
• Freshwater lakes	100.0	0.0072
• Inland seas	105.0	0.0076
• Soil water	70.0	0.0051
• Groundwater	8 200.0	0.592
• Ice caps/glaciers	27 500.0	1.984
• Biota	1.1	0.00008
Annual exchange	**Values (km^3 × 10^3)**	**Values (km^3 × 10^3)**
Evaporation	496.0	
Of which		
• Ocean		425.0
• Land		71.0
Precipitation	496.0	
Of which		
• Ocean		385.0
• Land		111.0
Runoff to oceans	41.5	
Of which		
• Rivers		27.0
• Groundwater		12.0
• Glacial meltwater		2.5

Table 5.2 Global water reservoirs and exchanges

TOK Link

What is the maximum sustainable yield?

The sustainable yield (SY) may be calculated as the rate of increased use of a natural resource, that is, that which can be exploited without depleting the original stock or its potential for replenishment. Thus, **maximum sustainable yield (MSY)** is the largest yield that can be taken from a resource over an indefinite period. MSY aims to maintain the resource size at the point of maximum growth rate by harvesting the amount that would normally be replenished, allowing the resource to continue to be productive indefinitely. MSY is often difficult to determine.

To research

Discuss the causes and consequences of changes in the balance of water stored in oceans and ice.

Maximum sustainable yield (MSY) – the maximum level of extraction of water that can be maintained indefinitely for a region.

To do:

a Comment on the stores of freshwater as shown in Table 5.2. What are the implications for human use of water resources?

b Identify and explain two ways in which water can be temporarily stored on the surface.

c Identify and explain three ways in which vegetation influences the hydrological cycle.

d Explain the difference in value between evaporation from oceans and precipitation into oceans.

e Explain how sea levels may rise without any melting of ice sheets and ice caps.

Water law principles	
A1	It is necessary to recognize the **unity of the water cycle** and the interdependence of its elements.
A2	The variable, uneven and unpredictable distribution of water in the water cycle should be acknowledged.
B1	All water is a **resource common to all**, the use of which should be subject to national control.
B2	There shall be **no ownership of water** but only a right to its use.
C1	The objective of managing the nation's water resources is to achieve **optimum long-term social and economic benefit** for our society from their use, recognizing that water allocations may have to change over time.
C2	The water required to meet people's **basic domestic needs** should be reserved.
C3	International water resources, specifically shared river systems, should be managed in a manner that will optimize the benefits for all parties in a spirit of mutual cooperation. Allocations agreed for downstream countries should be respected.
D1	The **national government has ultimate responsibility** for, and authority over, water resource management, the equitable allocation and usage of water, the transfer of water between catchments and international water matters.
D2	The development, apportionment and management of water resources should be carried out using the criteria of **public interest, sustainability, equity and efficiency** of use in a manner which reflects the value of water to society while ensuring that basic domestic needs, the requirements of the environment and international obligations are met.
D3	As far as is physically possible, water resources should be managed in such a manner as to enable all user sectors to gain equitable access to the desired quantity, quality and reliability of water, using conservation and other measures to **manage demand** where this is required.
D4	Water quality and quantity are interdependent and should be managed in an integrated manner, which is consistent with broader environmental management approaches.
D5	Water-quality management options should include the use of **economic incentives and penalties** to reduce pollution; the possibility of irretrievable environmental degradation as a result of pollution should be prevented.
D6	The **regulation of land use** should, where appropriate, be used as an instrument to manage water resources.
E1	The institutional framework for water management should be self-driven, minimize the necessity for state intervention and provide for a right of appeal.
E2	Responsibility should, where possible, be delegated to a **catchment or regional level** in such a manner as to enable interested parties to participate and reach consensus.
E3	**Beneficiaries** of the water management system **should contribute to the cost** of its establishment and maintenance.
F1	Where existing rights are taken away, compensation should be paid.
G1	The **right of all citizens to have access to basic water services** (the provision of potable water supply and the removal and disposal of human *excreta* and wastewater) necessary to afford them a healthy environment on an equitable and economically and environmentally sustainable basis should be supported.
G2	Water services should be provided in a manner consistent with the goals of water resource management.
G3	Where water services are provided in a monopoly situation, the interests of the individual consumer and the wider public must be protected and the broad goals of public policy promoted.

(Source: *Water law principles* 1996. Department of Water Affairs and Forestry, South Africa)

Table 5.3 Water law principles of South Africa

Be a critical thinker

How might water principles vary from place to place and over time? Give examples to support your answer. How might they change in the future? Again, give examples to support your views. What can be done to manage changing water principles?

The drainage basin hydrological cycle

In studying rivers, use is made of the drainage basin hydrological cycle (Figure 5.4). In this the drainage basin is taken as the unit of study rather than the global system (Figure 5.1). The basin cycle is an open system: the main input is precipitation that is regulated by various means of storage. The outputs include channel runoff, evapotranspiration and groundwater flow.

Drainage basin – the area drained by a river and its tributaries.

Drainage divide – also known as a watershed, it is the line defining the boundary of a river or stream drainage basin separating it from adjacent basin(s).

Water balance – the relationship between the inputs and outputs of a drainage basin.

Case study: *Water balance in Australia*

Australia's water balance assessment is based on studies in 51 catchments, ranging from the Great Artesian Basin and the Murray–Darling Basin to smaller basins such as Kangaroo Island. Many of the monitored areas are state capital cities, and the eastern, southern and southwestern coastlines.

Total input is nearly 13 million litres (2.8 million gallons). Over 11 million litres (2.5 million gallons) return to the atmosphere via evapotranspiration. Nearly 227 000 litres (50 000 gallons) make up aquifer recharge, while runoff accounts for over a million litres (240 000 gallons).

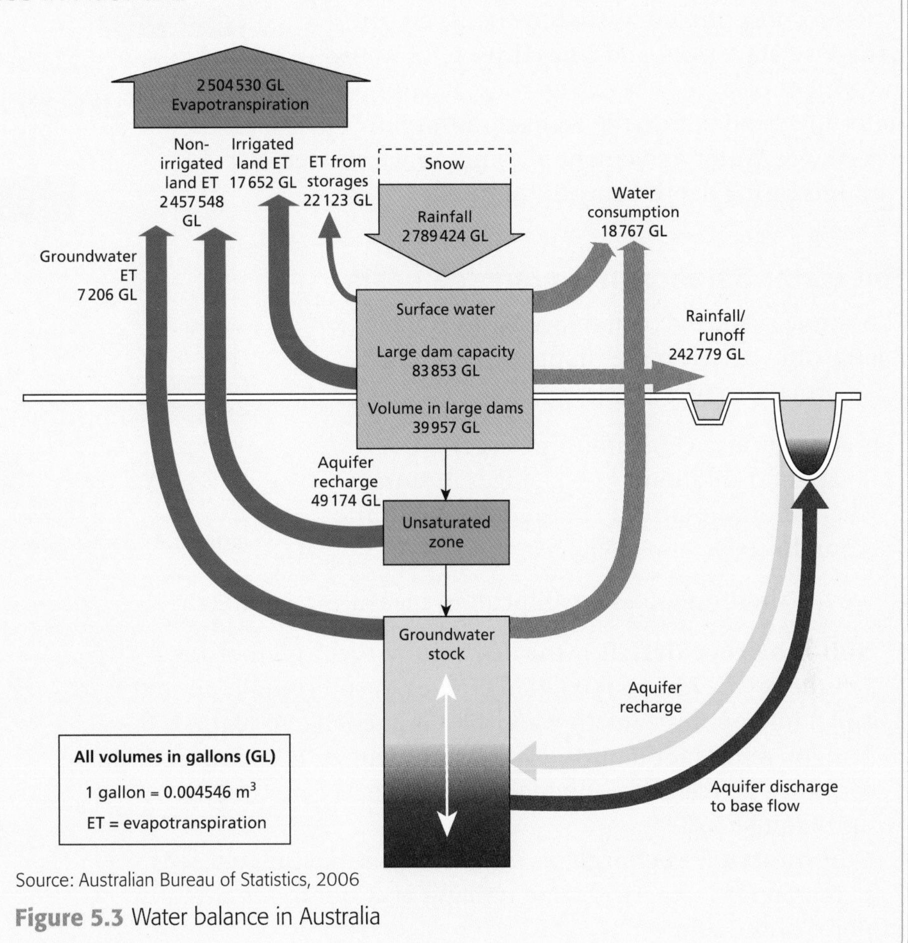

Source: Australian Bureau of Statistics, 2006

Figure 5.3 Water balance in Australia

Infiltration is the process by which water sinks into the ground. **Infiltration capacity** refers to the amount of moisture that a soil can hold. By contrast, the **infiltration rate** refers to the speed with which water can enter the soil. **Percolation** refers to water moving deep into the groundwater zone. **Overland runoff** occurs when precipitation intensity exceeds the infiltration rate, or when the infiltration capacity is reached and the soil is saturated.

The **groundwater** zone is normally divided into a zone of saturation, in which the underground water fills all the spaces in the rock, and a zone of aeration above it, in which the water does not fully saturate the pores. The **water table** divides one zone from the other. **Aquifers** are rocks that hold water. They provide the most important store of water, regulate the hydrological cycle and maintain river flow.

The **zone of aeration** is a transitional zone in which water is passed upwards or downwards through the soil. Soil moisture varies with porosity (the amount of pore spaces) in a soil, and with permeability (the ability to transmit water).

Water flowing through the basin can do so in a number of ways. **Overland flow** occurs in two main ways:

- when precipitation exceeds the infiltration rate
- when the soil is saturated (all the pore spaces are filled with water).

To do:

a Make a copy of Figure 5.3 and replace the absolute values with percentages. The input of rainfall, 13 million litres (2 789 424 gallons), represents 100%. All other figures are relative to this figure.

b Write a paragraph summarizing the main points raised by the South African Department of Water Affairs and Forestry in Table 5.3.

c State the three principles that you think are the most important. Give reasons to support your answer.

By contrast, **throughflow** refers to water flowing through the soil in natural pipes and **percolines** (lines of concentrated water flow between soil horizons). **Baseflow** and **interflow** refer to the movement of water within the zone of aeration (interflow) and within the zone of saturation (baseflow). Water movement becomes much slower with increasing depth beneath the surface.

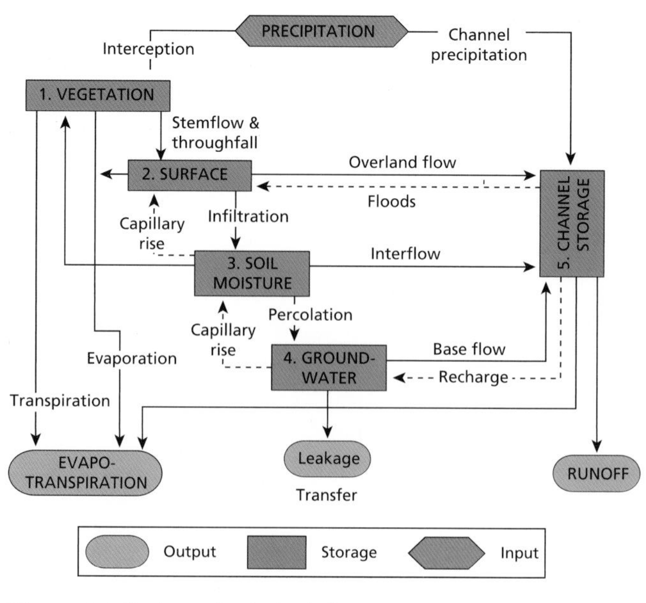

Figure 5.4 Drainage basin hydrology

The water balance in a drainage basin

The water balance in a drainage basin is the relationship between the inputs and outputs. It is normally expressed as

> precipitation = Q (runoff/discharge) = E (evapotranspiration) +/− changes in storage (such as on the surface, in the soil, and in the groundwater).

However, these variables are difficult to measure accurately.

- **Soil moisture deficit** is the degree to which soil moisture falls below **field capacity** (large pores in soil contain air, but small micropores contain water that is available to plants). In London and Madrid during late winter and early spring, soil moisture deficit is low, owing to high levels of precipitation and limited EVT.
- **Soil moisture recharge** occurs when precipitation exceeds potential EVT – there is some refilling of water in the dried-up pores of the soil.

To do:

Study Figure 5.4. Identify the inputs, stores and outputs from the drainage basin hydrological cycle. On a copy of Figure 5.4 add a feedback loop that operates within the hydrological cycle.

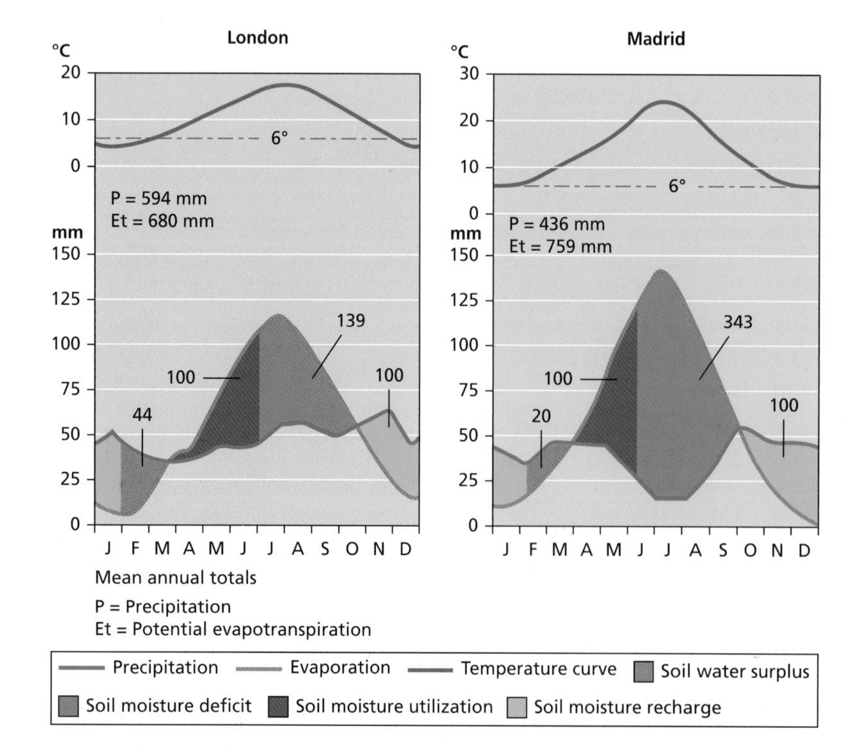

Figure 5.5 Water balance graphs for London and Madrid

To research

Visit www.nwlg.org for an animation on the hydrological cycle.

- **Soil moisture surplus** is the period when soil is saturated and water cannot enter, and so flows over the surface.
- **Soil moisture utilization** is the process by which water is drawn to the surface through capillary action and may be used by plants or evaporated.

Temporal changes

On a local scale, Figure 5.5 shows the water balance for London and Madrid. Both locations are in the northern hemisphere, where precipitation exceeds evapotranspiration in winter. The reverse is true in summer. However, there is enough surplus water to be evaporated in late spring and early summer, whereas in late summer and early autumn there is a water deficit. During this period, it may be important for farmers to irrigate their crops.

In addition, there are long-term changes. For example, the water balance in most places would have been very different during the last glacial period, and we would expect it to change during this next century as global warming leads to an increase in temperature and changes in rainfall pattern.

Spatial changes

The water balance varies considerably between countries and continents. Table 5.4 shows the average precipitation, evapotranspiration and runoff for each of the continents. Such data is hugely generalized. Nevertheless, runoff is divided into base (groundwater) flow and surface (flood) flow. This is an important distinction, because in some places there are severe seasonal differences in flood runoff (as in a monsoonal area). Thus there may be insufficient amounts of water at certain times of the year, and too much at other times. The base flow represents the water that generally can be relied upon.

To do:

a Study Figure 5.5.

i Describe and explain the variations in precipitation and evapotranspiration for London and Madrid.

b Study Table 5.4, which shows the water balance components for the world's continents. This could also be considered in terms of inputs, outputs and potential maximum sustainable yield (MSY).

i Complete the table (NB precipitation – evapotranspiration = total, and total = surface flow + base flow)

ii Calculate the percentage of runoff that is (i) surface runoff and (ii) base flow for each continent. Describe your results.

iii Work out the percentage of precipitation that becomes base flow for each continent. Describe your results.

Continent	Precipitation (inputs)	Evapotranspiration (outputs)	Total (MSY)	Surface runoff (flood flow)	Base flow
Europe	657	375	282	185	97
Asia	696	420		205	
Africa	696	582			40
Australia (and Oceania)		534	269		64
North America	645	403		171	
South America	1,564		618		223

Table 5.4 Water balance components for the world's continents

The concept of water deficiency and surplus varies between and within countries. It depends upon on how the water is obtained and used, and the problems it creates in terms of supply and demand.

Discharge, stream flow and channel shape

Discharge is found by multiplying the cross-sectional area of a river or stream by the mean velocity of the water. Steeper slopes should lead to higher velocities because of the influence of gravity. Velocity

Discharge – the volume of water passing a given point over a set time.

also increases as a stream moves from pools of low gradient to rapids. Discharge is normally expressed in cubic metres per second, or m^3/sec (cumecs).

Discharge (Q) normally increases downstream, as does width, depth and velocity. By contrast, channel roughness decreases (Figure 5.7). The increase downstream in channel width is normally greater than that of channel depth. Large rivers with a higher width/depth (w/d) ratio are more efficient than smaller rivers with a lower w/d ratio, since less energy is spent in overcoming friction. Thus the carrying capacity increases and a lower gradient is required to transport the load. Although river gradients decrease downstream, the load carried is smaller and therefore easier to transport.

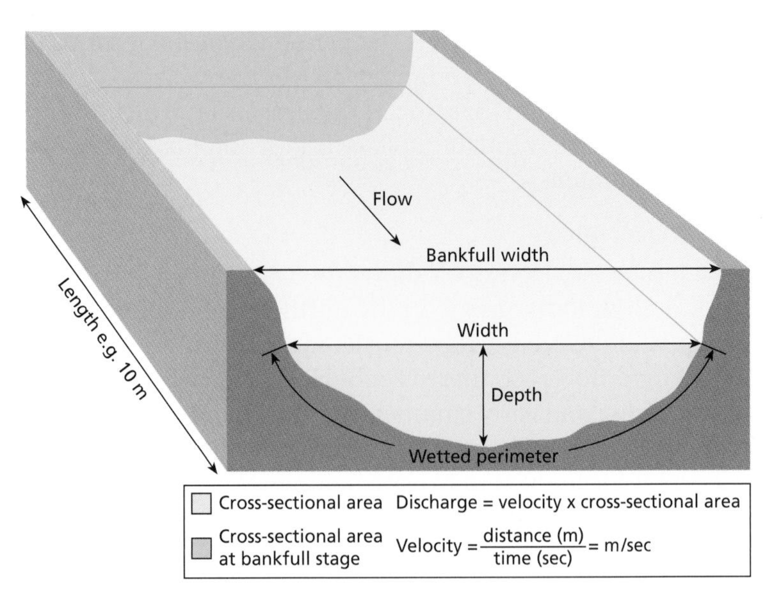

Figure 5.6 Discharge in a river

Velocity

Velocity varies with a number of factors: volume of water, roughness of bed, gradient of stream, and width, depth and shape of channel.

The effect of friction is to create an uneven distribution of velocity in a stream. Water closest to the bed and bank travels slowest, while water nearest the centre travels fastest. The highest velocity is thus midstream about a third of the way down (that at the surface is affected by surface resistance). In asymmetric channels maximum velocity is nearer the deep bank and slightly under the surface. This has important implications for erosion and deposition.

Channel shape

The efficiency of a stream's shape is measured by its **hydraulic radius**, the cross-sectional area divided by wetted perimeter (Figure 5.8). The higher the ratio the more efficient the stream is, and the smaller the frictional loss is. The ideal form is semicircular.

There is a close relationship between velocity, discharge and the characteristics of the channel in which the water is flowing. These include depth, width, channel roughness and hydraulic geometry. The w/d ratio is a good measure of comparison. The shape of the channel is also determined by the material forming the channel and river forces. Solid rock allows only slow changes, whereas alluvium allows rapid changes. Silt and clay produce steep, deep, narrow valleys (the fine material being cohesive and stable), whereas sand and gravel promote wide, shallow channels.

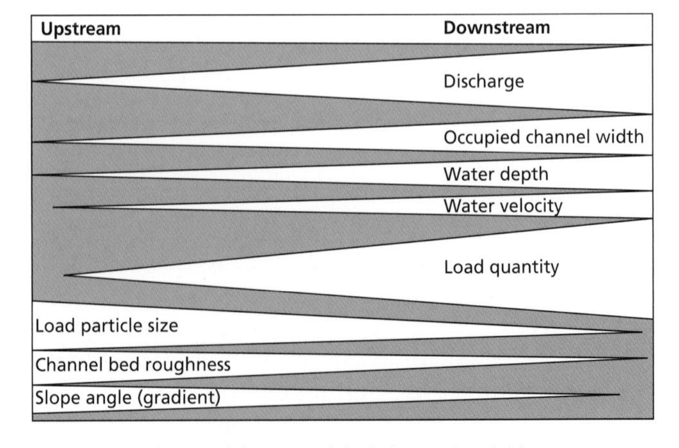

Figure 5.7 The Bradshaw model of channel variables

Channel roughness

Channel roughness causes friction, which slows down the velocity of the water. Friction is caused by irregularities in the riverbed, boulders, trees and vegetation, and contact between the water and the bed and bank. Manning's "n" is a formula that describes the relationship between channel roughness and velocity:

$$v = \frac{R2/3S1/2}{\text{"n"}}$$

where v = velocity, R = hydraulic radius, S = slope and "n" = roughness. The higher value of "n" the rougher the bed, as shown below:

Bed profile	Sand and gravel	Coarse gravel	Boulders
Uniform	0.02	0.03	0.05
Undulating	0.05	0.06	0.07
Irregular	0.08	0.09	0.10

Table 5.5 Channel roughness and velocity

River level

③ Flood – high friction

② Bankfull – maximum efficiency (low friction)

① Below bankfull – high friction

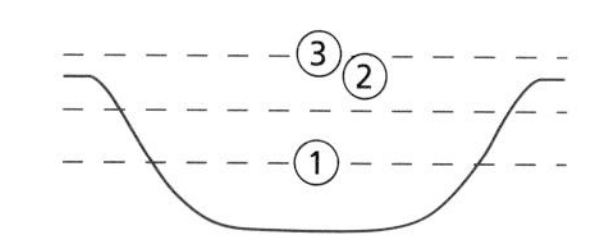

Shape

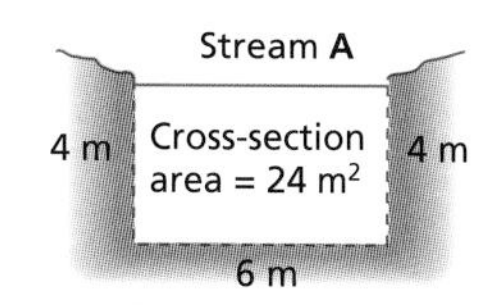

Very efficient (low relative friction)

Stream B

2 m

Cross-section area = 24 m²

2 m

12 m

Inefficient (high relative friction)

- - - wetted perimeter

Wetted perimeters	Hydraulic radius
Stream **A**: $4 + 4 + 6 = 14$ m	Stream **A**: $\frac{24}{14} = 1.71$ m
Stream **B**: $2 + 2 + 12 = 16$ m	Stream **B**: $\frac{24}{16} = 1.5$ m

Figure 5.8 Hydraulic radii

Mean annual discharge

A useful statistic is the mean annual discharge divided by the drainage basin area. This gives a depth-equivalent discharge (i.e. how much water runs off the surface for each area). The values range from over 1,000 millimetres for the Amazon River to 31 millimetres for the Colorado. In terms of absolute discharge, the Amazon is highest, at 230 000 cubic metres per second (just 700 metres upstream from its mouth it is already 2.5 kilometres wide and 60 metres deep). Second is the Zaire, at 40 000 cubic metres a second, while the Mississippi is just 18 000 cubic metres a second. Even in flood, the discharge of the Mississippi has only once ever reached 57 000 cubic metres a second.

River regimes

The river regime is the seasonal variation in the flow of a river. Arctic streams have maximum flow in spring, following snowmelt, whereas monsoonal rivers have maximum flows following the summer floods. Variations in a river's flow depend on many factors:

- the amount and nature of precipitation
- seasonal variations in temperature and evapotranspiration
- changes in vegetation cover
- variations in rock types, soil types and the shape and size of the drainage basin.

Of these, seasonal changes in climate generally have the greatest impact on changes in river flow.

To research

For a list of rivers arranged by discharge, visit http://en.wikipedia.org and search for "List of rivers by average discharge". Comment on the distribution by continent. Suggest why there are so few rivers with a large discharge in Europe.

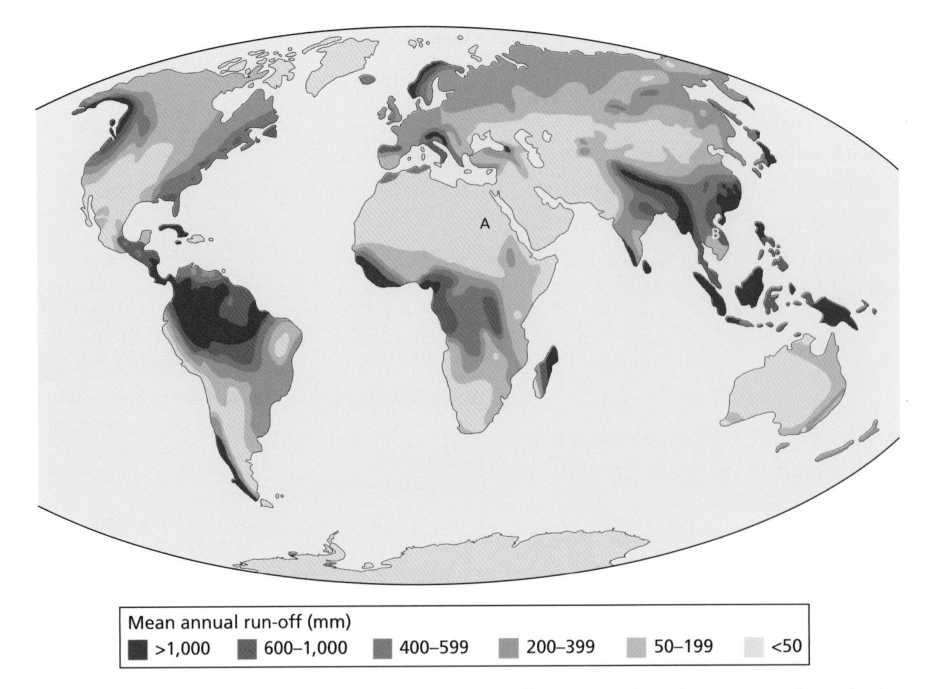

Figure 5.9 Global variations in mean annual runoff expressed as depth-equivalent discharge

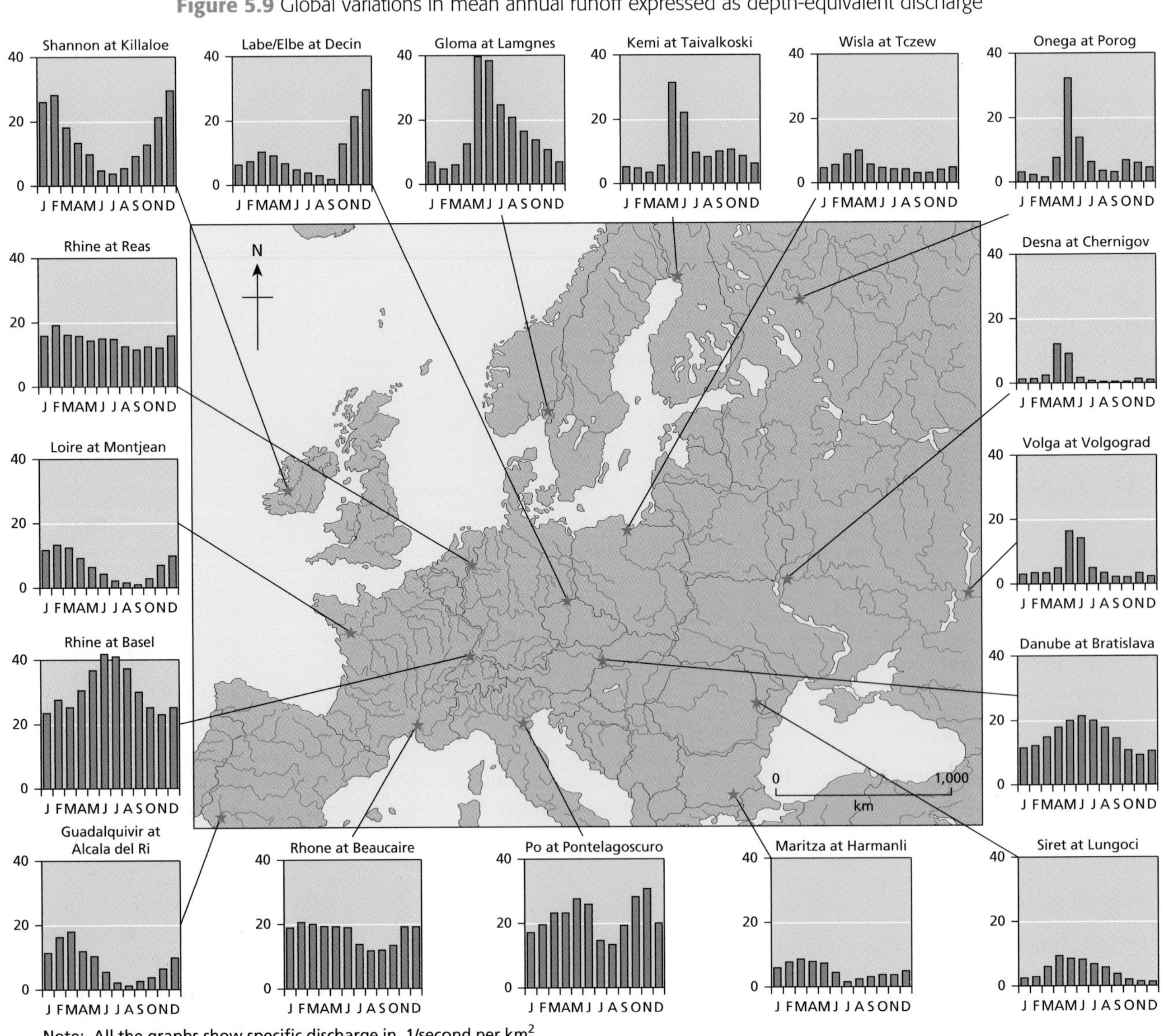

Note: All the graphs show specific discharge in. 1/second per km^2

Source Guinness P. and Nagle G. AS Geography Cases and Concepts, Hodder, 2000

Figure 5.10 River regimes in Europe

Storm hydrographs

A **storm** or **flood hydrograph** shows how a river channel responds to the key processes of the hydrological cycle. It measures the speed at which rain falling on a drainage basin reaches the river channel. It is a graph on which river discharge during a storm or runoff event is plotted against time.

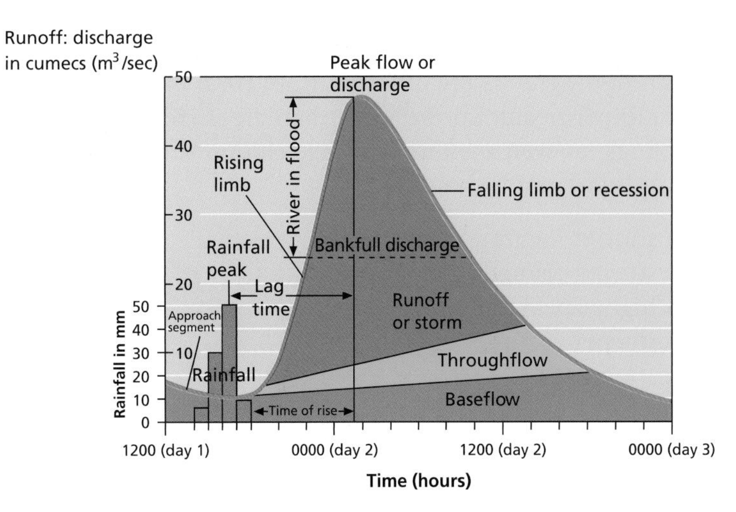

Figure 5.11 A storm hydrograph

Rising limb

- indicates the amount of discharge and the speed at which it is increasing
- very steep in a flash flood or in small drainage basins where the response is rapid
- generally steep in urbanized catchments

Peak flow or discharge

- higher in larger basins
- steep catchments have lower infiltration rates
- flat catchments have high infiltration rates, so more throughflow and lower peaks

Lag time

- time interval between peak rainfall and peak discharge
- influenced by basin shape, steepness, stream order

Runoff curve

- reveals the relationship between overland flow and throughflow
- where infiltration is low, antecedent moisture high, surface impermeable and rainfall strong, overland flow will dominate

Normal (base) flow

- the seepage of groundwater into the channel – very important where rocks have high pore space
- a slow movement, and the main, long-term supplier of the river's discharge

Recessional limb

- influenced by geological composition and behaviour of local aquifers
- larger catchments have less steep recessional limbs, likewise flatter areas

Hydrograph size (area under the graph)

- the higher the rainfall, the greater the discharge
- the larger the basin size, the greater the discharge

To do:

a Define the terms (a) discharge and (b) hydraulic radius.

b Study Figure 5.9.

  i Explain what is meant by the depth-equivalent discharge.

  ii Describe the patterns shown in Figure 5.9. Using an atlas, suggest reasons for (i) the value at location A and (ii) the value at location B.

c Study Figure 5.10.

  i Describe the regimes for Shannon (Ireland), Gloma (Norway) and the Po (Italy).

  ii Using an atlas, explain the differences you have noted.

To research

Visit www.s-cool.co.uk for a simple hydrograph (go to GCSE geography/rivers/hydrology), and to see differences in urban and rural hydrographs (go to A-level geography/river profiles/ storm hydrographs and river discharge). You may find the s-cool revision site quite useful.

Urban hydrology and the storm hydrograph

Urban hydrographs are different from rural ones because there are more impermeable surfaces in urban areas (roofs, pavements, roads, buildings) as well as more drainage channels (gutters, drains, sewers). Urban hydrographs have:

- a shorter lag time
- a steeper rising limb
- a higher peak flow (discharge)
- a steeper recessional limb.

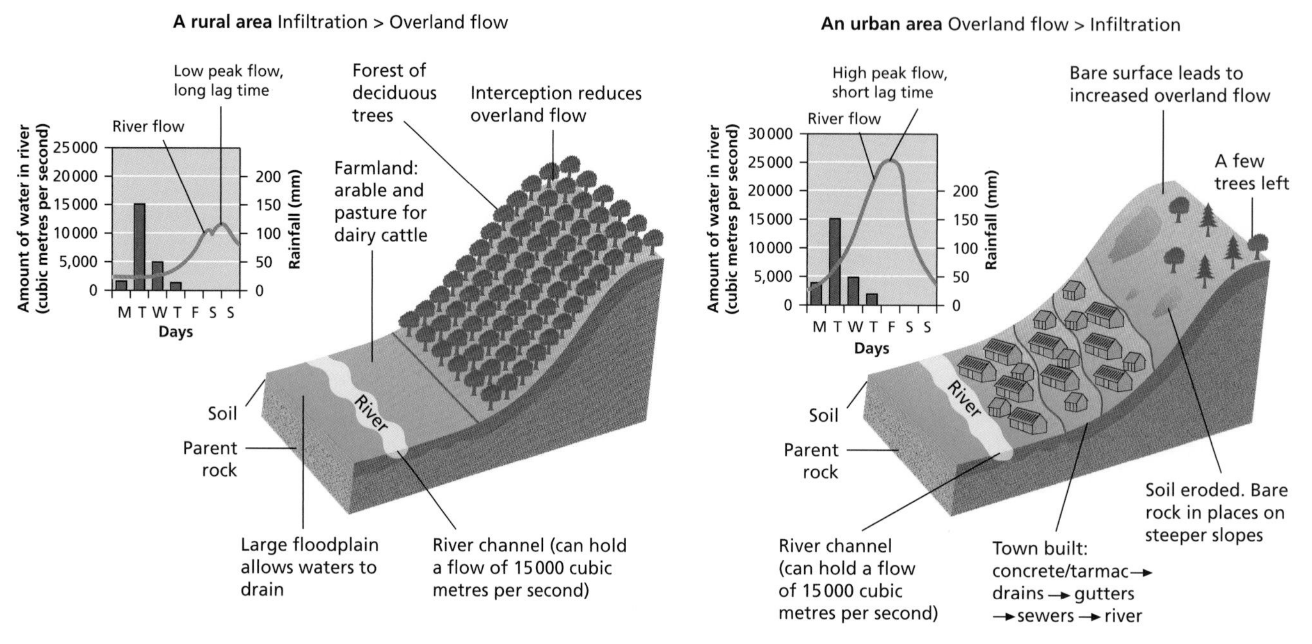

Figure 5.12 Urban and rural hydrographs

Variation in hydrographs

A number of factors affect flood hydrographs:

- climate (rainfall total, intensity, seasonality) – more intense rainfall leads to a shorter time lag, while a greater total of rainfall should lead to a higher peak flow
- soils – impermeable clay soils create more flooding, whereas permeable, sandy soils allow more infiltration
- vegetation – vegetation intercepts rainfall and so flooding is less likely
- infiltration capacity – soils with a low infiltration capacity cause much overland flow, therefore a higher flood peak and a shorter lag time
- rock type – permeable rocks will allow water to infiltrate, thereby reducing the flood peak
- slope angle – on steeper slopes there is greater runoff causing a greater peak flow and a shorter lag time
- drainage density – the more stream channels there are the more water that gets into rivers, thereby reducing the time it takes for the peak flood, and increasing the amount of water in the flood (Figure 5.13)
- human impact – creating impermeable surfaces and additional drainage channels increases the risk of flooding; dams disrupt the flow of water; afforestation schemes increase interception

To research

Visit http://www.nohrsc.noaa.gov/technology/gis/uhg_manual.html for the US National Weather Service's National Operational Remote Sensing Center's Hydrologic Technical Bulletin on Hydrographs. Some of their tables have been reproduced here.

- the drainage area – this increases over time (the partial area contribution model) and the peak of the flood increases as more drainage channels contribute, and drainage density increases (Figure 5.13)
- basin size, shape, and relief – small, steep basins reduce lag time, while basin shape influences where the bulk of the floodwaters arrive (Figure 5.14).

General description	Peaking factor	Limb ratio (recession to rising)
Urban areas; steep slopes	575	1.25
Mixed urban/rural	400	2.25
Rural, rolling hills	300	3.33
Rural, slight slopes	200	5.5
Rural, very flat	100	12.0

Table 5.6 Hydrograph peaking factors and recessional limb ratios

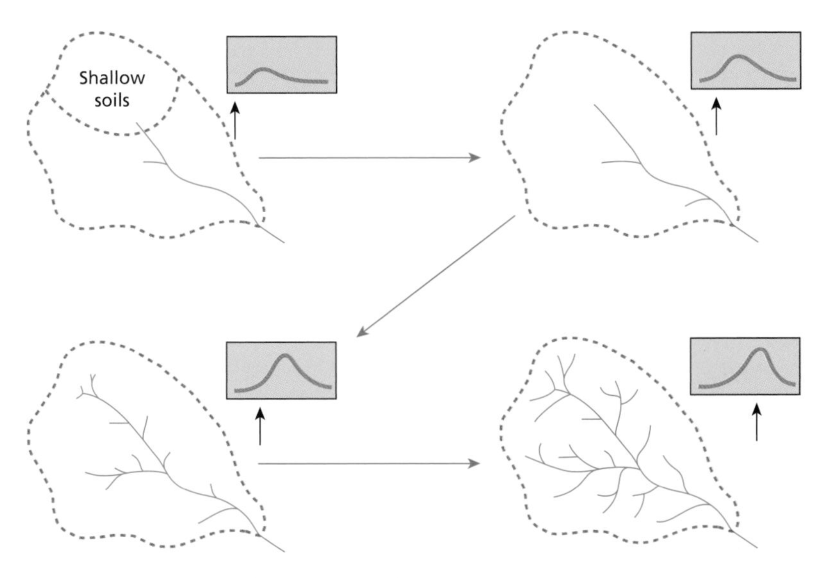

Figure 5.13 The partial area contribution model

To do:

a Describe the differences between the rural flood hydrograph and the urban flood hydrograph, as shown in Figure 5.12. Use terms such as peak flow, time lag, rising limb and recessional limb. Suggest reasons for the differences that you have noted.

b What impact has drainage basin shape (Figure 5.14) and drainage density (Figure 5.13) on storm hydrographs? Suggest reasons for their impacts.

c Study Figure 5.15. Compare the flood hydrographs of moorland that has been drained and burned with moorland that is undrained and/or unburned. What conclusions can you make about the effects of drainage and burning on flood hydrology? Suggest reasons to support your answer.

d Study the data in Table 5.6. State the meaning of "peaking factor" and "limb ratio". Using the data in the table, comment on the relationship between different types of land use and gradient on "peaking factor" and "limb ratio".

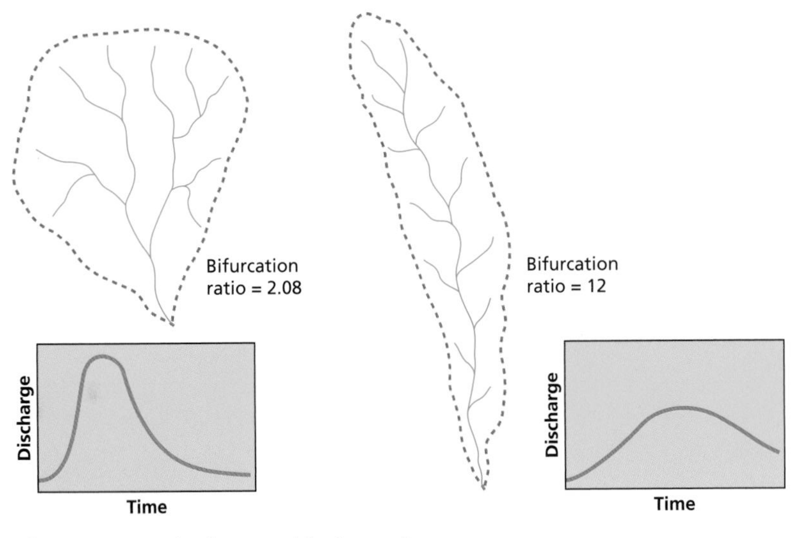

Figure 5.14 Basin shape and hydrographs

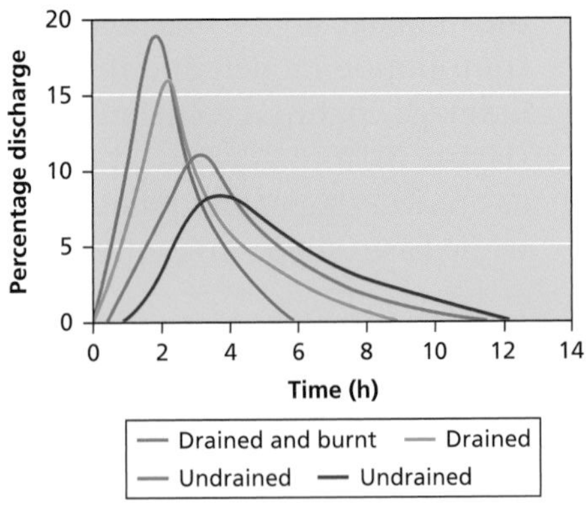

Figure 5.15 The effect of drainage and moorland burning on hydrographs

The causes and impacts of floods

Case study: *Flooding in Bangladesh, 1998*

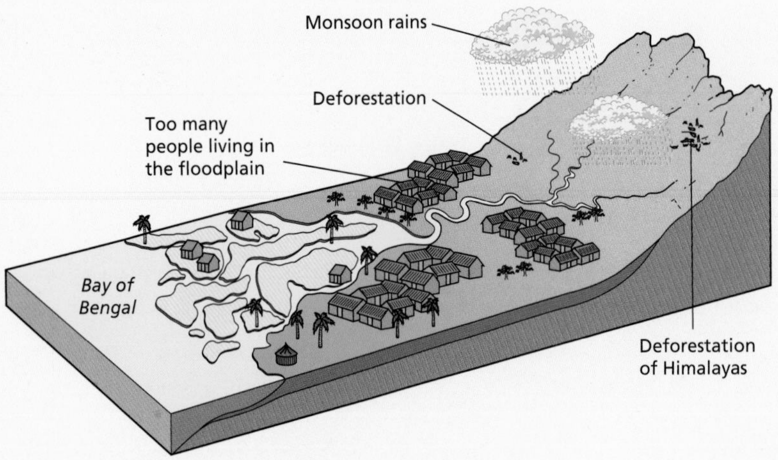

Figure 5.16 Some causes of flooding in Bangladesh

Bangladesh is one of the world's poorest and most densely populated countries (over 1,000 people per square kilometre), and it is experiencing rapid population growth (nearly 2.7% per annum). Every year the country experiences flooding and tropical storms, and the worst of these cause widespread damage and suffering.

Much of the country has been formed by deposition from three of the world's major rivers: the Brahmaputra, the Ganges and the Meghna. The sediment from these and over 50 other rivers forms one of the largest deltas in the world, and up to 80% of the country is located on the delta. As a result much of the country is just a few metres above sea level.

Almost all of Bangladesh's rivers have their source outside the country. For example, the drainage basin of the Ganges and Brahmaputra covers 1.75 million square kilometres, and includes the Himalayas, the Tibetan Plateau and much of northern India. Total rainfall within the Brahmaputra-Ganges-Meghna catchment is very high and seasonal: 75% of annual rainfall occurs in the monsoon between June and September. Moreover, the Ganges and Brahmaputra carry snowmelt waters from the Himalayas. Peak discharges of the rivers are immense – up to 100000 cumecs in the Brahmaputra, for example. In addition to water, the rivers carry vast quantities of sediment. This is deposited annually to form temporary islands and sandbanks, and helps to fertilize the land.

The advantages of flooding to Bangladesh

During the monsoon, between 30% and 50% of the entire country is flooded. Outside the monsoon season, heavy rainfall also causes extensive flooding. The floodwaters have certain advantages. They:

- replenish ground water reserves
- provide nutrient-rich sediment for agriculture in the dry season
- provide fish (fish supply 75% of dietary protein and over 10% of annual export earnings)
- reduce the need for artificial fertilizers
- flush pollutants and pathogens away from domestic areas.

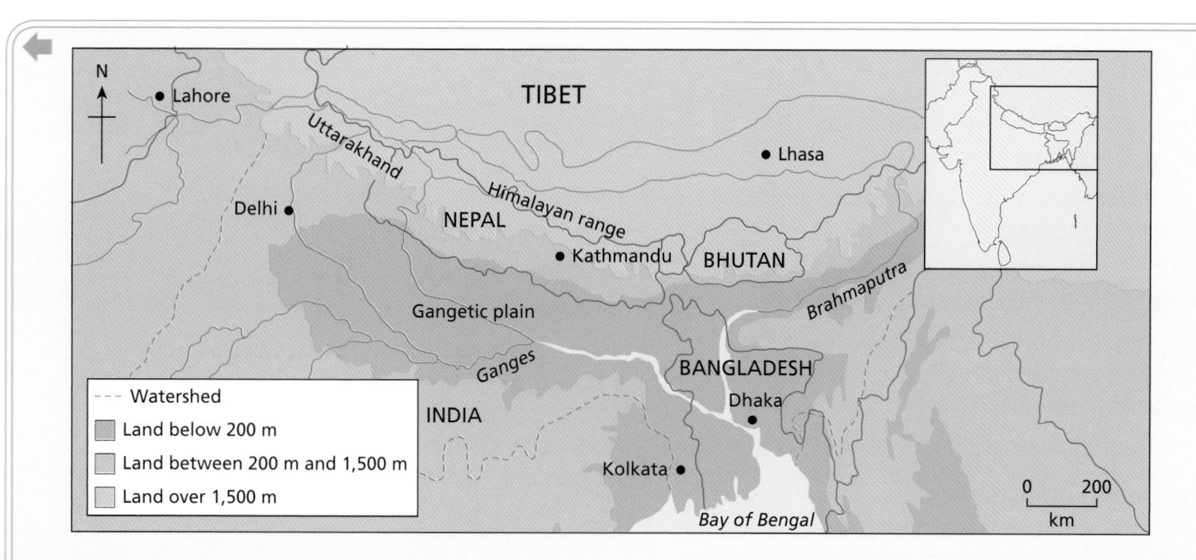

Figure 5.17 The Ganges catchment

The causes of the 1998 floods in Bangladesh

Large-scale flooding in Bangladesh is due to a variety of factors:

- discharge peaks of the big rivers
- high runoff from the Meghalaya Hills
- heavy rainfall associated with the monsoon
- high groundwater tables
- spring tides
- lateral river embankments
- the disappearance of natural water storage areas in the lowlands.
- snow melt in the Himalayas

There are five main types of flooding in Bangladesh: river floods, overland runoff, flash floods, "back-flooding" and storm surges. Snowmelt in the Himalayas, combined with heavy monsoonal rain, causes peak discharges in all the major rivers during June and July. This leads to serious flooding and destruction of agricultural land. (It is small-scale flooding that is advantageous to agricultural production, since the sediment deposited is a source of new nutrients and the water irrigates the rice paddies). In addition, the effects of flash floods, caused by heavy rainfall in northern India, have been intensified by the destruction of forest, which reduces interception, decreases water retention and increases the rate of surface runoff.

Human activity in Bangladesh has also increased the problem. Attempts to reduce flooding by building embankments and dikes have prevented the backflow of floodwater into the river. This leads to a ponding of water (also known as "drainage congestion") and back-flooding. In this way, embankments have sometimes led to an increase in deposition in drainage channels, and this can cause large-scale deep flooding.

The country is also subject to coastal flooding, when storm surges caused by intense low-pressure systems are funnelled up the Bay of Bengal to the delta.

Bangladesh is widely recognized as one of the countries most vulnerable to climate change, as sea levels rise and extreme weather events become more common.

Impacts

The worst flooding in Bangladesh's history occurred in 1998, when over two-thirds of the country was inundated. In the 1998 floods:

- 4,750 people were killed
- 23 million people were made homeless
- 130 000 cattle died
- 660 000 hectares of crops were damaged
- 400 factories were closed
- 11 000 km of roads were damaged
- 1,000 schools were damaged or destroyed.

To do:

Outline (i) the natural and (ii) human-induced causes of floods in Bangladesh.

Describe and explain (i) the advantages and (ii) the disadvantages of flooding in Bangladesh.

To research

Try to get hold of a copy of *Floods in Bangladesh – History, Dynamics and Rethinking the Role of the Himalayas* (Hofer, Messerli 2006) for a detailed account of flooding in Bangladesh, and excellent comparisons with floods elsewhere (the Mississippi, Yangtze and Rhine).

Be a critical thinker

Is deforestation in the Himalayas the cause of Bangladesh's floods?

There are at least three viewpoints regarding deforestation in the Himalayas:

- the belief that human activity is the cause of the problem (the environmentalists' view)
- the belief that human activity has little impact, and that the causes of rapid erosion are entirely natural (the naturalists' view)
- the view of local geographers (the local viewpoint).

The environmentalists' view

According to environmental scientists, for a balanced ecology forests should cover about 25% of a country's land area, but only 6% of Bangladesh's land is forested. The decline of forest cover in Bangladesh dates back to the 19th century, and is associated with commercial exploitation during British colonial rule, when forests were cut down for:

- railway sleepers
- road building
- urban development and house building in expanding villages
- fuelwood
- paper, timber and other forest products.

In addition, much forest was cleared to make way for agriculture. Half the country's forests have been destroyed in the last 20 years to meet the demands of an increasing population, and most recently to make way for tourist-related facilities. This reduction of forest cover has had a number of important effects:

- It traps less rain.
- The reduced litter layer intercepts less rainfall.
- The proportion of bare ground increases.
- Raindrop impact compacts the soil.

As forests become thinner, more light gets through to the ground layer and vegetation can grow there. This encourages grazing animals, which eat the buds of growing saplings. This favours vegetation that grows from the base, such as grasses, over vegetation that grows from buds, such as trees. In addition, the grazers compact the soil and increase its bulk density.

Thinner soils store and transport less moisture, leading to increased surface runoff and sediment discharge. Moreover, since the thin soils of the thinned forest are more exposed to direct sunlight, they lose more moisture through evaporation, leading to a reduction in the amount of groundwater. By contrast, soils under a complete forest canopy are shaded, and so evaporation losses are less, and they provide more water to groundwater stores.

Thus deforestation is associated with reduced infiltration rates, reduced soil water storage, and increased rates of surface runoff and soil erosion. There are also changes in stream morphology (shape and size), leading to increased flooding. Landslides also become more frequent, since whereas on a forested slope tree roots bind the soil, on deforested slopes there are no tree roots and so there is less anchorage of soil.

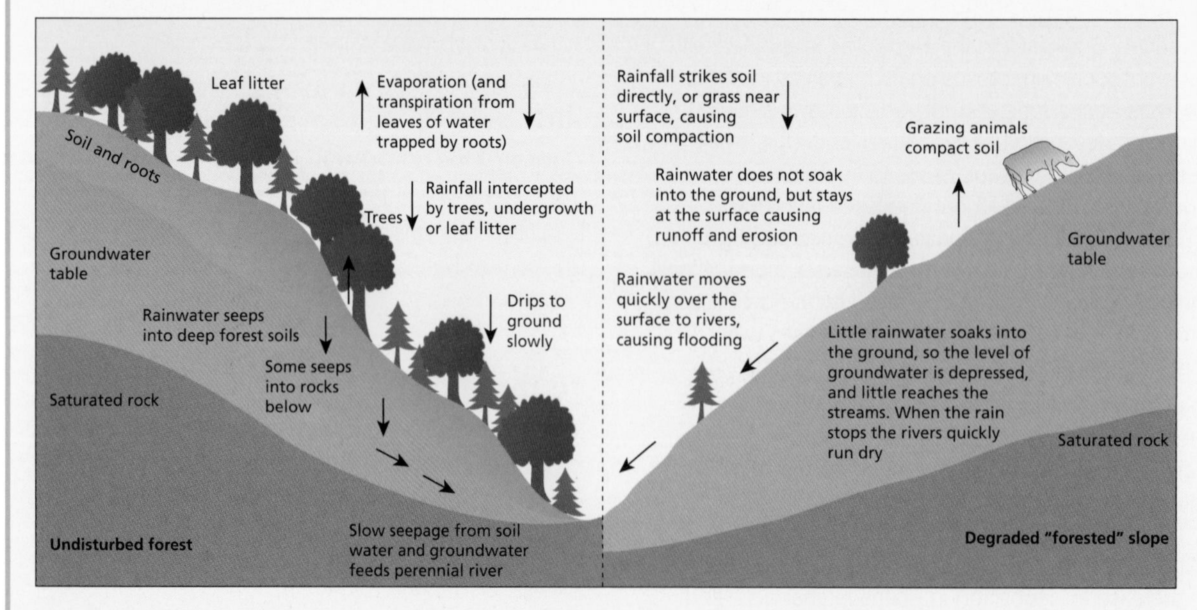

Figure 5.18 Natural and deforested slopes

Rainfall intensity (mm/hr)	Average soil loss (kg/ha) by % area forested			
	20–30	40–50	60–70	80–90
0-9	6.1	4.0	2.9	2.6
10-19	19.1	19.2	9.8	10.6
>20	43.6	25.2	28.1	16.9

Table 5.7 Soil erosion and deforestation in the Himalayas

Thus, according to environmentalists, deforestation in Nepal is causing erosion, floods, landslides, and declining groundwater (and a drying up of springs), with effects that reach as far away as the Bay of Bengal.

The naturalists' view

Other geographers do not accept that an environmental crisis is causing the high rates of erosion, in particular deforestation. They state that there is a great deal of uncertainty. For example, the estimates of fuelwood consumption per capita vary by a factor of 26. In addition, there is little agreement over the definition of deforestation. For instance, does it include the removal of a single branch? Does it include the sustainable harvesting of forest products? Does it include areas where replanting and regeneration are important features of forestry management?

The question of scale is also important. The effect of forest thinning is proportional to the amount of the catchment affected, and in many places this is very small.

Deforestation may have little impact on severe floods, since the ability of forest soils to absorb storm rainfall is limited. The area experiences high and seasonal rainfall due to the monsoon. Once the soil is saturated the forest behaves like any other impermeable, saturated surface. Moreover, the area is tectonically active, with rapid uplift of the mountains and rapid erosion by mountain streams. The rivers undercut the already steep Himalayan slopes, increasing their instability and the risk of landslides.

According to these geographers, deforestation causes surface erosion, gulleying and shallow landslides (less than 3 metres), which are important on a local scale. They also argue that the increased flood risk in the lower floodplains of the Ganges and Bramaphutra is due to increased human habitation of the floodplain, channel modifications, and the construction of ditches and drains that speed up the flow of water.

The local viewpoint

The two opposing viewpoints above are based on the work (and views) of western observers. Local geographers have also studied the causes and impacts of deforestation in the hill districts of Uttar Pradesh, India, south of Nepal (Figure 5.17). The area covers over 50 000 square kilometres, and is between 300 and 8,000 metres above sea level. The population of 5 million is growing rapidly, at about 2.3% per annum.

These researchers suggest that the agroforestry is increasingly unstable and imbalanced. The forest is used for fodder, manure, fuelwood and building materials. Population growth has increased the demand for forest land, but the government has seized much of this land, thereby reducing the amount available to local people. The land available to the local community is also the most severely eroded.

Forest cover and erosion in Uttarakhand, India

Forest cover in the area has fallen from about 60% to about 35%. Agricultural areas are increasing at about 1.5% per annum at the expense of the forest. However, up to 85% of the farmland now suffers from accelerated erosion, especially when it is close to roads. In addition to clearance during construction, landslides are triggered by the increase in slope instability.

Most of the work has been carried out on the Himalayan foothills and there are very few results from the central Himalayas. Much of the increase in erosion in the Himalayas is associated with new developments in the region. Most of these are concentrated along the foothills and the fringes of the area. Although the environmental change has been described as "fast, furious and negative", most of it is occurring alongside roads. Away from the roads there appears to be much less change.

To do:

a Explain why the Himalayas is naturally active in terms of geomorphological processes.

b Explain how human activities have affected the rates of geomorphological processes in the Himalayas. Use data to support your answer.

c Describe and suggest reasons why human impact varies spatially.

d Study Figure 5.18, which compares the processes on a natural and a degraded slope in the Himalayas. Describe and explain the differences in slope processes between the two slopes.

e To what extent do you think deforestation in the Himalayas is a cause of flooding in Bangladesh?

f Discuss the views of environmentalists, naturalists and locals regarding flooding in Bangladesh. Whose view do you agree with? Give reasons for your choice.

Dams and reservoirs

Be a critical thinker

Are large dams good or bad for the environment and for people?

The number of large dams (more than 15 metres high) being built is increasingly rapidly and is reaching a level of almost two completions every day. Examples of such megadams include the Akosombo (Ghana), Tucuruí (Brazil), Hoover (USA), and Kariba (Zimbabwe).

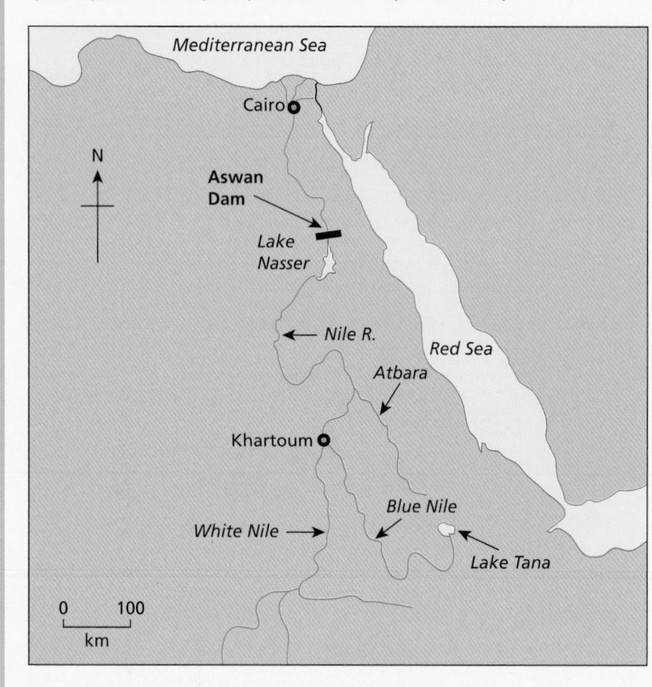

Figure 5.19 Location of the Aswan Dam

The advantages of dams are numerous. In the case of the Aswan High Dam on the River Nile, Egypt, they include:

- flood and drought control: dams allow good crops in dry years as, for example, in Egypt in 1972 and 1973
- irrigation: 60% of water from the Aswan Dam is used for irrigation and up to 3.4 million hectares (4,000 square kilometres) of the desert are irrigated
- hydroelectric power: this accounts for 7,000 million kilowatt hours each year
- improved navigation
- recreation and tourism.

It is estimated that the value of the Aswan High Dam to the Egyptian economy is about $500 million each year.

On the other hand, there are numerous costs. For example, in the case of the Aswan High Dam:

- water losses: the dam provides less than half the amount of water expected
- salinization: crop yields have been reduced on up to a third of the area irrigated by water from the dam due to salinization - this is due to poor water management
- groundwater changes: seepage leads to increased groundwater levels and may cause secondary salinization
- displacement of population: up to 100 000 Nubian people have been removed from their ancestral homes
- an increase in the humidity of the area has led to increased weathering of ancient monuments
- increased storage of water in the reservoir
- increased evaporative losses
- seismic stress: the earthquake of November 1981 is believed to have been caused by the Aswan Dam; as water levels in the dam decrease, so too does seismic activity
- deposition within the lake: infilling is taking place at about 100 million tonnes each year
- channel erosion (clear water erosion) beneath the channel: lowering the channel by 25 millimetres over 18 years, a modest amount
- erosion of the Nile delta: this is taking place at a rate of about 25 millimetres a year
- loss of nutrients – it is estimated that it costs $100 million a year to buy commercial fertilizers to make up for the lack of nutrients
- decreased fish catches: sardine yields are down 95% and 3,000 jobs in Egyptian fisheries have been lost
- spread of diseases such as schistosomiasis (bilharzia), due to increased stagnant water.

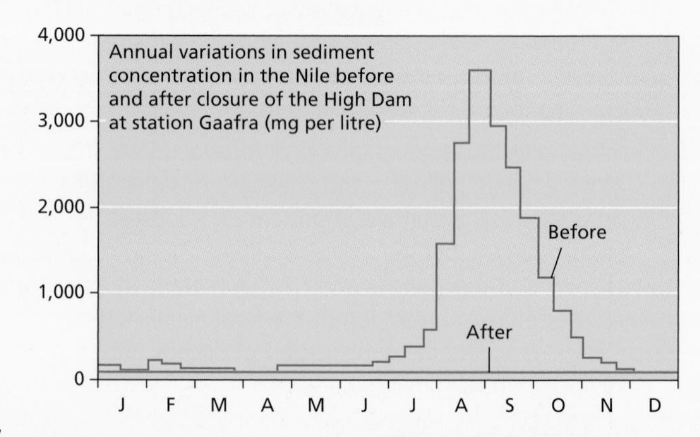

Figure 5.20 Changes in sedimentation along the Nile Valley

Logistics	Benefits	Side-effects	General assessment and prospects
Land development was a necessity to cope with the huge imbalance between the country's population growth and agricultural production. Agriculture is basic to the Egyptian economy.	It controls high floods and supplements low ones, saving Egypt from the costs of the damage from both high and low floods.	Water loss through evaporation and seepage is likely to affect the water supply needed for development plans; studies show that the water loss is within the predicted volume.	The Aswan High Dam is solid engineering work; more importantly, it is fulfilling a vital need for 40 million people.
Egypt had no option but to increase the water supply for land development policies of both horizontal and vertical extension.	Land reclamation allowed for increase in cultivated area, and increased the crop production of the existing land through perennial irrigation.	Loss of the Nile silt requires costly use of fertilizers; it has also caused riverbed degradation and coastal erosion of the northern delta.	All dams have problems; some are recognized while others are unforeseen at the time of planning.
Building a dam and forming a water reservoir in an Egyptian territory mimimizes the risks of water control politics on the part of riparian countries.	Nile navigation improved, and changed from seasonal to year-round.	Soil salinity is increasing and land in most areas is becoming waterlogged, due to delays in implementing drainage schemes.	A lesson learned from the Aswan project is that dams may be built with missionary zeal but little careful planning and monitoring of side-effects.
Egypt perceived the Aswan High Dam as a multipurpose project, basic to national development plans.	Electric power generated by the dam now supplies 50% of Egypt's consumption, but the dam was built primarily not for power generation but for water conservation.	Increased contact with water through irrigation extension schemes was expected to increase schistosomiasis rates; evidence exists that rates have not increased.	As a result of the new semi-capitalist policies and also the aid that Egypt received from several western countries, it is expected that several of the dam's problems will be efficiently controlled.
	The new lake has economic benefits, including land cultivation and settlement, fishing and tourist industries.	The Nile water quality has been deteriorating, but does not yet constitute a health hazard.	Dam-related studies have recognized problems and provided possible solutions. The dam will meet expectations providing that research findings are utilized.
			The development potential and economic returns of this water project are expected to be rewarding in the long run, if the project's developments are systematically studied and monitored.

Table 5.8 The benefits and costs of the Aswan Dam

Figure 5.21 Paphos Dam, Cyprus

To do:

Evaluate the effectiveness of large dams.

Floodplain management: stream channel processes

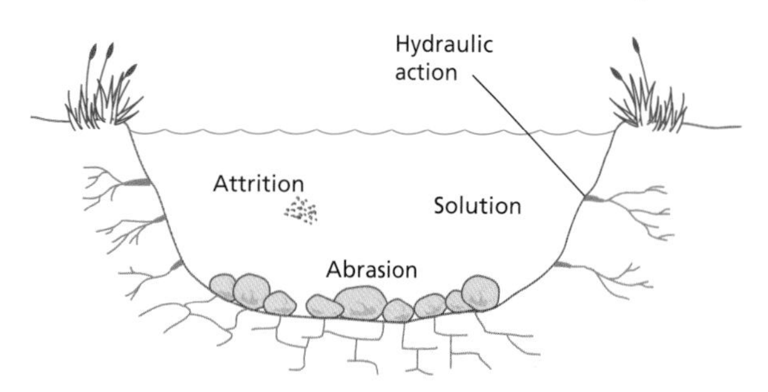

Figure 5.22 Types of erosion in a river

The main types of erosion

Abrasion (or **corrasion**) – the wearing away of the bed and bank by the load carried by a river.

Attrition – the wearing away of the load carried by a river. It creates smaller, rounder particles.

Hydraulic action – the force of air and water on the sides of rivers and in cracks.

Solution (or **corrosion**) – the removal of chemical ions, especially calcium, which causes rocks to dissolve.

Factors affecting erosion

As water drains downhill and enters the floor of a valley, it typically flows into a river or stream. This body of water moves with tremendous force, gradually wearing away, or abrading, its bed and bank. Its load of silt and debris is also gradually broken down and worn away in a process of attrition. Many factors affect this erosion process:

- **load** – the heavier and sharper the load the greater the potential for erosion
- **velocity and discharge** – the greater the velocity and discharge the greater the potential for erosion
- **gradient** – increased gradient increases the rate of erosion
- **geology** – soft, unconsolidated rocks, such as sand and gravel, are easily eroded
- **pH** – rates of solution are increased when the water is more acidic
- **human impact** – deforestation, dams, and bridges interfere with the natural flow of a river and frequently increase the rate of erosion.

To research

Visit the International Rivers Network at www.internationalrivers.org, and go to "The way forward" section for a video slideshow called "We all live downstream".

River-channel load theory

The **capacity** of a stream refers to the largest amount of debris that a stream can carry; its **competence** refers to the diameter of the largest particle that can be carried. The **critical erosion velocity** is the lowest velocity at which grains of a given size can be moved. The relationship between these variables is shown by means of a **Hjulström curve**.

There are three important features on Hjulström curves:

- The smallest and largest particles require high velocities to lift them.
- Higher velocities are required for entrainment than for transport.
- When velocity falls below a certain level (**settling** or **fall velocity**), particles are deposited.

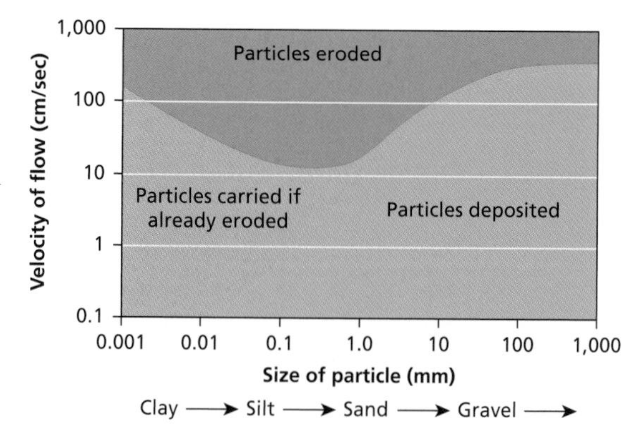

Figure 5.23 A Hjulström curve

To do:

Study Figure 5.23, which shows the relationship between water velocity and the size of material carried. Using this diagram, describe the sequence of geomorphological processes and their likely consequences in a stream channel with a wide range of sediment sizes, as discharge changes during a flood event from a low flow velocity of 0.05 m/sec-1, to 0.5m/sec-1, then to peak velocity of 1.5m/sec-1.

Features of erosion

Oxbow lakes are the result of erosion and deposition. Lateral erosion, caused by helicoidal flow, is concentrated on the outer, deeper bank of a **meander**. During times of flooding, erosion increases. The river breaks through and creates a new, steeper channel. In time, the old meander is closed off by deposition to form an oxbow lake.

Rapids frequently occur on horizontally bedded rocks. Layers of hard rock resists erosion and form shallow areas of turbulent flow within rivers – the Nile cataracts are a good example.

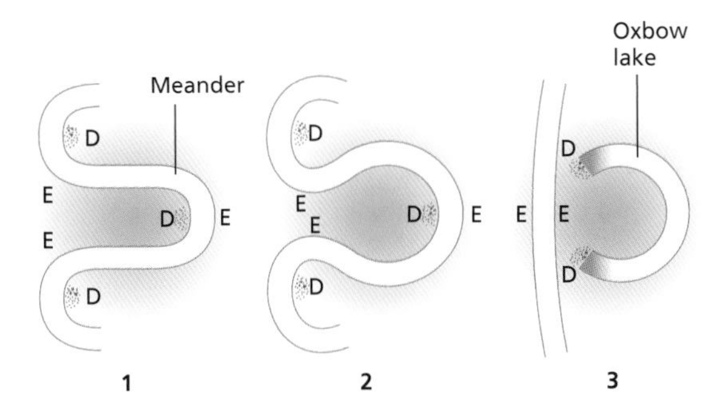

1 Eroson (E) and deposition (D) around a meander (a bend in a river).
2 Increased erosion during flood conditions. The meander becomes exaggerated.
3 The river breaks through during a flood. Further deposition causes the old meander to become an oxbow lake.

Figure 5.24 Formation of an oxbow lake

Floodplain management: deposition

Deposition occurs as a river slows down and it loses its energy. Typically, this occurs as a river floods across a floodplain, or enters the sea behind a dam. It is also more likely during low-flow conditions (such as in a drought) than during high-flow (flood) conditions, as long as the river is carrying sediment. The larger, heavier particles are deposited first, the smaller, lighter ones later. Features of deposition include deltas, levées, slip-off slopes (point bars), oxbow lakes, braided channels and floodplains.

The main types of transportation

Suspension – small particles held up by turbulent flow in the river.

Saltation – heavier particles bounced or bumped along the bed of the river.

Solution – the chemical load carried dissolved in the water.

Traction – the heaviest material dragged or rolled along the bed of the river.

Flotation – leaves and twigs carried on the surface of the river.

Levées

When a river floods its speed is reduced, slowed down by friction caused by contact with the floodplain. As its velocity is reduced, the river has to deposit some of its load. It drops the coarser, heavier material first to form raised banks, or levées, at the edge of the river. This means that over centuries the levées are built up of coarse material such as sand and gravel, while the floodplain consists of fine silt and clay.

Meanders

Meandering is the normal behaviour of fluids and gases in motion. Meanders can occur on a variety of materials, from ice to solid rock.

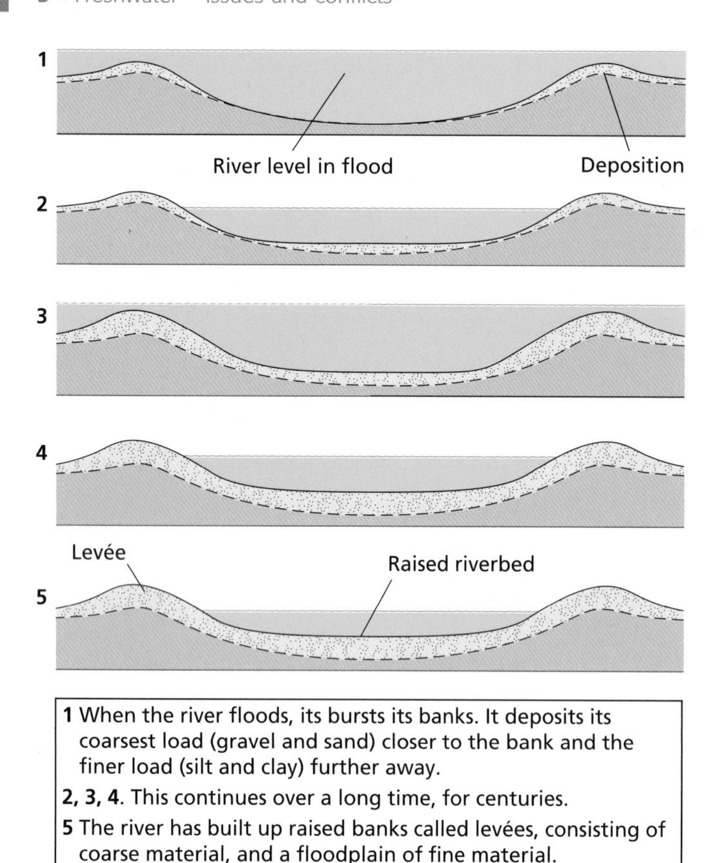

Figure 5.25 Levées

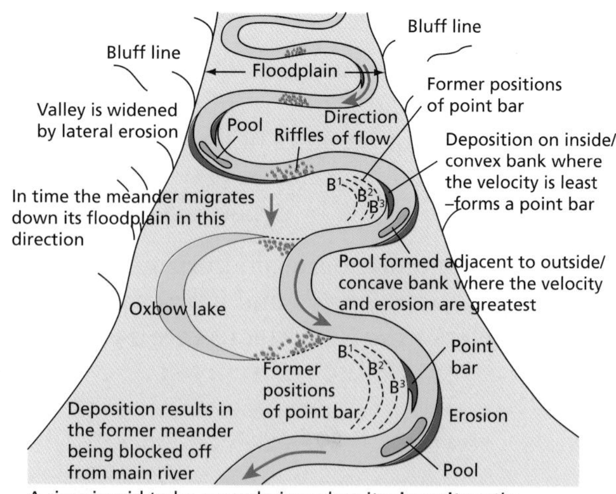

A river is said to be meandering when its **sinuosity** ratio exceeds 1.5. The **wavelength** of meanders is dependent on three major factors: channel width, discharge, and the nature of the bed and banks.

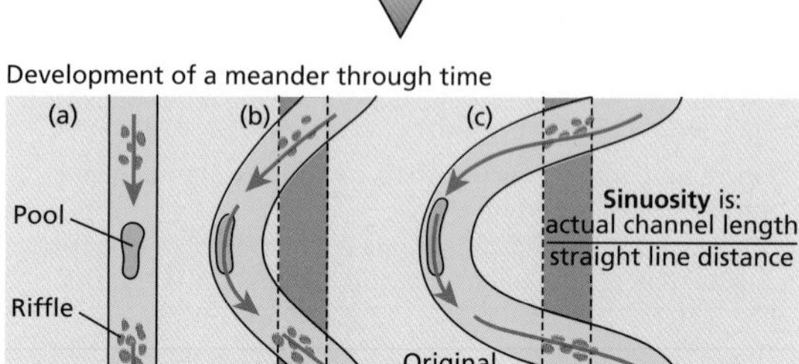

Figure 5.26 Meanders

Meander development occurs in conditions where channel slope, discharge and load combine to create a situation where meandering is the only way that the stream can use up the energy it possesses equally throughout the channel reach.

River terraces

A river terrace is an eroded floodplain, generally separated from the new floodplain by a steep slope. It is formed by changes in gradient, sediment load, climate change or human activity, or, indeed, any combination of these. It is the result of both deposition and erosion.

- Many terraces are formed by changes in base level (sea level).
- Changes in fluvial erosion and deposition, due to alternating cold and warm phases, are associated with the formation of terraces.
- Human activity can also lead to the formation of terraces. Deforestation for agricultural land reduces vegetation cover. As interception decreases, overland runoff increases, and there is accelerated erosion of part of the floodplain. This can lead to the formation of terraces upstream, as well as increased deposition downstream.

To do:

Comment on the factors that affect a river's ability to erode.

Draw a fully labelled diagram to show the formation of (i) a waterfall and (ii) a levée.

Human modification of floodplains

Urban hydrology

Urbanization changes the hydrology of a drainage basin. Roads and other artificial impermeable surfaces cut down infiltration and storage, increasing runoff and the risk of flooding. Various other factors such as building activity, encroachment on the river channel and the specific climatology of built-up areas mean that urbanized floodplains pose a range of challenges. When natural land is altered, storm sewers must collect the rainfall that would have otherwise been absorbed into the ground. These sewers speed up the flow of water into rivers, and if the river channel has also been restricted a storm event is likely to increase mean annual flood levels. In addition, the storm water washed off roads and other surfaces can contain contaminants, increasing pollution of rivers.

Stormwater sewers reduce the distance that storm water must travel before reaching a channel. Because sewers are smoother than natural channels the water drains away quickly, reducing storage.

Encroachment on the river channel occurs as a result of embankments, reclamation and riverside roads. The resulting reduction in channel width leads to higher floods. Bridges can also restrict free discharge of floods and increase levels upstream.

Replacement of vegetated soils with impermeable surfaces reduces infiltration, percolation and storage and so increases runoff and the velocity of overland flow. Evapotranspiration decreases because urban surfaces are usually dry.

Building activity clears vegetation, which exposes soil and increases overland flow, and disturbs and dumps the soil, increasing erodability. This lead to an increase in the likely flood peak.

Dams such as the three Gorges Dam on the Aswan – help reduce flood peaks, by storing water in a reservoir. They therefore make flood peaks lower and more predictable.

To research

Visit http://www.slideshare.net/maliadamit/river-channel-processes-landforms-1026801 for a slide show on general river processes.

For videos and animations of stream processes, visit http://serc.carleton.edu; go to the geoscience section and look for geomorphology videos.

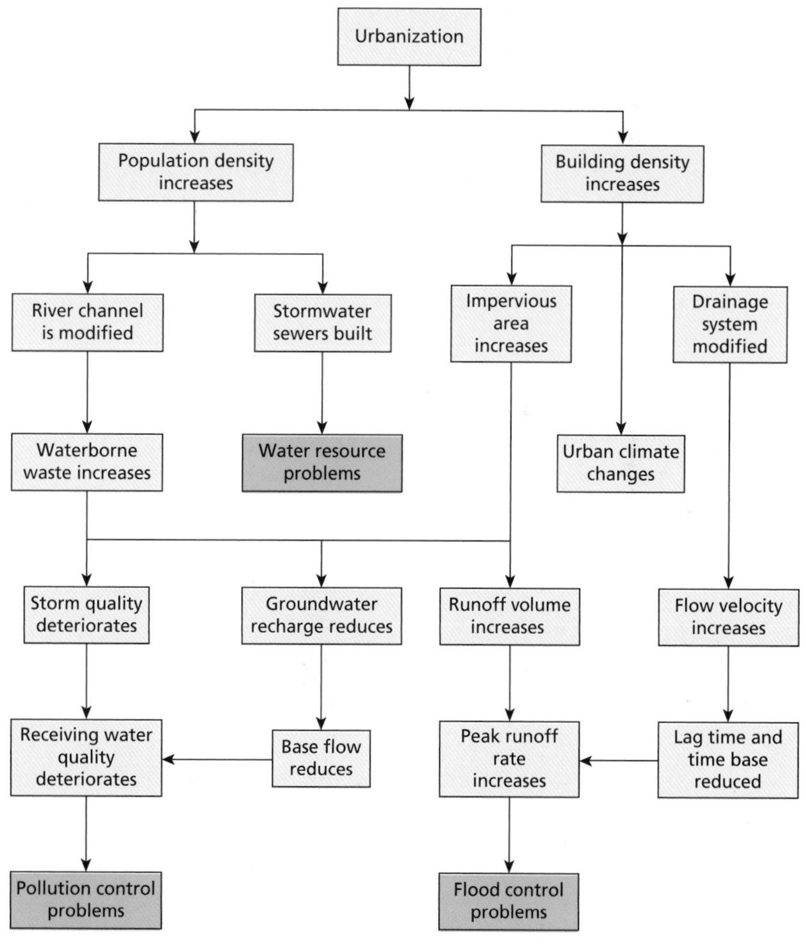

Figure 5.27 Urban hydrology

To research

For animations of deposition, visit http://highered.mcgraw-hill.com/sites/0072402466/student_view0/chapter10/animations_and_movies.html#

The Aswan High Dam reduces the high flood water and supplements the rivers when flows are low. It has saved Egypt from the monetary cost to cover damage from high flows and low flows.

Water resource problems in urban areas include reduced groundwater recharge, caused by sewers bypassing the mechanisms of percolation and seepage. Groundwater abstraction through wells may also reduce the store locally. Irrigation can draw on water resources, leading not only to depletion but also pollution.

Flood control problems arise because urbanization increases the peak of the mean annual flood, especially in moderate conditions. For example, a 243% increase in flood levels resulted from the building of Stevenage New Town in England. However, during heavy prolonged rainfall, saturated soil behaves in a similar way to urban surfaces.

Type of surface	Impermeability (%)
Watertight roof surfaces	70–95
Asphalt paving in good order	85–90
Stone, brick and wooden block pavements	
with tightly cemented joints	75–85
with open or uncertain joints	50–70
Inferior block pavements with open joints	40–50
Tarmacadam roads and paths	25–60
Gravel roads and paths	15–30
Unpaved surfaces, railway yards, vacant lots	10–30
Park, gardens, lawns, meadows	5–25

Table 5.9 Impermeability of urban surfaces

To do:

a Study Table 5.9. Identify which parts of the urban area have (a) the highest levels of impermeable surfaces and (b) the lowest amount of impermeable surfaces. How might impermeability and urban hydrology vary with distance from a city centre? Give reasons for your answer.

b Figure 5.28 shows the mean annual flood increments with increasing urban cover and sewerage facilities. The ratio R shows the factor by which the flood level has increased (for instance, when R = 3 flood levels have tripled). Describe what happens to flood levels as the proportion of impervious land increases and the number of storm sewers increases.

c Using the data shown in Figure 5.28, describe and explain the changes that occur in the magnitude and frequency of flood discharge as urbanization increases.

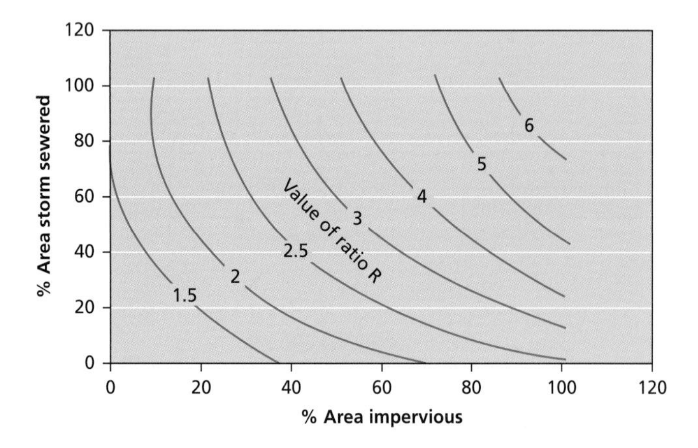

Figure 5.28 Annual flood increments with increasing urban and sewer cover

Figure 5.29 Impermeability and urban hydrology

Alternative stream management strategies

Perception of flooding is in part related to the frequency and the magnitude of floods. The responses to flooding are the result of knowledge, perception, money, technology and the characteristics of the flood and the success of the prediction. Responses include

- bearing the loss
- emergency action
- flood-proofing
- flood control
- land-use zoning
- flood insurance.

Emergency action includes the removal of people and property, and flood-fighting techniques such as sandbags. Much depends on the efficiency of forecasting and the time available to warn people and clear the area. Flood-proofing includes sealing walls, sewer adjustment by the use of valves, and covering buildings and machinery.

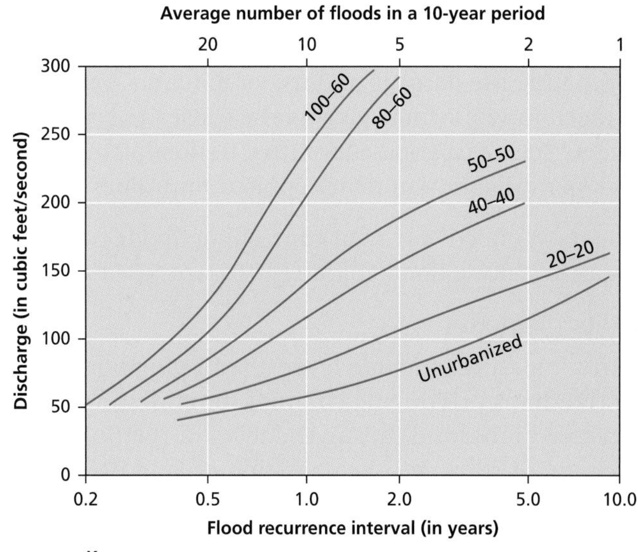

Figure 5.30 Flood frequency curves for different levels of urbanization

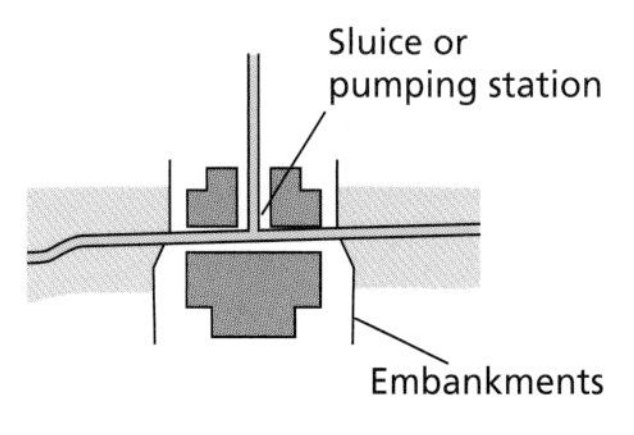

1 Flood embankments with sluice gates. The main problem with this is it may raise flood levels up- and downstream.

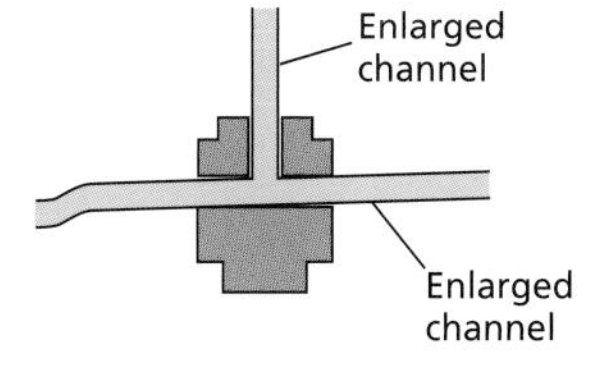

2 Channel enlargement to accommodate larger discharges. One problem with such schemes is that as the enlarged channel is only rarely used it becomes clogged with weed.

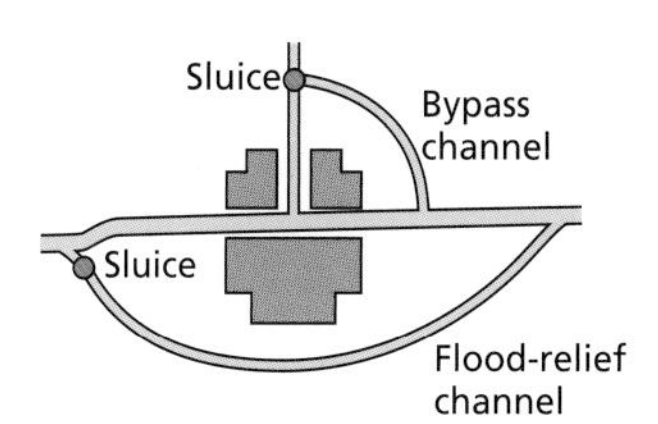

3 Flood relief channel. This is appropriate where it is impossible to modify the original channel as it tends to be rather expensive, e.g. the flood relief channels around Oxford.

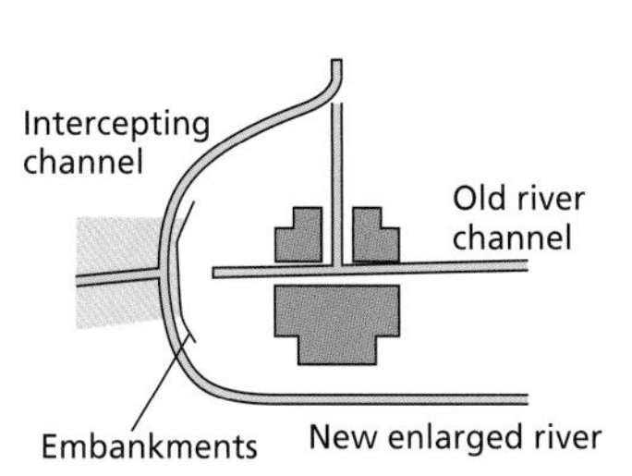

4 Intercepting channels. These divert only part of the flow away, allowing flow for town and agricultural use, e.g. the Great Ouse Protection Scheme in England's Fenlands.

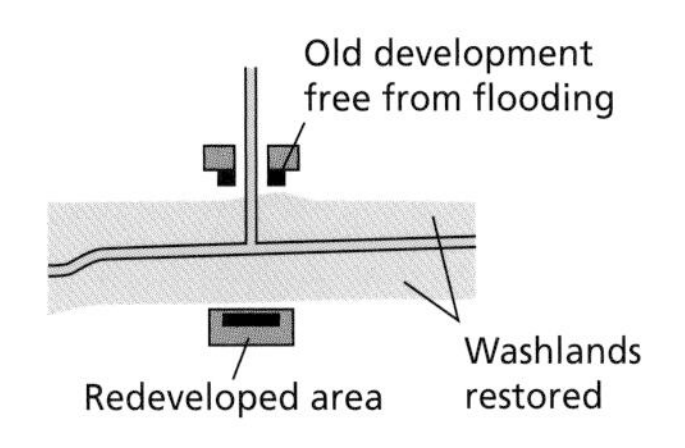

5 Removal of settlements. This is rarely used because of cost, although many communities, e.g. the village of Valmeyer, Illinois, USA, were forced to leave as a result of the 1993 Mississippi floods.

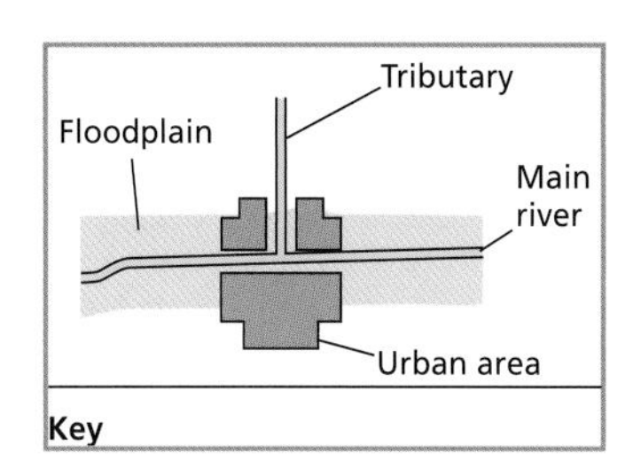

Figure 5.31 Flood control measures

Land-use management is a further way of limiting the damage. However, there are practical problems, such as the difficulty of estimating the damage and use of potential land. Moreover, protection works may give a false sense of security. Flood insurance is widely seen as a good alternative to floodplain management, but its lack of availability in many poor communities makes it of limited use.

The most effective way of controlling floods is through protective measures along flood channels. (Figure 5.31).

To research

Visit www.apfm.info for a case study about flood management on the Mississippi River Basin. Find out about the nature of floods, flood management and mitigation strategies. Comment on the main lessons that have been learnt in the management of the Mississippi.

Other methods

Other measures include levées, removing boulders from riverbeds to riverbanks (reducing channel roughness and protecting banks from erosion), and raising the level of the floodplain. Flood abatement (through the changing of land use in the drainage basin) tackles the problem by slowing down the rate at which water from storms reaches the river channel. There are several methods, including:

- afforestation, to increase interception and evapotranspiration
- terracing of farmland, contour ploughing and strip cultivation, to enable control of overland flow.

Groundwater management

The over-exploitation of ground water loads to falling ground water levels. In areas where nitrate fertilizers are used, groundwater is at risk of nutrient enrichment. In hot areas, salinization may be a problem, and in coastal zones saline intrusion can affect groundwater quality.

Groundwater – subsurface water.
Aquifers – porous rocks such as sandstone or limestone that contain significant quantities of water.

Groundwater pollution in Bangladesh

Since the 1970s, following the lead of UNICEF, Bangladesh has sunk millions of tube wells into groundwater, providing a convenient supply of drinking water free from the bacterial contamination of surface water that was killing a quarter of a million children a year. As a result of this action, infant mortality fell by half, but it has now been found that about one in five of the wells is contaminated with arsenic.

It is projected that one in 10 people who drink the water containing arsenic will ultimately die of lung, bladder or skin cancer. Arsenic poisoning is a slow disease. Skin cancer typically occurs 20 years after people start ingesting the poison. The real danger is internal cancers, especially of the bladder and lungs, which are usually fatal. Bangladeshi doctors have been warned to expect an epidemic of cancers by 2010. The victims will be people in their 30s and 40s who have been drinking the water all their lives – people in their most productive years.

One solution to the problem is a concrete butt, collecting surface water by pipe from gutters. Other possible solutions include a filter system. Neither is as convenient as the tube well it is designed to replace. Tube wells are easy to sink in the delta's soft alluvial soil, and for tens of millions of peasants the wells have revolutionized access to water. However, some experts believe that better watershed management would make it possible to use surface water as a safe source of drinking water.

Be a critical thinker

To what extent is groundwater a renewable resource?

Most **groundwater** is found within a few hundred metres of the surface, but has been found at depths of up to 4 kilometres. The permanently saturated zone within solid rocks and sediments is known as the **phreatic zone**, and here nearly all the pore spaces are filled with water. The upper layer of this is known as the water table. The water table varies seasonally: it is higher in winter following increased levels of precipitation. The zone that is seasonally wetted and seasonally dries out is known as the **aeration zone** or the **vadose zone**. Groundwater may take as long as 20 000 years to be recycled. Hence, in some places, groundwater is considered a non-renewable resource.

Aquifers can provide a great reservoir of water into which wells can be sunk. This water moves very slowly and acts as a natural regulator in the hydrological cycle by absorbing rainfall which otherwise would reach streams rapidly. In addition, aquifers maintain stream flow during long, dry periods. A rock that will not hold water is known as an **aquiclude** or **aquifuge**. These are impermeable rocks, such as clay, which prevent large-scale storage and transmission of water.

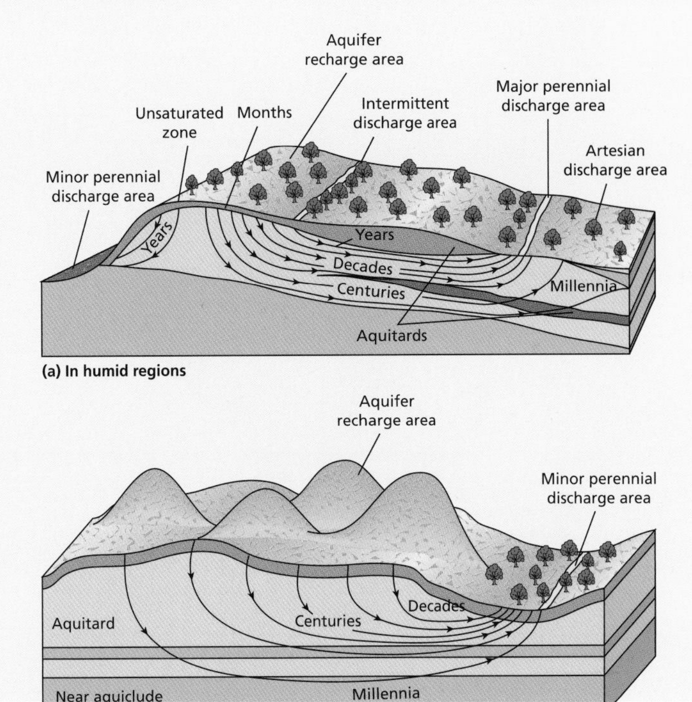

Figure 5.32 Groundwater sources

There are a number of methods of recharging groundwater reservoirs. Where the materials containing the aquifer are permeable water spreading is used. By contrast, in sediments with unpermeable layers pumping water into deep pits or wells is used. This method is used extensively on the heavily settled coastal plain of Israel to replenish the groundwater reservoirs. When surplus irrigation water is available, and to reduce the problems associated with salt water intrusion from the Mediterranean.

Groundwater recharge also occurs as a result of:

- infiltration of part of the total precipitation at the ground surface
- seepage through the banks and bed of surface water bodies such as rivers, lakes and oceans
- groundwater leakage and inflow from adjacent aquicludes and aquifers
- artificial recharge from irrigation, reservoirs, etc.

Losses of groundwater result from:

- evapotranspiration, particularly in low-lying areas where the water table is close to the ground surface
- natural discharge by means of spring flow and seepage into surface water bodies
- leakage and outflow through aquicludes and into adjacent aquifers
- artificial abstraction.

Groundwater depletion in India

In India, groundwater irrigation has transformed the lives of millions. It has also rectified problems of waterlogging and salinization caused by canals. But in many places, including the productive Punjab and Haryana, whose rather well-off farmers also get free or cut-price electricity, the rate of groundwater extraction is unsustainable. Nearly a third of India's groundwater blocks were defined in 2004 as "critical, semi-critical or over-exploited". The World Bank reckons that 15% of India's food is produced by "mining" (unrenewable extraction) of groundwater. In the Punjab region the rate of depletion is accelerating in 18 of its 20 districts, where groundwater levels have dropped 10 metres since 1979.

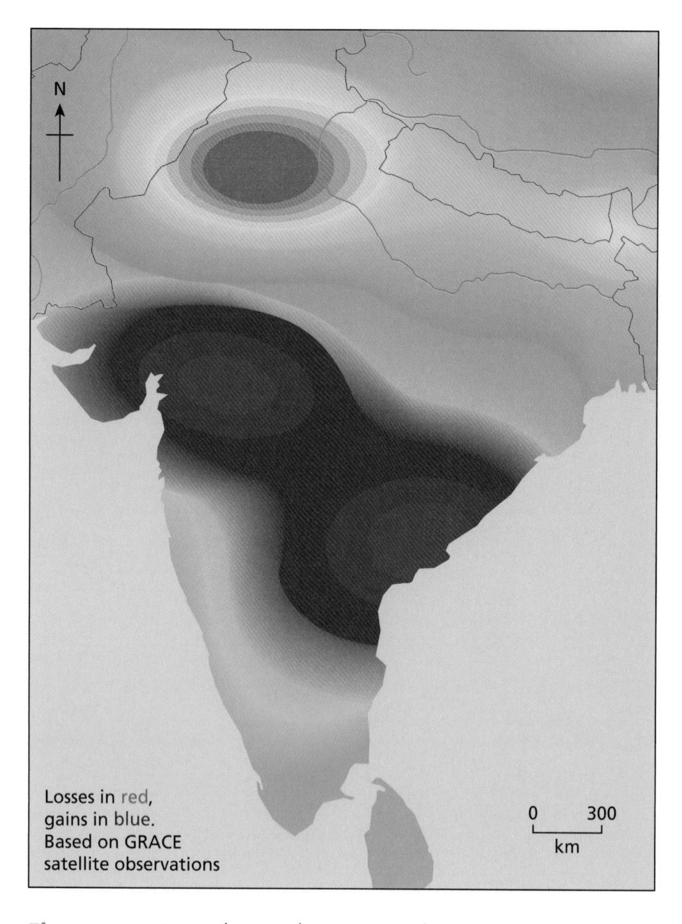

Figure 5.33 Groundwater changes in India, 2002–8

To do:

a Define in your own words the terms groundwater and aquifer.

b Explain why the use of groundwater can be unsustainable.

c Describe the changes to groundwater in India, as shown in Figure 5.33. Suggest reasons for (i) falling levels of groundwater and (ii) rising levels of groundwater.

d Suggest why the pollution of groundwater in Bangladesh is serious.

Freshwater wetland management

The Ramsar Convention, an international treaty to conserve wetlands, defines wetlands as "areas of marsh, fen, peatland or water, whether natural or artificial, permanent or temporary, with water that is static or flowing, fresh, brackish or salt." Thus, according to the Ramsar classification, there are marine, coastal and inland types, subdivided into 30 categories of natural wetland and nine human-made ones, such as reservoirs, barrages and gravel pits. Wetlands now represent only 6% of the earth's surface, of which 30% are bogs, 26% are fens, 20% are swamps, 15% are floodplains and 2% are lakes. It is estimated that there was twice as much wetland area in 1900 as in 2000. As well as being an important water resource, wetlands provide many important social, economic and environmental benefits (see Table 5.10).

Wetland – land with soils that are permanently flooded.

Functions	Products	Attributes
Flood control	Fisheries	Biological diversity
Sediment accretion and deposition	Game	Culture and heritage
Groundwater recharge	Forage	
Groundwater discharge	Timber	
Water purification	Water	
Storage of organic matter		
Food-chain support/cycling		
Water transport		
Tourism/recreation		

Table 5.10 Benefits of wetlands

Loss and degradation

The loss and degradation of wetlands is caused by several factors, including:

- increased demand for agricultural land
- population growth
- infrastructure development
- river flow regulation
- invasion of non-native species and pollution.

Case study: *Changing river management – the Kissimmee River*

The 165-kilometre Kissimmee River once meandered through central Florida. Its floodplain, reaching up to 5 kilometres wide, was inundated for long periods by heavy seasonal rains. Wetland plants, wading birds and fish thrived there, but the frequent, prolonged flooding caused severe impacts to people.

Between 1962 and 1971 engineering changes were made to deepen, straighten and widen the river, which was transformed into a 90-kilometre, 10-metre deep drainage canal. The river was **channelized** to provide an outlet canal for draining floodwaters from the upper Kissimmee lakes basin, and to provide flood protection.

Impacts of channelization

The channelization of the Kissimee River had several unintended impacts (see Figure 5.35). Concerns about the **sustainability** of existing ecosystems led to a massive restoration project, on a scale unmatched elsewhere.

The Kissimmee River Restoration Project

The aim is to restore over 100 square kilometres of river and associated floodplain wetlands by 2015. The project was started in 1999 and will benefit over 320 species, including the bald eagle, wood stork and snail kite.

Restoration of the river and its associated natural resources requires **dechannelization**. This entails backfilling approximately half of the flood-control channel and reestablishing the flow of water through the natural river channel. In residential areas the flood-control channel will remain in place. Seasonal rains and flows now inundate the floodplain in the restored areas.

The costs of restoration

- It is estimated the project will cost $414 million (initial channelization cost $20 million), a bill being shared by Florida and the federal government.
- Restoration of the river's floodplain could result in higher losses of water due to evapotranspiration during wet periods. In extremely dry spells, navigation may be impeded, but it is expected that navigable depths will be maintained at least 90% of the time.

Benefits of restoration

- Higher water levels should ultimately support a natural river ecosystem again.
- Reestablishment of floodplain wetlands is expected to result in decreased nutrient loads to Lake Okeechobee.
- Populations of key avian species such as wading birds and waterfowl have returned to the restored area, and numbers have increased.
- Dissolved oxygen levels have doubled, which is critical for the survival of aquatic species.
- Potential revenue associated with increased recreational usage on the restored river could significantly enhance local and regional economies.

To do:

a Outline the benefits of wetlands.

b Describe how the channelization of the Kissimmee affected local ecosystems.

c Explain the term "river restoration". What are the (i) aims and (ii) benefits of river restoration?

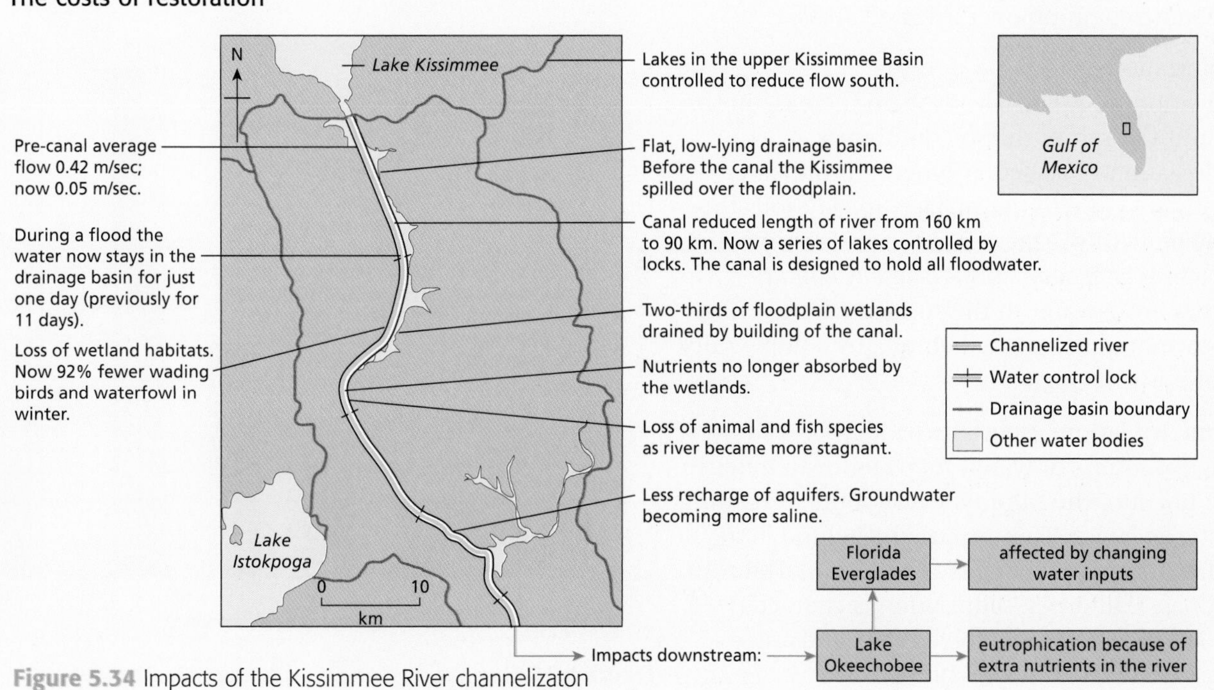

Figure 5.34 Impacts of the Kissimmee River channelizaton

Irrigation and agriculture

Irrigation – the artificial addition of water to soil in areas where there is insufficient for adequate crop growth.

Salinization – the build-up of salt in the soil, sometimes caused by poor irrigation techniques.

People have irrigated crops since ancient times: there is evidence for irrigation in Egypt for nearly 6,000 years. Water for irrigation can be taken from surface stores, such as lakes, dams, reservoirs and rivers, or from groundwater. Types of irrigation range from total flooding, as in the case of paddy fields, to drip irrigation, where precise amounts are measured out to each individual plant.

Impacts of irrigation

Irrigation occurs in developed as well as developing countries. For example, large parts of the USA and Australia are irrigated. The advent of diesel and electric motors in the mid-20th century led for the first time to systems that could pump groundwater out of aquifers faster than it was recharged. This has led in some regions to loss of aquifer capacity, decreased water quality and other problems. In Texas, irrigation has reduced the water table by as much as 50 metres. By contrast, in the fertile Indus Plain in Pakistan, irrigation has raised the water table by as much as 6 metres since 1922.

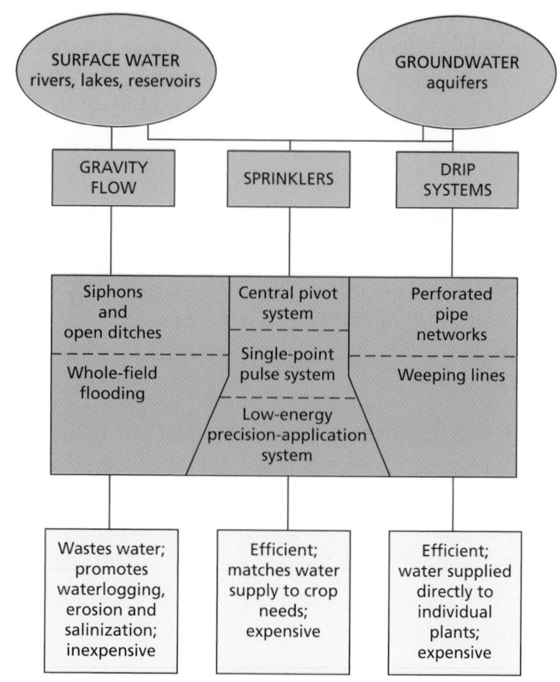

Figure 5.35 Types of irrigation

Irrigation can reduce the earth's albedo (reflectivity) by as much as 10%. This is because a reflective sandy surface may be replaced by one with dark green crops. Irrigation can also cause changes in precipitation. Large-scale irrigation in semi-arid areas, such as the High Plains of Texas, has been linked with increased rainfall, hailstorms and tornadoes. Under natural conditions semi-arid areas have sparse vegetation and dry soils in summer. However, when irrigated these areas have moist soils in summer and a complete vegetation cover. Evapotranspiration rates increase, resulting in greater amounts of summer rainfall across Kansas, Nebraska, Colorado and the Texas Panhandle. In addition, hailstorms and tornadoes are more common over irrigated areas.

Salinization

Irrigation frequently leads to an increase in the amount of salt in the soil, known as salinization. This occurs when groundwater levels are close to the surface. In clay soils this may be within 3 metres of the surface, whereas on sandy and silty soils it is less. Capillary forces bring water to the surface where it may be evaporated, leaving behind any soluble salts that it is carrying.

Some irrigation, especially paddy rice, requires huge amounts of water. As water evaporates in the hot sun, the salinity levels of the remaining water increase. This also occurs behind large dams. Chemical changes are also important. In Salinas, California, salinization is characterized by an increased in dissolved salts and an increase in the ratio of chlorides to bicarbonates.

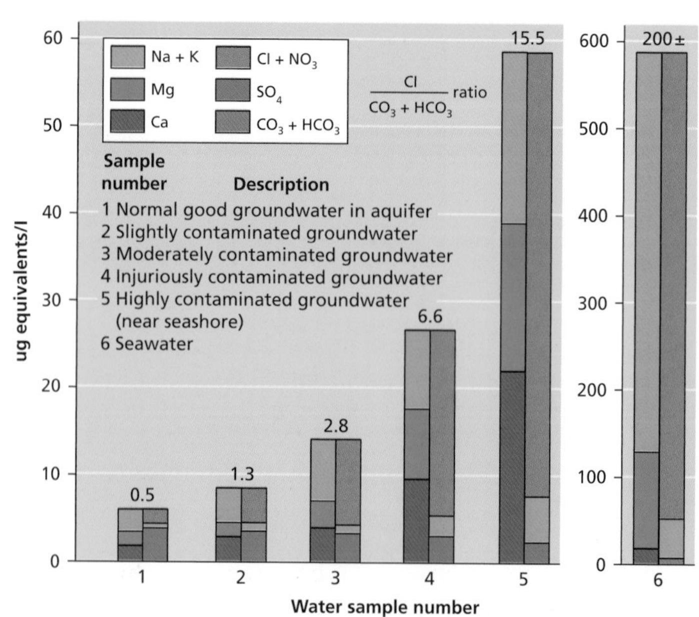

Figure 5.36 Changes in groundwater quality with distance from an aquifer

To do:

- **a** Explain how irrigation can both (i) raise groundwater levels and (ii) lower groundwater levels.
- **b** Study Figure 5.37, which shows changes in the quality of groundwater with distance from the centre of an aquifer in California. Describe the differences in water quality between "normal" groundwater in the aquifer (site 1) with that nearest the coast (site 5).
- **c** Compare and contrast the composition of (i) Cl and NO_3 and (ii) CO_3 and HCO_3 with distance from the centre of the aquifer. Suggest reasons for these changes.

Eutrophication – the nutrient enrichment of streams, ponds and groundwater.

Eutrophication

In eutrophication, increased amounts of nitrogen and/or phosphorus are carried in streams, lakes and groundwater, causing nutrient enrichment. This is an example of agro-chemical runoff. This leads to an increase in algal blooms as plants respond to the increased nutrient availability. This is an example of positive feedback. However, the increase in algae and plankton shade the water below, cutting off the light supply for submerged plants. The prolific growth of algae, especially in autumn as a result of increased levels of nutrients in the water and higher temperatures, results in anoxia (oxygen starvation in the water).

Figure 5.37 Eutrophication: algal bloom caused by excessively high nitrate levels in freshwater

There are three main reasons why the high concentrations of nitrogen in rivers and groundwater are a problem. First, nitrogen compounds can cause undesirable effects in the aquatic ecosystems, especially excessive growth of algae. Second, the loss of fertilizer is an economic loss to the farmer. Third, high nitrate concentrations in drinking water may affect human health.

Percentage of lakes and reservoirs suffering eutrophication	
Asia and the Pacific	54%
Europe	53%
Africa	28%
North America	48%
South America	41%

Table 5.11 Global variations in eutrophication (UNEP/ILEC surveys)

Dealing with eutrophication

There are three main ways of dealing with eutrophication. These include:

- altering the human activities that produce pollution, for example by using alternative types of fertilizer and detergent
- regulating and reducing pollutants at the point of emission, for example in sewage treatment plants that remove nitrates and phosphates from waste
- clean-up and restoration of polluted water by pumping mud from eutrophic lakes.

Case study:

Eutrophication in Kunming City, China

Eutrophication is the most widespread water-quality problem in China. In Dianchi Lake near Kunming City in Yannan Province, blue/green algae have killed over 90% of native water weed, fish and molluscs, and destroyed the fish culture industry. Because alternative water supplies have run short, water from Dianchi Lake has been used since 1992 to supply Kunming's growing population of 1.2 million residents. The city opened its first sewage treatment plant in 1993, but this copes with only 10% of the city's sewage.

Case study: *Managing the Murray–Darling basin*

Although the Murray – Darling is Australia's longest river system, draining a basin the size of France and Spain combined (Figure 5.39), it no longer carries enough water to carve its own path to the sea. After 10 years of drought, and many more years of overexploitation and pollution, the only hope of restoring the river to health lies in a complete overhaul of how it is managed.

Droughts have long plagued the Murray–Darling. The region is affected by the periodic El Niño weather pattern. As a result, the flow of the Darling, the longest tributary of the Murray, varies wildly, from as little as 0.04% of the long-term average to as much as 911%. A region that accounts for 40% of Australia's agriculture, and 85% of its irrigation, was on the verge of ruin.

As Australia's population continues to grow, so does demand for water in the cities and for the crops that grow in the river basin. By 1994 human activity was consuming 77% of the river's average annual flow, even though the actual flow falls far below the average in dry years. Thanks to a combination of reduced flow and increased runoff from saline soils churned up by agriculture, the water was becoming unhealthily salty, especially in its lower reaches. The tap water in Adelaide, which draws 40% of its municipal supplies from the river and up to 90% when other reserves dry up, was beginning to taste saline. The number of indigenous fish was falling, since the floods that induce them to spawn were becoming rarer. Toxic algae flourished in the warmer, more sluggish waters.

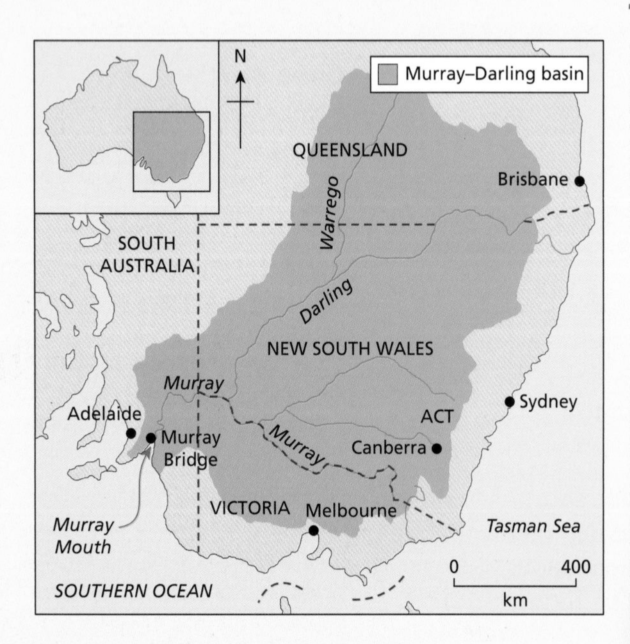

Figure 5.38 The Murray–Darling river basin

Australia embarked on a series of reforms that in many ways serve as a model for the management of big, heavily exploited rivers. New South Wales, Victoria and South Australia agreed to cap the amount of water they took from the river. They also made plans to reduce salinity and increase "environmental flows". They agreed on the following reforms:

- There would be no more subsidies for irrigation.
- Farmers would be responsible for the maintenance of channels and dams.
- For each river and tributary, scientists would calculate the maximum sustainable allocations of water, and states would make sure that extractions did not exceed that figure.

Guided by these principles, the states have made much progress. By 1999 the average salinity of the river in South Australia had fallen by over 20%. The construction of fish ladders around dams and weirs, and the release of extra water into important breeding grounds, has spawned a recovery in native species.

However, many problems remain. Farmers in certain states can still drill wells to suck up groundwater, and tree plantations absorb much of the rainwater that would otherwise find its way into the river. Small dams on farms, which block small streams or trap runoff from rain or flooding, also prevent water reaching the river.

To do:

a Describe the natural and human causes of water shortages in Australia.

b Comment on two of the problems caused by water shortages in Australia.

c Evaluate one or more measures to tackle the problem of water shortages in Australia.

To research

Visit www.wwf.org.au for the WWF Australia's report "Pipe dreams?" on inter-basin transfers and water shortages.

Possible measures to reduce nitrate loss are as follows (based on the northern hemisphere):

1 Avoid using nitrogen fertilizers between mid-September and mid-February when soils are wet.
2 Prefer autumn-sown crops – their roots conserve nitrogen in the soil and use up nitrogen.
3 Sow autumn-sown crops as early as possible, and maintain crop cover through autumn and winter.
4 Don't apply nitrogen to fields next to a stream or lake.
5 Don't apply nitrogen just before heavy rain is forecast.
6 Use less nitrogen after a dry year because less will have been lost. (This is difficult to assess precisely.)
7 Do not plough up grass, as this releases nitrogen.
8 Incorporate barley straw to prevent the growth of green algae; it uses nitrogen as it decays (Figure 5.39).

Figure 5.39 Using barley bales to tackle eutrophication

To do:

- **a** Choose an appropriate method to show the data in Table 5.11. Comment on your results.
- **b** Describe the human causes of eutrophication.
- **c** Outline the ways in which it is possible to manage eutrophication.

To research

Visit http://sciencebitz.com for a case study on managing eutrophication in Lago Paranoa, Brazil.

Demand for water: local/national scale

The Middle East

The problem of water supply is widespread throughout the Middle East region, with Jordan, Israel and Palestine suffering the most acute water shortages. The problem has created great friction between Arabs and Jews; the example of Israeli-Palestinian tensions illustrates the problem clearly.

For decades Israel has obtained up to 80% of the water provided by the mountain aquifer under the West Bank (Figure 5.49). Occupied by the Israelis since 1967, the West Bank has 120 000 Jewish settlers who use about 60 million cubic metres annually, compared with the 137 million cubic metres used by 1.5 million West Bank Arabs. In addition, the West Bank settlers irrigate 70% of their cultivated land, compared with just 6% of Palestinian land.

In the Palestinian Gaza Strip, most of the area's meagre domestic water supply comes from an aquifer so badly depleted that the water is saline.

As part of the Israeli-Jordan peace process in 1995, Israel agreed to provide Jordan with 150 million cubic metres of water per annum, by diverting water, building new dams and desalinization. Other possibilities for improving the region's water supply include using the Litani and Awali rivers in Lebanon, cutting back on agriculture, and creating a regional water market whereby people pay for the water they use.

To do:

- **a** Outline the reasons for water shortages in the Middle East.
- **b** In what ways may it be possible to manage water in the region?

More material available:

www.oxfordsecondary.co.uk/ibgeography

Water shortages in California and the Mekong Basin

To research

Visit http://www.unu.edu/unupress. Look up the publication *Managing water for peace in the Middle East* for a detailed account of the problems of managing the Jordan River.

6 Oceans and their coastal margins

By the end of this chapter you should be able to:

- understand the features and distribution of oceans and ocean currents, and their effect on climate
- explain the importance of oceans as a resource and the geopolitical issues arising from their use
- describe the characteristics of coastal margins, processes and landforms
- describe the conflicts and management strategies relating to the coastal zone.

The distribution of the oceans and ocean currents

Covering 70% of the earth's surface, oceans are of great importance to humans in a number of ways. One of the most important ways is through the atmosphere–ocean link, through which oceans regulate

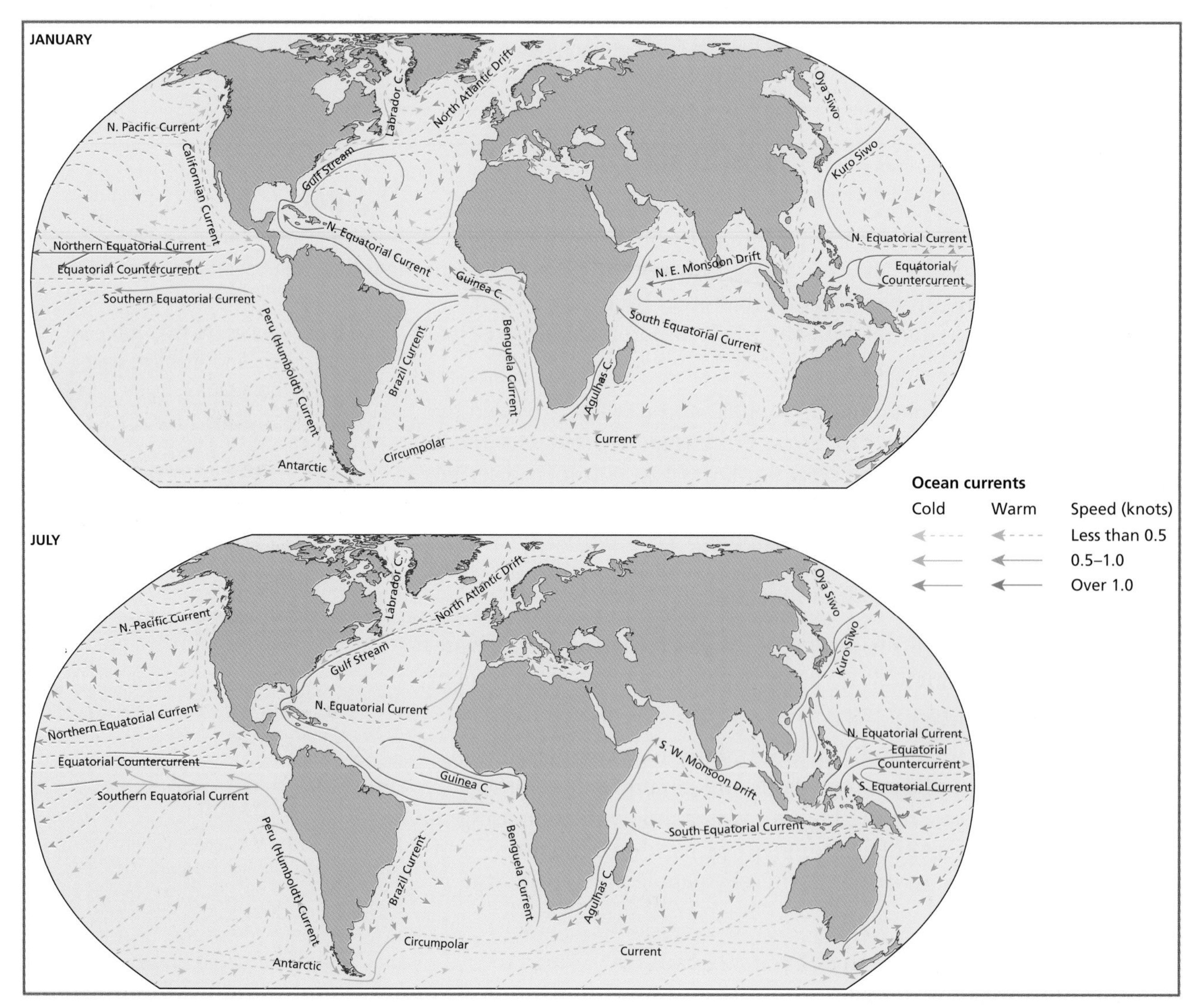

Figure 6.1 Ocean currents

climatic conditions. Warm ocean currents move water away from the equator, whereas cold ocean currents move water away from the cold regions towards the equator. The major currents move huge masses of water over long distances. The warm Gulf Stream, for instance, transports 55 million cubic metres per second. Without it, the temperate lands of northwestern Europe would be more like the sub-Arctic. The cold Peru Current and the Benguela Current of southwest Africa bring in nutrient-rich waters dragged to the surface by offshore winds.

In addition, there is the Great Ocean Conveyor Belt (see page 139). This deep, grand-scale circulation of the ocean's waters effectively transfers heat from the tropics to colder regions, such as northern Europe.

TOK Link

How maps can shape our perceptions

Oceans cover about 50% of the earth's surface in the northern hemisphere and about 90% in the southern hemisphere. This is not always clear when looking at world maps. See Figure 6.2.

Ocean or sea	Area in 1,000 km^2
Pacific Ocean	166 229
Atlantic Ocean	86 551
Indian Ocean	73 442
Arctic Ocean	13 223
South China Sea	2,975
Caribbean Sea	2,516
Mediterranean Sea	2,509
Bering Sea	2,261

Table 6.1 The world's largest oceans and seas

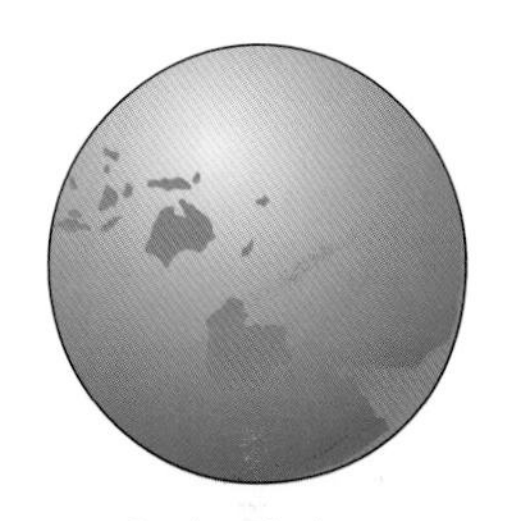

Figure 6.2 Land and sea hemispheres

Ocean morphology

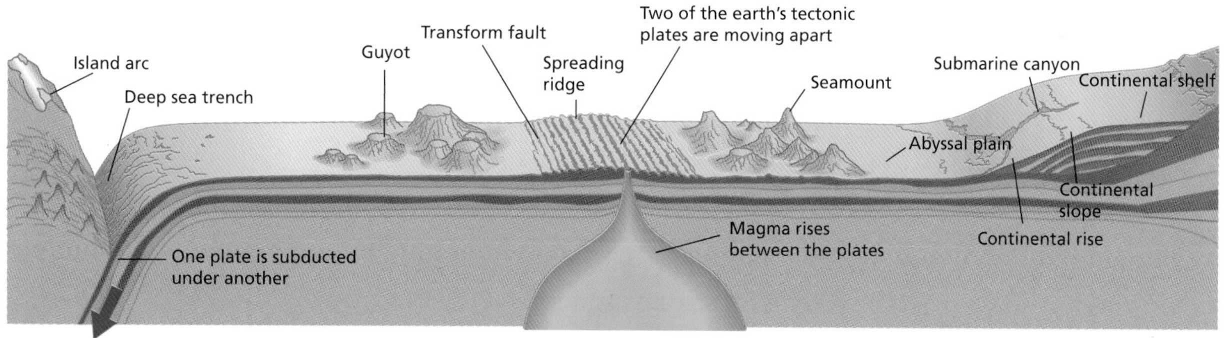

Figure 6.3 Features of the ocean floor

The ocean floor consists of many features such as deep sea trenches, mid-ocean ridges, transform faults, rift valleys, deep abyssal plains, continental slopes and continental shelves. Smaller features include submarine canyons and submarine volcanoes or seamounts.

The **continental shelf** is a relatively flat area of seabed, stretching from the land to the edge of the continental slope. The continental shelf is less than 250 metres deep and may be up to 70 kilometres wide.

The **continental slope** is the steeply sloping area of the seabed that stretches from the continental shelf to the abyssal plain. The continental slope may contain **submarine canyons** eroded by fast-flowing currents of water and sediment. In addition, sediment slumps

To do:

a Calculate the total surface area of the world's oceans and seas.

b Work out the percentage of the total surface area that each ocean or sea accounts for.

c Comment on the distribution of the world's oceans as shown by your calculations.

d Describe the changes in the ocean currents found in the Indian Ocean in January and July.

down the canyon to form a steep, narrow valley on the continental slope. Submarine canyons are located close to the point where a large river flows into the sea.

The **abyssal plain** is at the edge of the continental slope. These plains cover large areas of the sea floor at depths of between 4,000 and 6,000 metres. They are generally flat and featureless.

Seamounts are extinct volcanic cones that lie below the surface. A **guyot** is is a flat-topped volcano that reached the surface but later subsided.

Mid-ocean ridges are the largest feature of the ocean floor. They are essentially a linear belt of submarine mountains. They occur at divergent (spreading or constructive) plate boundaries. New magma forces its way up between two plates and pushes them apart. In slow-spreading ridges, such as in the Mid-Atlantic, the rate of spreading is up to 5 centimetres per year. The ridges are characterized by a wide **rift valley** at their centre. This rift valley can be up to 20 kilometres wide. In contrast, where the rate of spreading is rapid, as in the case of the East Pacific Rise, which spreads at a rate of about 17–18 centimetres a year, there are no rift valleys. **Transform faults** are also a feature of ocean ridges. They are fracture zones usually at right angles to the main ridge, caused by the plates moving apart.

Ocean trenches are the deepest parts of the oceans. These are arc-shaped depressions, formed at subduction zones where one tectonic plate (usually an oceanic one) plunges under a less dense continental one. The Mariana Trench in the Pacific Ocean is over 11 000 metres deep.

Oceanic water

Salinity

Oceanic water varies in salinity and temperature. Average salinity is about 35 parts per 1,000 (ppt). Concentrations of salt are higher in warm seas, owing to the high evaporation rates of the water. In tropical seas, salinity decreases sharply with depth. Another factor that can change the salinity of seawater is a very large river discharging into it. The runoff from most small streams and rivers is quickly mixed with ocean water by the currents, and has little effect on reducing salinity. However, a large river such as the Amazon in South America may result in the ocean having little or no salt content for over a kilometre or more out to sea.

The freezing and thawing of ice also affects salinity. The thawing of large icebergs (made of frozen freshwater and lacking any salt) will decrease the salinity, while the actual freezing of seawater will increase the salinity temporarily. This temporary increase happens in the early stages of the freezing of seawater as small ice crystals form at about −2° C. These tiny, needle-like ice crystals are frozen freshwater and the salts are not part of them, so the liquid between these crystals becomes increasingly salty. Eventually, though, as seawater freezes, the ice crystals trap salts and the entire large piece of frozen seawater (ice floe) is salty. Salinity levels increase with depth.

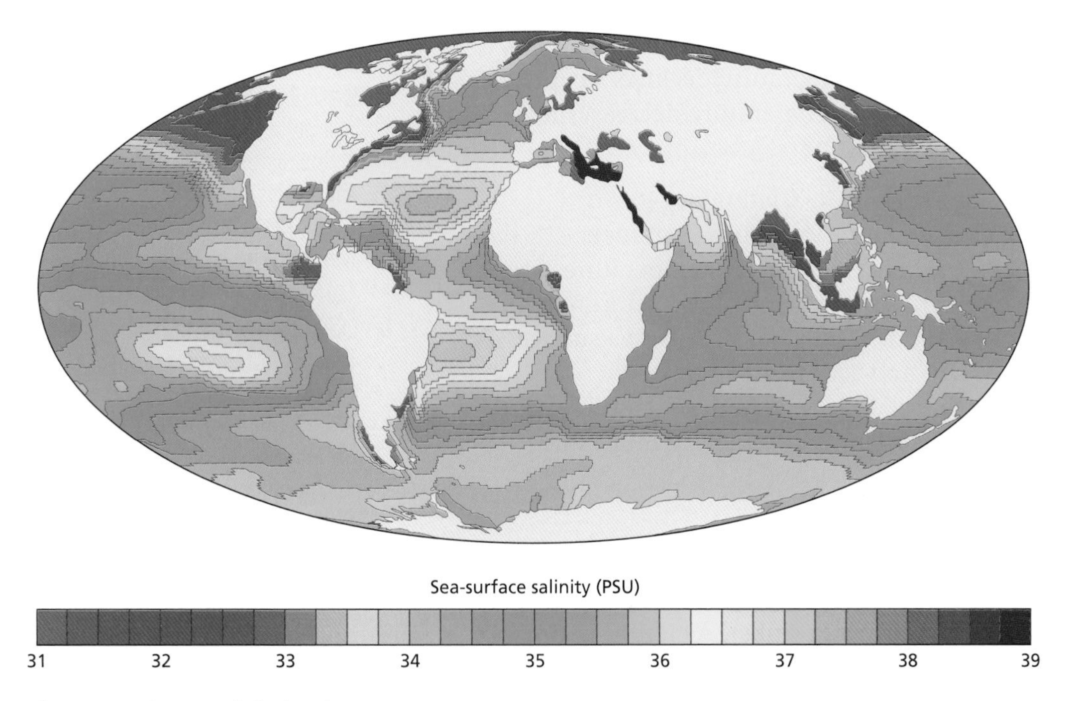

Figure 6.4 Ocean salinity levels

The predominant minerals in seawater are chloride (54.3%) and sodium (30.2%), which combine to form salt. Other important minerals in the sea include magnesium and sulphate ions.

Temperature

Temperature varies considerably at the surface of the ocean, but there is little variation at depth. In tropical and subtropical areas, sea

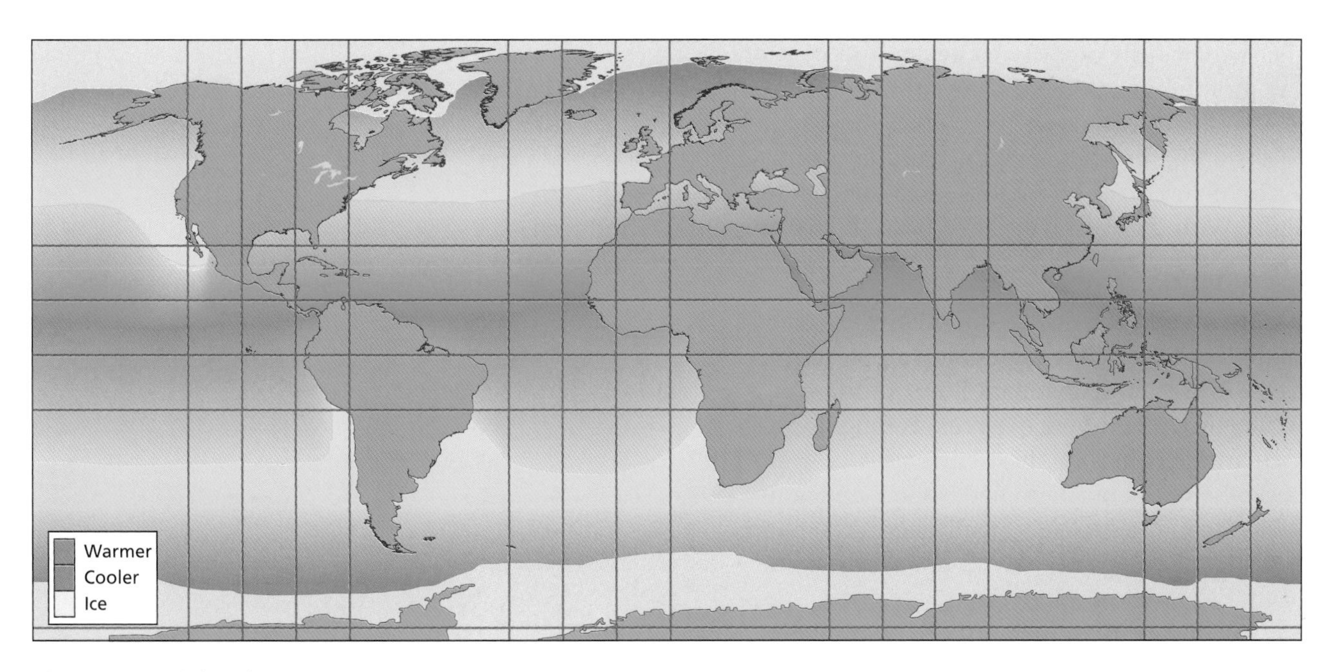

Figure 6.5 Variations in temperature

surface temperatures in excess of 25°C are caused by insolation. From about 300 to 1,000 metres the temperature declines steeply to about 8–10°C. Below 1,000 metres the temperature decreases to a more uniform 2°C in the ocean depths.

The temperature profile is similar in the mid-latitudes (40–50°N and S), although there are clear seasonal variations. Summer temperatures may reach 17°C, whereas winter sea temperatures are closer to 10°C. There is a more gradual decrease in temperature with depth (thermocline). In high latitudes and polar oceans, sea surface temperatures range between 0°C and 5°C. In some cases the temperature may be below freezing, but the water does not freeze because of its salinity. Below the surface, it reaches the uniform temperature of 2°C in the deep ocean.

Density

Temperature, salinity and pressure affect the density of seawater. Large water masses of different densities are important in the layering of the ocean water (denser water sinks). As temperature increases, water becomes less dense. As salinity increases, water becomes more dense. As pressure increases, water becomes more dense. A cold, highly saline, deep mass of water is very dense, whereas a warm, less saline, surface water mass is less dense. When large water masses with different densities meet, the denser water mass slips under the less dense mass. These responses to density are the reason for some of the deep ocean circulation patterns.

To do:

- **a** Describe the pattern of salinity as shown in Figure 6.4. Suggest reasons for the levels of salinity in (a) Arctic areas and (b) the Mediterranean region and (c) off the Amazon (Brazil).
- **b** Describe the pattern of temperatures shown in Figure 6.5. Identify the areas where there appears to be an extension of cold water into areas of warm water. Suggest why this happens.

Oceans and climate

Specific heat capacity

The specific heat capacity is the amount of energy it takes to raise the temperature of a body. It takes more energy to heat up water than it does to heat land. However, it takes longer for water to lose heat. This is why the land is hotter than the sea by day, but colder than the sea by night. Places close to the sea are cool by day, but mild by night. With increasing distance from the sea this effect is reduced.

Sea currents

Surface ocean currents are caused by the influence of prevailing winds blowing steadily across the sea. The dominant flow of surface ocean currents (known as **gyres**) is roughly circular; the pattern of these currents is clockwise in the northern hemisphere and anti-clockwise in the southern hemisphere. The main exception is the circumpolar current that flows around Antarctica from west to east. There is no equivalent current in the northern hemisphere because of the distribution of land and sea. Within the circulation of the gyres, water piles up into a dome. The effect of the rotation of the earth is to cause water in the oceans to push westward; this piles up water on the western edge of ocean basins, rather like water slopping

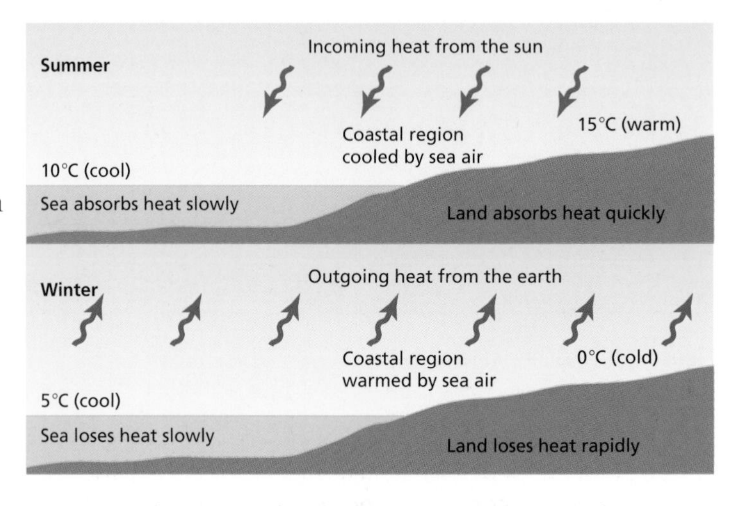

Figure 6.6 The effect of the sea on temperatures in coastal margins

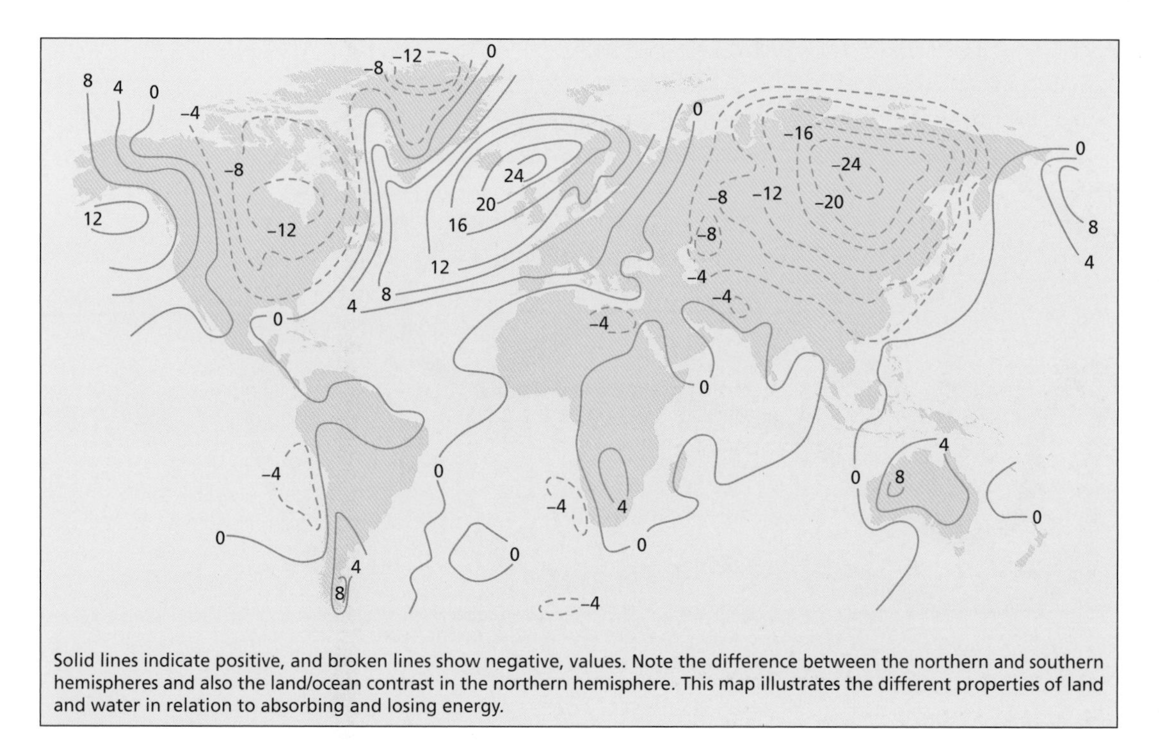

Solid lines indicate positive, and broken lines show negative, values. Note the difference between the northern and southern hemispheres and also the land/ocean contrast in the northern hemisphere. This map illustrates the different properties of land and water in relation to absorbing and losing energy.

Figure 6.7 Global temperature anomalies

in a bucket. The return flow is often narrow, fast-flowing currents such as the Gulf Stream. The Gulf Stream in particular transports heat northwards and then eastwards across the North Atlantic; it is the main reason why the British Isles have mild winters and relatively cool summers.

The effect of ocean currents on temperatures depends upon whether the current is cold or warm. Warm currents from equatorial regions raise the temperatures of polar areas (with the aid of prevailing westerly winds). However, the effect is only noticeable in winter. For example, the North Atlantic Drift raises the winter temperatures of northwestern Europe. Some areas are more than 24°C warmer than the average for their line of latitude (Figure 6.7). By contrast, there are other areas which are made colder by ocean currents. Cold currents such as the Labrador Current off the northeast coast of North America may reduce summer temperature, but only if the wind blows from the sea to the land.

In the Pacific Ocean there are two main atmospheric states: the first is warm surface water in the west with cold surface water in the east; the other is warm surface water in the east with cold in the west. In whichever case, the warm surface causes low pressure. As air blows from high pressure to low pressure there is a movement of water from the colder to the warmer area. These winds push warm surface water into the warm region, exposing colder deep water behind them.

The Great Ocean Conveyor Belt

Oceanic conveyor belt – a global thermohaline circulation, driven by the formation and sinking of deep water and responsible for the large flow of upper ocean water.

In addition to the transfer of energy by wind and the transfer of energy by ocean currents, there is also a transfer of energy by deep sea currents. Oceanic convection occurs from the polar regions, where cold, salty water sinks into the depths and makes its way towards the

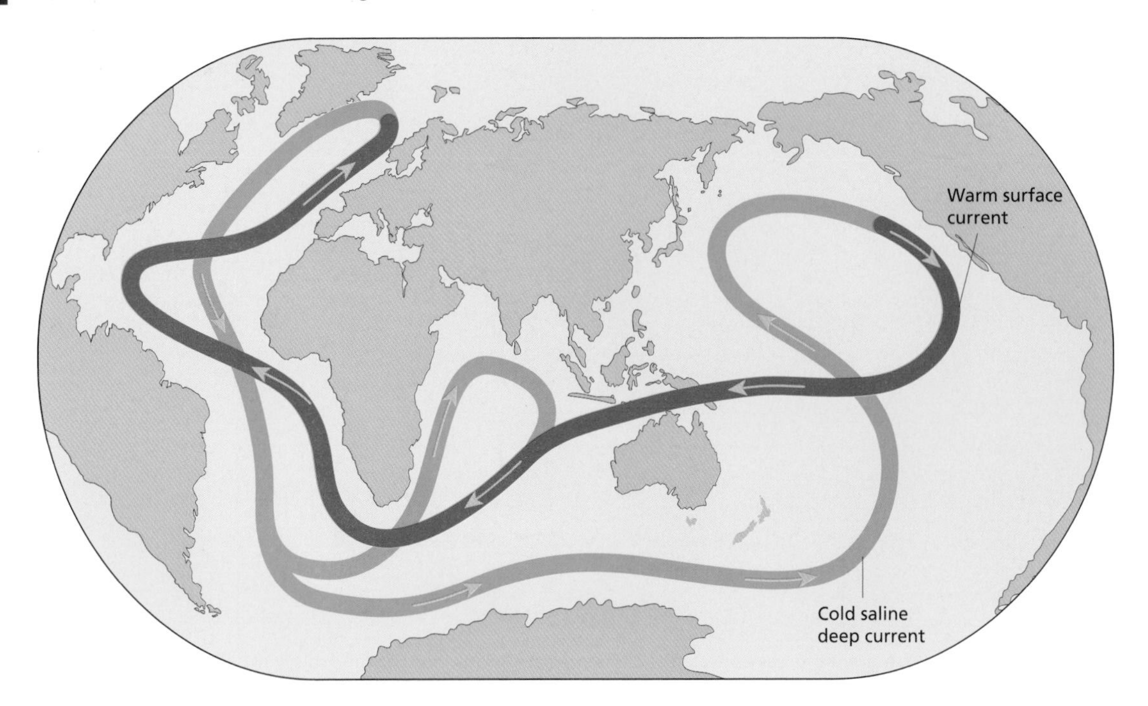

Figure 6.8 The Great Ocean Conveyor Belt

equator. The densest water is found in the Antarctic area; here seawater freezes to form ice at a temperature of around about −2°C. The ice is fresh water; hence the seawater left behind is much saltier and therefore denser. This cold, dense water sweeps round Antarctica at a depth of about 4 kilometres. It then spreads into the deep basins of the Atlantic, the Pacific and the Indian Oceans. Surface currents bring warm water to the North Atlantic from the Indian and Pacific Oceans. These waters give up their heat to cold winds which blow from Canada across the North Atlantic. This water then sinks and starts the reverse convection of the deep ocean current. The amount of heat given up is about a third of the energy received from the sun.

Because the conveyor operates in this way, the North Atlantic is warmer than the North Pacific, so there is proportionally more evaporation there. The water left behind by evaporation is saltier and therefore much denser, which causes it to sink. Eventually the water is transported into the Pacific where it picks up more water and its density is reduced.

To do:

a Describe the temperature anomalies shown in Figure 6.7:

i across the equator

ii at about 60°N.

b How do you account for these anomalies?

c Identify two areas where the temperatures over oceans are lower than those over adjacent land. How do you account for these temperatures?

El Niño and La Niña

El Niño – the "Christ Child" – is a warming of the eastern Pacific that occurs at intervals between two and 10 years, and lasts for up to two years. Originally, El Niño referred to a warm current that appeared off the coast of Peru, but we now know that this current is part of a much larger system.

Normal conditions in the Pacific Ocean

The Walker circulation is the east–west circulation that occurs in low latitudes. Near South America, winds blow offshore, causing upwelling of the cold, rich waters. By contrast, warm surface water is pushed into the western Pacific. Normally sea surface temperatures (SSTs) in the western Pacific are over 28°C, causing

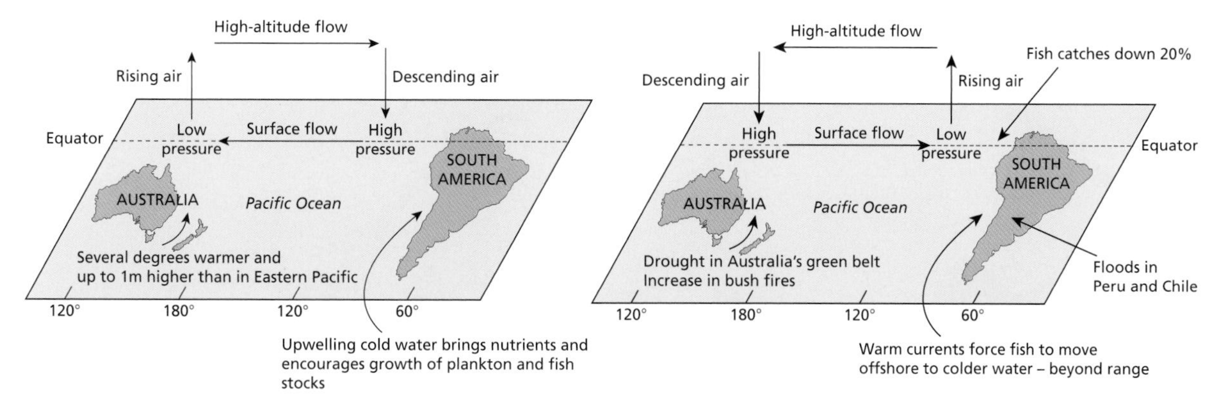

Figure 6.9 Normal conditions (left) and El Niño conditions in the South Pacific Ocean

an area of low pressure and high rainfall. Over coastal South America, SSTs are lower, high pressure exists and conditions are dry.

El Niño conditions in the Pacific Ocean

During El Niño episodes, the pattern is reversed. Water temperatures in the eastern Pacific rise as warm water from the western Pacific flows into the east Pacific. During ENSO (El Niño Southern Oscillation) events, SSTs of over 28°C extend much further across the Pacific. Low pressure develops over the eastern Pacific, high pressure over the west. Consequently, heavy rainfall occurs over coastal South America, whereas Indonesia and the western Pacific experience warm, dry conditions. These events can cause extreme weather events, which may be disastrous.

La Niña

La Niña is an intermittent cold current that flows from the east across the equatorial Pacific Ocean. It is an intensification of normal conditions, whereby strong easterly winds push cold upwelling water off the coast of South America into the western Pacific. Its impact extends beyond the Pacific and has been linked with unusual rainfall patterns in the Sahel and in India, and with unusual temperature patterns in Canada.

Managing the impacts of El Niño and La Niña

In the past, El Niño events could not be predicted with much accuracy. Now sensors across the Pacific predict El Niño months in advance: the last one was predicted so far in advance that Peru was supplied with food and people moved from vulnerable areas. Nevertheless, managing the impacts is difficult for many reasons:

- They affect large parts of the globe, not just the Pacific.
- Some of the countries affected do not have the resources to cope.
- There are indirect impacts on other parts of the world though trade and aid (teleconnections).

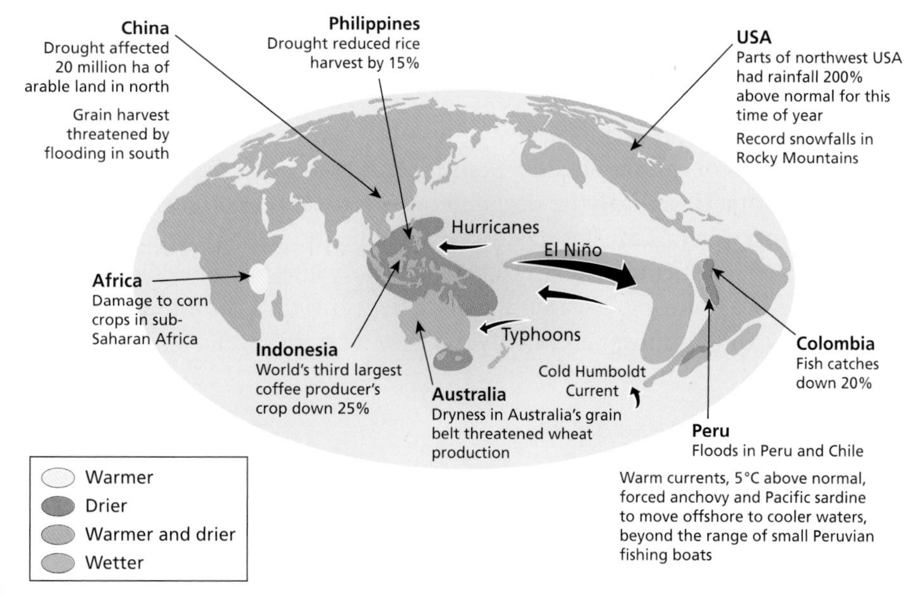

Figure 6.10 The effects of the 1997–8 El Niño

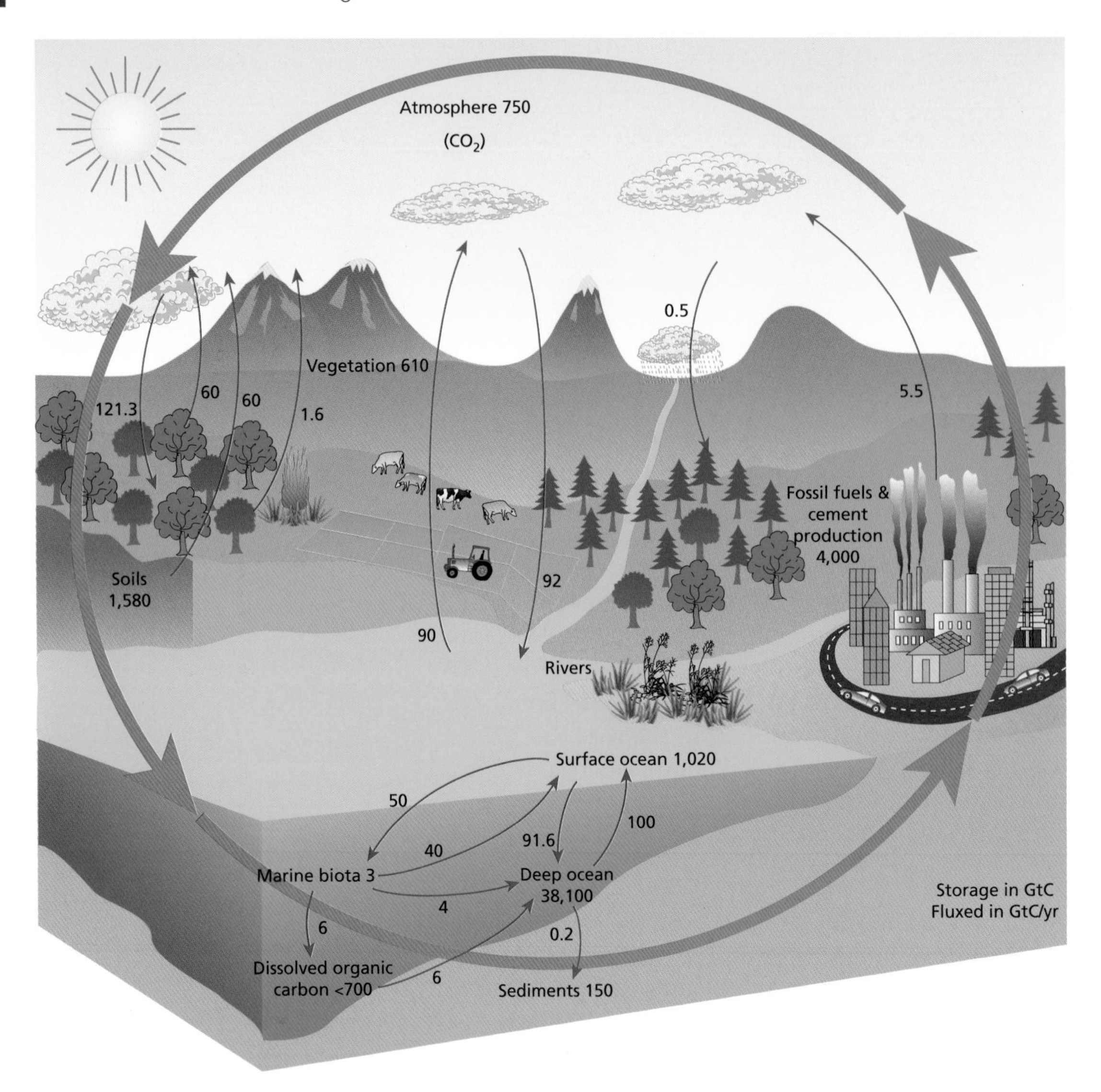

Figure 6.11 The global carbon cycle (gigatonnes of carbon)

Oceans as a store and source of carbon dioxide

The major reservoirs of carbon dioxide are fossil fuels (10 000 · 10^{12} kilograms of carbon), the atmosphere (750 x 10^{12} kilograms of carbon) and the oceans (38,000 · 10^{12} kilograms of carbon). Oceans play a key role in the carbon cycle. Photosynthesis by plankton generates organic compounds of carbon dioxide. Some of this material passes through the food chain and sinks to the ocean floor, where it is decomposed into sediments. Eventually it is destroyed at subduction zones, where ocean crusts are subducted beneath the continental plates. Carbon dioxide is later released during volcanic activity. The transfer of carbon dioxide from ocean to atmosphere involves a very long timescale.

To research

Visit http://news.bbc.co.uk and look up the 2007 article "Oceans are soaking up less CO_2" to explore the implications of the world's oceans now absorbing less CO_2.

Oceans and resources

Oceans contain a rich variety of resources. Biotic resources are living things such as fish; abiotic resources are non-living things such as oil. Saltwater contains nutrients and minerals.

On the **continental shelf** are oil and gas deposits. The Persian Gulf accounts for 66% of the world's proven oil reserves and 33% of the

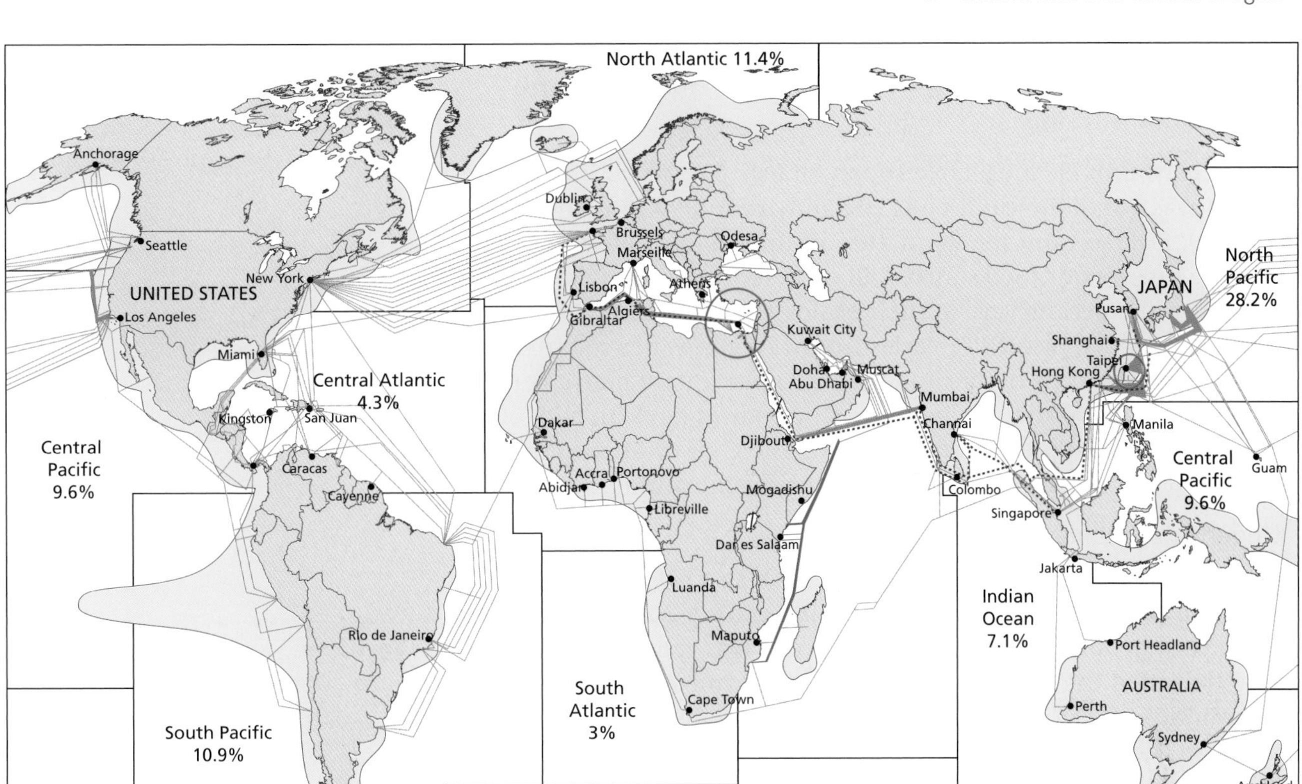

Figure 6.12 The world's main fishing grounds, showing percentage of global catch (2001) and the location of undersea cables

world's proven gas reserves. The continental shelf area of the Gulf of Mexico has been explored and developed since the 1940s. The continental shelf also contains sediments such as gravel, sand and mud. These come from the erosion of rocks and are transported by rivers to the sea. Diamonds can be found in the continental shelf areas off Africa and Indonesia.

On the **ocean floor** are found gold and manganese. Ocean floor sediments are formed of sand, mud and silt. Deep ocean floors are covered in ooze. Ocean sediments have a varied source: some are fine silts carried by turbidity currents; others come from sands and dust blown by wind off the continents; and some heavier material is carried by icebergs, and deposited as the ice gradually melts.

On the **abyssal plain** there are authigenic sediments, which are precipitates of chemicals such as iron oxide, from seawater, in forms such as fist-sized manganese nodules. To date, no economic way has been developed for mining these chemicals. Biogenic ooze is the skeletal remains of microscopic organisms that once lived in the ocean.

Near **ocean ridges and rift valleys** there are rich deposits of sulphur, sometimes associated with hydrothermal vents ("black smokers"). In the future, the biological riches of the "black smokers" face threats from deep-sea mining.

The **mid-ocean hot springs** spew out potentially valuable metal sulfides such as gold, silver and copper. In the cold water they are deposited in thick crusts, attracting exploitation. Rights have

To do:

a Define the terms El Niño and La Niña.

b Compare the conditions in the south Pacific Ocean (a) under normal conditions and (b) during El Niño conditions.

c Describe the impacts of the 1997–8 El Niño event.

d Study Figure 6.11, which shows the global carbon cycle.

- **i** Write a paragraph describing the carbon cycle.
- **ii** To what extent are the oceans an important part of the carbon cycle?

already been given to one company to prospect for metals on 4,000 square kilometres of the bed of the Bismarck Sea, north of Papua New Guinea.

The oceans provide a valuable supply of fish. The worldwide harvest of fish was 5 million tonnes in 1900 and around 90 million tonnes in 2000. Fish account for about 10% of the protein eaten by people. It is the only major food source still gathered from the wild. However, the oceans vary in their ecological productivity. Net primary productivity (NPP) varies from 120 grams per square metre per year in the open oceans to 360 grams per square metre per year in the continental shelves. In contrast, estuaries have an NPP of 1,500 grams per square metre per year. The Gulf of Mexico has a very large fishing industry, especially for shrimp and red snapper.

Underwater cables

The continental shelf and open ocean have also been used for the laying of cables. An internet blackout in January 2008, which left 75 million people with only limited access, was caused by a single ship that tried to moor off the coast of Egypt in bad weather. Telephone and internet traffic was severely reduced across a huge swath of the region, including India, Egypt and Dubai.

The incident highlighted the fragility of our global communications network. The impact of the blackout spread wide, with economies across Asia and the Middle East struggling to cope.

Despite the clean, hi-tech image of the online world, much of our planet still relies on real-world connections put in place through massive physical effort. The expensive fibre-optic cables are laid at great cost in huge lines around the globe, directing traffic backwards and forwards across continents and streaming millions of conversations simultaneously from one country to another.

Overfishing

The decline of fish stocks

Fishing fleets now catch fewer large, **predatory fish** but more, smaller fish further down the food chain. The most prized food fish, such as cod, which tend to be top-level predators, are declining, leaving smaller, less desirable fish. This not only affects the type of fish available for human consumption, but could change marine ecosystems for ever.

Larger, predatory fish feed need to eat large quantities of smaller fish. As their numbers fall, the numbers of smaller fish increases. This is why, despite **overfishing** of cod and other important species, total fish catches have remained high. However, the type of fish being caught is changing.

Even with larger boats and better technology, catches of species like cod are falling. World **fish stocks** have rapidly declined, and some species have even become extinct. More and more ships are chasing fewer fish, and prices have risen quickly. Despite many attempts to save the fish industry, for example through quotas and bans, there has been little success.

To research

Visit the National Maritime Museum's website at www.nmm.ac.uk and go to the "researchers" section, then "research areas and projects" for Planet Ocean, which has news items about ocean resources.

To do:

a Distinguish between biotic and abiotic resources, giving examples.

b Using examples, outline two ways in which ocean resources are important to people.

c Describe the pattern of underwater cables as shown in Figure 6.12.

d Suggest reasons for the high density of cables between (a) Japan and China and (b) northeast USA and Europe. Suggest reasons for the low density of cables off Africa.

Visit http://en.wikipedia.org and find the article "Submarine communications cable".

i Identify the continent that is not connected to submarine cables.

ii Study the 1901 map. State two similarities and two differences compared with the map on page 143.

iii Comment on the proportion of "international traffic" carried by submarine cables.

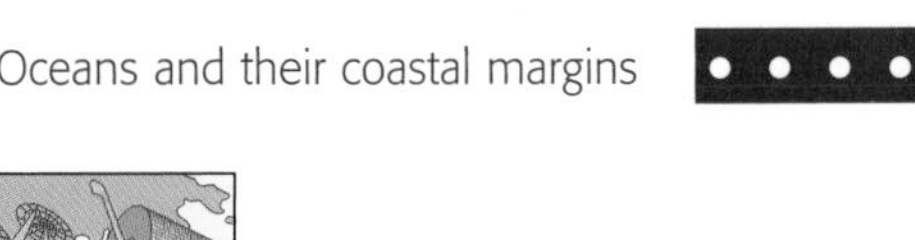

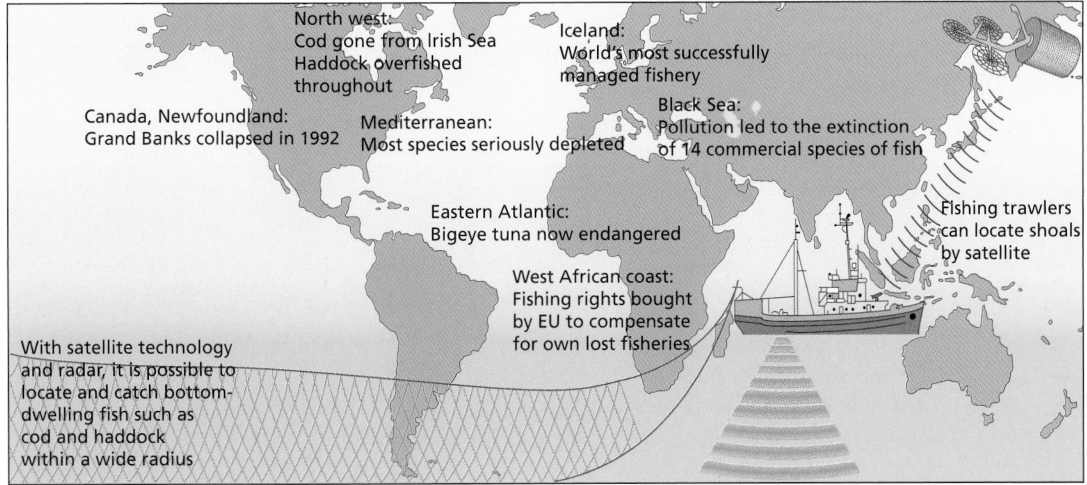

Figure 6.13 Fishing grounds under threat

Nearly 70% of the world's stocks are in need of management. Cod stocks in the North Sea are less than 10% of 1970 levels. Fishing boats from the EU now regularly fish in other parts of the world such as Africa and South America, to make up for the shortage of fish in EU waters. More than half the fish consumed in Europe is now imported.

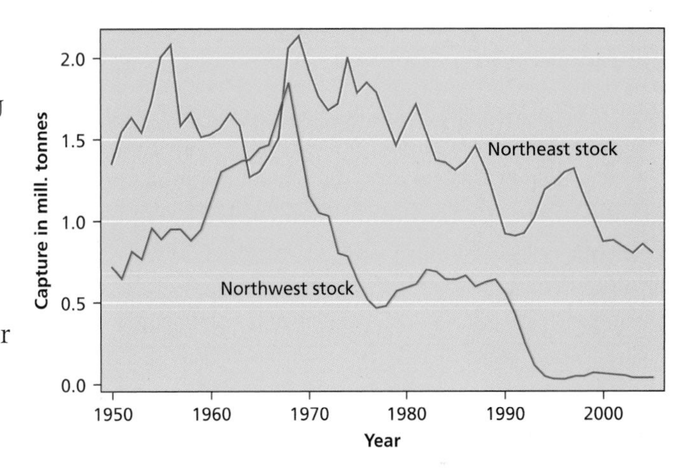

Figure 6.14 The decline of Atlantic cod

The closure of Grand Banks

Once a fish stock is overfished, it is very difficult for it to recover. The Grand Banks off Newfoundland were once the world's richest fishery. In 1992 the area had to be closed to allow stocks to recover. It was expected to be closed for three years, but fish numbers, especially cod, have not yet recovered and it is still closed. The cod's niche in the ecosystem has been taken by other species, such as shrimp and langoustines.

Decade	No. of species
1890s	2
1900s	1
1910s	0
1920s	4
1930s	2
1940s	3
1950s	4
1960s	1
1970s	8
1980s	53
1990s	3

Table 6.2 Fish extinctions

Too many fishermen, too few fish

Many argue that measures such as quotas, bans and the closing of fishing areas still fail to address the real problems of the European fishing industry: too many fishermen are chasing too few fish and too many immature fish are being caught. For the fisheries to be protected and for the industry to be competitive on a world scale, the number of boats and the number of people employed in fishing must be reduced. At the same time, the efficiencies which come from improved technology must be embraced.

A World Bank and FAO report in 2008 showed that up to $50 billion per year is lost in poor management, inefficiency and overfishing in world fisheries. The report puts the total loss over the last 30 years at $2.2 trillion. The industry's fishing capacity continues to increase. The number of vessels is increasing slowly. However, each boat has greater capacity due to improved technology. Due to overcapacity, much of the investment in new technology is wasted. The amount of fish caught at sea has barely changed in the last decade. Fish stocks are depleted, so the effort to catch the ones remaining is greater.

To research

Visit http://ec.europa.eu/fisheries/cfp_en.htm and research fish quotas (by country and by species).

Strategies for the European fishing industry

The table below suggests some possible strategies for the future, but there are clearly no simple solutions to the problems associated with such a politically, economically and environmentally sensitive industry.

Action	Type of measure	Objectives
Conservation of resources		
Technical measures	Small meshed nets, minimum landing sizes, boxes	Protect juveniles and encourage breeding; discourage marketing of illegal catches
Restrict catches	TACs (total allowable catches) and quotas	Match supply to demand; plan quota uptake throughout the season; protect sensitive stocks
Limit number of vessels	Fishing permits (which could be traded inter- or intra-nationally)	System applicable to EU vessels and other countries' vessels fishing in EU waters
Surveillance	To check landings by EU and third-country vessels (log books, computer/satellite surveillance)	Apply penalties for overfishing and illegal landings
Structural aid	Structural aid to the fleet	Finance investment in fleet modernization (although commissioning of new vessels must be closely controlled), while providing reimbursement for scrapping, transfer and conversion
Reduction in unemployment leading to an increase in productivity	Inclusion of zones dependent on fishing in Objectives 1, 2 and 5b of Structural Funds	Facilitate restructuring of the industry, to finance alternative local development initiatives to encourage voluntary/early retirement schemes
Markets		
Tariff policy	Minimum import prices, restrictions on imports	Ensure EU preference (although still bound under WTO)
Other measures		
Restrict number of vessels	Fishing licences	Large licence fees to discourage small, inefficient boats
Increase accountability of fishermen	Rights to fisheries	Auction off sections of the seabed where fish stay put (e.g. shellfish)
		Where a whole fishery is controlled, trade quotas, which would allow some to cash in and leave the sea

Table 6.3 Strategies for the future

Threats to the world's oceans

The oceans, like the atmosphere, are fundamental to the health of our planet. They dominate many of its cycling processes, as well as being the ultimate sink for a variety of pollutants. They absorb about 2 billion tonnes of carbon – in the form of carbon dioxide (CO_2) – and disperse an estimated 3 million tonnes of oil spilt annually from ships and, predominantly, from sources on land.

The oceans store a thousand times more heat than the atmosphere, and transport enormous amounts of it around the globe. In consequence, they are largely responsible for determining climate on land. The warm Gulf Stream washing up from the tropics in the Atlantic Ocean keeps Europe many degrees warmer in winter than Hudson Bay on the opposite shore. The oscillation between El Niño and La Niña currents in the tropical Pacific Ocean fundamentally changes the weather across the ocean, flipping Indonesia, Australia and coastal South America into and out of droughts and floods.

To do:

a Define the term "predatory fish".

b Describe the trend in extinctions as shown in Table 6.2.

c Compare the capture of the northeast Atlantic cod with that of the northwest Atlantic cod, as shown in Figure 6.14.

All these processes now face disruption from the global scale of human activity, particularly climate change. Currently, the oceans moderate climate change by absorbing a third of the CO_2 emitted into the air by human activity. But several studies suggest that global warming will stratify the oceans and reduce their capacity to act as a CO_2 "sink" by 10–20% over the next century, accelerating warming.

The impacts of waste on the marine environment

Less than a tenth of the sea floor has ever been explored; even so the human hand is increasingly evident. Humanity has used the oceans as a dustbin for far too long, and the sea is now extremely polluted. It is now known that accumulation of waste in the ocean is detrimental to marine and human health. When waste is dumped it is often close to the coast and very concentrated. Alternatives to ocean dumping include recycling, producing less wasteful products, saving energy and changing the dangerous material into more benign waste.

To research

Visit the EU facts page of www.civitas.org.uk and find out about the Common Fisheries Policy of the European Union (FSPOL).

- When was it introduced?
- What happened to North Sea cod stocks in 2008?
- What are the arguments for and against a common fisheries policy?

Over 80% of marine pollution comes from land-based activities. The most toxic waste material dumped into the ocean includes dredged material, industrial waste, sewage sludge, and radioactive waste. Dredging contributes about 80% of all waste dumped into the ocean. Rivers, canals, and harbours are dredged to remove silt and sand buildup or to establish new waterways. About 20–22% of dredged material is dumped into the ocean. About 10% of all dredged material is polluted with heavy metals such as cadmium, mercury and chromium, hydrocarbons such as heavy oils, nutrients including phosphorous and nitrogen, and organochlorines from pesticides. When "pure" dredged material is dumped into the ocean, fisheries suffer adverse affects, such as unsuccessful spawning in herring and lobster populations where the sea floor is covered in silt.

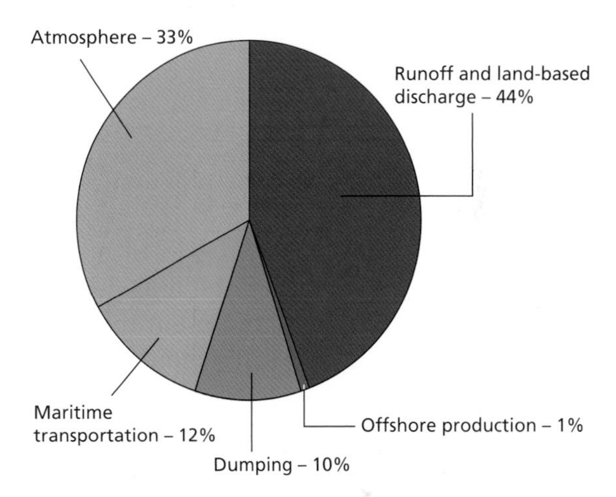

Figure 6.15 Sources of waste entering the sea

Over 60 million litres of oil run off America's roads and, via rivers and drains, find their way into the oceans each year. Through sewage and medical waste, antibiotics and hormones enter the systems of seabirds and marine mammals. Mercury and other metals turn up in tuna, orange roughy, seals, polar bears and other long-lived animals. In the 1970s, 17 million tonnes of industrial waste were legally dumped into the ocean. In the 1980s, 8 million tonnes were dumped, including acids, alkaline waste, scrap metals, waste from fish processing, flue desulphurization, sludge and coal ash. The peak of sewage dumping was 18 million tonnes in 1980, a number that fell to 12 million tonnes in the 1990s.

To do:

a Outline the main sources of ocean pollution, as shown in Figure 6.15.

b Which areas of the world's oceans have (a) the highest impact of human activity and (b) the lowest impact of human activity? Suggest reasons for these differences.

c Discuss the implications of the pollution of oceans by the disposal of waste materials.

Oil pollution

All over the world, oil spills regularly contaminate coasts. Oil exploration is a major activity in such regions as the Gulf of Mexico, the South China Sea and the North Sea. The threats vary. For example, there is evidence of widespread toxic effects on benthic (deep-sea) communities on the floor of the North Sea in the vicinity of the 500-plus oil production platforms in British and Norwegian waters. Meanwhile, oil exploration in the deep waters of the North Atlantic, northwest of Scotland, threatens endangered deep-sea corals. There is

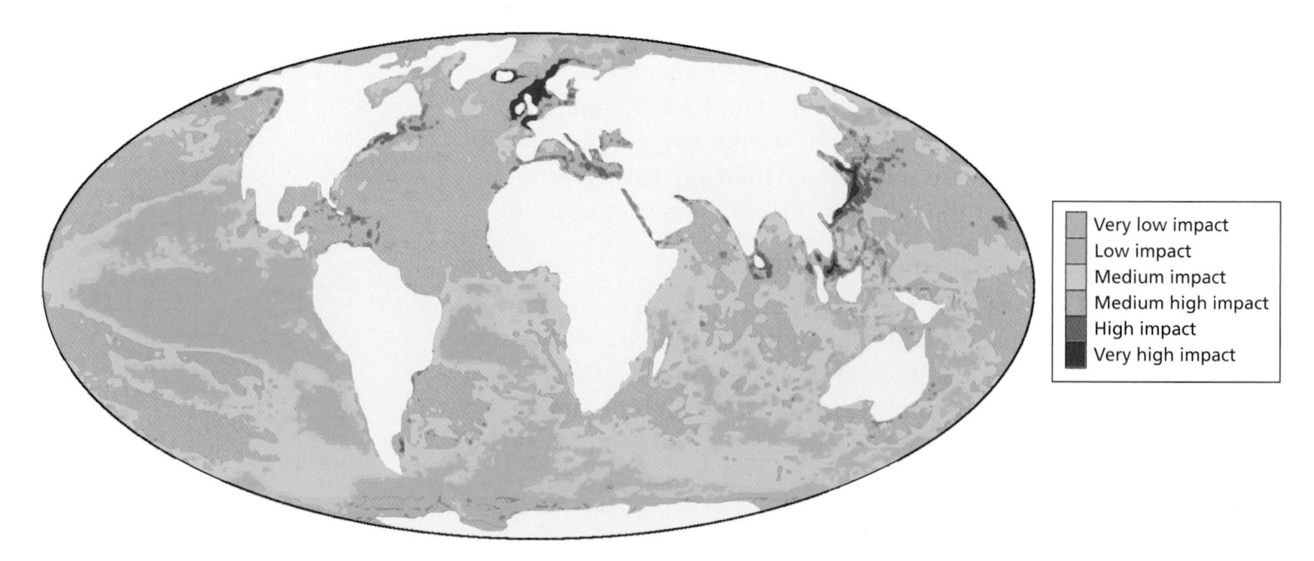

Figure 6.16 The human impact on the oceans

evidence, too, that acoustic prospecting for hydrocarbons in these waters may deter or disorientate some marine mammals.

Shipping itself is a huge cause of pollution. Since ships burn bunker oil, the dirtiest of fuels, that means not just more CO_2 but also more "particulate matter", which may be responsible for about 60 000 deaths each year from chest and lung diseases, including cancer. Most of these occur near coastlines in Europe, East and South Asia. Some action is being taken. Oil spills should become rarer after 2010, when all single-hulled ships were banned.

Efforts are also being made to prevent the spread of invasive species through the taking on and discharging of ships' ballast water. Similarly, a UN convention may soon ban the use of tributyltin, a highly toxic chemical once added to the paint used on almost all ships' hulls, in order to kill algae and barnacles.

See also chapter 4, page 87, for more on oil's environmental impact.

Radioactive waste

Radioactive effluent also makes its way into the oceans. Between 1958 and 1992, the Arctic Ocean was used by the Soviet Union, or its Russian successor, as the resting place for 18 unwanted nuclear reactors, several still containing their nuclear fuel. Radioactive waste is also dumped in the oceans, and usually comes from the nuclear power process, medical and research use of radioisotopes, and industrial uses. Nuclear waste usually remains radioactive for decades. The dumping of radioactive material has reached a total of about 84 000 terabecquerels (TBq), a unit of radioactivity equal to 10^{12} atomic disintegrations per second, or 27.027 curies.

Plastic

More alarming still is the plague of plastic. The UN Environment Programme reckoned in 2006 that every square kilometre of sea held nearly 18 000 pieces of floating plastic. Much of it was, and is, in the central Pacific, where scientists believe as much as 100 million tonnes of plastic jetsam are suspended in two separate gyres of garbage over an area twice the size of the United States. About 90% of the plastic in the sea has been carried there by wind or water from land. It takes decades to decompose or sink. Turtles, seals and birds inadvertently eat it, and not just in the Pacific. A Dutch study of 560 fulmars picked up dead in countries around the North Sea found that 19 out of 20 had plastic in their stomach – an average of 44 pieces in each. Moreover, when plastic breaks up it attracts toxins, which become concentrated in barnacles and other tiny organisms and thus enter the marine food chain.

Dead zones and red tides

Dead zones, red tides and their associated plagues of jellyfish seem to have occurred naturally for centuries, but their appearance is becoming increasingly frequent. Red tides, for example, regularly form off the Cape coast of South Africa, fed by nutrients brought up from the deep, and off Kerguelen Island in the Southern Ocean. Nowadays, though, most are associated with a combination of phenomena including overfishing, warmer waters and the washing into the sea of farm fertilizers and sewage.

Most of the larger fish in shallow coastal waters have already been caught. As the larger species disappear, so the smaller ones thrive. These smaller organisms are also stimulated by nitrogen and phosphorous nutrients running off the land. The result is an explosion of growth among phytoplankton and other algae, some of which die, sink to the bottom and decompose, combining with dissolved oxygen as they rot. Warmer conditions, and sometimes the loss of mangroves and marshes, which once acted as filters, encourage the growth of bacteria in these oxygen-depleted waters.

The result may be a sludge-like soup, apparently lifeless—hence the name dead zones—but in fact teeming with simple, and often toxic, organisms. In such places red tides tend to form, some producing toxins that get into the food chain through shellfish, and rise up to kill bigger fish (if there are any left), birds and even seals and manatees.

Response to threats

Governments world-wide were urged by the 1972 Stockholm Convention to control the dumping of waste in their oceans by implementing new laws. The International Maritime Organization was given responsibility for this convention, and a protocol was finally adopted in 1996, a major step in the regulation of ocean dumping.

There have been some successes in the international handling of the marine environment. The International Whaling Commission's moratorium introduced in the mid-1980s has helped revive whale stocks. The United Nations Convention on the Law of the Sea, signed in 1982 but not enforced until 1994, established a framework of law for the oceans, including rules for deep-sea mining and economic exclusion zones extending 200 nautical miles around nation states.

A series of international laws have effectively eliminated the discharge of toxic materials – from drums of radioactive waste to sewage sludge and air pollution from incinerator ships – into the

To research

Visit http://marinebio.org, an excellent site for information about ocean pollution (including radioactive waste and ocean dumping).

Visit http://oceans.greenpeace.org and look for an ocean pollution animation showing the journey of trash in the Pacific, and how rubbish becomes concentrated in two areas.

Visit www.panda.org and look in the "About our earth" section to find information about marine pollution.

Describe and explain the distribution of pollution in the Pacific Ocean, as shown in the animation.

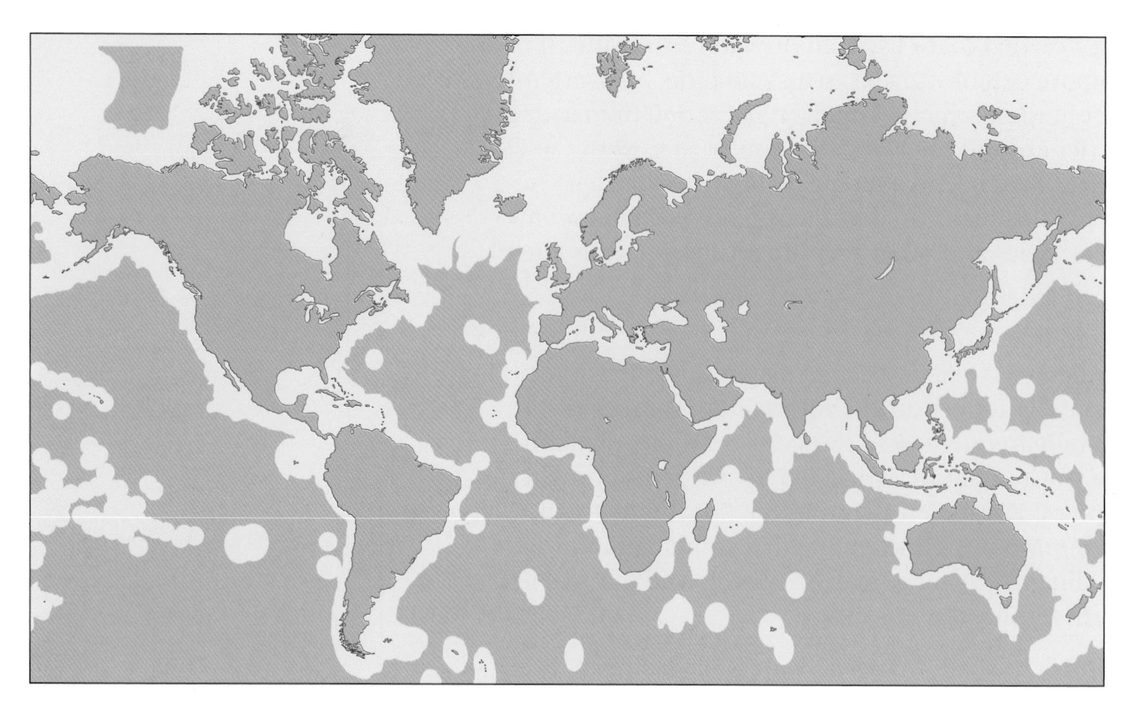

Figure 6.17 The international seabed (dark areas of ocean)

waters around Europe. International public pressure in the mid-1990s forced the reversal by a major oil company of plans to scuttle the Brent Spar, a large structure from the North Sea offshore oil industry, into deep water west of Scotland. European agreements since then have indicated that all production platforms and other structures should be removed from the oil fields at the end of their lives, wherever possible.

The geopolitics of oceans

Be a critical thinker

What legal rights do nations have over the oceans?

Geopolitics is short for geographical politics. It refers to the geographical relationships between different states. An economic exclusive zone (EEZ) is an area in which a coastal nation has sovereign rights over all the economic resources of the sea, seabed and subsoil, extending up to 200 nautical miles from the coast (the international nautical mile is 1.852 kilometres). The term "sovereign" means having independent authority over a territory.

Many people have for years believed that the seabed was paved with many minerals. Roughly a quarter of the ocean floor is strewn with "manganese nodules", usually about the size of an apple, which contain not just manganese but also cobalt, copper and nickel. Gathering the nodules is technically difficult, chiefly because most of them lie under 4 kilometres of water. The process is not popular with environmentalists either, since the necessary dredging stirs up quantities of sediment that kills everything nearby.

To research

Visit http://aipsg.org for an article on the geopolitics of South Asia and the threat of war. It examines the changing role of the Indian Ocean.

Visit http://www.lse.ac.uk/collections/alcoa/pdfs/berkmanpresentation.pdf for a presentation on the geopolitics of the Arctic Ocean.

To research

Visit the International Seabed Authority at http://www.isa.org.jm/en/home and find out about the Mining Code for the seabed.

Visit http://www.nautilusminerals.com/s/Home.asp to find out about Nautilus Minerals operations off Papua New Guinea. There is a link to their environmental impact statement.

However, the economics of mining has changed recently, as industrial commodity prices have risen much higher than they were in the 1970s, and as technology has advanced. This means that it may now become profitable to exploit the manganese crusts and other minerals recently discovered. Since 2004 China, France, Germany, India, Japan, Russia, South Korea and a consortium of east European countries have all been awarded licences by the International Seabed Authority to explore mining possibilities on the deep-ocean seabed, and the Canadian company Nautilus Minerals hopes to be the first deep-water mining company to start production. Its plan is to bring up ore containing copper and gold from the bottom of the Bismarck Sea north of Papua New Guinea, using technology developed by the offshore oil industry.

The scramble for the Arctic

Countries around the Arctic Ocean are rushing to stake claims on the Polar Basin seabed and its oil and gas reserves. Resolving territorial disputes in the Arctic has gained urgency because scientists believe rising temperatures could leave most of the Arctic ice free in summer months in a few decades' time. This would improve drilling access and open up the Northwest Passage, a route through the Arctic Ocean linking the Atlantic and Pacific that would reduce the sea journey from New York to Singapore by thousands of miles.

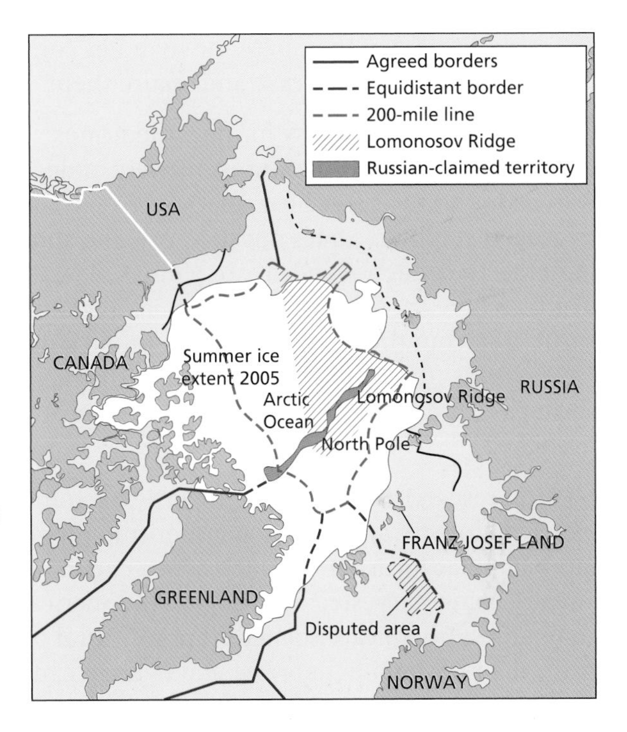

Figure 6.18 The geopolitics of the Arctic

Canada is establishing a year-round Arctic presence on land and sea as well as in the air. Denmark is trying to prove that a detached part of the underwater Lomonosov ridge is an extension of Greenland, which is Danish territory. Russia has staked a claim by sending a submersible to plant a corrosion-resistant titanium flag some 4 kilometres below the North Pole. In 2008 Canada, Denmark, Norway, Russia and the United States met in Greenland to discuss how to divide up the resources of the Arctic Ocean.

According to the US Geographical Survey, the Arctic could hold a quarter of the world's undiscovered gas and oil reserves. This amounts to 90 billion barrels of oil and vast amounts of natural gas. Nearly 85% of these deposits, they believe, are offshore. The five countries are racing to establish the limits of their territory, stretching far beyond their land borders.

Environmental groups have criticized the scramble for the Arctic, saying it will damage unique animal habitats, and have called for a treaty similar to that regulating the Antarctic, which bans military activity and mineral mining.

Under the 1982 UN Law of the Sea Convention, coastal states own the seabed beyond existing 370-kilometre zones if it is part of a continental shelf of shallower waters. While the rules aim to fix shelves' outer limits on a clear geological basis, they have created a tangle of overlapping Arctic claims.

Deep-water discoveries

Hydrates for energy

Hydrates are compounds that usually consist of methane molecules trapped in a cage of water. They were first found in permafrost in the 1960s and then, in the 1970s, on the slopes of continental shelves deep beneath the ocean floor. Some scientists believe these hydrates together contain more energy than all the known deposits of fossil fuels, a possibility that makes them highly attractive to countries such as Japan and India, with little or no oil or gas.

The oil companies, though, are cautious. Hydrates occur naturally in pipelines, and are unpopular because they clog the flow of oil. Extracting them would be intensely difficult. And methane, though it burns more cleanly than coal, absorbs a wider range of wavelengths of the earth's outgoing radiation than CO_2. It therefore traps more heat, making it an even more pernicious greenhouse gas. Some scientists also argue that methane hydrates contain less energy than the energy needed to release and secure them.

Exploiting genetic diversity for new medicines

A second discovery is the strange variety of deep-sea life. Some of it is to be found near so-called "black smokers", caused when dissolved hydrogen sulphide comes out of a rock and suddenly cools, causing minerals to condense and create plumes of "smoke". Life at the dark depths where these vents occur gets its food and energy from the earth, not from the sun. Here can be found sulphur-eating bacteria, scale worms that thrive in hot water at the tips of the smokers, and shrimps that can mend their DNA even after it has been highly irradiated.

Medical researchers hope the properties of some of these creatures may lead to new anti-carcinogens, or tumour-reducing drugs. Some more familiar forms of life, such as sea cucumbers, for instance, are already being harvested by French and American pharmaceutical companies. Chemicals isolated from soft coral off the coast of western Australia may help fight breast and ovarian cancers.

Most of the compounds found so far, say the doubters, come from volcanic deposits. Some scientists say there are fewer nodules than people generally believe and most of them are in very deep water. In addition, everything done at sea is becoming more expensive, so the economics of mining may not have improved much.

Exclusive economic zone (EEZs)

Exclusive economic zones have a profound impact on the management and conservation of ocean resources, since they recognize the right of coastal states to control over 98 million square kilometres of ocean space. Coastal states are free to "exploit, develop and manage and conserve all resources – fish or oil, gas or gravel, nodules or sulphur – to be found in the waters, on the ocean floor and in the subsoil of an area, extending almost 200 nautical miles from its shore." Almost 90% of all known oil reserves under the sea fall under some country's EEZ. So too do the rich fishing areas – up to 98% of the world's fishing regions fall within an EEZ.

To research

Visit http://www.isa.org.jm/files/images/maps/Crusts_Pacific_3D.jpg for a map of cobalt-rich crusts on seamounts and guyots.

Visit http://www.isa.org.jm/files/images/maps//Sulphides_Global.jpg for a map of polymetallic sulphides deposits.

Describe the distribution of polymetallic sulphide deposits on the sea floor. Suggest reasons to account for their distribution. In what ways does the map of the international seabed (Figure 6.17) influence the development of polymetallic sulphides?

Ascension – a British EEZ in the South Atlantic?

The UK has claimed 200 000 square kilometres of the Atlantic seabed surrounding Ascension Island. The mountainous ocean floor up to 560 kilometres from the isolated island in the South Atlantic is believed to contain extensive mineral deposits. With no near neighbours, other states are unlikely to challenge the claim.

Ascension Island has a land area of around 100 square kilometres but, due to its isolated location, it generates an EEZ with an area of more than 440 000 square kilometres. As mineral and energy prices have soared, there has been growing international interest in exploring the seabed for increasingly scarce reserves. The first deep-sea mining project – aiming to extract gold, silver, copper and zinc from extinct volcanic vents – started operating in the waters off Papua New Guinea in 2009. Britain has also lodged, or is preparing, claims to underwater territories around Antarctica, the Falklands, Rockall in the north Atlantic and in the Bay of Biscay.

To do:

a Discuss the sovereignty rights of nations in relation to territorial limits and exclusive economic zones (EEZs).

b Examine one geopolitical conflict in relation to an oceanic resource other than fishing.

Features of coastal margins

Coastal environments are influenced by many factors, including physical and human processes. As a result there is a great variety in coastal landscapes. For example, landscapes vary on account of:

- **geology properties** (rock) – hard rocks such as granite and basalt give rugged landscapes, e.g. the Giant's Causeway in Northern Ireland, whereas soft rocks such as sands and gravels produce low, flat landscapes, e.g. around Poole Harbour on the south coast of England
- **geological structure** – **concordant** or **accordant** (Pacific) coastlines occur where the geological strata lie parallel to the coastline, e.g. the south coast of Ireland or California, whereas **discordant** (Atlantic-type) coastlines occur where the geological strata are at right angles to the shoreline, e.g. on the southwest coast of Ireland
- **processes** – erosional landscapes, e.g. the east coast of England, contain many rapidly retreating cliffs, whereas areas of rapid deposition, e.g. the Netherlands, contain many sand dunes and coastal flats
- **sea-level changes –** these interact with erosional and depositional processes to produce advancing coasts (those growing either due to deposition and/or a relative fall in sea level) or retreating coasts (those being eroded and/or drowned by a relative rise in sea level)
- **human impacts** – these are increasingly common: some coasts, e.g. in Florida, are extensively modified, whereas others are more natural, e.g. southwest Ireland
- **ecosystem type –** mangrove, coral, sand dune, salt marsh and rocky shore add further variety to the coastline.

Waves

Most waves are generated by wind blowing over the sea surface, although some waves (**tsunami**) are caused by tides and earthquakes. Tsunami waves can reach heights of up to 15 metres,

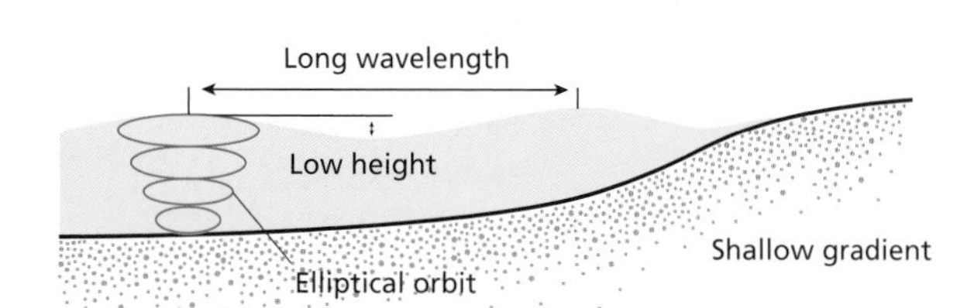

- Depositional waves
- Also called "spilling" waves
- Long wavelength, low height
- Low frequency (6–8 per minute)
- High period (one every 8–10 seconds)
- Swash greater than backwash
- Low gradient
- Low energy

Figure 6.19 Constructive, or swell waves

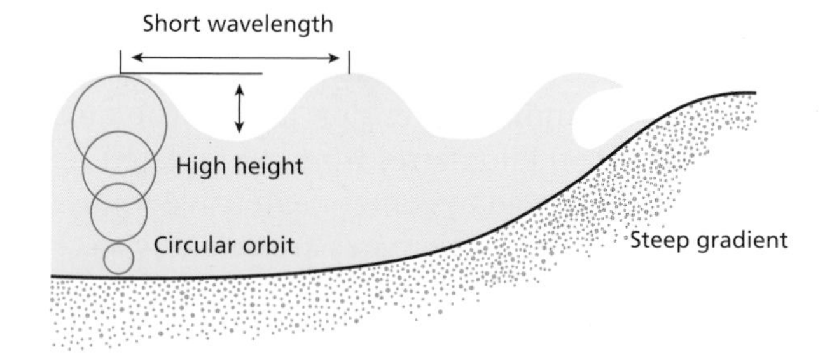

- Erosional waves
- Also called "surging" or "plunging" waves
- Short wavelength, high height
- High frequency (10–12 per minute)
- Low period (one every 5–6 seconds)
- Backwash greater than swash
- Steep gradient
- High energy

Figure 6.20 Destructive, or storm waves

and can travel at speeds of up to 600 kilometres per hour. The wave energy is controlled by

- wind strength and duration
- fetch, or distance of open water
- depth of the seabed.

Waves in open water can reach heights of up to 15 metres, and are capable of travelling huge distances as constructive waves, or **swell waves**. These are characterized by very long wavelengths and a reduced height (Figure 6.19). **Storm waves**, by contrast, are generated by local winds and travel only a short distance (Figure 6.20). They are also known as destructive waves.

As the wave breaks, its energy is transferred to the shore. The movement of water up the beach is known as the **swash**. The effectiveness of this is controlled by wave energy, size of beach material and beach gradient. The **backwash** is the movement of water down the beach under the force of gravity.

Wave refraction

Wave refraction occurs when waves approach an irregular coastline or are at an oblique angle. Refraction reduces wave velocity and, if complete, causes wave fronts to break parallel to each other (Figure 6.21a). Wave refraction concentrates energy on the flanks of headlands and dissipates energy in bays (Figure 6.21b).

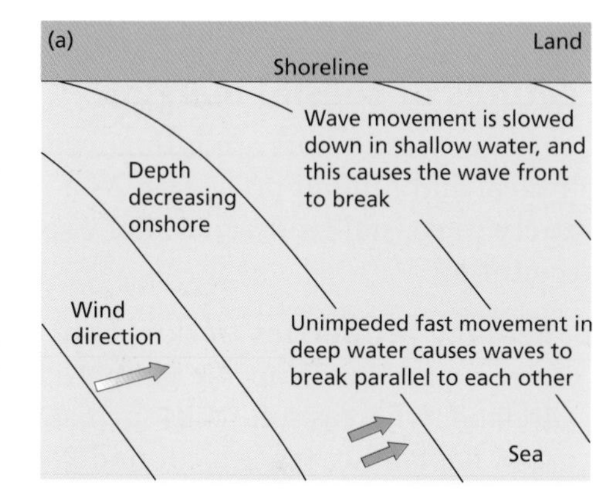

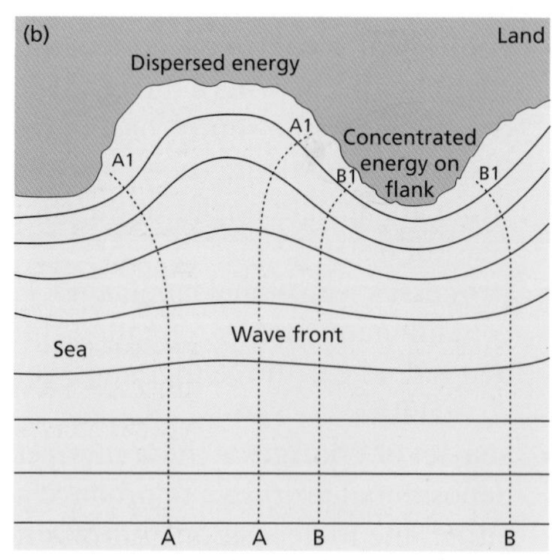

Figure 6.21 Wave refraction

Longshore drift

Refraction is rarely complete and consequently longshore drift occurs. For drifting to occur, the swash carries material up the beach in a direction parallel to the prevailing wind, whereas the backwash operates at right angles to the shore owing to the steepness of the beach slope (Figure 6.22). The result is a net transfer of sediment along a beach. Drifting is not always straightforward. Secondary drifting may be caused by less regular secondary waves, and refraction around an island may cause drifting to occur in converging directions.

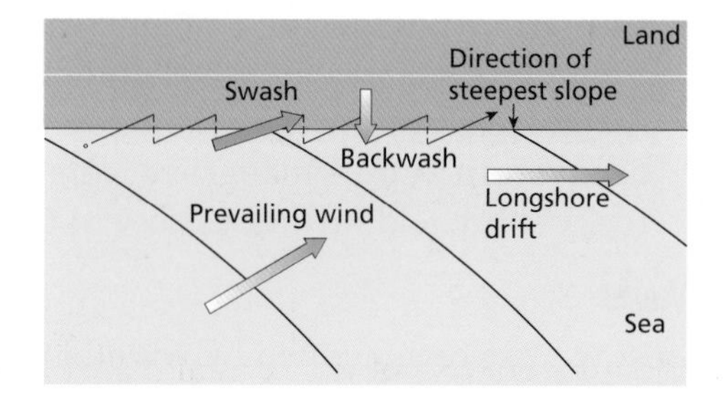

Figure 6.22 Longshore drift

Tides

Tides are regular movements in the sea's surface, caused by the gravitational pull of the moon and sun on the oceans. Tides are influenced by the size and shape of ocean basins, the characteristics of the shoreline, the Coriolis force and meteorological conditions. In general:

- tides are greatest in bays and funnel-shaped coastlines
- in the northern hemisphere, water is deflected to the right of its path
- during low-pressure weather systems, water levels rise 10 centimetres for every decrease of 10 millibars.

The difference between high and low tide is called the **tidal range**. This varies from almost nothing in enclosed seas such as the Mediterranean to almost 15 metres in the Bay of Fundy, Canada. The tidal range has important influences on coastal processes:

- It controls the vertical range of erosion and deposition.
- Weathering and biological activity are affected by the time between tides.
- Velocity is influenced by the tidal range and has an important scouring effect.

Coastal processes and landforms

Coastal erosion

Coasts are eroded in a number of ways

- **Abrasion** is the wearing away of the shoreline by material carried by the waves.
- **Hydraulic impact,** or **quarrying,** is the force of water and air on rocks (up to 30 000 kilograms per square metre in severe storms).
- **Solution** is the wearing away of base-rich rocks, especially limestones, by acidic water. Organic acids aid the process.
- **Attrition** is the rounding and reduction of particles carried by waves.

In addition, many forms of weathering cause erosion:

- **Salt weathering** is the process by which sodium and magnesium compounds expand in joints and cracks, thereby weakening rock structures.
- **Freeze-thaw weathering** is the process whereby water freezes, expands and degrades jointed rocks.
- **Water-layer weathering** refers to the tidal cycle of wetting and drying (**hydration**).
- **Biological weathering** is carried out by molluscs, sponges and urchins and is important in low-energy coasts.

Mass movements of the coastal margin are also a common type of erosion.

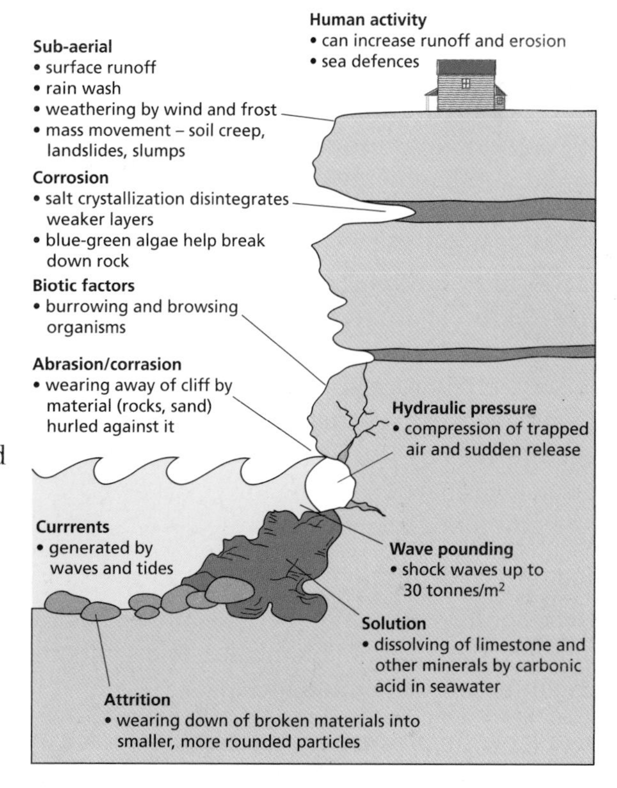

Figure 6.23 Types of erosion

Features of coastal erosion

Rapid erosion occurs on rocks where there is a weakness. Faults in the rock may be eroded to form sea caves. If two sea caves meet, an arch is formed. If the roof of the arch collapses, a stack is formed, and if this is eroded a stump is formed (Figure 6.24). These features are largely determined by the location of weaknesses in the rock, and variations in wave energy caused by refraction on the flanks of the headland.

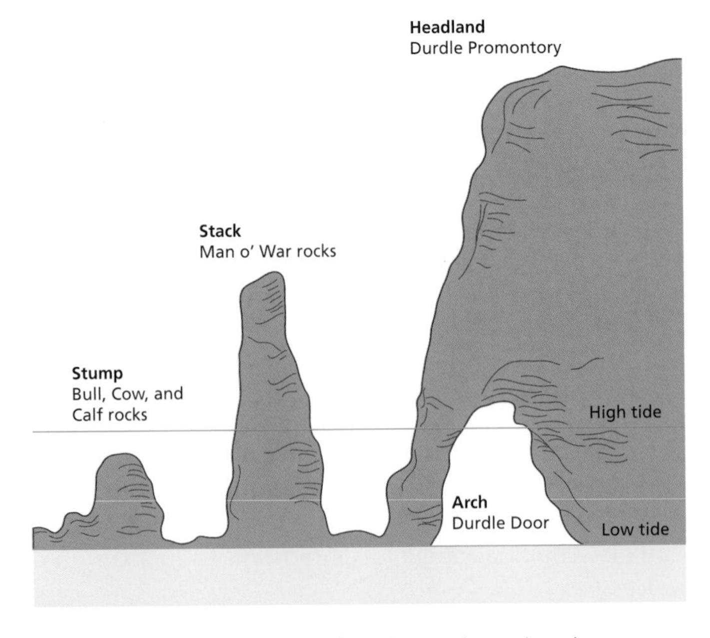

Figure 6.24 Features of coastal erosion: arch, stack and stump

Cliffs

The profile of a cliff depends upon a number of factors, including:

- geological structure
- sub-aerial and marine processes
- amount of undercutting
- rates of removal
- stage of development.

Rocks of low resistance are easily eroded and unable to support an overhang. Jointing may determine the location of weaknesses in the rock, just as the angle of dip may control the shape of the cliff (Figure 6.25). Past processes are also important. In areas affected by periglaciation, bevelled cliffs are found.

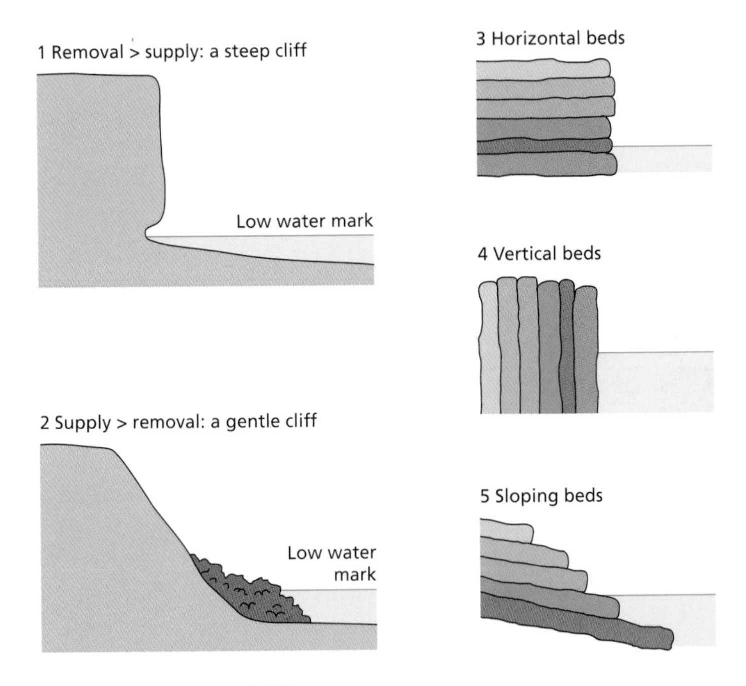

Figure 6.25 Types of cliff

Shore platforms

Shore platforms include **intertidal platforms** (**wave-cut platforms**), **high-tide platforms** and **low-tide platforms**. Wave-cut platforms are most frequently found in high-energy environments and are typically less than 500 metres wide with an angle of about 1°. A model of cliff- and shore-platform evolution (Figure 6.26) shows how steep cliffs (1) are replaced by a lengthening platform and lower-angle cliffs (5), subjected to sub-aerial processes rather than marine forces. Alternatively, platforms might have been formed by frost action, salt weathering, or biological action during lower sea levels and different climates.

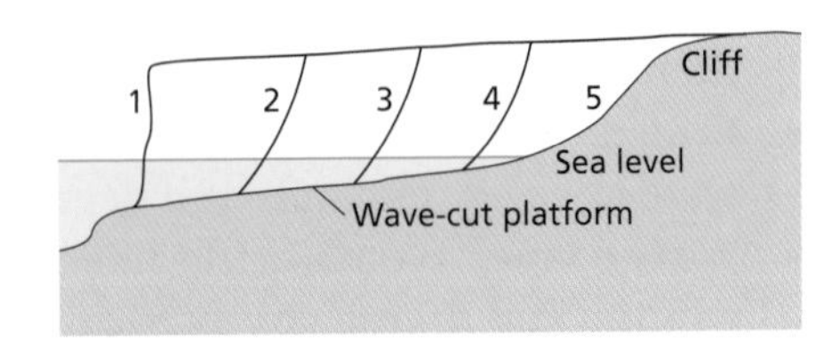

Figure 6.26 Shore platforms

Features of deposition

Deposition creates a number of features, including beaches, spits, bars and tombolos, and sand dunes. Essential conditions for deposition include:

- a large supply of material (beach deposits, offshore deposits, river sediments, material eroded from cliffs and headlands)
- longshore drift
- an irregular, indented coastline, e.g. river mouths
- low-energy coastlines
- bioconstruction, namely the work of plants.

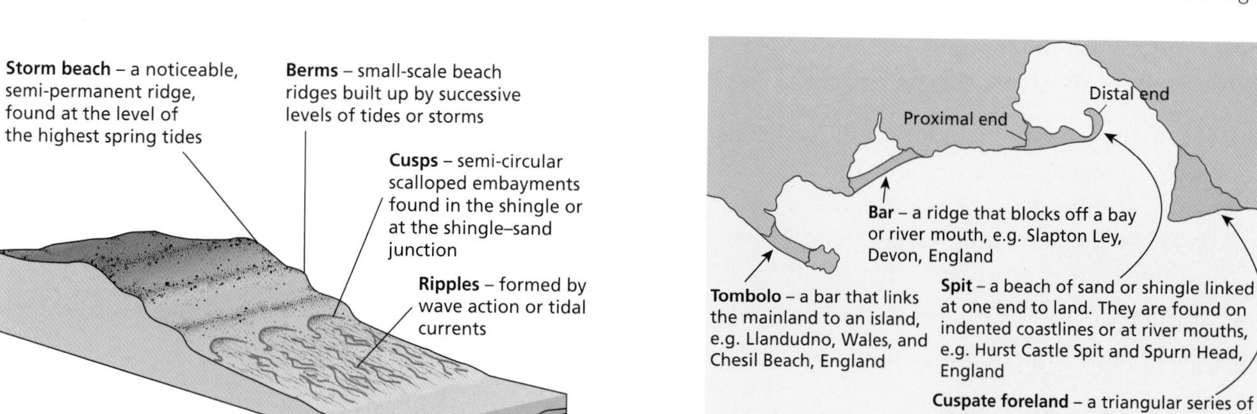

Figure 6.27 Beach profile

Figure 6.28 Features of deposition

Sand dunes

The largest **sand dunes** are found in mid-latitude storm wave environments. Their formation is favoured by:

- a large supply of sand on the beach
- high onshore wind speeds
- low precipitation
- low humidity.

A large tidal range will expose more sand, especially if the beach gradient is low. Large, fresh supplies of sand are provided by steep, actively eroding rivers.

Two types of dune are commonly described, **free** and **impeded dunes**. Free dunes are formed in the absence of vegetation and occur largely along desert margins. Impeded dunes are common in humid areas where vegetation is essential for trapping sand. A large tidal range, strong prevailing wind and large supply of sediment are important in the development of extensive dune systems.

Material	Diameter (mm)	Beach angle°
Cobbles	32	24
Pebbles	4	17
Coarse sand	2	7
Medium sand	0.2	5
Fine sand	0.02	3
Very fine sand	0.002	1

Table 6.4 Beach profiles and particle size

Figure 6.29 Sand dunes

Be a critical thinker

How do levels of land and sea vary?

Sea levels change in connection with the growth and decay of ice sheets. **Eustatic** change refers to a global change in sea level. At the height of glacial advance, 18 000 years ago, sea level was 100–150 metres below current sea level.

The level of the land also varies in relation to the sea. Land may rise as a result of tectonic uplift or following the removal of an ice sheet. The localized change in the level of the land relative to the level of the sea is known as **isostatic adjustment** or **isostasy**. Parts of Scandinavia and Canada are continuing to rise at rates of up to 20 millimetres a year.

A simple sequence of sea level change can be described:

1. Temperatures decrease, glaciers and ice sheets advance and sea levels fall eustatically.
2. Ice thickness increases and the land is lowered isostatically.
3. Temperatures rise, ice melts, and sea levels rise eustatically.
4. Continued melting releases pressure on the land and the land rises isostatically.

As a result of global warming (the enhanced greenhouse effect), rising sea levels are impacting especially on low-lying communities (see page 159).

Short- and long-term changes

According to **Valentin's classification** (1952):

- **Retreating coasts** include submerged coasts and coasts where the rate of erosion exceeds the rate of emergence.
- **Advancing coasts** include emerged coastlines and coasts where deposition is rapid (Figure 6.30).

Features of emerged coastlines include:

- raised beaches, such as the Portland raised beach, Dorset
- coastal plains
- relict cliffs such as those along the Fall Line in eastern USA
- raised mudflats, for example the Carselands of the River Forth, Scotland.

- **Retreating coasts** include submerged coasts and coasts where the rate of erosion is greater than the rate of emergence
- **Advancing coasts** include emerged coastlines and coasts where the deposition is rapid.

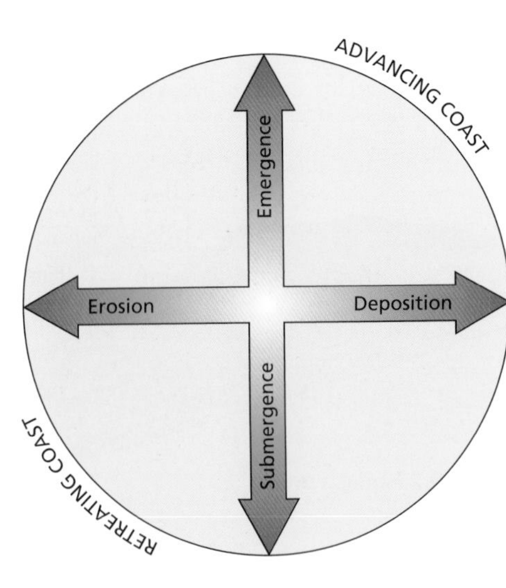

Figure 6.30 Advancing and retreating coasts

Submerged coastlines include:

- **rias**, such as the River Fal, which are drowned river valleys caused by rising sea levels during the Flandarin Transgression or due to sinking of the land
- **fjords**, such as Loch Torridon and the Oslo Fjord, which are glacial troughs occupied by the sea. They are common in uplifted mid-latitude coasts, notably in Norway, Greenland and Chile. An early view was that they were tectonic in origin, but now it is thought that they are drowned U-shaped valleys.
- **fjards,** or drowned glacial lowlands.

To do:

a Distinguish between advancing and retreating coastlines.

b Explain how (i) stacks and (ii) spits are formed.

c Examine the relationship between coastal processes, sub-aerial processes, lithology and landform, in two contrasting coastal environments.

Conflicts and management strategies

Coastal conflicts: tourism in St Lucia

Those who wish to develop tourism in St Lucia are in conflict with environmentalists, fishermen and locals. Most of the hotels built on the island are along the beachfront, and the clearing of land has led to slope instability, erosion and sedimentation of the shallow offshore environment. These developments have had several other negative impacts:

Overexploitation of the sand has led to significant increases in beach erosion and environmental degradation, and threatened the nesting grounds of endangered turtles. Although many hotels are artificially replenishing the beaches, the introduced sand is rapidly eroded and causing problems for offshore coral reefs, which require clear water.

Over-abstraction of water can lead to a decline in the amount of water available, as well as a decline in its quality. As freshwater supplies become reduced, saline water intrudes into the aquifers, thereby contaminating supplies. Tourists are great consumers of water (and energy), which has an impact on water supply for local people.

Altered wave and tide patterns occurred as a result of the building of the Pointe Seraphine cruise facility, to meet the needs of the many tourists who visit the island by cruise ship. Wave and swell

Figure 6.31 Tourist development at Rodney Bay, St Lucia

patterns changed significantly in the harbour. At Gros Inlet, wetlands were destroyed in 1969 to make way for building land and an artificial lagoon to expand the Rodney Bay resort. The results of the reclamation were unforeseen. The ebb and flood tide patterns were modified, creating stronger currents, which in turn increased erosion on nearby beaches. Local fisheries declined as the offshore waters became murkier, and the problem of sandflies remained.

Human activity	Agents/consequences	Coastal zone problems
Urbanization and transport	Land-use changes (e.g. for ports, airports); road, rail, and air congestion; dredging and disposal of harbour sediments; water abstraction; wastewater and waste disposal	Loss of habitats and species diversity; visual intrusion; lowering of groundwater table; saltwater intrusion; water pollution; human health risks; eutrophication; introduction of alien species
Agriculture	Land reclamation; fertilizer and pesticide use; livestock densities; water abstraction	Loss of habitats and species diversity; water pollution; eutrophication; river channelization
Tourism, recreation and hunting	Development and land-use changes (e.g. golf courses); road, rail and air congestion; ports and marinas; water abstraction; wastewater and waste disposal	Loss of habitats and species diversity; disturbance; visual intrusion; lowering of groundwater table; saltwater intrusion in aquifers; water pollution; eutrophication; human health risks
Fisheries and aquaculture	Port construction; fish processing facilities; fishing gear; fish farm effluents	Overfishing; impacts on non-target species; litter and oil on beaches; water pollution; eutrophication; introduction of alien species; habitat damage and change in marine communities
Industry (including energy production)	Land-use changes; power stations; extraction of natural resources; process effluents; cooling water; windmills; river impoundment; tidal barrages	Loss of habitats and species diversity; water pollution; eutrophication; thermal pollution; visual intrusion; decreased input of freshwater and sediment to coastal zones; coastal erosion

Table 6.5 Relationships between human activities and coastal zone problems

To research

The USGS coastal change site has useful explanation and examples, at http://pubs.usgs.gov. Look for "Coasts in crisis".

S-cool has information on coastal processes in the A-level geography section at www.s-cool.co.uk

Case study: *Increasing problems for people living in Australia's coastal zone*

Up to 80% of Australians live by the coast. However, the beach culture, as much part of the Australian identity as the bush and barbecues, is under threat, according to a government report on climate change that raises the prospect of banning its citizens from coastal regions at risk of rising sea levels.

The report says that property worth AUS $150 billion (£84 billion) is at risk. If sea levels rise 80 centimetres by 2100, some 711 000 homes, businesses and properties, which sit less than 6 metres above sea level and lie within 3 kilometres of the coast, will be vulnerable to flooding, erosion, high tides and more frequent storms. The report argues that Australia needs a national policy to respond to sea-level rise brought on by global warming, which could see people forced to abandon their homes.

On the far north coast of New South Wales, the state government has intervened to allow residents in the Byron Bay area to build special walls to protect their homes from rising sea levels. The government could consider "forced retreats", and prohibiting the "continued occupation of the land or future building development on the property due to sea hazard".

No one knows exactly how much sea levels will rise this century, but some estimates say it could be at least 60 centimetres; some predict a 1–2 metre rise by 2100. The subtropical state of Queensland is the most at risk, with almost 250 000 buildings vulnerable. Next is the most populous state, New South Wales (NSW), with more than 200 000. Coastal flooding and erosion already cost NSW around AUS $200 million (£112 million) a year.

Increased solid and liquid waste and its disposal are among the greatest environmental challenges facing St Lucia. The tourism industry generates more solid waste per capita than any other sector. This reduces the attractiveness of the tourist experience and raises issues about standards of health, the freshwater and marine environment, and the aesthetics of the island.

Marine pollution arising from the discharge of poorly treated wastewater into coastal waters poses environmental and health risks. Nutrient loading has led to the loss of coral productivity. The discharge of sewage by yachts also contributes to nearshore marine pollution. (Cruise ships generate a considerably higher amount of waste in total and per capita, but in general this is not disposed of in St Lucia, as port reception facilities are limited.)

Figure 6.32 Refuelling capability at Rodney Bay, St Lucia

Coastal management

Type of management	Aims/methods	Strengths	Weaknesses
Hard engineering	**To control natural processes**		
Cliff-base management	*To stop cliff or beach erosion*		
Sea walls	Large-scale concrete curved walls designed to reflect wave energy	Easily made; good in areas of high density	Expensive; lifespan about 30–40 years; foundations may be undermined
Revetments	Porous design to absorb wave energy	Easily made; cheaper than sea walls	Lifespan limited
Gabions	Rocks held in wire cages absorb wave energy	Cheaper than sea walls and revetments	Small scale
Groynes	To prevent longshore drift	Relatively low costs; easily repaired	Cause erosion on downdrift side; interrupt sediment flow
Rock armour	Large rocks at base of cliff to absorb wave energy	Cheap	Unattractive; small scale; may be removed in heavy storms
Offshore breakwaters	To reduce wave power offshore	Cheap to build	Disrupt local ecology
Rock strongpoints	To reduce longshore drift	Relatively low cost; easily repaired	Disrupt longshore drift; erosion downdrift
Cliff-face strategies	*To reduce the impacts of sub-aerial processes*		
Cliff drainage	Removal of water from rocks in the cliff	Cost-effective	Drains may become new lines of weakness; dry cliffs may produce rockfalls
Vegetating	To increase interception and reduce overland runoff	Relatively cheap	May increase moisture content of soil and lead to landslides
Cliff regrading	Lowering of slope angle to make cliff safer	Useful on clay (most other measures are not)	Uses large amounts of land – impractical in heavily populated areas
Soft engineering	**Working with nature**		
Offshore reefs	Waste materials, e.g. old tyres weighted down, to reduce speed of incoming wave	Low technology and relatively cost-effective	Long-term impacts unknown
Beach nourishment	Sand pumped from seabed to replace eroded sand	Looks natural	Expensive; short-term solution
Managed retreat	Coastline allowed to retreat in certain places	Cost-effective; maintains a natural coastline	Unpopular; political implications
"Do nothing"	Accept that nature will win	Cost-effective!	Unpopular; political implications
Red-lining	Planning permission withdrawn; new line of defences set back from existing coastline	Cost-effective	Unpopular; political implications

Table 6.6 Types of coastal management

To do:

a Define the term coastal management.

b Distinguish between hard engineering and soft engineering.

c **i** Explain how a groyne works.

ii Using a sketch diagram, suggest the likely distribution of sediment around groynes 50 years after the groynes' construction. Suggest reasons to support your answer.

d **i** Identify the benefits of sea walls.

ii Outline some of the disadvantages of using sea walls as a form of coastal management.

e For a coastal area you have studied, describe how the coastline is being protected, and comment on the effectiveness of the measures used.

The management of erosion on the West African coast

The Guinea Current along the coast of West Africa is one of the strongest in the world, and its longshore drift is removing huge amounts of coastline between Ghana and Nigeria. Ghana, Benin and Togo have been worst affected: their coasts are losing around 1.5 million cubic metres of sand each year. Hundreds of houses and beach hotels along this coast have crumbled into the sea, and the port cities are also under threat.

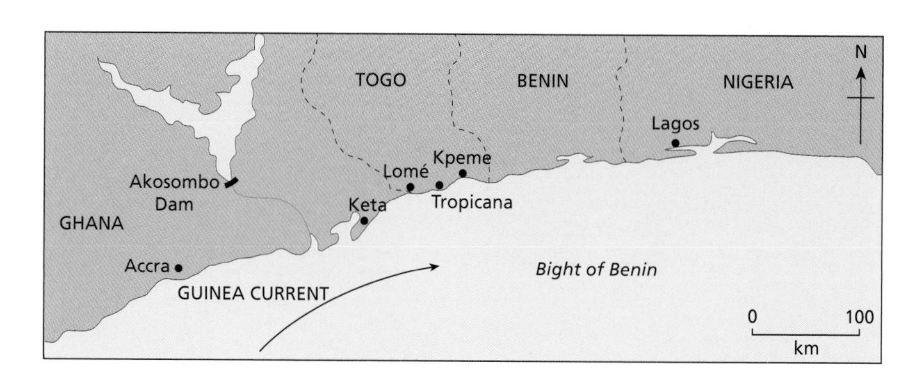

Figure 6.33 Longshore drift in West Africa

It is thought that human activities over the last few decades are partly responsible for the coastline's rapid retreat. In the past, sediments washed out to sea from rivers would be deposited along the coast, replacing the sand washed away by the ocean current. Environmentalists say the building of new ports 40 years ago, the construction of dams on rivers near the coast, the removal of sand from beaches to make cement, and other human activities, have caused the sea to advance, in some places at a rate of 20 metres a year.

The Akosombo Dam

Much of the increase in coastal retreat has been blamed on the construction of the Akosombo Dam on the Volta River in Ghana. Just 110 kilometres from the coast, it disrupts the flow of sediment from the river and stops it reaching the shore. Thus there is less sand to replace what has already been washed away, and so the coastline retreats as it is eroded by the Guinea Current. Towns such as Keta, 30 kilometres east of the Volta estuary, have been destroyed as their protective beach has been removed.

Erosion in Southern Nigeria

In southern Nigeria, the port city of Lagos is located at a break in the coast, and developed rapidly in the 19th and early 20th century. Dredging started in 1907 and the harbour was begun in 1908. Breakwaters and jetties provide a deepwater channel for large ships. These developments interrupted the west-east longshore drift, resulting in increased deposition on Lighthouse Beach (on the western, updrift side of the jetty) and on the eastern, downdrift side of the jetty. Across the harbour, Victoria Beach has been eroded by almost 70 metres per year, and 2 kilometres of beach have been lost.

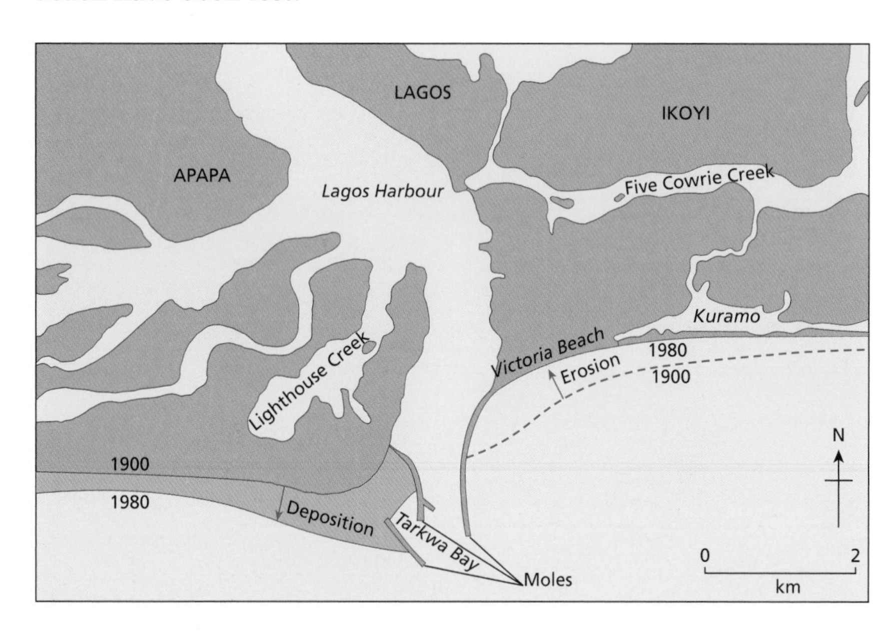

Figure 6.34 Erosion near Lagos, Nigeria

Victoria Beach is widely used as a recreational area for Lagos's population, and beach replenishment has taken place since 1976, partly for recreation and partly to protect the land from the sea. The problems of Victoria Beach are difficult to solve, however, since Lagos must be kept open for shipping. The jetties on the west side of the harbour capture the sediment to stop it drifting eastwards and blocking the harbour mouth. The unfortunate effect of this is to promote erosion of the eastern beach. However, continued deposition on Lighthouse Beach will eventually lead to deposition beyond the jetties of Tarkwa Bay, thus cutting off the deep channel of Lagos Harbour.

Storm surges

During storm surge periods, the sea level has often risen by up to 2 metres above normal. Government response has been in the form of regular beach replenishment and monitoring. Recent experiences show that this strategy is no longer enough, as storm surges become more frequent. In August 1995, for example, waves over 4 metres high flooded over half of Victoria Island, causing severe damage to roads, electric power lines, drainage systems and houses.

The current thinking of Nigeria's federal government is to construct breakwaters some distance away from the coastline. This should dampen the effect of waves reaching the shore, and consequently reduce the problem of flooding and erosion on the island. However, the lack of long-term data on sea-level values, waves, surges and meteorology will bring serious design problems.

Managing tsunamis

Coastal populations are on the increase in many parts of the world. Adequate mitigation measures from tsunami hazard can save lives, property, and the livelihoods of those living on the coast. A wide range of approaches can be used, including warning systems, education, building code standards, land-use planning, and other engineering solutions. The territories and coastlines most frequently affected by damaging tsunamis are Japan and the Hawaiian Islands.

Education is one of the easiest ways to reduce tsunami life loss. Most tsunami hazard is concentrated along coastlines close to the earthquake fault rupture. The duration of ground shaking is a good indicator of an impending tsunami: anything lasting a minute or more is a sign of a great earthquake with the potential to cause a life-threatening tsunami. In many (but not all) tsunamis, the first movement of the sea is a withdrawal. Any occasion when the sea shore recedes rapidly and inexplicably should be taken as a signal for immediate flight to higher ground. In addition, the first tsunami wave is not necessarily the most devastating one.

To research

To find out more about how Japan and Hawaii manage the risk of tsunamis, visit Hawaii's Department of Emergency Management at www.honolulu.gov/dem/tsunami2.htm and http://civil.eng.monash.edu.au/drc/symposium-papers/drokn1-takahasi.pdf for management in Japan.

Fish farming (aquaculture)

Fish farming was first introduced when overfishing of wild Atlantic salmon in the North Atlantic and Baltic seas caused their populations to crash. Aquaculture involves raising fish commercially, usually for food. (In contrast, a fish hatchery releases juvenile fish into the wild for recreational fishing or to supplement a species' natural numbers.) The most important fish species raised by fish farms are salmon, carp, tilapia, catfish and cod. Salmon make up 85% of the total sale of Norwegian fish farming, but most global aquaculture production now uses non-carnivorous fish species, such as tilapia and catfish.

Figure 6.35 A fish farm in Brunei

Technological costs are high, and include using drugs, such as antibiotics, to keep fish healthy, and steroids to improve growth. Breeding programmes are also expensive. Outputs are high per hectare and per farmer, and efficiency is also high. Environmental effects can be damaging, especially with salmon.

Salmon are carnivores and so need to be fed pellets made from other fish. It is possible that farmed salmon actually represent a net loss of protein in the global food supply, as it takes 2–5 kilograms of wild fish to grow a kilogram of salmon. Fish like herring, mackerel, sardines and anchovy are used to produce the feed for farmed salmon, and so the production of salmon leads to the depletion of other fish species on a global scale.

Other environmental costs include the sea lice and disease that spread from farmed fish into wild stocks, and pollution (created by uneaten food, faeces, and chemicals used to treat them) contaminating surrounding waters. Accidental escape of fish can affect local wild fish gene pools, when escaped fish interbreed with wild populations, reducing their genetic diversity, and potentially introducing non-natural genetic variation. In some parts of the world, escapes from farmed salmon threaten native wild fish, as it may be an alien species (e.g. the British Columbia salmon farming industry has inadvertently introduced a non-native species – Atlantic salmon – into the Pacific ocean).

Figure 6.36 Dune restoration

More material available:

www.oxfordsecondary.co.uk/ibgeography

Local-scale conflict in Studland beach and sand dunes

The environmental benefits of not removing fish from wild stocks, but growing them in farms, are great. Wild populations are allowed to breed and maintain stocks, while the farmed variety provides food.

Habitat restoration

Sand dunes are vulnerable ecosystems. However, their restoration is not always straightforward. For example, the Sefton dunes in northwest England have recently been partially stabilized by maintaining their natural vegetation. Pine trees have been planted, further stabilizing the dunes.

However, the pinewood plantations have resulted in the loss of dune habitat and associated species, and limit the ability of the coastal system to respond to environmental change. Once a woodland canopy is formed, the pine trees shade out the light so other plants are unable to grow. Huge areas of natural dune landscape with specialized sand dune plants and animals have been lost in this way.

The pine plantations were colonized by red squirrels, a rare, declining and protected species in Great Britain. The pine woodlands are now managed to maximize the conservation of the red squirrels, and welcomed as a feature of the local landscape.

Be a critical thinker

Should we preserve and protect native vegetation and native animal species rather than introduced ones?

To research

Visit the Restoration Centre at www.nmfs.noaa.gov to find out about coastal habitat restoration in the USA. View the projects and track their progress using the Restoration Atlas, a new interactive web-based map. You'll find information about ecological impacts.

Visit the Sands of Time website for management of the Sefton dunes, at www.sandsoftime.hope.ac.uk

Coral reefs and mangroves

Coral reefs and mangroves are fundamentally connected ecosystems. Mangroves protect coral reefs from sedimentation from land-based sources, as well as helping to keep the water clear of particles and nutrients. Both of these functions are necessary to maintain reef health. Mangroves also provide spawning and nursery areas for many animal species that spend their adult lives on the reefs. In return, the coral reefs provide shelter from the impact of waves and storms for the mangroves and their inhabitants, while the calcium carbonate eroded from the reef provides nutritious sediment from which the mangroves grow.

Coral – the limestone skeleton secreted by certain marine polyps, often deposited in extensive masses to form a reef.

Coral reef – a ridge in a relatively shallow, tropical sea, consisting of colonies of coral and other organic matter.

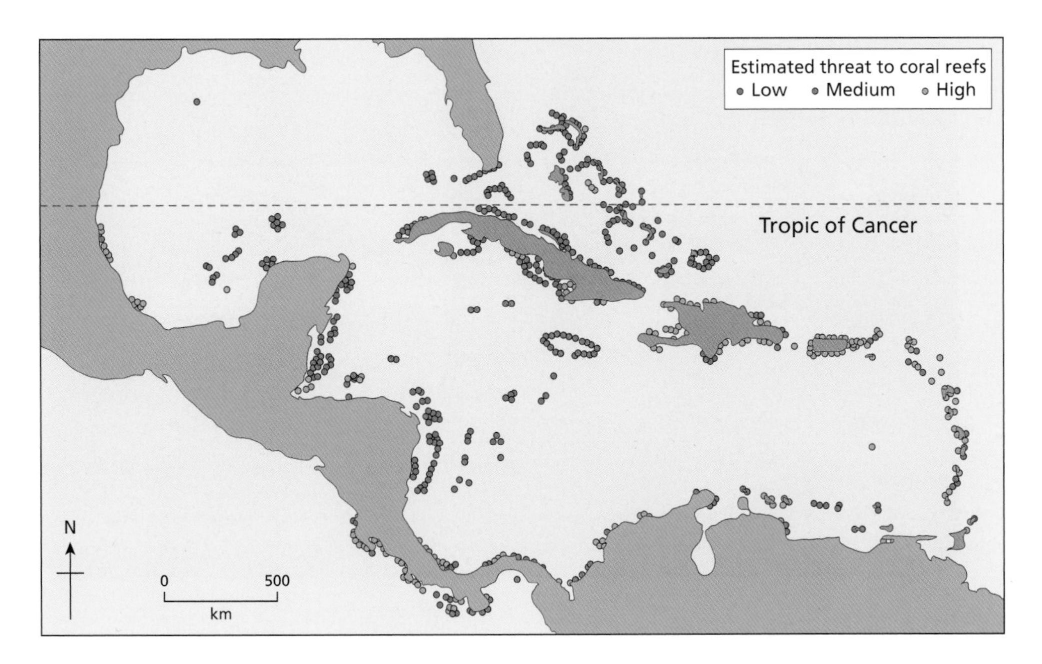

Figure 6.37 Coral reefs in the Caribbean

Coral reefs

Coral reefs are often described as the "rainforests of the sea" on account of their rich biodiversity. Coral reefs resemble tropical rainforests in two ways: both thrive under nutrient-poor conditions (where nutrients are largely tied up in living matter), yet support rich communities through highly efficient recycling processes. Some coral is believed to be 2 million years old, although most is less than 10 000 years old. Coral reefs contain nearly a million species of plants and animals, and about 25% of the world's sea fish breed, grow, spawn and evade predators in them. Some of the world's best coral reefs include Australia's Great Barrier Reef and many of the reefs around the Philippines and Indonesia, Tanzania and the Comoros, and the Lesser Antilles in the Caribbean.

Figure 6.38 Coral reef in Antigua

Pressures on coral reefs

Nearly two-thirds of the world's coral reefs are at risk from human activity. According to the World Resources Institute, 58% of the world's coral reefs are at high or medium risk of degradation, with more than 80% of South East Asia's extensive reef systems under threat.

Indirect pressures include dust storms from the Sahara, which have introduced bacteria into Caribbean coral, and global warming, which may cause coral bleaching. Bleaching occurs when high temperatures cause the corals to expel their algal symbionts, removing their colour so that the coral appears bleached (the algae are responsible for the colour of the coral). Many areas of coral in the Indian Ocean were destroyed by the 2004 tsunami.

Reef building necessitates warm, clear water. Unfortunately, pollution, sedimentation and global climate change, in addition to

To research

Research the types of human activity that may lead to the destruction of the coral reefs.

other pressures, threaten this need by impeding light penetration in the water, effectively halting the photosynthesis of the living part of the coral reefs.

Overfishing, destruction of coastal habitat and pollution from industry, farms and households are endangering not only fish – the leading individual source of animal protein in the human diet – but also marine biodiversity and even the global climate.

Figure 6.39 Coral and tourists

More material available:

www.oxfordsecondary.co.uk/ibgeography
Managing the Great Barrier Reef

Be a critical thinker

In what ways is coral useful to people?

Coral reefs are of major biological and economic importance. About 4,000 species of fish and 800 species of reef-building corals have been described. The reefs around Barbados, the Seychelles and the Maldives are important for helping to attract the tourists on which their economies rely. Florida's reefs attract tourism worth $1.6 billion annually. The global value of coral reefs in terms of fisheries, tourism and coastal protection is estimated to be $375 billion.

Biodiversity The NPP (net primary productivity) of coral reefs is 2 kilograms per square metre per year, representing an important component of the carbon cycle. Coral reefs are among the most diverse marine communities; one in four marine species is found on or near reefs. Coral reefs are important not only for biodiversity, but also to the human population as well. There are many advantages.

Seafood In developing countries, coral reefs contribute about a quarter of the total fish catch, providing food for one up to a billion people in Asia alone. Properly managed, reefs can yield, on average, 15 tonnes of fish and other seafood per square kilometre per year.

New medicines Coral reef species produce an array of chemicals for their self-protection. These chemicals are now being exploited for medical purposes, and corals are being used for bone grafts, and for treating viruses, leukaemia, skin cancer, and other tumours.

Other products Reef ecosystems yield a host of other economic goods, ranging from corals and shells made into jewellery and tourism curios, to live fish and corals used in aquariums, to sand and limestone used by the construction industry.

Coral reefs offer a wide range of environmental benefits that are of enormous importance to the nearby inhabitants. These benefits are sometimes difficult to quantify, but they include:

- **recreational value:** the tourism industry is one of the fastest-growing sectors of the global economy. Coral reefs are a major draw for snorkellers, scuba divers and recreational fishers
- **coastal protection:** the benefits from this protection are widespread, and range from maintenance of highly productive mangrove fisheries and wetlands to supporting local economies built around ports and harbours, where, as is often the case in the tropics, these are sheltered by nearby reefs.

Mangroves

Mangroves are salt-tolerant forests of trees and shrubs that grow in the tidal estuaries and coastal zones of tropical areas. The muddy waters, rich in nutrients from decaying leaves and wood, are home to a great variety of fish, sponges, worms, crustaceans, molluscs and algae. Mangroves cover about 25% of the tropical coastline, the largest being the 570 000 hectares of the Sundarbans in Bangladesh.

Figure 6.40 Mangrove trees

Mangrove – tropical evergreen tree or shrub with intertwined roots and stems, which grows in dense groves in swamps or along tidal coasts.

To research

Explore the Reefs at Risk website at www.wri.org/wri/reefsatrisk

The value of mangroves

Mangroves have many uses, such as providing large quantities of food and fuel, building materials and medicine. One hectare of mangrove in the Philippines can yield 400 kilograms of fish and 75 kilograms of shrimp. Mangroves also protect coastlines by absorbing the force of hurricanes and storms. They also act as natural filters, absorbing nutrients from farming and sewage disposal.

Pressures on mangroves

Despite their value, many mangrove areas have been lost to rice paddies and shrimp farms. As population growth in coastal areas is set to increase, the fate of mangroves looks bleak. Already most Caribbean and South Pacific mangroves have disappeared, while India, West Africa and South-East Asia have lost half of theirs.

Thailand	185 000 ha (1960–91) to shrimp ponds
Malaysia	235 000 ha (1980 and 1990) to shrimp ponds and farming
Indonesia	269 000 ha (1960–90) to shrimp ponds
Vietnam	104 000 ha (1960–74) due to US army
Philippines	170 000 ha (1967–76) mostly to shrimp ponds
Bangladesh	74 000 ha (since 1975) largely to shrimp ponds
Guatemala	9,500 ha (1965–84) to shrimp ponds and salt farming

Table 6.7 Mangrove losses

7 Extreme environments

By the end of this chapter you should be able to:

- describe the global distribution of extreme environments
- understand the different strategies used by people for coping with life in an extreme environment
- explain the processes that produce the different landforms of different types of extreme environment
- understand the challenges posed to agriculture in arid and semi-arid areas, and to mineral extraction in periglacial areas
- describe resource development in extreme environments, with reference to tourism and sustainability.

This option covers two different kinds of extreme environment:

- cold and high-altitude environments (polar, glacial and periglacial areas, and high mountains in non-tropical latitudes)
- hot, arid environments (hot desert and semi-arid areas).

This chapter describes the distribution and characteristics of each type of environment, and the natural processes and physical features of these distinctive landscapes. It also describes the ways in which people have responded to the opportunities that extreme environments offer, despite their rugged terrain, harsh climates and unproductive soils. Such extreme conditions mean that settlement is limited, but new technologies are providing opportunities for exploiting the resource potential of these environments. Activities such as mining, oil drilling, irrigated agriculture and tourism make significant contributions to the economies of these regions, but the physical conditions present challenges to all human activities here and are often a source of conflict.

The issue of maintaining sustainability is common to all extreme environments because they share the same problem of low carrying capacity. Their fragile landscapes and ecosystems are sensitive to change and recover slowly. Human activity of all kinds can have serious negative impacts, and the careful management of these environments and their resources is essential.

The option is presented by subtopic rather than by environment, to enable you to make comparisons between them. Reference to other chapters such as Leisure, sport and tourism (chapter 9) may provide additional material and case studies.

The global distribution of extreme environments

Cold and high-altitude environments

The distribution of cold environments is very uneven. Polar environments are located towards the North Pole and the South Pole, where levels of **insolation** (solar radiation) are very low. In the northern hemisphere, there is a belt of **periglacial** environments. This zone is generally not found in the southern hemisphere except in small areas, given the relative lack of land mass at around 60°– 65°S.

Arid – having less than 250 millimetres of precipitation per year.
Active layer – the highly mobile, often saturated surface layer of permafrost that melts in summer and freezes in winter. Its depth can vary from a few centimetres to 5 metres.
Albedo – the reflectivity of a surface. Snow-covered landscapes are highly reflective and have a high albedo, whereas dark surfaces (tarmac) reflect less and retain heat.
Carrying capacity – the maximum number of individuals that a given environment can support with the resources available and without detrimental effects.
Glacial – characterized by the presence of ice masses, either at high altitudes or at the poles.
Periglacial – snow and ice cover on the fringe of glaciated areas ("peri" = on the edge of), usually associated with permafrost or ground that remains frozen for at least two years. These regions include high mountain and tundra areas of northern Europe and North America.
Semi-arid – having 250–500 millimetres of precipitation per year.

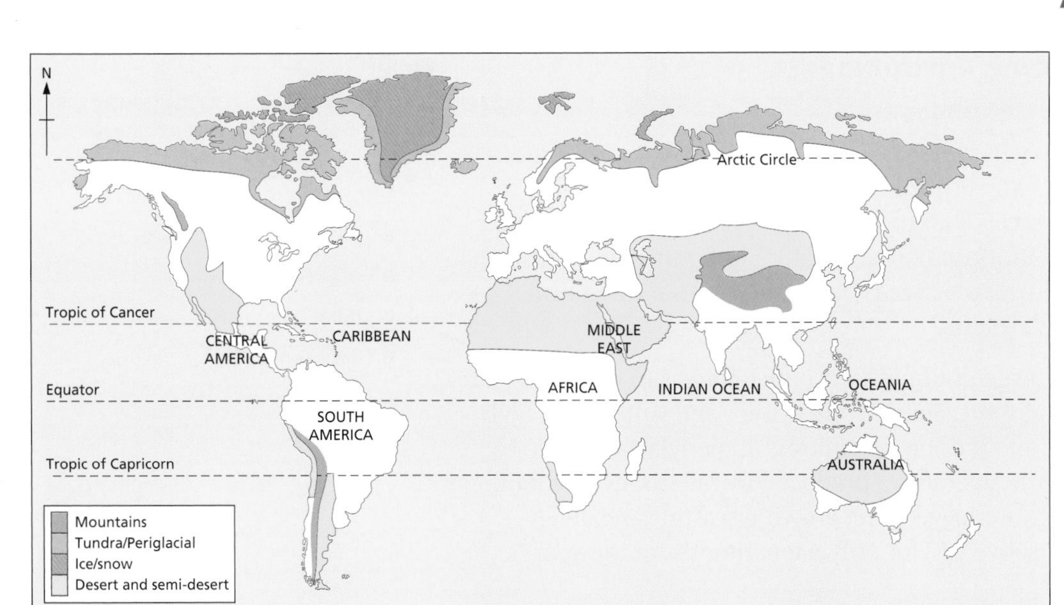

Figure 7.1 The distribution of extreme environments

Other cold environments are associated with high mountains. There are extensive areas of high ground in Asia, associated with the Himalayas; other high-altitude areas include the Andes and the Rockies.

Desert and semi-arid environments

Desert and semi-desert areas cover as much as a third of the earth's surface. They are generally located around the tropics and are associated with permanent high-pressure systems which limit rain formation. Four main factors determine the location of the world's main deserts. They include:

- the location of stable, **high-pressure conditions** at the tropics, e.g. the Sahara and the Great Australian deserts
- large distance from the sea (known as **continentality**), such as the central parts of the Sahara and Australia and also parts of the southwest USA
- **rain-shadow effects**, as in Patagonia (South America) and the Gobi Desert in central Asia
- proximity to **cold upwelling currents**, which limit the amount of moisture held in the air, e.g. off the west coast of South America, helping to form the Atacama Desert, and off the west coast of southern Africa, helping to form the Namib desert (see Figure 7.2).

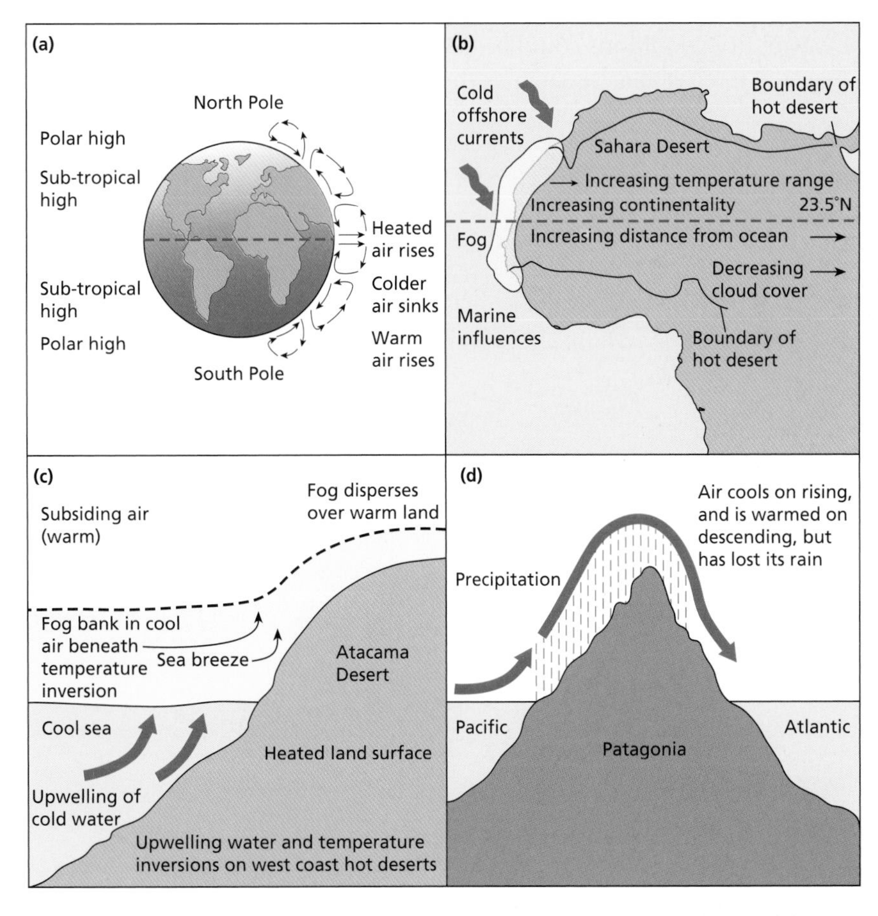

Figure 7.2 Causes of aridity: (a) subtropical high pressure, (b) continentality, (c) cold offshore current and (d) rain shadow

Conditions in extreme environments

Cold and high-altitude environments

Cold environments are very varied in their characteristics. Mountain environments can be characterized by warm days and very cold nights. They may also receive large amounts of rainfall due to relief rain. In contrast, some mountain areas receive low rainfall because they are in a rain shadow. Polar areas generally receive low rainfall. They are, in effect, cold deserts.

Owing to their steep nature, mountains are difficult areas to build on, and they act as barriers to transport. Soils are often thin, and suffer from high rates of overland runoff and erosion. In contrast, in periglacial areas – or tundra regions – the low temperatures produce low rates of evaporation, and soils are frequently waterlogged. The growing season is relatively short: temperatures are above 6°C for only a few months of the year.

Study Figure 7.2. Explain how the following factors influence the development of arid and semi-arid areas:

- **a** atmospheric pressure
- **b** ocean currents
- **c** relief
- **d** continentality.

Desert and semi-arid environments

In desert areas, such as Aden, the lack of water acts as a major constraint for development. Temperatures are hot throughout the year but, in the absence of freshwater, farming is almost impossible. In semi-arid areas, annual rainfall varies between 250 millimetres and 500 millimetres, so there is some possibility for farming, especially where water conservation methods are used. On the other hand, the guarantee of warm, dry conditions could be excellent for tourism developments, especially in coastal areas such as the Red Sea coast of Egypt.

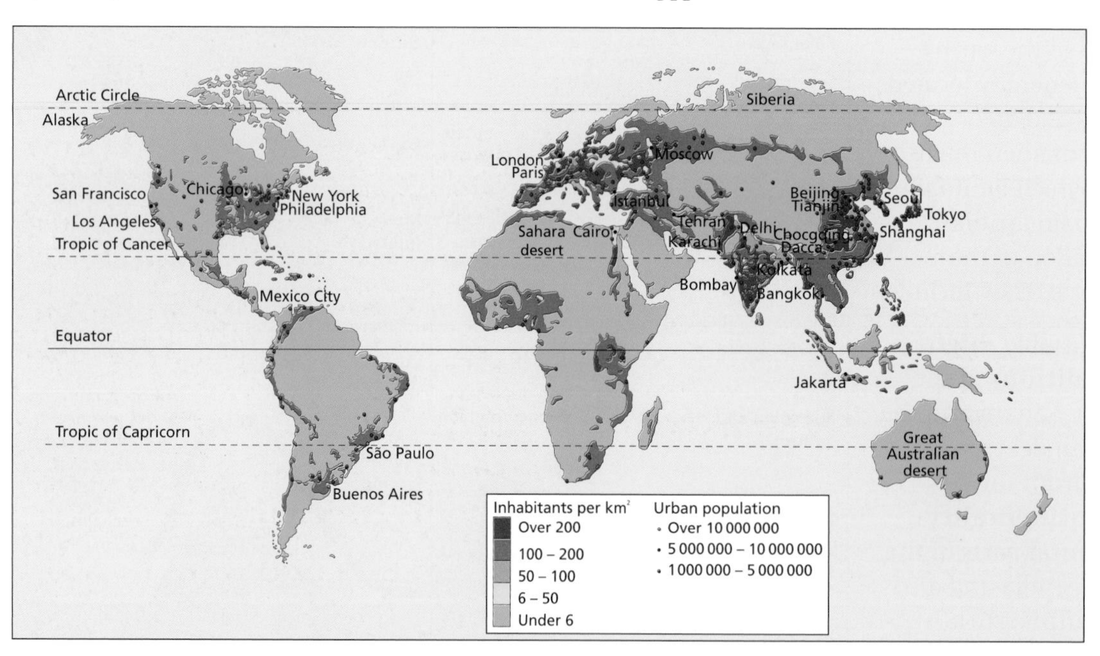

Figure 7.3 Extreme environments have very low population densities

People in extreme environments

Extreme environments are characterized by low population densities. Examples include densities of three people per square kilometre in central Australia, Iceland and northern Canada, two people per square kilometre in Namibia and just one per square kilometre in the western Sahara. Much of this can be put down to the extremes of climate: insufficient heat in Iceland and Canada, and insufficient water in the other three areas are largely to blame.

None of these environments is particularly "comfortable"; they all fall a long way outside the recognized "comfort zones for human habitation".

Other factors are important, too. Iceland is relatively remote and isolated. This makes communications costly, if not difficult. It also increases the cost of materials, which have to be imported, such as timber for building. Similarly, Namibia is a long way from the economic core of southern Africa, and this increases the costs of imports and exports. Coastal areas are better off than inland areas but are still relatively undeveloped.

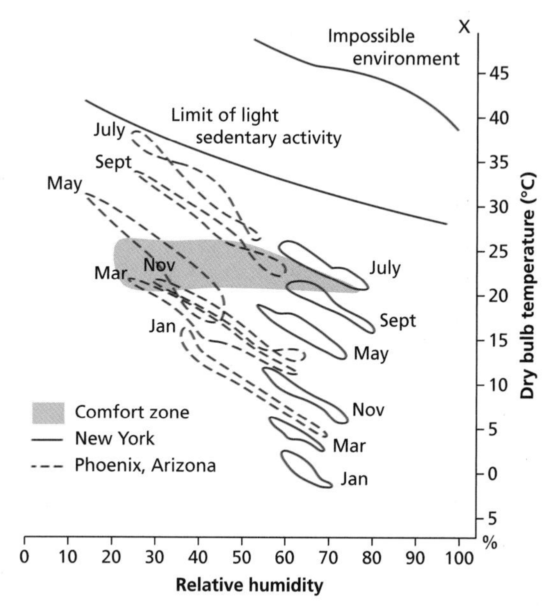

Note how the climates of New York and Phoenix fall outside the comfort zone in most of the six months plotted.

Figure 7.4 Comfort zones

Coping with periglacial environments

Traditionally, the Inuit peoples of the Arctic regions of Alaska, Canada and Greenland have used periglacial pastures for herding or hunting caribou. The Inuit tend to be migratory, moving north into the tundra during the brief months of summer and heading southwards to the forest margins in winter. The Sami of Scandinavia also follow this pattern.

To make up for the lack of good pasture on land, many indigenous peoples have turned to rivers and the oceans. Fishing is extremely important in periglacial environments. For the Nyetski of the Yamal Peninsula in Siberia, fish are an important supplement to their diet. At the other end of the scale, fishing and fish-related products accounted for up to 70% of Iceland's GDP. To cope with the cold conditions Inuit populations have evolved a layer of fat that protects them from the extreme cold.

To do:

a Describe the conditions of the comfort zone shown on Figure 7.4.

b Explain why the conditions are impossibly uncomfortable for humans at X.

Building on permafrost

Building on permafrost is challenging. Heated buildings can thaw the permafrost underneath them. Once the permafrost thaws, it sinks, damaging the building it supports. Engineers solve this problem by raising buildings above the ground on a steel frame, so that cold air can flow under them and stop the permafrost from thawing. Another way to stop damage from thawing permafrost is to thaw the ground first. This method makes the ground more stable to build on, because the structure keeps the ground from freezing.

Water supplies

In places with large, continuous stretches of permafrost, finding water takes a lot of effort. People can sometimes get water from nearby lakes and rivers or by melting ice or snow, but they cannot get liquid water directly out of the ground in the winter. Water pipes from the water supply to the buildings have to be protected so that water inside the pipe does not freeze and the ground around the pipe does not thaw.

Transport

Transport infrastructure such as roads and bridges may be damaged by frost heave, and need constant repair to keep them safe. Soil under roads can be replaced with gravel so that water drains better and there is less frost heave, or roads may be painted white to reflect more heat and keep them cooler.

To research

Research environment and human adaptation at http://ameeta.tiwanas.com/files/presentations/physical/Environment_and_Human_adaptation.pdf

Research Agriculture in the Himalayas of Nepal at http://www.colby.edu/personal/p/pccabot/Agr.htm

Be a critical thinker

Is human activity no longer restricted by physical geography?

Is it true to say that harsh climate, high altitudes and difficult terrain are no longer barriers to human habitation because of technological advances?

A cooler road surface helps prevent frozen ground from thawing underneath. In some places with permafrost, the top layer of ground thaws during the summer, creating marshland, so people can only drive on these areas during the winter, on ice roads at least a metre thick built on top of the frozen marshes. Trucks weighing up to 64 tonnes can then drive across them and haul supplies to mines and drill sites in northern Canada and Alaska. Ice roads are also built on frozen lakes for winter travel. In summer, the roads melt and must be rebuilt each winter.

Coping with arid environments

Like the traditional peoples of periglacial areas, desert inhabitants are also migratory. The Bedouin of the Arabian Peninsula and the Fulani of Africa, for example, have learned to cope with the extreme temperatures by avoiding the direct sun and taking a rest during the middle of the day. They tend to travel in early morning and late afternoon. Their clothing – long, loose-fitting garments – also helps them to cope with high temperatures. It reduces sweating and allows them to remain reasonably fresh.

Glacial environments

Glacial systems

A glacial system is the balance between inputs, storage and outputs. Inputs include **accumulation** of snow, avalanches, debris, heat and meltwater. The main store is that of ice, but the glacier also carries debris, **moraine** and meltwater. The outputs are the losses due to **ablation**, the melting of snow and ice, and sublimation of ice to vapour, as well as sediment.

Glacier – a large body of ice and compacted snow moving slowly down a valley.
Moraine – a line of loose rocks, weathered from the valley sides and carried by the ice of the glacier.
Ablation – losses from the glaciar system such as melting.

The **regime** of the glacier refers to whether the glacier is advancing or retreating:

- if accumulation > ablation the glacier advances
- if accumulation < ablation the glacier retreats
- if accumulation = ablation the glacier is steady.

Glacial systems can be studied on an annual basis or on a much longer timescale. The size of a glacier depends on its regime, i.e. the balance between the rate and amount of supply of ice and the amount and rate of ice loss. The glacier will have a positive regime when the supply is greater than loss by ablation (melting, evaporation, calving, wind erosion, avalanche, etc.) and so the glacier will thicken and advance. A negative regime will occur when the wasting is greater than the supply (e.g. the Rhône glacier today) and so the glacier will thin and retreat. Any glacier can be divided into two sections: an area of accumulation at high altitudes generally, and an area of ablation at the snout.

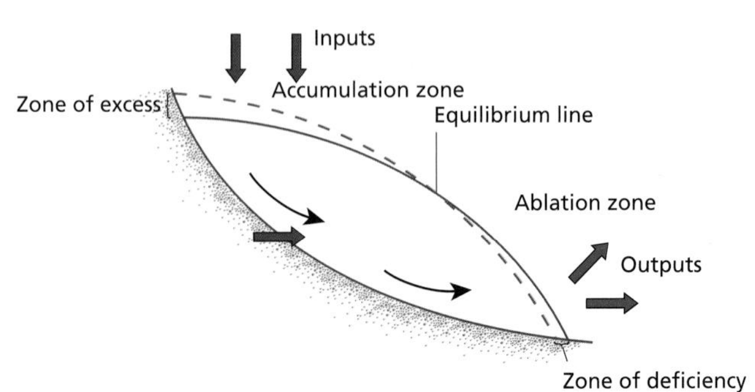

Figure 7.5 Glacier movement

Glacial erosion

The amount and rate of erosion depend on local geology, the velocity of the glacier, the weight and thickness of the ice, and the amount and character of the load carried. The methods of glacial erosion include plucking and abrasion.

Plucking

This occurs mostly at the base of the glacier and to an extent at the side. It is most effective in jointed rocks or those weakened by freeze–thaw. As the ice moves, meltwater seeps into the joints and freezes to the rock, which is then ripped out by the moving glacier.

Abrasion

The debris carried by the glacier scrapes and scratches the rock, leaving striations, or grooves, in the rock.

Other mechanisms

Other mechanisms include meltwater, freeze–thaw weathering and pressure release. Although neither strictly glacial nor erosional, these processes are crucial in the development of glacial scenery.

Figure 7.6 Glacial landscape, Austria

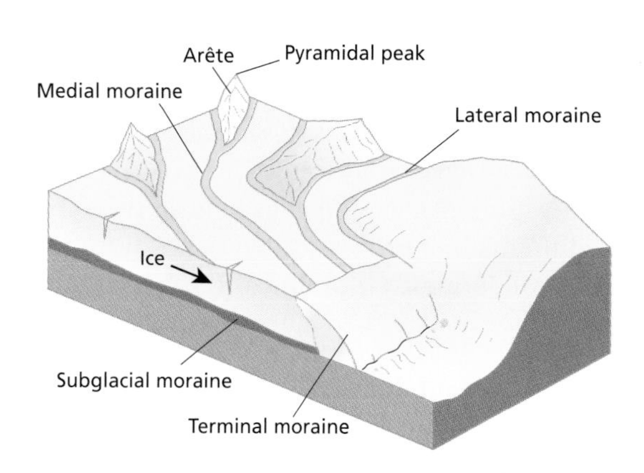

Figure 7.7 During glaciation

Landforms produced by glacial erosion

Cirques

A cirque (or corrie) is an amphitheatre-shaped valley that has been scooped out by erosion at the head of a glacier. In the northern hemisphere cirques are generally found on north- or east-facing slopes, where accumulation is highest and ablation is lowest. They are formed in stages:

1. A preglacial hollow is enlarged by **nivation** (freeze–thaw and removal by snowmelt).
2. Ice accumulates in the hollow.
3. Having reached a critical weight and depth, the ice moves out in a rotational manner, eroding the floor by plucking and abrasion.
4. Meltwater trickles down the *bergschrund* (a crevasse that forms when the moving glacier ice separates from the non-moving ice above), allowing the cirque to grow by freeze–thaw.

After glaciation, an armchair-shaped hollow remains, frequently filled with a lake, e.g. Blue Lake cirque, New South Wales, Australia, and the Cirque de Gavarnie in the Central Pyrenees, France.

Arêtes, peaks, troughs, basins and hanging valleys

Other features of glacial erosion include **arêtes** and **pyramidal peaks** (horns) caused by the headward recession (cutting back) of two or more cirques. Glacial **troughs** (or U-shaped valleys) have steep sides and flat floors. In plan view they are straight, since they have **truncated** the interlocking spurs of the preglacial valley. The ice may also carve deep **rock basins** frequently filled with **ribbon lakes**. **Hanging valleys**

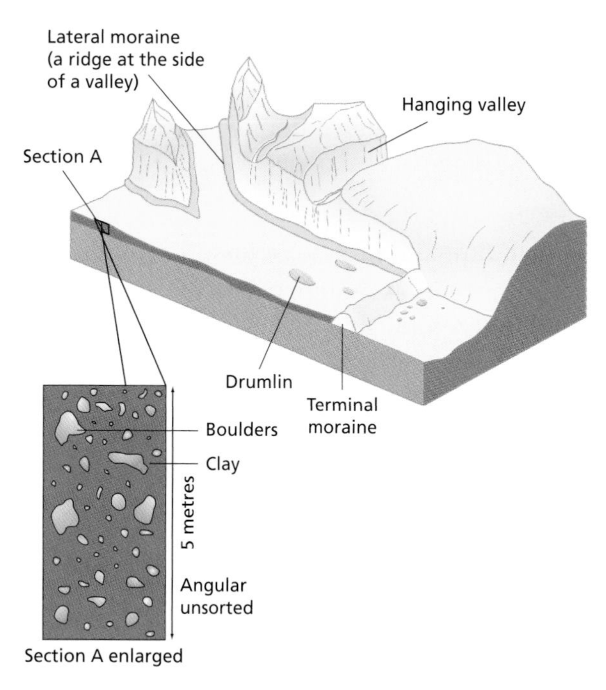

Figure 7.8 After glaciation

are formed by tributary glaciers, which, unlike rivers, do not cut down to the level of the main valley but are left suspended above, e.g. Stickle Beck in the Lake District, UK. They are usually marked by waterfalls.

A **roche moutonnée** (or sheepback) is a bare mound of rock that can vary in size from a few metres to hundreds of metres high. These rocks are smoothed and polished on their up-valley side (stoss) by abrasion but plucked on their down-valley side (lee) as ice accelerates. They can be several kilometres long. There are examples in the Cairngorms in Scotland and Glacier Bay, Alaska.

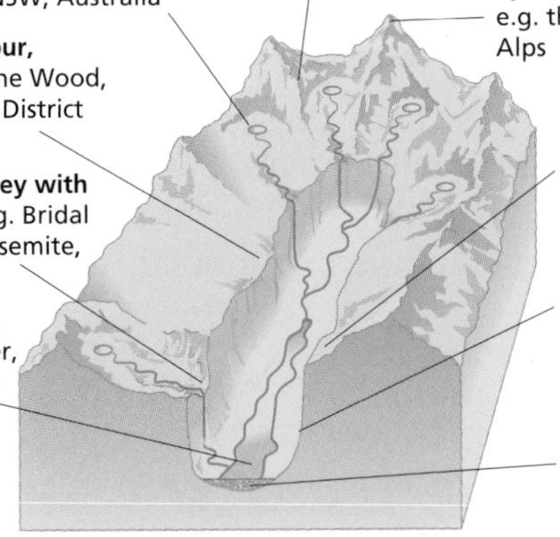

Figure 7.9 Landforms of a glaciated valley

Glacial deposition

Although glaciers move slowly – usually at a rate of a few centimetres a year – the force of gravity and the sheer weight of their ice give them tremendous power to erode the landscape. As it moves downhill, the ice picks up and transports rock fragments produced by freeze–thaw weathering. As the glacier melts and retreats, it leaves behind these rock fragments and sediments in a process called deposition. The rocks and sediments form till, erratics, moraine and other depositional features.

Drift – glacial and fluvioglacial deposits left after the ice has melted.
Till – angular and unsorted glacial deposits that include erratics, drumlins and moraines.
Erratic – a large boulder foreign to the local geology.
Drumln – a small egg-shaped hill on the floor of a glacial trough.

The characteristics of till

Till is often subdivided into **lodgment till**, material dropped by actively moving glaciers, and **ablation till**, deposits dropped by stagnant or retreating ice. Till has the following characteristics:

- poor sorting – till contains a large range of grain sizes, e.g. boulders, pebbles, clay
- poor stratification – no regular sorting by size
- mixture of rock types – from a variety of sources
- striated and subangular particles
- long axis orientated in the direction of glacier flow
- some compaction of deposits.

The characteristics of erratics

Erratics are glacier-transported rock fragments that differ from the local bedrock, and may be embedded in till or on the ground surface. They range in size from pebbles to huge boulders: one example is the Madison Boulder in New Hampshire, USA, estimated to weigh over 4,600 tonnes. Erratics that have moved over long distances – some more than 800 kilometres – generally consist of rock resistant to the shattering and grinding effects of glacial transport. Erratics composed of distinctive rock types can be traced to their place of origin, indicating the direction of glacial movement that brought them to their resting place.

The characteristics of moraine

Moraines are loose rocks, weathered from the valley sides and carried by glaciers. At the snout of the glacier is a crescent-shaped mound or ridge of **terminal moraine**. It represents the maximum advance of a glacier, and its character is determined by the amount of load the glacier was carrying, the speed of movement and the rate of retreat. The **ice-contact slope** (up-valley) is always steeper than the down-valley slope. Cape Cod in Massachusetts, USA, is a fine example of a terminal moraine.

Lateral moraine is a ridge of loose rocks and sediment running along the edge of a glacier where it meets the valley side. The lateral moraines on the Gorner Glacier in Switzerland are good examples. Where two glaciers merge and the two touching lateral moraines merge in the middle of the enlarged glacier, the ridge is known as a **medial moraine**. Again, the Gorner Glacier contains many examples of medial moraines.

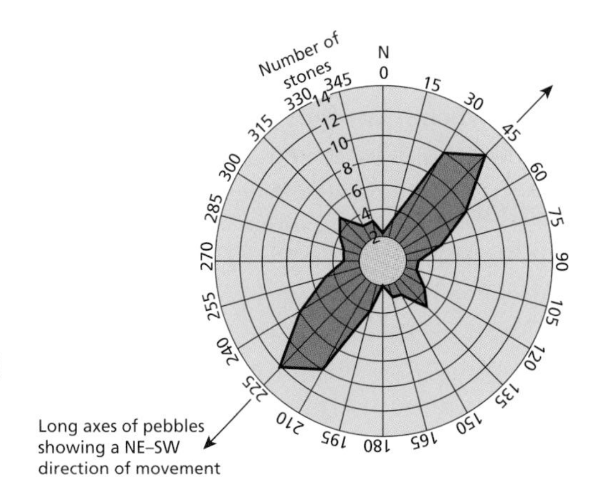

Figure 7.10 Determining the direction of glacier movement

The characteristics of drumlins

Drumlins are small oval mounds up to 1.5 kilometres long and 100 metres high, e.g. the drowned drumlins of Clew Bay in Co. Mayo, Ireland. One of the largest concentrations is in New York State, where there are over 10 000 drumlins. They are deposited as a result of friction between the ice and the underlying geology, causing the glacier to drop its load. As the glacier continues to advance it streamlines the mounds.

More material available:

www.oxfordsecondary.co.uk/ibgeography

Case study and skills exercise using topographical maps and air photographs of the Aletsch glacier Switzerland.

Direction of glacier movement

These features can be used to determine the **direction of glacier movement**. Erratics pinpoint the origin of the material; drumlins and the long axes of pebbles in glacial till are orientated in the direction of glacier movement.

Periglacial environments

Periglacial – relating to or found in a region that borders on a glacier.
Permafrost – impermeable, permanently frozen ground.
Freeze–thaw action – the action of water as it freezes and thaws in cracks in rock, causing the rock to shatter.

Periglacial areas are found on the edge of glaciers or ice masses, and more than a third of the earth's land surface was once subject to periglacial conditions at some time. Periglacial environments are characterized by permafrost, usually hundreds of metres deep, and freeze–thaw action, which continuously alters the ground surface so that large quantities of angular, fractured rock (frost-shattered rock) are common. Summer temperatures in these environments briefly rise above freezing, so ice in the soil near the surface melts.

Three types of periglacial region can be identified: Arctic continental, Alpine and Arctic Maritime. These vary in terms of mean annual temperature and therefore the frequency and intensity with which periglacial processes operate.

Landforms associated with periglacial environments

Periglacial environments show a wide range of different processes that relate to the action of permafrost and freeze–thaw weathering. These processes include mass

Figure 7.11 Periglacial environment, Iceland

movement, a common phenomenon during the warmer seasons, which usually occurs in four forms: solifluction, gelifluction, frost creep and rockfalls.

Figure 7.12 Periglacial environment, Italian Dolomites

Permafrost

Approximately 20% of the world's surface is underlain by permafrost. Three types of permafrost exist: continuous, discontinuous and sporadic, and these are associated with mean annual temperatures of −5° to −50°C, −1.5° to −5°C and 0° to −1.5°C respectively. Above the permafrost is the active layer, a highly mobile layer which seasonally thaws out and is associated with intense mass movements. The depth of the active layer depends on the amount of heat it receives, and varies in Siberia from 0.2–1.6 metres at 70°N to between 0.7–4 metres at 50°N.

Mass movement

Solifluction literally means flowing soil. In winter, water freezes in the soil, causing expansion of the soil and segregation of individual soil particles. In spring the ice melts and water flows downhill. It cannot infiltrate the soil because of the impermeable permafrost. As it moves over the permafrost it carries segregated soil particles (peds) and deposits them further downslope as a U-shaped solifluction lobe or terracette.

In periglacial environments, solifluction occurs only when temperatures are well above zero and free liquid water is available in the active layer. Solifluction is common when surface sediments are poorly drained and saturated with water. **Gelifluction** is a form of solifluction where soil particles slide in water over a permafrost layer.

Frost creep is a type of solifluction that occurs because of frost heaving and thawing. The process begins with the freezing of the ground surface elevating particles at right angles to the slope. The particles rise up because cold temperatures causes water in between particles to freeze and expand. As the ice thaws in the warm season, turning back to water, the contracting surface drops the particles in elevation. This drop, however, is influenced by gravity causing the particles to move slightly downslope.

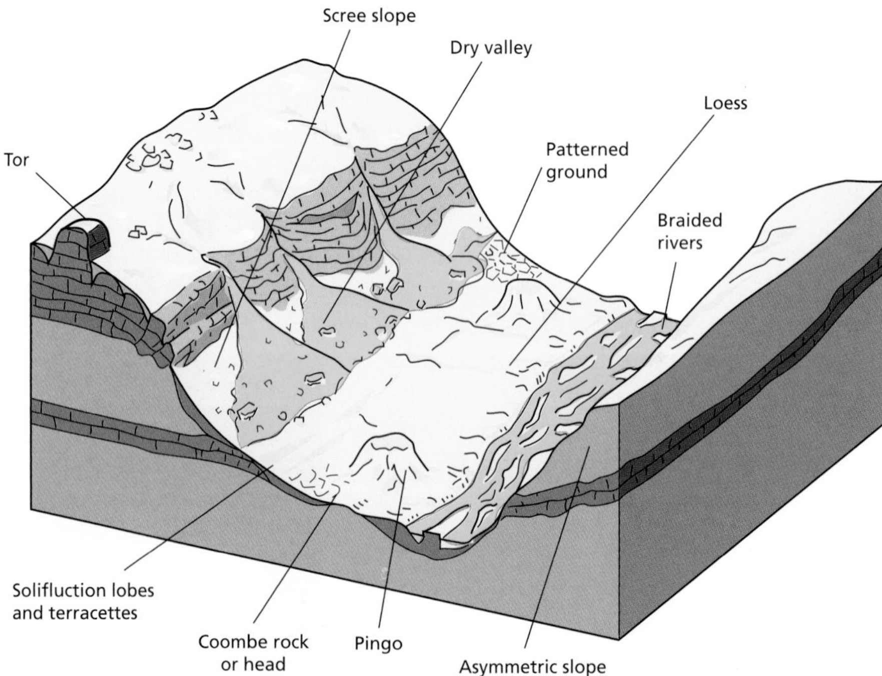

Figure 7.13 Features of the periglacial environment

Rockfalls occur when fragments of rock break away from a cliff face, often as a result of freeze–thaw weathering.

Patterned ground is a general term describing the stone-circles, polygons and stripes that are found in soils subjected to intense frost action, e.g. on the slopes of Kerio crater, southern Iceland. On steeper slopes, stone stripes replace stone circles and polygons. Their exact mode of formation is unclear, although ice sorting, differential frost heave, solifluction and the effect of vegetation are widely held to be responsible.

Pingos

A **pingo** is an isolated, conical hill up to 90 metres high and 800 metres wide, which can only develop in periglacial areas. Pingos form as a result of the movement and freezing of water under pressure. Two types are generally identified: **open-system** and **closed-system** pingos. Where the source of the water is from a distant elevated source, open-system pingos form, whereas if the supply of water is local, and the pingo arises as a result of the expansion of permafrost, closed-system pingos form. Nearly 1,500 pingos are found in the Mackenzie Delta of Canada. When a pingo collapses, ramparts and ponds are left.

Thermokarst

Thermokarst refers to a landscape of hummocks and wet hollows, resulting from subsidence caused by the melting of permafrost. This may be because of broad climatic changes or local environmental changes.

To research

Research Alpine protected areas at Alparc, http://www.alparc.org/the-alps/a-sensitive-area

To do:

Figure 7.14 shows the cumulative change in the length of the Aletsch Glacier, the largest and one of the most rapidly retreating glaciers in the Alps.

- **a** What does cumulative change mean, and how is it calculated?
- **b** Describe the rate of change over this 120-year period.
- **c** Research the methods used to monitor glacial changes.
- **d** Examine the factors causing variation in the rate of glacial retreat.
- **e** Assess the consequences of glacial retreat for a country such as Switzerland.

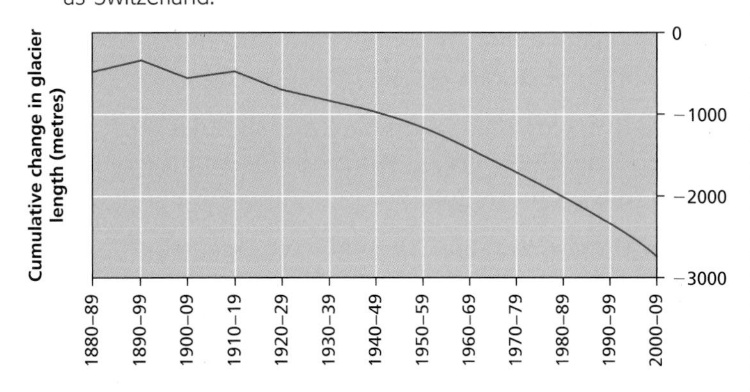

Figure 7.14 Retreat of the Aletsch Glacier, Swiss Alps, 1880–2009

Hot, arid environments

Various processes of erosion, weathering and deposition shape desert landscapes. Their distinctive topography is carved by water: although rainfall is low, it is an important climatic factor in the formation of desert erosion features. As a result of the dryness, wind and mechanical weathering also play an important part in desert erosion. The lack of moisture and scant vegetation make the wind a more powerful agent of erosion in deserts than in humid environments. Sand and sediments are blown along the surface, eroding rocks and other objects with which they come in contact. However, wind lifts the sand only up to a metre above the ground, so higher features have been eroded by water.

The work of water in deserts

Not only is water vital for the development of many desert landforms, but it is also important for the operation of mechanical and chemical weathering in deserts. There are a number of sources of water in deserts: rainfall, the water table (which may be exposed by deflation to produce an oasis) and rivers.

More material available:

www.oxfordsecondary.co.uk/ibgeography

Desert rains may fall as short torrential downpours over the normally dry ground, and there is a high amount of runoff. The flash floods rush down desert mountains in networks of channels, gullies or canyons, carrying large amounts of rock fragments that have great erosive power. At the bottom of the desert mountain the streams spread out, depositing the transported material to build an alluvial fan. Channels in the fan carry any water that has not sunk into the ground or evaporated, along with finer silt, to the toe of the fan. It then spreads out still more as it washes down to the lowest part of the basin, where it may form a playa lake, which then evaporates over a few days to form a salt pan.

Rivers that flow through deserts can be classified as exotic (exogenous), endoreic or ephemeral.

Exotic or **exogenous** rivers are those that have their source in another, wetter environment and then flow through a desert. The Nile in Egypt is an exotic river, being fed by the White Nile, which rises in the equatorial Lake Victoria, as are the Blue Nile and Atbara, which rise in monsoonal Ethiopia.

Endoreic rivers are those that drain into an inland lake or sea. The River Jordan, which drains into the Dead Sea, is a good example.

Ephemeral rivers are those that flow seasonally or after storms, and often have high discharges and high sediment levels. Even on slopes as gentle as 2°, overland flow can generate considerable discharges. This is a result of factors including:

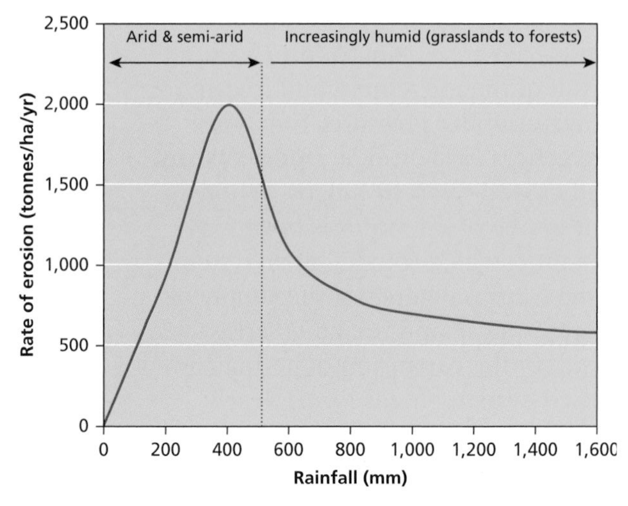

Figure 7.15 Rainfall and soil erosion

- an impermeable surface (in places)
- limited interception (lack of vegetation)
- rainsplash erosion displacing fine particles, which in turn seal off the surface and make it impermeable.

To do:

Referring to figure 7.15:

a Describe the relationship between the rate of soil erosion and the amount of rainfall.

b Suggest reasons for the relationship you have described.

Weathering in deserts

- **Salt crystallization** causes the decomposition of rock by solutions of salt. There are two main types of **salt crystal growth**. First, in areas where temperatures fluctuate around 26–28°C, sodium sulphate (Na_2SO_4) and sodium carbonate (Na_2CO_3) expand by about 300%. This creates pressure on joints, forcing them to crack. Second, when water evaporates, salt crystals may be left behind. As the temperature rises, the salts expand and exert pressure on rock. Both mechanisms are frequent in hot desert regions where low rainfall and high temperatures cause salts to accumulate just below the surface.
- **Disintegration** is found in hot desert areas where there is a large diurnal temperature range. In many desert areas, daytime temperatures exceed 40°C, whereas night-time ones are little above freezing. Rocks heat up by day and contract by night. As rock is a poor conductor of heat, stresses occur only in the outer layers. This causes peeling or **exfoliation** to occur. Griggs (1936) showed that moisture is essential for

Figure 7.16 Desert landscape, Tenerife

this to happen. In the absence of moisture, temperature change alone does not cause rocks to break down. It is possible that the expansion of many salts such as sodium, calcium, potassium and magnesium can be linked with the exfoliation.

Figure 7.17 Desert landscape showing weathering

Arid	Egypt	0.0001–2.0
Semi-arid	Australia	0.6–1.0

Table 7.1 Rates of weathering (mm/yr)

Weathering produces regolith, a superficial and unconsolidated layer above the solid rock. This material is easily transported and eroded, and may be used to erode other materials.

Wind action in deserts

Many of the world's great deserts are dominated by subtropical high-pressure systems. Large areas are affected by trade winds, while local winds are important too. Wind action is important in areas where winds:

- are strong (over 20 kilometres per hour)
- are turbulent
- come largely from a constant direction
- blow for a long period of time.

	Annual temp. (°C)	Annual rainfall (mm)	Processes
Semi-arid	5–30	250–600	Strong wind action, running water
Arid	15–30	0–350	Strong wind action, slight water action

Table 7.2 Peltier's classification of regions and their distinctive processes

Near the surface, wind speed is reduced by friction (but the rougher the ground the more turbulent it becomes).

Sediment is more likely to be moved if there is a lack of vegetation, and it is dry, loose and small. Movement of sediment is induced by **drag** and **lift** forces, but reduced by **particle size** and **friction**. Drag results from differences in pressure on the windward and leeward sides of grains in an airflow.

There are two types of wind erosion.

- **Deflation** is the progressive removal of small material, leaving behind larger materials. This forms a stony desert, or reg. In some cases, deflation may remove sand to form a deflation hollow. One of the best known is the Qattara Depression in Egypt, which reaches a depth of over 130 metres below sea level.
- **Abrasion** is the erosion carried out by wind-borne particles. They act like sandpaper, smoothing surfaces and exploiting weaker rocks.

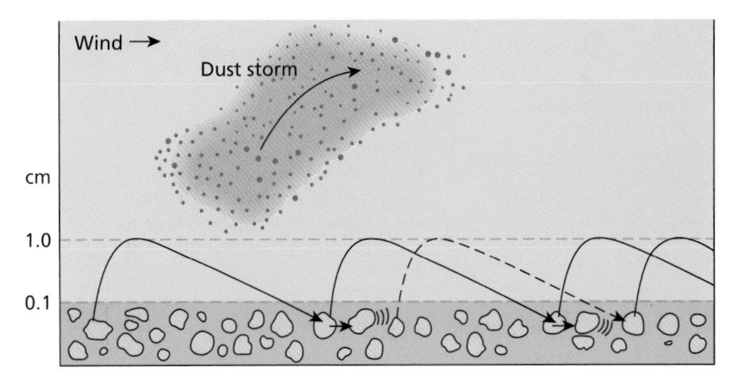

Figure 7.18 Wind action in deserts

Features of the arid landscape

What differentiates deserts and arid landscapes from other ecosystems is not only the extreme climate but also the landforms scattered across their surfaces. The popular assumption is that deserts are largely dunescape, but only a fifth of deserts are covered with sand. Pediments, wadis, canyons, salt pans, alluvial fans and plateaux are among other typical features of arid environments.

To do:

a Distinguish between the processes of weathering and erosion in arid and semi-arid areas. Refer to specific landform features in your answer.

b Distinguish between the terms **exotic**, **endoreic** and **ephemeral** in the context of rivers, and give two examples in each case.

(Note that making a distinction means picking out the differences, not just defining the terms.)

Sand dunes may cover thousands of square kilometres and be up to 500 metres high. They all have a gentle slope on the windward side and a steep slope on the leeward side, but are classified into many types according to shape. Their shape and size depends on the supply of sand, direction of wind, nature of the ground surface, and presence of vegetation. The most common types are barchan or crescent dunes, which are U-shaped. They form around shrubs or large rocks, which hold the main part of the dune in place.

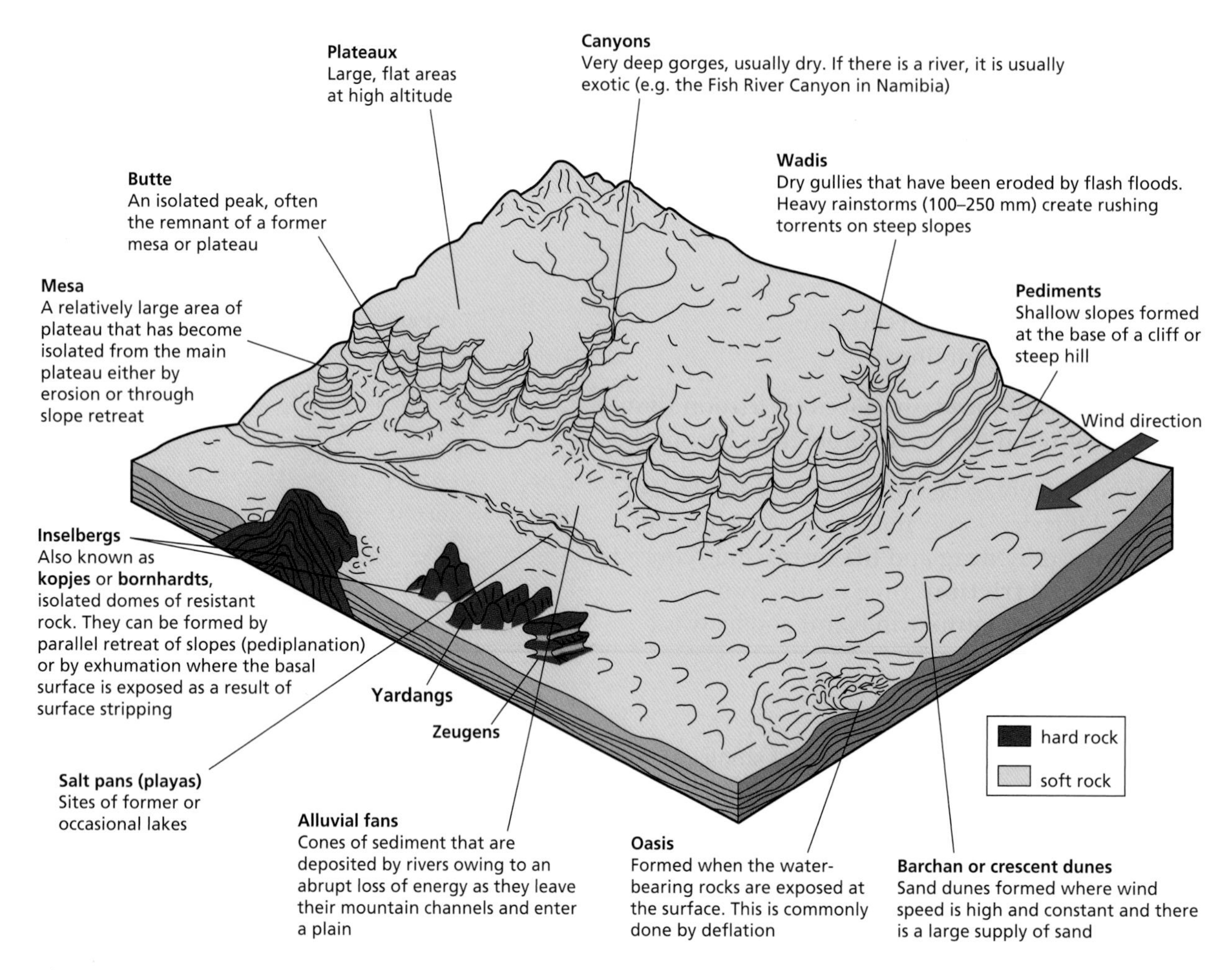

Figure 7.19 Landforms in the arid landscape

Yardangs and **zeugens** are wind-eroded landforms where the softer rock strata are removed, leaving the more resistant layers to form either mushroom-shaped features (zeugens) where strata are horizontal, or long ridges (yardangs) where the strata are vertical. These ridges could be as high as 100 metres and stretch for many kilometres.

Agriculture in arid and semi-arid areas

Agriculture in arid and semi-arid areas is dominated by lack of freshwater, an abundance of heat and sunlight, very low rainfall or a short rainy season, saline soil or water, strong dry winds, poor soil structure, overgrazing, and limited technological development.

The shortage of water and the high temperatures determine many of the characteristic processes in arid and semi-arid areas, as well as many of the characteristics of their soils and ecosystems. All arid and semi-arid areas have a **negative water balance**. That means the outputs from evapotranspiration and stores of water exceed the input from precipitation (pEVT > ppt). The shortage of water can be made up by using irrigation water – that is, by artificially increasing the amount of water that places receive through pipes and other watering systems (centre-pivot irrigation, drip irrigation).

Desert soils are arid (dry) and infertile, due to:

- a low organic content because of the low levels of biomass
- being generally very thin with few minerals
- lack of clay (the amount increases with rainfall)
- not generally being leached because of the low rainfall; hence soluble salts remain in the soil and could be toxic to plants.

Salinization may occur in areas where annual precipitation is less than 250 millimetres. In poorly drained locations like valley floors, and basins in the continental interior, surface runoff evaporates and leaves behind large amounts of bicarbonates. The pH of soils affected by salinization is usually below 8.5. The saline soils adversely affect the growth of most crop plants by reducing the rate of water uptake by roots, and plants die as a result of wilting.

Desertification

Desertification occurs when already fragile land in arid and semi-arid areas is overexploited. Many areas at risk of desertification are in less developed countries, where people depend on farming the land to make a living. Poor management of livestock and other unsustainable farming practices may often be the direct causes of the problem, but the indirect factors leading to desertification may include a combination of population pressure and economic and political issues, as well as global climate change leading to drought. See Figure 7.20.

More material available:

www.oxfordsecondary.co.uk/ibgeography

Case study on desertification in central and southern Europe

TOK Link

How does classification help our understanding of geography?

Figure 7.20 categorizes the causes of desertification into natural processes and human activities. There are many ways of classifying the causes of desertification, and you might think of alternatives. Do we need to classify at all?

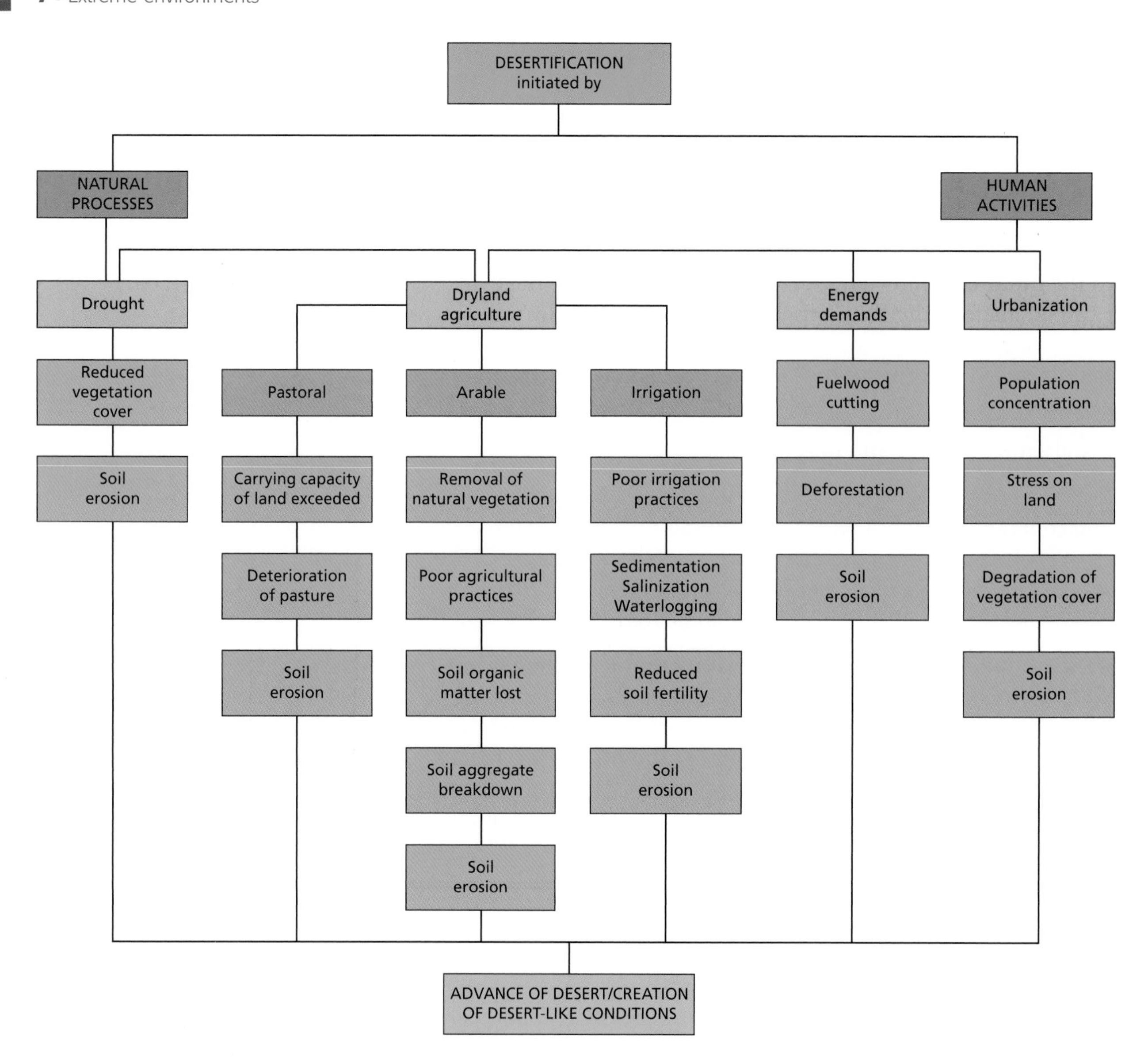

Figure 7.20 The causes and development of desertification

Consequences of desertification

The consequences of desertification may be environmental, economic, and social and cultural.

The **environmental** consequences are:

- soil nutrients lost through wind and water erosion
- changed composition of vegetation and loss of biodiversity, as vegetation is removed
- reduced land available for cropping and pasture
- increased sedimentation of streams because of soil erosion; sediment accumulations in reservoirs
- expanding areas of sand dunes.

The **economic** consequences are:

- reduced income from traditional economy (pastoralism and cultivation of food crops)
- decreased availability of fuelwood, necessitating purchase of oil/kerosene
- increased dependence on food aid
- increased rural poverty.

To do:

a Referring to Figure 7.20, define the terms:

(i) carrying capacity

(ii) soil organic matter

(iii) soil aggregate breakdown

(iv) salinization.

b Explain the relationship between urbanization and desertification.

The **social and cultural** consequences are:

- loss of traditional knowledge and skills
- forced migration due to food scarcity
- social tensions in reception areas for migrants.

The exploitation of periglacial areas

The exploitation of periglacial areas for their mineral and fossil fuel resources creates both opportunities and challenges. Resource development can improve the economies of these regions, but can also put their fragile environment under pressure and create conflicts among local communities.

Fragility of periglacial areas

Periglacial areas are fragile for three reasons: first, the ecosystem is highly susceptible to interference, because of the limited number of species involved. Secondly, the extremely low temperatures limit decomposition, and hence **pollution**, especially from oil, has a very long-lasting effect on periglacial ecosystems. Thirdly, **permafrost** is easily disrupted, posing significant problems. Heat from buildings and pipelines, and changes in the vegetation cover, rapidly destroy it. Thawing of the permafrost increases the active layer, and subsequent settlement of the soil causes subsidence. Consequently, engineers have had to build special structures to cope.

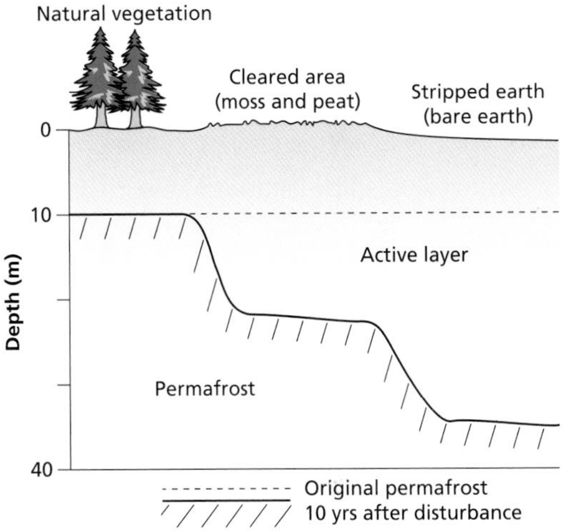

Figure 7.21 Permafrost disrupted

Frost heave

Close to rivers, owing to an abundant supply of water, frost heave is very significant and can lift piles and structures out of the ground. Piles for carrying oil pipelines have therefore needed to be embedded deep in the permafrost to overcome mass movement in the active layer. In Prudhoe Bay, Alaska, they are 11 metres deep. However, this is extremely expensive.

The human impact on periglacial areas

The hazards associated with the use of periglacial areas are diverse and may be intensified by human impact. Problems include mass movements, flooding, thermokarst subsidence, low temperatures, poor soils, a short growing season and a lack of light.

For example, the Nyenski tribe in the Yamal Peninsula of Siberia has suffered as a result of the exploitation of oil and gas. Oil leaks, subsidence of railway lines, destruction of vegetation, decreased fish stocks, pollution of breeding grounds and reduced caribou numbers have all happened directly or indirectly as a result of human attempts to exploit this remote and inhospitable environment.

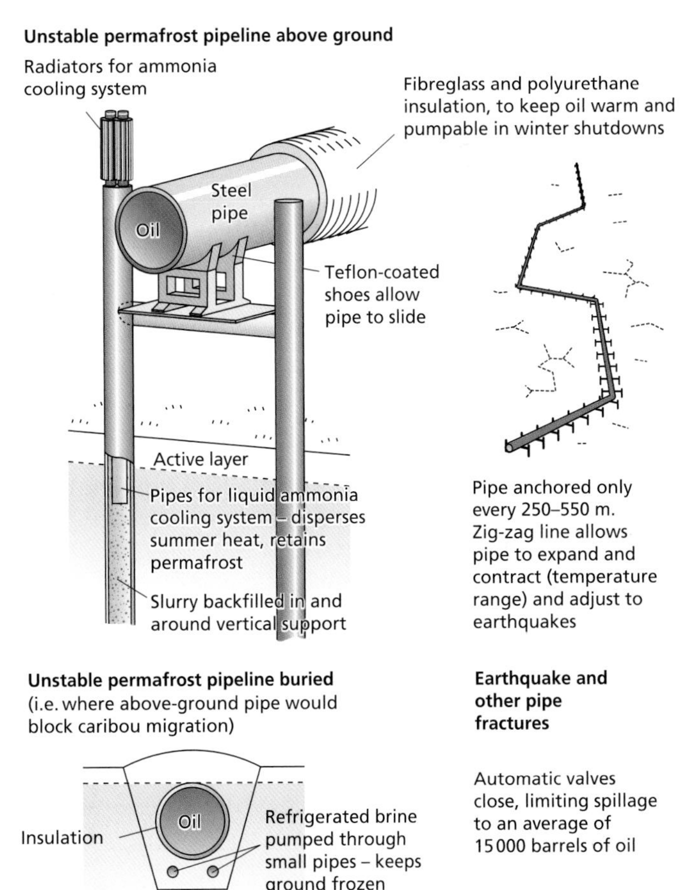

Figure 7.22 Problems with piplines

Case study: Oil drilling in Alaska

Alaska, at 60° north and covering 20° of latitude, has some remarkable periglacial landforms. Climatic conditions range from mild, maritime and wet in the south to arid and very cold in the north. The region covers 40% of the USA's surface and is rich in primary resources; it possesses 20% of the USA's oil, as well as gold and abundant fish and timber reserves. However, the exploitation of these resources is technically difficult and compounded by the possibility of tectonic activity. The cold climate, extensive permafrost and inaccessibility have created difficulties for developers, but exploitation of these resources is justified by the economic benefits.

Heat generated by infrastructure such as central heating, hot water and sewerage systems causes melting of the permafrost below it, disturbing the active layer and causing subsidence. Dark materials with low albedo such as tarmac have the same effect. Drilling for oil generates frictional heat, which also melts the permafrost, and drilling vibration leads to permafrost breakage.

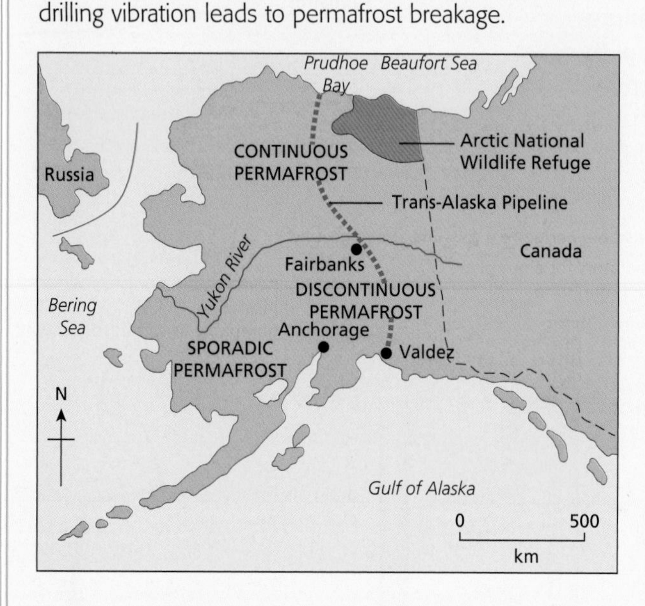

Figure 7.23 Periglaciation and the primary economy

The trans-Alaska pipeline (The Aleyska Pipeline)

Construction began in 1975 and was completed in 1977 at a cost of $8 billion. The 1,300-kilometre pipeline crosses the Brooks and Alaska mountain ranges and 800 rivers and streams. Oil is pumped through the pipeline at 80 °C because air temperatures drop as low as minus 50 °C and the oil needs to flow. To prevent melting of the permafrost, the pipeline sits on top of an insulated support about 3 metres above the ground. The pipeline is also built on sleepers to allow for displacement of 6 metres horizontally and 1.5 metres vertically during an

Figure 7.24 The trans-Alaskan pipeline, on the north slope of the Brooks mountain range

earthquake. Any disruption or oil leakage would be environmentally and ecologically damaging. In Alaska the tundra ecosystem to the northwest and the boreal forest to the east have limited biodiversity and short food chains. These ecosystems are fragile and slow to recover from any kind of trauma, and oil spillage is particularly destructive and long-lasting.

The case of ANWR

The Arctic National Wildlife Refuge (ANWR) is the largest protected wildlife area in the USA, with 9 million hectares in the remote northeast of Alaska. It includes a range of natural habitats such as tundra, boreal forest, barrier islands and coastal lagoons, and is home to many rare birds and animals. Long-running political controversy has arisen over the future of ANWR, because oil companies active in Prudoe Bay to the west want to drill for oil farther east along the northern coastal plain, within the protected area.

To do:

Identify the different arguments both for and against the exploitation of oil and or other minerals in cold environments. Present your arguments through discussion or formal debate by adopting different roles, such as indigenous people, transnational corporation, government and conservationist. Write a summary of the arguments and draw a conclusion.

The exploitation of hot, arid areas

Hot, arid areas are some of the least developed and fragile environments in the world, where most people survive by subsistence farming. Many countries in hot, arid regions want to develop by exploiting their mineral and other resources, but this poses challenges as well as opportunities. Increasing awareness about the sensitivity of arid lands has led to increased monitoring of impacts of human activities. Satellite images allow arid areas to be monitored. Programmes to reverse land degradation have been introduced, and many arid areas are now protected.

The growth of desert tourism

Remote destinations such as deserts are becoming more accessible due to the expansion of global air transport networks and information technology. Deserts are like open-air museums, displaying the artefacts and monuments of ancient civilizations, striking scenery, impressive wildlife and indigenous tribes. All these resources are attracting an ever-increasing number of tourists who are bored with the sun-sea package holiday and are looking for an authentic experience.

Tourism benefits to the host country

Tourism offers many countries with arid climates and areas of desert the opportunity to develop in several respects:

Economy – Tourism generates valuable foreign revenue and can become a springboard for economic development. The tourist industry employs people directly in a range of jobs such as drivers, guides, park rangers and cooks. It also provides indirect employment in other businesses, causing a multiplier effect and revitalizing local economies.

Social welfare – Many arid and semi-arid areas suffer from land degradation and failing agriculture. Alternative employment in desert tourism can stem the outflow of migrants to cities, and relieve rural poverty and urban congestion. Revenue gained from tourism can be invested in local infrastructure and services such as roads, schools and medical services.

Culture – Tourism can also preserve traditions, customs and heritage sites that might otherwise disappear.

Environment – Strict regulation of hunting and the development of game reserves and conservation areas have prevented the disappearance of some species and preserved the biodiversity and uniqueness of desert ecosystems.

International relations – A well-managed tourist industry that involves local communities, conserves local resources and adopts sustainable practices will receive international recognition and respect.

To research

Research pro-poor tourism, and summarize one case study in an extreme environment.

http://www.propoortourism.org.uk/16_stats.pdf

More material available:

www.oxfordsecondary.co.uk/ibgeography

Mining in Botswana and weapons testing in Australia

To do:

Explain what is meant by the terms fragility and resilience in the context of arid or semi-arid areas.

TOK Link

Are deserts becoming more fragile and less productive?

First, find out how fragility and productivity might be assessed or measured. Research the history of desert areas to find out past civilizations, economies and physical characteristics. On the basis of one or more case studies, assess their current state in terms of fragility and productivity relative to that of the past.

To do:

Analyse the benefits of desert tourism from the perspective of:

a the tourist

b the indigenous population

c the government.

TOK Link

Does tourism exceed the carrying capacity of most extreme environments?

This question requires you to investigate the concept of carrying capacity, how it is quantified or measured, and to judge the validity of this question with reference to all three environments studied. Reference to changing capacity may be found in chapters 9 and 11.

Case study: Sustainable tourism in Namibia

Namibia is the most arid country south of the Sahara Desert, receiving an average of only 258 millimetres of rain a year. With a population density of only 2 people per kilometre, it is one of the least populated countries in the world. Namibia is divided into three topographical regions: the western coastal zone; with its intensely dry Namib Desert; the eastern desert zone, and the semi-arid areas of the central plateau, where desertification is a concern and a barrier to economic progress (Figure 7.25). Drought dominates the climate, although some rain occurs during the summer months. Temperatures are highest in summer when they can reach up to 40° C in the daytime. Temperatures along the coast are cooled by the Benguela ocean current (Figure 7.25). With over 300 days of sunshine a year, tourism potential is great. (Figure 2.6).

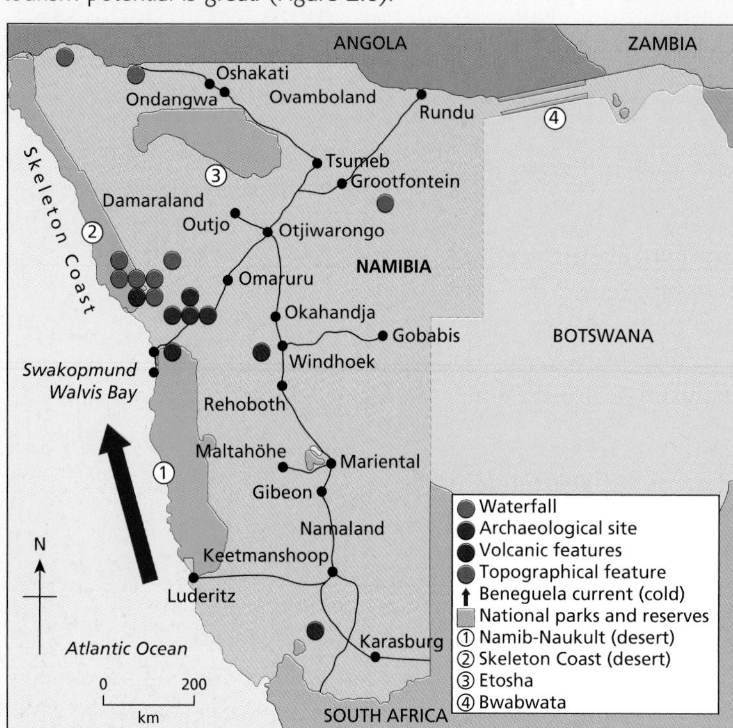

Figure 7.25 Features of Namibia

The importance of tourism to the economy

2009	Namibia	Sub-Saharan Africa	World
% contribution to GDP	13.9	6.7	9.4
% of people employed	17.8	1.7	7.6
% annual growth	7.9	4.6	4

Table 7.3 The importance of tourism in the Namibian economy

The Namibian economy is based on primary activities such as mining (diamonds), agriculture and fishing, which contribute around 40% of GDP in Table 7.3. Tourism now makes a significant contribution to the Namibian economy and has grown steadily since independence in 1990. Namibia has 21 parks and reserves which cover 14% of its land, containing 32 government-owned resorts. The main attractions of Namibia are its natural environment, diverse cultures and archaeological sites. Protected and communal areas each hold about 40% of the wildlife, while commercial farms on private land contain the majority. The privately owned semi-arid to arid range lands have established a new multimillion-pound industry based on both the viewing of game animals on safari and their controlled destruction (known as trophy hunting).

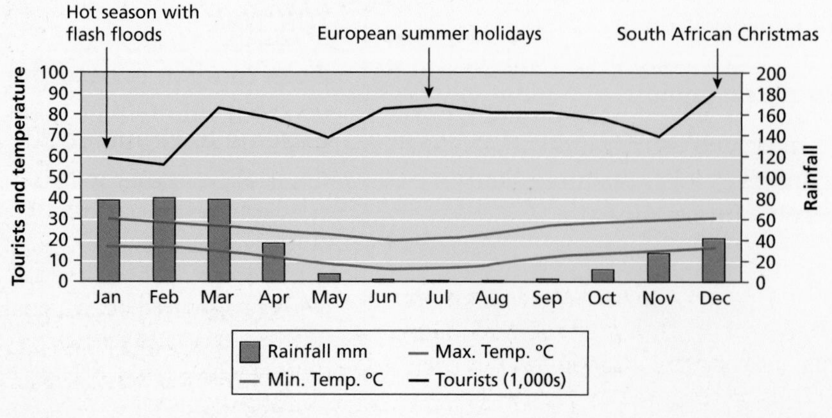

Figure 7.26 Climate in Namibia and the seasonal pattern of international tourist arrivals

To do:

a Referring to Figure 7.26, suggest reasons for the seasonal pattern of international tourist arrivals in Namibia.

b How resilient is the tourist industry to external events? Refer to Figure 7.26 in your answer.

Three regulatory bodies are responsible for the development of tourism in Namibia:

- **MEP** (Ministry for the Environment and Tourism): monitors the impact of tourism on the environment. Current issues concern land degradation and desertification, aridity and water scarcity, and threats to biodiversity
- **NTB** (Namibia Tourism Board): responsible for coordinating and regulating all aspects of the tourist industry including accommodation, transport and catering
- **NACOBTA** (Namibia Community Based Tourism Association): helps to develop local tourist enterprises by providing funding for new ventures, ensuring that the income generated from foreign tourists reaches the local community.

Sustainable management of tourism in Namibia

Tourism has been a relatively recent development in Namibia, and the government has adopted the principles of sustainable management from the start. It is following the recommendations issued by UNEP (United Nations Environment Programme) in 2006.

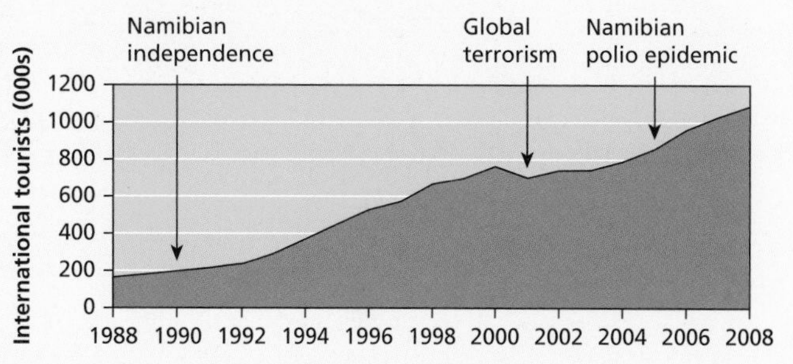

Figure 7.27 Tourism in Namibia, 1988–2008

Issue	UNEP recommendation
Resource overconsumption	
The ecological footprint (resource consumption and waste output) of tourists relative to local people is high; they consume up to three times the amount of water and their demand for hot water may deplete reserves of fuelwood.	Tourists should be encouraged to use alternatives to fuelwood, and use purified water rather than mineral water in plastic bottles. They should also use dry toilets and burn toilet paper with a lighter. Tourists should not use well water, which must be reserved for the nomadic population.
The environment and wildlife	
Desert environments are fragile and exposed, and have a low carrying capacity. Tourists have higher levels of resource consumption (energy and water) and waste output (sewage seepage, vehicle fuels and noise through hot-air balloons, desert vehicles and fly-in aircraft) Refuse disposal systems are lacking and aridity makes decomposition time very slow: • a sheet of paper: 4–9 months • a banana skin: 6–12 months • an aluminium can: 2,000–4,000 years	To prevent the contamination of the water supply, biodegradable detergents should be used instead of chemicals and soap. Tourists should take their non-biodegradable rubbish home, e.g. ointment tubes, aerosols, photographic film packaging. Tourists should be accompanied by guides when touring game reserves, and avoid intruding on animals; and prevent trafficking in protected species. Feeding of game animals is forbidden.
Archaeological sites	
Looters and collectors have used hacksaws to remove paintings and engravings. Religious fanatics and vandals have caused damage.	Group size must be restricted to minimize environmental impact. The impact of tourist numbers at any one site should be restricted by price and timing. Flash photography, the wetting of the pictures to enhance their colour, graffiti, engraving and looting are forbidden. Vehicle-free zones should exist near tourist sites. Qualified guides with specialist local knowledge and expertise should run desert tours.

The indigenous population	
Disrespectful behaviour by tourists through inappropriate dress, offensive gestures and intrusive photography may contribute to changing traditional values. Giving inappropriate "gifts" such as T-shirts and money to children can encourage begging. Tourism may also marginalize local populations and restrict their access to grazing land and water in an environment where population is growing and these resources are under pressure.	Produce should be bought from local communities to maintain their income and avoid leakage. Respect for the natural and cultural heritage of local communities should be given through photos, dress and behaviour. In return, intercultural understanding should be encouraged, and local populations should accommodate tourist culture by adopting hygienic practices and preparing suitable food. Tour operators should provide language training, employ women and draw on the expertise and local knowledge of these communities.

Table 7.4 UNEP principles of sustainable management

Responsible tourism at the local scale

The Himba of Kaokoland

Kaokoland, in remote northern Namibia, is an area measuring roughly 40 000 square kilometres. It has a very low population of under 30 000 inhabitants, who include the Himba people, a semi-nomadic pastoral population. Like the Masai in East Africa, their main food is cow milk and meat, and their cattle are an important symbol of wealth.

The Himbas' beehive huts are made from tree saplings mixed with mud and dung and surrounded by a pen where the animals are kept. These pastoralists maintain traditional dress, language and behavioural codes, which have attracted anthropologists and tourists with a romantic notion of Africa. The preservation of traditional lifestyles is probably a result of the inhospitable landscape in which they live. The Himba plaster their skin and hair with butter and ash to give protection against the sun and keep their skin looking younger. They also wear elaborate jewellery and weave their hair in intricate styles.

Today the Himba way of life is threatened by the proposed Epupa dam scheme, which will impose a sedentary lifestyle upon them if it is constructed. In addition, traders from Angola and Namibia have caused corruption by exchanging Himba cattle for alcohol, and tourists have persuaded some to accept tobacco, cigarettes and sweets in exchange for taking photos of them.

Model tourist accommodation at Damaraland Camp

In 2005 the World Travel and Tourism Council (WTTC) named Damaraland Camp the overall winner of the Tourism for Tomorrow Conservation Award. This award seeks to recognize and promote the world's leading examples of best practice in responsible tourism.

Damaraland Camp is an example of community-based tourism run by the local Damara people and the western operator Wilderness Safaris. The camp was set up in 1996 in the 800-square-kilometre Torra conservancy, 90 kilometres from the Skeleton Coast. The camp consists of en-suite tents built on wooden platforms, shaded and open to the breeze. A verandah overlooks the Haub River Valley to the distant mountains. After the morning's game drive into the valley system, food is served in a bar/dining area constructed from local stone and canvas. The mini-gorge behind the camp has a plunge pool for guests.

This camp is rated as one of the best community projects to sustain itself without donor funding. Profits are now being directed towards mobile clinics, educational materials, running water and anti-poaching patrols. Most of the staff are from nearby communities and trained by Wilderness Safaris. One woman who was once a local goatherd is now the manager of the camp. In addition, Wilderness Safaris puts 10% of its

Figure 7.28 Damaraland, Namibia

income straight back into these communities. Since the project was started in 1996, wildlife populations in the conservancy have doubled.

To do:

- **a** Referring to Figure 7.28, identify the aspects of this tourist accommodation that conform to principles of responsible practice.
- **b** Referring to Figure 7.29:
  - **i** Name and describe the natural features in this landscape.
  - **ii** Explain how the impact of tourists on this landscape can be minimized.

Figure 7.29 Sossusvlei Dunes in the Namib Naukluft Park, Namibia

Issues of sustainability in extreme environments

The exposure of remote and once inaccessible extreme environments has brought significant economic benefits, but environmental costs. All economic activity impacts upon the environment, and tourism more than most. Climate change exacerbates these problems and confronts the governments of these regions with further management challenges. The following case studies are drawn from varied types of extreme environment with different development paths and levels of environmental resilience.

Mountain regions – tourism

Clean air, breathtaking scenery, rare species, cultural interest and heroic potential make mountain zones the second most popular international tourist destination, after coasts. They have become increasingly accessible and attractive to tourists looking for adventure holidays in pristine landscapes. Whereas trekking and skiing have been the traditional attractions, the tourist economy of many mountain areas has diversified to include other activities such as mountain biking, snowboarding, paragliding and whitewater rafting. Indigenous populations and archaeological sites also generate interest.

Tourism is a means of economic development for some of the world's poorest countries, and it can generate growth and reduce regional disparity through the **multiplier effect** and social disparity through the **trickle-down** effect. However, the economic and social success of tourism depends on careful management of the primary resources that the tourists come to see. Mountain environments have a low **carrying capacity** and are sometimes referred to as fragile. This means that the environment is easily damaged by human impact because of steep slopes, thin erodible soil and vegetation that does not regenerate easily.

The tourism industry in many landlocked mountainous countries such as those in the Himalayas has grown dramatically in the last 50 years, The development of mountain tourism therefore brings benefits, but presents some contentious environmental and social issues which are difficult to resolve.

Case study: *The Nepalese Himalayas and the tourist industry*

Nepal's development has been restricted by its remote and landlocked location, difficult terrain and shortage of natural resources for industrialization. Traditionally, its economy has depended on agriculture, with little participation in world trade until tourism began to develop as a major industry in the 1980s. Tourism's annual contribution to GDP is now 6%, it employs 5% of the working population and is predicted to grow at 4% per year. Nepal's unique physical features have attracted increasing numbers of tourists since the 1960s. These include the 10 highest mountains in the world, such as Mount Everest. It is also culturally diverse, with two major religions – Hinduism and Buddhism – existing side by side, and 12 major ethnic groups.

Tourism in Nepal has increased unsteadily since the 1950s; in 1961 4,000 tourists were recorded in Nepal, almost all of them trekkers. It was only after 1990 that Nepal's tourism increased dramatically, until the hijacking of an Indian Airlines flight from Kathmandu in 1999 caused it to decline. The Royal massacre of 2001 and the growing Maoist insurgency pushed it to the brink. Following the signing of a peace agreement in 2006 between the government and the Maoists, tourist demand has been recovering.

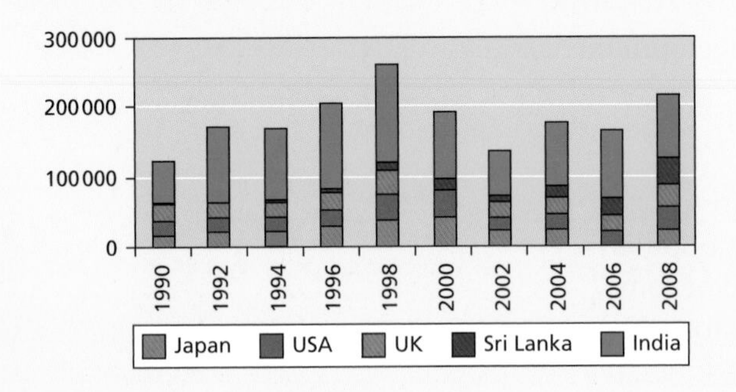

Figure 7.30 International tourist arrivals in Nepal by country of origin, 1990–2008

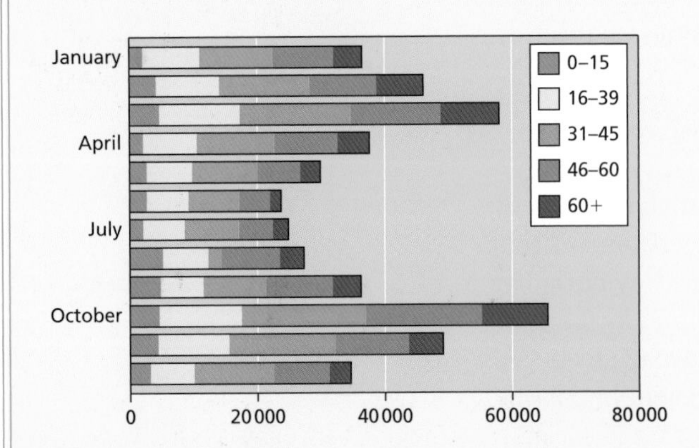

Figure 7.31 Age structure and seasonal trends in tourist arrivals for Nepal, 2008

To do:

a Explain the changes in the origins of international tourist arrivals countries using Figure 7.30.

b Describe the trend shown by Figure 7.31. (Remember to quantify your answer by referring to the data given.)

c Using the information given in both graphs, describe and explain the characteristics of tourists visiting Nepal.

d Suggest how the characteristics of international tourists to Nepal may change in the future.

The benefits of tourism for Nepal

Tourism presents Nepal with many opportunities:

- international recognition and integration into the global economy
- a rise in foreign revenue to help clear national debts
- improvement in local income and employment, generating a multiplier effect, especially in big towns and cities such as Kathmandu and Pokhara.
- improved cultural understanding through the locals' learning of new skills.

The costs of tourism in Nepal

Increasing tourist demand for Nepal presents the following problem:

- Deforestation and road construction disrupt drainage patterns and may cause excessive runoff and flooding downstream.
- Landslides, mudslides and rock falls result from the destabilization of soil and regolith on steep slopes. This is compounded by torrential monsoon rain between June and September, when slopes become unstable.
- Mountain trekking can destroy vegetation and wildlife habitats.
- Trails of litter now line most of the mountaineering, tramping and trekking routes, especially around mountain base camps. Inadequately covered toilet pits and makeshift toilets pollute streams and rivers, and have become a serious health issue.

Case study: Environmental degradation around Mount Everest

The Khumbu region of Nepal, containing Kathmandu and Mt Everest, can comfortably hold about 40 000 people, but during peak tourist season in the lower valley there can be as many as 700 000. Supplying local people with electricity and water is problematic in this mountainous country, but the demands of tourists for hot water and showers put this under further pressure. Increasing resource consumption is matched by increasing waste generation, and the infrastructure is inadequate. The tonnes of rubbish discarded on Everest include climbing equipment, foods, plastics, cans, glass, clothes, papers, tents and even electronic equipment.

In 2008 a geological team, sponsored by the United Nations Environment Programme (UNEP), found signs that the landscape of Mount Everest had changed significantly since Sir Edmund Hillary and Tenzing Norgay first conquered the peak in 1953. Tourism was found to be a major cause of environmental problems, but global warming is likely to have even more far-reaching effects. The government is now willing to consider a strategy to manage trekking in the Everest region now that a 10-year era of political unrest is no longer distracting attention from the environmental issues.

Closing down the mountain is no solution as far as the local Sherpa population is concerned. A Sherpa reaching the summit of Everest will make about $1,600, five times the per capita GNI. Trekking provides this population with the most significant part of their income, and so closing the area around Everest would be devastating to their economy. Critics say it is no surprise that the Nepali authorities have no plans to scale back tourism in the region: even to set foot on the slopes of Everest, each team of seven climbers must pay a royalty of £50,000 to the Nepalese government.

Figure 7.32 Pollution on Mount Everest

Be a critical thinker

Should Mount Everest be closed?

- How do we decide on priorities—the livelihood of the Sherpa v the preservation of the natural environment?
- Is the conquest of Everest more to do with macho neocolonialism than conservation?

The impact of climate change in the Himalayas

If temperatures increase by 1.2 °C by 2050 and 3° by the end of the century, there may be far-reaching consequences for mountainous regions and countries adjoining them or fed by their meltwaters.

The Himalayas have the largest concentration of glaciers outside the polar region. These glaciers are a freshwater reserve; they provide the headwaters for nine major river systems in Asia – a lifeline for almost a third of humanity. There is clear evidence that Himalayan glaciers have been melting at an unprecedented rate in recent decades; this trend causes major changes in freshwater flow regimes and is likely to have a dramatic impact on drinking water supplies, biodiversity, hydropower, industry and agriculture.

One result of glacial retreat has been an increase in the number and size of glacial lakes forming, impounded behind terminal moraines. These in turn give rise to an increase in the potential threat of **glacial lake outburst floods (glofs)** occurring. Such disasters often cross boundaries: the water from a lake in one country threatens the lives and properties of people in another.

Regional cooperation is needed to formulate a strategy to deal effectively both with the risk of outburst floods and with water management issues. Long-term shrinkage of Himalayan glaciers may severely reduce runoff in major rivers such as the Indus, Yellow River and Mekong, with impacts on water supplies.

Experts have identified 25 potentially dangerous glacier lakes in Bhutan, and predict a worst-case scenario of massive flooding at any time down the Punakha-Wangdi valley. UNDP is working to establish an early warning system for glacier lake outburst floods, strengthen disaster risk-management plans, and improve the preparedness of local communities.

In an unprecedented effort, a project has been launched to lower the water level of Lake Thorthormi, a glacial lake ranked as one of the most dangerous in the country. Such a controlled drainage effort requires detailed surveys, an evacuation plan in case the lake collapses, a sound engineering plan and training and evaluation. Once completed, the project will provide valuable experiences to countries with similar problems, such as China, Nepal, Pakistan, India and Chile.

To research

There is convincing evidence for the melting of glaciers in the Himalayan region, but some climatologists dispute the rate of ice-melt in mountainous districts.

Research the following websites to present an argument supporting or rejecting the notions that:

- ice-melt is occurring at an increasing rate, and the implications for adjoining counties could be catastrophic
- the current quantity of glacial ice in the Himalayas is so great that there is no cause for concern.

Visit Times online, "World misled over glacier meltdown", at http://www.timesonline.co.uk/tol/news/environment/and Reuters AlertNet, "UN panel re-examines Himalayan thaw report", at http://mobile.alertnet.org/thenews/newsdesk

See also "Climate change forming dangerous high-altitude lakes" at http://mobile.alertnet.org

Case study: *Coping in semi–arid regions – the African Sahel*

The indigenous people of the semi-arid Sahel region, located between the Sahara desert and the humid subtropics of Africa, have adapted to the extreme environmental conditions by a combination of strategies. As pastoralists, they make use of the limited resources of the Sahel, and combat overgrazing by migrating to areas of seasonal growth while there is an opportunity. In doing so, they tend to leave vegetation around more permanent water sources for times when they will need it later. Such migration patterns also utilize arid areas that are not suitable for cultivation. The livestock herds are diversified – cattle are kept for income in the meat market, sheep and goats for milk, and meat for internal consumption. Herd diversification also allows pastoralists to make use of a greater variety of the available vegetation resources because the animals have different grazing patterns. The diet of the indigenous people varies with conditions. More milk is consumed in the wetter periods, with meat being more common in the drier periods. Their animals are bartered with sedentary farmers for grain.

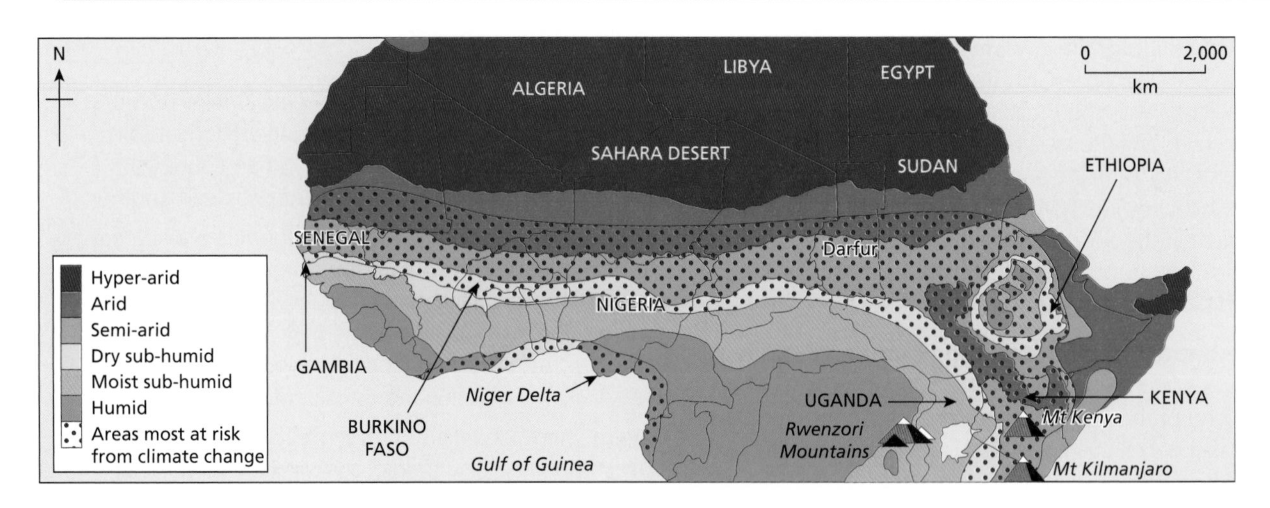

Figure 7.33 Areas or the Sahel at risk of desertification

Semi-arid regions and climate change

Unreliable rainfall and drought are normal features of the climate of semi-arid regions such as the countries of the Sahel. However, the pattern has changed since the 1970s, and droughts have become more frequent, sometimes, as during the 1980s, bringing devastation and famine. These changes in precipitation may be a result of climate change and specifically global warming. Current climate projections predict increases in temperatures that are likely to have environmental and socio-economic consequences, both in the Sahel and beyond. The implications of climate change are serious and far-reaching, and

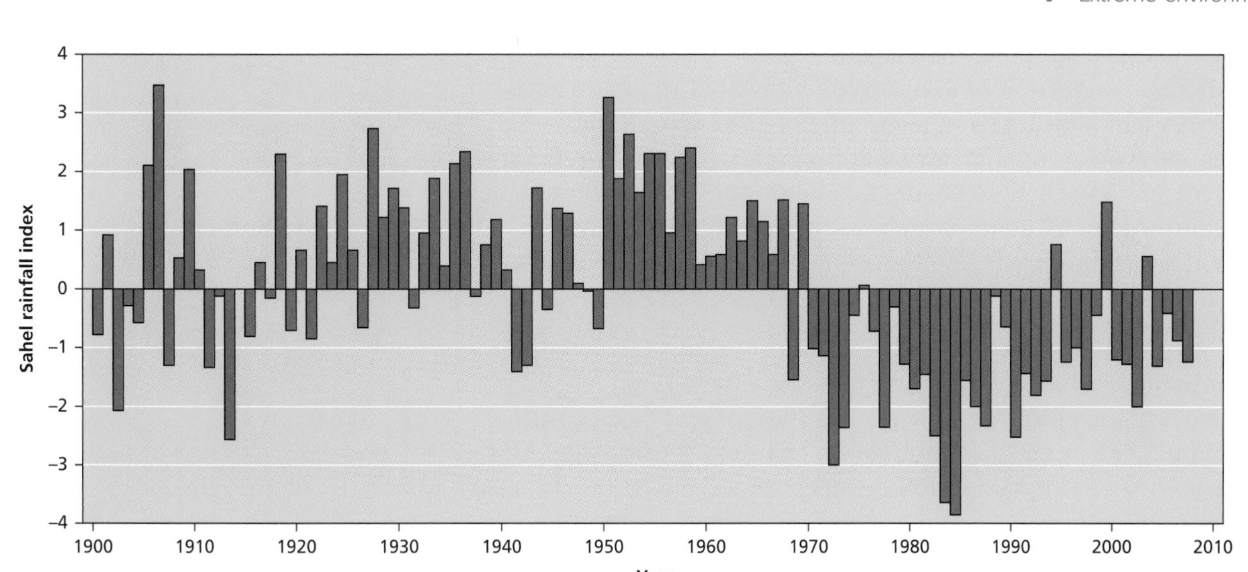

Figure 7.34 Rainfall in the Sahel, 1900–2010

can threaten the lives of individuals and households but also international security.

The primary **environmental effects** of climate change may include:

- reduced agricultural production and increased food insecurity
- additional pressure on water availability, water accessibility and water demand
- flooding of low-lying areas
- changes in humidity likely to encourage the spread of disease, through mosquitoes and other vectors.

The secondary **socio-political effects** of climate change may include:

- conflict over resources such as water and pasture
- loss of territory and border disputes due to transborder migration and pressure on resources, e.g. the conflict between Darfuri tribes in Sudan in 2007
- environmentally induced migration, often resulting in urban overcrowding
- tension over energy supplies resulting from deforestation and competition for charcoal
- international political tension which cannot be resolved without global arbitration.

Climate change and food security

The predicted rise in temperature of the Sahel is likely to be 1–2.75°C, and arid conditions will therefore worsen. Even where precipitation is likely to increase, higher temperatures will raise evaporation levels and reduce its effectiveness. Environmental hazards such as floods and droughts will become more common in the Sahel, causing disruption to agricultural systems and marketing. These hazards are likely to reduce **food production** and availability and increase prices, food insecurity and hunger. Poverty will therefore exacerbate the impacts of climate change in the Sahel region of Africa.

To do:

Study Figure 7.34.

a Describe the trends in rainfall variability shown.

b The rainfall as an average for the period is represented as 0 on the graph. Explain the value of using an index in this way.

c "The extent to which countries accommodate climate change depends more upon human than physical factors." Discuss this view.

TOK Link

- How reliable is statistical data?
- How can we draw conclusions from a limited sample?
- Who collected the data and how reliable was their method?
- Can we draw conclusions from such a short timespan?

To research

Visit http://www.psi.org.uk/ehb/projectskeenan.html for information on the effect of climate change on the Taureg people of the Sahel.

Coping strategies adopted in Senegal

Senegal has a projection of a 20% reduction in rainfall and a 4°C increase in temperature from a baseline of 1961–90. Climate change will worsen food yields by approximately 25% in relation to the current situation, exposing an additional 1–4 million people to food insecurity by 2050. Overgrazing, overpopulation of marginal lands and natural soil erosion are causing desertification. Senegal is investigating various coping strategies that can be applied at the household or local level. These include:

- improving soil fertility by the careful use of fertilizers
- improving the efficiency of irrigation systems and pest control
- adopting water and soil conservation techniques using diguettes (stone rows) to impede runoff on steep slopes.

Coping with water shortages in dry areas

Some solutions are "natural" and require farmers to adapt to the natural environment. Adaptations to water shortages, include:

- increased mobility (the traditional way)
- management of size and composition of herds
- exchange of livestock and livestock products
- increased use of drought-tolerant species
- utilization of wild species and tree crops
- windbreaks to reduce wind erosion of bare soil
- irrigating with silt-laden river water to restore soil
- dune stabilization using straw checkerboards and xerophytes
- land enclosure to reduce wind erosion.

More material available:

www.oxfordsecondary.co.uk/ibgeography

Desertification in southern and central Europe

The Canadian tundra

Until recently, Canada paid little attention to its northern region. Only 104 000 of the country's 33 million people live north of the 60th parallel. Two things are now forcing them to pay attention to the north:

- climate change: the warming climate has made accessible minerals once locked in the ice, just when their prices are high, unleashing an exploration boom
- the people: those who live in the north are demanding and getting more of a say in their future.

There is no dispute that the Arctic is warming. Arctic temperatures have increased at almost twice the global average rate in the last 100 years; 70 000 square kilometres of sea ice (an area about the size of Ireland) are disappearing annually.

A warming climate brings many problems for the Inuit. Unpredictable sea ice can be fatal. Life is becoming more expensive: snowmobiles must take longer routes, buildings are weakened by melting permafrost and, ironically, the local council in Kujuaq felt obliged to buy 10 air-conditioners in 2006 after temperatures reached 31°C.

The effects of climate change – more shipping, mining, and oil and gas exploration – may threaten the environment and with it the Inuit's traditional life, based on hunting and fishing. Some want development, but on their terms. In 2006 Nunavut's economy grew by 5.8%, second only to that of oil-rich Alberta. Much of the boost came from the opening of the territory's first diamond mine. Of the 130 companies exploring in Nunavut in 2009, 32 were looking for uranium. Others seek gold, diamonds, silver, zinc, nickel, copper, iron ore and sapphires.

Case study: *Antarctica – wilderness under threat*

The continent of Antarctica is a unique environment of mountains, plains, ice shelves and ice caps, consisting of some 13.8 million square kilometres of land, of which 68% is covered with ice. Antarctica has 90% of the world's ice, which in many places may be up to 5 kilometres thick. It is the coldest place on earth, with the lowest recorded temperature of –89.2° at Vostok. The average temperatures at the South Pole are –27.5°C in summer and –62°C degrees in winter.

Antarctica possesses valuable resources that belong to no one. Antarctica has no indigenous population, no laws, no government and no economy. While there are no permanent residents in Antarctica, several nations have made overlapping territorial claims. They include Chile, Great Britain, Argentina, Norway, Australia, France and New Zealand. Regulations concerning their rights are set out in the Antarctic Treaty of 1959, which allowed all signatory nations to use the continent for peaceful or scientific purposes only, in a spirit of cooperation.

Scientific research in Antarctica

Antarctica offers scientists the opportunity of working in a pristine environment, which enables them to monitor long-term and current changes in environments that are normally obscured by human activity. Antarctica is a sensitive indicator of global change. The polar ice cap holds within it a record of past atmospheres, which indicate past and present climatic change. Another focus of research is the thinning of the ozone layer and the amount of UV radiation that penetrates the atmosphere. Biodiversity, pollution geology and mineral resources are other topics of scientific interest and research.

In 1994 a further Antarctic Treaty was drawn up in order to prevent international disputes over research activities and resource exploitation, and to make sure that the environment of Antarctica was protected. It reiterated the role of Antarctica as a rural reserve devoted to peace and science, but emphasized the importance of conservation with clauses about waste management and minimizing pollution. Bans on military use are in place, and no mineral resource activity is allowed except for the purpose of scientific research. There are now 60 research bases managed by various nations, and they are making efforts to minimize environmental impact and control their waste output by recycling.

Tourism in Antarctica

Considered as a remote location for so long, Antarctica is becoming more accessible to outsiders. During the 2007–08 tourism season more than 46 000 tourists, mostly from Europe and the USA, made the journey, more than 20 times the number in 1983. Tourists on cruise ships arrive every few days during summer. The ships carry between 40 and 450 passengers and come from South America, Australia and New Zealand. The most popular destinations include Half Moon Island and the Antarctic Peninsula, with their relatively mild climate and abundant wildlife, including penguins and seals. Other attractions include McMurdo Sound and the Commonwealth Bay Sectors of the continent.

Tourism management in Antarctica

Tourism in Antarctica is regulated by the International Association of Antarctic Tour Operators (IAATO). This organization applies strict guidelines to its members,

Guidelines for tourists adopted at the Antarctica Treaty Meeting, Kyoto 1994

These guidelines were devised by IAATO tour operators to help visitors comply with the Protocol on Environmental Protection.

- Protecting Antarctic wildlife – disturbing, feeding, handling of animals is forbidden. Damage to plants, moss beds and lichens is forbidden.
- No guns, explosives, poultry, pet dogs and cats or houseplants allowed.
- Respect for protected areas of ecological, scientific or historic value.
- Respect for scientific research:
  - Camp supplies and equipment to be left untouched.
- Keeping Antarctica pristine:
  - No litter, burning, pollution of lakes and streams, painting, engraving names, graffiti. No geological, biological or historical items to be taken away.
- Safety:
  - Keep a safe distance away from wildlife, both on land and sea.
  - Follow all leaders' instructions.
  - Do not walk on to glaciers or large snowfield without proper equipment, due to the danger of falling in crevasses.
  - Respect smoking restrictions and avoid causing fires.

limiting the size of the ships that may cruise Antarctic waters and the number of people who can be landed at sites around Antarctica at any one time.

The real potential threat from tourists is from non-IAATO ships and tour operators who run cruises with larger ships and greater numbers of people landing.

Figure 7.36 Tourism in Antarctica

Be a critical thinker

Should Antarctica become a world wildlife reserve?

There is concern over the rate of increase in the number of tourists and the impact that they may have. Moss and lichen grow very slowly in the harsh climate and are therefore highly sensitive to trampling. The tourist season during the short summer coincides with the penguins' breeding season, raising concern that visitors may disturb the penguin rookeries. While tourists may only spend a relatively short time on landings, it is by its nature relatively "high-impact" time compared to that of the scientists on a permanent or semi-permanent research station.

8 Hazards and disasters – risk assessment and response

By the end of this chapter you should be able to:

- define and characterize the different types of hazard, including earthquakes, volcanoes, hurricanes, droughts and human-induced technological hazards, and their impact on human populations
- understand why people live in hazardous environments, and the factors that influence their vulnerability to hazard events
- describe how people cope with hazards and disasters, and the different short-term, mid-term and long-term responses to those events.

Hazard – a threat (whether natural or human) that has the potential to cause loss of life, injury, property damage, socio-economic disruption or environmental degradation.
Hazard event – the occurrence (realization) of a hazard, the effects of which change demographic, economic and/or environmental conditions.
Disaster – a major hazard event that causes widespread disruption to a community or region, with significant demographic, economic and/or environmental losses, and which the affected community is unable to deal with adequately without outside help.
Vulnerability – the geographic conditions that increase the susceptibility of a community to a hazard or to the impacts of a hazard event.
Risk – the probability of a hazard event causing harmful consequences (expected losses in terms of death, injuries, property damage, economy and environment).

Environmental hazards exist at the interface of physical geography and human geography. Natural hazard events are often made worse by human activities. In addition, natural environmental conditions affect the outcomes of human-induced hazard events. The principles involved in studying human-induced hazard events are the same as those involved in studying natural hazard events.

This chapter examines the range of human adjustments and responses to hazards and disasters at a variety of scales. The term "natural disaster" is not used, as it does not convey the complex underlying factors that expose populations to risk and subsequently the conditions necessary for a disaster to occur.

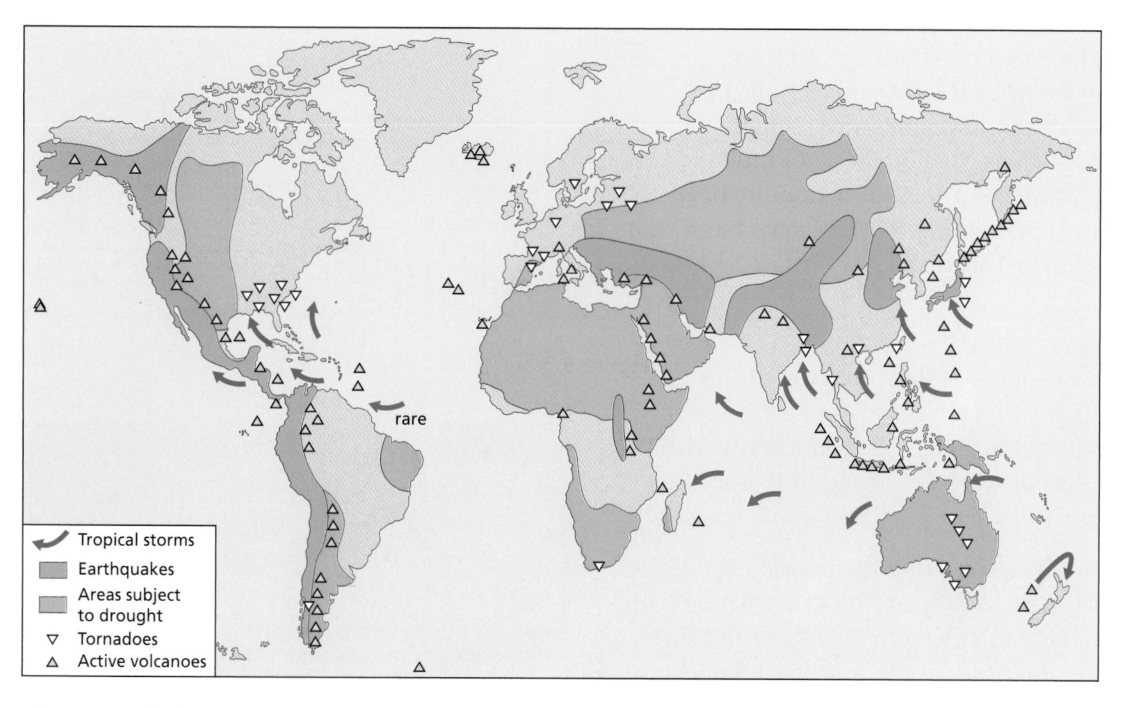

Figure 8.1 The location of natural hazards

Characterizing hazards and disasters

It is possible to characterize hazards and disasters in a number of ways.

1 **Magnitude:** the size of the event, e.g. Force 10 on the Beaufort Scale, which measures wind speed, the maximum height or discharge of a flood, or the size of an earthquake on the Richter scale.
2 **Frequency:** how often an event of a certain size occurs. For example, a flood 1 metre high may occur, on average, every year. By contrast, in the same stream a flood of 2 metres in height might occur only once every 10 years. The frequency is sometimes called the recurrence interval. According to Gumbel's laws of extremes, the larger the event, the less frequently it occurs. Earthquakes with a magnitude of over 8 occur, on average, once a year, whereas those with a magnitude of 3 or 4 occur far more frequently. However, it is the very large events that do most of the damage (to the physical environment and to people, properties and livelihoods, as with the 2004 South Asian tsunami).
3 **Duration:** the length of time that an environmental hazard exists. This varies from a matter of hours, such as with urban smog, to decades, in the case of drought, for example.
4 **Areal extent:** the size of the area covered by the hazard. This can range from very small scale, such as an avalanche chute, to continental, as in the case of drought.
5 **Spatial concentration:** the distribution of hazards over space; whether they are concentrated in certain areas, such as tectonic plate boundaries, coastal locations, valleys and so on.
6 **Speed of onset:** this is rather like the lag time in a flood hydrograph. It is the time difference between the start of the event and the peak of the event. It varies from rapid events, such as the Kobe earthquake, to slow timescale events such as drought in the Sahel of Africa.
7 **Regularity (or temporal spacing):** some hazards, such as tropical cyclones, are regular, whereas others, such as earthquakes, are much more random. Some volcanoes, for example, have a very long return period (e.g. Mt St Helens has a return period of over 120 years).

To do:

a Describe the distribution of earthquakes and volcanoes, as shown in Figure 8.1.

b Compare the locations of areas subject to drought with those that experience tropical storms.

Figure 8.2 Lava flow on Etna

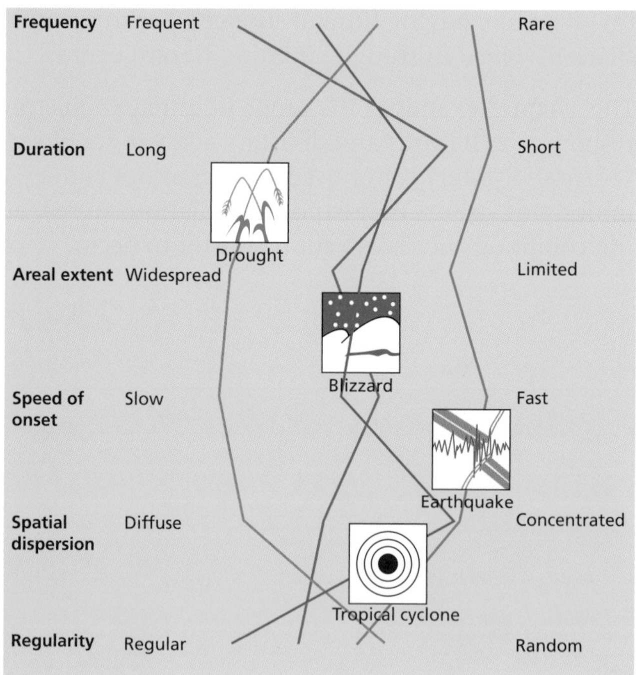

Figure 8.3 The characteristics of hazards

Earthquakes

An earthquake is a sudden, violent shaking of the earth's surface. An earthquake may occur without much warning, and last for tens of seconds. Powerful aftershocks can affect an area for a long time after the main earthquake.

Earthquakes generally occur along the boundaries of the tectonic plates that make up the earth's crust. These plates are continually moving relative to each other, and sometimes pressure builds up at the boundaries and causes rocks and other materials to give way. Earthquakes are

To research

Visit www.intute.ac.uk/sciences/hazards/ for a general introduction to world hazards. This site also provides access to interactive maps of the world (through "World guide"), and allows you to focus on a country of interest to you.

associated with all types of plate boundaries. The **focus** refers to the place beneath the ground where the earthquake takes place. **Deep-focus earthquakes** are associated with plate margins where the oceanic plate is forced under the continental plate, in a process known as subduction. **Shallow-focus earthquakes** are generally located along constructive and conservative boundaries. The **epicentre** is the point on the ground surface immediately above the focus.

Some earthquakes are caused by human activity such as:

- nuclear testing
- building large dams
- drilling for oil
- coal mining.

To do:

a Make a copy of Figure 8.3. Draw a hazards-characteristics curve for a volcano and one for a technological hazard.

b Describe the characteristics of the natural hazard shown in Figure 8.2.

Spatial distribution

Most of the world's earthquakes occur in clearly defined linear patterns. These linear chains generally follow plate boundaries. For example, there is a clear line of earthquake along the centre of the Atlantic Ocean between the African and American plates, and around the Pacific Ocean at the edge of the Pacific plate. In some cases these linear chains are quite broad, particularly along the west coast of South America and around the western Pacific. Broad belts of earthquakes are associated with **destructive plate boundaries** and **subduction zones** (where a dense ocean plate plunges beneath a less dense continental plate), whereas narrower belts of earthquakes are associated with **constructive plate boundaries**, where new material is formed and the plates are moving apart. **Collision boundaries** occur when two plates collide, such as in the Himalayas, and are also associated with broad belts of earthquakes, whereas **conservative plate boundaries**, where the plates are sliding past each other, such as at Califonia's San Andreas fault line, give a relatively narrow belt of earthquakes (this can still be over 100 kilometres wide). In addition, there appear to be isolated occurrences of earthquakes. These may be due to human activities, or isolated plumes of tectonic activity known as hotspots.

Factors affecting earthquake damage

The extent of earthquake damage is influenced by the following:

- **strength and depth of earthquake and number of aftershocks**
  The stronger the earthquake, the more damage it can do, e.g. an earthquake of 6.0 on the Richter Scale is 100 times more powerful than one of 4.0; the more aftershocks there are the greater the damage that is done. Earthquakes that occur close to the surface (shallow-focus earthquakes) are potentially more damaging than earthquakes deep underground (deep-focus earthquakes), since overlying rocks will absorb more of the energy of the latter.
- **population density**
  An earthquake that hits an area of high population density, such as in the Tokyo area of Japan, could inflict far more damage than one which hits an area of low population and building density
- **type of buildings**
  MEDCs generally have better-quality buildings that have been built to be earthquake resistant. (People in MEDCs are also more likely to have insurance cover than those in LEDCs.)

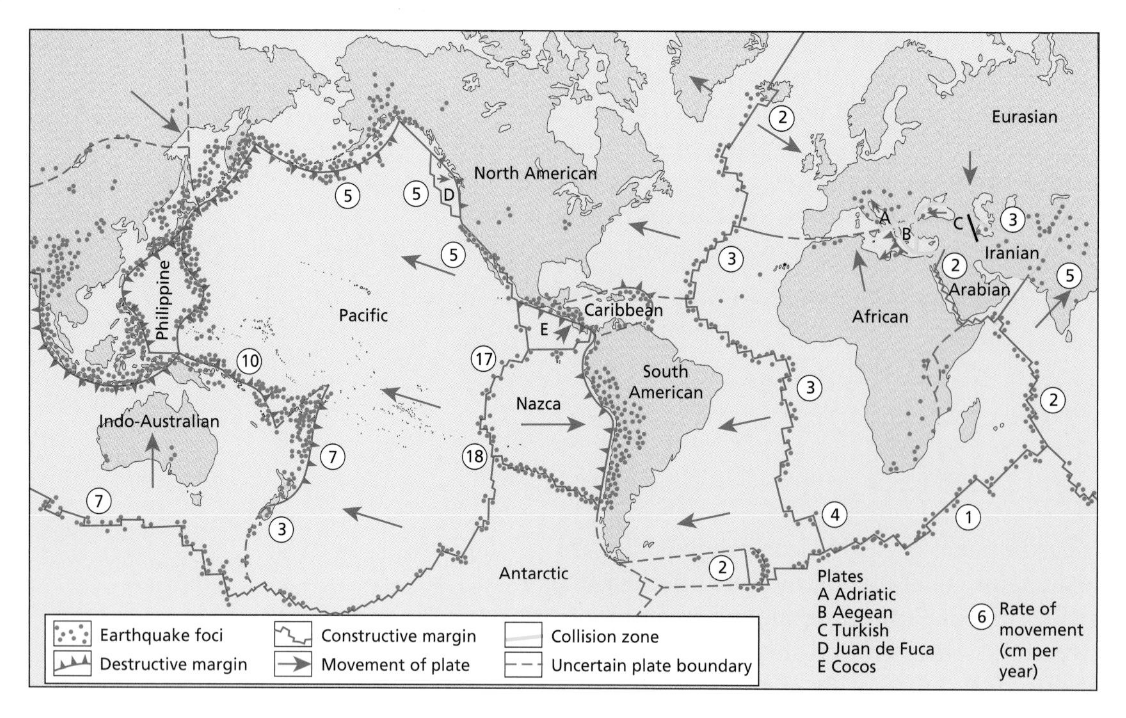

Figure 8.4 The distribution of earthquakes

- **time of day**
  An earthquake during a busy time, such as rush hour, may cause more deaths than one at a quiet time. There are fewer people in industrial and commercial areas on Sundays; at night more people are in their homes.
- **distance from the centre of the earthquake**
  The closer a place is to the centre (epicentre) of the earthquake, the greater the damage that is done.
- **types of rocks and sediments**
  Loose materials may act like liquid when shaken, a process known as **liquefaction**; solid rock is much safer, and buildings built on flat areas of solid rock are more earthquake resistant.
- **secondary hazards**
  These include mudslides and tsunamis (high sea waves), fires, contaminated water, disease, hunger and hypothermia.
- **economic development**
  MEDCs will generally have a better level of preparedness and more effective emergency response services, better access to technology and better health services. The funds to cope with disasters will be greater.

Country	Year	Death toll (est.)	Richter scale
Southeast Asia	2004	248 000	9.1
Haiti	2010	230 000	7.0
Kashmir, Pakistan	2005	86 000	7.6
Bam, Iran	2003	30 000	6.6
Chengdu, China	2008	78 000	7.9

Table 8.1 The world's worst earthquakes by death toll in the 21st century

Dealing with earthquakes

The main options for dealing with the risks of living in an earthquake-prone region include:

- improving forecasting and warning systems
- improving building design, building locations and emergency procedures.

Ways of predicting and monitoring earthquakes include making observations of:

- crustal movement – small-scale movement of plates
- changes in electrical conductivity
- strange and unusual animal behaviour, especially of carp fish
- historic evidence – whether there are trends in the timing of earthquakes in a region.

Case study: *Two contrasting earthquakes – Kobe and Kashmir*

Kobe, Japan, January 1995

The Kobe earthquake was responsible for over 5,000 deaths, 30 000 injuries and for making over 300 000 people homeless. It was caused by the oceanic Pacific plate plunging under the continental Eurasian plate. The earthquake, which registered 7.2 on the Richter scale, struck at 5.46 a.m., when most people were still asleep. Many people were crushed in their beds, although the number of people killed by collapsing motorways was relatively low. Many important buildings, such as the City Hall and public hospitals, were destroyed. Up to 80% of the schools, museums and sports facilities were also destroyed.

The earthquake came as a surprise to Japanese scientists, since the area was considered to be one of the safest for earthquake activity. Conditions for the survivors worsened as rain, strong winds and lightning increased the risk of landslides. Doctors were faced with outbreaks of disease due to the damp, unhygienic conditions. Over 1,300 aftershocks were recorded, and these toppled many buildings. Gas and water pipes were broken and there were 175 separate fires in Kobe. Water supplies to deal with the fire were badly disrupted. Transport and communications were badly affected. A 1-kilometre stretch of elevated highway collapsed, and Japan's bullet train was closed. The damage was estimated at $160 billion.

Muzaffarabad, Kashmir, October 2005

The Kashmir earthquake recorded 7.7 on the Richter scale. There were over 22 aftershocks in the 24 hours after the main earthquake, some of which measured over 6.0 on the Richter scale. The quake and its aftershocks were felt from central Afghanistan to western Bangladesh. The cause of the earthquake was the Indo-Australian plate moving against the Eurasian and Iranian plates.

The epicentre was at the city of Muzaffarabad (Pakistan Kashmir), but buildings were wrecked in an area spanning at least 400 kilometres across, from Jalalabad in Afghanistan to Srinagar in Indian Kashmir. The death toll was over 86 000, at least 100 000 were seriously injured, and more than 3 million people were left homeless. Of the homeless, nearly a million had to sleep in the open. In built-up areas, water and sanitation systems were broken, giving a high risk of an outbreak of disease. There was a desperate shortage of tents capable of withstanding the Kashmiri winter, as well as a lack of blankets, sleeping bags, warm clothes, medicine and food.

Some charities expressed concern that the public might be suffering compassion fatigue. The UN had received just 12% of the $312 million pledged to its emergency appeal, in contrast with 80% of pledges at the same stage after the south Asian tsunami of the previous year. Whatever the reason, the earthquake did not provoke the response from the rest of the world that it needed. Perhaps the tsunami, killing hundreds of thousands, had deadened compassion for the victims of another disaster.

To do:

a Suggest reasons why the Kobe earthquake was so devastating.

b Briefly explain why the Kashmir earthquake had a greater impact than the Kobe earthquake.

Volcanoes

A volcano is an opening or fissure through the earth's crust through which hot molten magma (lava), gases, molten rock and ash erupt. Eruptions happen when pressure beneath the crust builds up, and can cause lateral blasts, flows of lava and hot ash, mudslides, avalanches and floods. An erupting volcano can trigger tsunamis, flash floods, earthquakes and rockfalls. Some eruptions let out so much material that the world's climate is affected for years.

- The greatest volcanic eruption in human history was Tambora in Indonesia in 1815. Some 50–80 cubic kilometres of material were blasted into the atmosphere.
- In 1883 the explosion of Krakatoa was heard as far as 4,776 kilometres away.
- The largest active volcano is Mauna Loa in Hawaii, 120 kilometres long and over 100 kilometres wide.

The distribution of volcanoes

Some volcanoes are on land but many are submarine, and most – but not all – occur at plate boundaries. For example, the volcanoes of Hawaii occur over hotspots, i.e. isolated plumes of rising magma. About three-quarters of the earth's 550 historically active volcanoes lie along the so-called Pacific "Ring of Fire", where the Pacific plate collides with its neighbours. Many of these volcanoes are among the world's most recent eruptions, such as Mt Pinatubo (the Philippines), Mt Unzen (Japan), Mt Agung (Java), Mt Chichon (Mexico), Mt St Helens (USA) and Nevado del Ruiz (Colombia). Other areas of active volcanicity include Iceland, Montserrat in the Caribbean, and Mt Nyiragongo in the Democratic Republic of Congo. Most volcanoes that are studied are above land, but some submarine volcanoes, such as Kick'em Jenny, off Grenada in the Caribbean, are also monitored closely.

To research

Visit http://earthquake.usgs.gov/earthquakes to research current earthquake activity (in the last seven days).

The site is also an excellent starting point for investigating significant earthquakes, including the 2010 Haiti earthquake. Click on the links to Summary and Additional Information.

Also on this site, go to Earthquakes lists and maps and click on "+ 1,000 deaths since 1900". Identify and comment on any pattern in the death rate associated with earthquakes over time.

Magma – molten material inside the earth's interior.

Lava – the magma ejected at the earth's surface through a volcano or crack at the surface.

Chamber – the reservoir of magma deep inside a volcano.

Crater – the depression at the top of a volcano following a volcanic eruption. It may contain a lake.

Vent – the channel through which magma within a volcano reaches the surface during a volcanic eruption.

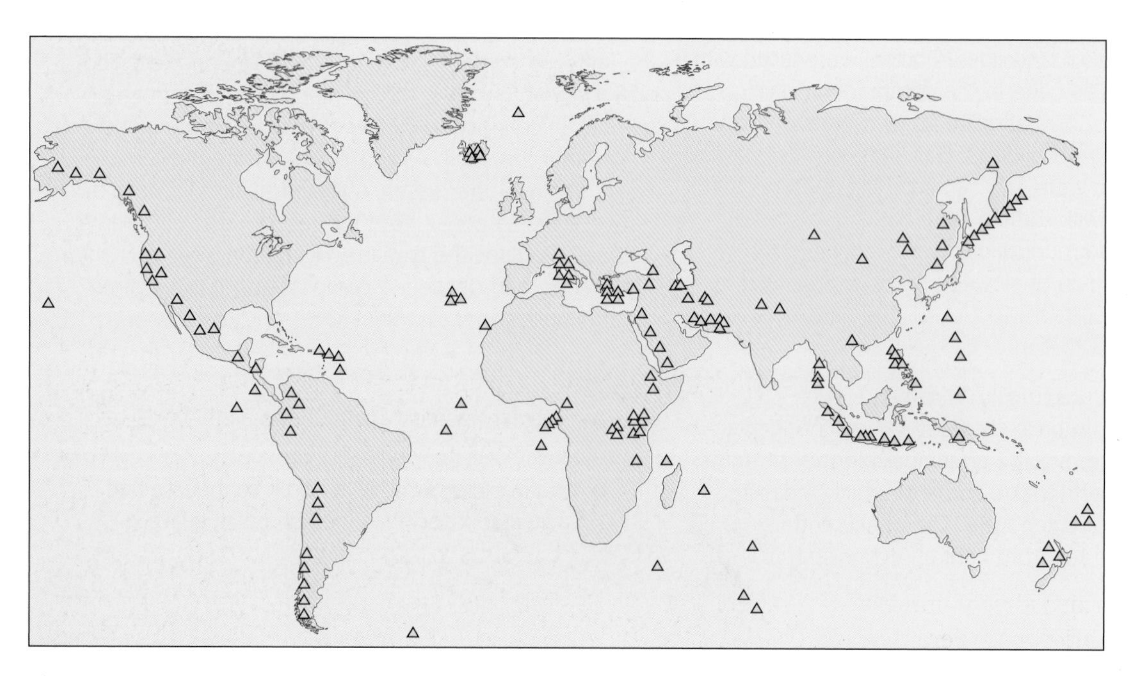

Figure 8.5 The distribution of volcanoes

Predicting volcanic eruptions

Volcanic eruptions are easier to predict than earthquakes, since there are certain signs. The main ways of predicting volcanoes include:

- seismometers, to record swarms of tiny earthquakes that occur as the magma rises
- chemical sensors, to measure increased sulphur levels
- lasers, to detect the physical swelling of the volcano
- ultrasound, to monitor low-frequency waves in the magma resulting from the surge of gas and molten rock, as happened at Pinatubo, El Chichon and St Helens.
- observations, such as at Gunung Agung (Java).

However, it is not always possible to state exactly when volcanic eruption will happen. The USGS successfully predicted the eruption of Mt Pinatubo in 1991, and evacuated the area. However, they unsuccessfully predicted a volcanic eruption at Mammoth Mountain Ski Area in California, USA, and the prediction reduced visitor numbers to the resort and caused economic distress to local business people.

It is also difficult to predict the timescale of an eruption. Some volcanoes may erupt for days, while others go on erupting for years. Mt Pinatubo erupted for a few weeks, while the Soufrière Hills volcano on the island of Montserrat erupted continuously between 1997 and 2005, and still vents ash in small quantities. Mt Etna in Sicily has been erupting intermittently for hundreds of years. In general, volcanoes at destructive plate boundaries tend to produce more explosive volcanoes, whereas those at hotspots, such as Hawaii, produce more frequent but less explosive eruptions.

To do:

a Describe the distribution of volcanoes as shown in Figure 8.5.

b Outline two ways in which a volcano can be considered both a resource and a hazard.

Hawaiian type
Runny basaltic lava which travels down sides in lava flows. Gases escape easily

Plinian type
Gas rushes up through sticky lava and blasts ash and fragments into sky in huge explosion. Gas clouds and lava can also rush down slopes. Part of volcano may be blasted away during eruption

Figure 8.6 Types of volcano

Types of volcano

The shape of a volcano depends on the type of lava it contains. Very hot, runny lava produces gently sloping **shield volcanoes (Hawaiian type)**, while thick material produces **cone-shaped volcanoes (Plinian type)**. These may be the result of many volcanic eruptions over a long period of time. Part of the volcano may be blasted away during eruption. The shape of the volcano also depends on the amount of change there has been since the volcanic eruption. Cone volcanoes are associated with destructive plate boundaries, whereas shield volcanoes are characteristic of constructive boundaries and hotspots (areas of weakness within the middle of a plate).

Volcanic eruptions eject many different types of material. **Pyroclastic flows** are super-hot (700°C) flows of ash and pumice (volcanic rock) moving at speeds of over 500 kilometres per hour. In contrast, a more explosive eruption produces a plume of **ash**, a very fine-grained but very sharp volcanic material that is often blasted far up into the air. **Cinders** are small-sized rocks and coarser ash-like material, also produced in explosive eruptions. The volume of material ejected varies considerably from volcano to volcano.

Eruption	Date	Volume of material ejected
Mt St Helens, USA	1980	1 km-3
Mt Vesuvius, Italy	AD79	3 km-3
Mt Katmai, USA	1912	12 km-3
Mt Krakatoa, Indonesia	1883	18 km-3
Mt Tambora, Indonesia	1815	80 km-3

Table 8.2 The biggest volcanic eruptions

Measuring volcanic strength

The strength of a volcano is measured by the Volcanic Explosive Index (VEI). This is based on the amount of material ejected in the explosion, the height of the resulting cloud, and the amount of damage caused. Any explosion above level 5 is considered to be very large and violent. A VEI 8, or supervolcano, ejects more than 1,000 cubic kilometres of material, 100 times more than a VEI 7. (The last eruption of a supervolcano was 75 000 years ago.)

Living with the volcano

People often choose to live in volcanic areas because they are useful in a variety of ways.

- Some countries such as Iceland and the Philippines were created by volcanic activity.
- Some volcanic soils are rich, deep and fertile, and allow intensive agriculture to take place, for example in Java. (Other countries, such as Sumatra and Iceland, have poor soils, Iceland because the climate is too cool to allow chemical weathering of the lava flows, and Sumatra because its soils are highly leached.)
- Volcanic areas are important for tourism, for example St Lucia and Iceland.
- Some volcanic areas have symbolic value and are part of the national identity, for example Mount Fuji in Japan.

Direct hazards	Indirect hazards	Socio-economic impacts
Pyroclastic flows	Atmospheric ash fallout	Destruction of settlements
Volcanic bombs (projectiles)	Landslides	Loss of life
Lava flows	Tsunamis	Loss of farmland and forests
Ash fallout	Acid rainfall	Destruction of infrastructure – roads, airstrips and port facilities
Volcanic gases		Disruption of communications
Lahars (mudflows)		
Earthquakes		

Table 8.3 Hazards associated with volcanic activity

The ash fallout from the Eyjafjalljokull glacier in Iceland (April 2010) caused widespread disruption to European air travel. No one was killed in the eruption, but the economic cost was great. It had a truly global impact, as countries that traded with the EU were badly affected. (See chapter 15 for a description of the impact of the fallout on the Kenyan flower industry.)

Hurricanes

Hurricanes are intense hazards that bring heavy rainfall, strong winds and high waves, and cause other hazards such as flooding and mudslides. Hurricanes are also characterized by enormous quantities of water. This is due to their origin over moist tropical seas. High-intensity rainfall with large totals of up to 500 millimetres in 24 hours invariably cause flooding. The path of a hurricane is erratic; hence it is not always possible to give more than 12 hours' notice. This is insufficient for proper evacuation measures.

Hurricanes develop as intense low-pressure systems over tropical oceans. Winds spiral rapidly around a calm central area known as the eye. The diameter of the whole hurricane may be as much as 800 kilometres, although the very strong winds that cause most of the damage are found in a narrower belt, up to 300 kilometres wide. In a mature hurricane, pressure may fall to as low as 880 millibars. This very low pressure, and the strong contrast in pressure between the eye and outer part of the hurricane, leads to strong gale-force winds.

To research

Visit Volcano World at http://volcano.oregonstate.edu/ for information on current volcanic activity.

See also the films *Dante's Peak* and *Supervolcano*, which provide excellent examples of some of the potential impacts of volcanic eruptions.

Supervolcano uses footage of many tectonic hazards such as lava flows, lahars, CO_2 clouds from eruptions such as Pinatubo, Montserrat, Nevado del Ruiz, and Lake Nyos in the Cameroon.

Visit www.associateofcontend.com and use the search button to find out which were the worst volcanic eruptions in human history.

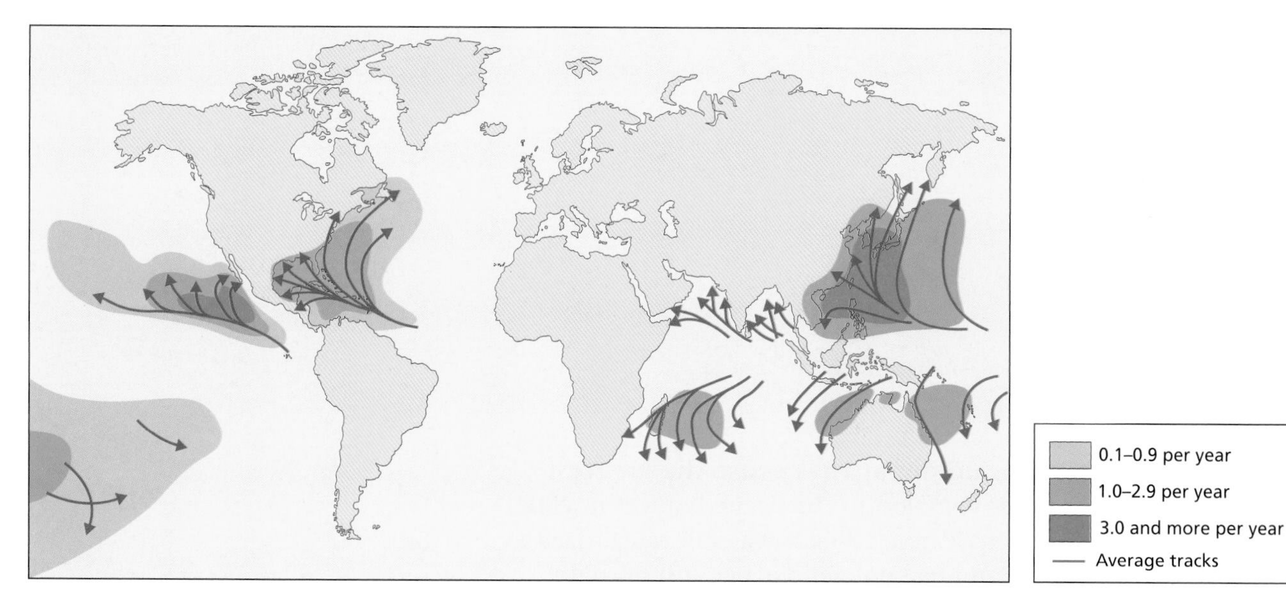

Figure 8.7 The distribution of hurricanes

Hurricanes move excess heat from low latitudes to higher latitudes. They normally develop in the westward-flowing air just north of the equator (known as an easterly wave). They begin life as small-scale tropical depressions, localized areas of low pressure that cause warm air to rise. These trigger thunderstorms that persist for at least 24 hours and may develop into tropical storms, which have greater wind speeds of up to 118 kilometres per hour (74 mph). Only about 10% of tropical disturbances become hurricanes, storms with wind speeds above 118 kilometres per hour (above 74 mph).

For hurricanes to form, a number of conditions are needed.

- Sea temperatures must be over 27°C to a depth of 60 metres (warm water gives off large quantities of heat when it is condensed; this is the heat that drives the hurricane).
- The low pressure area has to be far enough away from the equator so that the Coriolis force (the force caused by the rotation of the earth) creates rotation in the rising air mass; if it is too close to the equator there is insufficient rotation and a hurricane would not develop.
- Conditions must be unstable: some tropical low-pressure systems develop into hurricanes, but not all of them, and scientists are unsure why some do but others do not.

Impacts of hurricanes

The Saffir–Simpson scale, developed by the National Oceanic and Atmospheric Administration, assigns hurricanes to one of five categories of potential disaster. The categories are based on wind intensity. The classification is used for hurricanes forming in the Atlantic and northern Pacific; other areas use different scales.

- **The unpredictability of hurricane paths makes the effective management of hurricanes difficult.** It was fortunate for Jamaica that Hurricane Ivan (2004) suddenly changed course away from the most densely populated parts of the island, where it had been expected to hit. In contrast, it was unfortunate for Florida's Punta Gorda when Hurricane Charley (2004) moved away from its predicted path.

To do:

- **a** Describe the distribution of hurricanes as shown in Figure 8.7.
- **b** Outline the main changes that occur as hurricane intensity increases.
- **c** Suggest reasons why the impacts of hurricanes vary from place to place.

Type	Hurricane category	Damage	Pressure (mb)	Windspeed (km/h)	Storm surge (metres above normal)
Depression	–	–	–	<56	–
Tropical storm	–	–	–	57–118	–
Hurricane	1	Minimal	>980	119–53	1.2–1.5
Hurricane	2	Moderate	965–79	154–77	1.6–2.5
Hurricane	3	Extensive	945–64	178–209	2.6–3.6
Hurricane	4	Extreme	920–44	210–49	3.7–5.5
Hurricane	5	Catastrophic	<920	>250	>5.5

Table 8.4 The Saffir–Simpson scale

- **The strongest storms do not always cause the greatest damage.** Only six lives were lost to Hurricane Frances in 2004, but 2,000 were taken by Jeanne when it was still categorized as just a "tropical storm" and had not yet reached full strength.
- **The distribution of the population throughout the Caribbean islands increases the risk associated with hurricanes.** Much of the population lives in coastal settlements and is exposed to increased sea levels and the risk of flooding.
- **Hazard mitigation depends on the effectiveness of the human response to natural events.** This includes urban planning laws, emergency planning, evacuation measures and relief operations, such as rehousing schemes and the distribution of food aid and clean water.
- **LEDCs continue to lose more lives to natural hazards, as a result of inadequate planning and preparation.** By contrast, insurance costs continue to be greatest in American states such as Florida, where multi-million-pound waterfront homes proliferate.

Case study: Hurricane Katrina

Hurricane Katrina was the USA's worst natural disaster in living memory. The storm hit land near New Orleans on 29 August 2005 at a speed of some 225 kilometres per hour. Katrina was a category 4 hurricane, but what set it apart from other hurricanes was the way it lingered rather than passing through. Unlike most hurricanes of this intensity, which are over in a relatively brief time, Katrina continued to build. The battering winds were not the only danger: the low pressure at the centre of the hurricane and the high winds made the ocean rise up by as much as 9 metres in places. (The maximum height of the Asian tsunami was 10 metres.)

New Orleans is located on the Mississippi delta in the Gulf of Mexico, and most of the city is below sea level. The hurricane flooded 80% of the city, killing over 1,800 people and damaging 204000 homes, making over 800000 people homeless. Economists suggest that Hurricane Katrina cost the US economy $80 billion. The rescue operation was criticized for not doing enough to help the most vulnerable members of the population (see also page 214).

New Orleans can no longer be protected from hurricane storm surges, according to the US army general in charge of the city's defences. In 2010 he claimed that although it was possible to develop better early-warning systems, better evacuation plans and better levées to hold back most of the water, it is not possible to prevent levées being overtopped and the city flooded. He declined to say whether this meant the city should be abandoned altogether and relocated inland.

The vast Mississippi delta is sinking by a centimetre a year. Sea levels are rising at an accelerating rate, and will be 2 metres higher by the year 2100. Much of the delta is less than a metre above sea level, so most communities will be submerged. In the last 50 years 6,000 hectares of marsh and swamp have been lost because of salt-water intrusion. In the four-month hurricane season, land disappears at the rate of 0.5 hectares every six minutes, or 6,500 to 10 000 hectares a year. The Office of Coastal Protection is responsible for integrating hurricane protection, storm damage reduction, flood control, coastal protection and restoration in the state of Louisiana. The Office is spending $1.5 billion (about £915 million) over four years on wetland restoration. Another $14.3 billion is being spent on new levées and defences for New Orleans. The cost of diverting the Mississippi to save the delta's wetlands and its settlements from sinking would be $200 billion.

Case study: *Cyclone Nargis*

In May 2008 Cyclone Nargis struck Burma. Winds exceeding 190 kilometres per hour (118 mph) and torrential rain devastated the area, killing some 134 000 people. As many as 95% of all buildings in the affected area were demolished by the cyclone and the resulting floods. The Burmese government identified 15 townships in the Irrawaddy delta that had suffered the worst. Seven of them had lost 90–95% of their homes, with 70% of their population dead or missing. International frustration mounted as disaster management experts failed to get the necessary visas to enter the country.

The low-lying land in the Irrawaddy delta is home to an estimated seven million of Burma's 53 million people. Nearly two million of the densely packed area's inhabitants live on land that is less than five metres above sea level, leaving them extremely vulnerable to flooding. As well as the cost in lives and homes is the cost of losing valuable agricultural land in the fertile delta, considered Burma's rice bowl.

To research

Visit www.nhc.noaa.gov for the National Hurricane Centre (USA).

http://www.nasa.gov/mission_pages has excellent information on Cyclone Nargis as well as satellite images, maps and photographs. (Go to Hurricanes, then Archives 2008.)

http://news.bbc.co.uk/1/shared/spl/hi/americas/05/katrina/html/ for a case study on Katrina, including animations on how hurricanes form, and the passage of Katrina over New Orleans.

Drought

A drought is an extended period of dry weather leading to extremely dry conditions. The definition of drought depends on the culture defining it. In the UK, for example, **absolute drought** is often defined as a period of at least 15 consecutive days with less than 0.2 millimetres of rainfall. **Partial drought** is defined as a period of at least 29 consecutive days during which average daily rainfall does not exceed 0.2 millimetres. In the longer term, drought in the UK is also defined as a 50% deficit over three months, or a 15% shortfall over two years. According to UNEP, a drought occurs when there are two or more years with rainfall substantially below the mean.

Figure 8.8 The impact of drought

A large proportion of the world's surface experiences dry conditions. Semi-arid areas are commonly defined as having a rainfall of less than 500 millimetres per annum, while arid areas have less than 250 millimetres, and extremely arid areas less than 125 millimetres per annum. Millet is the only crop that will grow in arid areas (less than 250 millimetres of annual rainfall); below that, no sedentary farming is possible. In addition to low rainfall, dry areas have **variable rainfall**. As rainfall total decreases, variability increases. For example, areas with an annual rainfall of 500 millimetres have an variability of about 33%. This means that in such areas rainfall could range from 330 to 670 millimetres. This variability has important consequences for vegetation cover, farming and the risk of flooding.

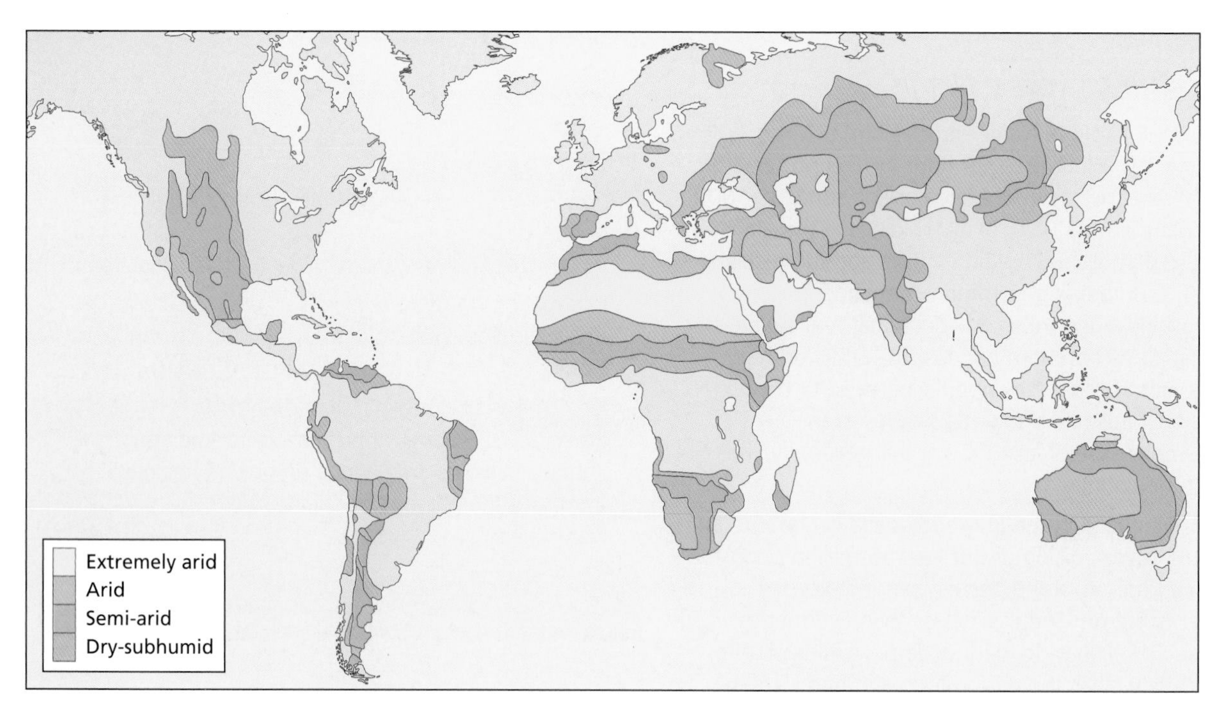

Figure 8.9 The distribution of dry regions

The causes of arid conditions

Permanently arid conditions are caused by a number of factors.

- The main cause is the global atmospheric circulation. Dry, descending air associated with the **subtropical high-pressure belt** is the main cause of aridity around 20°–30° N.
- In addition, distance from sea, **continentality**, limits the amount of water carried across by winds.
- In other areas, such as the Atacama and Namib deserts, **cold offshore currents** limit the amount of condensation into the overlying air.
- Others are caused by intense **rain-shadow effects**, as air passes over mountains.
- Human activities may also cause drought by practices such as deforestation leading to erosion, followed by the spread of desert conditions into areas previously fit for agriculture. This is known as desertification, and is an increasing problem.

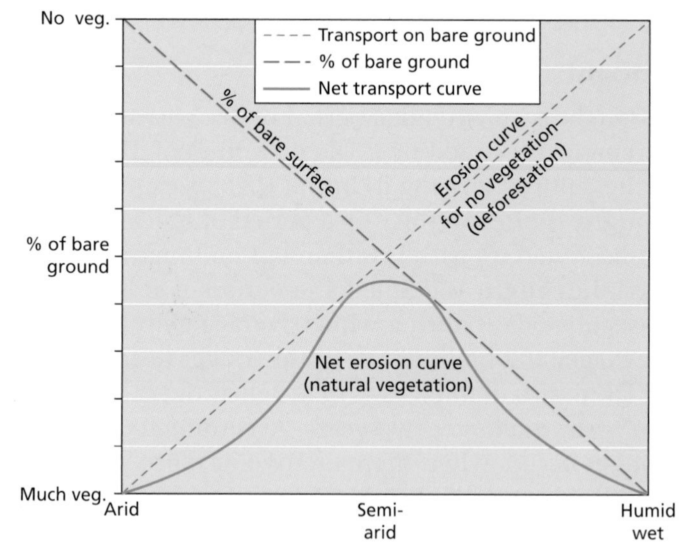

Rates of transport (and therefore denudation) tend to be highest in semi-arid areas and especially in more humid areas where vegetation is removed

Figure 8.10 Rates of transport in relation to aridity

Short-term droughts can be caused by El Niño events (see page 140), particularly in Australia and Indonesia. China was also affected by prolonged drought in 1997.

Case study: Europe's drought of 2003

The 2003 European heatwave was one of the hottest on record in Europe. Estimates for the death toll were as high as 30 000, nearly half of them in France. In France, harvests were down by 30–50% on 2002. France's electricity grid was affected, as demand for electricity soared as the population turned up air conditioning and fridges. At the same time, nuclear power stations, which generate around 75% of France's electricity, were operating at a much reduced capacity because less water was available for cooling.

Portugal declared a state of emergency after the worst forest fires for 30 years. Temperatures reached 43°C in Lisbon in August 2003: 15°C hotter than the average for the month. Over 1,300 deaths occurred in the first half of August, and up to 35 000 hectares of forest, farmland and scrub were burned. Some fires were deliberately started by arsonists seeking insurance or compensation money, and over 70 people were arrested.

The prolonged heatwave left some countries facing their worst harvests since the end of the Second World War. Some countries that usually export food were forced to import it for the first time in decades. Across the EU, wheat production was down 10 million tonnes, about 10%.

Case study: Drought in Africa

In 2003 parts of southern Ethiopia experienced the longest drought anyone had known. The world's largest emergency food aid programme was in operation, but it proved inadequate. Because of a sixth poor rainy season in three years, 20 million people needed help. The situation was worse than the 1984 famine, when 10 million people lacked food.

People under threat of famine in Africa	
Ethiopia	20 million
Zimbabwe	7 million
Malawi	3.2 million
Sudan	2.9 million
Zambia	2.7 million
Angola	1.9 million
Eritrea	1 million
plus around 7.3 million across Swaziland, Congo, Uganda, Congo-Brazzaville, Lesotho and Mozambique	

Table 8.5 Africa's "at risk" population

Hazards related to drought

A variety of hazards affect dry areas. These include:

- declining water resources and food shortages
- flooding of valleys, alluvial fans and plains (playas)
- increased soil erosion and gulleying
- surface subsidence due to water abstraction
- sedimentation or deposition of river sediments
- landslides and rockfalls
- weathering.

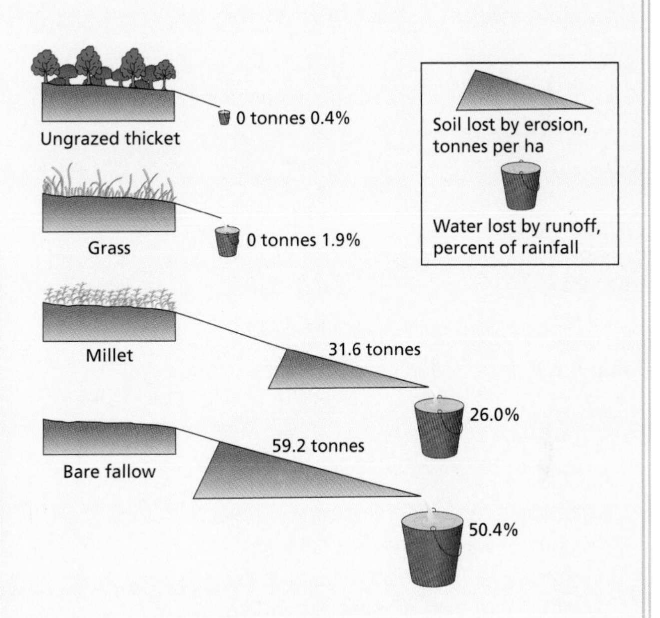

Figure 8.11 Soil and water losses

According to some geographers, erosion is most likely in semi-arid rather than arid or extremely arid areas. The relative lack of vegetation in a semi-arid region means that when it does rain, a high proportion of rain will hit bare ground, compact it, and lead to high rates of overland runoff. By contrast, in much wetter areas such as rainforests, the vegetation intercepts much of the rainfall and reduces the impact of rainsplash. At the other extreme, areas that are completely dry do not receive enough rain to produce much runoff. Hence, it is the areas with more variable rainfall (and variable vegetation cover) that experience the highest rates of erosion and runoff. Moreover, as the type of agriculture changes, the rate of erosion and overland runoff change. Under intensely dry conditions this causes gulley formation.

To do:

- **a** Define the main types of dry regions as shown in Figure 8.9. Describe the distribution of arid and extremely arid areas.
- **b** Explain why soil erosion may be high in semi-arid regions.
- **c** Comment on the relationship between land-use cover and soil erosion.
- **d** Outline the main hazards associated with drought.

Technological hazards

Technological hazards are extremely wide-ranging, and include war, nuclear (radioactive) material, oil spills, industrial accidents, and contamination of water and soil. In these hazards, the misuse of technology endangers lives and property. It is generally people, not the technology, that have caused the disaster.

Class	Examples
Multiple extreme hazards	Nuclear war (radiation), recombinant DNA, pesticides
Extreme hazards	
Intentional biocides	Antibiotics, vaccines
Persistent teratogens	Uranium mining, rubber manufacture
Rare catastrophes	LNG explosions, commercial aviation (crashes)
Common killers	Auto crashes, coal mining (black lung)
Diffuse global threats	Fossil fuel (CO_2 release), sea surface temperatures (ozone depletion)
Hazards	Saccharin, aspirin, appliances, skateboards, bicycles

(Source Smith, K. *Environmental hazards.* Routledge 1992)

Table 8.6 A classification of technological hazards

To research

Visit http://www.guardian.co.uk/environment/2010/jan/22/kenya-drought-insurance to see how Kenyan farmers are being helped during recent droughts.

Visit http://ochaonline.un.org/OchaLinkClick.aspx?link=ocha&docId=1087247 for a map of drought in Africa, 1980–2001.

How does the map on this site compare with the map of the world's dry regions? What does this tell you about the distribution of drought in Africa?

More material available:

www.oxfordsecondary.co.uk/ibgeography

Industrial pollution in Bhopal

Case study: Pollution in Iraq

According to an official Iraqi study, three decades of war and neglect have left environmental ruin in large parts of the country. More than 40 sites across Iraq are believed to be contaminated with high levels or radiation and dioxins. Areas in and near Iraq's largest towns and cities, including Basra and Falluja, account for around 25% of the contaminated sites, which appear to coincide with communities that have seen increased rates of cancer and birth defects since 2005. Scrap-metal yards in and around Baghdad and Basra contain high levels of ionizing radiation, which is thought to be a legacy of depleted uranium used in munitions during the first Gulf War and since the 2003 invasion.

High levels of dioxins on agricultural land in southern Iraq are thought to be a key factor in a general decline in the health of people living in the poorest parts of the

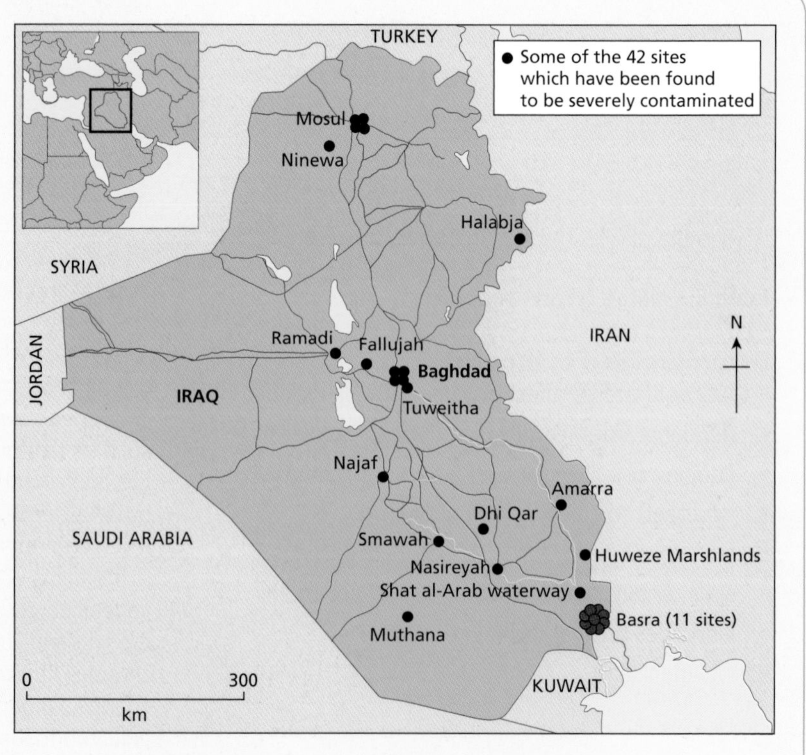

Figure 8.12 Pollution in Iraq

country. For example, Basra has some heavily polluted areas. It has been a battlefield for two wars, the Gulf War and the Iran–Iraq War, where many kinds of bombs were used. In addition, oil pipelines were bombed and most of the contamination settled in and around Basra.

In Falluja there are claims of a large rise in birth defects, notably neural tube defects which affect the spinal cord and brain. Although depleted uranium was used in the area, the city also lacks a sewerage system, which means that household waste and sewage leaks directly on to the streets or into storm drains discharging into the river Euphrates, used for drinking water. Clear problems of contamination create sickness that may have directly affected genetics, but proving causation is difficult.

Ten areas have been classified by Iraq's nuclear decommissioning body as having high levels of radiation. They include the sites of three former nuclear reactors on the southeastern outskirts of Baghdad, as well as former research centres around the capital that were either bombed or dismantled between the two Gulf wars.

Scrap sites remain a prime concern. Wastelands of rusting cars and war damage dot Baghdad and other cities between the capital and Basra, offering unchecked access to both children and scavengers.

Iraq's environmental degradation is being intensified by an acute drought and water shortages across the country, which have seen a 70% decrease in the volume of water flowing through the Euphrates and Tigris rivers. Much of the water reaching Iraq has first been used by Turkey and Syria for power generation. When it reaches Iraq it is of poor quality and often contaminated. This water is used for agriculture as well as for drinking water.

Vulnerable populations: why people live in hazardous areas

A hazard is a perceived natural event that threatens both life and property, whereas a disaster is the realization of this hazard. A distinction can therefore be made between extreme events in nature, which are not environmental hazards (because people and/or property are not at risk), and environmental hazards, in which people and/or property are at risk.

Environmental hazards are caused by people's use of dangerous environments. A large part of an environmental hazard is therefore human behaviour, namely the failure to recognize the potential hazard and act accordingly. Hence the term "natural hazard" is not a precise description, as natural hazards are not just the result of "natural" events.

Why do the poor often live in hazardous environments?

Environmental hazards occur only when people and property are at risk. Although the cause of the hazard may be geophysical or biological, this is only part of the explanation. It is because people live in hazardous areas that hazards occur. So why do they live in such places?

The **behavioural** school of thought considers that environmental hazards are the result of natural events. People put themselves at risk by, for example, living in floodplains. By contrast, the **structuralist** school of thought stresses the constraints placed on (poor) people by the prevailing social and political system of the country. Poor people live in unsafe areas – such as steep slopes or floodplains – because they are prevented from living in better areas. This school of thought provides a link between environmental hazards and the underdevelopment and economic dependency of many developing countries. Even in rich countries such as the USA, certain populations may be more at risk than others; for example, the impacts of Hurricane Katrina were greater on the poor, black population of the affected region than on other sections of society.

To research

Visit www.lenntech.com and type "environmental effects of warfare" into the search button. This site also has excellent links to other sites (in the bibliography).

Visit www.bhopal.com for the Union Carbide Bhopal website, and http://bhopal.org for the view of those affected by the disaster.

There are many video clips on You Tube; go to www.youtube.com and search for "Bhopal".

To do:

a Compare and contrast the causes of pollution in Bhopal with those of Iraq.

b Briefly explain who you think is responsible for the pollution in (i) Bhopal and (ii) Iraq. Outline the implications of your answer to parts (i) and (ii).

Resource or hazard?

People choose to live in certain environments because of the resources they offer. Millions of people live in river floodplains, since during a "normal" year the river does not flood, and it provides many resources, such as water, silt and opportunities for transport and recreation (Figure 8.13). River deltas are particularly rich in water, silt and fertile soils, and the potential for trade and communications. Flood events on the scale of the 2008 Burma floods in the Irrawady delta (see page 207) and those of 2005 in the Mississippi delta caused by Hurricane Katrina (see page 206) are rare. Most of the time, water levels operate at a level where they can be considered an important resource.

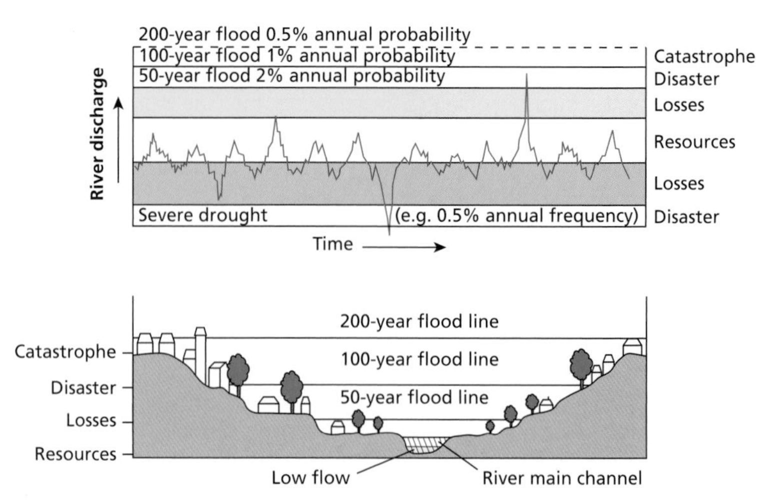

Figure 8.13 Flood recurrence intervals

Too little water is just as much of a problem as too much water, and when rivers dry up competition increases for the water that is available, and this can lead to conflict, desertification or famine. The more extreme the hazard event, the greater the problem will be. For example, a 1:200-year event is worse than a 1:50-year event, and if there are people living in the area affected, the event is likely to be disastrous.

To research

Visit http://usgs.gov and search for "The 100-year flood" to find out about 100-year floods and recurrence intervals.

The same is true for volcanic environments. These may provide rich, productive soils, minerals to mine, and an attraction for tourists. Some of the earth's most fertile land is formed from the breakdown of volcanic material around the base of volcanoes. However, when a volcano is erupting, lives are threatened and it may be necessary to evacuate, as in the case of Plymouth in Montserrat (1997) and Chaiten, Chile (2008).

Changing patterns?

In some locations, the sheer number of people triggers hazards. For example, in megacities the volume of vehicles on roads almost inevitably causes air quality to decline. The concentration of manufacturing industry in certain regions (such as southeast China and southeast India) is also linked to a decline in air quality, increasing water pollution and increased acidification. As more people move into urban areas – whether into slums or formal housing – the risk of hazards increases, since rapid urbanization tends to be poorly managed, with more alteration of the natural habitat.

In some areas, changing climate patterns are putting people at risk. For example, in southern Spain and Portugal, increasingly dry years are turning large areas into desert. This natural process is compounded by overuse of water for golf courses and recreational facilities. Consequently, groundwater levels are declining, soils are drying, vegetation is dying and the region is becoming desertified. This leads to increased risk of wind and water erosion, and further declines in productivity.

To do:

a Outline the differences between the behavioural school of thought and the structuralist school of thought in relation to hazards.

b Outline the ways in which rivers can be considered (i) resources and (ii) hazards.

Vulnerability

The concept of vulnerability encompasses not only the physical effects of a natural hazard but also the status of people and property in the affected area. Many factors can increase one's vulnerability to natural hazards, especially catastrophic events. Aside from the simple fact of living in a hazardous area, vulnerability depends on:

Vulnerability – the conditions that increase the susceptibility of a community to a hazard or to the impacts of a hazard event.

- **population density:** many rapidly growing cities are in hazardous areas; large urban areas such as New Orleans are especially vulnerable to natural hazards
- **understanding of the area:** recent migrants into shanty towns may be unaware of some of the natural hazards posed by that environment
- **public education:** educational programmes in Japan have helped reduce the number of deaths in earthquakes
- **awareness of hazards:** the 2004 tsunami in South Asia alerted many people to the dangers of tsunamis
- **the existence of an early-warning system:** the number of deaths from hurricanes in the USA is usually low partly because of an effective early-warning system
- **effectiveness of lines of communication:** the earthquake in Sichuan (China) in 2008 brought a swift response from the government, which mobilized 100 000 troops and allowed overseas aid into the country
- **availability and readiness of emergency personnel:** there were many deaths following Cyclone Nargis in Burma due to a shortage of trained personnel

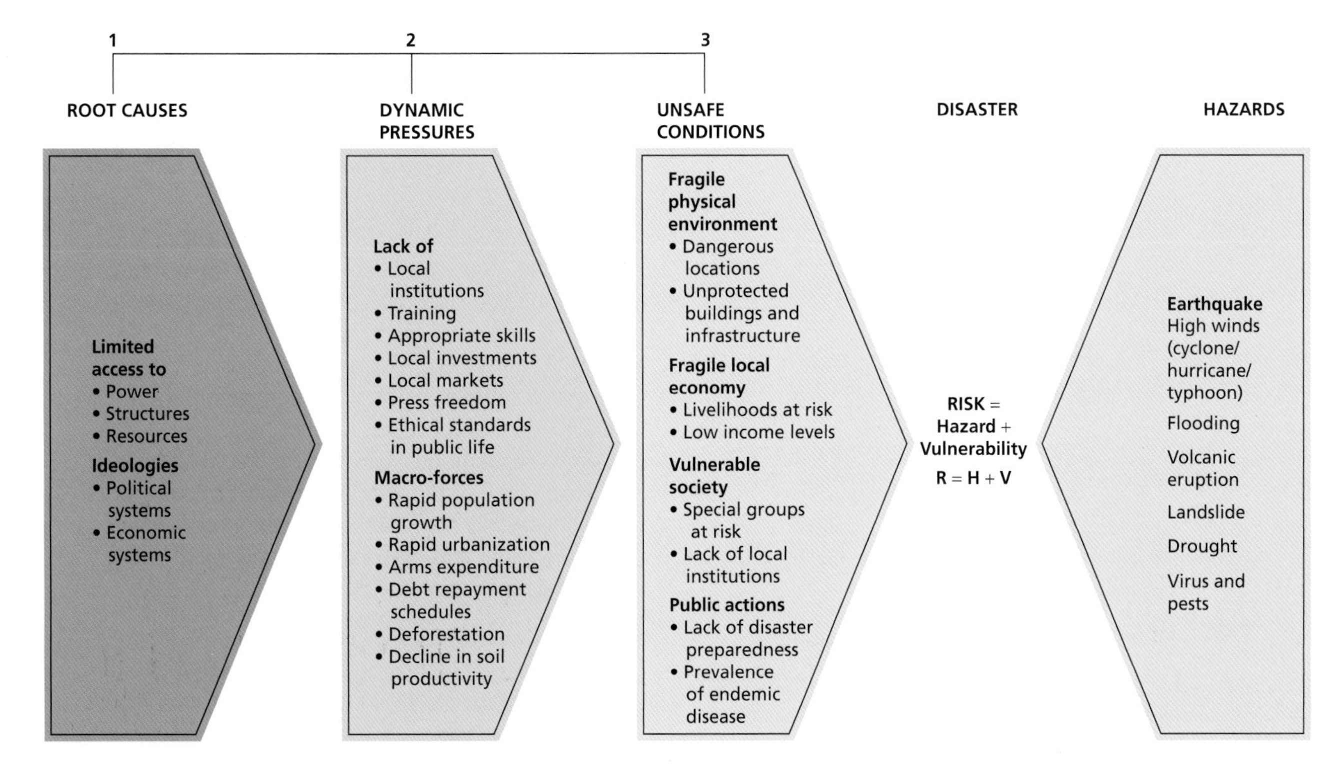

Figure 8.14 The progression of vulnerability

- **insurance cover:** generally it is the poor who have no insurance cover and they are most likely to be affected in a natural hazard as their housing quality is poor
- **construction styles and building codes:** there was criticism during the Sichuan earthquake that many schools were destroyed (by implication, poorly built), whereas government buildings remained standing
- **the nature of society:** the failure of the Burmese government to allow aid to the victims of Cyclone Nargis in 2008 increased the number of deaths from disease and malnutrition
- **cultural factors:** some of the cultural factors that influence public response to warnings are: the extent of trust in government, scientists or other authority figures; the extent and success of social networks; the amount of control or autonomy a community feels it has; and perceived hazard level.

Many of these factors help explain the fact that less developed countries are much more vulnerable to natural hazards than industrialized countries.

To research

Visit the publications page of the resources section at www.fhrc.mdx.ac.uk for free downloads on flooding and impacts/responses.

To do:

a Define the term "vulnerability".

b Outline the factors that contribute to vulnerability.

To research

To find out more, visit http://webra.cas.sc.edu/hvri/, the Hazards and Vulnerability Research Institute from the University of South Carolina. The site includes the social vulnerability index (SOVI).

Describe the distribution of social vulnerability to natural hazards in the USA.

http://webra.cas.sc.edu/hvri/docs/SoVIRecipe.pdf gives an outline of the factors considered in determining social vulnerability.

Suggest reasons why these factors are considered to be important.

Case study: *Social vulnerability and Hurricane Katrina*

Hurricane Katrina (see also page 206) was a particular threat to New Orleans, which is built on land below sea level. When Katrina made landfall, it flooded the streets, wrecked the power grid, tore roofs and walls off historic buildings and brought down many trees. Many homes in New Orleans were submerged by the surge of floodwater brought on by the storm. New Orleans' levées gave way under the pressure of the storm surge. Most of the city went under water, with some sections as deep as 6 metres. One of the areas worst affected was the Ninth Ward, a poor district to the east of the city's famous French Quarter.

The floods brought with them poisonous snakes, waterborne diseases, carcasses of livestock and abandoned pets and grotesquely swollen human corpses. This was a shocking sight for an MEDC society like the USA. There were also health dangers arising from fallen power lines and sewage-tainted water. The floodwaters in New Orleans became 10 times more toxic than is considered safe.

People with cars were able to evacuate the city before Katrina hit, but others were left stranded. Of the more than 1,800 people killed and 800 000 made homeless, the great majority were from the poorer ethnic minorities. The rescue operation was criticized for not doing enough to help the poorest members of the population. Most of those left without help were from the poor neighbourhoods, which were the worst hit by the hurricane. The disproportionate impact of the disaster on the poor, mainly black sections of the population served to highlight and exacerbate existing racial and class inequalities within the USA.

Impact and social status of damage

- Median household income in the most devastated neighbourhood was $32,000, or $10,000 less than the national average.

- In the disaster area 20% of households had no car, compared with 10% nationwide.
- Nearly 25% of those living in the hardest-hit areas were below the poverty line, double the national average.
- About 60% of the 700 000 people in the 36 neighbourhoods affected (in the states of Louisiana, Mississippi and Alabama) were from an ethnic minority. Nationwide, about one in three US citizens is from a racial minority.

To research

Visit http://www.youtube.com for You Tube video clips on Katrina.

http://dsc.discovery.com/convergence/katrina/facts/facts.html has some basic information on Katrina.

Case study: *Vulnerability to tropical storms in Haiti*

Haiti is a Caribbean country characterized by poverty, environmental degradation, corruption and violence. Since 2000 more than $4 billion has gone to rebuild communities and infrastructure devastated by hurricanes, floods and landslides.

Haiti has long been vulnerable to tropical storms and hurricanes; in recent years, the country has been afflicted by a significant increase in severe natural disasters. Loss of human life from tropical storms in Haiti is due primarily to severe flash floods in eroded watersheds that wash down on poor riverine and coastal floodplain communities. Haiti's disastrous floods of 2004 in Gonaïves and other areas serve as a warning of major threats to densely populated districts of Port-au-Prince and other major coastal cities.

Haiti has a youthful and rapidly growing population that is increasingly clustered in urban areas. Like other countries in the region, Haiti is experiencing rapid urban growth. Haiti's overall rate of urban population growth is 3.63% compared to 0.92% in rural areas. Port-au-Prince alone is growing by 5% annually, and 40% of Haiti's population lives in urban settlements, including shanty towns in coastal floodplains such as Cité Soleil in Port-au-Prince, Raboteau in Gonaïves, and La Faucette in Cap-Haïtien. The Port-au-Prince metropolitan area now comprises a quarter of Haiti's entire population. Given the sheer scale of settlement in coastal loodplains, predicted deaths due to catastrophic flooding in Port-au-Prince would far surpass all other disasters in Haiti's meteorological record.

The high rate of population growth and rapid urban expansion do not allow aquifers and floodplains to function as natural storage and filters, particularly during flood conditions. Due to unplanned urbanization, hard surfaces caused by anarchic construction methods prevent the infiltration of surface water required to recharge the country's most important aquifers, located in the major plains of Cul-de-Sac, Gonaïves, Léogane, Les Cayes and Cap-Haïtien. There is virtually no chance of diminishing Haiti's vulnerability to severe flooding without mitigation efforts that target densely populated urban neighbourhoods.

As the earthquake of 2010 demonstrated (see page 222), the root causes of environmental disaster in Haiti are acute poverty, rapid population growth and unplanned urbanization. The earthquake itself was shallow and severe, registering 7.0 on the Richter scale, but what made it so devastating was that its epicentre was close to the capital, Port-au-Prince, where a million people live in densely built housing that is not earthquake resistant. Prospects for reduced vulnerability to natural disaster in Haiti are very limited in the absence of broad-based economic development.

To research

Visit http://www.cnn.com/SPECIALS/2010/haiti.quake/ for a CNN special on Haiti.

Risk and risk relationships

Every year, natural hazards cause thousands of deaths, hundreds of thousands of injuries, and billions of dollars in economic losses. Disasters appear to be increasing in frequency, and they represent a major source of risk for the poor, wiping out development gains and accumulated wealth in developing countries. A hazard's destructive

More material available:

www.oxfordsecondary.co.uk/ibgeography
Vulnerability and the Guatemala earthquake

potential is a function of the magnitude, duration, location and timing of the event.

- About 25 million square kilometres (about 19% of the earth's land area) and 3.4 billion people are relatively highly exposed to at least one hazard.
- Some 3.8 million square kilometres and 790 million people are relatively highly exposed to at least two hazards.
- About 0.5 million square kilometres and 105 million people are relatively highly exposed to three or more hazards.
- In some countries, large percentages of the population reside in hazard-prone areas.

The world's geophysical hazards tend to cluster along fault boundaries characterized by mountainous terrain. Hazards driven mainly by hydro-meteorological processes – floods, cyclones, and landslides – strongly affect the eastern coastal regions of the major continents as well as some interior regions of North and South America, Europe and Asia. Drought is more widely dispersed across the semi-arid tropics. The areas subject to both geophysically and hydro-meteorologically driven hazards fall primarily in East and South Asia, Central America and western South America. Large areas of the eastern Mediterranean and Middle East also appear at high risk of loss from multiple hazards.

Many of these areas are more densely populated and developed than average, leading to high potential for casualties and economic losses. The areas at high risk of economic losses are more widely distributed in industrial and lower-middle-income countries than areas of high mortality risk.

Factors affecting the perception of risk

Be a critical thinker

How do people cope with hazards?

At an individual level, there are three important influences on an individual's response to hazard:

- experience – the more experience of environmental hazards, the greater the adjustment to the hazard
- material well-being – those who are better off have more choice
- personality – is the person a leader or a follower, a risk-taker or a risk-minimizer?

Ultimately, in terms of response there are just the three choices: do nothing and accept the hazard; adjust to the situation of living in a hazardous environment; leave the area. It is the adjustment to the hazard that we are interested in.

The level of adjustment will depend, in part, on the risks caused by the hazard. This includes:

- identification of the hazard
- estimation of the risk (probability) of the environmental hazard
- evaluation of the cost (loss) caused by the environmental hazard.

A number of factors influence the perception of risk.

To do:

a Discuss the view that "vulnerability is a function of demographic and socio-economic factors".

b Suggest why some sectors of a population are more vulnerable than others.

c In what ways are some communities more prepared and able to deal with a natural hazard than others?

Risk – the probability of a hazard event causing harmful consequences (expected losses in terms of death, injuries, property damage, economy and environment).

Analysis of risk

$R = H \cdot Pop \cdot Vul$

Where:

R = Risk, that is, the number of expected human impacts (killed)

H = Annual hazard occurrence probability

Pop = Population living in a given exposed area

Vul = Vulnerability: depends on sociopolitico-economic context.

More material available:

www.oxfordsecondary.co.uk/ibgeography
Populations at risk

Factors tending to increase risk perception	Factors tending to reduce risk perception
Involuntary hazard (radioactive fallout, e.g. Chernobyl, 1986)	Voluntary hazard (professional mountaineers)
Immediate impact (Cyclone Nargis, Burma, 2008)	Delayed impact (drought in Ethiopia, 2003, 2008)
Direct impact (Sichuan earthquake, 2008; Haiti earthquake, 2010)	Indirect impact (drought in Spain and Portugal and the effect on tourism)
Dreaded impact (cancer; AIDS)	Common accident (car crash)
Many fatalities per disaster (Hurricane Katrina, 2005; Haiti earthquake, 2010)	Few fatalities per disaster (UK floods, 2007)
Deaths grouped in space or time (Bhuj earthquake, India, 2000)	Deaths random in space and time (stomach cancer)
Identifiable victims (chemical plant workers, Bhopal, 1984)	Statistical victims (cigarette smokers)
Processes not well understood (nuclear accident, e.g. Chernobyl)	Processes well understood (flooding)
Uncontrollable hazard (Hurricane Katrina, 2005)	Controllable hazard (ice on motorway)
Unfamiliar hazard (tsunami, Indonesia, 2004)	Familiar hazard (river flood)
Lack of belief in authority (young population)	Belief in authority (university scientist)
Much media attention (nuclear hazards, e.g. Chernobyl; Mozambique floods, 2000)	Little media attention (factory discharge in water or atmosphere)

Table 8.7 Factors influencing public risk perception, with examples

Hazard event prediction

Predicting a hurricane

Although hurricanes are very difficult to predict, there are certain patterns that can be seen (Table 8.7). Hurricane movement is generally westward and northwards in the Caribbean. Other than basic knowledge of general hurricane occurrence, there are no atmospheric conditions that can be measured and combined in order to predict where a hurricane will develop. Therefore we can only forecast its path once it has formed.

Hurricanes form between 5° and 30° latitude and initially move westward (owing to easterly winds) and slightly towards the poles. Many hurricanes eventually drift far enough north or south to move into areas dominated by westerly winds (found in the middle latitudes). These winds tend to reverse the direction of the hurricane to an eastward path. As the hurricane moves poleward, it picks up speed and may reach between 30 and 60 kilometres per hour. An average hurricane can travel up to about 650 kilometres in a day, and up to 4,800 kilometres before it dies out. Hurricanes normally die out when they leave the tropics or start moving over land.

Hurricanes occur between July and October in the Atlantic, eastern Pacific and the western Pacific north of the equator. South of the equator, off Australia and in the Indian Ocean, they occur between November and March.

On average, 10–25% of the rain experienced in the Caribbean comes from hurricanes. Areas within 80 kilometres of the centre of a hurricane receive 100–250 millimetres of rain over a period of up to four days. Even places up to 320 kilometres from the eye will receive rainfall from the hurricane.

More material available:

www.oxfordsecondary.co.uk/ibgeography

Skills: locating places where disasters have struck.

To do:

a Outline the factors that influence a person's perception of risk.

b Suggest why some individuals or communities may underestimate the probability of hazard events occurring. (See also pages 222–3.)

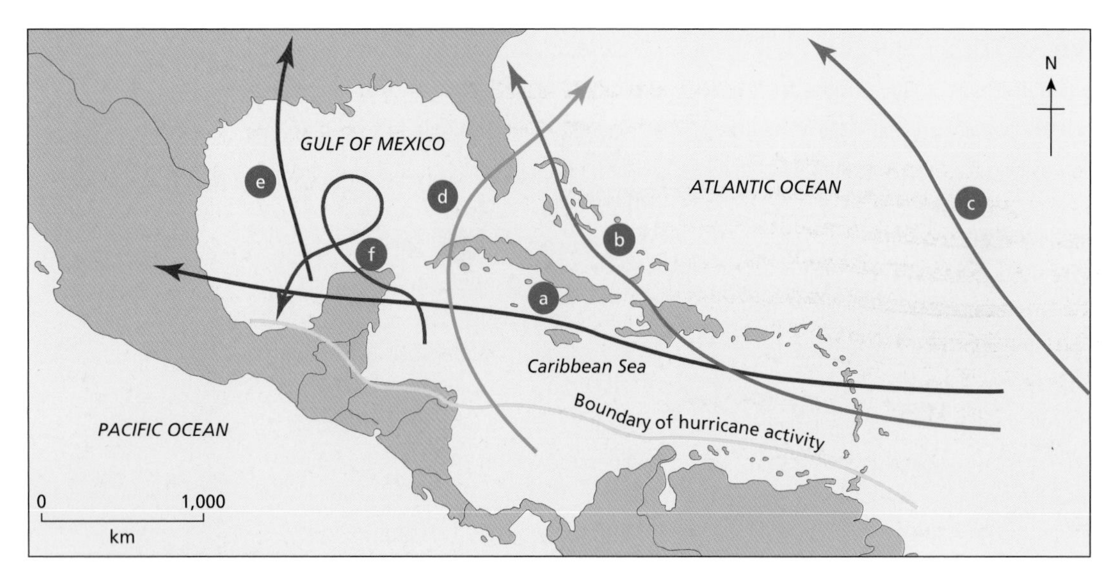

- Some hurricanes originate in the Atlantic between 10°N and 20°N. Of these (a), some generally travel westwards towards the USA.
- Others (b) travel westwards and then north or northeast towards the Bahamas and the USA.
- Some (c) avoid the Caribbean and head northwest towards Bermuda and the USA.
- Some hurricanes originate in the Caribbean and strengthen as they travel north towards Cuba and the USA (d).
- Others originate in the Gulf of Mexico, usually late in the season, and head westwards or northwards towards the densely populated US and Mexican coasts (e).
- Other hurricanes in the region, such as Hurricane Mitch in 1999, are very unpredictable (f).

Figure 8.15 Hurricane pathways in the Caribbean

Satellites detect hurricanes in their early stages of development and can help provide early warning of imminent hurricanes. Reinforced aircraft fitted with instruments fly through and over hurricanes, and weather radar can locate storms within 320 kilometres of the radar station.

Predicting the size and impact of a hurricane

Most of the damage caused by a hurricane is related to wind speeds and driving rain. The Saffir-Simpson scale (see also page 206) describes the likely damage. The scale rates the hurricane's intensity on a scale of 1 to 5. This is used to give an estimate of the potential property damage and flooding expected along the coast from a hurricane landfall. Wind speed is the determining factor in the scale, as storm surge values are highly dependent on the slope of the continental shelf and the shape of the coastline in the landfall region.

To research

Visit www.nhc.noaa.gap and go to the Saffir-Simpson hurricane wind scale table, to find out how increasing levels of hurricane activity impact on
(i) people, livestock and pets
(ii) mobile names, and
(iii) power and water.

Case study: *Monitoring Soufrière Hills volcano, Montserrat*

In July 1995 the small Caribbean island of Montserrat was devastated by the eruption of its previously dormant Soufrière Hills volcano. Volcanic activity ever since has made over 60% of southern and central parts of the island uninhabitable. Plymouth, the island's capital, was evacuated three times in 1995 and 1996, and two-thirds of the population was forced to flee abroad. An exclusion zone is still in place around the volcano, and Plymouth had to be abandoned. The volcano was responsible for 19 deaths – all of them farmers – caught out by an eruption during their return to the exclusion zone in 1997. Volcanic dust is another hazard, and is a potential cause of silicosis and aggravating asthma. There are many hazards around Plymouth (Figure 8.16).

Much of the northern third of the island has seen considerable development. Redevelopment has been centred around the Davy Hills, where new housing, schools, a hospital and infrastructure have been built as the islanders attempt to learn to live with the volcano in the south.

Volcanic management includes monitoring and prediction. GPS is used to monitor changes in the surface of the volcano – volcanoes typically bulge and swell before an eruption. Hazard risk maps can be used to good effect (Figure 8.17). There are risks on other Caribbean islands, too: St Vincent and St Kitts are high-risk islands, whereas St Lucia, Grenada and Nevis are lower risk.

Figure 8.16 Hazards at Plymouth, Montserrat

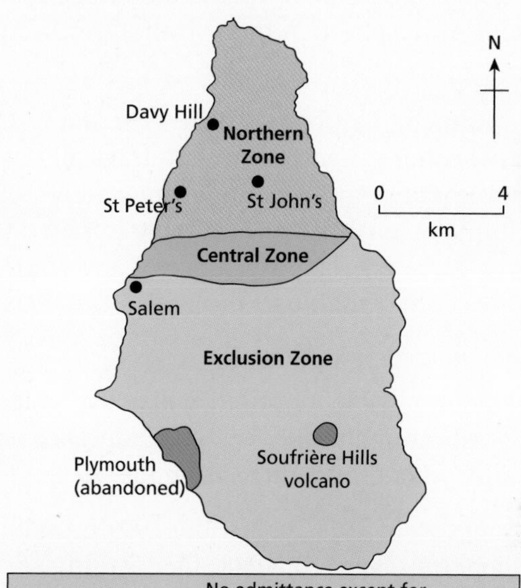

Exclusion Zone	No admittance except for scientific monitoring and national security matters.
Central Zone	Residential area only. All residents in state of alert. All have rapid means of exit 24 hours a day. All residents must have hard hats and dust masks.
Northern Zone	Significantly lower-risk, suitable for residential and commercial occupation.

Figure 8.17 Volcanic hazard risk in Montserrat

To do:

a Outline the main hazards related to volcanic activity in Montserrat (Figure 8.16).

b Describe the distribution of risk as shown in Figure 8.17.

To research

Visit the Montserrat Volcano Observatory at www.montserratvolcanoobservatory.info/ and find out about the volcano's current activity.

Predicting a tsunami

At present it is impossible to predict precisely where and when a tsunami will happen. In most cases it is only possible to raise the alarm once a tsunami has started. In theory there is time to issue warnings: a tsunami off the coast of Ecuador will take 12 hours to reach Hawaii and 20 hours to reach Japan; a tsunami from the Aleutians will take five hours to reach Hawaii. However, the impacts will vary with shoreline morphology.

It is, however, possible to monitor submarine volcanoes to predict the risk of tsunami. For example, Kick'em Jenny, the underwater volcano north of Grenada, has erupted 10 times since the late 1970s and grown by 50 metres. Volcanologists believe it could cause a tsunami and threaten Venezuela.

The first effective tsunami warning system was developed in 1948 in the Pacific, following the 1946 tsunami that hit Alaska and Hawaii. The system consisted of over 50 tidal stations and 31 seismographic stations, spread over the ocean between Alaska, Hong Kong and Cape Horn. When water passes a critical threshold, a warning is automatically sent to Honolulu (Hawaii). In addition, the earthquake epicentre is plotted and its magnitude investigated. Satellites have improved the effectiveness of the system, which is now operated by the US National Oceanic and Atmospheric Administration (NOAA).

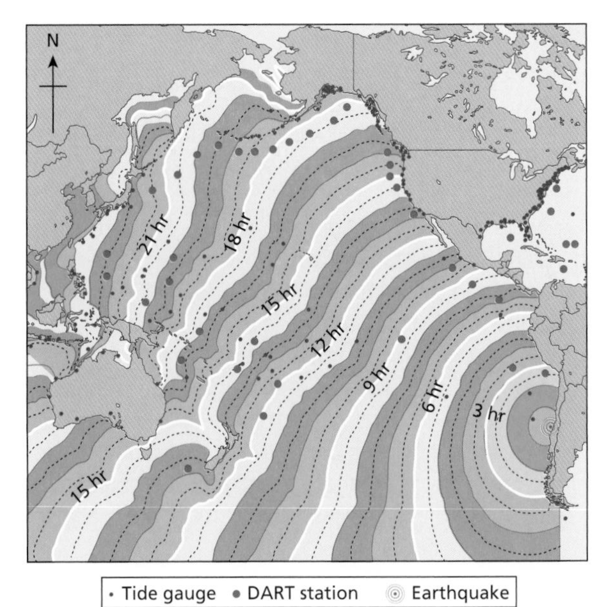

Figure 8.18 Tsunami early-warning system (travel times) following the 2010 Chile earthquake

Other tsunami early-warning systems include those in Japan and Kamchatka (Russia). However, many LEDCs lack early-warning systems, as was so tragically exposed in the 2004 Boxing Day tsunami. Immediately following the 2010 Chile earthquake, a tsunami warning was issued (Figure 8.18). Fortunately, there was little evidence of any large-sized waves affecting areas other than part of the Chilean coast.

Predicting a volcanic eruption

It is virtually impossible to monitor all active volcanoes. Satellites offer the prospect of global coverage from space and are being developed for remote warning systems.

Early warnings were issued before the 1991 eruption of Mount Pinatubo in the Philippines. Over 320 people died, mostly after the collapse of their ash-covered roofs, but many more lives were saved, and at least 58000 people were evacuated from the high-risk areas. Management of the eruption seems to have been well coordinated and effective:

- State-of-the-art volcano monitoring techniques and instruments were applied.
- The eruption was accurately predicted.
- Hazard zonation maps were prepared and circulated a month before the violent explosions.
- An alert and warning system was designed and implemented.
- The disaster response machinery was mobilized on time.

Predicting an earthquake

A number of techniques can be used to help predict the risk of an earthquake, although scientists are still not able to say precisely when and where an earthquake will occur. The most reliable predictions of earthquake risk focus on:

- measurement of small-scale ground surface changes
- small-scale uplift or subsidence
- ground tilt
- changes in rock stress

- microearthquake activity (clusters of small quakes)
- anomalies in the earth's magnetic field
- changes in radon gas concentration
- changes in electrical resistivity of rocks.

One intensively studied site is Parkfield in California, on the San Andreas fault. Parkfield, with a population of less than 50 people, claims to be the earthquake capital of the world. Parkfield is heavily instrumented: strain meters measure deformation at a single point; two-colour laser geodimeters measure the slightest movement between tectonic plates; and magnetometers detect alterations in the earth's magnetic field, caused by stress changes in the crust. Nevertheless, the 1994 Northridge earthquake was not predicted and it occurred on a fault that scientists did not know existed. Technology helps, but not all the time.

More material available:

www.oxfordsecondary.co.uk/ibgeography

Average annual earthquakes based on observations since 1930 and selected criteria for Volcanic Explosive Index (VEI)

Disasters

Stages in a disaster

I Preconditions	
Phase 1	*Everyday life* (years, decades, centuries)
	"Lifestyle" risks, routine safety measures, social construction of vulnerability, planned developments and emergency preparedness.
Phase 2	*Premonitory developments* (weeks, months, years)
	"Incubation period" – erosion of safety measures, heightened vulnerability, signs and problems misread or ignored.
II The disaster	
Phase 3	*Triggering event or threshold* (seconds, hours, days)
	Beginning of crisis; "threat" period: impending or arriving flood, fire, explosion; danger seen clearly; may allow warnings, flight or evacuation and other pre-impact measures. May not, but merging with:
Phase 4	*Impact and collapse* (instant, seconds, days, months)
	The disaster proper. Concentrated death, injury, devastation. Impaired or destroyed security arrangements. Individual and small-group coping by isolated survivors. Followed by or merging with:
Phase 5	*Secondary and tertiary damage* (days, weeks)
	Exposure of survivors, post-impact hazards, delayed deaths.
Phase 6	*Outside emergency aid* (weeks, months)
	Rescue, relief, evacuation, shelter provision, clearing dangerous wreckage. "Organized response". National and international humanitarian efforts.
III Recovery and reconstruction	
Phase 7	*Clean-up and "emergency communities"* (weeks, years)
	Relief camps, emergency housing. Residents and outsiders clear wreckage, salvage items. Blame and reconstruction debates begin. Disaster reports, evaluations, commissions of enquiry.
Phase 8	*Reconstruction and restoration* (months, years)
	Reintegration of damaged community with larger society. Re-establishment of "everyday life", possibly similar to, possibly different from pre-disaster. Continuing private and recurring communal grief. Disaster-related development and hazard-reducing measures.

Table 8.8 Phases of a disaster, with reported durations and selected features of each phase

To research

Visit http://news.bbc.co.uk/1/hi/world/americas/8510900.stm, which has an interesting set of graphics comparing the Haiti earthquake, the Sichuan (China) earthquake and that of Italy (2009).

Responses to disaster: two events of 2008 contrasted

In May 2008 an earthquake registering 7.9 on the Richter scale devastated the Chinese province of Sichuan. The earthquake was caused by the Indian plate pushing northwards against the Eurasian plate. Over 69 000 people were killed and nearly 18 000 people were missing as a result of the earthquake. A further 4.8 million people were made homeless. Many rivers were blocked by landslides, and formed 34 "quake lakes". The risk of landslides was increased by the arrival of the summer rains. The Chinese government received praise for its swift rescue attempts and its willingness and openness to receive foreign aid.

In contrast, the Burmese government received considerable criticism for the way it dealt with Cyclone Nargis (see also page 207). The cyclone hit Burma just weeks before the Sichuan earthquake occurred. Over 134 000 people were killed and a further 56 000 people were missing. The disaster cost an estimated $10 billion in damage. However, the event was also a human-made disaster. The country's military rulers seemed unaware of the scale of the disaster and refused international aid at first, only accepting a small amount of aid after two weeks had gone by. Aid workers suggested that this response of the Burmese government would result in a final death toll of over a million, far higher than it need have been, as a result of lack of clean water, food, medicine and shelter rather than the cyclone itself.

Case study: *Haiti's earthquake, 2010*

In 2008 four tropical storms killed 800 people in Haiti, left a million of the 9 million population homeless, and wiped out 15% of the economy (see page 215). However, a yet crueller blow to this small nation was the earthquake of 2010 that devastated the country. On 12 January an earthquake measuring 7.0 on the Richter scale occurred just 25 kilometres west of the capital Port-au-Prince, only 13 kilometres below the surface. Aftershocks occurred just 9 kilometres below the surface, 56 kilometres southwest of the city. A third of the population was affected: about 230 000 people were killed, 250 000 more were injured and around a million made homeless.

The island of Hispaniola (shared by Haiti and the Dominican Republic) sits on the Gonave microplate, a small strip of the earth's crust squeezed between the North American and Caribbean tectonic plates. This makes it vulnerable to rare but violent earthquakes. The Dominican Republic suffered a serious quake in 1946. But the Enriquillo-Plantain Garden fault, which separates the plates on the Haitian side of the border, had been accumulating stress during more than a century of inactivity. What magnified its destructive power was the position of its epicentre, so close to the densely populated capital and so shallow.

The city and the region around it are mainly shanty settlements of overcrowded, badly constructed buildings, hopelessly ill suited to withstanding a shaking. Most of Port-au-Prince's 2 million residents live in tin-roofed shacks perched on unstable, steep ravines. After a school collapsed in the suburb of Pétionville in 2008, the capital's mayor said that 60% of its buildings were unsafe.

Emergency aid

Seven days after the earthquake, the United Nations had got food to only 200 000 people. Help was pledged from Mexico, Venezuela, China, Britain, France, Germany, Canada and Cuba. Crews of Dominicans, including engineers, telecoms technicians and the Red Cross, were among the first to join the relief effort.

Financial assistance also poured in. The World Bank led with a $100 million commitment, pending the approval of its board. The UN released $10 million from its emergency fund and European countries pledged $13.7 million. Yet most of this aid arrived too late for the thousands who were trapped in rubble or awaiting treatment for their injuries.

One of the world's poorest countries, Haiti was overpopulated and vulnerable even before the disaster. The country had only two fire stations and no army – the Haitian army was abolished in 1995 – and was powerless to do anything for itself. The earthquake degraded an already feeble health service by destroying many hospitals and clinics, including all three aid centres run by Médecins sans Frontières, an NGO. Crowded Haiti has long suffered from squabbling politicians, extreme inequality and ecological stress.

A month after the disaster, hospitals were working again. The World Food Programme handed rice to 2.5 million people in the capital and nearby areas. Most streets in Port-au-Prince were cleared of rubble.

After the rescue and relief phase, the focus of aid changed to providing shelter robust enough to withstand the rains (and landslides) that normally begin in May, and the hurricanes that may follow. Around 550 000 people gathered in makeshift camps; almost as many were sleeping rough. With aftershocks continuing, many were too scared to venture back into their houses even when these survived. Some were issued with tents, but relief workers reckoned that simple plastic tarpaulins, suspended on poles, were a more durable option. Months after the disaster the camps were still crowded, and for most people proper housing is years away.

The 1989 earthquake near San Francisco in California was of similar magnitude to Haiti's but killed just 63 people, mainly because most buildings there are designed to withstand the shock. There are plenty of ideas for cheap earthquake-proofing: one is to fit rubber pads from recycled tyres between concrete blocks as shock-absorbers. But the Haitian government has never enforced building codes.

Rebuilding Haiti's homes, schools, roads and other infrastructure will take between $8 and $14 billion. Many Dominicans fear a flood of illegal migrants into their country unless reconstruction is swift and effective.

Long-term rebuilding

A long-term strategy for rebuilding Haiti is vital. Even before the earthquake, Haiti was environmentally degraded and had few basic services. "Building back better" must be more than just a slogan.

Fortunately, a blueprint drawn up by Haiti's government was presented to donors in 2009. It calls for investment to be targeted on infrastructure, basic services and combating soil erosion to make farmers more productive and the country less vulnerable to hurricanes. The country needs a strong government to put it to rights. A temporary development authority with wide powers to act might be the best way forward.

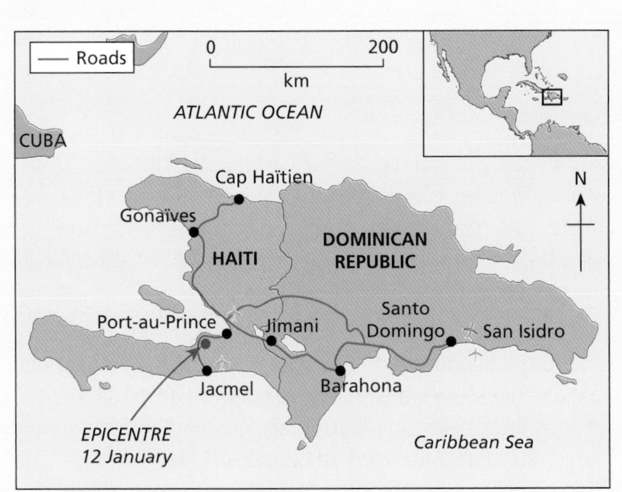

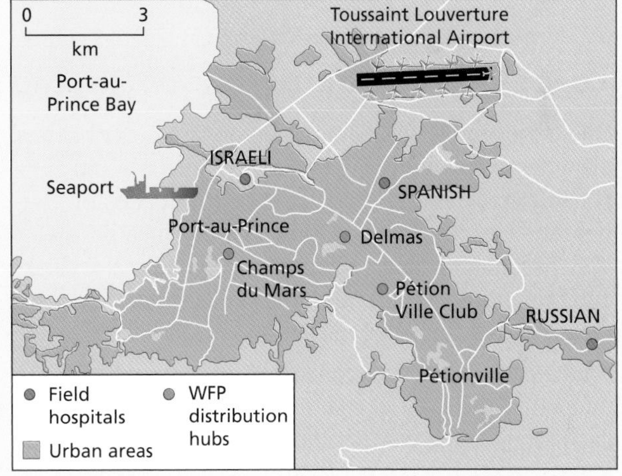

Figure 8.19 Location of Haiti and the 2010 earthquake.

Be a critical thinker

What are the advantages and disadvantages of child resettlement?

Is resettlement the same as displacement?

After the Haiti earthquake, the Joint Council on International Children's Services (JCICS), a US advocacy organization, received 150 enquiries in just three days about Haitian adoption. Usually there are 10 a month. Children's groups in the US asked people to wait before trying to adopt Haitian orphans, warning that mass adoptions or airlifts could break up families and open the door to fraud, abuse and trafficking

The Council also dismissed a plan from the Catholic archdiocese of Miami to airlift thousands of displaced Haitian children to Florida, in an echo of an initiative that saw 14 000 unaccompanied Cuban children begin new lives in the US in the early 1960s. Just like that two-year programme, which was designed to remove children from the control of Fidel Castro's government, the new arrivals would live in temporary shelters in south Florida until foster homes were found or they were reunited with family members.

Chile, 2010

The 2010 Chilean earthquake, which occurred off the coast of the Maule region of Chile on 27 February, at 03:34 local time, had a magnitude of 8.8 on the Richter scale and lasted 90 seconds. The cities experiencing the strongest shaking were Arauco and Coronel. The earthquake triggered a tsunami, which devastated several coastal towns in south-central Chile and damaged the port at Talcahuano. However, the death toll was just 521 victims. This low figure has been put down to Chile's building standards, adequate emergency preparations, and a high standard of living (compared with, for example, Haiti).

To research

Visit http://news.bbc.co.uk to read articles about the Chile earthquake, access a map, and find other links. Look for the article called "Why did the Haiti earthquake have a greater impact than the Chile earthquake?", and answer that question.

Case study: *Benzene pollution on the Songhua River, China, 2005*

On 13 November 2005, an explosion at a petrochemical plant owned by the China National Petroleum Corp (CNPC) in Jilin City released at least 100 tonnes of benzene into the Songhua River. The Songhua is 1,927-kilometres long, and runs through 30 major cities and numerous villages. Benzene is a dangerous toxin that can cause anaemia, cancer and blood disorders.

An 80-kilometre slick of contaminated water passed through Harbin 230 kilometres downstream (population 10 million), on its way to other cities. On the day after the explosion, China's environmental watchdog agency found benzene levels in the river near the plant to be 108 times the safe limit. When the contaminated slick reached the border of Jilin and Heilongjiang, pollution levels were still 29 times above national safety levels.

Many farmers and towns depend on the river for drinking water, washing and irrigation. Before reaching Harbin, the slick had gone through the town of Songyuan without causing any problems, because local water supplies come from underground water. However, there was concern that many farmers might not have heard or read public announcements, and might still be using river water.

Immediately after the blast, the Jilin authorities denied that any pollution had taken place. National TV channels repeatedly reported that "the incident had not caused any serious pollution", explaining that the benzene had been burnt and reduced to harmless carbon dioxide. The real situation became clear only when water supplies to the city of Harbin were cut off without warning, leaving 3.8 million residents in Heilongjiang province with no access to clean water for several days. This caused immediate panic in the population, and led to the following effects:

- Residents attempted to buy up as much food and water as they could.

- Many tried to flee, jamming roads out of the city.
- The price of water more than doubled.
- Schools closed, and many parents sent their children to relatives in other cities.
- The provincial government shipped 16 000 tonnes of water from neighbouring cities (but Harbin requires on average 18 600 tonnes per day)
- People living within 1,000 metres of the river were evacuated, and walking by the river was banned.

Had the information been made public in a timely fashion, the rush for food and water and other panic responses were unlikely to have happened.
Jilin Petrochemical, a subsidiary of CNPC, was ordered to pay one million yuan (US$ 128 000) for polluting the river, the maximum possible fine but never imposed before on corporate polluters.

When the slick of contaminated water eventually reached the Amur River in Siberia (of which the Songhua is a tributary) and the Russian city of Khabarovsk (population 650 000), there was an international protest. The fears of possible contamination of the Amur River highlighted China's worsening pollution problems. In the wake of the Songhua incident, in 2007 Beijing adopted a plan that includes spending 13.4 billion yuan (US$ 1.7 billion) to clean up the Songhua River and put in place pollution controls by 2010.

To research

Visit http://news.bbc.co.uk and look for the article "Toxic leak threat to Chinese city" for a link to a video clip, timeline and quotes from people living in the region.

1. Outline the potential impacts of the slick of contaminated water.
2. In what ways could the impacts of the slick have been managed differently?

Research the Cuyahoga River in northeast Ohio, USA, which has caught fire at least 13 times since 1868. Why has the river caught fire? When did it last do so?

Adjustment and response to hazards

People cope with hazards in a number of ways. At an individual level, three important factors affect how a person copes:

- experience: people with more experience of hazards are better able to adjust to them
- levels of wealth: people with more money have more choices open to them
- personality: is the person a leader or a follower, a risk-taker or very cautious?

The three basic options from which they can choose are:

- Do nothing and accept the hazard.
- Adjust to living in a hazardous environment.
- Leave the area.

In most cases, people take the middle option. How people adjust to hazards depends on:

- the **type of hazard**
- the **risk (probability) of the hazard** – several factors influence how people view the risk
- the **likely cost (loss)** caused by the hazard.

Ways of managing the consequences of a hazard include:

- **modifying the hazard event,** through building design, building location and emergency procedures
- **improved forecasting and warning**
- **sharing the cost** of loss, through insurance or disaster relief.

Class of adjustments	*Earthquakes*	*Volcanoes*	*Hurricanes*
Affect the cause	No known way of altering the earthquake mechanism	No known way of stopping volcanic eruptions	No known way of preventing hurricanes
Modify the hazard	Stable site selection: soil and slope stabilization; sea wave barriers; fire protection	Diversion channels for lava (e.g. Etna); spray water on lava (e.g. Heimaey)	Have wide belts of forest to reduce impact (especially wind speeds); build back from the coast
Modify loss potential	Warning systems; emergency evacuation and preparation; building design; land-use change; permanent evacuation	Warning systems; emergency evacuation and preparation; mobile facilities, e.g. mobile retailers on Etna	Forecasting; warning systems; emergency evacuation and preparation
Adjust to losses			
Spread the losses	Public relief; subsidized insurance	Public relief; subsidized insurance	Public relief; subsidized insurance
Plan for losses	Insurance and reserve funds	Insurance and reserve funds	Insurance and reserve funds
Bear the losses	Individual loss bearing	Individual loss bearing	Individual loss bearing

Table 8.9 An early version of alternative adjustments to natural hazards

Northern Nigeria		Tanzania
Nothing permanent	*Change location*	Nothing permanent
Nothing	*Change use*	Drought-resistant crops; irrigation
Store food for next year; seek work elsewhere temporarily; seek income by selling firewood, crafts, or grass; expand fishing activity; plant late cassava; plant additional crop	*Prevent effects*	More thorough weeding; cultivate larger areas; work elsewhere; tie ridging; planting on wet places; send cattle to other areas; sell cattle to buy food; staggered planting
Consult medicine men; pray for end of drought	*Modify events*	Employ rainmakers; pray
Turn to relatives; possible government relief	*Share*	Send children to kinsmen; store crops; government relief; move to relative's farm; use savings
Suffer and starve; pray for support	*Bear*	Do nothing

Table 8.10 Adjustments to drought suggested by peasant farmers in Nigeria and Tanzania

Hurricane watches and warnings

A hurricane watch is issued when there is a threat of hurricane conditions within 24–36 hours. A hurricane warning is issued when hurricane conditions are expected in 24 hours or less.

During a hurricane watch

- Listen to the radio or television for hurricane progress reports.
- Check emergency supplies.
- Fuel car.
- Bring in outdoor objects such as lawn furniture, toys and garden tools, and anchor objects that cannot be brought inside.
- Secure buildings by closing and boarding up windows.
- Remove outside antennas and satellite dishes.
- Turn refrigerator and freezer to coldest settings. Open only when absolutely necessary and close quickly.
- Store drinking water in clean jugs, bottles and cooking utensils.
- Know where you are going to shelter if the need arises.

More material available:

www.oxfordsecondary.co.uk/ibgeography
Hurricane preparedness for home

During a hurricane warning

- If you need to evacuate your home, lock up and go to the nearest shelter.
- Take blankets and sleeping bags to shelter.
- Listen constantly to a radio or television for official instructions.
- Store valuables and personal papers in a waterproof container on the highest level of your home.
- Stay inside, away from windows, skylights and glass doors.
- Keep a supply of flashlights and extra batteries handy. Avoid open flames, such as candles and kerosene lamps, as a source of light.
- If power is lost, turn off major appliances to reduce power "surge" when electricity is restored.
- Feed animals and pets and move indoors or let loose.

After the hurricane

- Assist in search and rescue.
- Seek medical attention for persons injured.
- Clean up debris and effect temporary repairs.
- Report damage to utilities.
- Assist in road clearance.
- Watch out for secondary hazards, fire, flooding, etc.
- Assist in community response efforts.
- Avoid sightseeing.
- Cooperate with damage assessors.

Prevention and Mitigation Measures

Risk Assessment

The evaluation of risks of tropical cyclones should be undertaken and illustrated in a hazard map. The following information may be used to estimate the probability of cyclones which may strike a country:

- analysis of climatological records to determine how often cyclones have struck, their intensity and locations
- history of wind speeds, frequencies of flooding, height location or storm surges
- information about the last 50–100 years of cyclone activity.

Land-use control

This is designed to control land use so that the least critical facilities are placed in most vulnerable areas. Policies regarding future development may regulate land use and enforce building codes for areas vulnerable to the effects of tropical cyclones.

Floodplain management

A master plan for floodplain management should be developed to protect critical assets from flash, riverine and coastal flooding.

Reducing vulnerability of structures and infrastructures

- Design new buildings to be wind and water resistant. Design standards are usually contained in building codes.
- Locate communication and utility lines away from the coastal area, or install underground.
- Improve building sites by raising the ground level to protect against flood and storm surges.

- Regularly inspect protective river embankments, levées and coastal dikes for breaches due to erosion, and take opportunities to plant mangroves to reduce breaking wave energy.
- Improve vegetation cover to help reduce the impact of soil erosion and landslides, and to facilitate the absorption of rainfall to reduce flooding.

Building design

A single-storey building has a quick response to earthquake forces. A high-rise building responds slowly, and shock waves increase as they move up the building. If the buildings are too close together, vibrations may be amplified between buildings and increase damage. The weakest part of a building is where different elements meet. Elevated motorways are therefore vulnerable in earthquakes because they have many connecting parts. Certain areas are very much at risk from earthquake damage: areas with weak rocks, faulted (broken) rocks, and soft soils. Many oil and water pipelines in tectonically active areas are built on rollers, so that they can move with an earthquake rather than fracture.

Figure 8.20 A hurricane shelter in Cavalla, Montserrat

Short-term, mid-term and long-term responses after an event

In the immediate aftermath of a disaster the main priority is to **rescue** people. This may involve the use of search and rescue teams and sniffer dogs. Thermal sensors may be used to find people alive among the wreckage. The number of survivors decreases very quickly. Few survive after 72 hours, although there were reports from Sichuan of people surviving for nearly 20 days: the number is extremely low, however.

Rehabilitation refers to people being able to make safe their homes and live in them again. Following the UK floods of 2007, some people were unable to return to their homes for over a year. For some residents in New Orleans, rehabilitation was not possible after Katrina, and so **reconstruction** (rebuilding) was necessary. This can be a very long, drawn-out process, taking up to a decade for major construction projects. The timescales involved are shown in the model of disaster recovery (Figure 8.22).

Figure 8.21 (a and b) Steel shutters for hurricane management at the Casuarina Hotel, Barbados

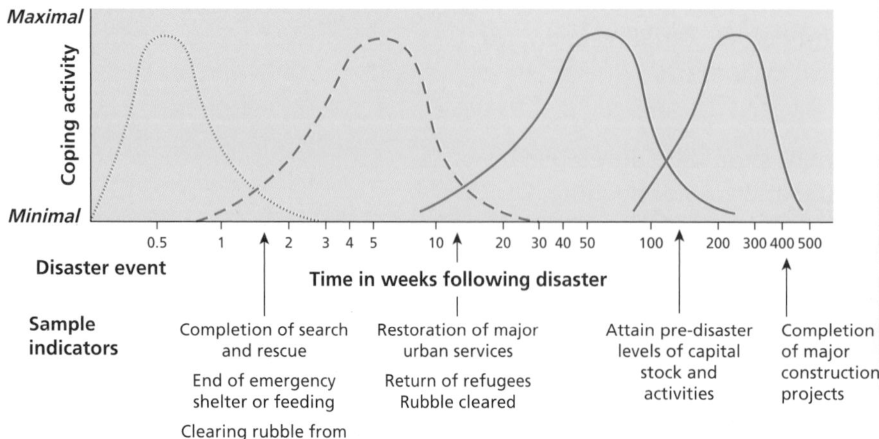

Figure 8.22 A model of disaster recovery for urban areas

As well as dealing with the aftermath of a disaster, governments try to plan to reduce impacts in future events. This was seen after the South Asian tsunami of 2004. Before the event, a tsunami early-warning system was not in place in the Indian Ocean. Following the event, as well as emergency rescue, rehabilitation and reconstruction, governments and aid agencies in the region developed a system to reduce the impacts of future tsunamis. It is just part of the progress needed to reduce the impact of hazards and to improve safety in the region.

Rescue, rehabilitation and reconstruction in practice

Following the 2004 South Asian tsunami, the Indonesian government produced a master plan for the rehabilitation and reconstruction of the region affected by the tsunami. They defined rescue, rehabilitation and reconstruction in the following ways:

- **rescue** – saving people so they can survive despite having only minimum life necessities
- **rehabilitation** – restoring the functions of public services, a process that needs one or two years
- **reconstruction** – rebuilding the public system, economic system, infrastructure and governance functions, predicted to take two to five years.

In practice, the Indonesian government provided the following rescue services:

(a) immediately helping the disaster survivors

(b) immediately burying the victims' dead bodies

(c) immediately enhancing basic facilities and infrastructure to be able to provide adequate services for the victims.

Reconstructing Haiti

Following the earthquake (see page 222), plans were discussed for the rescue, rehabilitation and reconstruction of Haiti. Reconstructing Haiti is a challenge to the international community, which has failed over decades to lift the island state out of poverty, corruption and violence. In the past 10 years more than $4 billion has gone to rebuild communities and infrastructure devastated by hurricanes, floods and landslides, but mismanagement, lack of coordination and attempts by global institutions to use Haiti as an economic testbed are believed to have frustrated all efforts. A foreign debt of $1.5 billion has weighed down the economy.

Stage 1: rescue, 1–10 weeks

The first step was emergency teams working with government and communities to rescue trapped people, clear rubble, and restore water and sanitation to prevent disease. Haiti has a number of self-help groups, NGOs, unions, faith groups, and youth brigades that helped mobilize the emergency effort.

Stage 2: assessment, 6–10 weeks

The UN, government donors, the World Bank and charities needed to know what capability remained. Ports, hospitals, houses, schools and roads were destroyed, and there was little piped water or electricity

To research

Visit http://www.erra.pk/ for the Earthquake Reconstruction and Rehabilitation Authority of Pakistan. Click on the link to the seismic zoning hazard map for Muzaffarabad.

What is the Earthquake Reconstruction and Rehabilitation Authority doing to help Pakistan following the Kashmir earthquake of 2005?

Visit http://www.usindo.org/publications/Blue_Plan_Aceh.pdf and research, in detail, the plans of the Indonesian Government for the rehabilitation and reconstruction of areas affected by the tsunami.

outside the business area in the capital. Pumps and pipes would have probably been severely damaged, government records would have been lost, and teachers, doctors, engineers and professionals may have been injured or dead. The extent of devastation needed to be fully understood in order for recovery to begin. There needed to be a pause, since planning was needed rather than reaction.

Attention has been on the capital, Port-au-Prince, but cities like Carrefour and Jacmel were also damaged, as well as hundreds of rural communities.

Stage 3: coordination, 1–10 weeks

The lesson of the 2004 tsunami and other disasters is that there was a real risk of wasting aid in the race to help. Coordination was needed at national and NGO level. The danger was that institutions could rush in, impose their own ideas, and duplicate efforts. The key, even at this emergency stage, was to think long term. Expectations also needed to be managed: people living in slums sometimes expect to get palaces after a hazard event has destroyed their home. When they don't, this can lead to problems.

Stage 4: rehabilitation, 1–52 weeks

If, as the Red Cross suggests, 3 million people need to be rehoused, the infrastructure of the country has to be rebuilt. Haiti's record of handling money is poor, and it is one of the world's most aid-dependent states. Hampering its recovery have been deep corruption, a poor civil service, and mistrust between the donor community and the government. Aid agencies have increasingly bypassed government, adding to administrative chaos.

Stage 5: reconstruction, 1–40 years

Many countries and aid groups already have major reconstruction programmes in Haiti. Realistically, it will take decades to rebuild the country. It is likely that the international community will now follow the example of Aceh, where a government agency was set up exclusively to coordinate the reconstruction; it set up a multi-donor fund to coordinate aid efforts.

Responses are affected by a number of factors. These include:

- the magnitude of the hazard – the greater the event, the greater the reaction
- the predictability of the event – hurricanes are annual events, whereas earthquakes are more random in time
- the level of wealth – how much the individual household, national government and international organizations can raise
- the perceived level of risk – whether a volcano is likely to erupt or not, and its level on the VEI
- the level of information provided in the media – probability of event, size of event, measures to be taken
- the degree of hazard event preparation – building codes, land-use zoning, drills
- personal factors – awareness of alternatives, ability to afford such alternatives, etc.

9 Leisure, sport and tourism

By the end of this chapter you should be able to:

- Understand the factors that have led to the growth and changing patterns of international tourism
- Describe leisure and sport at various scales and their significance for urban regeneration, using examples
- Explain the characteristics of a national tourist industry, ecotourism, and tourism as a development strategy, with reference to the principles of sustainable tourism.

Leisure – any freely chosen activity or experience that takes place in non-work time.
Recreation – a leisure-time activity undertaken voluntarily and for enjoyment. It includes individual pursuits, organized outings and events, and non-paid (non-professional) sports.
Sport – a physical activity involving events and competitions at the national and international scale with professional participants.
Tourism – travel away from home for at least one night for the purpose of leisure. This definition excludes day trips, some of which may be international trips.

This option is designed to cover a range of topics relating to the leisure industry: its diversity, opportunities and constraints. The topics are discussed under the headings of sport and tourism, and categorized according to scale. Each one has a conceptual base and focuses on one or more contemporary and sometimes controversial issues such as cultural conflict between visitors and hosts, fair participation in sport, environmental responsibility and leakage of tourist revenue. Case studies illustrate the impact of a specific leisure activity in different environments. The case studies assess reasons for its development, its economic, social and environmental impacts and the issue of future management. In this syllabus the theme of tourism is not confined to this option, and other relevant case studies can be found in chapter 7, Extreme environments, and chapter 11, Urban environments.

It is sometimes difficult to define the terms leisure, sport and tourism, since they often overlap, and participation in them may be simultaneous. For example, you may play golf or go swimming or skiing while on holiday. In addition, there are many possible subdivisions of tourism. **Ecotourism** is tourism focusing on the natural environment and local communities; **heritage tourism** is tourism based on a historic legacy such as a landscape feature, historic building or event as its major attraction; and **sustainable tourism** is tourism that conserves primary tourist resources and supports the livelihoods and culture of local people.

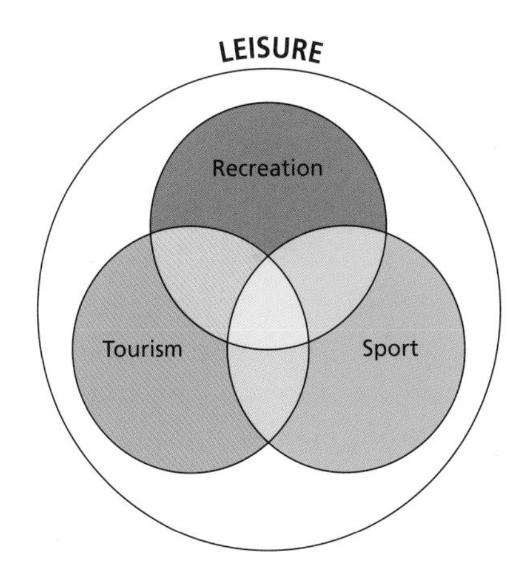

Figure 9.1 The leisure domain

To do:

Explain how leisure activities vary according to the age, gender and level of wealth of the participating population in your local area.

The growth of tourism

Tourism is the world's most popular leisure activity. It generates 10% of global GDP and provides 8.55% of all jobs. It is a dynamic industry, influenced by the interaction of three major factors: tourist **demand**, the organization of **access** by tour operators, and the **supply** of destinations.

Increased demand

There are three main reasons for the growth of tourism since the late 20th century. Firstly, dual incomes and paid holidays have increased the amount of disposable income of many households, especially in

MEDCs, but also in middle-income countries such as China and Brazil, where the demand for both international and domestic tourism is expanding rapidly. Secondly, higher levels of education have allowed people to acquire languages and knowledge of other cultures. This has reduced apprehension about communication in foreign countries. Lastly, 50% of the world's population is urbanized and there is a growing desire by its wealthier and better-educated population to escape the pressures of urban living, with its noise, overcrowding, congestion, air pollution and garbage.

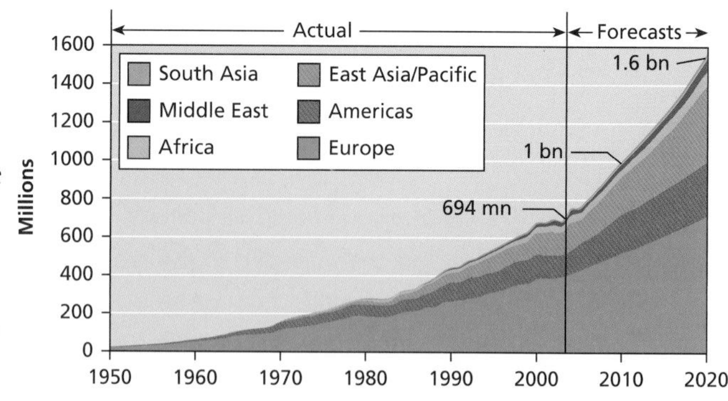

Figure 9.2 The growth of global tourism: international arrivals by destination

Improved accessibility

The Internet has facilitated research and booking operations, and credit cards have made advance payment much easier for the customer and the tour operator. Wide-bodied jet planes, high-speed trains and large cruise ships have allowed economies of scale and compressed space and time. The global flight network has expanded, allowing access to remote destinations on the "pleasure periphery".

Supply: economic growth and stability at the destination

Across large areas of the world, general levels of prosperity have been rising since the 1950s. Political stability is important, too. From the end of the Second World War until the late 1990s in western Europe, there was almost a complete absence of major political and military conflict. This is not the case in eastern Europe, however, and tourism there is less important.

Changing patterns of international tourism

Global tourism is a thriving business, with a predicted 130% increase in the number of international tourists between 2000 and 2020 (Figure 9.2). Europe continues to receive the greatest share, but tourist preference is shifting in favour of the Asia–Pacific region, Africa and the Middle East. Not surprisingly, these areas have also been experiencing very rapid economic development, and tourism has assisted this to some extent. These destinations have improved their infrastructure, levels of hygiene and knowledge of languages to a standard acceptable to the affluent tourist.

For an area to grow, it must have primary and secondary resources:

- **Primary tourist/recreational resources** are the pre-existing attractions for tourism or recreation (that is, those not built specifically for the purpose), including climate, scenery, wildlife, cultural and heritage sites, and indigenous people.
- **Secondary tourist/recreational resources** include accommodation, catering, entertainment and shopping.

A number of factors affect international tourism, either positively as attractions or negatively as deterrents (shocks). The effect of shocks may be long or short term.

Tourist attractions at the destination	
Climate	Winter snow in the Alps; all-year sunshine in the Maldives
Landscape	Spectacular desert in Namibia; tropical rainforest in Belize
Culture	Tribal customs in the Andaman Islands; historic architecture in Oxford
Sport	Olympic Games, Beijing, 2008; World Cup, Cape Town, 2010
Tourist deterrents at the destination	
Hazards	Earthquake in Haiti, 2010; oil slick in the Gulf of Mexico, 2010
Political unrest	Sri Lanka, 2009; Greece, 2010
Disease	Avian flu (Egypt, 2010); cholera (Zimbabwe, 2010)
Exchange rate	May reduce the spending power of tourists

Table 9.1 Tourist attractions and deterrents at the destination

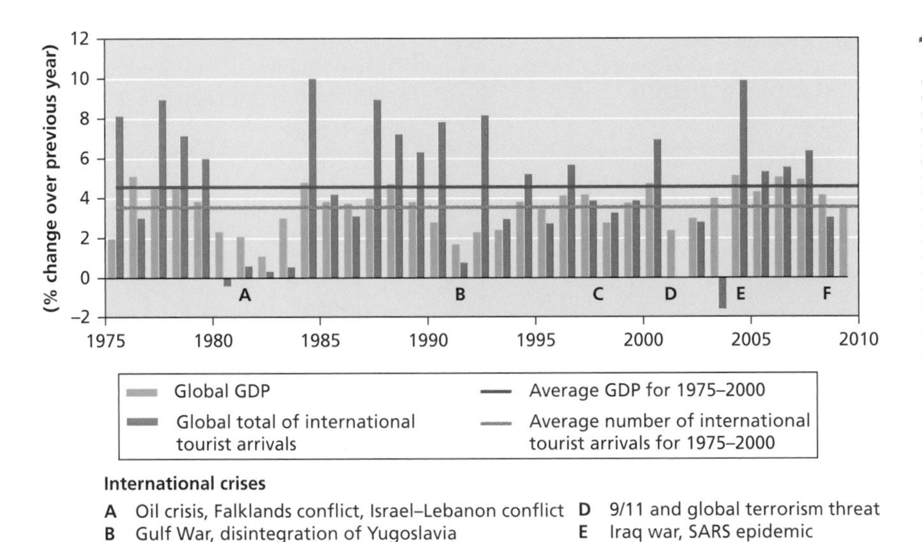

Figure 9.3 Economic growth (GDP) and international tourist arrivals

To do:

Study Figure 9.3.

a Describe the growth rate for international tourist arrivals in the years when the growth rate for GDP is (a) greater than 4%, and (b) less than 2%.

b Explain the relationship between global tourist demand and GDP, as shown by the graph.

c Classify the different types of crises shown on the graph.

d Explain the differences in the level of impact that these crises have had on tourist demand.

e Globalization has made the tourism industry more volatile. Discuss this statement with reference to specific events, places and dates.

Changing fashions in tourism

Figure 9.4 shows a classification of tourism types. The tourism industry has grown in terms of its global extent and the volume of tourists involved. It is a dynamic industry that has developed to meet an ever-growing demand for new and exotic experiences. Mass tourism, which involves large numbers of tourists and has significant impact upon destinations, has lost favour with the more affluent tourist, bored with the sun, sea and sand experience.

Niche tourism has evolved out of a desire for diversity and something new. It offers specialized tourism and operates on a smaller scale than mass tourism. It is more likely to uphold the principle of sustainability because tourist numbers are controlled, and their impacts on the host culture and environment are minimized.

Why does the tourism industry always eventually recover from shocks and global crises?

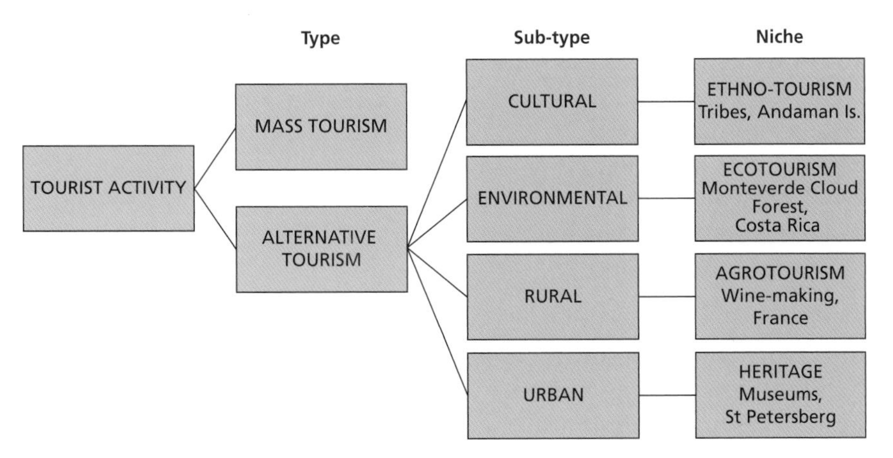

Figure 9.4 Types of tourism

To do:

a Critically evaluate the classification of niche tourism used in Figure 9.4.

b Redesign this diagram using an alternative classification. Justify the changes you have made.

c For each of the niche activities, suggest two additional types and state a location for each one.

Leisure at the international scale: sport

There are significant spatial variations in the participation of sport, and in international sporting success.

Participation rate – the proportion of a population that takes part in a specific sporting activity.

1 Cross-country running

Rank	Country	Number of participants in global top 200	Per capita GNP (2007) US$
1	Kenya	38	1,600
2	Ethiopia	24	700
3	USA	13	46 000
4	Japan	12	33 800
5	Morocco	10	3,800
6	France	10	33 800
7	Spain	9	33 700
8	Portugal	7	21 800
9	UK	6	35 300
10	Russia	6	14 600

2 Golf

Rank	Country	Number of participants in global top 200	Per capita GNP (2007) US$
1	USA	96	46 000
2	UK	27	35 300
3	Japan	16	33 800
4	Australia	16	37 500
5	Sweden	13	36 900
6	South Africa	5	10 600
7	Argentina	4	13 000
8	Spain	3	33 700
9	Ireland	2	45 600
10	Germany	2	34 400

Table 9.2 Global participation in two sports

Factors affecting participation in sport

There are many variations in sporting activity by nations. The USA, for example, is strong in baseball, American football and basketball. In contrast, cricket is largely played in parts of the former British Empire. Gaelic football and hurling are played almost exclusively in Ireland (and sometimes in the UK and the USA among expatriots and their descendants), and Australian Rules football is played mostly in Australia. Global sports include football and athletics.

Physical factors

Physical factors have an impact on sporting participation and success. For example:

- Areas with regular and reliable snow in winter, such as the Alps, favour skiing and winter sports.
- Coastal areas with large plunging breakers produce ideal conditions for surfing, such as in Hawaii and California.
- High-altitude regions, such as those in Kenya and Ethiopia, produce people with a higher concentration of red blood cells, which is said to favour endurance sports such as long-distance running.

Human factors

Most sports take place in sporting venues, such as tennis courts, football pitches and swimming pools. Thus, physical, geographical factors might not be as important as human factors. There is a strong correlation between **economic wealth** and provision of sporting facilities. Most golf courses are found in MEDCs and NICs, although a large number of golf courses have now been created in LEDCs for the benefit of tourists. The development of golf courses in Vietnam is a good example. On the other hand, many children will play ball games on any area of open space they can find.

To do:

Using the data in Table 9.2, showing participation in cross-country running and golf, use Spearman's Rank correlation to test the hypothesis that there is no relationship between wealth and participation in sport.

a Comment on the type and strength of the correlations, and suggest reasons for them.

b Represent this data on two scatter graphs, and plot the best-fit lines.

c Referring to your correlation coefficients and scatter graphs, draw conclusions on the hypothesis that there is no relationship between wealth and participation in sport on a global scale.

d Comment on the strengths and weaknesses of each technique used to show the relationships between two variables.

There is also evidence that **political factors** influence the provision of sporting infrastructure. Geographers have identified different "models" of national sports systems. For example:

- The Eastern Bloc model uses sports to show how successful the communist system is.
- The New World democratic model uses sport to forge an identity for the new nation.
- The Emergent Nation model for Southeast Asia and Africa often uses organizations such as the police force and the army to develop its sporting talent.
- The American model is based on competition and rewarding success.

In the UK, the government has attempted to use sport to develop underprivileged areas and rural areas. In some developing countries a lack of funding means that sports resources are limited. This is important in explaining variations in success in sport.

Social factors are also important. Some people cannot afford the membership fees associated with certain sports. Golf clubs are generally expensive. Boxing is a sport usually associated with a working-class population (although Oxford and Cambridge University each have boxing clubs). Polo is another sport that is largely the preserve of the wealthy.

Cultural factors also influence participation in sport. A good example is the low participation of Muslim women in athletics and swimming. The convention for Muslim women to remain robed means that successful Muslim athletes, such as the Moroccan middle-distance runner Hasna Benhassi, receive much criticism at home.

Case study: *The 2008 Beijing Olympic Games*

The choice of Beijing in China as the host for the 2008 Olympic Games was due to the country's economic progress from a largely rural society towards a more open, wealthy and mainly urban society. Nevertheless, some politicians and NGOs criticized the choice on account of China's disappointing human rights record. China had much to gain from hosting the Games: aside from the prestige and potential economic benefits, the initial outlay on buildings, transport and infrastructure would be a permanent **legacy**, the costs being offset by the profits made from hosting the event.

Economic costs

US$43 billion was spent on infrastructure, energy, transportation and water supply projects, which made it the most expensive Olympic Games by a wide margin (Figure 9.5). The Olympic Park covers 1,215 hectares, and includes an 80 000-seat stadium, 14 gymnasia, an athletes' village and an international exhibition centre, surrounded by 260 hectares of green belt. Other building works include the decentralization of 200 industrial enterprises and the Capital iron and steel works. New hotels were also built and old buildings refurbished.

Beijing's transport infrastructure was expanded significantly. A third airport terminal was constructed, the 120-kilometre Tianjin intercity railway running high-speed trains (350 kilometres per hour) was opened, and the capacity of the subway network was doubled.

Environmental benefits

Measures taken before the Games to reduce Beijing's notoriously high levels of air pollution included:

- the relocation of 150 central factories
- the exclusion of 45% of Beijing's 3.3 million cars and 300 000 heavy-polluting vehicles from the city
- a reduction in coal consumption to generate electricity and replacing it with renewable energy.

The UNEP report, *Beijing 2008 Olympic Games – Final Environmental Assessment*, identified several environmental benefits for Beijing, which justified the $17 billion cost of environmental projects. The successes included:

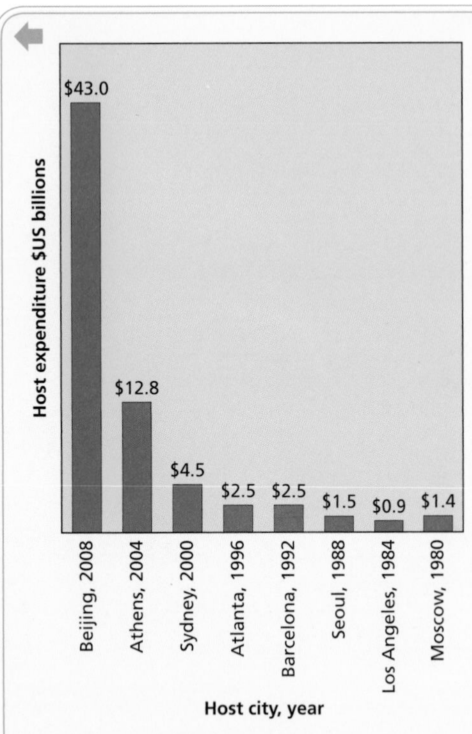

Figure 9.5 Expenditure at the Olympics, in billions

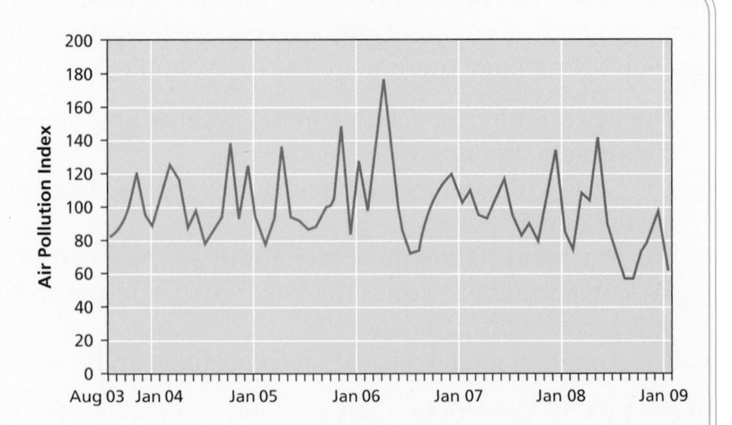

Figure 9.6 Air pollution in Beijing, August 2003–January 2009

- the creation of 720 green spaces in Beijing.
- a waste recycling rate in the Olympic venues 23% higher than targets
- a 47% drop in carbon monoxide concentration, 38% in nitrogen dioxide, 30% in volatile organic compounds, 20% in particulate matter and 14% in SO^2
- hydrofluoric fluorocarbons (ozone-damaging greenhouse gases) phased out ahead of the 2030 target
- more than 20% of the electricity consumed in the venues supplied by renewable energy
- the number of "blue sky" days, when the air pollution index was 100 or below, rose from under 180 in 2000 to 274 in 2008 (Figure 9.6)
- average air pollution reduced by over 45%.

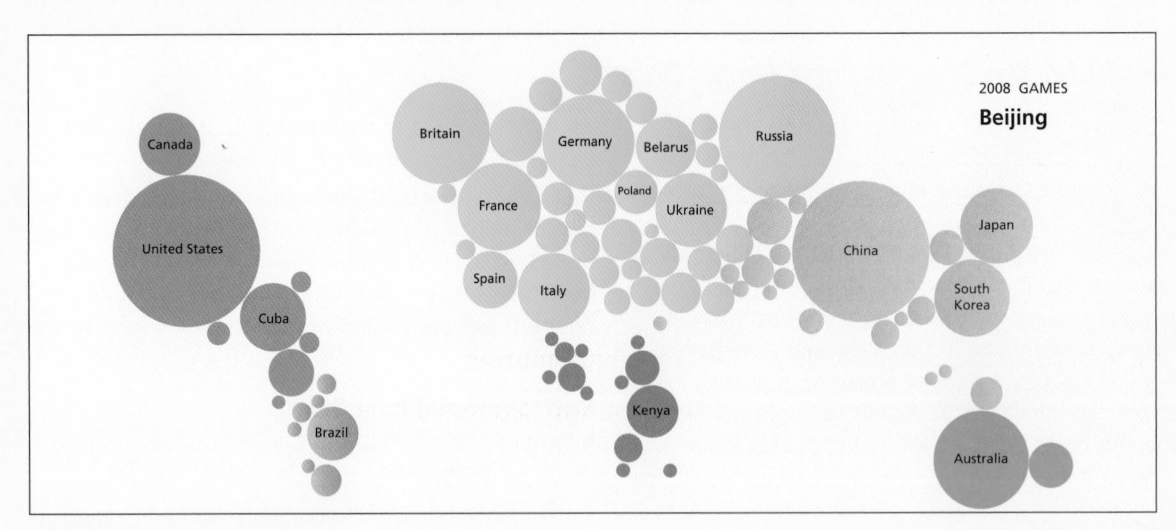

2008 Top 10 medal-winning countries

Country	Gold	Silver	Bronze	Total
United States	36	38	36	110
China	51	21	28	100
Russia	23	21	28	72
Britain	19	13	15	47
Australia	14	15	17	46

Country	Gold	Silver	Bronze	Total
Germany	16	10	15	41
France	7	16	18	41
South Korea	13	10	8	31
Japan	9	6	10	25
Italy	8	9	10	27

Source: Science of Coaching 4 August 2008

Figure 9.7 Beijing medals: the circles represent the total number of medals awarded to each country.

To do:

The map (Figure 9.7) shows the global pattern of medals awarded at the 2008 Olympic Games.

a Describe and evaluate this technique as a means of making international comparisons.

b Suggest two alternative methods of showing this data spatially (the global distribution of medals).

c Describe and explain the distribution of medals shown on the map.

To research

Research the medals awarded by country for previous Olympic Games events, and track the changes in national participation over time. Comment on any changes.

See Olympic Medals Interactive Graphic New York Times

Advantages	Disadvantages
Prestige – it is considered an honour to host the event, and if the games are a success the host city gains in reputation	Possible financial loss – Montreal made a loss of over $1 billion in 1976 and the debt took years to pay off
Economic spin-offs – the event boosts both trade and tourism	Terrorism – the shooting of Israeli athletes at the 1972 Munich Olympics is an example
National unity – such an event unites the country and gives a sense of pride	Overcrowding – a large number of visitors puts a strain on hotels, transport, water supplies, etc.
Improved sports and other facilities – cities build or improve their permanent facilities to host events	Increased security risks – due to the international television coverage, large events are now prime targets for terrorists
Profit – the event may make money through sales of radio and TV rights, tickets and merchandise, as well as spending in hotels, restaurants, etc.	Loss of esteem – if an event does not do well, the host country's image suffers. The host will have difficulty attracting other events

Table 9.3 The advantages and disadvantages of hosting the Olympic Games

Leisure at the International scale: tourism

Butler's model of how tourist areas evolve

This model says that any tourist area develops in seven main stages:

1 Exploration: A small number of tourists is attracted to a new location or to exotic adventurous travel. There is minimal impact, since local people have not developed services exclusively for tourists.

2 Involvement: If tourists are accepted by the locals, the destination becomes better known. There are improvements in the tourist infrastructure, and some local involvement in tourism.

3 Development: Inward investment takes place. Tourism becomes a big business. Firms from MEDCs control, manage and organize tourism, with package tours and less local involvement.

4 Consolidation: Tourism becomes an important industry in an area or region. Former agricultural land is used for hotels. Facilities such as beaches and pools may become reserved for tourists. Resentment begins and there is a decelerating growth rate.

5 Stagnation: There is increased local opposition to tourism and an awareness of the problems it creates. Fewer new tourists arrive.

6 Decline: The area decreases in popularity. International operators move out and local involvement may resume. Local operators may be underfunded, but it is possible for the industry to be rejuvenated, such as in the modernization of UK coastal resorts in the 1990s.

Case study: *Spain, a national tourist industry*

Spain is a classic example of post-1945 growth in tourism, with over 34 million tourists annually. Spain illustrates many of the problems that resort areas encounter as they reach capacity, and the tendency for tourist places to drift downmarket, setting in motion a downward spiral. The key factors that led to the rise of mass tourism to Spain include: its attractive climate, long coastline, distinctive Spanish culture, its accessibility to countries in northwest Europe, and the competitive price of package holidays, especially for the under 30s and over 60s age groups.

Over half the foreign visitors to Spain come from France, Germany and the UK. Most of the travellers head for the south coast, for holidays based on sun, sea and sand. Over 70% of tourists are concentrated into just six regions, which are all in the coastal areas and the Mediterranean islands. The rapid growth of tourism has led to many unforeseen developments. For example, Torremolinos has changed dramatically – before 1960 it was a small fishing village and a tourist resort for only select tourists. However, the town became popular as a centre for package tours, and rapid, uncontrolled developments led to the area being swamped by characterless buildings, a lack of open space, limited car parking and inaccessible sea frontage. Overcommercialization, crowding of facilities such as bars, beaches and streets, and pollution of the sea and beach also occurred.

Linear development on the Costa del Sol

The Spanish Costa del Sol is an excellent example of the linear development of a tourist resort. Before the era of cheap package tours, the area was relatively underdeveloped. From the 1960s, however, rapid and unchecked expansion created a long linear development along the coast. Initially expansion was focused on Torremolinos, Marbella, Fuengirola and Malaga, the main point of entry to the region. However, as tourism increased, tourist developments diffused from larger centres to secondary centres, and there was also infilling between the centres. This infilling included campsites, golf courses and *urbanizaciones*, villa developments that include second or retirement homes belonging to both Spanish and foreign owners, as well as houses and apartments for rent by tourists.

Impacts of tourism on the Spanish Costas

Tourists from the UK to Spain	1960: 0.4 million	1971: 3.0 million	1984: 6.2 million 1988: 7.5 million	1990: 7.0 million	2009: 8.5 million
State of, and changes in, tourism	Very few tourists	Rapid increase in tourism. Government encouragement	Carrying capacity reached: tourists outstrip resources, e.g. water supply and sewerage	Decline: world recession, prices too high, cheaper upmarket hotels elsewhere	Attracting more affluent visitors
Local employment	Mainly in farming and fishing	Construction work. Jobs in cafés, hotels, shops. Decline in farming and fishing	Mainly in tourism - up to 70% in some areas	Unemployment increases as tourism decreases (20%). Farmers use irrigation	Decrease in unemployment
Holiday accommodation	Limited accommodation, very few hotels and apartments, some holiday cottages	Large blocks built (using breeze block and concrete), more apartment blocks and villas	More large hotels built, also apartments and timeshare, luxury villas	Older hotels looking dirty and run-down. Fall in house prices. Only high-class hotels allowed to be built	Development of up-market quality accommodation
Infrastructure (amenities and services)	Limited access and few amenities. Poor roads. Limited street lighting and electricity	Some road improvements but congestion in towns. Bars, discos, restaurants and shops added	E340 opened, "the highway of death". More congestion in towns. Marinas and golf courses built	Bars/cafés closing. Malaga bypass and new air terminal opened.	Upgrading of infrastructure that has deteriorated
Landscape and environment	Clean, unspoilt beaches. Warm sea with relatively little pollution. Pleasant villages. Quiet. Little visual pollution	Farmland built on. Wildlife frightened away. Beaches and seas less clean	Mountains hidden behind hotels. Litter on beaches. Polluted seas (sewerage). Crime (drugs, vandalism and mugging). Noise pollution	Attempts to clean up beaches and seas (EU Blue Flag beaches). New public parks and gardens opened. Nature reserves	20% of Spanish golf courses are found in Andalucia – 8% annual growth

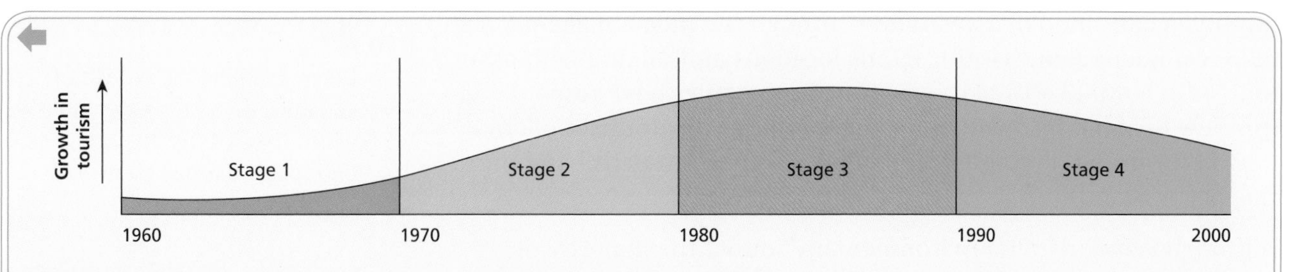

Table 9.4 Changes in tourism on Spanish Costa del sol, according to Butler

To research

a Choose a country (not Spain) with a long-established tourist industry (more than 30 years), and undertake research into the development of this industry. This will involve finding data on international arrivals over 30 years. Other useful supporting data would include:

- tourism as a % of GDP
- % employment in the tourist economy
- tourist exports as a % of total exports.

b Review the development of this country's tourist industry and identify stages according to the Butler model. This will provide a useful case study.

Visit the World Tourism Organization at www.world-tourism.org and the World Travel and Tourism Council at www.wttc.org for information.

Ecotourism

Ecotourism is the responsible development and management of tourism, with the aim of having minimal impact on the environment and the local community. It generally occurs in remote areas with a low density of tourists, and operates at a fairly basic level. Ecotourism includes tourism that is related to ecology and ecosystems. These include game parks, nature reserves, coral reefs and forest parks. It aims to give people a first-hand experience of natural environments and to show them the importance of conservation. Its characteristics include:

- planning and control of tourist developments so that they fit in with local conditions
- increasing involvement and control by local or regional communities
- appropriateness to the local area
- a balance between conservation and development, and between environment and economics.

However, in areas where ecotourism occurs, there is often a conflict between allowing total access and provision of tourist amenities, and conservation of the landscape, plants and animals of the area. Another conflict arises when local people wish to use the resource for their own benefit rather than for the benefit of animals or conservation.

Ecotourism – a "green" or "alternative" form of tourism that aims to preserve the environment by managing it sustainably.

Leakage – economic loss of tourist money, by tourists using companies not owned by the host country, and spending money outside the host country (e.g. on a cruise ship).

The growth of ecotourism

Awareness of the cultural, economic and environmental costs of mass tourism, with its high consumption and disregard for local communities, has been a major impetus to the growth of ecotourism. Consumers are also beginning to translate their concerns about the environment into appropriate action, by choosing holidays that are environmentally and ecologically benign. Ecotourism has been growing dramatically since the 1980s, and now accounts for around 5% of the global tourism market.

Improved education and awareness through the media make people more confident about visiting exotic locations and mixing with other tourists in a multicultural group. Travel companies have responded to this new demand by opening up new locations. Ecotourist holidays now occur in the remotest destinations, such as Antarctica and the Galapagos Islands.

Although ecotourism is environmentally sound, the majority of tourists will have undertaken a long-haul flight to reach a remote destination. The environmental impacts of aircraft noise and emissions are damaging, especially at take-off and landing. At cruise level aircraft emissions can affect the concentration of greenhouse gases, including carbon dioxide, ozone and methane. They also trigger the formation of contrails and may increase cirrus cloudiness.

To do:

Essay: Examine the growth of ecotourism and the extent to which it maintains the principles of sustainability.

Case study: *Ecotourism in Belize*

Belize in Central America, formerly known as British Honduras, was one of the first countries to promote itself as an ecotourism destination. It is English speaking and only two hours flying time from Miami in the USA. Belize covers only 23 000 square kilometres, but has a range of primary tourist resources: mangrove swamps, wetlands, savannah, mountain pine forests and tropical rainforests. It can also offer a coral reef, the second largest in the world after the Australian Barrier Reef, and several notable archaeological sites including those of the Mayan civilization. It has a number of reserves including the Coxcomb Basin Wildlife Sanctuary, and forest which was once exploited for its cedar and mahogany and is now preserved. Visitors to Coxcomb Basin can see a variety of animals and birds, including black howler monkeys, tapir, keel-billed toucans and macaws. It can also offer a range of activities such as excellent fishing, scuba-diving, whitewater rafting, kayaking, and also nature tourism – studying the exotic species of plants and animals of local ecosystems.

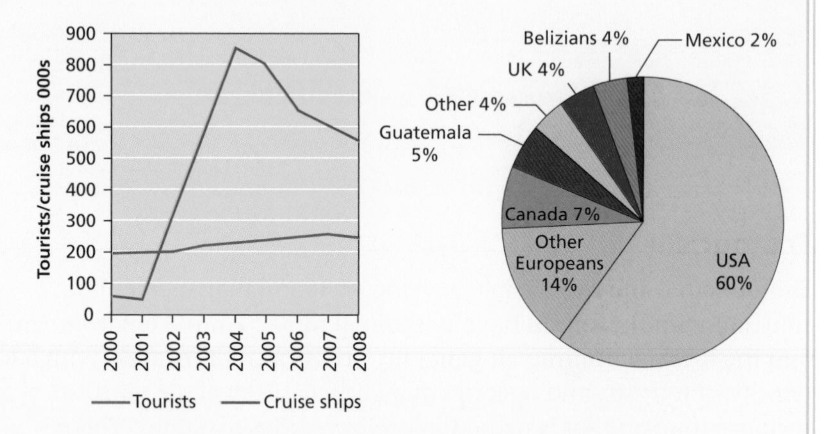

Figure 9.8 Tourist arrivals in Belize and their countries of origin

Tourism is managed by the Belize Tourist Board, which works closely with the government's Ministry of Tourism and the private sector. Tourism has traditionally been regarded as a viable route towards economic development, and in Belize the tourism industry is now an important part of the economy. In 2007 over 25% of all jobs were in tourism and it made up over 18% of the GDP. It has now overtaken cash crop farming as the principal economic sector.

The Community Baboon Sanctuary: successful sustainable tourism

One of the most successful ecotourism projects in Belize has been the Community Baboon Sanctuary at Bermudian Landing, just 43 kilometres from Belize City and along the length of the Belize River. It is a cooperative venture of private landowners (many of whom are subsistence farmers), conservationists and biological researchers. Their common aim is to preserve extensive forest habitats of howler monkeys (known locally as baboons).

Each landowner agrees to adopt more sustainable practices in farming, which will conserve both species and soil as well as increase yields. In return, the farmers agree to preserve areas of forest along the riverside between properties, where the monkeys can survive and swing from tree to tree. More than 100 families living in the villages along a 32-kilometre stretch of the river are involved in this project, and many provide services to tourists such as guided tours, restaurants and accommodation. Some do not have running water and the facilities are often basic, but the hospitality is good.

The principles of ecotourism such as water and energy conservation and waste minimization are practised. This project has been successful in many respects: the howler monkey population has increased threefold since the project began in the mid-1980s, habitats and ecosystems are preserved for scientific research, local people are able to supplement their income, tourist

Figure 9.9 A howler monkey in the Community Baboon Sanctuary, Belize

revenue is kept within the community and leakages minimized, biotic resources are conserved for the scientists, and farming yields have increased.

An evaluation of ecotourism in Belize

Ecotourism, in theory, should sustain the resources on which tourism depends, but once the carrying capacity of an area is exceeded, environmental degradation is almost inevitable. Belize is still not equipped to cope with the high levels of consumption and waste left by tourism. In order to generate funding to improve local infrastructure, the government has allowed much private and unregulated development to occur along the coastline and a massive increase in cruise ships. In one week a typical large cruiser with 5,000 passengers might disgorge up to 800 000 litres of raw sewage, 40 million litres of effluent (from washing machines, basins, showers), 400 litres of hazardous waste including cleaning fluids and acids, and 10 tonnes of semi-decomposed solid food waste. The unregulated dumping of waste along the coastline can be devastating to ecosystems and repulsive to tourists. The future of ecotourism is seriously threatened by the cruise ships.

Cruise tourism fails also to benefit local people because these visitors spend less than half of those on a stopover. Passengers do most of their spending on board ship and that money is repatriated as **leakage**. Investment in tourism has also disadvantaged local people where property development is concerned. The extensive development of the coastline has been financed and managed by foreigners, mostly Americans, and the escalating price of land excludes local people from the market.

The environmental consequences of the arrival of 245 000 tourists per year are now readily visible. Some of the coral of the Hol Chan marine reserve, established in 1987, has become infected with black band disease, a form of algae which kills damaged coral. There has also been a decline in the number of conches owing to overfishing, partly to satisfy the tourism demand for shellfish. Other environmental insults include eutrophication of freshwater resulting from fertilizer runoff.

The following objectives for the sustainable management of the coastline of Belize, including the barrier reef, have been drawn up but not yet implemented. The government aims to:

- conserve the world heritage site of the barrier reef by forming the World Heritage Alliance
- increase knowledge and understanding of the functioning of the country's ecosystems by promoting training programmes
- persuade cruise tourists to spend more time on land
- diversify into two options, offering quality products outside the world heritage site and an alternative area inside, in order to reduce the concentration of tourists in specific areas
- support planning and development of a buffer zone
- impose stricter regulations on cruise ships to make them take their waste back home.

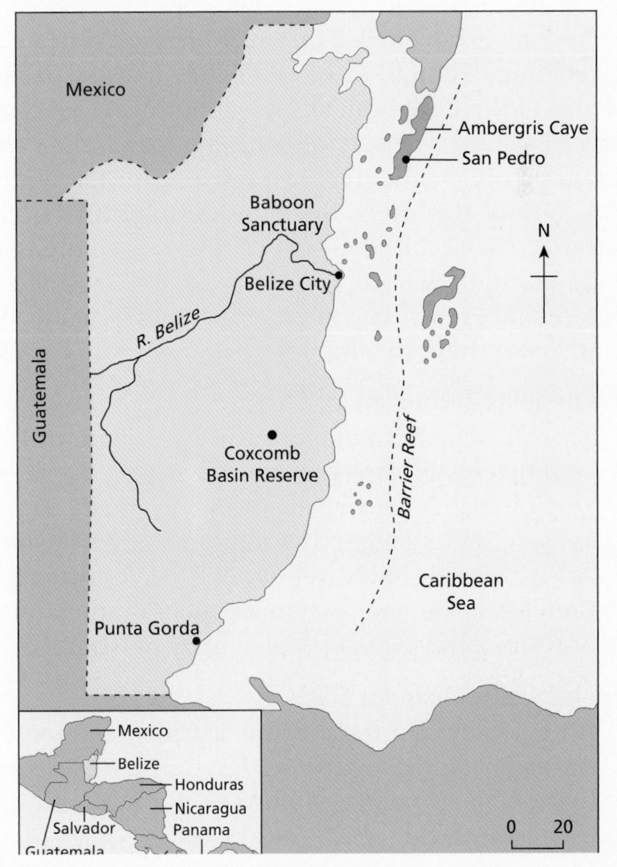

Figure 9.10 Location of Belize

Tourism as a development strategy

Tourism as a development option has many attractions. Tourism is a positive route towards economic development for poor countries, especially when they lack the raw materials for manufacturing. Tourism can be regarded as an export, but goods and services are not subject to the price fluctuation of commodities sold on the world market. Tourism can be an effective way for a country to overcome its problems of balance of payments.

From the perspective of the **host country** and its population, tourism is labour-intensive and can overcome the problem of unemployment both in rural and urban environments. The tourist economy provides jobs directly (tourist welfare, catering, transport, guiding and accommodation) or indirectly (construction, engineering and food production). Tourism also provides opportunities to acquire new skills, for example in languages, catering and entertainment.

Tourism can create a **multiplier effect**, which means that income gained by local people is circulated through the economy by their purchasing of products within the host area. Tourism can redistribute wealth globally, nationally and locally, provided that leakage is not allowed to drain the economy (as when foreign-owned companies manage the business, reap the profits and repatriate them).

Tourism adds diversity to the export base of the country and thereby helps to stabilize its foreign exchange earnings.

Multiplier effect – when an initial amount of spending (usually by the government) leads to increased spending by tourists and so results in an increase in national income greater than the initial amount of spending.

To do:

- **a** Explain the terms "multiplier effect" and "leakage", giving three named and located examples of each. (Named and located means referring to specific tourist destinations.)
- **b** Define "income elasticity" in the context of tourism.

Tourism in small island developing states (SIDS)

Small islands have relied on tourism as a lever to economic development, and in many cases they have been successful. The islands of the Caribbean and Mediterranean are characterized by large-scale, high-density resort complexes, relatively short visitor stays and the gradual replacement of artificial attractions with natural and cultural amenities. Since the 1990s, tourist preference has shifted towards a more authentic experience on remote islands such as St Lucia in the Caribbean, the Maldives and Seychelles in the Indian Ocean, and French Polynesia in the South Pacific. As a development strategy for these islands, tourism brings both benefits and risks.

The benefits of tourism for SIDS

- A small land area and narrow resource base makes manufacturing an unlikely development strategy.
- Tropical islands are well endowed with many natural attractions such as coasts, mountains, ecosystems, heritage sites and indigenous tribes.
- As an exporting industry, tourism is not restricted by quotas or tariffs.
- Both direct and indirect employment in the tourist industry provides jobs for local people, many of whom are untrained.

The risks of tourism for SIDS

- After decades of mass tourism, isolation and remoteness are two of the major tourist attractions, but transport costs are high and access to the core economies of Europe, North America and Japan is usually limited.
- Global shocks will depress international tourist demand, but domestic demand is too weak to compensate for lost international revenue (see Figure 9.16).

Case study: Health tourism as a development strategy in Tunisia

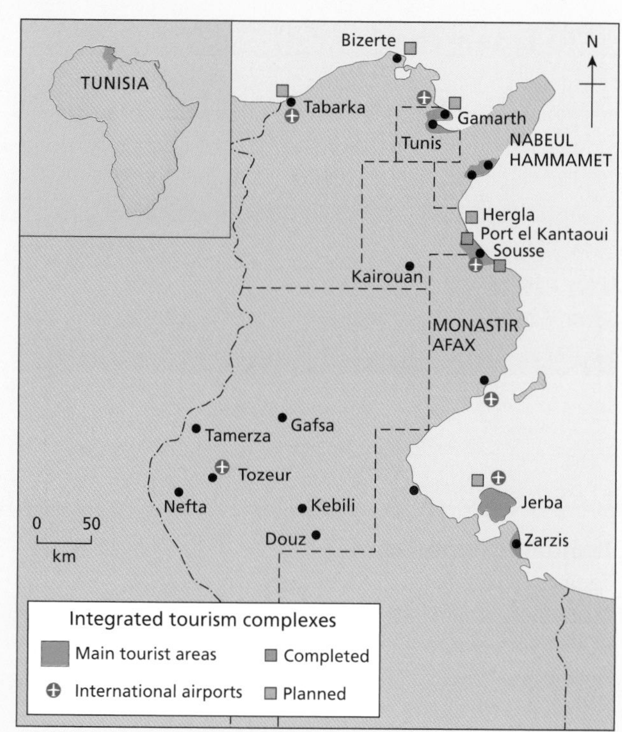

Figure 9.11 Tourist developments in Tunisia

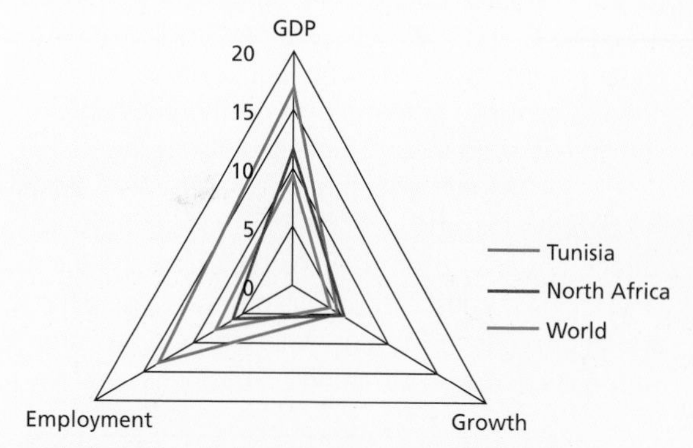

Figure 9.12 The relative importance of tourism in Tunisia, 2009

The tourist industry of Tunisia grew rapidly during the 1990s, but began to stagnate from 2000, when tourism growth in other North African counties such as Morocco and Egypt began to exceed that of Tunisia. After a setback in 2002, when a synagogue in Djerba was bombed, the tourists began to return, but growth rates were still well below those of the rest of North Africa. Since 2005 the Tunisian government has focused on the niche tourist market in an attempt to diversify the offer beyond the traditional mass beach tourism, and is developing all types of tourism that promote well-being. Ecotourism and golf tourism have both become popular, but the demand for health tourism has grown most dramatically, from 42 000 visitors in 2003 to 111 207 in 2008, according to the Ministry of Tourism.

Health and therapeutic tourism is based on the concept of helping the mind and body together, where body ailments are treated in a relaxing atmosphere that takes away the stress of daily life. Natural elements – mineral water, sea, sand and all-year Mediterranean sunshine – provide the supplementary therapies that may be combined with cosmetic surgery or dental treatment. The country has important reserves of mineral water in springs, which are either cold (with water less than 25° C) or hot. The Carthaginians knew about the benefits of thalassotherapy – bathing in warm sea water – even before the Romans built their thermal baths in Tunisia's Cap Bon region. Tunisian cities are rich in water springs running from north to south, with more than 51 therapeutic mineral water centres, which aim to help pain, stress and fatigue. The government's strategy is to encourage the construction of comprehensive health cities, complete with specialized centres, hospitals, pharmacies and pharmaceutical production plants. In partnership with the private sector, 17 centres will be constructed.

Health tourism, and especially cosmetic surgery, is approved and regulated by the Ministry of Tourism, and health specialists oversee every aspect of patient treatment. This includes arranging the flight and transfers, the medical treatment and convalescence at a 5-star hotel overlooking the beach for five days.

Figure 9.13 Thalassotherapy pool, Tunisia

Tunisia appeals to those who are short of time and want to save money. For instance, the cost of a facelift in

Tunisia is around €3,200. The same operation in France would be €5,000 or more. The next challenge for the country is to change customer perception: doctors still need to convince potential patients that they are competent, even though they are working in a developing country. With this aim, government strategies are directed towards expanding the market beyond Europe to Canada, the USA, Russia, China and the Gulf. Developing domestic tourism is proving to be difficult, and the majority of customers continue to be European women with money to spend on cosmetic surgery.

To research

Study your syllabus (Section Two, Option E), and decide how this case study on Tunisia might be relevant to different topics within this option. Do the same for other case studies in this chapter. You will find that some can also be applied to other options.

Case study: Tourism as a development strategy in the Maldives

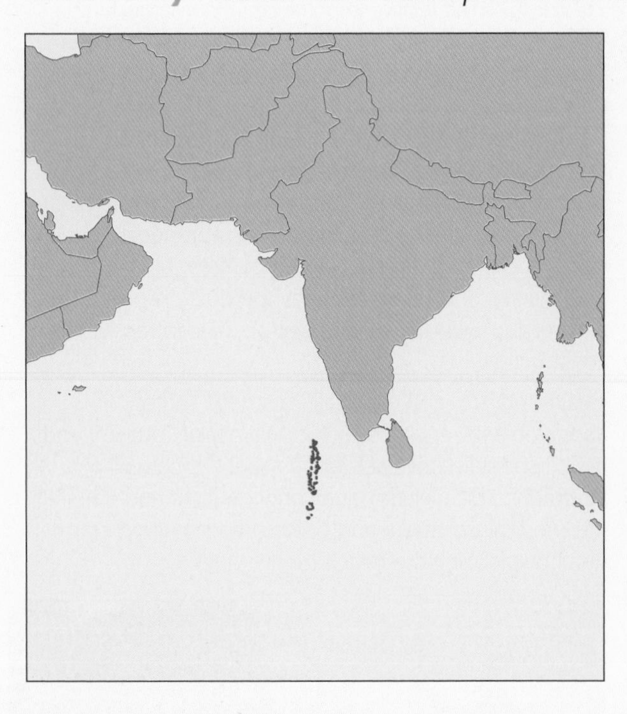

Figure 9.14 Location of the Maldives

The Maldives consist of an archipelago of 1,190 tiny islands, only 200 of which are inhabited by its population of 300 000. Located at latitude 3° north, the islands experience a tropical monsoon climate with hot (26–31°C) and humid conditions most of the year and a dry season from October to April. For the majority of the 400 000 tourists who visit the islands annually, the sun-sea-sand combination makes this an ideal holiday destination.

Economic importance

Tourism accounts for 28% of the Maldives' GDP and more than 60% of its foreign exchange receipts. Over 90% of government tax revenue comes from import duties. The development of tourism has fostered the overall growth of the country's economy. It has created direct and indirect employment and income generation opportunities in other related industries, the so-called multiplier effect. Fishing was once the prime industry here, but more recently tourism has assumed much greater importance. Agriculture and manufacturing continue to play a lesser role in the economy, constrained by the limited availability of cultivable land and the shortage of domestic labour. Most staple foods must be imported. Industry, which consists mainly of garment production, boat building and handicrafts, accounted for only about 7% of the Maldives' GDP in 2009.

Vulnerability to external shocks

Tourist demand and revenue grew rapidly between 1990 and 2009, at an annual rate of 5%, but the rate has fallen and is predicted to average 2% between 2010 and 2020. The Maldives, like many SIDS, are vulnerable to external shocks, whether from natural causes or human activity, and both are beyond their control.

Sea-level rise is a real threat to the Maldives, where the average ground level is only 1.5 metres above sea level; it is the lowest country on the planet. Over the last century sea level has risen globally by about 20 centimetres, and current estimates suggest that it will rise by a further 59 centimetres by 2100.

Tsunamis are another threat. These islands are in a tectonically active zone and were badly affected by the

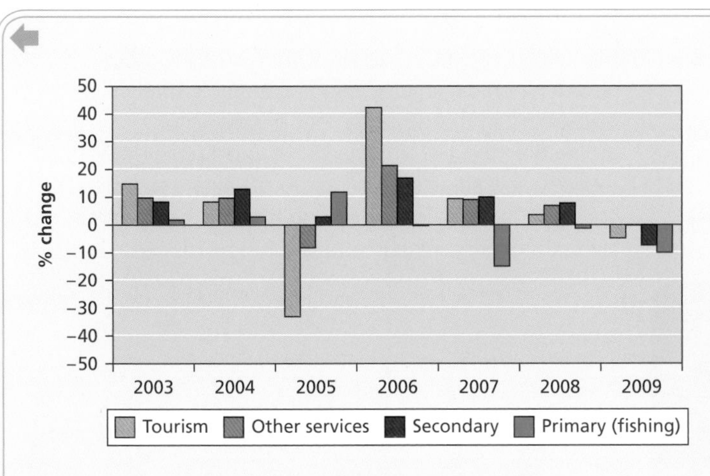

Figure 9.15 Changes in the contribution to GDP by economic sector in the Maldives

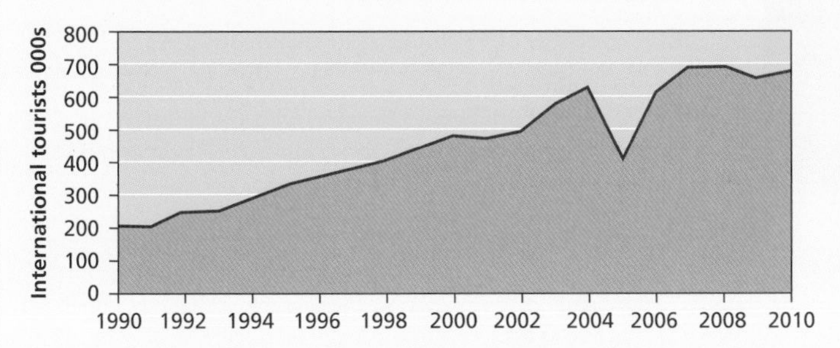

Figure 9.16 The growth of tourism in the Maldives

Asian tsunami of December 2004, which left 100 dead, displaced 12 000 and caused $300 million of property damage. As a result of the tsunami tourist demand fell, and GDP contracted by 4.6% in 2005. Although reconstruction was rapid and tourist numbers recovered, the trade deficit grew as a result of high oil prices and the import of construction materials required to rebuild resorts.

Since the world economic recession of 2008, growth has been stagnating for all sectors. The national economy has a narrow base that relies heavily on tourism, with a limited source area supplying 70% of all tourists. Any threats to this limited market from changes in the world economy will be serious.

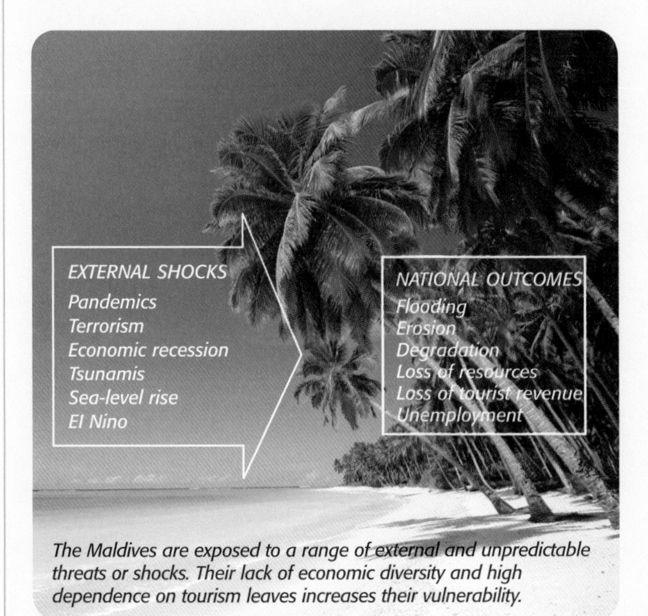

Figure 9.17 The vulnerability of the Maldives

Vulnerability to internal limitations

Depletion of resources

The average tourist consumes more water and energy than the average Maldivian, and both are expensive to produce. Almost all water is provided by desalination. Groundwater supplies are running short, and often contaminated by sewage or saltwater incursion. All energy is produced by generator except on Male and Hulhulé Islands. The Maldives have no economic minerals and agricultural potential is very low.

Pollution

Solid and liquid waste production by tourists is also high, and matches their consumption. Solid waste is either dumped in poorly managed landfill sites, incinerated or dumped at sea. All three processes are unsustainable.

How the Maldives is responding to threats

The environment

All new tourism development sites must undergo an environmental impact assessment to measure potential construction carrying capacity for the area. Developers must produce a mitigation plan against flooding. This must include a 40-metre setback from the high-water mark, a maximum of 20% land coverage by buildings, and building height restriction to tree-top level. Building materials such as sand and aggregates must be imported mainly from India, but there is scope for recycling demolition and building waste.

The problem of waste management is addressed through the compulsory installation of incinerators, bottle crushers and compactors in all resorts. Sewage disposal through soak-pits into the aquifer is discouraged, and all new resort developments must now have their own wastewater treatment plants. The installation of desalination plants for the provision of desalinated water in tourist resorts has also substantially reduced the stress on the natural aquifer.

President Mohamed Nasheed pledged to make the Maldives carbon neutral within a decade, which means encouraging the development of solar and wind energy. In 2009 he held an underwater cabinet meeting to emphasize his commitment to addressing the problem of climate change and consequent sea-level rise.

The economy

The Maldivian government aims to diversify the economy beyond tourism and fishing, and encourage linkage between tourism and other sectors such as construction, manufacturing and transport. Other aims include reforming public finance and encouraging foreign investment in the development of new resorts, broadening the tourism market by promoting domestic tourism, and attracting visitors from China and India. The goal is to increase employment, which is problematic given the limited resource base of the Maldives.

Figure 9.18 Underwater cabinet meeting in the Maldives

Be a critical thinker

Are the Maldives a realistic long-term tourism destination?

Consider the advantages and disadvantages of tourism development in the Maldives, covering a range of perspectives. For example, consider who benefits, who loses, and the long- and short-term impacts.

Sport at the national level

Case study: *A national sports league: rugby in South Africa*

Rugby is one of South Africa's big three sports, alongside soccer and cricket. The country has fared extremely well on the world stage.

For the disadvantaged people of the old apartheid South Africa, rugby was the white person's game, and even more so the game of the Afrikaner. Traditionally, most communities of colour played soccer while, for white communities, rugby was the winter sport of choice.

The Super 14

The Super 14 competition features 14 regional teams from South Africa, New Zealand

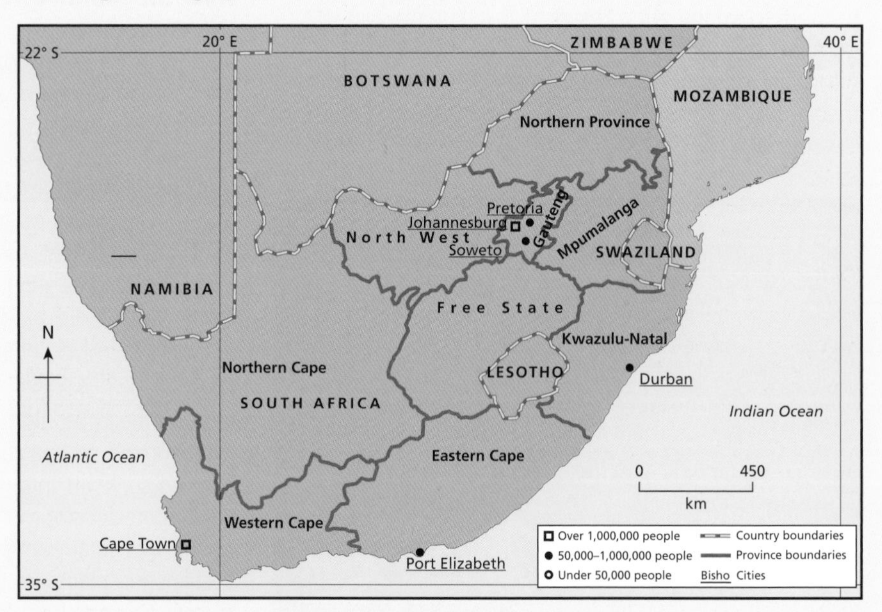

Figure 9.19 South Africa and its provinces

and Australia, with South Africa providing five teams, New Zealand five and Australia four.

The South African teams in the competition are:

- the Sharks – made up of players from the Natal Sharks (based in Durban, Natal and KwaZulu)
- the Stormers – made up of players from provincial teams the Western Province and Boland Cavaliers (Cape Winelands and the west coast of the Western Province)
- the Central Cheetahs – made up of players from provincial teams the Cheetahs, Griquas and Griffons (largely northern Free State)
- the Cats – made up of players from provincial teams the Lions, Pumas and Leopards (Johannesburg, North West and Mpumalanga)
- the Bulls – made up of players from provincial teams the Blue Bulls and Falcons (Northern Transvaal, East Rand and Pretoria).

The Currie Cup

The Currie Cup is the premier provincial rugby competition in South Africa, and was first contested in 1892. The format of the Currie Cup varied from year to year, and finals were held intermittently until 1968, after which the final became an annual event.

Up to and including 2007, the most successful province in the history of the Currie Cup is Western Province (Western Cape), with 32 titles (four shared), followed by the Blue Bulls (Northern Transvaal) with 22 (four shared), the Lions (Transvaal) with nine (one shared), the Natal Sharks (Natal) with four, and the Cheetahs (Free State) with four (one shared). Other teams that have lifted the trophy include Griquas (Northern Cape) (three times) and Border (Eastern Cape) (twice, both shared).

For many years the biggest rivalry in South African rugby was between Western Province and the Blue Bulls. During the early to mid-1990s this was superseded by a three-way rivalry between Natal, the Lions and Western Province. The Blue Bulls have returned to Currie Cup prominence, however, while the Free State Cheetahs won three titles in succession, from 2005 to 2007, including sharing the Currie Cup with the Blue Bulls in 2006.

The Currie Cup takes place roughly between July and October. The format divides 14 teams into eight Premier Division and six First Division teams. The teams, in alphabetical order, are: Blue Bulls, Boland Cavaliers, Border Bulldogs (East London), Eagles, Falcons (East Rand and Gauteng), Free State Cheetahs, Golden Lions (Johannesburg), Griffons (Welkom), Griquas (Northern Cape), Leopards (Mpumalanga), Mighty Elephants (Port Elizabeth), Natal Sharks, Pumas (North West) and Western Province (Western Cape).

The Vodacom Cup

The Vodacom Cup has become an important competition on the South African rugby calendar. It takes place at the same time as the Super 14 competition – starting in late February and finishing in mid-May – and thus creates a platform for talented young players who might otherwise not get a chance to make their mark.

It has also been a fertile breeding ground for strong players from previously disadvantaged backgrounds, thanks to the enforcement of quotas. Quotas, successfully implemented lower down, now extend through the higher levels of South African rugby, including the Super 14.

The Vodacom Cup is divided into two sections – North and South – with the top two teams advancing to the semi-finals and playing cross-section matches of one-versus-two for a place in the final.

The North is made up of the Golden Lions, Griffons, Leopards, Pumas, Falcons, Blue Bulls and Griquas. The South's teams are the Mighty Elephants, Boland Cavaliers, Border Bulldogs, Free State Cheetahs, Eagles, Western Province, and KZN Wildebeests (KwaZulu Natal).

Leisure at the local scale

In the context of leisure, we need to distinguish between **environmental carrying capacity** (the maximum number before the local environment becomes damaged) and **perceptual carrying capacity** (the maximum number before a specific group of visitors considers the levels of impact, such as noise, to be excessive). For example, young mountain bikers may be more crowd-tolerant than elderly walkers.

Carrying capacity – the maximum number of visitors or participants that a site or event can satisfy at one time.

Case study: Carrying capacity in Venice

Figure 9.20 Flooding in Venice, 2001

The historic centre of Venice comprises 700 hectares, with buildings protected from alterations by government legislation. Environmental degradation threatens the city, which is sinking into the lagoon and floods several times a year. There is a conflict of interest between those employed in the tourist industry (and who seek to increase the number of tourists) and those not employed in the tourist industry (and who wish to keep visitor numbers down). However, the pattern of tourism is not even. There are clear seasonal variations, with an increase in visitor numbers in summer and at weekends. Research has estimated that an average of 37 500 day-trippers a day visit Venice in August. A ceiling of 25 000 visitors a day has been suggested as the maximum carrying capacity for the city.

Exceeding the carrying capacity has important implications for the environment and its long-term preservation. The environmental carrying capacity (concerned with preservation) and the economic carrying capacity (concerned with economic gain) have different values, but the 25 000 figure is a useful benchmark. In 2000 the carrying capacity of 25 000 visitors was exceeded on over 200 days, and on seven of those days visitor numbers exceeded 100 000.

The large volume of visitors travelling to Venice creates a range of social and economic problems for planners. The negative externalities of overpopulation stagnate the centre's economy and society through congestion and competition for scarce resources. This in turn has resulted in a vicious circle of decline, as day-trippers, who spend little or nothing in hotels and restaurants and thus contribute less to the local economy than resident visitors, replace the resident visitors as it becomes less attractive to stay in the city. Thus the local benefits of tourism are declining.

A number of measures have been made to control the huge number of day-trippers. These include:

- denying access to the city by unauthorized tour coaches via the main coach terminal
- withdrawing Venice and Veneto region's bid for EXPO 2000, since a study concluded that it would cause irreparable ecological damage to the city.

There has also been a proposal to charge for entrance to the city in order to control numbers. Nevertheless, the city continues to market the destination, thereby alienating the local population, who are increasingly leaving the city to live on the mainland.

The excessive numbers of day-trippers have also led to a deterioration in the quality of the tourist experience. This is significant in that it highlights problems affecting many historic cities around the world, especially those in Europe.

Leisure in urban areas

A simple hierarchy of leisure can be established for urban areas, depending on population size and number of people needed to support a sporting activity. In most small settlements few sports facilities are available. However, as settlement size increases and the threshold population increases, settlements are able to offer a greater variety of sports facilities with increasingly specialist functions.

Community size	Recommended facilities	Activities offered
Village (pop. 500–1,500)	Community hall, community open space, mobile library	Badminton, keep fit, yoga, football, cricket
Small country town (pop. 2,500–6,000)	As above, plus tennis courts, sports hall, swimming pool	As above, plus tennis, netball, gym, hockey
Town	As above, plus specialist sports venues, golf courses, skateboard parks, bowling green	As above, plus bowling, golf, skateboarding, judo, karate
City	Sports stadiums, athletics grounds	As above, plus home grounds of sports clubs (football, rugby, hockey, athletics grounds)
Capital city	National sports centre for selected sports	As above plus international facilities

Table 9.5 The leisure hierarchy

Intra-urban spatial patterns

Figure 9.21 shows the distribution of leisure facilities around a typical large town or small city in the UK. In most large towns there is a concentration of leisure facilities and tourist attractions in the central area, while on the periphery there are sports and leisure centres, garden centres and country parks. The central area contains the main concentration of restaurants, cinemas, theatres and other facilities that do not require much space. Finally, there may be some other leisure facilities dispersed into neighbourhoods, such as parks, recreation grounds and community centres.

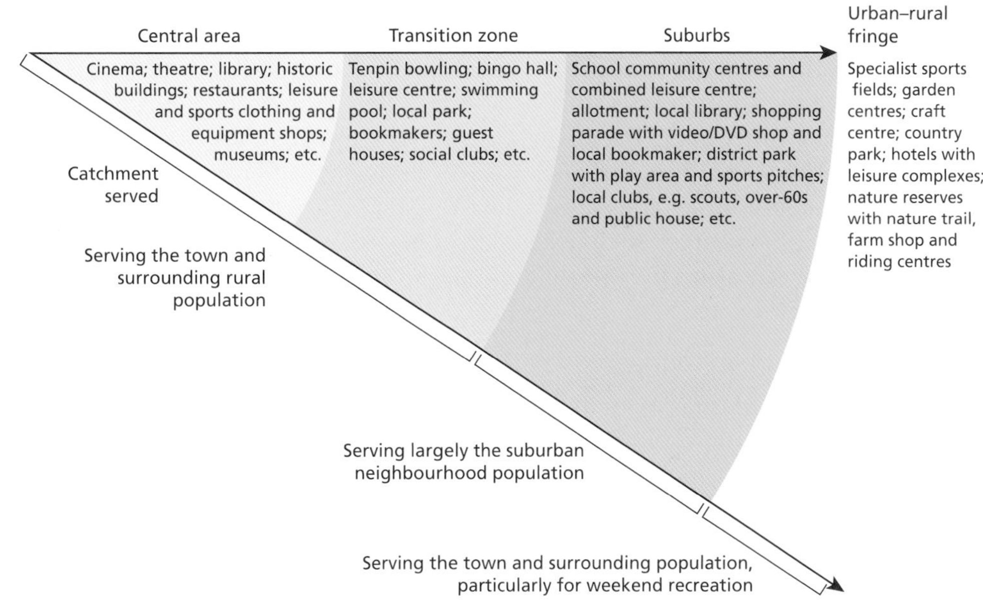

Figure 9.21 A model of leisure activities and land uses in and around a typical UK town or city.

To do:

- **a** Draw an annotated map of one segment of land stretching from the CBD to the outer edge of the suburbs of your local town.
- **b** Plot leisure facilities from the centre to the edge.
- **c** Compare this segment with that of the model (Figure 9.21) and explain any major differences.

Tourist facilities in urban areas

Urban areas are important for tourism because they are:

- destinations in their own right
- gateways for tourist entry
- centres of accommodation
- bases for excursions.

The tourist business district

In most urban areas there is a distinct pattern in the distribution of tourist activities and facilities. The tourist centre of the city is often referred to as the RBD (recreational business district) or the TBD (tourist business district). In many cities, the TBD and the CBD (central business district) coincide.

Tourist facilities in urban areas include accommodation, catering and shopping. Most tourist-related accommodation is found in urban areas, and urban infrastructure and accessibility are vital in the location of hotels and guesthouses.

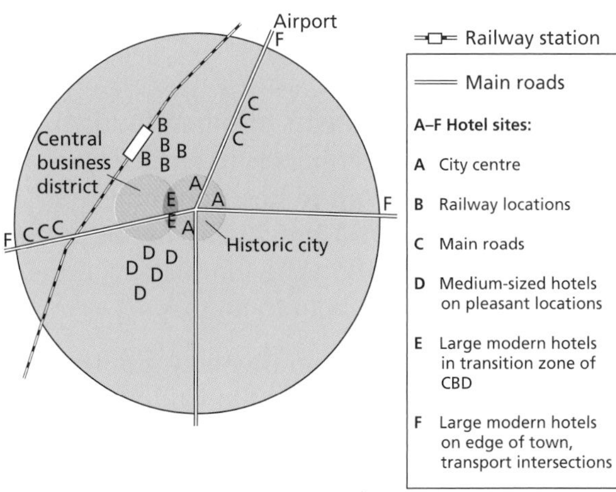

Figure 9.22 The tourist business district of an urban area

Urban tourism

Urban tourism is nothing new, and cities continue to attract more visitors than rural areas. Cities aiming to attract tourists present many challenges for planners. They have plenty of potential, especially if they are old, which partly accounts for the popularity of many European cities as a destination.

Several old industrial cities in Britain have had a facelift, investing money in conserving old buildings and turning parts of the city into heritage zones. Cultural and ethnic variations are also dimensions of a city that can be developed for tourists; many cities have Chinatowns such as London, San Francisco, New York and Vancouver. The biggest attractions in urban areas are often heritage monuments such as cathedrals, art galleries and museums, but large cities also have the advantage of a greater range of resources such as concert halls, theatres and stadia, so they can also offer music, drama and sporting events such as the World Cup.

The benefits of urban tourism

- Urban tourism has created employment through the regeneration of old buildings.
- It is labour-intensive and produces a multiplier effect, which are especially important in areas of high unemployment.
- It revives old inner-city areas.
- It may improve infrastructure (roads, water, telecom).
- It may help preserve aspects of the local culture (food, clothes, festivals and history)
- It generates resources through increased foreign exchange.
- It may promote development of healthcare and other services for tourists that then become available to local people.

The disadvantages of urban tourism

- Jobs are often poorly paid, unskilled and seasonal.
- Councils spend money on affluent tourists rather than poor locals.
- Tourism produces additional environmental stresses of litter, congestion and traffic.
- Conflicts may arise between residents and visitors because shops stock gifts and souvenirs rather than food, and hotels occupy land needed for housing.
- Application of visitor management techniques might impact on local residents, reducing access to certain areas.
- It may cause cultural destruction through modernization and homogenization.

Solutions to the problems of urban tourism

The challenge for planners is to ensure that visitors can support the values of the host community in a way that sustains (or improves) the quality of life in the community. It has been suggested that the following actions by both visitors and residents would help to do this:

- Act to support cultural diversity and legacy.
- Engage in activities that add value to the community.
- Avoid activities that degrade the environment.
- Act to disperse the benefits.
- Patronize locally owned enterprises.

To do:

Urban field work techniques

a Explain the difficulties in classifying the shops and other types of outlet in the CBD.

b Explain how you would distinguish tourist from non-tourist outlets.

c Explain how you would determine the edge of the TBD in order to map it.

To do:

Ideas for fieldwork

a Assess the impact of a local recreational event such as a sports competition, festival, agricultural show, or display.

b Assess carring capacity of a rural honeypot site, and evaluate management strategies.

c Conduct a survey of tourist/resident conflict in an urban tourist centre.

d Conduct a survey of recreational provision in socially contrasting urban areas.

More material available:

www.oxfordsecondary.co.uk/ibgeography
Urban tourism in Oxford

Sport and urban regeneration

Sport has great potential for tourism and economic development. For example, the 1991 World Student Games were held in Sheffield, UK, and attracted over 5,500 competitors from 110 countries. Four new sports arenas were built specially for the Games, costing almost £150 million, including the 25 000-capacity Don Valley Stadium. These facilities remain important not only for Sheffield but for a large region surrounding the city.

The Olympic Games and regeneration

Barcelona 1992

An example of a city that has benefited from the Olympics is Barcelona. The 1992 Olympics marked a watershed in the city's economic development, acting as a springboard in its drive to become one of Europe's leading cities.

The Olympics were more than just an expensive marketing campaign. They brought big investments, not just in sports facilities but in major infrastructure projects such as telecommunications and roads. These in turn attracted investment badly needed by a city which had seen its long-established and old-fashioned industrial base decline in the 1970s and 1980s.

Transport improvements were crucial. Over £6.5 billion, raised from the public and private sector, was spent on enlarging the port, building an additional runway, establishing a high-speed rail link and improving public transport. This infrastructure, coupled with no less than five university campuses, has helped bring high-technology industries. Sony, Sharp, Hewlett Packard, Pioneer, Panasonic and Samsung are just some of the electronics groups represented in or near Barcelona.

Tourism has also benefited. Barcelona attracts an increasing number of tourists. More and more cruise ships either start or end their voyages at Barcelona's port. Many of their passengers choose to stay a night or two in town, either at the beginning or the end of the cruise.

London 2012

One important element in bringing the Games to London was the prospect of regeneration in the East End of London. The Games should bring an economic boom to the capital and to the UK as a whole.

The main effect of staging the games in 2012 will be the complete transformation of the Lower Lea Valley, the largest remaining regeneration opportunity in inner London. The area is home to one of the most deprived communities in the UK, with high unemployment. Key Olympic venues will be built there, including an 80 000-seat stadium, a world-class aquatic centre, a velodrome and BMX track and a three-arena sports complex. There will also be an Olympic village to accommodate up to 17 800 athletes and officials.

Up to 12 000 permanent jobs will be created in the area of the Olympic park alone, as well as thousands of temporary ones. At least 7,000 jobs will be created in the construction sector. The games should also provide a boost for the tourism industry as more than

Be a critical thinker

Who benefits from hosting an international sporting event?

Consider the meaning of "benefit" in its broadest context, to include the type of benefit, its extent, and the beneficiaries. This means considering:

- benefit type – which might be economic, environmental or social
- extent – the area affected, which might be the county, region, city or local area.
- The people affected who gains and who loses art.
- benefits over time – in the long, medium and short term.

You will need to use case study evidence.

half a million visitors head to the UK. Analysts believe Olympics-generated income from tourism that year could reach £2 billion.

The Commonwealth Games, Manchester 2002

In 1995 Manchester was awarded the 2002 Commonwealth Games. It also had an unsuccessful bid for the 2000 Olympic Games. Despite losing the bid, Manchester has benefited from the construction of new facilities and a considerably improved infrastructure.

Greater Manchester's motorway box – the equivalent of the M25 around London – was completed in 1997. Its tram network has been extended to the stadium site, and road approaches spiral into the stadium parking area, making parking relatively easy.

The central theme of Manchester's approach has been urban regeneration. The east side of Manchester has not enjoyed the economic buoyancy of the city centre or south Manchester, nor did it benefit from major projects such as the redevelopment of Trafford Park and Salford Quays, the old Ship Canal docklands, in west Manchester. However, regeneration of this side of the city is now under way.

The stadium is located within the city's inner ring road and is on a derelict site once contaminated by heavy engineering waste and gasworks. Since the area was decontaminated and cleaned up, there have a number of new developments such as the national velodrome, a world-class cycling stadium, and a national indoor tennis centre.

To research

Visit www.london2012.com, the official website of the London Olympics, which includes information on the regeneration of the area, and www.leevalleypark.org.uk/ for more information on the regeneration of the Lee Valley.

Visit www.neweastmanchester.com for news about the regeneration projects for this part of the city.

Sustainable tourism

Sustainable development has been defined as "development that meets the needs of the present without compromising the ability of future generations to meet their own needs" (Brundtland Report, 1987). The concept of sustainable tourism has often used the idea of carrying capacity. **Carrying capacity** may be physical, ecological or perceptual.

The principles of sustainable tourism

Sustainable tourism is that which

- operates within natural capacities for the regeneration and future productivity of natural resources
- recognizes the contribution of people in the communities, customs and lifestyles linked to the tourism experience
- accepts that people must have an equitable share in the economic benefits of tourism.

The key objectives for sustainable tourism are to maintain the quality of the environment while maximizing the economic benefits. This entails:

- **using resources sustainably:** the sustainable use of natural, social and cultural resources is crucial and makes long-term business sense
- **reducing overconsumption and waste:** this avoids the cost of restoring long-term environmental damage and contributes to the quality of tourism
- **maintaining biodiversity:** maintaining and promoting natural, social and cultural diversities is essential for long-term sustainable tourism and creates a resilient base for industry

Sustainable – within the limits of our resources so that human needs can be met indefinitely.
Physical carrying capacity – the measure of absolute space, for example the number of spaces within a car park.
Ecological carrying capacity – the level of use that an environment can sustain before environmental damage occurs.
Perceptual carrying capacity – the level of crowding that a tourist will tolerate before deciding that a location is too full.

- **supporting local economies:** tourism that supports a wide range of local economic activities and which takes environmental costs and values into account both protects these economies and avoids environmental damage
- **involving local communities:** the full involvement of local communities in the tourism sector not only benefits them and the environment in general but also improves the quality of the tourist experience
- **training staff:** staff training which integrates sustainable tourism into work practices, along with recruitment of local personnel of all levels, improve the quality of the tourism product
- **marketing tourism responsibly:** encouraging tourists to visit sites during off-peak periods reduces visitor numbers at times when ecosystems are less robust; marketing provides tourists with full and responsible information, increases respect for the natural social and cultural environments of destination areas and enhances customer satisfaction
- **undertaking research:** ongoing monitoring by the industry, using effective data collection analysis, is essential to help solve problems and to bring benefits to destinations, the industry, tourists, and the local community
- **integrating tourism into planning:** this entails including plans for tourism and development into national and local planning policies, as well as management plans which undertake environmental impact assessments, to increase the long-term viability of tourism
- **providing better information:** giving tourists information in advance and in situ (such as through visitor centres) about tourist destinations.

Managing tourists–environmental impact

The usual ways of controlling tourists are to use:

- spatial zoning
- spatial concentration or dispersal
- restrictive entry or pricing.

Spatial zoning defines areas of land that have different suitabilities or capacities for tourists. Honeypot sites are commonly protected. These are locations that attract tourists by virtue of their promotion and provision of information, refreshment and parking, and then prevent further penetration of tourists into more fragile environments. The Grand Canyon is a good example. Elsewhere, restrictions on tourists may be achieved through pricing. In the USA national parks charge an entry fee whereas in the UK entry to national parks is free.

Be a critical thinker

How can we assess carying capacity?

Constantly changing environmental conditions and human behaviours are both impredictable.

Case study: *The impact of tourism in the Swiss Alps*

Ten million tourists visit the Swiss Alps every year. Slopes and skies are exposed to hundreds of flights, freight and holiday traffic. New ski lifts replace old ones, which may be left abandoned and obsolete: cables, pylons and deserted construction sites litter the mountains, while the human traffic destroys vegetation. The region is a water reservoir for both the Po (Italy) and the Rhône (France), and for Switzerland itself. Rising temperatures are melting glaciers, reducing snow cover and accelerating rates of rock weathering.

The OECD predicts that ski resorts in the Alps below 1,050 metres will be unviable within 20 years, and by the end of the 21st century only ski resorts over 2,000 metres will have guaranteed snow; and by 2030 50 of Switzerland's 230 ski resorts will be redundant.

A predicted increase in the number of Alpine skiers combined with a reduction in the number of ski slopes will result in over-intensive use and environmental degradation such as erosion, loss of species diversity, disruption of ecosystems and pollution. Many communities in the Alps are dependent on tourism, which provides both direct employment (hotel staff, restaurant owners, ski lift operators) and indirect employment (farmers, builders, mortgage companies). Global warming will reduce the tourist season and the income of those dependent on the skiing industry.

Monitoring climate change in the Swiss Alps

Switzerland is one of the world's richest economies, which allows it to invest in research, monitoring and hazard mitigation programmes to reduce the threat to its population.

The **Swiss Glacier Monitoring Network** has kept records of glacier characteristics over a number of decades. They have monitored their length, area, volume, mass change (the balance between snow accumulation and ice ablation) through remote sensing, the use of high-resolution topographical maps and fieldwork research. These characteristics are monitored twice yearly, once after the spring snowmelt and once after the summer ablation.

The purpose of the research is to identify high-risk glaciers, those that are melting more quickly than others and experiencing high rates of mass change. Local communities can set up warning and preparedness programmes in the event of an impending hazard event.

Mitigation techniques include monitoring meltwater flows, averting valley blockage by driftwood, and constructing escape channels.

Managing mountain tourism in the Alps

Water conservation

The Alps provide regular supplies of fresh water through river systems and glacier melt. The flow may be regulated by dams and reservoirs to protect lowland populations from flooding and provide them with HEP.

Watershed management

Maintaining existing tree cover or reafforestation reduces water and soil loss upstream and the likelihood of destructive landslides, flooding and sedimentation downstream. Both soil and vegetation are slow to regenerate under cold conditions, and careful monitoring of mountain hydrology is essential.

Maintaining biodiversity

High-altitude species may be restricted to small mountain areas, and have nowhere to grow if their habitat is damaged. Conservation areas aim to minimize visitor impacts by zoning them in time and space. This means restricting access to certain areas and controlling opening times.

Minimizing the impact of skiing

Mountain Riders, a Grenoble-based green campaign group, carried out the first carbon audit of a ski resort. It showed that 75% of its greenhouse-gas emissions arose from transporting skiers and boarders to their destination, as well as from servicing the resort with everything from beer to bed linen. Mountain Riders now produces an annual environmental record of 250 ski resorts around the world. Information is provided on more than 40 different environmental issues, ranging from a resort's policy on recycling to climate change.

To research

For further information on glacier monitoring in Switzerland, visit the Swiss Glacier monitoring Network at Eidergösslische Technische Hochschule, Zurich. Go to www.glaciology.ethz.ch and click on the button.

10 The geography of food and health

By the end of this chapter you should be able to:

- understand and calculate the different ways of measuring the health of a population
- explain variations in health across the world, with reference to the epidemiological transition model
- describe global patterns of food intake, types of food sufficiency and food shortage and the reasons for them
- explain the factors that affect food production and markets and ways of alleviating food shortages, with reference to sustainable agriculture
- describe global patterns of disease, the spread of disease and the geographic factors relating to disease, with reference to malaria, AIDS and the "diseases of affluence".

This optional theme explores the ways in which food availability and susceptibility to disease determine the health of a population. The problem of food shortage is discussed in terms of lack of access and entitlement, which lead to long-term insecurity. The case study of Ethiopia is used to examine national and local causes and consequences of **chronic** and **acute** food shortage. The globalization of food production and issues of trade, aid and market access are considered. Environmental management and the adoption of sustainable agricultural practices are recognized as fundamental aims in global food production, and essential for meeting the growing global demand for food.

Poverty underlies the problems of food insecurity and disease susceptibility, and can be both a cause and a consequence of both these predicaments. This option examines several examples of diseases associated with poverty or affluence: their prevalence, impacts and the strategies designed to control them. It also examines the shift from infectious to chronic disease experienced by countries moving through the epidemiological transition as they develop. The burden of chronic disease that confronts them now poses difficult management challenges for the future.

Chronic/acute – occurring over a long/short period of time.

Infant mortality rate (IMR) – $\frac{\text{total no. of deaths of children} < 1 \text{ year old}}{\text{total no. of live births}} \times$ 1,000 per year

Life expectancy (E_0) – average number of years that a person can be expected to live if demographic factors remain unchanged.

Access to safe water – access to water that is affordable, in sufficient quantity and available without excessive effort and time.

Access to health services – usually measured in the number of people per doctor, health worker or hospital.

YLDs – years lived with disability.

DALYs (disability-adjusted life years) – the sum of years of potential life lost due to premature mortality and the years of productive life lost due to disability.

Variations in health: key indicators

Various key indicators are used to measure the health of a population or a country, and allow us to make comparisons between them.

Infant mortality rate (IMR) and child mortality rate

Infant mortality represents an important component of under-five mortality. Both the IMR and child mortality indicators reflect the social, economic and environmental conditions in which children

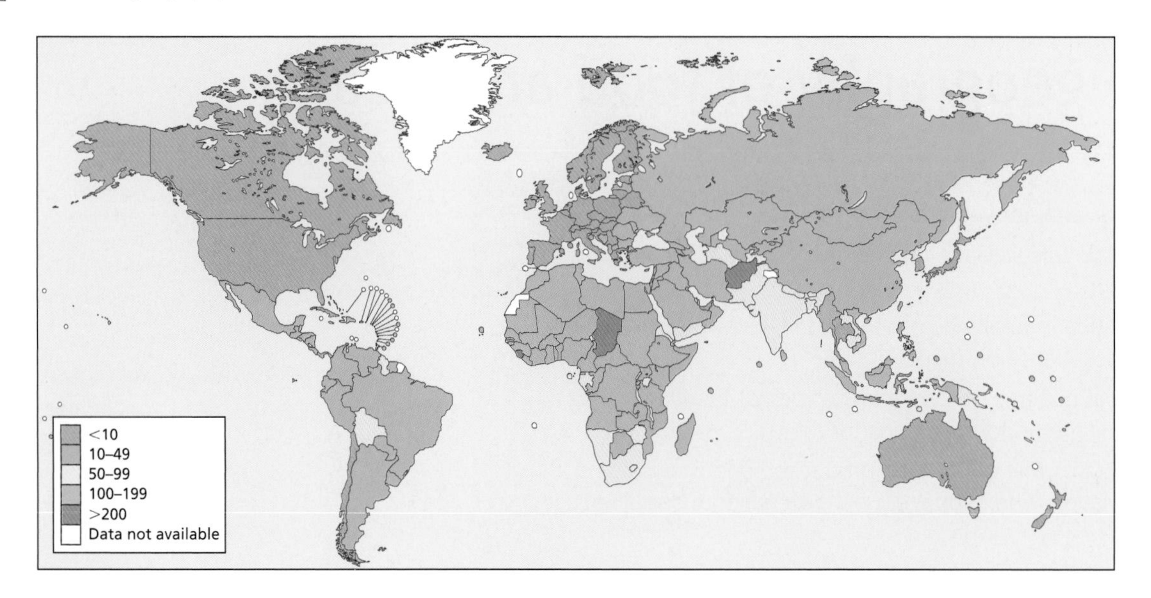

Figure 10.1 Under-five mortality rate by country, 2007

(and others in society) live, including their healthcare. Infant mortality rate is an MDG indicator.

The infant mortality rate is a good measure of human welfare because it reflects household income, nutrition, maternal age and education, housing condition, sanitation and other factors. IMR ranges from 2 per 1,000 in Sweden to 178 per 1,000 in Angola, and the world average is 44. However, the data used to calculate both infant and child mortality may be unreliable if it is collected by household survey, and national birth and death registration systems may also be unreliable. In some African countries the AIDS epidemic, civil war and migration are likely to have led to further inaccuracy, which the World Health Organization (WHO) and other organizations have to account for in estimating probabilities for infant and child mortality rates.

Life expectancy

Life expectancy varies from over 80 years in a number of rich countries such as Australia, Canada and Japan to under 45 in parts of sub-Saharan Africa. Spatial variation can also be seen within countries. For example, the life expectancy of people in LEDCs is longer in urban than rural areas. Similar variations exist between social and ethnic groups. For example, black minority groups in US cities have a life expectancy five years shorter than the white population.

Male life expectancy is universally shorter than that of females. This has been linked to the higher incidence of degenerative illness in men caused by smoking, heavy alcohol consumption and higher exposure to pollutants. Lower life expectancy in men can also be attributed to accidental death by violence, road accidents and suicide, particularly in the 19–24 age group.

Life expectancy has increased significantly in all parts of the world since 1950. Reasons for the rise include greater food production, greater availability of clean water, better living

To do:

a Describe the global pattern of child mortality shown on the map (Figure 10.1).

b Identify one country with a very high rate (over 200). Research this country and suggest reasons for the high levels of child mortality there. You will need to support your explanation with recent indicators of the country's level of economic development.

conditions and better healthcare, especially for the young and the old. However, the AIDS epidemic has had a significant impact on sub-Saharan Africa, where life expectancy fell in countries such as Botswana between 1995 and 2005. Efforts to control the spread of the disease are evident from recovery in life expectancy seen on the graph (Figure 10.2).

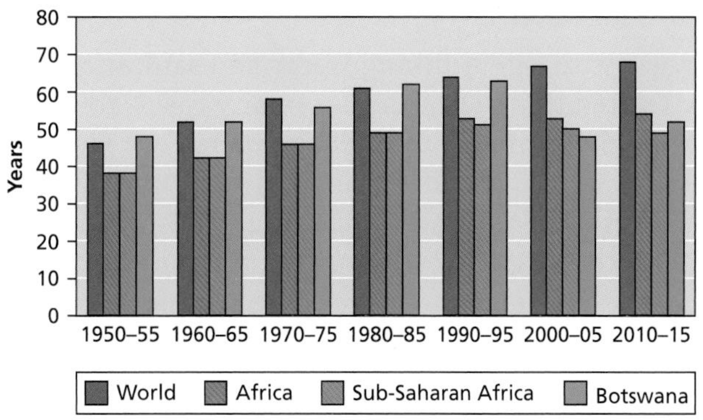

Figure 10.2 Changes in life expectancy, world and Africa, 1950–2015

Although life expectancy is regularly used as an indicator of health, it has some limitations. For example, it considers only length of life, and overlooks the number of years spent living in a state of ill health. Another limitation is that it provides an average impression of mortality in a country. In order to explain regional, international and local variations as well as changes over time, age-specific mortality rates need to be studied. Unreliable data can make comparisons difficult. For example, countries may adopt different techniques of data collection and representation; in some countries life expectancy is not taken from birth.

Malnutrition – a state of poor nutrition, resulting from a deficiency or imbalance of proteins, energy and minerals. Mineral deficiency may lead to diseases such as kwashiorkor, and calorie/energy deficiency to marasmus. Calorie/energy excess may result in **obesity**.

Obesity – an unhealthy condition where excess body fat has accumulated and the body mass index (BMI) exceeds 30 (calculated by weight in kilograms divided by height in metres squared). BMI has limitations, however, because of racial variation in human physique. Obesity may be linked to diseases of offence such as cardiovascular disease and cancer.

Calorie intake

The intake of food varies, from a low of just over 1,500 calories per person per day in Afghanistan and Eritrea. In contrast, the largest intakes are twice that amount, seen in countries such as the USA (3,774 calories), Portugal (3,740 calories) and Greece (3,721 calories). Newly industrializing countries (NICs) such as China and India are associated with rising calorie intakes: 2,951 and 2,459 respectively.

Calorie intake is not totally reliable as an indicator of well-being and diet. First, it does not take nutrient consumption into account. Secondly, it needs to be linked to calorie requirement. For example, an adult male lumberjack working on a cold winter day in northern Canada will require 3,500 calories, whereas a female textile worker in subtropical Mumbai will require less than 2,000 per day.

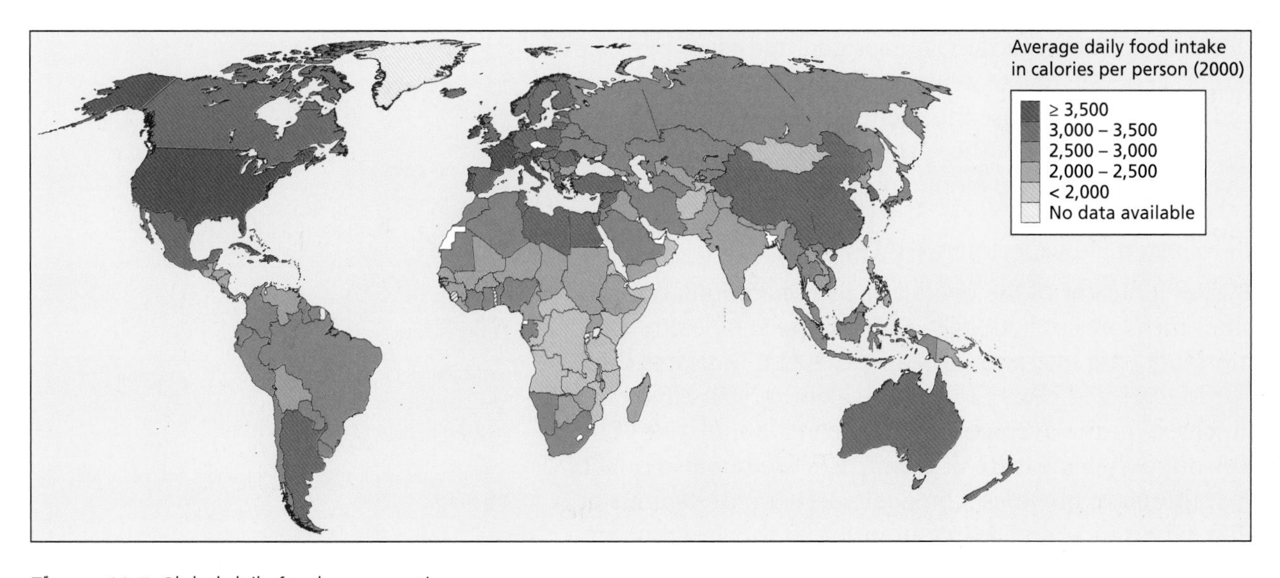

Figure 10.3 Global daily food consumption

Access to safe water

Lack of access to safe drinking water and sanitation is directly related to poverty, and, in many cases, to the inability of governments to finance satisfactory water and sanitation systems. The direct and indirect human costs of these failings are enormous, including widespread health problems, excessive use of labour (particularly for women, who are forced to travel long distances to obtain water for their families), and severe limitations for economic development. Globally, the percentage of the world's population with access to safe drinking water increased from 77% to 87%, between 1990 and 2006 which is sufficient to reach the Millennium Development Goal (MDG) target if the rate of improvement is maintained. In low-income countries, however, the annual rate of increase needs to double in order to reach the target, and a gap persists between urban and rural areas in many countries.

Figure 10.4 UNICEF's WASH programme for helping to reduce child mortality

In 2008, 2.6 billion people had no access to a hygienic toilet or latrine, and 1.1 billion were defecating in the open. The slowest improvement has been in the WHO African region, where the percentage of the population using toilets or latrines increased from 30% in 1990 to only 34% in 2008. Inadequate sewerage spreads infections such as schistosomiasis, viral hepatitis and cholera.

UNICEF's water, sanitation and hygiene (WASH) programme has helped to improve the provision of clean drinking water and sanitation and to educate communities about hygiene. These are major steps towards reducing the incidence of diarrhoeal disease and child mortality. Providing separate toilets for girls has also helped to improve their attendance at school. (Figure 10.4).

Access to health services

Access to health services might be measured as number of doctors, hospital beds or health expenditure per 10 000 population. It has been estimated, in the World Health Report 2006, that countries with fewer than 23 physicians, nurses and midwives per 10 000 population generally fail to achieve adequate coverage rates for selected primary healthcare interventions as prioritized by the MDGs. The data represent three countries at high, medium and low levels of economic development. (Figure 10.5).

Country	Switzerland	Brazil	Ethiopia
Physicians per 10 000 population	40	17	0.5
Hospital beds per 10 000 population	55	24	2
Health expenditure $ per capita *	4,417	674	26

*Using average exchange rate for International dollar with PPP adjustment

Table 10.1 Access to health services for selected countries

Health-adjusted life expectancy (HALE)

HALE is an indicator of the overall health of a population. It combines measures of both age- and sex-specific health data and mortality data into a single statistic. HALE indicates the number of expected years of life equivalent to years lived in full health, based on the average experience in a population. Thus, HALE is not only a measure of quantity of life but also quality of life. It therefore provides a broader spectrum of health status than life expectancy, and draws attention to the growing influence of chronic illness on the quality of life.

Calculating HALE

The World Health Organization (WHO) uses life expectancy tables and Sullivan's method (the number of remaining years, at a particular age, which an individual can expect to live in a healthy state) to compute the HALE for countries. The calculation method also includes a weight assigned to each type of disability, adjusted for the severity of the disability.

Mortality data for calculation of life tables are obtained from death registration data reported annually to the WHO. For countries without such data, available survey and census sources containing information on child and adult mortality are analysed and used to estimate life-expectancy tables.

The HALE indicator has some limitations, however: a major challenge is the lack of reliable data on mortality and morbidity, especially from low-income countries. Other problems with the indicator include the lack of comparability of self-reported data from health interviews.

Region	Life expectancy	HALE
Europe	76	70
North America	78	69
Oceania	76	68
Sub-Saharan Africa	52	41

Table 10.2 Life expectancy relative to HALE

To research

Visit www.euphix.org and enter the search term HALE for tables on life expectancy (HALE) at birth for men, women and the total population in Iceland, Norway, Switzerland and the EU-27, 2002.

Go to http://www.phac-aspc.gc.ca/publicat/cdic-mcc/24-4/gfx/cdic244-arte-fig2e.gif for a graph showing HALE by disease (loss of life in years by disease).

NB: IHD = heart disease;
COPD = cardio or pulmonary disorders (heart and lungs)

TOK Link

- How do indicators help us understand geography?
- How reliable are these indicators – were the data collection techniques standardized, and how were the standards set?

Prevention relative to treatment

The provision of healthcare and its range of services varies internationally, and is a reflection of the funds available and the public spending priorities set by governments. There are three key aspects of medical care to which patients may have varying degrees of access: health professionals, medication and other therapies. Patient access may depend on ability to pay, as well as the accessibility of these services. In many developing countries both these limitations tend to exclude the poorer sectors of society, thereby reinforcing their disadvantage and making them even more vulnerable to disease. Such differences in access are evident on all scales, from global to local.

Providing adequate medical care in many developing countries is therefore difficult once an illness is established. A more effective and far less costly way of containing disease is to prevent its occurrence and present barriers to its spread. In order to minimize the spread of infection, various improvements in domestic hygiene can be adopted, but these will only be effective where there is also a safe water supply and adequate sanitation. The success of such practices also depends on the provision of health education, with a focus on childcare. For example, long-term breastfeeding improves a child's resistance to disease and delays the return of maternal fertility. The promotion of contraception and avoidance of HIV and other infections are also priorities. In the case of non-infectious degenerative diseases such as heart disease and cancer, education and public awareness of risks can significantly reduce the incidence of disease.

Healthcare programmes need to consider the big picture and ensure that they are implemented with other objectives in place, such as better access to food, which will improve resistance to disease and safe water which will limit its trasmission.

The global availability of food

Food security and food shortage

A commonly used definition of food security comes from the United Nation's Food and Agriculture Organization (FAO):

"Food security exists when all people, at all times, have access to sufficient, safe and nutritious food to meet their dietary needs and food preferences for an active and healthy life."

Starvation – a state of extreme hunger, resulting from lack of food over a prolonged period of time.
Temporary hunger – a short-term physiological need for food, resulting from deprivation.
Famine – an extreme shortage of food, resulting in mortality. Famine can occur where food is available, but people lack the means to buy it.

FAD and FED

Much of the early literature on famine and hunger contained reports on climate and its effect on food supplies, and on the problems of transport, storage and relief organizations. Such studies often used the umbrella term **food availability deficit (FAD)**, which implied that food deficiencies were caused by local shortages due to physical factors such as drought or flood.

Amartya Sen (1981) observed that hunger could be found in areas where food was not only available, but production was, in fact, increasing. This has been the case in India, Ethiopia and Sudan. In the analysis of the population "at risk" of malnutrition, it became clear that it was important to encompass the political and economic system in which food is produced, distributed and consumed. Access to food may be severely limited by barriers such as its rising cost relative to wages. When wages dry up, hungry households may have to acquire food by selling their "exchange entitlements", or assets such as livestock, for food. Consequently the household would experience a **food entitlement deficit (FED)**. Sen's work has generally been accepted, although the causes of famine are now considered to be much more complex, and it is important to consider physical factors such as precipitation and environmental degradation as potential triggers of famines.

Areas of food sufficiency

In spite of the gloomy predictions of Malthus and his followers, the neo-Malthusians, world food production has managed to keep pace with population growth. This is been achieved by the adoption of new technology, which enables farmers to increase output per hectare as well as putting more land under cultivation.

Increasing food output

Ways of increasing food output are well known.

Higher crop yields are achieved by using genetically engineered **high-yielding varieties** (HYVs) of staple crops such as wheat, maize and rice. This means that productivity per hectare is higher, and there can be several harvests in one year. By using HYVs, India feeds twice as many people as Africa on just 13% of the land area.

In order to maintain this yield, **artificial fertilizers** (nitrates and phosphates) must be applied, along with **pesticides**, which destroy competitors such as insects, slugs, fungi and weeds.

Irrigation is essential if maximum yield is to be sustained, and it has allowed agriculture to take place in arid and semi-arid areas. For example, the North Sinai Canal development, which runs from the

River Nile delta to the Sinai peninsula, irrigates 62 000 square kilometres of desert.

Increasing the **scale of operations** – by bringing together **fragmented land holdings** – forms more manageable and productive land areas. This process occurred in France during the 1960s and 1970s.

Biotechnology has the capacity to create another green revolution, but much of the agricultural research and development is carried out by large-scale companies (agribusinesses) in developed countries concerned with providing food for those markets rather than for developing countries.

Two mechanisms that have had a powerful influence on farming are markets and human productivity. Farmers will increase output in response to guaranteed prices and guaranteed markets. In part, this was the cause of the "food mountains" and "milk lakes" of Europe during the 1980s, when surplus output had to be sold to developing nations at lower prices that had been paid to the EU farmers.

In order to increase production, it is necessary to pay farmers properly. Nowhere is this more needed than in developing countries, where agricultural progress has stagnated relative to industrialization. To keep the better-educated, more skilled labour in rural areas, better pay and working conditions are needed, otherwise migration will continue to have the same effect as soil erosion – reducing the ability of the land to feed the population.

Case study: *The Green Revolution*

The Green Revolution was the application of science and technology to increase food productivity. It includes a variety of techniques such as genetic engineering to produce high-yielding varieties (HYVs) of crops, mechanization, pesticides, herbicides, chemical fertilizers and irrigation water. Some LEDCS, such as Mexico, India and the Philippines, initially adopted it during the 1960s and 1970s.

HYVs are the flagship of the Green Revolution. During 1967–8 India adopted Mexican rice IR8, which had a short stalk and a larger head than traditional varieties, and yielded twice as much grain as them. However, it required considerable amounts of water and nitrogen. Up to 55% of India's crops are HYVs and 85% of those in the Philippines. By contrast, only 13% of Thailand's crops are HYVs.

The consequences of the Green Revolution

The main benefit is that more food can be produced to meet the ever-growing demand of the population in countries such as India. This will reduce the balance of payments deficit in many LEDCs, helping to reduce food imports and boost the productivity of commercial crops.

However, population growth is more rapid than the increase in food production. In India, for example, by 2020 the population will reach 1.3 billion and food production will need to increase by 50% to match demand. But much of India's land is of limited potential.

In India, grain production has increased substantially, but the benefits have been greatest in the richer states. For example, in West Bengal, a rice-growing state where centuries of land division have left tiny fragmented holdings, few farmers have been able to generate sufficient reserves to adopt the new technology. Even in the rich state of Punjab, known as the "breadbasket of India", disparities in wealth have resulted in further division between rich and poor. Mechanization has reduced the demand for labour, and rural–urban migration is occurring at an ever-increasing rate. Many poorer farmers have been forced into debt.

The environmental costs have also been high. Mismanaged irrigation systems have led to salinization, which affects 20% of Pakistan's and 25% of central Asia's irrigated land. One of the biggest worries is the impact of agricultural fertilizers and pesticides, both of which are essential for the growth of HYVs of cereal crops, which use up nutrients from the soil at a faster rate than other varieties. In many parts of Asia groundwater contamination is a problem, and in southwest Punjab the incidence of neonatal illness and adult cancer is marked. The health costs of the Green Revolution cannot be ignored, despite its success in averting food shortage in many LEDCs.

Who gained and who lost through the Green Revolution?

The Green Revolution did not benefit all, even though it averted a chronic food crisis in the Indian subcontinent during the 1970s. However, a study has found that cumulative global emissions since 1850 would have been much higher without the Green Revolution's higher yields. In addition, although modern farming uses more energy and chemicals, feeding the world at current levels without the new crops would use more than twice as much land as is currently used for agriculture. Emissions from the extra land clearance, releasing carbon stored in trees and soil, would have been a third as much again as those arising from growing the HYVs.

"Converting a forest or some scrubland to an agricultural area causes a lot of natural carbon in that ecosystem to be oxidized and lost to the atmosphere," said Steven Davis, from the Carnegie Institution's Department of Global Ecology at Stanford University in California. "What our study shows is that these indirect impacts from converting land to agriculture outweigh the direct emissions that come from the modern, intensive style of agriculture." (The study is published in the US *Proceedings of the National Academy of Sciences*.)

Areas of food deficiency

A number of environmental, demographic, political, social and economic factors cause food shortages around the world.

Soaring oil and energy prices can push up the cost of food production dramatically. For example, in 2007 rising oil prices caused the cost of fertilizer to increase by more than 70%, and fuel for tractors and farm machinery by 30%. Other oil-related increases included pesticide manufacture and labour. Oil-driven inflation was also the underlying factor in the rice crisis.

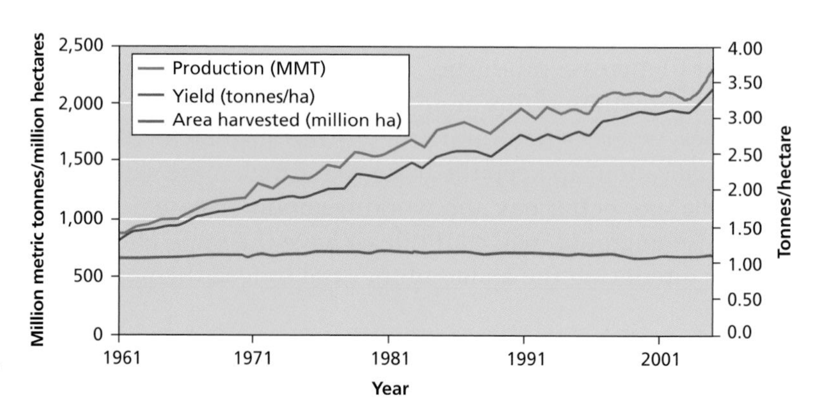

Figure 10.5 World cereals production and yields, 1961–2005

Rising demand results from global population growth and from people in emerging economies, such as China and India, using their increasing affluence to buy more meat, eggs and dairy products. Over 30% of the world's grain now goes to feeding animals rather than people directly. Farming one hectare of decent land can produce 60 kilograms of protein from grain, but only 20 kilograms of beef protein.

Natural hazards such as drought and floods can severely disrupt agricultural production. For example, droughts in grain-producing areas of the world have hit harvests in the last few years. Grain stocks are at a historic low. Cyclones in Burma (2008) and Bangladesh have also reduced the world supply of grain.

Biofuels are competing with food for arable land, with both the US and the EU mandating their use. Since 2008 about 30% of the US corn crop has been diverted to biofuels.

To do:

Study Figure 10.5, which shows global cereal production.

a Define the terms productivity and yield.

b Explain the three trends shown on the graph.

c Explain the meaning of "chronic food shortage".

Underinvestment in agriculture over many years has, as food experts have warned, brought disaster to many developing countries. In 1986, 20% of foreign aid spent by rich countries was devoted to agriculture in the developing world. By 2006, that share had shrunk to less than 3%. African governments, more wary of the political clout of their urban citizens, now spend less than 5% of their budgets on supporting farming and the rural communities, which are home to the poorest two-thirds of their populations.

Speculative trading in agricultural commodities has grown dramatically as investment banks look for new areas to make profits in following the credit crunch. The result has been enormous fluctuations in market prices that do not appear to relate to changes in fundamentals such as supply and demand. Four years ago, $10–15 billion was invested in these funds – now that figure is more than $150 billion. Wall Street investment funds own 40% of US wheat futures and more than a fifth of US corn futures.

With **climate change**, some areas will become drier and face water shortages; others may experience more extreme weather conditions. One estimate is that by 2050, half the arable land in the world might no longer be suitable for production. By then the global population is expected to have grown from today's 6.3 billion to 9 billion.

Different experts give different weight to each of these factors, but agree that their coincidence has led to the current turbulence.

Be a critical thinker

How far are economic factors responsible for food shortages?

Remember to consider this issue on different scales. Figure 10.6, "Against the grain", shows price rises for essential commodities. Such fluctuations make farmers' incomes uncertain and can harm the economies of countries still dependent on trading their raw materials and food on the world market.

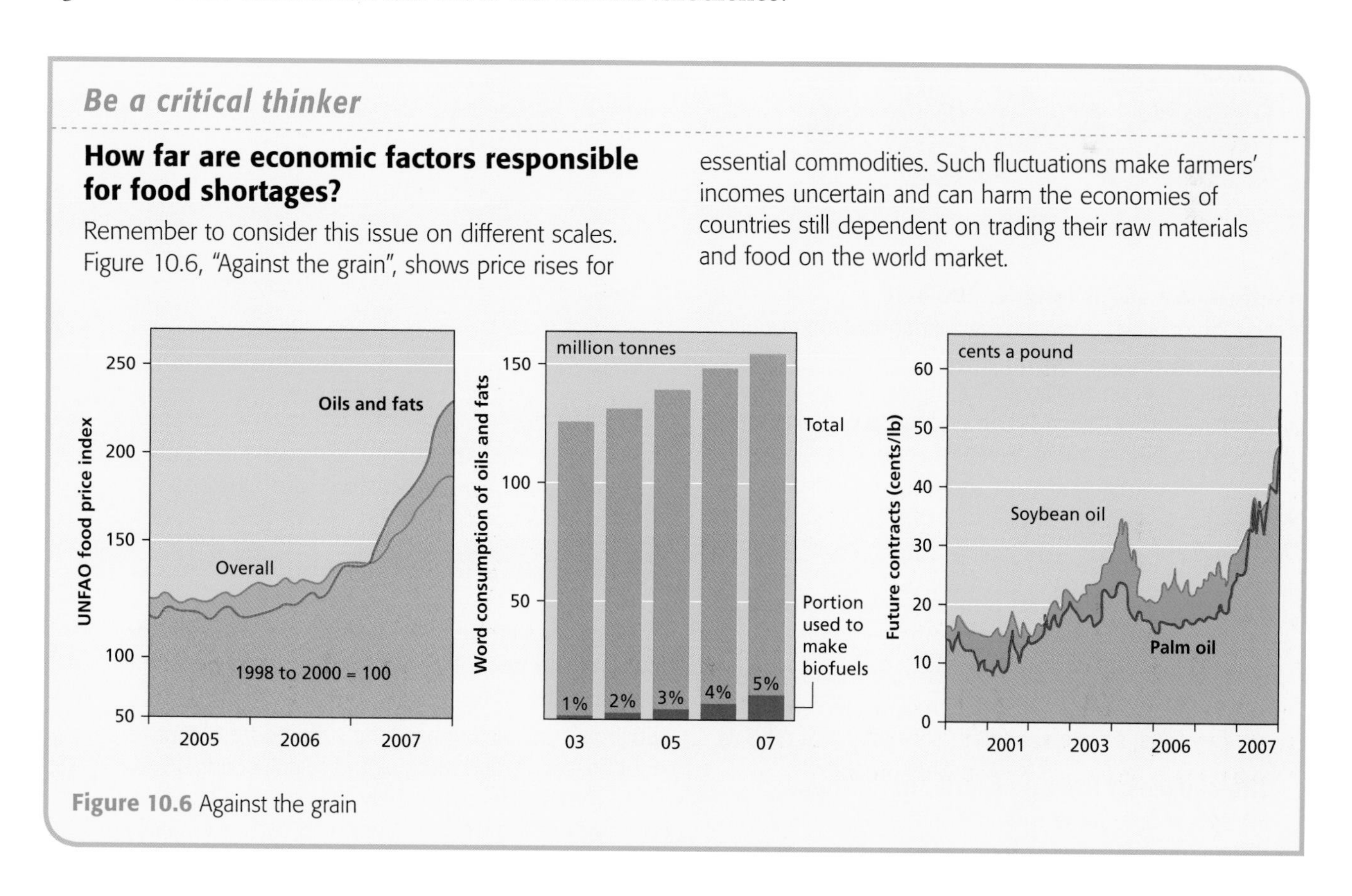

Figure 10.6 Against the grain

Case study: Food insecurity and famine in Ethiopia

Climate	Tropical monsoon with wide topographic-induced variation. Seasonal rainfall	
Terrain	High plateau with central mountain range divided by Great Rift Valley	
Food production	Cereals, pulses, oilseed, sugar cane, potatoes, quat, hides, cattle, sheep, and goats	
Current environmental issues	Deforestation; overgrazing; soil erosion; desertification; water shortages	
GDP per capita	(PPP) $900 (2009 est.).	World rank 214/228
Human Development Index	0.37	World rank 171/178
% of children underweight aged < 5	38.4	World rank 13/178
Infant mortality rate per 1,000	82.6	World rank 20/226

(Source: Human Development Report, 2009)

Table 10.3 Ethiopia at a glance

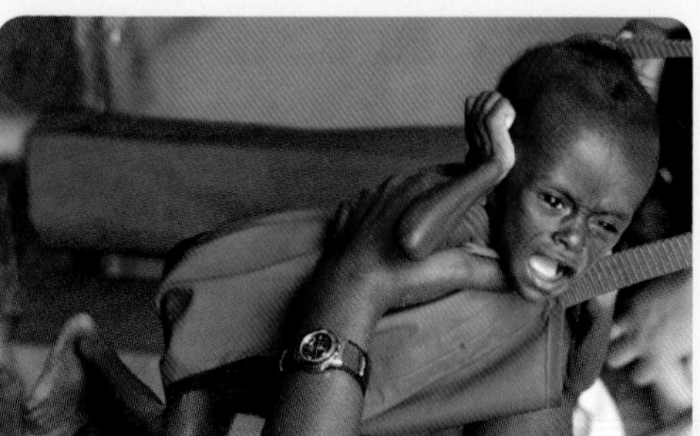

Figure 10.7 Starvation in Ethiopia

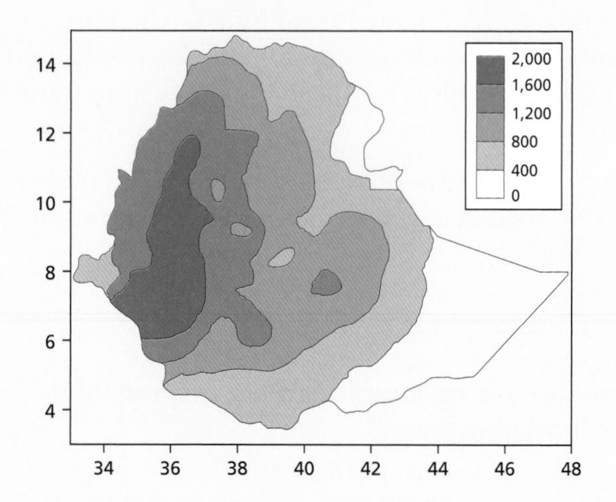

Figure 10.8 Mean annual rainfall, Ethiopia (mm)

Ethiopia's silent famine, 2009–10

The Ethiopian government has hidden the extent of its food crisis and dismissed the word "famine", despite admitting that an additional 6.2 million Ethiopians have needed food aid since 2009. Wishing to play down the food crisis before an election, the Zenawi government has made access to outside agencies very difficult.

Restrictions include:

- limiting work permits and visas to foreigners
- limiting access to relief centres to avoid exposure of the food crisis to the foreign press
- preventing foreign aid agencies from making independent assessments of malnutrition; instead, they must be accompanied by government officials.

The causes of food insecurity in Ethiopia

Ethiopia is one of the countries of the Horn of sub-Saharan Africa, the world's most food-insecure region. Average food intake in the region barely exceeds the daily requirement of 2,100 calories, and is by far the lowest in the world.

Ethiopia's population has long suffered from chronic food insecurity and famine, particularly during 1984–5. The causes of food insecurity are common to all countries of the Horn of Africa, and each event will be

To research

Using an atlas, explain the spatial and seasonal pattern of annual precipitation over Ethiopia.

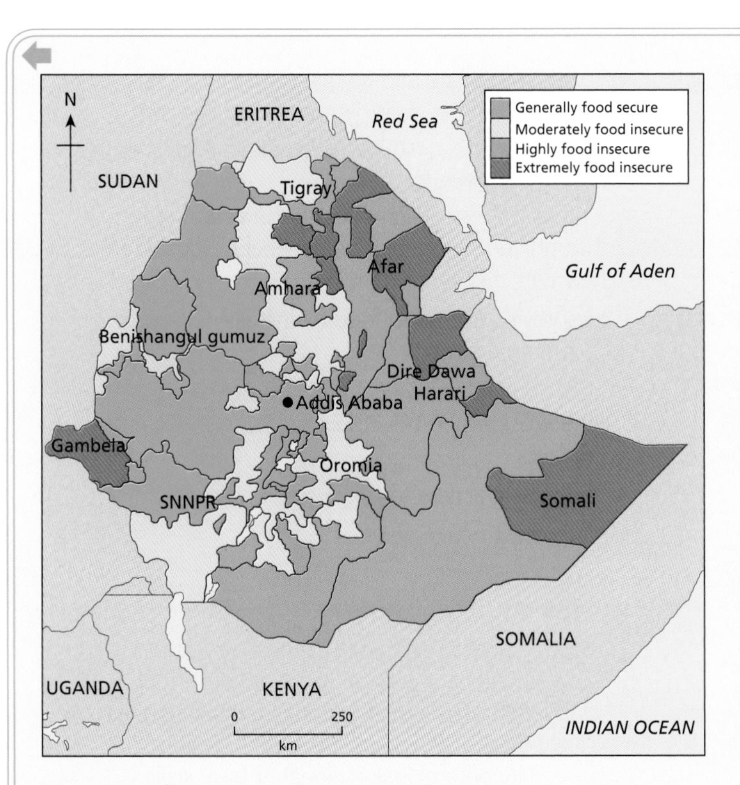

Figure 10.9 Food insecurity in Ethopia, January–June 2010

triggered by one or more of the following causes.

Drought and flooding

Most of Ethiopia experiences one main wet season, the "Kiremt", from mid-June to mid-September when the ITCZ is at its northernmost position. During this season 90% of the country's rain falls. Flooding, sometimes caused by El Niño, can result in soil erosion and crop damage. The dry season extends from October to May, but in 2008 and 2009 it was prolonged, and rain was late to arrive and low in quantity. Ethiopian communities are resilient, and cope with drought by building up food reserves in a good year, but recurrent and prolonged drought makes famine a real threat.

Shortage of land

Rapid population growth has resulted in subdivision and redistribution of land holdings to an inadequate size. There is little incentive for farmers to improve such small plots, which they do not own.

Land degradation

Overpopulaton leads to overgrazing and overcultivation, which in turn leads to soil exposure, followed by wind and water erosion. Around 28% of Ethiopia's land is classified by the FAO as either degraded or severely degraded. In some areas of the uplands, 50–100 tonnes of topsoil are eroded each year. In some parts of the country up to 40 sheep are grazed on 0.1 hectares of land. Traditionally, soil structure and fertility were maintained through practices such as leaving land fallow and applying animal dung as fertilizer. Nowadays, a lack of fuelwood means that the dung often has to be used for fuel instead. The unfertilized soil particles lack cohesion and are easily eroded by wind or water.

Political conflict

In the Somali region, an ongoing conflict with Eritrea, combined with drought, has disrupted infrastructure and access to markets and food. Political conflict has reduced agricultural productivity, and diverted national government money away from agricultural programmes that provide relief to areas of food shortage.

Food aid

Much of the aid that Ethiopia receives is emergency aid rather than the assistance aid that could help the country work its way out of poverty. The USA is the largest donor of food aid, but much of this comes in sacks of grain, leading to dependency. This grain may also undercut the market for local farmers and reduce their income.

Cash rather than food aid would allow the people to buy what is most appropriate for them, and invigorate the economy from the bottom upwards. In addition, the food rations are not necessarily suited to real nutritional needs. Emergency food aid is both visible and costly, and eats into other forms of international assistance. As a result, Ethiopia has one of the world's largest food-aid programmes, but the lowest rate of official aid per capita in sub-Saharan Africa.

Population growth

Ethiopia is Africa's second most populous country and is home to more than 85 million people. The Ethiopian population has doubled in the last 25 years (a Malthusian increase). The latest UN assessments foresee the addition of another 50 million people by 2025. The average fertility rate is one of the highest in the world, with Ethiopian women giving birth to six children each. At the start of 2010, 6.2 million people were threatened by hunger and malnutrition, and required urgent food assistance.

Rise in world food prices

The World Food Programme (WFP) has calculated that across Ethiopia the price of maize has doubled and wheat has gone up by 40% since the end of 2008, with prices set to keep rising. Recent price hikes mean that after the crops failed again in 2009, families are now unable to afford to buy the staple foods they need to keep going.

Terms of trade

Ethiopia's economy is based on agriculture, which accounts for 50% of GDP, 85% of exports and 80% of total employment. The major agricultural export crop is coffee, providing approximately 35% of Ethiopia's foreign exchange earnings, down from 65% a decade ago because of the slump in coffee prices since the mid-1990s. The IMF and World Bank's trade liberalization policies have had a catastrophic effect on the Ethiopian coffee industry, by exposing it to unfair competition.

Poverty

Low income and lack of exchange entitlements (assets) mean that Ethiopians may be unable to get food even if it is available. This situation is exacerbated by rising food prices on the world market.

In the case of food emergencies, the UN World Food Programme (WFP) and the Productive Safety Net Programme are implemented but otherwise food shortage is the concern of national governments.

Government policies are aimed at reducing the risk of temporary shortfalls and providing security, by:

- improving crop and animal production through irrigation (only 3% of cultivated land is under irrigation) and water points in pastoral areas, and by promoting soil conservation and fodder plantations
- diversifying crop and animal production, encouraging specialization in accordance with ecological characteristics (promoting vegetable production in dryland areas)
- intensifying cropping through the use of inputs (improved seeds, fertilizer, animal or mechanical traction, and pest control)
- improving access to food by increasing farm and off-farm income
- developing "safety-net" programmes to limit the erosion of productive capacity in the event of crisis
- improving access to micro-credit mechanisms
- improving health services.

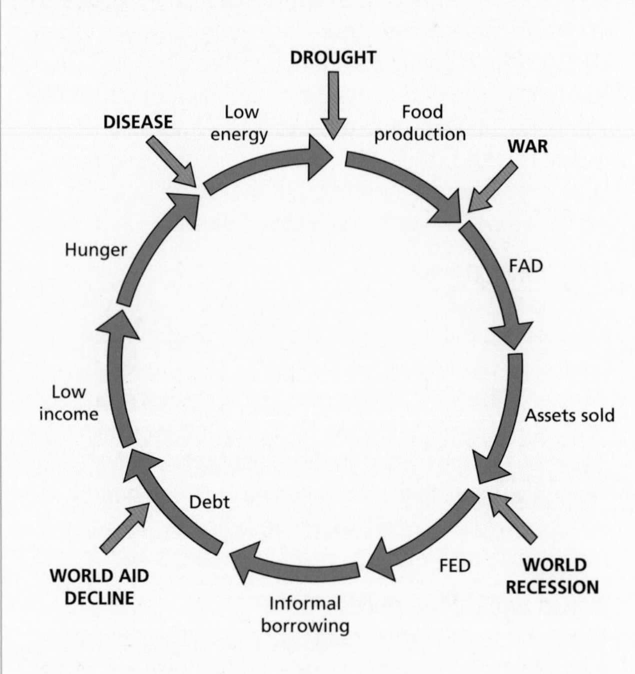

Figure 10.10 The hunger trap in Ethiopia

To do:

Study Figure 10.10, which shows a cycle of adverse events that may occur at household level and lead to famine. The shocks are factors that might initiate or worsen the state of food insecurity on a global, national or local scale.

a Briefly explain the cycle of events shown.

b Select three types of shock and explain how they are linked to food insecurity.

c Explain the advantages and disadvantages of emergency and long-term food aid.

To research

Visit the UN's World Food Programme website at www.wfp.org, for more information on how it is responding to the situation in Ethiopa.

Visit www.odi.org.uk for more about the Overseas Development Institute's Productive Safety Net Programme.

Food production and markets

Many factors affect the production and availability of food. Among the political factors are trade barriers, agricultural subsidies, and bilateral and multilateral agreements. Large farming companies – or agribusinesses – are often part of transnational companies (TNCs), and they have a major impact on trade and trading arrangements.

Trading blocs and farm subsidies

A trading bloc is an arrangement among a number of countries to allow free trade among member countries, but to impose tariffs (charges) on countries that may wish to trade with them. The European Union (EU) is an example of a trading bloc, and it has a major impact on global food production and trade.

To increase farm productivity, the EU introduced the Common Agricultural Policy (CAP). At the centre of the CAP was the system of **guaranteed prices** for unlimited production. This encouraged farmers to maximize their production as it provided a **guaranteed market**. Imports were subjected to duties or levies, and export subsidies were introduced to make EU products more competitive on the world market.

Overproduction became a problem both in Europe and the USA during the 1980s and early 1990s. Surplus food was dumped on the world market, which depressed the price and made it impossible for farmers outside the European or US protection zones to compete. Farm support schemes in the West cost poor families in developing countries $100 billion a year in lost income.

Free trade

Free trade allows a country to trade **competitively** with another country. There are no restrictions regarding what can be exported or imported. By contrast, **protectionism** creates restrictions to trade. It creates barriers to imports as well as to exports.

To research

Visit the World Trade Organization at www.wto.org and investigate the advantages and disadvantages of free trade.

Multilateral arrangements

Multilateral arrangements occur when a number of countries (such as those in the European Union) agree to import goods from a number of other countries. For example, the ACP (African, Caribbean and Pacific nations) had an arrangement through the 1975 Lome (Trade) Convention that gave ACP farmers preferential access to the entire EU. Nevertheless, Caribbean bananas accounted for only 7–9% of EU banana imports.

Bilateral arrangements

A bilateral arrangement is when one consumer enters an agreement with one producer. For example, in 2007 the Caribbean island of St Lucia had cause for celebration when the British supermarket chain Sainsbury announced that all the bananas it would sell in future would be fairly traded bananas, and that nearly 100 million of these would come from St Lucia.

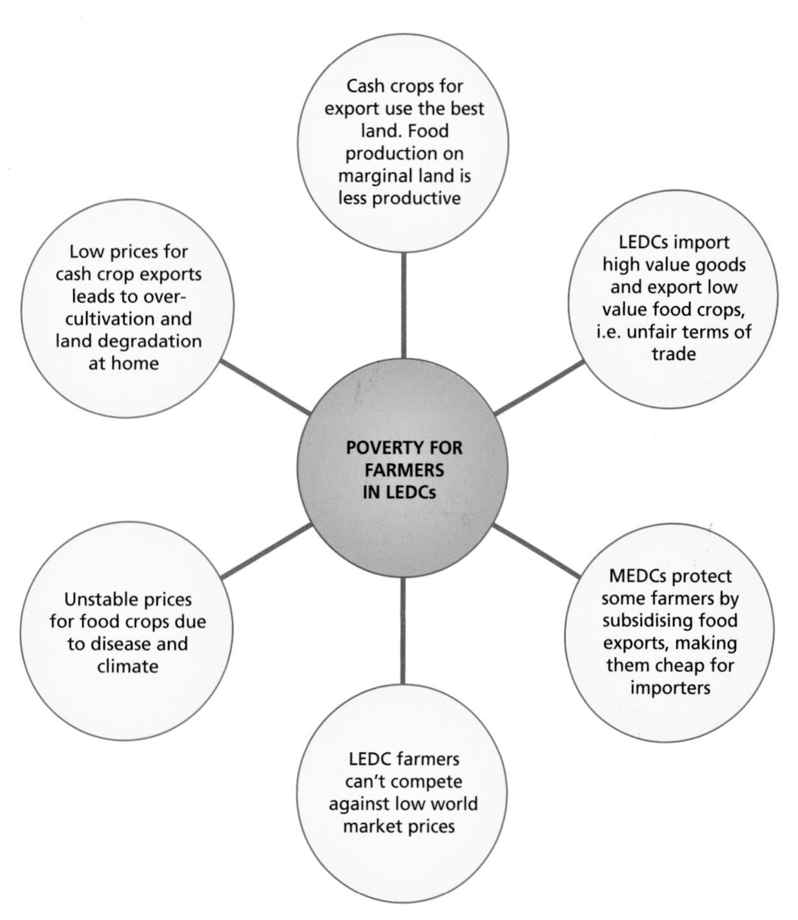

Figure 10.11 The unfair terms of global trade and the effect upon LEDC farmers.

Alleviating food shortages

By 2009, when G8 agriculture ministers convened in an effort to prevent a repeat of the food crisis of the previous year, the number of chronically hungry had surpassed 1 billion for the first time in history. A significant outcome of the summit was the first official admission that the world is "very far" from achieving the Millennium Development Goal of halving the proportion of the world's population facing undernourishment by the year 2015.

Solutions to food insecurity

Global trade and investment policies have driven instability in global food markets, increasing the number of people who are hungry and hindering the ability of countries to grow their own food. The global financial crisis is only compounding the issue of hunger. However, action is being taken and there have been some successes.

Technological solutions

GM crops: Agriculture experts at the UN and in LEDCs do not expect GM crops on their own radically to improve yields. The main trouble, they argue, is that almost all the research has been devoted to developing crops for rich countries in the northern hemisphere.

Expanding irrigation: The introduction of irrigation schemes includes perennial irrigation systems, where water is supplied all year in accordance with the crop requirement.

Appropriate technology: These include small-scale self-help schemes, such as diguette construction (stone lines placed along hillsides to trap runoff) in Burkina Faso.

Seeds and fertilizer: As well as needing food to survive, the rural poor urgently need help planting next season's crops if there is to be any end to the crisis. Millions have been forced to eat next season's seeds to survive, and the price of fertilizer (largely dependent on oil) has risen sixfold in some regions over the course of a year.

Sustainable practices: Using local knowledge and appropriate technology, and avoiding pollution, will help conserve resources (soil and biodiversity) for future generations.

Socio-economic solutions

Agricultural investment: Experts believe yields in Africa can be increased up to fourfold with the right help. Whereas 40% of Asian agriculture is irrigated, that figure is 4% in Africa. The average Asian farmer uses 110 kilograms of fertilizer a year, while the average African uses just 4 kilograms. At least a third of the crops in an average African season are lost after the harvest, largely because farmers cannot get them to markets on time.

Better credit: Women farmers in Bangladesh have developed cooperatives to share money to buy equipment. The Grameen Bank provides loans to those unable to find alternative sources of credit.

Food aid: The World Food Programme has managed to raise all its $755 million appeal to maintain its emergency feeding programmes, largely thanks to a surprise Saudi donation of $500 million. However, the WFP only reaches about 80 million

To research

Visit www.fairtrade.org.uk for more about fair trade.

Visit www.tradeobservatory.org for more on food security issues.

Who decides what is fair or unfair, and where is the distinction between the two?

Do the aims of equity and fairness only matter when resources are in short supply?

Research on the politics of food: www.pubs.sociolistreviewindex.org.uk *International Socialism Journal*, 2003, 'The Perks of Food,' by Carlo Morelli.

Figure 10.12 Food aid from Operation Hunger

of the most desperate, mostly refugees from conflicts and natural disasters. There are over 800 million more chronically hungry people scattered around the world. Food aid can alleviate emergency famine situations, but is not a long-term solution because it has a depressive effect on local market prices.

Land reform – the redistributor of land to individual farmers giving them the inculters to invest and make the land more productive.

Land reform: This has been successful in Kenya, but is often slow to occur, and the situation is particularly pressing in Bangladesh.

To research

www.thezimbabwean.co.uk. "Zimbabwes land refuses: challenging the myths" by Ian Scoones, 27 October 2010.

Improved infrastructure: roads, railway, electricity, and water supply would be suitable policy. This can be provided by agribusinesses.

Trade reform: Free trade, lowering farm subsidies in the USA and undoing some of the protectionism of Europe's Common Agricultural Policy should help poor farmers in the long term, but its direct impact could be to raise food prices in the developing world, as producers focus on western markets.

Fair trade: Fair or ethical trade can be defined as trade that attempts to be socially, economically and environmentally responsible. It is trade in which companies take responsibility for the wider impact of their business, and aims at sustainable development for excluded and disadvantaged producers. Ethical trading helps farmers obtain fairer prices for their products, as part of an attempt to address failings of the global trading system.

Case study: *Fair trade pineapples in Ghana*

Fair trade aims to ensure that producers in poor countries get a fair deal. A fair deal includes a fair price for goods and services, decent working conditions, and a commitment from buyers so that there is reasonable security for the producers.

Good examples of fair trading include Prudent Exports and Blue Skies, both pineapple-exporting companies in Ghana. Prudent Exports has introduced better working conditions for its farmers, including longer contracts and better wages. The company has its own farms but also buys pineapples from smallholders, and exports directly to European supermarkets. They have also responded to requests to cut back on the use of pesticides and chemical fertilizers.

The Blue Skies Organic Collective Association (BSOC) comprises 80 pineapple farmers, including four women, who are members of four village-level collectives. The collectives collaborate with Blue Skies Products Limited, supplying it with organic pineapples. Blue Skies Products, a large-scale fruit processor, wanted to consolidate the trading relationships with its small-scale local suppliers, and helped BSOC achieve organic and Fairtrade certification. Blue Skies also built collection points where pineapples would be taken from the field each day before being collected free of charge. Blue Skies markets the farmers' pineapples to European supermarkets on Fairtrade terms. Boreholes have now been constructed in the villages using funds raised by the Fairtrade premium, helping to solve the acute water supply problems within the four communities.

Some retailers appear to be the driving force behind fair trade as they seek out good practice in their suppliers in terms of health and safety at work, employment of children, pay and conditions, and even the freedom of association of workers.

Sustainable agriculture

Sustainable agriculture refers to the ability of a farm to produce food indefinitely, without causing irreversible damage to ecosystems. A sustainable farming system conserves resources and reduces or prevents environmental **degradation**, while at the same time ensuring farm profitability and a prosperous farming community.

Degradation – depletion of vegetation, loss of biodiversity, soil and water.

Sustainable farming practices are therefore those that maintain or enhance:

- the economic viability of agricultural production
- the environment's natural resources
- other ecosystems that are influenced by agricultural activities.

The environmental costs of increasing food production

Increasing food production has brought a greater reliance on mechanization and the use of artificial inputs such as fertilizer, pesticides and **energy subsidies**. The challenge is to satisfy human demand without destroying ecosystems, landscapes and resources.

Energy subsidies – sources of energy not directly received from the sun, e.g. fossil fuels.

A useful way of assessing the extent to which a particular agricultural system depends upon energy subsidies is to calculate its **energy efficiency ratio** (see Table 10.4). For example, battery hens and greenhouse lettuces are dependent on shelter, heating, automatic feeding and watering. Relative to other systems they consume more energy, are resource destructive and are non-sustainable.

Agribusiness

The basic principle of agricultural production is to maximize food production to meet the needs of the population. Profit maximizing is a driving force behind commercial agriculture, and efficiency requires **economies of scale**. Worldwide, the scale of farm units and their food yield has increased, and operations are often overseen by transnational corporations, from planting or breeding to harvesting or slaughtering, to processing, packaging, transport and marketing. Production on this scale has usually put profit before environmental issues.

Figure 10.13 Free-range hens. Low energy subsidies

Increasing crop yield

Maximum yield is achieved through the application of new technology and manufactured inputs. These include high-yielding, genetically modified seeds and artificial fertilizers such as phosphates and nitrates, which allow multiple cropping. Intensive production is also achieved by modifying the microclimate and reducing the risk of frost and drought by heated greenhouses and irrigation.

Increasing livestock yield

Maximum yield is achieved by selective breeding, the use of antibiotics, hormones and growth promoters. Animals such as pigs and poultry are reared indoors to restrict movement and metabolic heat loss. Intensive livestock production (factory farming) has received much criticism. It may cause **eutrophication** of nearby watercourses and lakes, and the low priority given to animal welfare is still a concern as the developing world adopts intensive practices.

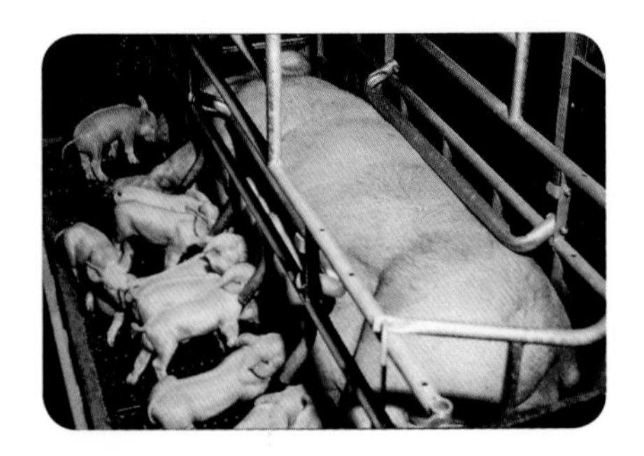

Figure 10.14 Intensively reared pigs. High energy subsidies in "factory farming"

Eliminating competitors

In all types of commercial agriculture, pests can reduce productivity. Insecticides, herbicides and fungicides have been remarkably successful in promoting crop growth by killing weeds and pests, but they can damage both the environment and its ecosystems. Many pesticides, such as organochlorines, create environmental problems because they can enter drainage systems and the soil, causing contamination. They are usually persistent and non-biodegradable, and have the tendency to **bioaccumulate** in food chains with serious implications for top consumers – humans.

Modifying the landscape and reducing biodiversity

Intensive commercial farming requires the area under cultivation or grazing to be increased by large-scale deforestation, hedgerow clearance, drainage of wetlands, irrigation of dry or marginal land and reclamation from the sea. Modern intensive agriculture may therefore be responsible for altering the natural hydrology and microclimate of an area. The removal of trees may reduce interception and increase soil exposure to erosion by water. Microclimatic modifications include increased **albedo** (reflectivity of the ground surface), evaporation and wind speed and higher **diurnal** (24-hour) temperature range. The resulting simplified **agroecosystems** lack the diversity and ecological resilience of natural ecosystems.

Food miles

Food miles refer to the distance that food travels from where it is produced to where it is consumed. It is a way of indicating the environmental impact of the food we eat. The global food industry has a massive impact on transport. Food distribution now accounts for between a third and 40% of all UK road freight. The food system has become almost completely dependent on crude oil. This means food supplies are vulnerable, inefficient and unsustainable.

Inputs	Borneo	Japan	California
Direct energy			
Labour	0.626	0.804	0.008
Axe and hoe	0.016		
Machinery		0.189	0.360
Vehicle fuel		0.910	3.921
Indirect energy			
Fertilizers		2.313	4.317
Seeds	0.392	0.813	1.140
Irrigation		0.910	1.299
Pesticides		1.047	1.490
Electricity		0.007	0.380
Transport		0.051	0.121
Total inputs			
Output – rice yield	**7.318**	**17.598**	**22.3698**
Energy efficiency ratio	**7.08**	**2.49**	**1.57**

(Source: Byrne, K. 2000. *Environmental Science, Bath Science.* Cheltenham UK. Nelson Thornes.)

Table 10.4 Energy efficiency ratios for rice production (Thousands of kilocalories per hectare)

Energy efficiency ratios

The energy efficiency ratio (EER) is a measure of the amount of energy input into a system compared with the output. In a traditional agroforestry system the inputs are very low relative to intensive pastoral farming or greenhouse cultivation, which have high energy subsidies. Direct energy subsidies may be involved with planting, cultivation, harvesting and marketing the crop, whereas indirect energy subsidies may not be obvious. Nevertheless, they should be counted. The output: input ratio is calculated by dividing total outputs by inputs. An efficient farming system has an EER equal to or greater than 1.

$$ER = \frac{\text{energy outputs}}{\text{energy inputs}}$$

To do:

a Referring to Table 10.4, calculate the total inputs and briefly explain the differences in energy efficiency ratios for Borneo, Japan and California.

b Referring to Figure 10.15, explain the differences in energy ratios for each type of food production.

Sustainable yield

The **sustainable yield** is the amount of food (yield) that can be taken from the land without reducing the ability of the land to produce the same amount of goods in the future, without any additional inputs. If the production of palm oil, for example, reduces the nutrient availability in the soil or moisture in the soil, it is not sustainable. Equally, if a particular type of farming leads to the build-up of salt in the soil (salinization) or nitrates in streams (eutrophication), the type of farming is not sustainable.

The principle can also be applied to fishing, where the tonnage of fish removed from the ocean in one year is the amount that allows fish stocks to recover and produce the same yield for years to come.

?

Is organic farm production efficient enough to feed future generations?

In rich countries there is a growing market for organic produce, and demand can be satisfied by producers in poorer countries, such as Kenya and Egypt, who can produce crops out of season and command a high price for their organic credentials.

Organic farming

In the rich world, the damaging environmental effects of intensive commercial farming have led to the promotion of more sustainable practices involving organic farming. These include:

- applying manure or compost rather than inorganic fertilizers
- using crop rotation to maintain soil nutrient status and allow recovery between harvests
- using biological controls rather than pesticides
- reducing energy subsidies, especially transport and manufacture
- allowing livestock to roam and graze freely in the open air. (Fig. 10.13)

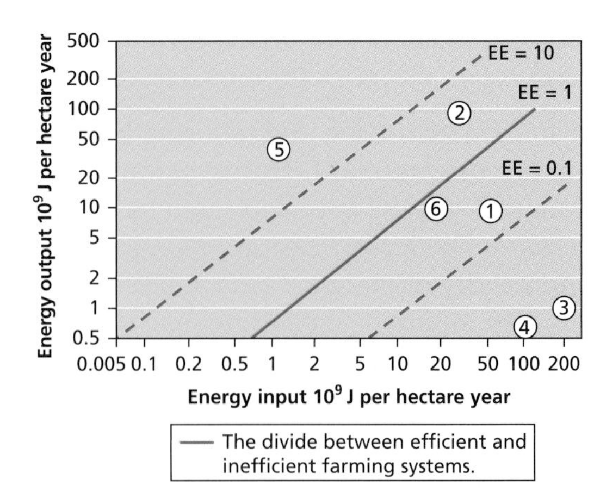

① Battery egg production, UK
② Organic farming, Austria
③ Greenhouse lettuces, France
④ Prawn fishing, Australia
⑤ Shifting cultivation, Belize
⑥ Dairy farming, UK

(Source Curriculum Press)

Figure 10.15 The energy efficiency of different farming systems

Global patterns of disease

Global disease varies in its pattern of **prevalence** (the number of cases per 10 000 population) and **incidence** (the number of confirmed cases annually). In general terms, prevalence rates for infectious or **communicable** diseases are higher in LEDCs than MEDCs, and the reverse is true for degenerative or **non-communicable** disease.

Are hunger and starvation determined more by politics than agricultural productivity?

Classifying diseases

Diseases may be classified in a variety of ways, as illustrated below.

Disease	Type	Means of transmission	Duration
Malaria	Infectious, parasitic	Plasmodium parasite via Anopheles mosquito (vector borne)	Rapid onset; long-term effects
Schistosomiasis	Infectious, parasitic	Bilharzia snail	Sudden onset; long-term debilitating effects
HIV-AIDS	Infectious, viral	Sexual contact and contaminated syringes and infected blood.	Gradual onset; slow deterioration
Heart disease	Non-communicative, degenerative	Ageing, smoking, alcohol and sedentary lifestyle	Gradual onset; surgery possible
Cholera	Infectious, bacterial	Contaminated water and food	Rapid onset, sometimes fatal

Table 10.5 Disease types

Infectious or communicable diseases (diseases of poverty)

These diseases may be contagious and transmitted through close human contact or by **vectors** (transmitters such as insects). They spread rapidly in the overcrowded and insanitary conditions associated with poverty. Examples include malaria, schistosomiasis, diarrhoeal disease and HIV-AIDS.

Non-communicable, chronic diseases (diseases of affluence)

The most common non-communicable diseases are heart disease, stroke, chronic respiratory failure and cancer. These are degenerative diseases associated with old age, but they now affect younger people, too. There are thought to be factors other than ageing which induce degenerative disease at a much earlier age, such as 40–50. These include smoking, high-energy diet, low levels of physical activity and high alcohol consumption. Exposure to air pollution, commonly

To do:

Refer to Figure 10.16.

a Explain what is meant by the epidemiological transition.

b Critically evaluate the use of the term "diseases of affluence".

found in cities, is responsible for some respiratory disease, and there is a known link between exposure to radiation (radon gas, x-rays, nuclear explosions and weapons testing) and cancer.

The epidemiological transition

As countries develop and become more urbanized, their standard of living and food supply improve and infectious diseases subside. Some people begin to adopt the habits of the more affluent world (see above). In time, these lifestyle changes become detrimental to health, causing chronic and degenerative diseases. These gradually come to replace infectious diseases as the major cause of death, a change known as the **epidemiological transition**. Today, many countries are going through this transition and although the death rates are falling and people are living longer, their quality of life may be reduced. Chronic disease can push individuals and their families into poverty because of the cost of treatment and caring for a sick person. It can also impose a heavy financial burden on the state. For example, the WHO estimates that between 2005 and 2010, national income loss for heart disease, stroke and diabetes was $18 billion in China, $9 billion in India and $3 billion in Brazil.

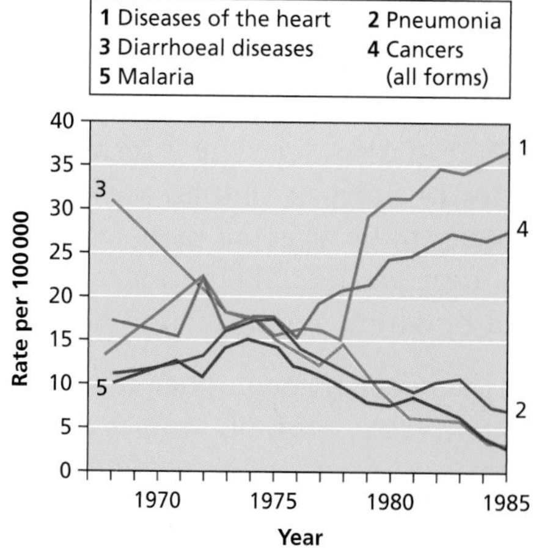

Figure 10.16 Epidemiological transition model for Thailand, 1968–85

More material available:

www.oxfordsecondary.co.uk/ibgeography
Re-emergent diseases and superbugs

Rank	Low-income countries	% of deaths	Middle-income countries	% of deaths	High-income countries	% of deaths
1	Lower respiratory infections	11.2	Stroke and other cerebrovascular disease	14.2	Coronary heart disease	16.3
2	Coronary heart disease	9.4	Coronary heart disease	13.9	Stroke and other cerebrovascular diseases	9.3
3	Diarrhoeal diseases	6.9	Chronic obstructive pulmonary disease	7.4	Trachea, bronchus, lung cancers	5.9
4	HIV/AIDS	5.7	Lower respiratory infection	3.8	Lower respiratory infections	3.8
5	Stroke and other cerebrovascular diseases	5.6	Trachea, bronchus, lung cancers	2.9	Chronic obstructive pulmonary disease	3.5
6	Chronic obstructive pulmonary disease	3.6	Road traffic accidents	2.8	Alzheimer's and other dementias	3.4
7	Tuberculosis	3.5	Hypertensive heart disease	2.5	Colon and rectum cancers	3.3
8	Neonatal infections	3.4	Stomach cancer	2.2	Diabetes mellitus	2.8
9	Malaria	3.3	Tuberculosis	2.2	Breast cancer	2.0
10	Prematurity and low birth weight	3.2	Diabetes mellitus	2.1	Stomach cancer	1.8

(Source: The Global Burden of Disease 2009. WHO)

Disease type

- Non-communicative, cancers
- Non-communicative, cardiovascular
- Communicative
- Non-communicative, degenerative
- Non-communicative, respiratory
- Other

Table 10.6 The top ten causes of death by global region

The spread of disease

Disease diffusion refers to the spread of a disease into new locations. It occurs when incidences of a disease spread out from an initial source. The **frictional effect of distance** or **distance decay** suggests that areas that are closer to the source are more likely to be affected by it, whereas areas further away from the source are less likely to be affected and/or will be affected at a later date.

The Swedish geographer Hägerstrand (1967) is known for his pioneering work on "waves of innovation". This has formed the basis for many medical geographers attempting to map the spatial diffusion of disease. Four main patterns of disease diffusion have been identified, namely expansion diffusion, contagious diffusion, hierarchal diffusion and relocation diffusion. There is also network diffusion and mixed diffusion. The diffusion of infectious disease, for example, tends to occur in a "wave" fashion, spreading from a central source.

Some physical features act as a barrier towards diffusion, including mountains and water bodies, while political and economic boundaries may also limit the spread of disease. The diffusion of disease can be identified as an S-shaped curve to show four phases: infusion (25th percentile), inflection (50th percentile), saturation (75th percentile), and waning to the upper limits.

Types of diffusion

- **Expansion diffusion** occurs when the expanding disease has a source and diffuses outwards into new areas.
- **Relocation diffusion** occurs when the spreading disease moves into new areas, leaving behind its origin or source of the disease, for example a person infected with HIV moving into a new location.
- **Contagious diffusion** is the spread of an infectious disease through the direct contact of individuals with those infected.
- **Hierarchal diffusion** occurs when a phenomenon spreads through an ordered sequence of classes or places, for example from cities to large urban areas to small urban areas.
- **Network diffusion** occurs when a disease spreads via transportation and social networks, for example the spread of HIV in southern Africa along transport routes.
- **Mixed diffusion** is a combination of contagious diffusion and hierarchal diffusion.

The pandemic risk index

The pandemic risk index takes the following into account:

- the risk of a particular disease emerging in a country
- the risk of disease spreading to and within one country
- the capacity of a country to contain the disease.

High-risk areas may be found in both high- and low-income countries. For example, in parts of Africa disease spreads easily and is hard to contain, owing to poor medical services and

To do:

Referring to Table 10.6, the top ten causes of death, analyse the evidense for the epidemiological transition.

To research

Prevention and control of cholera – a water borne disease of poverty.

More material available:

www.oxfordsecondary.co.uk/ibgeography

Case study of cholera epidemic in Haiti in 2010.

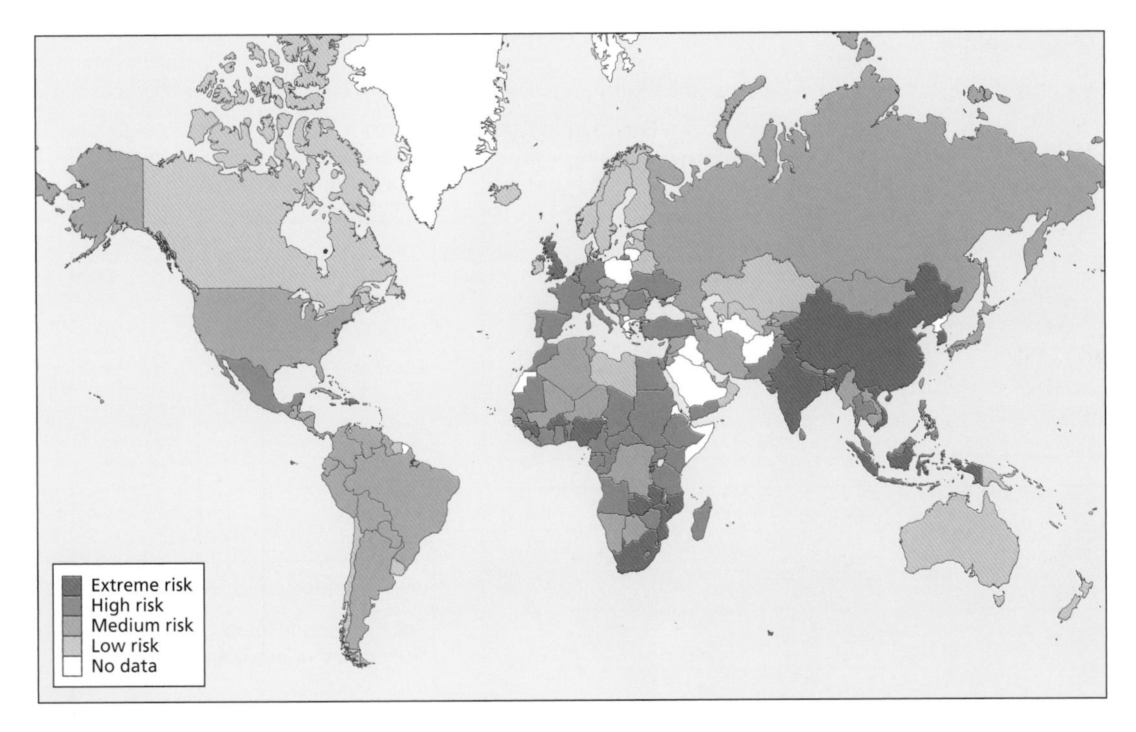

Figure 10.17 Global pandemic risk

infrastructure. Rich countries such as the UK are a hub for international air travel and therefore a focus for disease transmission.

Mexico, the source area for H1N1

In a startling measure of just how widely a new disease can spread, researchers accurately plotted swine flu's course around the world by tracking air travel from Mexico, where the disease originated. The research, based on an analysis of flight data from March and April 2009, showed that more than 2 million people flew from Mexico to more than 1,000 cities in 164 countries worldwide, four out of five of them to the USA. Swine flu began its spread in those countries from that time.

To do:

a Study the four maps of Figure 10.18.

- **i** Describe the diffusion of the H1N1 (swine flu) virus in 2009.
- **ii** Name the type of disease diffusion indicated by the maps.

b Visit the WHO site for statistical changes for each time period and for information about the spread of the disease, at http://www.who.int/csr/disease/swineflu/updates/en/index.html.

- **i** Represent the data as a cumulative frequency curve on a graph.

c Draw one or more maps showing the diffusion of the disease from Mexico, by flow lines.

d Suggest (i) the type of diffusion pattern shown by this disease, and (ii) reasons for the pattern of diffusion.

e Using the link above, describe and evaluate the methods used by governments to contain the spread of this disease.

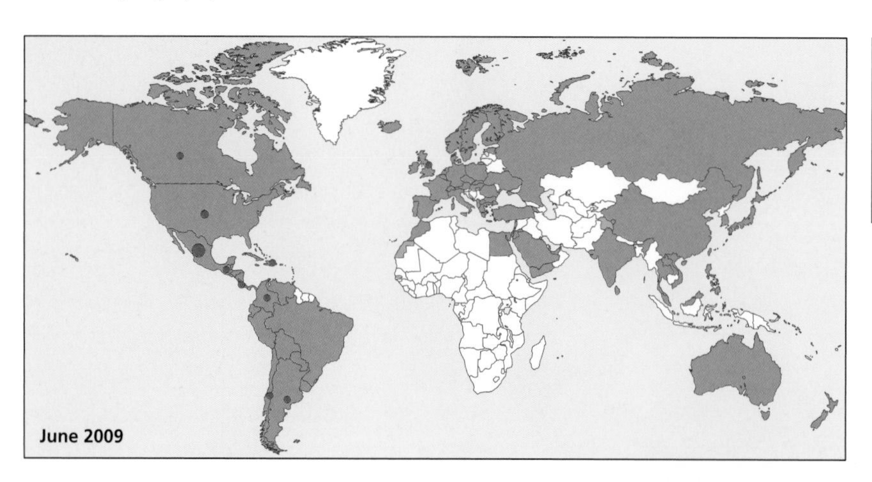

More material available:

www.oxfordsecondary.co.uk/ibgeography

See the website for information on 'Superbugs' or newly emerging infectious diseases

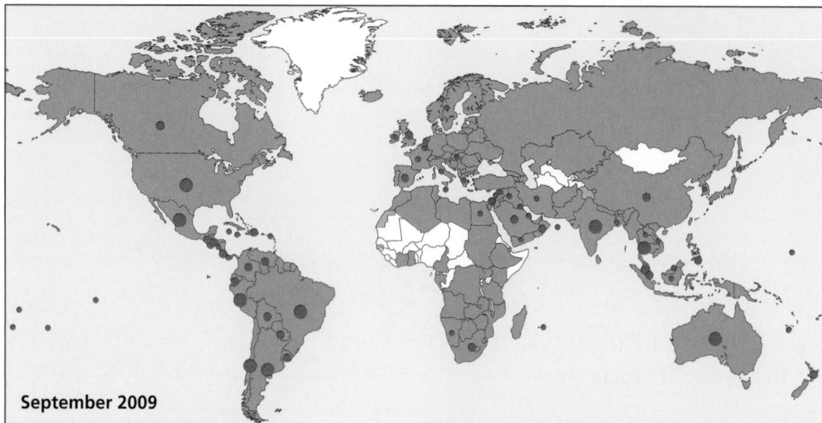

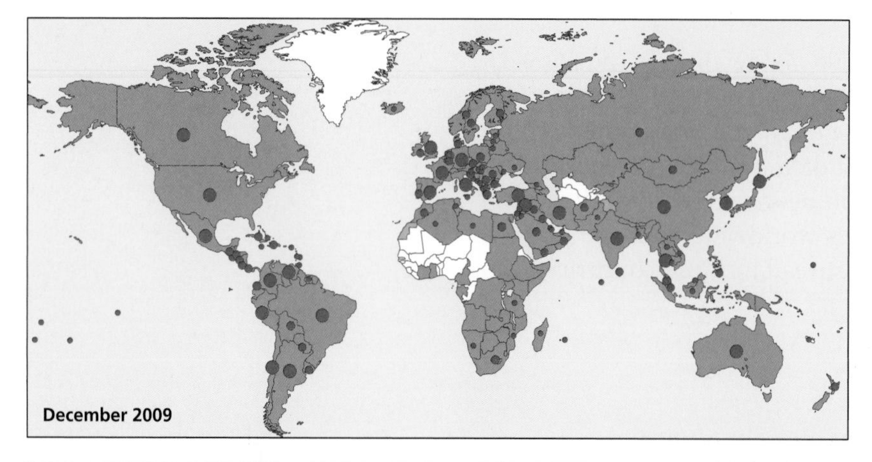

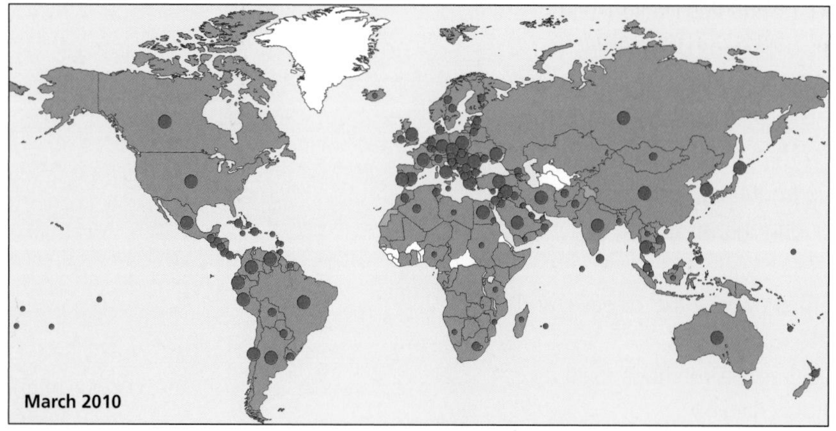

Figure 10.18 H1N1 diffusion, June 2009–March 2010

Geographic factors and impacts: malaria

Figure 10.19 Probing mosquito

Approximately 40% of the world's population is at risk of catching malaria, which kills up to 3 million people annually, mostly in sub-Saharan Africa where 80% of all cases occur. The plasmodium parasite that causes malaria is transmitted via the bites of infected mosquitoes. In the human body, the parasites multiply in the liver, and then infect red blood cells. Symptoms of malaria include fever, headache and vomiting, and usually appear between 10 and 15 days after the mosquito bite. If not treated, malaria can quickly become life threatening by disrupting the blood supply to vital organs.

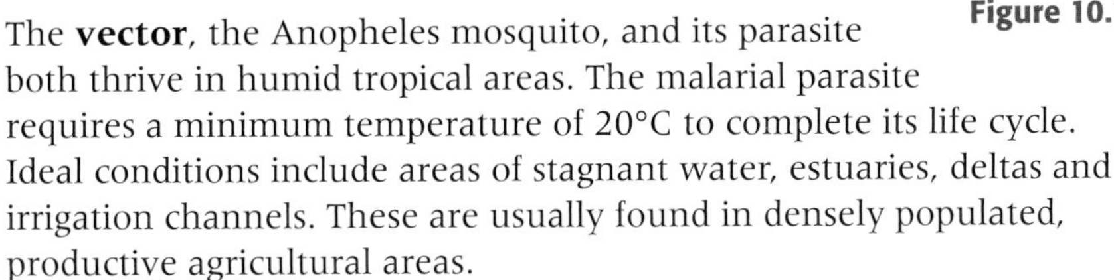

The **vector**, the Anopheles mosquito, and its parasite both thrive in humid tropical areas. The malarial parasite requires a minimum temperature of 20°C to complete its life cycle. Ideal conditions include areas of stagnant water, estuaries, deltas and irrigation channels. These are usually found in densely populated, productive agricultural areas.

Natural factors or human activities may precipitate a malaria epidemic. Natural triggers include climatic variations and natural disasters, while conflict and war, agricultural projects, dams, mining and logging are among the human factors. Most of these factors modify the physical environment, and increase the capacity of mosquitoes to transmit malaria. Some factors also result in massive population movements that expose non-immune populations to malaria infection. Populations with already depressed immune systems such as those infected with HIV are particularly vulnerable.

In southern Tanzania up to 80% of the children are infected with the disease by the age of six months. There, 4% of children under the age of five die as a result of malaria. Pregnant women, travellers and refugees are also especially vulnerable to the disease. Deaths from malaria – concentrated among African children – could be halved to 500 000 by spending another £600 million a year on known prevention and treatment measures.

The cost of malaria

The direct cost of malaria to individual households includes medication, doctor's fees and preventive measures such as bed nets, which help to reduce transmission. Individuals are unable to work during their attacks, which can reduce family incomes by up to 25% a year. Public health spending also increases: malaria treatment can cost up to 40% of all national health expenditure per year.

Indirect costs include loss of productivity for employers. The WHO estimates that malaria cost up around $200 million in lost productivity between 2005 and 2010. Malaria can also interrupt education, jeopardize national productivity and deter tourists and business investors from visiting high-risk areas.

Methods of control

There are two approaches to controlling the spread of malaria. The first is to protect humans from mosquito attack and the second is to reduce the mosquito population.

Protecting humans

Drug treatment

In many parts of the world, the parasites have developed resistance to a number of malaria medicines. Prompt and effective treatment with artemisinin-based combination therapies can relieve symptoms, but there is currently no vaccine for malaria, although 30 new drugs are at the preclinical trial stage.

The insecticide DDT has been used successfully to eradicate mosquitoes in many parts of the world, including Belize, Brazil, Ecuador, Ethiopia, India, Kenya, and Thailand. It is still used in areas of high risk, but there are serious environmental consequences: DDT is non-biodegradable, persistent, toxic to wildlife, and passes up the food chain. It is also known to be carcinogenic, with a possible link to cancer in humans.

The adverse consequences of using DDT have resulted in many countries prohibiting its use. Organophosphates have been used as an alternative method of killing the insect and although these are less persistent and biodegrade more easily, they are also toxic to humans. One further problem with the application of all insecticides is that the insect will build up resistance, as it has done with DDT. Pyrethrine is the most recent insecticide to be used, but is too expensive to make a major impact.

Bed nets

Insecticide-treated bed nets are an effective method of protecting humans from mosquitoes at night. They can reduce child deaths by over 20% and illness by 50%. However, the cost of around $3–$5 per net is prohibitive for most families in tropical Africa.

Killing mosquito larvae

The mosquito population can be reduced by interrupting its life cycle. The most effective methods are to cover open water tanks, drain areas of stagnant water and infill redundant irrigation channels. Ponds and lakes can be stocked with fish to eat the larvae. These methods are environmentally sound, but often impractical and slow to take effect.

TOK Link

The map is the message

How do geographers convey a meaning through maps?

How might maps distort reality?

What is reality?

Go to www.worldmappser.org and look for the map of malaria deaths (map 230).

Open the population map (map 2) for comparison. Use these maps to comment on these questions.

Geographical factors and impacts: HIV and AIDS

Case study: *The impact of HIV & AIDS in Africa*

Two-thirds of all people infected with HIV live in sub-Saharan Africa, although this region contains only10% of the world's population. AIDS has caused immense human suffering from the household to the national level. During 2008 alone, an estimated 1.4 million adults and children died as a result of AIDS in sub-Saharan Africa. Since the beginning of the epidemic, more than 15 million Africans have died from AIDS.

Although access to antiretroviral treatment is starting to lessen the toll of AIDS, fewer than half of Africans who need treatment are receiving it. The impact of AIDS will remain severe for many years to come.

The impact at the national level

Treatment costs

Ill-health and loss of life have had a severe impact on all sectors of the economy, and millions of key workers in the health sector, schools and farming have died or are too ill to work. In sub-Saharan Africa, the direct medical costs of AIDS (excluding antiretroviral therapy) have been estimated at about US$30 per year for every person infected, at a time when overall public health spending is less than US$10 per year for most African countries. Additional costs include hospitalization and training for health workers in the administration of antiretroviral treatment.

Depleted workforce

The biggest increase in deaths, has been among adults aged between 20 and 49: years; the economically active population. This reduces government revenue from lost income tax as well as the overall loss in productivity. Individual businesses struggle to survive with lower productivity, extra costs and absenteeism. By making labour more expensive and reducing profits, AIDS limits the ability of African countries to attract industries that depend on low-cost labour and makes investments in African businesses less desirable. HIV can thus threaten the foundations of economic development in Africa.

Loss of key workers in education and health can hold back progress in managing the disease. Many teachers live with AIDS, and their absence cannot be adequately covered. This may limit AIDS education programmes. Health workers are also in short supply. A study in one region of Zambia found that 40% of midwives were HIV-positive. Excessive workloads, poor pay and migration to richer countries also contribute to the shortage.

The AIDS epidemic adds to food insecurity in many areas, as agricultural work is neglected or abandoned due to household illness. It is thought that by 2020, Malawi's agricultural workforce will be 14% smaller than it would have been without HIV and AIDS. In other countries, such as Mozambique, Botswana, Namibia and Zimbabwe, the reduction is likely to be over 20%.

The impact at the household level

There has been a dramatic increase in destitute households where there are no income earners. Such a situation has repercussions for every member of the family. Children, especially girls, may have to leave school to work in farming, women may have to abandon domestic work to earn wages in male-dominated jobs such as carpentry. Prostitution provides an income for destitute households, increasing further the risk of contracting HIV. Poor family welfare may be an outcome of this social change.

Figure 10.20 Schoolchildren in South Africa, one in four of whom are living with HIV/AIDS

Rising costs

It is estimated that, on average, HIV-related care can absorb one-third of a household's monthly income. The financial burden of death can also be considerable, with some families in South Africa easily spending seven times their total household monthly income on a funeral.

Coping strategies

Savings are used up or assets sold; assistance is received from other households. Using up savings if available and taking on more debt are usually the first options chosen by households struggling to pay for medical treatment or funerals. As debts mount, precious assets such as bicycles, livestock and even land are sold. Once households are stripped of their productive assets, the chance of them recovering and rebuilding their livelihoods becomes more remote.

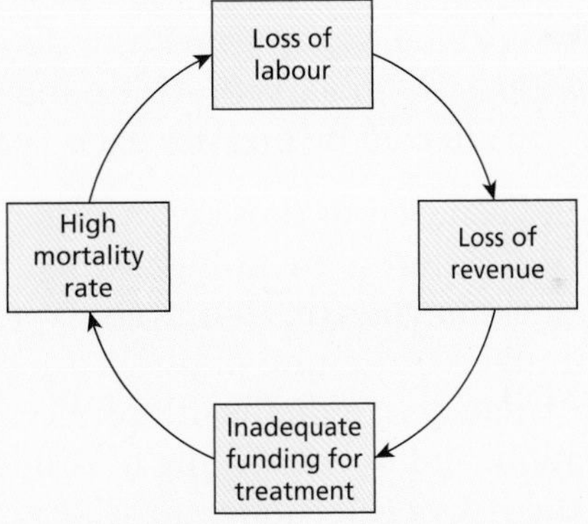

Source AVERT AIDS

Figure 10.21 The vicious circle of poverty and the AIDS epidemic

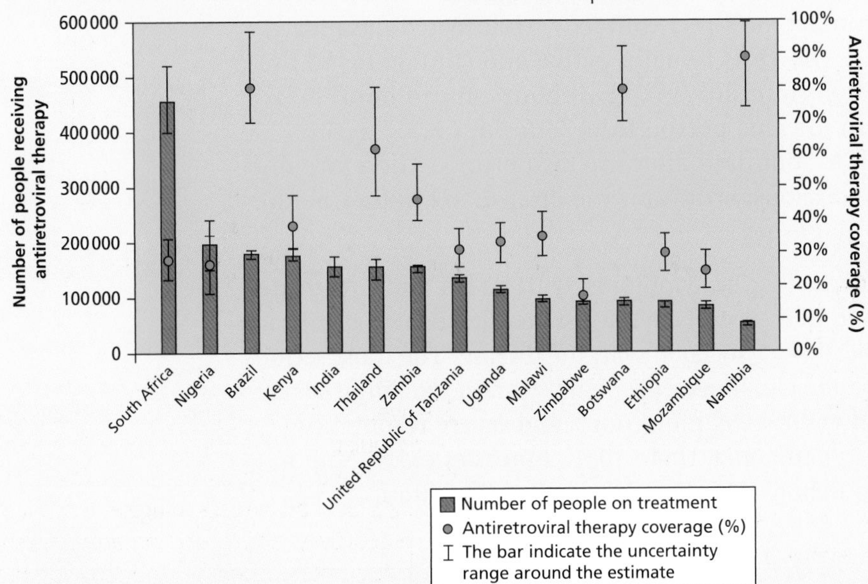

Figure 10.22 International variations in antiretroviral treatment

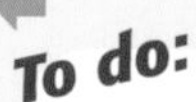

To do:

Figure 10.21 shows the vicious circle of consequences imposed on a country by the AIDS epidemic and poverty. The situation is self-perpetuating and recovery is difficult.

a Draw a similar circular diagram to show the consequences of AIDS at the household level.

b Draw a second diagram showing the reverse situation, where the household starts to recover.

c Referring to figure 10.22, analyse the international variations in the treatment of AIDS using antiretroviral medication.

The benefits of antiretroviral treatment

- More people are living with HIV than ever before, as people live longer due to the beneficial effects of antiretroviral therapy. The number of AIDS-related deaths declined by over 10% from 2005–10, as more people gained access to the life-saving treatment. UNAIDS and WHO estimate that since the availability of effective treatment in 1996, some 2.9 million lives have been saved.
- Since 2001 around 200 000 new infections in children have been prevented by antiretroviral therapy, as more HIV-positive mothers gain access to treatment preventing them from transmitting the virus to their children.

Geographic factors and impacts: measles

Measles is one of the leading causes of death worldwide among young children, even though a safe and cost-effective vaccine is available. An estimated 164 000 people died from measles in 2008: that is nearly 450 deaths every day, or 18 deaths every hour in children under the age of five. More than 95% of measles deaths occur in low-income countries with underdeveloped health services, particularly in parts of Africa and Asia. More than 20 million people are affected by measles each year.

Unvaccinated young children or those with suppressed immune systems such as AIDS sufferers are at highest risk of measles and its complications, including death. Countries experiencing or recovering from conflicts or natural disasters are particularly vulnerable, because damage to the health services interrupts routine immunization, and overcrowding in residential camps greatly increases the risk of infection.

Transmission, symptoms and treatment

The highly contagious virus is spread by coughing and sneezing, close personal contact, or direct contact with infected nasal or throat secretions. The virus remains active and contagious in the air or on infected surfaces for up to two hours. It can be transmitted by an infected person from four days prior to the onset of the rash to four days after the rash erupts. Therefore one sufferer can unconsciously transmit the disease to another before they have been diagnosed.

Measles causes severe flu-like symptoms, with a rash that covers the face and hands and lasts several days. Most measles-related deaths are caused by complications associated with the disease. The most serious complications include blindness, encephalitis (brain swelling), severe diarrhoea and dehydration, ear infections, and severe respiratory infections such as pneumonia. Up to 10% of measles cases result in death among populations with high levels of malnutrition.

Severe complications from measles can be avoided by means of good nutrition and adequate fluid oral rehydration. The rehydration solution replaces fluids and other essential minerals lost through diarrhoea or vomiting. Eye and ear infections and pneumonia need antibiotic treatment.

Prevention

Routine measles vaccination for children, combined with mass immunization campaigns in countries with high case and death rates, are key public health strategies to reduce global measles deaths. The measles vaccine has been in use for over 40 years. It is safe, effective and inexpensive. It costs less than one US dollar to immunize a child against measles.

In 2008 about 83% of the world's children received one dose of measles vaccine by their first birthday through routine health services – up from 72% in 2000. Two doses of the vaccine are recommended to ensure immunity, as about 15% of vaccinated children fail to develop immunity from the first dose.

Global health response

The global improvement in measles vaccination is recognized as a significant indicator of progress towards achieving the fourth Millennium Development Goal (MDG 4), which aims to reduce the under-five mortality rate by two-thirds between 1990 and 2015.

To do:

Identify the geographic factors responsible for the development and spread of measles.

To research

Visit the WHO or UNICEF website (www.who.int or www.unicef.org) and find out about the aims of Measles Initiative that they have developed.

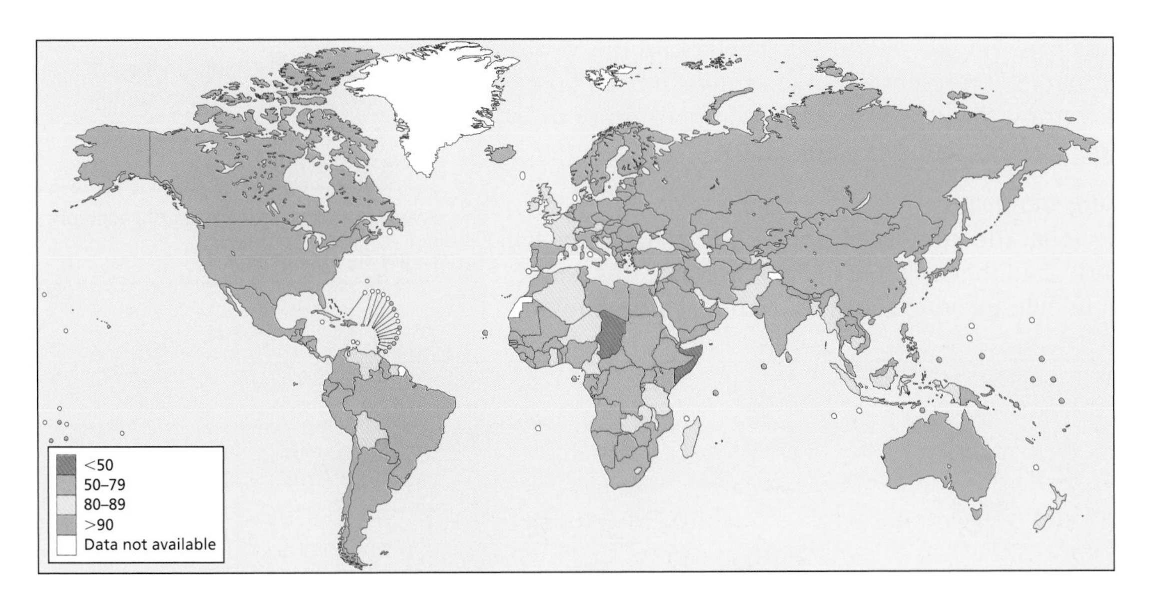

Figure 10.23 Percentage of one-year-olds immunized against measles, 2008

Geographic factors and impacts: diseases of affluence

The WHO cites the following risk factors for chronic degenerative diseases:

- **Age:** Chronic and degenerative diseases are more prevalent among the elderly. The burden of ill health increases as national populations age and pass through the demographic transition.
- **Wealth:** When economic development occurs, people gain wealth and change their lifestyles to become less active and less healthy.
- **Urban residence:** Easy access to transport and services in urban areas avoids the exertion of rural water collecting and farming.

The diet changes to include less dietary fibre and more saturated fat. Fast food outlets that tend to promote unhealthy eating are found mostly in urban areas.

Smoking: a preventable cause of early death

Smoking was one of the main causes of death in the 20th century in the rich world, but by 2010 72% of those dying from tobacco-related illnesses were in low- and middle-income countries. Today many high-income countries have implemented effective anti-smoking campaigns, and the number of smokers has fallen sharply. However, in low- and middle-income countries anti-smoking campaigns have been less successful, due to cultural resistance, poor education and lack of funding.

Economic impacts of smoking

In low- and middle-income countries, smokers spend a disproportionate part of their income on tobacco that could otherwise be spent on food, healthcare and other necessities. Tobacco's cancer connections are well known, but it may also cause malnutrition in several ways. In Bangladesh an estimated 10 million people are undernourished due to tobacco instead of food expenditure. Tobacco replaces potential food production on almost 4 million hectares of the world's agricultural land.

A quarter of smokers die and many more become ill during their most productive years, when income loss devastates families and communities. In 2006, about 600 billion smuggled cigarettes were sold on the global market, representing an enormous missed tax opportunity for governments, as well as a missed opportunity to prevent many people from starting to smoke.

WHO's anti-smoking campaign

The WHO estimates that the number of smoking-related deaths will increase significantly, and that by 2030 83% of these deaths will occur in low- and middle-income countries. National anti-smoking

To research

Visit http://www.mapsofworld.com/world-top-ten/maps/countries-by-highest-death-rate-from-lung-cancer.jpg for the top 10 countries with lung cancer.

Figure 10.24 WHO anti-smoking campaign directed at young women

More material available:

www.oxfordsecondary.co.uk/ibgeography
Smoking in China

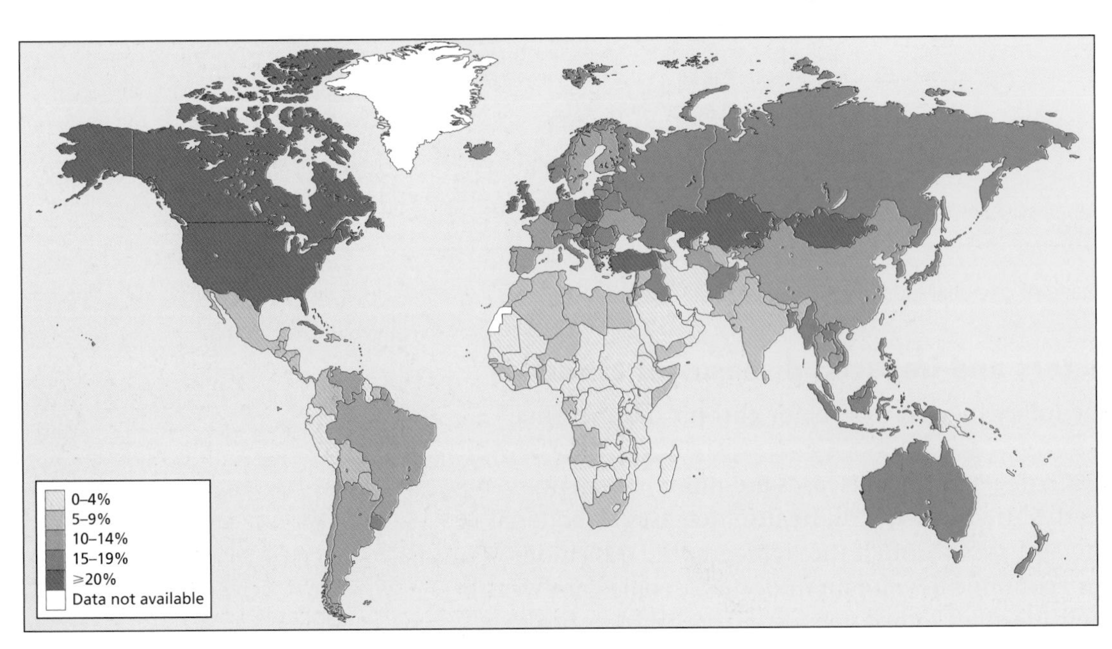

Figure 10.25 Tobacco-related deaths as % of total deaths, 2004

campaigns have had more success in richer countries, with measures that include tobacco taxes, advertising bans, smoke-free public places and health warnings on packs. Recent emphasis has been on the detrimental effects of passive smoking. The most effective campaign has been the Framework Convention on Tobacco Control (FCTC), a global treaty endorsed by more than 160 countries, and recommended by the WHO's MPOWER policy package.

Global obesity

Globally, 65% of the population lives in a country where overweight and obesity kill more people than underweight. This includes all high-income and middle-income countries. It is estimated that 44% of diabetes cases, 23% of heart disease cases and 7–41% of certain cancers are attributable to overweight and obesity.

Figure 10.26 Children in an exercise class

Social and economic development as well as policies in the areas of agriculture, transport, urban planning, environment, education, food processing, distribution and marketing influence children's dietary habits and preferences as well as their physical activity patterns. Increasingly, these influences are promoting unhealthy weight gain, leading to a steady rise in the prevalence of childhood obesity.

Be a critical thinker

Is obesity the worst form of malnutrition?

Malnutrition is a broad term that covers both under- and overfeeding. It reflects the quality as well as the quantity of food consumed. The consequences of obesity should be compared to those of other forms of malnutrition. Remember to consider the short- and long-term impacts, not just on individuals, but also on the nation.

Case study: Obesity in Mexico

Mexico is the second fattest nation in the world. By 2008 71% of Mexican women and 66% of Mexican men were considered overweight, compared with only 10% overall in 1989. A quarter of Mexican children aged 5–11 were overweight. Obesity is linked to poverty, and is common among the population with a mean daily per capita income of US$2.

It is cheaper to buy high-calorie junk or fast food such as burgers, chips and carbonated soft drinks than healthy food, and the number of fast-food outlets has increased dramatically. Mexico has 205 McDonald's restaurants, ranking it 16th in the world, and the country's consumption has increased by 60% since 2005. Obesity is also a consequence of an inactive urban lifestyle and the replacement of manual labour by technology. In elementary schools only 32% of children in south Mexico City get daily exercise, and 60% of those children are considered obese, overweight or undernourished according to the body mass index (BMI) scale. Three out of four schools do not have drinking water, and it is easier and cheaper to provide soft drinks.

Mexico is using several methods to try to reform its obese and overweight population, and began a new health campaign in 2008. PepsiCo launched a programme for children using a computer game to persuade them to take more exercise and eat more healthily.

The repercussions are immense. Obesity is linked to a range of degenerative and life-shortening diseases, notably heart disease and diabetes. Up to 10 million Mexicans (out of a total population of 110 million) have diabetes, and 70 000 of those die of the disease each year. Obesity, once a mark of wealth, is now a sign of poverty in the crowded inner cities of transitional economies such as Mexico. Despite this, in rural regions many are still under- rather than overnourished.

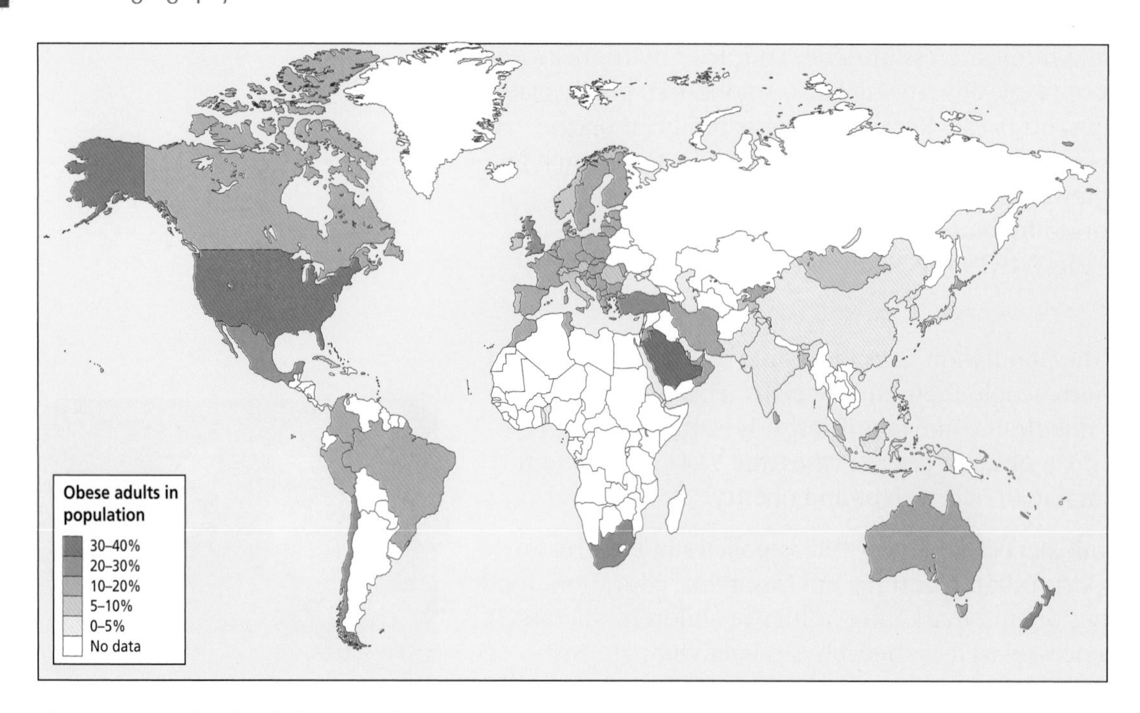

Figure 10.27 The global obesity problem

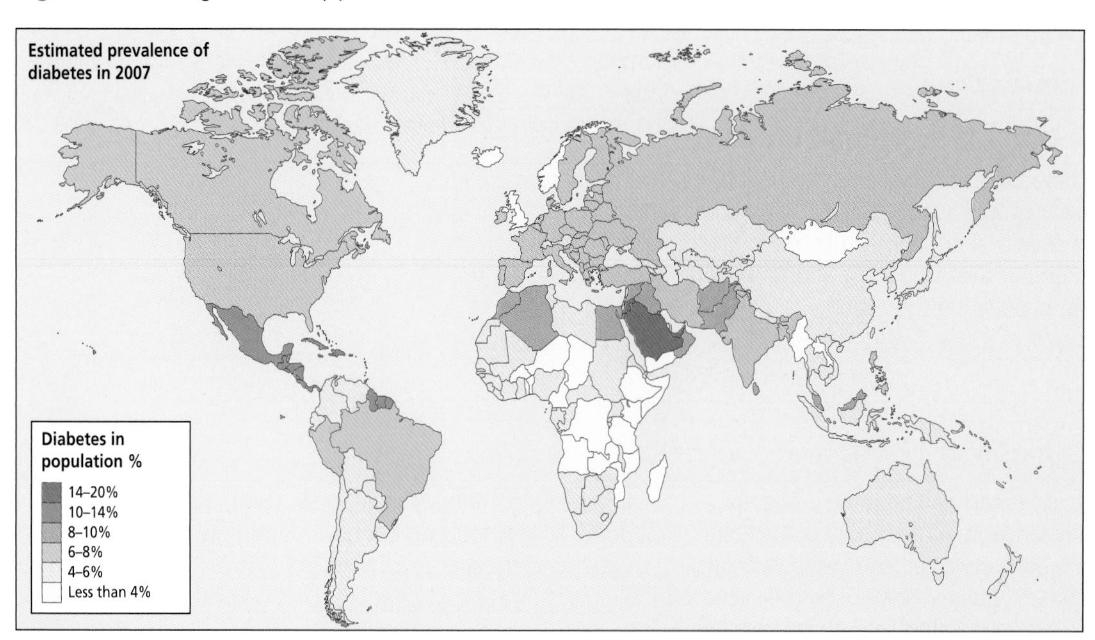

Figure 10.28 Prevalence of diabetes, 2007

Diabetes: the forgotten epidemic

Diabetes mellitus is an incurable condition affecting increasing numbers of people worldwide Almost 80% of diabetes deaths occur in low- and middle-income countries. More than 220 million people worldwide have diabetes, and the WHO predicts that deaths from the disease will double between 2005 and 2030. Diabetes and its complications – cardiovascular disease, neuropathy and blindness – can have a significant economic impact on individuals, families, health systems and countries. For example, WHO estimates that in the period 2006–15 China will lose $558 billion in foregone national income through heart disease, stroke and diabetes alone.

To do:

Compare the two maps (Figures 10.27 and 10.28), showing the prevalence of global obesity and diabetes, and describe the similarities and differences.

If the global pattern between these two conditions shows some similarities, is it valid to conclude that one causes the other?

Management and prevention

Diabetes is a metabolic disease involving pancreatic malfunction and insulin deficiency. The management of diabetes involves careful monitoring of blood sugar levels, medication and, for type 1, daily insulin injections. Simple lifestyle measures have been shown to be effective in preventing or delaying the onset of type 2 diabetes:

- Achieve and maintain healthy body weight.
- Be physically active, taking 30 minutes daily regular, moderate exercise.
- Eat a healthy diet with between three and five servings of fruit and vegetables a day, and reduce sugar and saturated fat intake.
- Avoid tobacco use: smoking increases the risk of cardiovascular disease.

Case study: *Diabetes management in India*

India's diabetic population is expected to increase by 150% to 90 million between 2000 and 2030; this is the highest **prevalence rate** (the number of cases as a proportion of total population) increase in the world. The comparable increase in India's total population will be only 40%.

Prevalence rates vary geographically and socio-economically. Rural rates in India are 3%, but in urban areas they are 9% and increasing relatively rapidly. The causes are related to the changing lifestyle that occurs with rural to urban migration. The usual risk factors apply: lack of exercise, high cholesterol consumption and the uptake of smoking. In 2005 the average annual cost of diabetes care was estimated to be 10 000 rupees for urban and 6,260 rupees for rural patients. For low-income families, caring for a diabetic member can consume up to 25% of household income. Access to and compliance with medication becomes a serious issue for poor families, especially those in rural areas. Around 30% of patients fail to take medication as prescribed, and 37% fail to change their diet. Delays in diagnosis can also jeopardize the success of treatment. There is a seven-year time lag between the highest educated and the least educated, and a four-year lag between the highest and lowest socio-economic groups. Delayed treatment is likely to worsen the prognosis and incur serious complications, such as blindness and heart disease.

National initiatives

India has an organization – diabetesindia.com – offering information and advice to professionals and patients, but lacks a national programme for diabetes. Although diabetes is actively researched, it is low on the government's current spending priorities and consumes only 1.2% of the annual budget. Non-communicable diseases such as diabetes receive far less attention and funding than acute infectious diseases, and diabetes receives less than AIDS even though its prevalence is higher.

There are a number of obstacles to the effective management of diabetes. India is a very large country with a diverse population, speaking 18 major languages in 200 different dialects. India has more cultural diversity than the whole of Europe. Effective management of diabetes in India depends upon education, self-help and the support of family members. This is essential where the caseload of doctors, especially in rural areas, is overburdened with acute infectious conditions. Religious and cultural differences as well as diet have to be taken into account with any programme, and educating patients and their families in blood-glucose monitoring, insulin and medication regimes is complex.

Be a critical thinker

To what extent is disease a consequence and/or a cause of poverty?

Consider the following:

- How can we define poverty on the global scale and apply a common term when geographical diversity is so great?
- How do we measure poverty, and who defines the standard?
- Is poverty always an undesirable state? Give reasons for your answer.
- How can we understand poverty if we have no personal experience of it?

11 Urban environments

By the end of this chapter you should be able to:

- explain the processes contributing to urbanization and the growth of cities and megacities
- describe the consequences of urbanization for countries at different stages of development
- understand the city as a system, with reference to residential patterns, urban poverty and deprivation, economic activity and environmental and social stress
- understand the challenges and problems facing cities and the sustainable strategies used to meet them.

This optional theme considers cities as places of intense human interaction. Urbanization promotes efficiency through economies of scale and concentration of the workforce. The rapid rate of urbanization, along with high levels of resource consumption and waste production, present challenges to planners. Cities are socially and economically distinct from one another, and variations in wealth and ethnicity can be seen in their internal structure. Their diversity makes cities focal points of multicultural discourse and ethnic assimilation, but segregation is also found. The future survival of cities depends on sustainable management of resources, the minimization of waste and the eradication of social inequality.

Centripetal movements

The process of urbanization

Urbanization is an increase in the proportion of a population living in towns and cities within a country or region. It is a relative term and should not be confused with urban growth, which is an increase in the absolute number of people living in urban areas and does not relate to rural areas or the country's population as a whole. **Centripetal** movements involve the migration of people into towns and cities.

Urbanization results from a positive migratory balance (more people moving into the city than moving out) and natural increase (where birth rate is higher than death rate). The urbanization process today is occurring very rapidly in less developed countries, and differs from the experience of European cities during the Industrial Revolution in Europe, when migration was the main cause. Another difference is that today's rural–urban migrants have a limited number of jobs available to them when they arrive in the city. This situation, where the volume of immigration far exceeds the provision of jobs, may be referred to as overurbanization, as in cities such as Lagos in Nigeria.

Push and pull factors

The decision to migrate from a rural to an urban area is triggered by both push and pull factors. Today's potential migrant from a developing country would need to justify the move by considering these factors:

Brownfield site – abandoned, derelict, or underused industrial buildings and land, which may be contaminated but have potential for redevelopment.
Counter-urbanization – a process involving the movement of population away from inner urban areas to a new town, new estate, commuter town or village on the edge or just beyond the city limits or rural–urban fringe.
Re-urbanization – the development of activities to increase residential population densities within the existing built-up area of a city. This may include the redevelopment of vacant land and the refurbishment of housing and the development of new business enterprises.
Suburb – a residential area within or just outside the boundaries of a city.
Suburbanization – the outward growth of towns and cities to engulf surrounding villages and rural areas. This may result from the out-migration of population from the inner urban areas to the suburbs, or from inward rural–urban movement.
Sustainable urban management strategy – an approach to urban management that seeks to maintain and improve the quality of life for current and future urban dwellers. Aspects of management may be social (housing quality, crime), economic (jobs, income) and environmental (air, water, land and resources).
Urbanization – the process by which an increasing percentage of a country's population comes to live in towns and cities. It may involve both rural–urban migration and natural increase.
Urban sprawl – the unplanned and uncontrolled physical expansion of an urban area into the surrounding countryside. It is closely linked to the process of suburbanization.

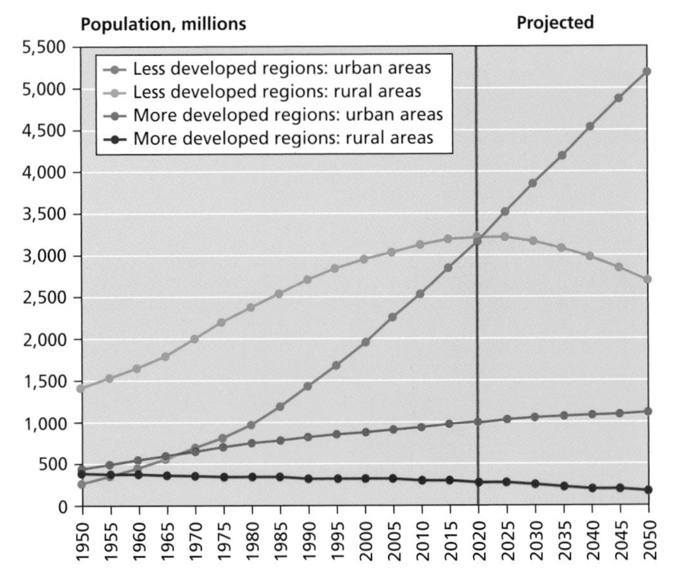

Figure 11.1 Changes in urban populations worldwide, millions

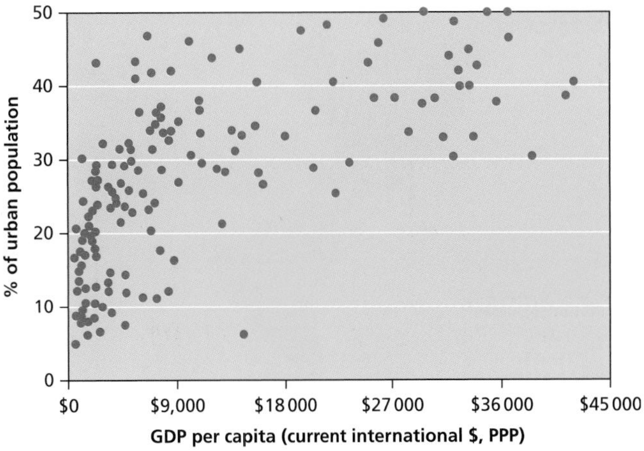

Figure 11.2 Urbanization versus income,

To do:

a Referring to Figure 11.1, describe the changes in the relationship between urban and rural populations for both more developed and less developed regions.

b Account for the differences.

Rural push factors

High rates of population growth have put pressure on natural resources such as water and energy and reduced the size of land holdings until they become unproductive. New farming technology has favoured the rich farmer but, for those at the other end of the socio-economic scale, unemployment or underemployment are typical. Migration for work is the only feasible solution for many who are supporting large families.

Urban pull factors

Higher wages and more varied employment and education opportunities make urban life attractive to the prospective migrant. Immigrants are willing to tolerate poor living conditions initially, in the hope of better prospects in future. The attraction of "bright lights" is often cited as a pull factor, but the reality is that many migrants are intimidated by the urban environment and would prefer the tranquillity and security of their original rural community.

The consequences of urbanization

Economic growth

Urban economies are almost always more productive than rural ones. For example, just 9% of Brazil's population is urban but it produces over 30% of the nation's gross domestic product. Most of the wealth is generated in **megacities**: 2% of China's population lives in Shanghai, a city that produces 12% of the country's output. Industrial productivity is higher in cities because of greater efficiency; there will be a large, concentrated, educated, accessible, skilled workforce.

Gentrification

Gentrification is the reinvestment of capital into inner-city areas. It refers mostly to an improvement of residential areas, although there is an economic dimension too. Thus, as well as residential rehabilitation and upgrading, there is also commercial redevelopment. Gentrification is a type of filtering that may lead to the social displacement of poor

TOK Link

Urbanization creates wealth. How true is this statement?

The scatter graph (Figure 11.2) suggests that there is a positive relationship between wealth and level of urbanization. In other words, countries with a lower level of urbanization are poor and those with a higher level of urbanization are rich. However, we need to be very cautious about making such a bold and simplistic statement. A number of aspects need to be checked before any assumptions can be made.

How was this data collected? Was it accurately measured? Was the sample involved fair and representative? How do you define urban? How accurate is GDP per capita as a measure of a country's level of wealth?

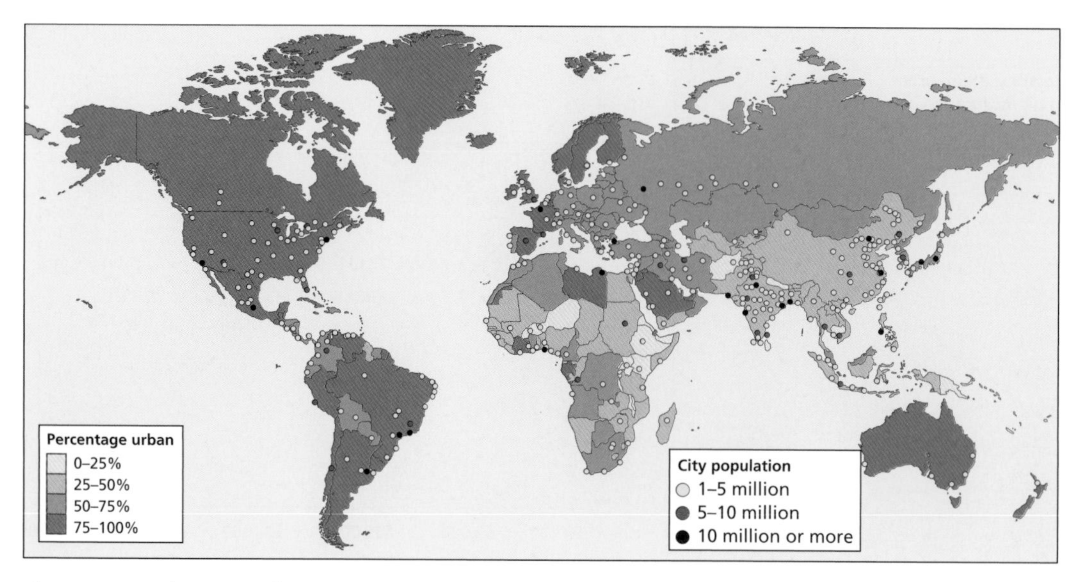

Figure 11.3 Urbanization by country

people – as an area becomes gentrified, house prices rise and the poor are unable to afford the increased prices. As they move out, young upwardly mobile populations take their place.

Gentrification has occurred in many large old cities throughout the world, such as in New York (Greenwich Village and Brooklyn Heights), Toronto (Riverdale) and London (Fulham and Chelsea). It has also been observed in cities as diverse as Johannesburg, Tokyo and Sao Paulo.

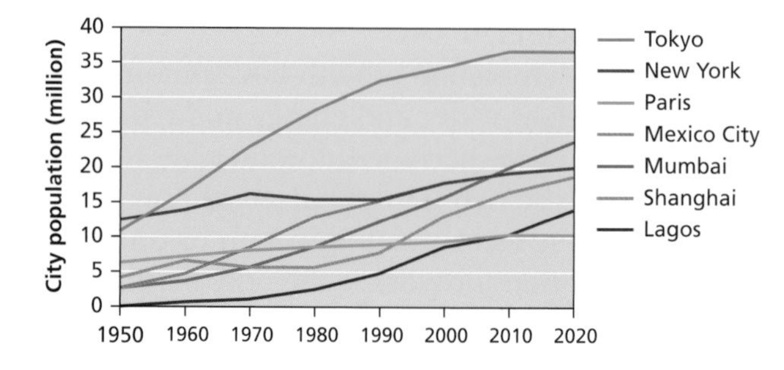

Figure 11.4 Growth of urban population in selected cities, 1950–2020 (projected)

Re-urbanization/urban renewal

Re-urbanization is a revitalization of urban areas and a movement of people back into these areas. It often reflects sound and sustainable policies that avoid the destruction of greenfield sites and thereby conserve open country and wildlife. It involves the reclamation of derelict land and the refurbishment of existing built-up areas, known as **brownfield sites**. Re-urbanization and urban renewal schemes are adopted in cities that have reached an advanced stage of development and have sufficient public and private funding to undertake ambitious schemes. Examples include London's Docklands, La Défense in Paris and the renewal of Manhattan in New York.

Figure 11.5 Gentrification of Spitalfields, London

Centrifugal movements

Centrifugal movements, or decentralization, are the outward movements of a population from the centre of a city towards its edge or periphery, resulting in an expansion of the city.

Suburbanization

Suburbanization is an example of centrifugal movement. It started in about 1900 in Europe and in the 1920s in the USA. It is now a worldwide phenomenon, even in developing cities, where increasing affluence has allowed urban residents to relocate. It is caused by the rapid growth of the urban population and a demand for better housing and more space. Rising disposable incomes have enabled people to meet both the cost of new housing and the associated transport costs of commuting back to the city centre for work. In some cases industry has also decentralized, providing employment outside the city centre.

Urban sprawl

Urban sprawl is the physical expansion of the city area and is closely associated with suburbanization. It is an inevitable consequence of unplanned urban growth and may now be found worldwide. Good examples of urban sprawl include Seoul, Mexico City and Riyadh. Since the discovery of the world's largest petroleum reserves in Saudi Arabia in the 1930s, the pace of the capital Riyadh's urbanization has been rapid. In the early 1970s, Riyadh, already the largest city in Saudi Arabia, had a population of 500 000. The first satellite image (Figure 11.6a) was taken in 1972. The second (Figure 11.6b) was taken in 2007, when the population had reached almost 6 million.

To do:

Refer to Figure 11.6.

a Describe and suggest reasons for the growth of Riyadh between 1972 and 2007.

b Describe and briefly explain the pattern of growth.

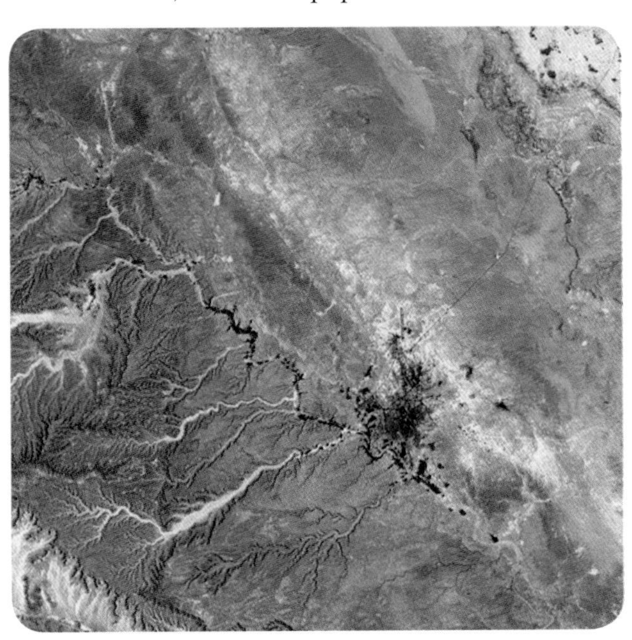

Figure 11.6 Satellite views of urbanization in Riyadh, Saudi Arabia, 1972 and 2007

Counter-urbanization

This is the movement of people from inner urban areas to areas beyond the city limits. It is a process of population decentralization, which often follows on from suburbanization and is characteristic of wealthy cities in MEDCs. Counter-urbanization, like suburbanization, is a response to the growth of the metropolitan area and population and the increasing stress of overcrowding, congestion, pollution, lack of community and declining services. The rural lifestyle is perceived to be free of these nuisances, and increased disposable income has allowed both workers and retirees to pursue the rural idyll. IT and the option of teleworking provide added incentives to move out of town.

To research

Explain the problems associated with urban growth in a hot desert environment such as Riyadh. Refer to chapter 7, Extreme environments.

The consequences of centrifugal movement

Centrifugal movements involve a shift of population and economic activity from the centre of the urban area to its periphery and beyond. The construction of roads and buildings destroys open space and increases air pollution. At the same time, loss of population and economic activity is detrimental to the centre. Planners have focused on ways of reviving the urban core and restricting new construction in urban **hinterlands** (the zone surrounding a city).

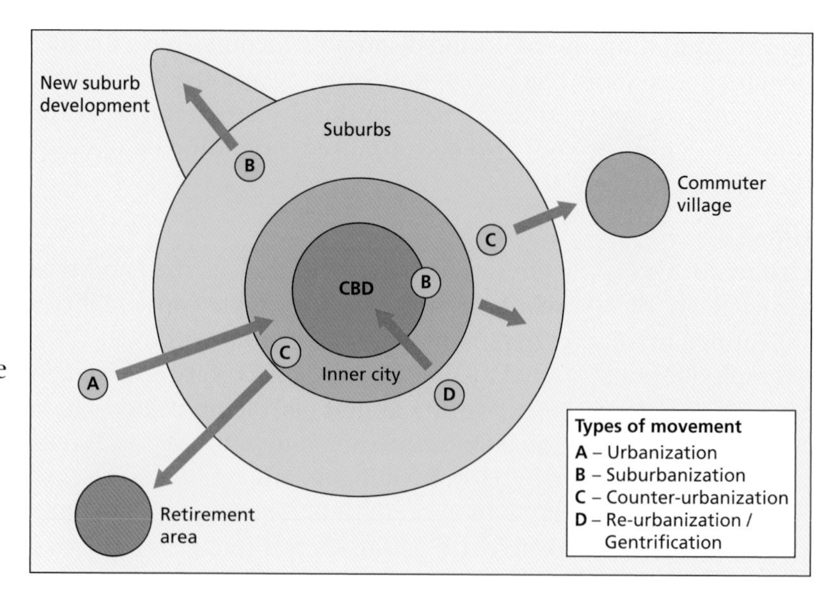

Figure 11.7 Urban residential relocation: types of movement

The family life cycle

Intra-urban population movement occurs within the city and may involve shifts of population during the family life cycle. A person is likely to move around different zones of a city, depending on their age and their need for a house of a certain size. This is true for those in rented accommodation as well as for homeowners. Residential patterns are influenced by banks, building societies, local authorities, housing associations and free choice. (Figure 11.8)

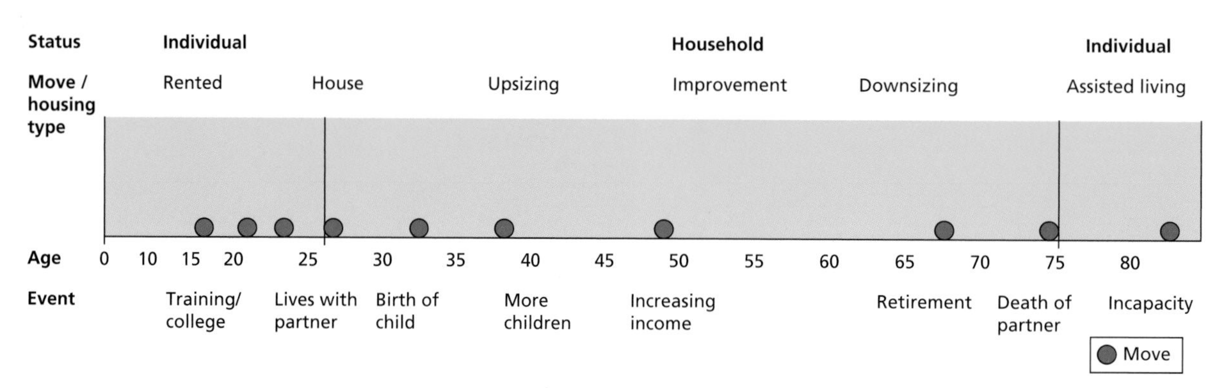

Figure 11.8 The effect of life cycle on residential location in a European country

> ***TOK Link***
>
> **How useful are models in geography?**
>
> Models are generalizations about the world which are made through empirical studies. For example, the life cycle model of urban residential change here was devised by research on the patterns of social changes in cities in the more developed world. To answer this question you will need to critically evaluate two urban models.

Megacities

Megacities are large metropolitan areas or urban **agglomerations** of 10 million inhabitants or more. (An agglomeration is the metropolitan area incorporating several large towns and cities, such as New York/Newark in Table 11.1). Changes in the rank order can be observed between the three time periods, with the greatest population growth occurring during the period 1975–2009. Whereas the top ranks were once held only by rich megacities, they are now dominated by cities in the less developed world such as Dhaka, Kolkata and Karachi, whose growth rates over the last few decades have exceeded 3% per annum, reflecting their national rates of natural change.

> **To do:**
>
> Study Figure 11.8, which shows the changes in residential location during the life cyle, and is representative of a European country.
>
> Suggest ways in which the residential changes might not follow this pattern in different part of the world.

1975			2009			2025		
Rank	Megacity	Pop. millions	Rank	Megacity	Pop. millions	Rank	Megacity	Pop. millions
1	Tokyo	26.6	1	Tokyo	36.5	1	Tokyo	37.1
2	NewYork/ Newark	15.9	2	Delhi	21.7	2	Delhi	28.6
3	Mexico City	16.7	3	Sao Paulo	20.0	3	Mumbai	25.8
			4	Mumbai	19.7	4	Sao Paulo	21.7
			5	Mexico City	19.3	5	Dhaka	20.9
			6	New York/ Newark	19.3	6	Mexico City	20.7
			7	Shanghai	16.3	7	New York/ Newark	20.6
			8	Kolkata	15.3	8	Kolkata	20.1
			9	Dhaka	14.3	9	Shanghai	20.0
			10	Buenos Aires	13.0	10	Karachi	18.7

(Source: World Urbanization Prospects 2009)

Table 11.1 The world's top 10 megacities ranked by population

Although the megacity has been an important feature of urbanization processes, it is usually the initial stage and is followed by the expansion of the **urban hierarchy**. The most rapidly growing cities are those with populations around 500 000, whose problems are not on the same scale and therefore less reported than those of the megacities.

Urban hierarchy – a national arrangement of urban areas, for example from one large city to many small villages.

Megacities are dynamic and vibrant centres of economic activity, social interaction and innovation. Many of those listed would classify as world cities, acting as hubs in the global network of economic activities such as trade and financial transfers. Megacities bring together people and resources, and are able to generate wealth efficiently through economies of scale.

Megacities also have a large number of specific problems and common characteristics. These include high population concentration and density, uncontrolled spatial expansion, severe infrastructural deficits such as inadequate water supply and sewerage, signs of ecological strain and overload, poor housing provision and increasing disparity between rich and poor. There are stark contrasts between megacities, as shown in Figure 11.2, and also within many of them.

Megacity	Population, millions	Persons per room	Percentage of houses with water and electricity	Infant mortality rate	Noise levels	Quality of Life Index
Tokyo	35	0.9	100	5	4	81
New York	17.4	0.5	99	10	8	70
Mexico City	19.4	1.9	94	36	6	38
Kolkata	12.8	3.0	57	46	4	34

Table 11.2 Some measures of quality of life in megacities

To do:

Refer to Table 11.2.

a Evaluate the indicators used to generate the Quality of Life Index.

b Suggest two additional indicators and justify your choice.

Case study: Shanghai – a rapidly growing megacity

Figure 11.9a The World Finance Center, Shanghai's tallest building of 101 floors

Figure 11.9b Low-income housing in the Old City, Shanghai

For 700 years Shanghai has been one of Asia's major ports, and it has a varied history. It thrived until 1949, when China closed itself to trade with the west. This changed Shanghai from an international centre of production and trade to an introverted city, serving the interests of a highly centralized socialist state.

During the 1970s, China began slowly reopening its economy to the world, and Shanghai was designated one of 14 open cities. The Shanghai Economic Zone was established in 1983, and in the early 1990s an ambitious major programme of redevelopment was started, especially in its eastern hinterland around Pu dong.

Having by 1990 established a strong industrial base, the city was well placed to take advantage of the new opportunities offered by globalization. It became a major centre for export manufacturing based on automobiles, biotechnology, chemicals and steel, and its service industry sector (trade, finance, real estate, tourism, e-commerce) helped to diversify its economy.

Between 1990 and 2000, Shanghai began to re-emerge as a world city. Changes in national laws, expenditure on local infrastructure and reforms of land management practice attracted foreign investment. More than half the world's top 500 transnational corporations and 57 of the largest industrial enterprises set up in Shanghai, contributing to an annual regional growth rate of over 20%, more than twice the national average. In 2009, Shanghai was ranked the seventh largest city in the world, with a population of 15 million.

Since 1990, the city's manufacturing sector has steadily contracted, shedding almost a million jobs, while the business services, finance and real estate sectors have expanded. Rising demand for highly skilled labour has led to further in-migration, resulting in increasing disparity in wealth between rich and poor. Shanghai's experience does lend support to the general hypothesis that world city status inevitably leads to a widening gap between rich and poor.

Today, Shanghai is a city-state within China. It is part of the Yangtze River Delta, the fastest growing urban area in the world, containing 16 megacities including Shanghai. This region has 75 million people and earns 25% of China's GNP, 50% of its foreign direct investment. The city has been described as the largest construction site in the world: 4,000 buildings with more than 24 storeys were under construction in 2010.

Managing Shanghai's rapid growth

Social and demographic issues

Housing shortages and overcrowding problems are acute. Almost half the population lives in less than 5% of the total land area, and in central Shanghai population density reaches 40 000–160 000 people per square kilometre. Population pressure is caused by in-migration, overcrowding, disparities in wealth and the social insecurity of Shanghai's poor "floating population".

The Shanghai government has a series of important policies to address these problems:

- a combination of widespread family planning and medical care, which has controlled fertility levels among the young immigrant population
- compulsory work permits
- educational initiatives to improve immigrant job opportunities.

These initiatives have reduced population density in the heart of the city and increased it in the suburban satellite cities such as Songjiang. This has been a successful strategy, although for many who have been moved out under the decentralization policy, the journey back to the centre for work is no advantage.

In 2004 the average life expectancy of Shanghai's population reached 80 years, and it is estimated that, by 2020, around 34% of the population will be aged over 60 and 28% will be over 65. This will put stress on the social security system.

Affluence and effluence

Water quality in Shanghai is a concern: less than 60% of waste water and storm water and less than 40% of sewage flows are intercepted, treated and disposed of. Waste disposal is also a major problem: the Huang pu River receives 4 million cubic metres of untreated human waste every day. The construction industry generates 30 000 tonnes of building waste per day, and municipal landfill sites have almost reached capacity. Nevertheless, since the 1990s there have been marked improvements in sanitation, and almost all households have access to piped water, electricity and waste disposal. Municipal organic waste is now used as fertilizer in the surrounding rural areas.

Shanghai has the highest cancer mortality rate in China, and until recently had the reputation of being the 10th most polluted city in the world. Industry generated over 72% of carbon dioxide emissions, 9% came from transport systems, and the remainder came from domestic use. Coal-fired power stations provide 75% of China's electricity, as well as emissions of suspended particulate matter, nitrous oxides and sulphur dioxide. Motor vehicle emissions are particularly harmful in the presence of strong sunlight, when photochemical smog is formed. A product of this is low-level ozone, a harmful irritant responsible for breathing difficulties. Efforts to reduce air pollution levels have been moderately successful: emissions of nitrous oxides have fallen from around 40 ppm to 32, and SO^2 emissions have fallen from 50 ppm to around 32 over the same time. Reducing PM 10s has proved to be a much more difficult task.

The government has responded to pollution problems by upgrading the city's transport systems and attempting to limit the growth in car ownership. Shanghai's underground system, with a daily capacity of 1.4 million, is now linked to Pu dong airport by the world's fastest commercial magnetic levitation train – MAGLEV – capable of reaching 431 kilometres per hour. Other strategies to improve safety have been pedestrianization and a reduction in the number of bicycles, currently estimated at 9 million and a cause of many road accidents.

Coastal flooding

Like many global ports, Shanghai is under threat of coastal flooding, partly due to its low elevation at only 4 metres above sea level, but also from monsoons and tropical cyclones. Future hazard events will be aggravated by climate change and the possibility of sea-level rise. The problem is compounded by subsidence, which has been caused by overabstraction of groundwater and the weight of high-rise buildings. Shanghai sank by 2.6 metres between 1921 and 1965, and in 2002 alone by 10.22 millimetres.

The rapid development of Shanghai has presented the government and planners with some challenging problems, only some of which have been solved. The question is whether Shanghai can maintain the principles of sustainability while growing at such a rapid pace.

Figure 11.10 Recycling by tricycle: collection from CBD businesses

To do:

- Examine the natural and artificial hazards that threaten the city of Shanghai and its population.
- Critically evaluate the government's policy to control air pollution in Shanghai.

Be a critical thinker

Should cities like Shanghai be allowed to continue to grow indefinitely?

The population of Shanghai is predicted to double by the year 2030. Review the problems that already exist in the city, and consider how more people can be accommodated and more jobs provided.

Visit websites that cover the Shanghai Urban Planning Exhibition, to see the plans for the city's development over the next two decades.

To do:

a Compare the population structures of Shanghai and New York (Figure 11.11) and discuss the socio-economic consequences for these two megacities.

b Compare the patterns of residential density shown in Figures 11.12 and 11.13, and discuss the environmental consequences for these two megacities.

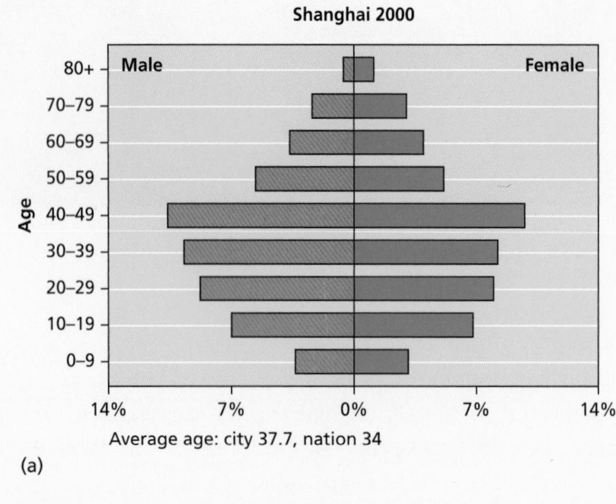

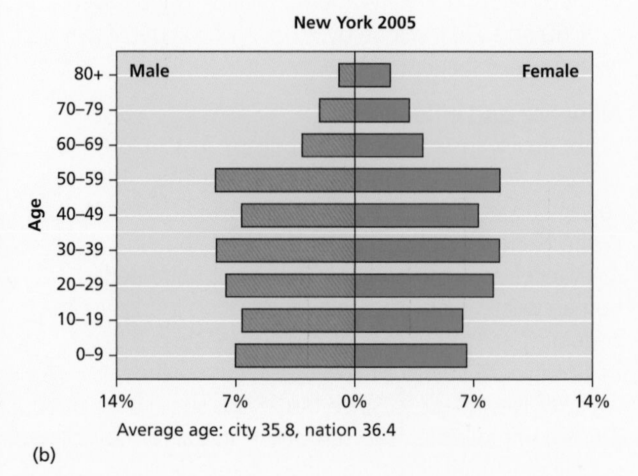

Figure 11.11 Population pyramids for Shanghai and New York

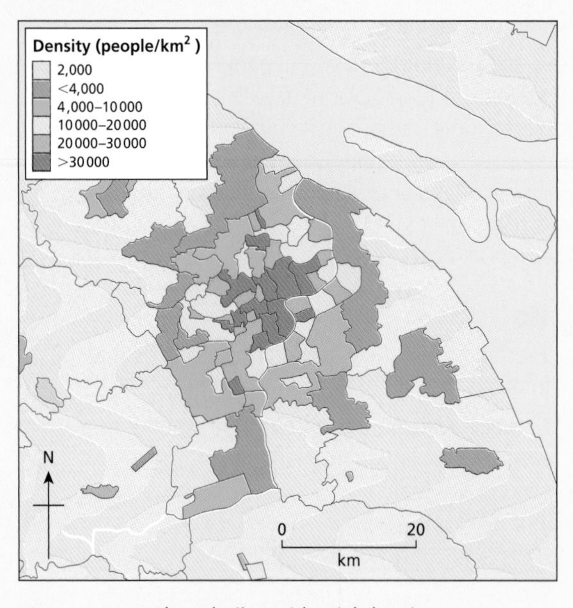

Figure 11.12 Shanghai's residential density

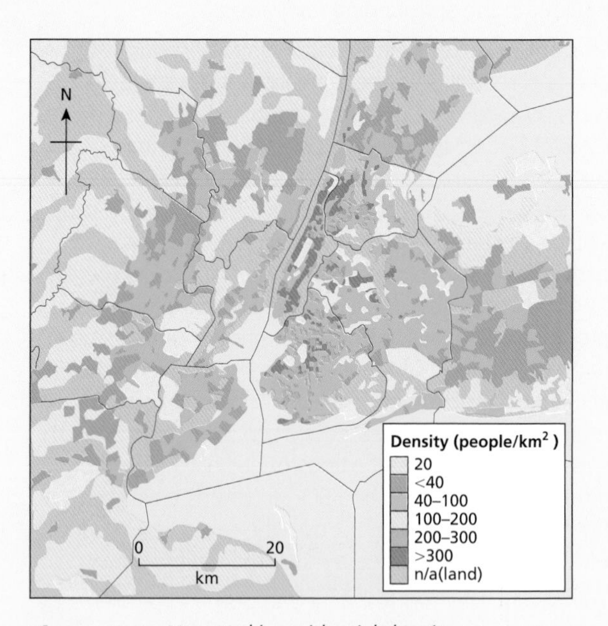

Figure 11.13 New York's residential density

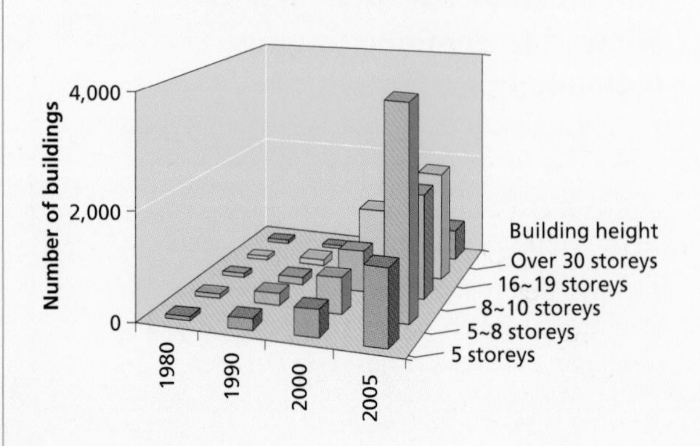

Figure 11.14 The evolution of Shanghai's skyline

Figure 11.15 The MAGLEV high-speed train in Shanghai

Residential patterns in rich countries

Residential segregation is common in all cities, irrespective of culture or economics. The causes of segregation are:

- **Socio-economic status** Whereas social contracts such as the caste system in India can determine social position in traditional societies, in western society individual socio-economic status is determined largely by employment and income.

Residential segregation – the physical separation of population by culture, income or other criteria.

Case study: *Ethnic segregation and diversity in London*

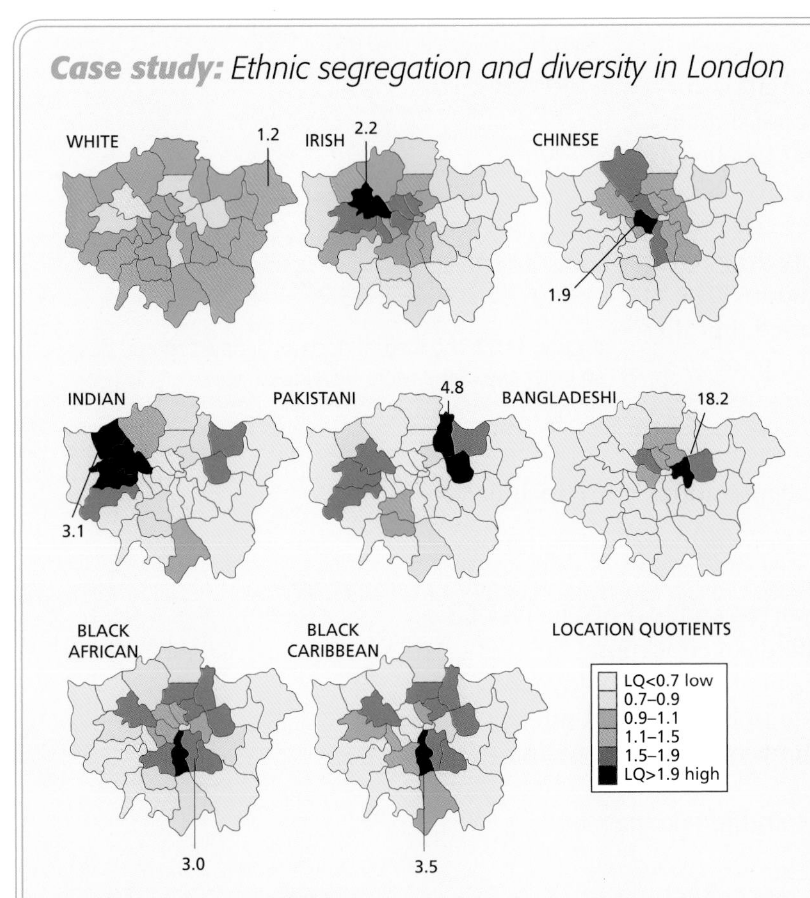

Figure 11.16 Location quotients for different ethnic groups in London

Measuring ethnic segregation

One way of measuring ethnic segregation is to use the **location quotient** (LQ). For example:

- Neighbourhood × has 25% Bangladeshi population and the city average is 10%.
- The LQ would be 2.5, a higher than average concentration.
- LQ < 1 means no concentration.
- LQ = 1 means average concentration.
- LQ > 1 means above average concentration.

The Bangladeshi population in London

Between 1960 and 1980 there was some decentralization of London's white population to new towns beyond the suburbs. The remaining urban housing stock was old, and later occupied by immigrant populations – Indians, Pakistanis, Black and Caribbean Africans and Bangladeshis. Today the Bangladeshis are concentrated in the London Borough of Tower Hamlets. The high level of segregation is due to their:

- late arrival
- inability to speak English on arrival
- wish to maintain cultural identity
- common origin in the Bengali province of Sylhet, so that segregation brought security.
- poverty, which leads to high concentration and overcrowding.

The Bangladesh community has established a network of businesses in Spitalfields, now a prosperous commercial zone in Tower Hamlets. This includes Brick Lane (often known as Banglatown) where restaurants and businesses take advantage of lunchtime workers from the City close by. This has now become a curry hot spot and tourist attraction. Around this district are small clothing factories originally owned by Jewish communities. Bangladeshis in the area are also employed in food retailing, wholesale clothing distribution, the travel industry and money lending.

To do:

To do:

a Referring to the maps, identify the ethnic groups with the highest and lowest levels of concentration and suggest reasons for the difference.

b Why do populations segregate themselves in cities? Are they:

- segregating themselves from threats or enemies?
- seeking to gain separate identity as a superior group?

c Using London or other examples, explain what is meant by ethnic diversity.

- **Ethnicity** The cultural differences between immigrants and existing residents often led to difficulties in communication, resulting in varying degrees of residential segregation. Over time many migrant groups have gradually been assimilated, while for others – in particular the more visible ethnic minorities – spatial segregation remains a fact of urban life.

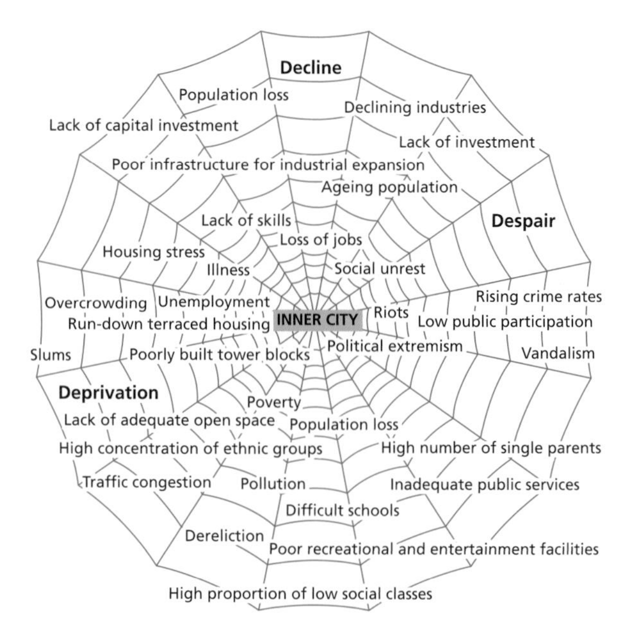

Figure 11.17 The web of decline, deprivation and despair in an inner city of the more developed world.

Urban poverty and deprivation

Within most cities there is considerable variation in the quality of life. This raises questions about equality of opportunity and social justice. The areas labelled as "poor" are zones of deprivation, poverty and exclusion. In MEDCs these are often inner-city areas or ghettos, whereas in LEDCs countries it is frequently shanty towns that exhibit the worst conditions. The factors associated with deprivation are varied, but they result in a cycle of urban deprivation and a poor quality of life.

Measuring deprivation

A number of indices are used to measure deprivation. These include:

- physical indicators – such as quality of housing, levels of pollution, incidence of crime, vandalism, graffiti
- social indicators – including crime (reported and fear of); levels of health and access to health care; standards of education; proportion of population on subsidized benefits (unemployment, disability, free school meals); proportion of lone-parent families
- economic indices – access to employment; unemployment and underemployment; levels of income
- political indices – opportunities to vote and to take part in community organization.

Slums and squatter settlements

The total number of slum dwellers in the world stood at around one billion in 2005 and is estimated to reach two billion by 2030. This represents about 32% of the world's urban population, but 78.2% of the urban population is in LEDCs. Slums are typically located in areas that planners do not want – steep slopes, floodplains, edge-of-town locations and/or close to major industrial complexes. See Dharavi case study on page 300.

Positive aspects	Negative aspects
They are points of assimilation for immigrants.	Security of tenure is often lacking.
Informal entrepreneurs can work here and have clienteles extending to the rest of the city.	Basic services are absent, especially water and sanitation
Informal employment, based at home, avoids commuting.	Overcrowding is common.
There is a strong sense of kinship and family support.	Sites are often hazardous.
Crime rates are relatively low.	Levels of hygiene and sanitation are poor, and disease is common.

Table 11.3 Living in a slum

Economic activities in cities

The formal and informal economies

Urban economies and labour forces may be classified as formal or informal. Both types exist together and universally. The contribution made by the latter to national economies is hard to measure, but an estimate is that it employs more than 60% of the urban population in South America and Asia and more than 70% in sub-Saharan Africa. In the majority of cases, employment in the informal economy is unskilled and poorly paid. However, in richer countries a new type of artisan vendor has emerged, selling handcrafted wares and overpriced bric-a-brac to affluent residents and tourists. Overall, the contribution to national economies made by informal activity is growing, and probably underestimated, especially in low- and middle-income countries.

Informal activity has no specific location in the city, although street vending and personal services tend to occur in and around the central business district, where pedestrian density is greatest. Other types of informal activity occur in residential areas, often on the ground floor of slum properties and squats, many of which are found in areas of disamenity such as railway sidings, river embankments and zones of discard.

Formal economy	Examples:	Informal economy	Examples:
• Qualifications and training required	Bank clerk	• No qualifications or training required	Fruit vendor
• Set hours of work and pay	Midwife	• Unregulated hours and pay	Rickshaw puller
• Job security and legal protection	Teacher	• No job security, no legal protection	Barber
• Pensions and unemployment benefits	Plumber	• No pensions, no job protection	Taxi driver
• Well-serviced purpose-built premises	Truck driver	• Small premises, sometimes domestic	Waste-picker
• Imported technology using non-local raw materials		• Adaptive technology using local raw materials	
• Capital intensive		• Labour intensive	
• Registered and documented business transactions		• Barter of cash transfers, no documentation	
• State-regulated business enterprise		• Some illegal business	

Table 11.4 The characteristics of formal and informal economies

The advantages of the informal economy

The informal economy plays a vital role in the developing urban economies of many low- and middle-income countries, and is an inevitable outcome of rapid urbanization. It has provided unskilled and semi-skilled migrants with casual, temporary, but immediate work.

Setting up a business formally is cumbersome and a deterrent to many potential entrepreneurs. For example, in Angola it takes 13 different procedures, 124 days and almost 500% of the average income of an Angolan. By contrast, in the USA the same process requires just five procedures, five days and 0.7% of the average annual income of an American. Confronted with these costs and delays, setting up an informal business is preferable in a low-income country.

The informal and formal economies are often interdependent; goods produced at minimum cost in small workshops are then passed on to be finished and sold within the formal economy on the national and international market.

To research

See the publication *Urban Age* published by the LSE for further information on Mumbai.

The informal economy makes a large contribution to urban wealth, and cannot be overlooked. It was the basis of the Industrial Revolution in 19th-century Europe, and may therefore be regarded as a transitional but essential stage in the evolution of city economies.

The disadvantages of the informal economy

It has been unfairly associated with such activities as drug pushing, prostitution, political corruption, bribery, and smuggling. This threatens the security of the residents, turns away potential visitors and downgrades the city's international image. Countries with a sophisticated legal and political system and stronger protection of physical and intellectual property rights experience higher economic well-being, Informally run businesses have none of these benefits, and the lack of legal property ownership limits access to credit.

Workers in the informal economy are often exposed to health and safety risks and deprived of the rights and benefits associated with law and regulation. Lack of protective clothing and adequate instruction cause contamination by toxic chemicals and heavy metals found in solvents and recycled waste.

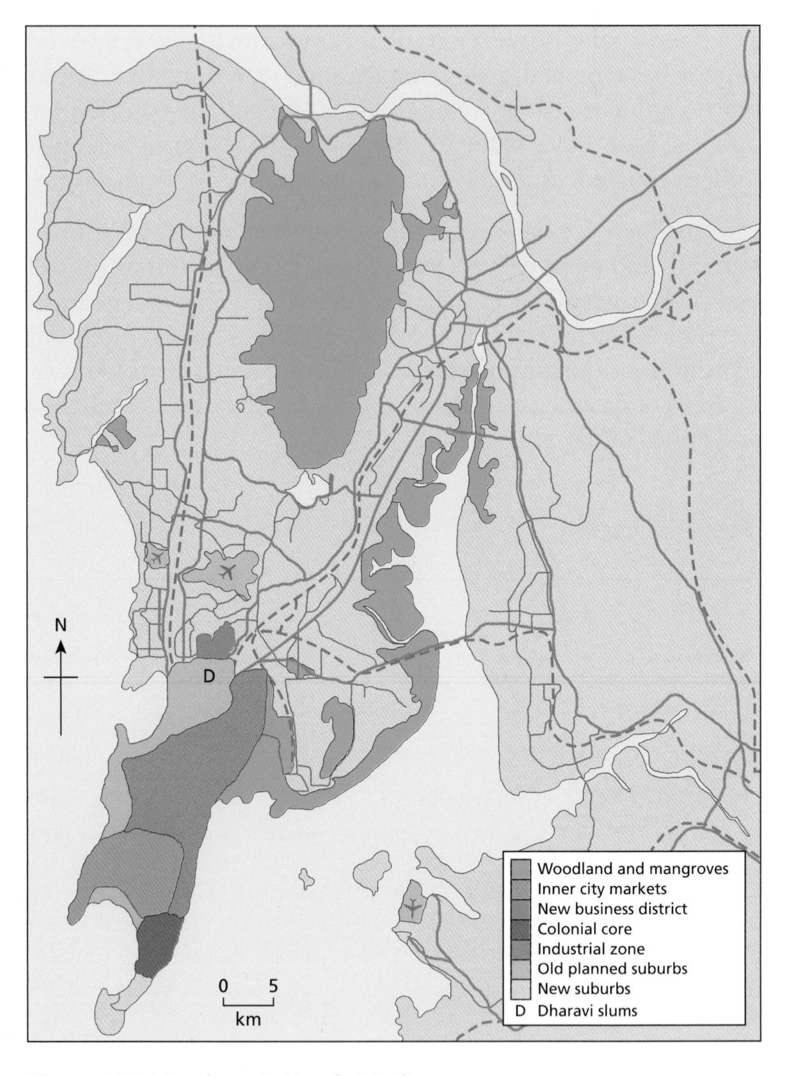

Figure 11.18 Land use in Mumbai, India

Case study: *The informal economy of Dharavi, Mumbai*

Dharavi is located in the heart of Mumbai, close to the fast-developing commercial centre and to Mumbai's domestic and international airports. Despite its plastic and tin shacks and lack of infrastructure, Dharavi is a unique, vibrant, and thriving "cottage" industry complex. Dharavi is the largest and oldest slum in Mumbai, the best-known in India, with a population of more than 600 000 people living in 100 000 makeshift homes. Its population density is the highest in the world, at over 100 000 people per square kilometre in an area barely half the size of New York City's Central Park. Although the concentration of businesses could easily deter consumers, the large scale at which informality occurs yields an estimated $665 million in annual revenues. However, the economic success of Dharavi has a high cost in human welfare.

The working environments and conditions are very poor, especially for the newly arrived casual worker.

In dark unventilated foundries, workers ladle molten metal into a belt-buckle mould held between their bare feet. In a warehouse, men smeared from head to toe in blue ink strip the casings from used ballpoint pens so they can be melted down and recycled; few wear gloves or other protective gear despite exposure to solvents and other chemicals. Environmental and health hazards are the realities that workers have to withstand to earn a living.

Its population has achieved a unique informal "self-help" urban development over the years, without any external aid. It is a humming economic engine. The residents, though bereft of housing amenities, have been able to lift themselves out of poverty by establishing thousands of successful businesses. A study by the Centre for Environmental Planning & Technology indicates that Dharavi currently has close to 5,000 industrial units, producing a wide range of goods including textiles, pottery, leather, plastics and poppadums. It also processes 80% of Mumbai's recyclable waste. Other residents are employed providing services such as transport for their own community, or for Mumbaikers in general.

Industrial activity takes place in nearly every home, with a workshop at street level and a living room above. Dharavi's informal economic activity is decentralized, home-based, low-tech and labour-intensive. It relies on close business networks within a strongly cohesive community. Entrepreneurs employing large numbers of cheap and flexible casual workers can compete in the national and international market, thereby indirectly contributing to the formal economy.

Threats to Dharavi

Unfortunately, the slum lacks residential infrastructure: one public toilet is shared by 1,400 residents, and water and electricity are available only once a day. The state government depicts Dharavi as a large junkyard, and wants to force its population to relocate into tiny cubbyhole apartments in high-rise towers, so that the vacated land can be commercially exploited by developers through the Dharavi Redevelopment Plan. At a conservative estimate, a development of this magnitude could fetch $460 million for a developer, a profit of at least 900%. Case studies all over the world have documented the inappropriateness of high-rise resettlement projects in poor areas. The social and economic networks on which the poor rely for subsistence in Dharavi cannot be sustained in high-rise structures.

The least that can be done in this redevelopment plan is to refurbish the workplaces of existing industries within the residential areas and remodel the project by providing low-rise, high-density row housing for existing families engaged in home-based occupations. This way, each house will have a ground floor and an additional storey, as well as a terrace and courtyard which can be used for these home-based business activities. Unfortunately, the formulation of Dharavi Redevelopment Plan as a profit-maximizing real-estate tool leaves no room for exploring such sustainable and economically viable approaches. It exposes the DRP as a weak cover-up for a land grab of the worst kind.

Figure 11.19 Dharavi the informal economy

Figure 11.20 Street in Dharavi

Housing strategies in Mumbai

Mumbai has an acute housing problem, with 8 million people living in slums and squatter settlements close to the centre, where land prices are some of the highest in India. A flat in the centre of the city costs $500 000, way beyond the reach of slum dwellers.

Two property companies have found a solution outside the city. Matheran Realty are building 15 000 ultra-low-cost flats in

Karjat, 90 kilometres east of Mumbai, costing 210 000 rupees ($4,500) for a unit of 19 square metres. Tata is building 1,300 basic units – "nano homes" – at Boisar, 100 kilometres north of the city, costing 390 000 to 670 000 rupees ($8,000 to $14,000). The simplest consist of a single room with a sink and a toilet behind a partition. The buildings have no more than three storeys so there is no need for expensive structural work. Instead of bricks and mortar, the walls are made of lightweight moulded concrete blocks. The concrete is made with foam, fly ash and other waste materials to make it lighter as well as cheaper. There are no lifts and just one staircase per block. All this means that the homes can be built quickly and with unskilled labour. Many slum dwellers are drivers, factory workers or tailors with incomes of around 90 000 rupees ($2,000) a year, enough to buy a flat costing 200 000 to 400 000 rupees ($4,200 to $8,200).

Obtaining credit was a problem until recently for slum dwellers with illegal tenure or no permanent address, but now the National Housing Bank and the National Bank for Agricultural Development have agreed to finance companies so they can offer mortgages to these people. To reduce risks, buyers must put down at least 25% of the purchase price and employers must confirm their income.

The construction of low-cost housing outside the city overcomes some of the problems of overcrowding and insanitary conditions found in the central slums such as Dharavi, but there are drawbacks:

- There is no accommodation for the informal businesses that were an integral part of a dwelling.
- Ex-slum dwellers will now have to commute back to the centre to work, causing further problems of traffic pollution and congestion.
- Displacement will disrupt extended families, their social networks and security.
- Close business linkages that existed within the informal economy of the slum will be lost.

There are no easy solutions to Mumbai's housing problems.

To research

Mumbai's Dharavi district has been controversial for some years and its survival is in the balance.

Visit http://ngm.nationalgeographic.com/2007/05/dharavi-mumbai-slum/dharavi-video-interactive for a slide show presentation on Dharavi, "Dharavi – Mumbai's shadow city".

Visit *The Economist*, www.economist.com, for an article on India's informal economy, "The tailors of Dharavi".

See also the film *Slumdog Millionaire*.

Use the above resources and this chapter to develop economic, social, political and environmental arguments both for and against the demolition of Dharavi.

To do:

Study Figure 11.18.

a Describe the pattern of land use for Mumbai, and suggest reasons for the location of specific land uses.

- **i** Name three urban processes involved in the development of these land-use zones.
- **ii** Explain the pressures that may result from the layout of the city for residents of Mumbai (Mumbaikers).

b Use the photos (Figures 11.19 and 11.20) to describe the features of the informal economy of Dharavi, Mumbai, in India.

c Find a case study of contamination, toxic leaks or air pollution incidents that have occurred in the poor areas of any city.

d Find examples of informal retailing in your area, e.g. market stalls and farmers' markets.

- **i** Describe and explain their location.
- **ii** Survey their customers to find out their place of origin and motives for using the service.
- **iii** Compare goods sold informally with those in your local supermarket, for price, quality and ecological footprint.
- **iv** Make a list of the benefits and costs of informal trading.

The location of industry in urban areas

Models of urban land use have located manufacturing industry in inner-city areas (Burgess's concentric model), along major routeways (Hoyt's sector theory) and in industrial suburbs (Harris and Ullman's multiple nuclei model). These reflect the variety of manufacturing industries and their differing locational requirements. In these models the location of industry is described but not explained in detail.

Industries found in cities include:

- those needing access to skilled labour, such as medical instruments; those needing access to the CBD, such as fashion accessories and clothes; and those which need the whole urban market for distribution, such as newspapers – these industries all having a central location
- port industries
- those located on radial routes, e.g. Samsung Electronics at Suwon, Korea
- those needing large amounts of land for the assembly, production or storage of goods, e.g. the Hyundai car works at Busan, Korea.

Large cities are attractive for industries for a number of reasons:

- Capital cities, such as Paris or Moscow, are often the largest manufacturing centres of a nation.
- Cities are large markets.
- Port cities have excellent access to international markets.
- Cities are major centres of innovation, ideas and fashion.
- A variety of labour is readily available, including skilled and unskilled workers, decision-makers and innovators.

Land use in New York

Industries, warehouses and factories occupy 4% of the city's total lot area. They are found primarily in the South Bronx, along either side of Newtown Creek in Brooklyn and Queens, and along the western shores of Brooklyn and Staten Island. River front locations are very important, especially for transportation and utility uses.

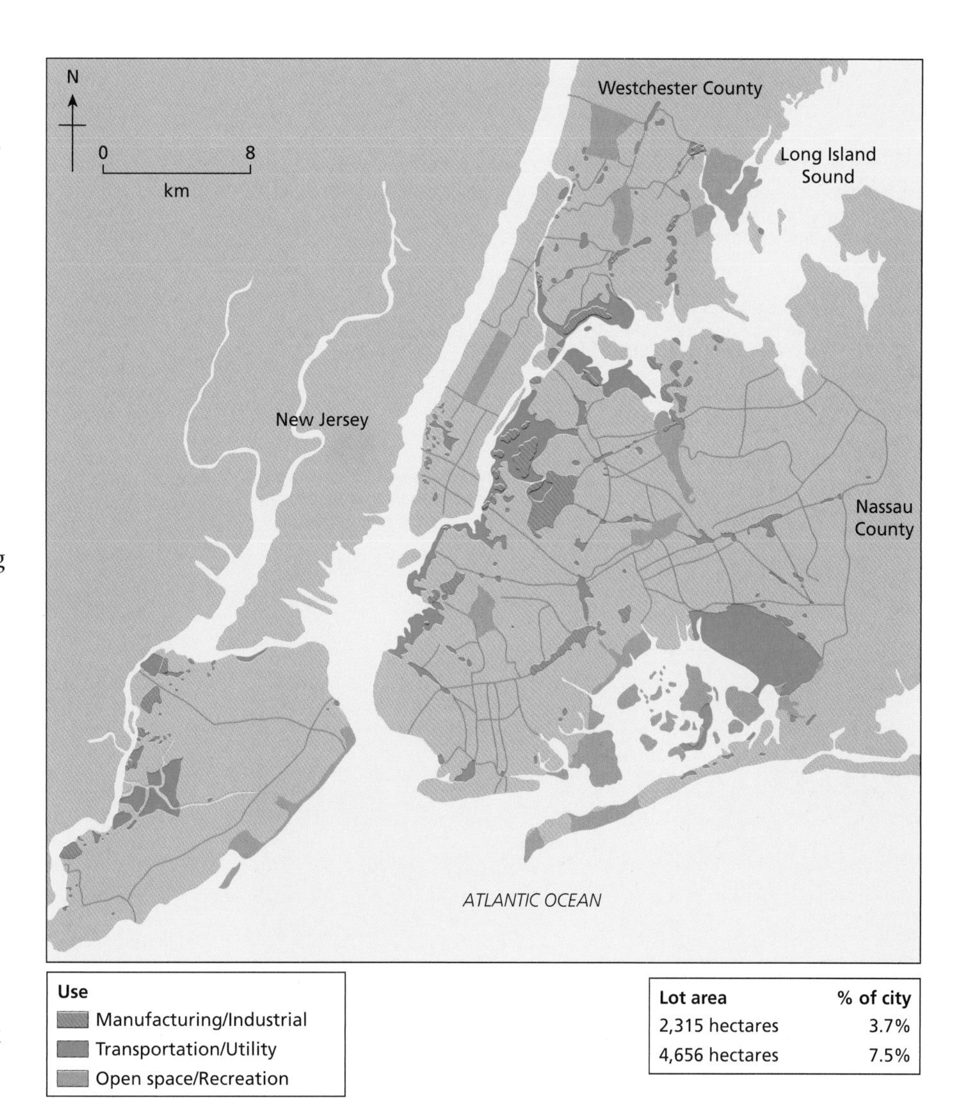

Figure 11.21 Land use in New York

The central business district (CBD)

The central business district (CBD) is the commercial and economic core of a city. It is the heart of the city, the area most accessible to public transport, and the location with the highest land values. It has a number of characteristic features and internal zoning (clustering of similar types of business). Figures 11.22 and 11.23 typify an MEDC city.

To do:

Describe the geographical features of the photo of Sao Paulo (Figure 11.24) by drawing an annotated sketch.

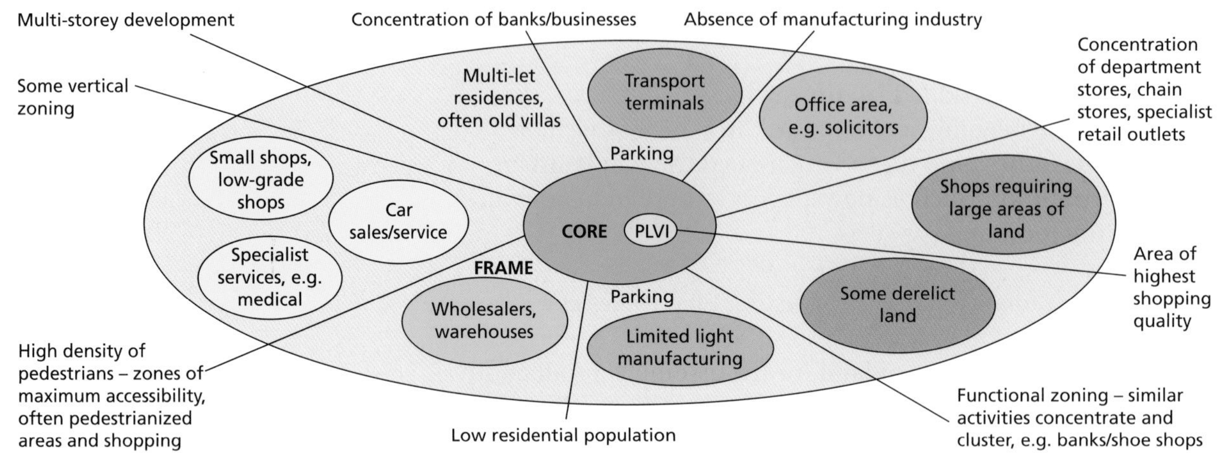

Figure 11.22 Core and frame elements of the CBD

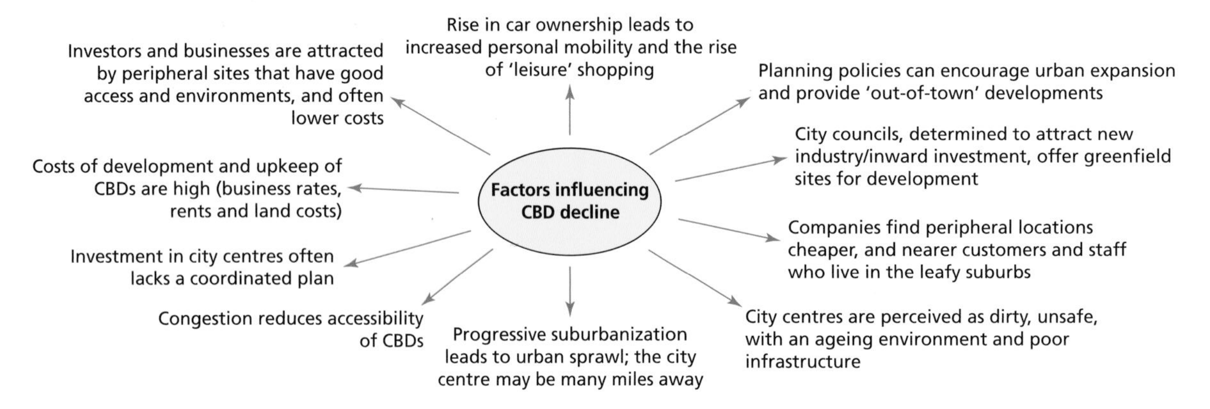

Figure 11.23 Factors affecting CBD decline

Figure 11.24 Sao Paulo, Brazil

To do:

Carry out the following fieldwork exercise:

- **a** Define the boundary of the CBD of your local town or city, using your own criteria. (Hints: pedestrians, parking restrictions.)
- **b** Identify its core using conventional criteria, and servey pedestrians to obtain a perceived core.
- **c** Analyse the real against the perceived CBD.
- **d** To what extent does the CBD of your local town shift over time, i.e. daily, weekly seasonally?

Case study: Changes in Reading's CBD, 1960–2010

Figure 11.25 The CBD of Reading, UK, 1967

Reading is a large town of 145 000 inhabitants situated 60 kilometres west of London. The population is relatively affluent, but there is marked inequality. The town's CBD has undergone many changes in its development that reflect contemporary aspects of culture and technology. The result of this evolution throughout the UK has been the loss of retail diversity and the increasing homogenization of the high street.

Before 1970

The area of Reading's CBD in 1960 was around half its current size. The map (Figure 11.25) shows that the core was dominated by comparison shops, mainly selling clothes and shoes, but also specialist shops offering goods and services that have since declined in popularity. Knitting and sewing shops, furriers and dancing schools were popular then. Land values were relatively low, and some peripheral parts of the CBD core were given over to car parks and storage, which would be considered unprofitable today.

The remainder was dominated by national and regional chain stores such as Heelas (John Lewis), BHS, Debenhams, Littlewoods, Woolworth's and the Co-op; all occupying prime sites close to the peak land value intersection (PLVI) at Heelas. Functional zoning (clustering of similar outlets) was evident, with banks and building societies concentrated at the eastern end of the CBD, closer to the start of the office zone. Other office-related businesses such as printers (before IT was widely used) were found close by.

Food shops were widely dispersed and varied. Union Street, in the central town, known as "Smelly Alley", had several small shops selling fish, meat and vegetables. Over the rest of town, grocers, greengrocers, bakers and butchers occupied central positions.

Figure 11.26 The CBD of Reading, UK, 2010

1970–90

Independent retailers who dominated Reading's CBD retail outlets began to experience some competition from shops in the Butts Centre (Broad Street Mall), where shopping could be done under cover with multistorey car parking provided. However, the main evolutionary force affecting Reading's CBD in the 1970s and 1980s was decentralization (the centrifugal or outward movement of people, manufacturing and services to suburban or rural locations).

This led to a spiral of falling demand for the CBD and the demise of the small in-town trader. One major out-of-town site was at Calcot, 6 kilometres west of the town centre. This site had several advantages over the town centre: it was accessible, land values and overhead costs were lower, congestion was less and car parking was free. These changes reflected the socio-economic shift towards more female employment during this period, and the resulting need for easier access to bulk buying from a superstore.

1990–2010

As competition from out-of-town retailing drew customers away from the congested and dispiriting CBD; Reading's town centre began to suffer from decay of its core – the "doughnut effect". The response to this decline was to refurbish the centre, and this took around a decade to complete. The initial refurbishment was designed to upgrade the environment and historic features of the CBD, and included the pedestrianization of Broad Street and nearby Cross Street. It improved the overall appearance of Broad Street by allowing market stalls, florists and street artists and musicians, and improving the shopping experience by providing street seating and some tree shade. Victorian buildings were refurbished, and Reading Station was rebuilt.

The Oracle

The major development to revitalize Reading's CBD has been the Oracle, a large in-town shopping centre completed in 1999. It is located between the CBD and the River Kennet on the south side of the town centre. In 1997, the 22-acre site consisted of derelict land once occupied by Simonds brewery and the telephone exchange. Hammerson, the private developer, aimed to create a high-quality retail centre including "anchor shops" such as House of Fraser to attract other retailers, in particular international clothes shops. The site offered potential for the development of riverside cafes and restaurants and scope for a three-storey car park with a capacity for 2,300 cars. The cost of the centre was £215 million, raised principally by private investors.

The Oracle has been successful in the following ways:

- In 2007 it was ranked 16th in a league table of best-performing retail centres in the UK compiled by economic analyst Experian.
- It has created 3,500 local jobs.
- It has allowed more flexible shopping, with opening hours until 8 pm.

However, the Oracle has also raised some controversy:

- It has created competition for trade in the CBD, notably in Friar Street to the north, where pedestrian densities have fallen and shop closures in the first five years were frequent.
- Reinvestment has involved the conversion of businesses to pubs or "superpubs", which attract large numbers of drinkers and make this zone crowded on Friday and Saturday evenings.
- Retailing in Friar Street continues to be depressed by the competition from Broad Street and the Oracle. The street is dominated by £1 shops, charity shops, fast food and "All you can eat" Chinese restaurants.
- It has exacerbated traffic congestion at peak times.
- Its outlets are exclusive and beyond the means of Reading's poorer population.
- Chain stores make up 90% of the outlets in the Oracle, thus reinforcing the homogeneity of the UK shopping centre.

a A "zone of discard", Friar Street

b A superpub, formerly a post office, Friar Street

c The River Kennet and the Oracle

d The Oracle in Broad Street

Figure 11.27 Reading's CBD, 2010

Managing environmental problems in urban areas

MEDCs	LEDCs
• Increased number of motor vehicles	• Private car ownership is lower
• Increased dependence on cars as public transport declines	• Less dependence on the car, but growing
• Major concentration of economic activities in CBDs	• Many cars are poorly maintained and are high polluters
• Inadequate provision of roads and parking	• Growing centralization and development of CBDs increases traffic in urban areas
• Frequent roadworks	• Heavy reliance on affordable public transport
• Roads overwhelmed by sheer volume of traffic	• Journeys are shorter but taking longer
• Urban sprawl results in low-density built-up areas, and increasingly long journeys to work	• Rapid growth has led to urban sprawl and longer journeys
• Development of out-of-town retail and employment leads to cross-city commuting	• Out-of-town developments are beginning as economic development occurs {e.g. Bogota, Colombia)

Table 11.5 Traffic problems in cities

Environmental problems in urban areas vary over time as economic development progresses. The greatest concentration of environmental problems occurs in cities experiencing rapid growth, such as Mexico City. This concentration of problems is referred to as the **Brown Agenda**. It has two main components:

- issues caused by limited availability of land, water and services
- problems such as toxic hazardous waste, pollution of water, air and soil, and industrial accidents such as at Bhopal in 1985.

Figure 11.28 Smog in Mexico City

Mexico City sits in a basin where cool air can remain trapped by a layer of warm air above it. This is called a temperature inversion. The air fails to mix and, if vehicle emissions are high, intense pollution events can occur. Mexico City also experiences photochemical smog in summer, due to the chemical reactions that occur with N0x on sunny days. Ozone is a biproduct, and is a strong respiratory irritant.

What is the urban heat island?

Urban areas are generally 2–4°C warmer than those of the surrounding countryside, depending upon the size and character of the city. This creates an urban heat island (UHI). Its intensity (UHII) is the temperature difference between the urban and the rural area.

The socio-economic impacts of the urban heat island

- Human health – high levels of ozone, suspended particulates and heat cause serious respiratory problems for the elderly, the long-term sick and young children.
- Human discomfort – high levels of humidity, atmospheric dust and poor air quality worsen the quality of life in cities for many.
- Disease – higher temperatures increase the likelihood of vector- and waterborne diseases in poor cities.

Be a critical thinker

Why does the urban heat island effect matter?

Consider:

- increasing rates of urbanization
- increasing affluence
- possible links to global warming
- positive feedback mechanisms, i.e. hotter cities create more heat.

Factor	Effect on UHII
Weather conditions	Clear and calm anticyclonic conditions, with little or no wind, intensify the UHI because there is maximum urban solar radiation in the daytime and maximum rural reradiation loss at night. Wind helps to disperse urban heat and reduce UHII.
Topography and hydrology	Inland cities far away from a water body have higher UHIIs, whereas coastal cities experience cooling onshore winds during the day. All urban water bodies cause local cooling through the release of latent heat by evaporation.
City population size and density	A large urban population consumes high levels of energy and generates both air pollutants and heat. Concentration of people, traffic and industrial activity increase the UHII.
Level of national economic development	Developing cities have high building densities, increasing levels of energy consumption, but weak emission controls. UHII will increase as population and heat generation grow.
Building and street design	Narrow canyon-like streets with a low sky-view angle are cooler by day but retain the heat at night. The overall roughness of the built urban environment obstructs wind and therefore retains heat more easily than open rural ground surfaces.
Land surface cover	Concrete and asphalt have a high thermal capacity and are able to absorb daytime solar radiation and release it as heat at night. However, urban vegetation, like water bodies, reduces UHII.
Anthropogenic heating	Moving or stationary traffic, heating and air-conditioning systems supplement solar radiation and increase UHI intensity.
Air pollution	Fossil fuel combustion by industrial activity, domestic burning and traffic emissions lead to the release of suspended particulate matter, causing heat retention in the lower atmosphere.

Table 11.6 Factors influencing urban heat island intensity

- Energy waste – high levels of energy expended through air-conditioning systems in richer cities are an increasing concern.
- Heat stress and illness – working days are lost and productivity lowered.
- Environmental degradation – intense heat causes the degradation of urban fabric such as rubber and tarmac, creating a long-term cost to city authorities.

Urban heatwaves

Urban heatwaves are a growing problem, and have affected cities irrespective of their latitude and climate type. They catch urban societies unawares and unprepared, causing higher mortality rates than many natural hazards. They are an increasing risk with urbanization and the rapid growth of megacities. The associated urban heat island effect, high-density residential living and high levels of vehicle and industrial emissions all warsen the problem.

Figure 11.29, the thermal profile of a temperate city in summer, shows the contrasts between urban and rural areas. The weather conditions are typically anticyclonic – clear and calm. The temperatures were recorded at ground level (surface temperature) and above the city at canopy level (air temperature). In the daytime intense solar radiation causes the ground surface to become much hotter than the rural area, or the canopy layer. At night, when there is no solar radiation, the surface and air temperatures are similar, and both are highest in the city centre.

To do:

Study Figure 11.29.

a Explain the fluctuation at the pond on the west side and the park on the east of the city.

b Explain how the intensity of the urban heat island might be affected by a change of weather in a city such as this one.

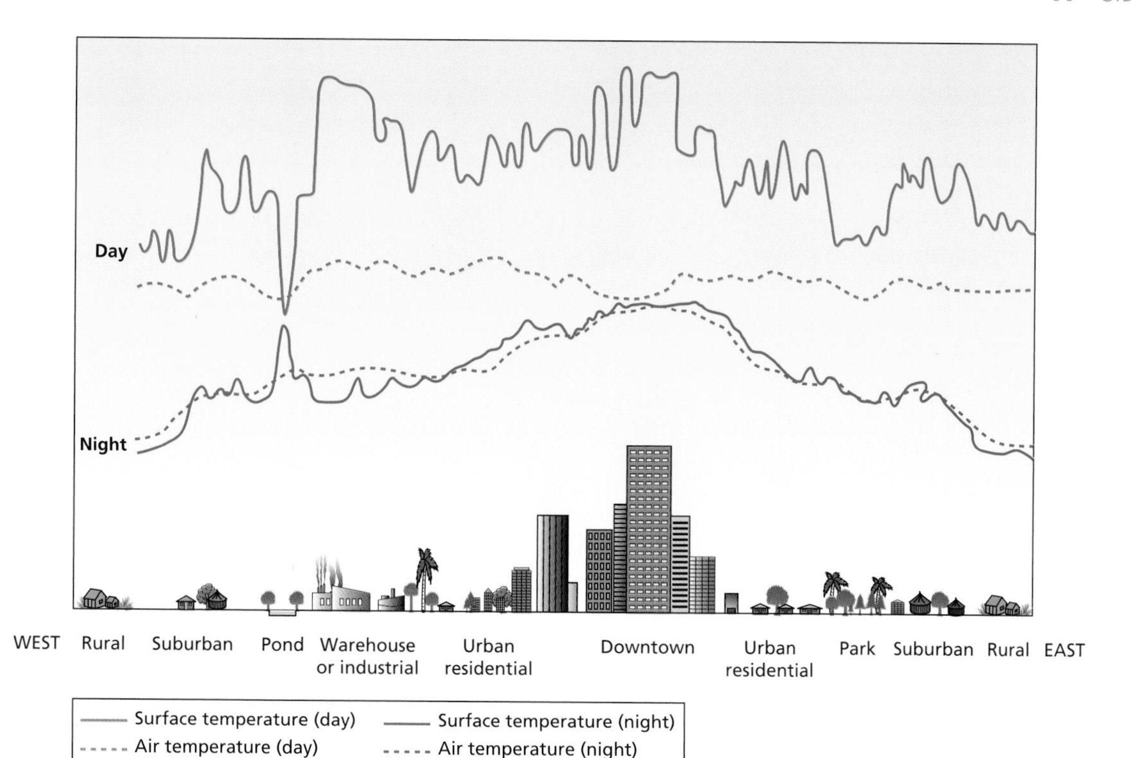

Figure 11.29 Thermal transect of a temperate city in summer

Case study: The Paris heatwave of 2003

In the summer of 2003, a persistent anticyclone over Europe blocked rain-bearing winds from the North Atlantic, allowing hot air from North Africa to penetrate. The resulting heatwave killed 30 000 people in Europe, most of them in cities. The August heatwave in the Paris Basin caused 4,867 deaths, an increase of 60% over normal mortality rates. In central and suburban Paris the seasonal mortality rate increased by 190%, and 1,294 people died.

The problems were most serious in Paris itself, where heating was most intense. For nine consecutive days in August the maximum daytime temperatures exceeded 38°C, which was shortly followed by a steep rise in mortality. (Figure 11.32)

Figures 11.30 and 11.31 are satellite images showing average land surface temperatures in the Paris Basin during the heatwave of August 2003, at midday and at midnight. The risk of dying increases at night, but in the daytime exceptionally high temperatures can be tolerated (see temperature keys).

The heat caused dehydration and cardiovascular illness, and the high levels of ground-level ozone and nitrous oxides

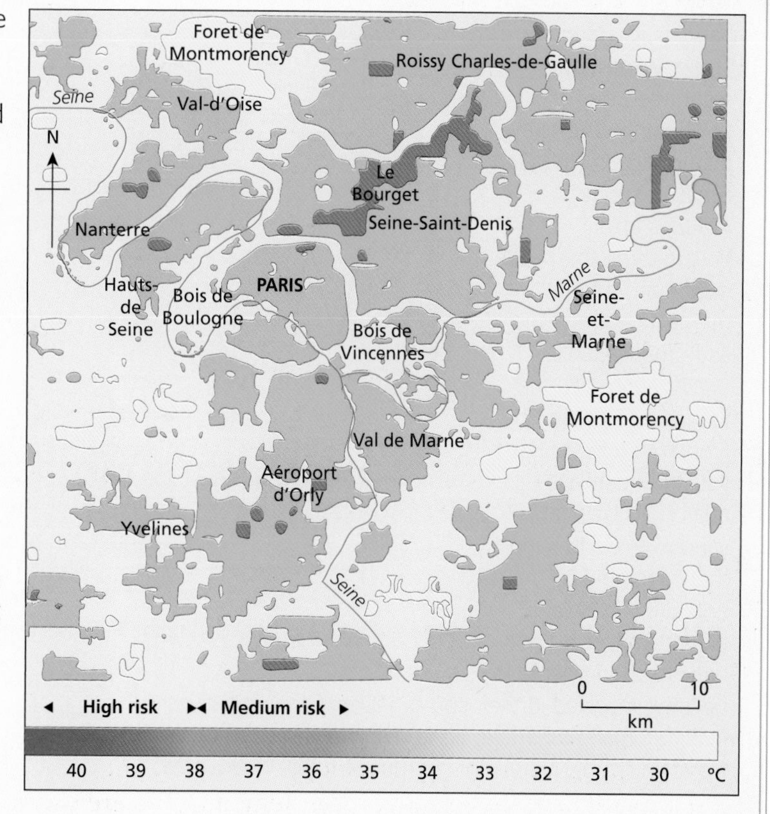

Figure 11.30 Midday temperatures, Paris, August 2003

induced or aggravated respiratory problems. Those most afflicted were the elderly and those with pre-existing health problems. Paris has a high proportion of elderly women living alone (60% of people aged over 80). In August 2003 many were unable to cope or adapt to the conditions. Immigrants to the city fared better, usually because they were younger and more resilient. Paris, like many northern European cities, has been unfamiliar with heatwaves until quite recently, and air conditioning, even in hospitals, is not regularly used.

Blame was put on the mass exodus of the able-bodied during August, leaving a shortage of medical staff and government ministers. Families were also held culpable for abandoning elderly relatives in the city while they went on holiday, and for failing to take responsibility for them on their return. By 3 September, 57 bodies still remained unclaimed in the municipal mortuary.

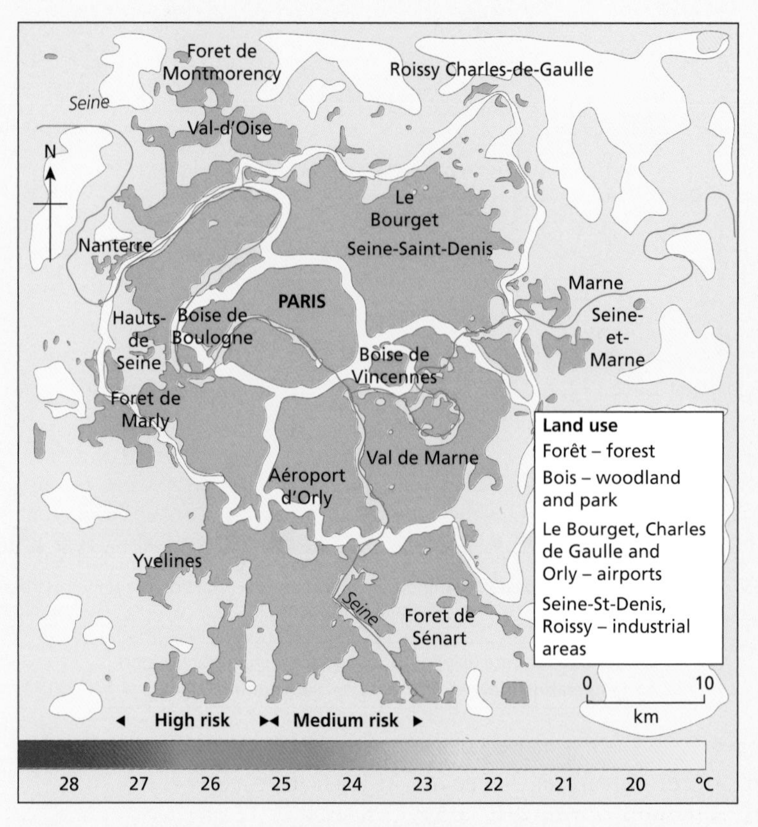

Figure 11.31 Midnight temperatures, Paris, August 2003

Responses to the Paris heatwave

When the scale of the disaster became clear, it led to a public outcry, and a debate about the way France handles its elderly care. The government responded by developing a heatwave plan of action (the *Plan Canicule*), which includes local government, the health sector and other relevant partners, such as humanitarian organizations like the French Red Cross. The French Red Cross has initiated heatwave-related activities at various levels to support the vulnerable by encouraging isolated elderly and/or handicapped people to become involved in their communities. Other initiatives include organizing a telephone helpline, a system of alerts and cool shelters for use during future heatwaves.

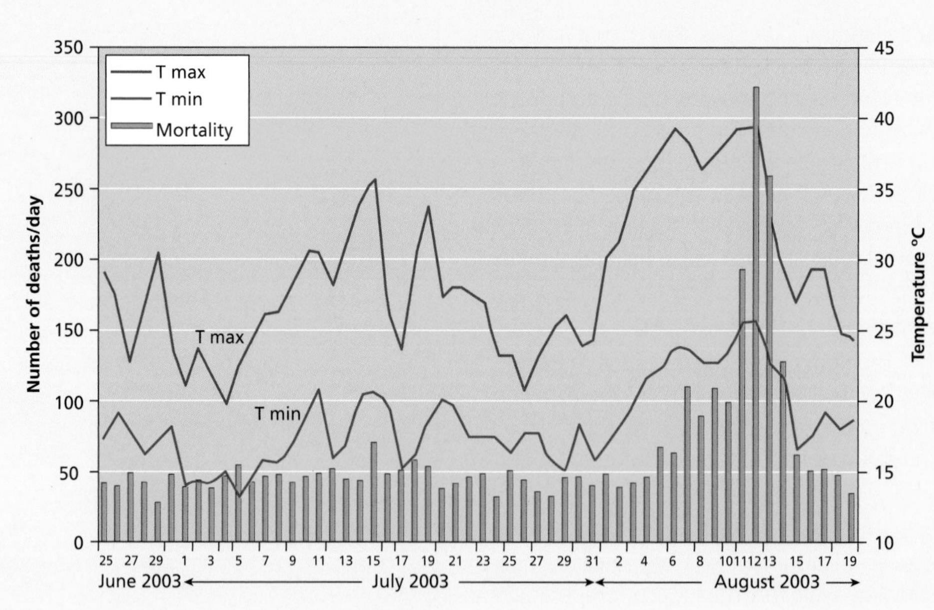

Figure 11.32 Mortality during the Paris heatwave of 2003

A subsequent heatwave in Paris in 2006, when temperatures again reached 40°C in parts of the city, resulted in only 42 deaths – testimony to the success of the *Plan Canicule*. However, the plan is designed to cope in an acute emergency situation, and does not address the long-term problem of overheating in cities.

To do:

a Refering to Figure 11.30 and 11.31, describe the patterns for day and night-time temperatures.

Remember to:

- describe the general pattern over the whole area of both maps.
- name and locate areas that have temperature anomalies (cooler or warmer patches).
- quote names, distances, directions and temperature data.

b Suggest reasons for the patterns you have found. Remember to look at all the evidence on both maps and the key.

c Explain why human tolerance to heat varies in time and from one place to another and between individuals.

Urban social stress

The rapid expansion of urban populations, particularly in developing countries, often overwhelms the ability of cities to provide adequate services, such as housing, sanitation, drinking water, education and employment. Rising inequalities greatly exacerbate these problems, as class differences tend to be more obvious in dense agglomerations than in rural settings. Noise, pollution, overcrowded living conditions and fear of crime all add to urban social stress.

Housing

Provision of enough good-quality housing is a major challenge for poorer countries. There are at least four aspects to the management of housing stock:

- quality of housing – with proper water, sanitation, electricity, and space
- quantity of housing – having enough units to meet demand
- availability and affordability of housing
- housing tenure (ownership or rental).

Much of the housing in many poorer cities is inadequate. Many homes lack access to water, sanitation, a reliable safe power supply, adequate roofs, solid foundations and secure **tenure**, i.e. the residents are not at risk of eviction.

Inequalities and social problems

Examples of the many social problems found in cities include:

- lack of access to services for the underclass
- problems related to crime
- ethnic and religious divisions causing social and **economic polarization** (The widening gap between rich and poor).

Issues of urban crime

Main actors	Organization	Impacts
Individual criminals	One-off acts of violent crime and delinquency, usually economically motivated	Murder, assault, gender-based violence, robbery and theft
Organized criminals	Drug cartels, human trafficking networks, arms smugglers, state security forces (intelligence) and police officers	Targeted killings, kidnaps, extortion, human trafficking and enslavement, small arms proliferation
Endemic community violence	Urban gangs, vigilante groups, ethnic militias, state security forces and police officers	High rates of gang/police/civilian casualties, unlawful killings, recruitment of urban child soldiers, social cleansing, gender-based violence, gang warfare and police shootouts, kidnaps
Open, armed conflict	Rebel groups, paramilitaries, state military forces	

Table 11.7 Types of urban crime and violence

The causes of urban violence and crime

- **Inequality:** the growing gap between rich and poor is far more likely to lead to crimes than poverty itself. Unequal access to employment, education, health and basic infrastructure is the underlying cause of crime.
- **Unemployment:** most research suggests that unemployed youths aged 15–24 are disproportionately more likely to be perpetrators as well as victims of crime and violence.
- **Limited police presence:** the speed of urbanization increases pressure on the ability of authorities to meet public security and safety demands through adequate policing. Large cities have lower levels of community cooperation with the police and require more police officers per inhabitant to effect an arrest.
- **Opportunity:** crime rates (number of crimes committed per head of population in a specific area) are higher in large cities because there is a greater concentration of wealthy potential victims, more opportunities to commit crime, and a more developed second-hand market for expensive stolen goods. Criminal networks can be persuasive and influential towards certain disadvantaged groups, luring them into criminal gangs.
- **Poor urban planning:** the design of urban environments may affect the security of citizens: for example, if streets have unlit or hidden areas and layouts with few escape routes, crime is more common.
- **Globalization and communications technology:** ease of communication through mobile phones and other devices assists the efficiency of organized crime.

Figure 11.33 Urban gangs create endemic community violence

City	% of population feeling unsafe after dark	Gini coefficient
Sao Paulo	72	0.61
Buenos Aires	66	0.52
Maputo	65	0.52
Johannesburg	57	0.75
Pnom Penh	48	0.36
Rome	44	0.34
Sydney	27	0.33
Hong Kong	22	0.53
Paris	22	0.38

Table 11.8 The safety of cities and their Gini coefficient

To do:

Research has shown a convincing link between homicide rates and inequality in cities.

a Using a statistical or graphical technique, correlate the two variables in Table 11.8 (Fear and the Gini coefficient).

b Draw a conclusion and evaluate your method by suggesting weaknesses and alternative techniques.

c Suggest other variables that might relate to crime and inequality in cities.

0.6 or above	Extremely high levels of inequality, not only among individuals but also among social groups (known as "horizontal inequality"); wealth concentrated among certain groups to the exclusion of the majority; high risk of social unrest and civil conflict
0.5–0.59	Relatively high levels of inequality reflecting institutional and structural failures in income distribution
0.45–0.49	Inequality approaching dangerously high levels; if no remedial actions taken, investment could be discouraged and lead to sporadic protests and riots; often denotes weak functioning of labour markets, inadequate investment in public services and lack of pro-poor social programmes
0.40	International alert line
0.3–0.39	Moderate levels of inequality; healthy economic expansion accompanied by political stability and civil participation (however, this could also mean that all groups are generally rich or poor, and that therefore disparities are not reflected in income or consumption levels)
0.25–0.29	Low levels of inequality; egalitarian society often characterized by universal access to public goods and services, political stability and social cohesion

(Source: UN-HABITAT Monitoring and Research Division)

Table 11.9 The Gini coefficient: what it means

The consequences of urban crime

Crime has a negative impact at city level, where endemic insecurity can generate a culture of fear, leading to segregation and social fragmentation and the creation of fortified spaces. At national and global levels, high rates of crime (or perception of crime) can impact negatively on economic development. At the international scale, the oldest cities often provide ideal environments for transnational criminal organizations dealing in drugs, arms and human beings.

The Latin American region has the highest homicide rate in the world, and widespread fear and insecurity are a fact of daily life. Cities such as outside this region, Lagos, Nairobi and Johannesburg are internationally notorious for crime and endemic violence, which can deter foreign investment and tourists.

Gated communities

While high levels of violence do occur, anxiety generated by the perception of crime and endemic violence often outstrips the actual level of danger. In Mexico City, for example, widespread acts of violence haunt the minds of its residents, forcing them to restrict their movements, avoid leaving home at night and retreat into private spaces. In this way fear and envy become self-perpetuating.

Gated communities predominate where inequalities are pronounced and public security is inadequate. Most are found in North American cities, but the phenomenon is widespread in countries such as Brazil and South Africa where socio-economic inequalities are marked. In South Africa alone, the number of private security guards has increased by 150% since the late 1990s.

Unfortunately, this desire to establish a safe refuge from violence often results in segregation along ethnic, religious or racial lines, resulting in further mistrust or antagonism between groups. In Indian cities spatial segregation between Hindus and Muslims is widespread, particularly in Mumbai and the Gujarati city of Ahmedabad. In 2002, communal violence between the two groups led to the deaths of over 1,000 people.

When cities gain a reputation for endemic crime and violence, their ability to generate wealth is undermined. In Brazil it is estimated that 10% of annual GDP is spent on private security and insurance. Transnational criminal networks gravitate toward cities by virtue of their infrastructure and the cover provided by the intricacy of small streets. These organizations thrive in many cities in Africa, central Asia and Latin America.

Spectacularly violent acts of international terrorism have been played out in various cities across the world. Notable examples include New York's World Trade Center collapse of 11 September 2001, with the loss of almost 3,000 lives, the 2004 detonation of bombs on packed commuter trains in Madrid, killing 191 and injuring over 1,800, and the London bombings of July 2005 which left 52 people dead and over 700 injured.

Insecurity in cities is common, and fear of crime is heightened by the scale and horror of international terrorism. At the other end of the scale is the small-town petty criminal working independently, but still capable of raising anxiety levels in the local community.

Although the gravity and scale of criminal activity vary, the common cause is usually urban inequality, and the consequence is fear and stress.

To research

The State of the World's Cities by UN-HABITAT provides comprehensive coverage of a range of urban issues including inequality, and uses the GINI Index as a measure of inequality. Visit www.unhabitat.org for information.

The city as a system

A system is a simplified way of looking at how things work. Systems generally include factors (inputs), processes (throughputs) and results (outputs). This systems approach can be applied to many aspects of geography, including cities.

Large cities are often considered to be unsustainable systems because they consume huge amounts of resources and produce vast amounts of waste. Sustainable urban development aims to meet the needs of the present generation without compromising the needs of future generations. The Rogers model (1997) compares a sustainable city with an unsustainable one. In the sustainable city, inputs and outputs are smaller and there is more recycling.

Compact cities minimize the amount of distance travelled, use less space, require less infrastructure (pipes, cables, roads, etc.), are easier to provide a public transport network for, and reduce urban sprawl. But if the compact city covers too large an area it becomes congested, overcrowded, overpriced and polluted. It then becomes unsustainable.

The following steps need to be taken to achieve sustainability:

1 Improve economic security
People should have access to employment and an adequate livelihood. If they become ill, permanently disabled or unemployed they should be entitled to economic security.

2 Meet social, cultural and health needs
Housing should be healthy, safe, secure, affordable and within a neighbourhood that provides piped water, sanitation, drainage,

transport, healthcare, education and child welfare. The home and workplace should be free from hazards and chemical pollution. There should be equality in service provision, so that it is available to all, irrespective of income.

3 Minimize the use of non-renewable resources

This means reducing consumption of fossil fuels in housing, commerce, industry and transport, and substituting renewable resources where possible. Since the car is a major source of pollution in cities, public transport should be promoted and urban car use discouraged. The practices of waste minimization, recycling, reuse and reclamation need to be carried out. Architectural, historical and natural assets within cities are irreplaceable and must be respected and conserved.

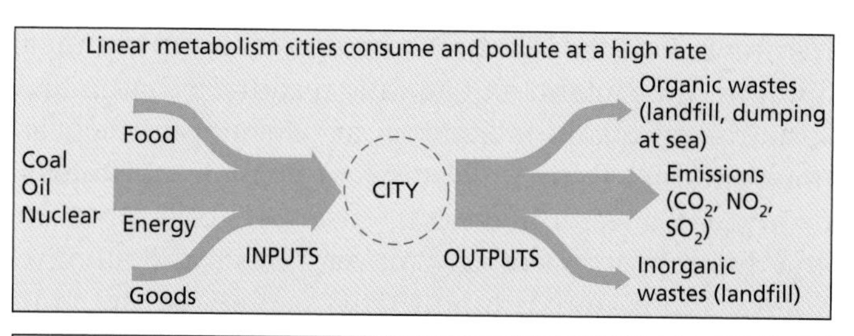

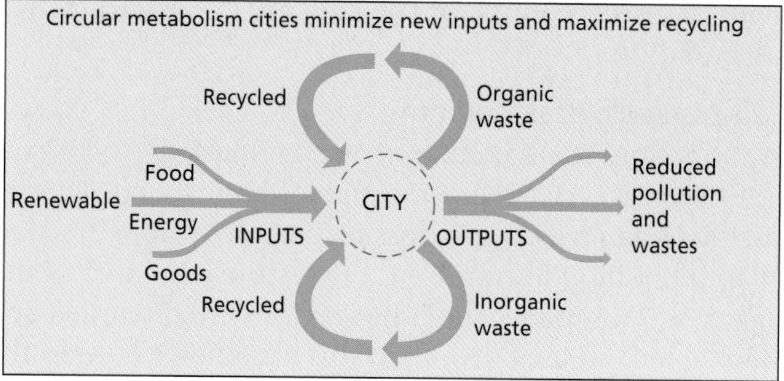

Figure 11.34 The Rogers model of city systems

4 Use finite renewable resources sustainably

Cities should draw on water resources up to their maximum sustainable yield, and not beyond. Waste minimization and recycling should be encouraged. City dwellers should be aware of their ecological footprint and its impact on the city and the land around it.

5 Preserve green space

"The green agenda" involves the provision and maintenance of green space such as parks and playgrounds. It also means reusing, wherever possible, existing inner urban sites or brownfield sites, and refurbishing or rebuilding there rather than on greenfield sites. Local communities should be actively involved in the processes of improving their local neighbourhoods.

The sustainable city

Rapid urbanization means that most cities are now huge consumers of resources and producers of waste, and their environmental impact is huge. A **sustainable city**, or **eco-city**, is a city designed to have minimal environmental impact. Its population wishes to create the smallest possible ecological footprint and produce the least amount of pollution and waste. A sustainable city uses land efficiently, and practises reuse and recycling as well as waste minimization.

The urban ecological footprint

According to the Global Development Research Centre, the urban ecological footprint is the land area required to sustain a population of any size. All the resources which people use for their daily needs, such as food, water and electricity, must be produced using raw natural resources. The urban ecological footprint measures the amount of farmland and aquatic resources that must be used to sustain a population, based on its consumption levels at a given point in time. To the fullest extent possible, this measurement incorporates water and energy use, uses of land for infrastructure and different forms of agriculture, forests and all other forms of energy and material "inputs" that people require in their day-to-day lives. It also accounts for the land area required for waste assimilation.

More material available:

www.oxfordsecondary.co.uk/ibgeography

Case study on sustainable city development a Curitiba, Brazil

Tokyo's ecological footprint

According to the Earth Council, a biologically productive area of 1.7 hectares is needed per capita for basic living. This means that, for sustainable living, the people in Tokyo alone need an area of more than 45 million hectares – 1.2 times the land area of the whole of Japan. If mountains and other regions are discarded and only habitable land included, then this becomes 3.6 times the land area of Japan.

Tokyo is a city where the land is used several times at several levels. The difference between high-density cities (compact cities with much vertical development) and extended cities (cities with suburban sprawl, like those in Australia and the US, for example) is that three or four times more land area is used in extended cities.

Compact cities such as Tokyo have a large population living in a very small and dense area of land, freeing land area for other purposes. They also require reduced amounts of infrastructure and resources – it is easier to provide services, utilities and infrastructure to a population concentrated in a small area than is the case when people are spread out over a large area.

Examples of sustainable strategies

Reducing pollution

- The *Hoy no circula* (day without a car) environmental programme, launched in Mexico City in 1989, saw air pollution fall 21% in the first year.
- In Cubatao, Brazil, local and national government, and some businesses, have combined to reduce air pollution and enforce stricter regulations.

Integrated transport and land use

- Curitiba in Brazil is a good example.

Recycling

- In Shanghai, a wide-ranging programme was established in 1957. It now employs 30 000 people retrieving and reselling reclaimed and recycled products, including 3,600 advisers working with factories on sorting and retrieving waste (see Figure 11.10).
- In Curitiba 70% of households separate recyclable rubbish, and in squatter settlements food and bus fares are exchanged for garbage.

Sustainable housing

Refer to the discussion of Mumbai on page 298.

To research

Check www.iied.org/eandu for up-to-date examples and case studies about sustainable housing.

Case study: *Solid waste management in Kathmandu, Nepal*

Solid waste management (SWM) has become a major issue in many countries experiencing rapid urbanization. In comparison to rural populations, urban residents consume more resources and produce more non-biodegradable waste such as food packaging, electronic products and hazardous medical waste. Until recently, waste was collected and dumped either in landfill sites or in open spaces such as riverbanks and unstable slopes. Unsorted organic waste produces serious problems of smell, leachate seepage and methane, and attracts vermin and other scavengers, posing a serious risk to health.

Kathmandu Metropolitan City (KMC) has a particularly acute problem of solid waste disposal. The average amount of solid waste produced is relatively low, at 0.37 kilograms per person per day (Japanese cities produce 1.1 and US cities 2.1 kilograms per person per day), but this is increasing rapidly by 12% per year, well above its population growth rate of 6%. Kathmandu possesses one of the greatest concentrations of architectural treasures in the world, but excessive waste left in unsorted dumps throughout the city is a serious deterrent to prospective tourists.

Composting – sustainable waste treatment

About 70% of household waste generated in the Nepalese municipalities consists of organic matter, 20% of recyclable materials such as paper, plastic and metal, and about 10% of other materials including hazardous hospital waste. Composting organic waste can significantly reduce the cost of environmental impacts of waste accumulation and makes landfill unnecessary. Many residents in Kathmandu practise community composting. Waste is gathered through door-to-door collection and then composted in piles or in large vessels or chambers. The compost is packed into bags and sold on the local market.

Kathmandu's environment department has a Community Mobilization Unit (CMU), whose major function is to promote composting. The municipality has designed 100-litre-capacity home-composting bins called "Saaga" and it is selling each one with 300 *Eisenia foetida* worms, the necessary accessories, gardening equipment and half a day's training at a subsidized total cost of 500 rupees. It is estimated that about 3,000 compost bins have so far been sold. Around 800 households are practising composting in their homes.

The CMU works with community groups, youth groups and schoolchildren to raise awareness and provide training in solid waste management. Sustainable solid waste management in Nepal aims to conserve resources by waste minimization, reuse and recycling and to support those currently working as rag pickers, sweepers, collectors and sorters. An integrated system, as seen in Figure 11.34, is planned and now partially achieved in Kathmandu.

Part 3 Global interactions

The study of global interactions has a broader perspective than a conventional study of globalization, which emphasizes the imposition of western culture and values on the world. Global interactions suggest a complex, two-way process, where commodities and cultural traits may be adopted, adapted, or resisted by societies. The process and outcomes are neither inevitable nor universal.

The theme focuses on the global interactions, flows and exchanges arising from the disparities that exist between places. It presents important geographic issues of change in space and time. Many of these changes can be questioned.

Global interactions has seven compulsory topics:

Measuring global interactions provides an introduction to the course by identifying the level and rate of global interactions. **Changing space – the shrinking world** identifies improved information and communications technology and transport as fundamental to all forms of global interaction. These first two sections provide a basis for further study by examining the pattern(s) and process(es) of global interactions and the technology that has enabled them.

The next four sections are **Economic interactions and flows, Environmental change, Sociocultural exchanges,** and **Political outcomes** of global interactions. They present an alternative perspective on these interactive processes and outcomes, and question their inevitability. These processes illustrate that globalization is not static but evolving. They examine the variations in its course and speed and how globalization results in different levels of involvement and acceptance. Globalization may be resisted and rejected in some countries or sub-regions, where local forces may reassert themselves as a reaction against the loss of distinctiveness and sovereignty.

The final topic, **Global interactions at the local level,** examines responses to the two-way global interactions operating at the more local scale. Global interactions may encounter local obstacles and resistance, which modify them and result in hybridized outcomes. This topic involves local investigation.

Measuring global interactions

Indices of globalization

The Kearney index

The Kearney globalization index tracks and assesses changes in four key components of global integration (see Figure 12.1). The 72 countries ranked in the 2007 globalization index account for 97% of the world's GDP and 88% of the world's population. The index covers major regions of the world, including developed and developing countries, to provide a comprehensive and comparative view of global integration.

Economic integration combines data on trade and foreign direct investment (FDI) inflows and outflows, international travel and tourism, international telephone calls, and cross-border remittances. Technological connectivity counts the number of Internet users and Internet hosts. Political engagement includes each country's memberships in a variety of representative international organizations.

The resulting data for each given variable are then "normalized" through a process that assigns the value of 1 to the highest data, and all other data points are valued as fractions of 1. The base year (1998 in this case) is assigned a value of 100. The given variable's scale factor for each subsequent year is the percentage growth or decline in the GDP – or population-weighted score of the highest data point, relative to 100. Globalization index scores for every country and year are derived by summing all the indicator scores.

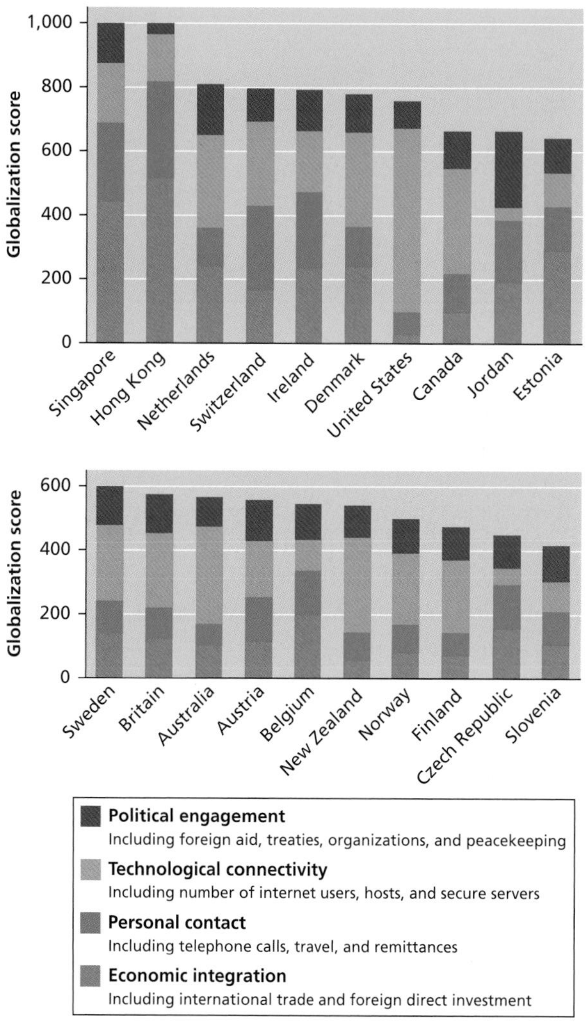

One characteristic that many of the most globalized countries have in common is size. Eight of the index's top 10 countries have small land areas, and seven have fewer than 8 million citizens. Canada and the United States are the only large countries that consistently rank in the top 10. Countries such as Singapore and the Netherlands lack natural resources. Countries like Denmark and Ireland have limited domestic markets. To be globally competitive, these countries have no choice but to open up and attract trade and foreign investment.

Political engagement Including foreign aid, treaties, organizations, and peacekeeping
Technological connectivity Including number of internet users, hosts, and secure servers
Personal contact Including telephone calls, travel, and remittances
Economic integration Including international trade and foreign direct investment

Figure 12.1 The Kearney globalization index

The KOF index

The KOF index of globalization, introduced in 2002, covers the economic, social and political dimensions of globalization. KOF defines globalization as: "the process of creating networks of connections among actors at multi-continental distances, mediated through a variety of flows including people, information and ideas, capital and goods. Globalization is conceptualized as a process that erodes national boundaries, integrates national economies, cultures,

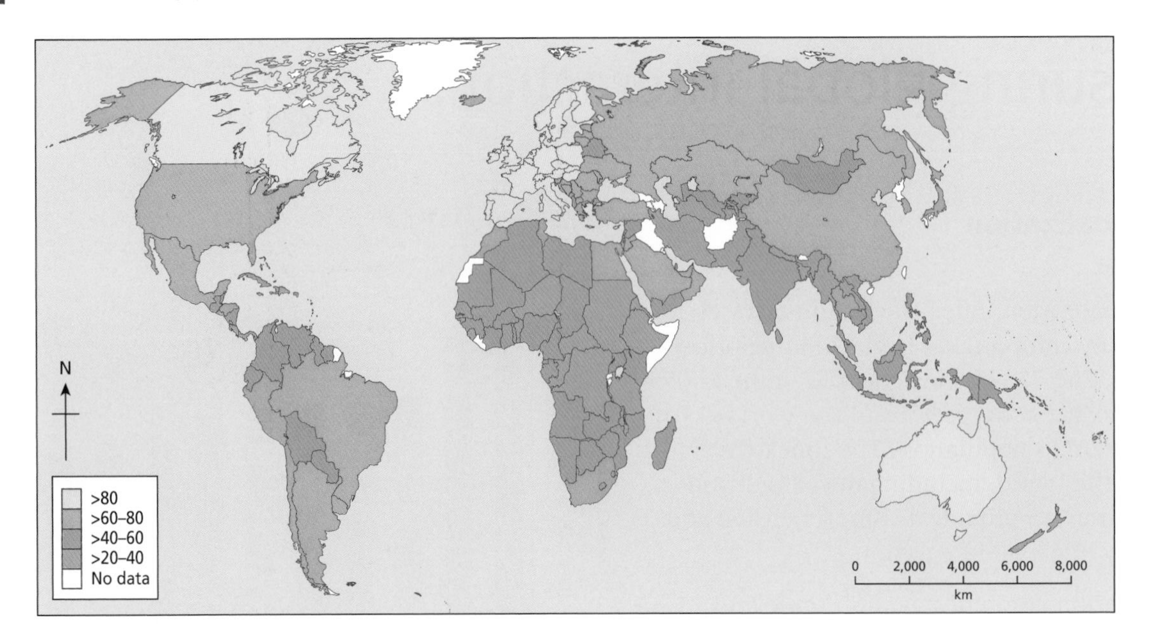

Figure 12.2 The KOF globalization index, 2010

technologies and governance and produces complex relations of mutual interdependence."

More specifically, the three dimensions of the KOF index are defined as:

- **economic globalization,** characterized as long-distance flows of goods, capital and services, as well as information and perceptions that accompany market exchanges (this accounts for 38% of the globalization index)
- **political globalization,** characterized by a diffusion of government policies (this accounts for 23% of the globalization index)
- **social globalization,** expressed as the spread of ideas, information, images and people (this accounts for the remaining 39% of the globalization index).

Globalization index – the Kearney index is one of several measures of globalization. It tracks changes in the four key components of global integration: trade and investment flows; movement of people across borders; volumes of international telephone traffic and Internet usage; participation in international organizations (A. T. Kearney. *Foreign Policy*)

In addition to the indices measuring these dimensions, KOF calculates an overall index of globalization and sub-indices referring to actual economic flows, economic restrictions, data on information flows, data on personal contact and data on cultural proximity. In 2009 Belgium had the highest index of globalization (91.5), followed by Ireland and the Netherlands. The country with the lowest index was Myanmar (Burma), at 23.7.

Economic globalization

Broadly speaking, economic globalization has two dimensions: first, actual economic flows, which are usually taken to be measures of globalization; and, second, restrictions to trade and capital. In 2009 Singapore had the highest level of economic globalization, followed by Luxembourg and Ireland.

Political globalization

Political globalization uses the number of embassies and high commissions in a country, the number of international organizations of which the country is a member and the number of UN peace missions a country has participated in. In 2009 political globalization was highest in France, followed by Italy and Belgium.

TOK Link

Which index?

Which index of globalization do you think is best – Kearney or KOF? How do they differ? How might their methodology affect their results?

Social globalization

The KOF index classifies social globalization in three categories. The first covers personal contacts, the second includes data on information flows and the third measures cultural proximity. In 2009 Switzerland had the highest level of social globalization, followed by Austria and Canada.

- **Personal contacts** include international telecom traffic (outgoing traffic in minutes per subscriber) and the degree of tourism (incoming and outgoing) a country's population is exposed to. Government and workers' transfers received and paid (as a percentage of GDP) measure whether and to what extent countries interact.
- **Information flows** include the number of Internet users, cable television subscribers, number of televisions (all per 1,000 people), and international newspapers traded (as a percentage of GDP).
- **Cultural proximity** is arguably the dimension of globalization most difficult to grasp. According to one geographer (Saich, 2000, *Globalization, governance and the authoritarian state*), cultural globalization mostly refers to the domination of US cultural products. However, there are many other companies and products that do not come from the USA: sushi, curry, Toyota and Sony spring to mind. KOF includes the number of McDonald's restaurants located in a country. In a similar vein, it also uses the number of Ikea stores per capita.

To research

Visit http://globalization.kof.ethz.ch/ for the main portal to the KOF index of globalization site.

http://globalization.kof.ethz.ch/static/pdf/rankings_2010.pdf has a detailed breakdown of the 2010 KOF index (note that this uses data from 2007).

1. Identify the most globalized countries in the 2010 index.
2. Outline the characteristics that these countries have in common.
3. Suggest reasons why these countries have a high level of globalization.
4. Study the list of countries that have a globalization index of > 20 − 40. Comment on their distribution and characteristics. What does this tell you about globalization?

	Indices and variables	Weights
A.	**Economic globalization**	**[38%]**
	i) Actual flows	(50%)
	Trade (percentage of GDP)	(19%)
	Foreign direct investment, flows (percentage of GDP)	(20%)
	Foreign direct investment, stocks (percentage of GDP)	(23%)
	Portfolio Investment (percentage of GDP)	(17%)
	Income payments to foreign nationals (percentage of GDP)	(21%)
	ii) Restrictions	(50%)
	Hidden import barriers	(21%)
	Mean tariff rate	(29%)
	Taxes on international trade (percentage of current revenue)	(25%)
	Capital account restrictions	(25%)
B.	**Social globalization**	**[39%]**
	i) Data on personal contact	(34%)
	Telephone traffic	(26%)
	Transfers (percentage of GDP)	(3%)
	International tourism	(26%)
	Foreign population (percentage of total population)	(20%)
	International letters (per capita)	(26%)
	ii) Data on information flows	(34%)
	Internet users (per 1,000 people)	(36%)
	Television (per 1,000 people)	(36%)
	Trade in newspapers (percentage of GDP)	(28%)
	iii) Data on cultural proximity	(32%)
	Number of McDonald's restaurants (per capita)	(37%)
	Number of Ikeas (per capita)	(39%)
	Trade in books (percentage of GDP)	(24%)
C.	**Political globalization**	**[23%]**
	Embassies in country	(25%)
	Membership of international organizations	(28%)
	Participation in UN Security Council missions	(22%)
	International treaties	(25%)

Table 12.1 KOF index of globalization

Global core and periphery

World systems analysis (see page 37) is a way of looking at economic, social and political development, where any analysis of development is seen as part of the overall capitalist world economy. According to the model, the capitalist world system has three main characteristics:

- a global market
- many countries, which allow political and economic competition
- three tiers of countries, defined as **core**, largely MEDCs, the **periphery**, which can be identified with LEDCs, and the **semi-periphery** (see page 38).

World cities and the global economy

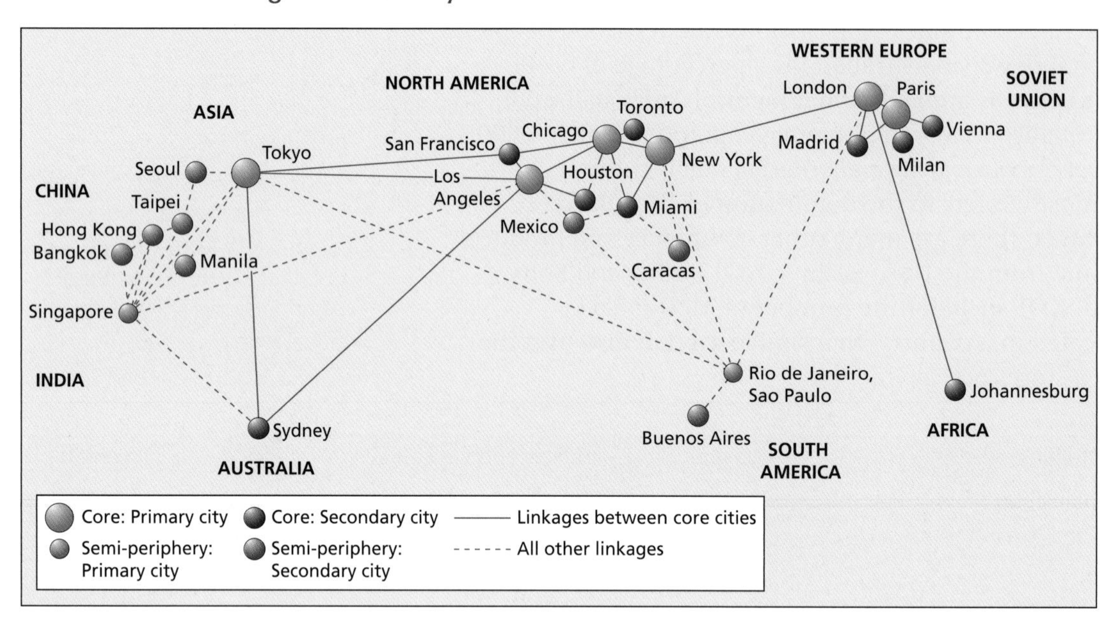

Figure 12.3 World cities

In 1986 John Friedmann developed the world cities hypothesis. It was an attempt to understand the major global cities of the world and their response to the shift from an international to a global economy. Friedmann defined world cities as:

- centres through which money, information and commodities flow
- large, urbanized regions defined by dense patterns of interaction
- hierarchical with respect to the economic power they command and their ability to attract global investment
- sites for the concentration and accumulation of capital.

World cities have been created by the shift from an international to a global economy. An **international economy** sees goods and services traded across national boundaries by individuals and firms from different countries under strict control of individual nations. However, in the **global economy** goods and services are produced by large MNCs who largely dictate the industrial policy of the nation state. They usually orchestrate their operations from world cities. The concept of world cities has been developed by other geographers. Appadurai (1990) has developed a list that describes the variety of functions which world cities offer, in terms of six cultural and economic landscapes:

1. **Ethnoscapes** are produced by flows of business personnel, guestworkers, tourists, immigrants and refugees.
2. **Technoscapes** are flows of machinery, technology and software from MNCs and government agencies.
3. **Finascapes** are flows of capital currency and securities.
4. **Mediascapes** are flows of images and information through newspapers, televisions and film.
5. **Ideoscapes** are flows of a western view of life, e.g. democracy, welfare rights and mass consumption.
6. **Commodityscapes** are flows of culture and style encompassing everything from architecture to interior design.

MNC Multinational corporation – an enterprise that has its headquarters in one country but operates in several other countries known as the host countries.

To research

Visit http://consultant-news.com/article_display.aspx?p=adp&id=5205 for the Kearney Index of Global Cities. Compare and contrast the five dimensions that the index measures: business activity, information exchange, cultural experience, political engagement and human capital.

Which city is the most globalized overall? Devise a ranking system (e.g. 10 points for being first, 9 for second, etc.) and work out the five most globalized countries in the world. Comment on your results.

a

b

Figure 12.4 Two world cities: New York (left) and Tokyo

Changing space – the shrinking world

Time–space convergence

The **frictional effect of distance** or **distance decay** suggests that areas that are closer together are more likely to interact, whereas areas further away are less likely to interact with each other. However, there has been a reduction in the frictional effect of distance as improvements in transport have allowed greater distances to be covered in the same amount of time. In addition, improvements in ICT have brought places on different sides of the world together almost instantaneously.

A basic requirement for the evolution of international trade and the development of transnational corporations (TNCs) is the development of technologies that overcome the frictional effect of distance and time. The most important of these *enabling* technologies are transport and communications. Neither of these technologies caused the development of international trade or of TNCs, but they allowed such developments to occur. Without them, today's complex global economic system would not exist.

Transport systems are the means by which materials, goods, and people are moved between places. Communication systems are the means by which information is transmitted between places. Before the development of electricity in the 19th century, information could move only at the same speed, and over the same distance, as the prevailing transport system would allow. Electricity broke that link, making it increasingly necessary to treat transport and communication as separate, though closely related, technologies. Developments in both have transformed the world, allowing unprecedented mobility of materials and goods and a globalization of markets.

Major developments in transport technology

The world has "shrunk" in the time it takes to get from one part of the world to another (Figure 13.1). For most of human history, the speed and efficiency of transport were low and the costs of overcoming the frictional effect of distance high. Movement over land was especially slow and difficult before the development of the railways. The major breakthrough came with the invention and application of steam power and the use of iron and steel for trains, railway tracks and ocean-going vessels. The railway and the steamship enabled a new, much enlarged, scale of economic activity.

The mid- to late 20th century saw an acceleration of this process of space–time convergence. The most important developments have been the introduction of commercial jet aircraft, the development of much larger ocean-going vessels (superfreighters) and the introduction of containerization, which greatly simplifies shipment from one mode of transport to another. Of these, the jet aircraft had much influence, on the development of TNCs. It is hardly coincidental that the take-off of TNC growth and the take-off of commercial jets both occurred during the 1950s.

Time–space convergence – the reduction in the time taken to travel between two places due to improvements in transportation or communication technology.
Transnational corporation (TNC) – a firm that owns or controls productive operations in more than one country through foreign direct investment.

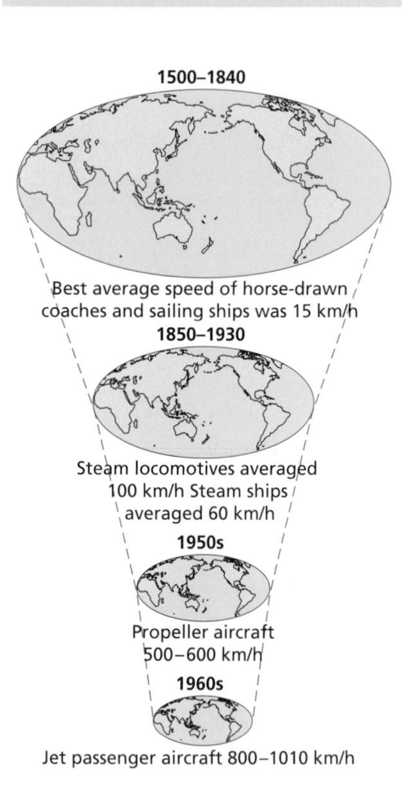

Figure 13.1 Time–space convergence

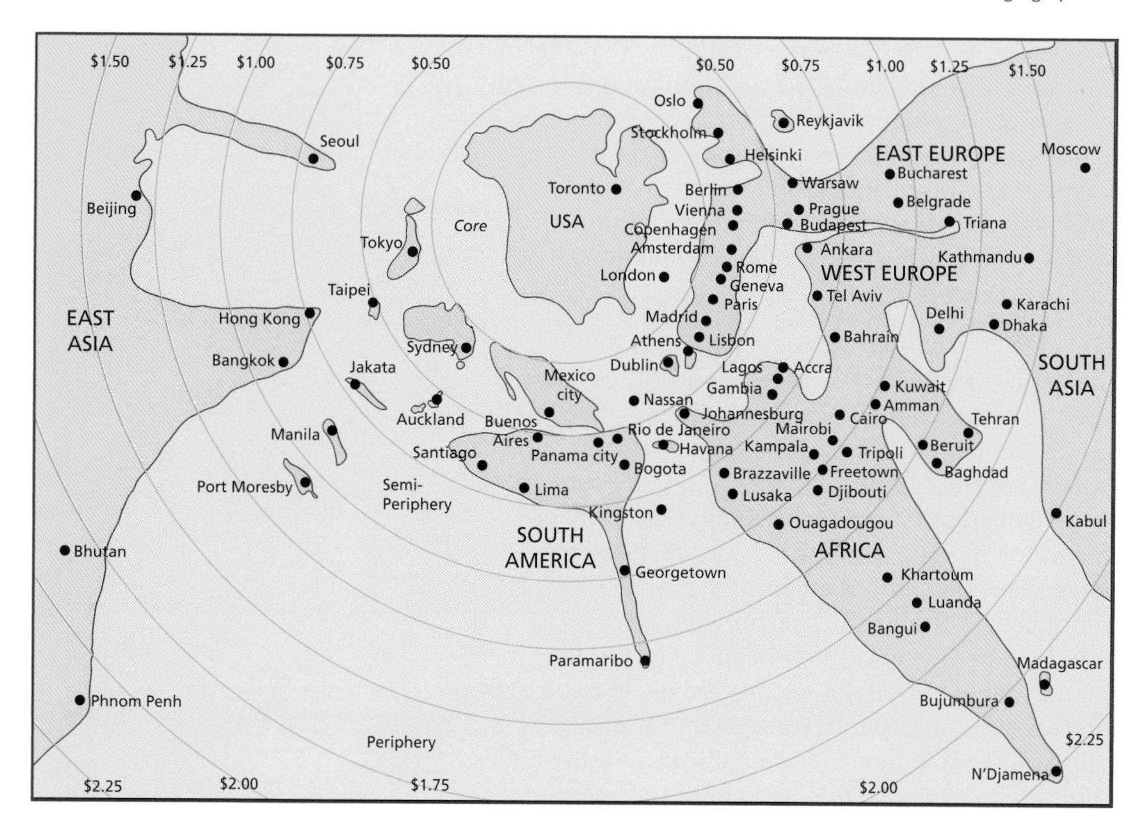

Figure 13.2 Time–space convergence – distance and cost of phone calls from the USA (landline)

Major developments in communications technology

Both the time needed and the relative cost of transporting materials, goods and people have fallen dramatically as the result of changes in transport technology. However, these developments have depended, to a large extent, on parallel developments in communications technology. In the 19th century, rail and ocean transport needed electric telegraph and, later, the oceanic cable, for their development. Similarly, the far more complex global transport system of the present depends fundamentally on telecommunications technology.

Communications technologies are the key technology transforming relationships at the global scale. Communications technologies are significant for all economic activities, but they are especially vital to those sectors and activities whose primary function is to collect, transform and transmit information. One of the most important catalysts to enhanced global communications has been the development of satellite technology. Satellite technology has made possible remarkable levels of global communication of conventional messages and also the transmission of data. Its key element is the linking together of computer technologies with information-transmission technologies over vast distances. Not only are transmission costs by satellite insensitive to distance but also the user costs have fallen dramatically. Satellite communications are now being challenged by optical fibre cables. These systems have a very large carrying capacity, and transmit information at very high speed and with a high signal strength.

However, only very large organizations, whether business or government, have the resources to utilize fully the new communications technologies. For a TNC, they have become essential to its operations. For example,

Texas Instruments, an electronics TNC, has approximately 50 plants in some 19 countries. It operates a satellite-based communications system to coordinate, on a global scale, production planning, cost accounting, financial planning, marketing, customer services and personnel management. The system consists of almost 300 remote job-entry terminals, 8,000 inquiry terminals and 140 distributed computers (computers that interact with one another to solve computational problems) connected to the network.

According to Dicken (2007), technological developments in communications have transformed time–space relationships between all parts of the world. However, not all places are equally affected. In general, the places that benefit most from innovations in the communications media are the "important" places, i.e. the core areas. In contrast, the peripheral areas have benefited move recently in terms of access to, and use of, new communications technologies. Recent developments in mobile phones are bringing major benefits to people who did not benefit from earlier technology (see page 332–333). New investments in communications technology are market-related; they go to where the returns are likely to be high. The cumulative effect is to reinforce both certain communications routes at the global scale and to enhance the significance of the nodes (cities/countries) on those routes (see the diagram of Internet users and the map of telephone lines, on pages 332 and 334).

Figure 13.2 shows the simultaneous time–space convergence and divergence of the world, measured in terms of the cost of a minute-long telephone call from the USA in 2000. In short, rather than making all places the same in terms of accessibility, globalization has increased spatial differentiation – i.e. some areas are more accessible (cheaper), whereas others are less so (more expensive).

Types of transport

Transport costs are made up of **operating costs** and the **profit rate** of the carrier. Operating costs include:

- **fuel and wages**
- **capital,** which includes equipment, terminal facilities, tracks and repairs
- **indirect costs,** such as insurance.

Some modes of transport are more competitive over a certain distance. For example, ocean transport is very competitive over long distances. This is due to low operating costs. However, over short distances it is not competitive. This is because of the high overhead costs of ports. By contrast, the operating costs of roads are high but their overhead costs are low. This makes them very competitive over short distances but not over long distances. To compete over longer distances they need to carry much greater loads. Articulated lorries are able to spread the cost over a greater load. The same feature can be seen in other forms of transport. Some aircraft, notably wide-bodied jumbos, are getting larger. Tankers have increased in size. Very large container ships are more competitive because they can carry a greater load. The use of standardized containers has greatly reduced transhipment costs (i.e. the costs of transferring from one type of transport to another).

Factors affecting the type of transport used include:

- the item to be transported
- the cost of transporting it
- the speed with which it needs to be transported.

Figure 13.3 Container ship

For example, perishable goods or high value goods such as flowers and fruit need to be transported rapidly, whereas bulky goods such as coal can be transported by the cheapest means possible. **Economies of scale** are also important. It is cheaper to carry bulk than small amounts, so **bulk carriers** are increasingly used. **Very large crude carriers** (VLCC) are bulk carriers built for transporting crude oil; other bulk carriers are designed to carry cargoes such as iron ore, coal or wheat. By contrast **container carriers** are ships designed to carry containers. They are equipped with specialized handling devices for carrying expensive freight such as machine parts or high-value manufactures such as electronic equipment as well as low value manufactured goods and non-perishable agricultural products.

	Advantages	Disadvantages
Sea	• Cheaper over long distances • No cost in building the route • Good for bulky, low-cost goods, e.g. coals, ores, grains • Costs spread over a large cargo	• Slow • Limited routes to deep-water ports • Ships expensive to build and maintain • Environmental problems, especially pollution • Ports take up great space
Air	• Fast over long distances • Limited congestion • Good for high-value transport such as people, high-tech industries and urgent cargo	• Large land area needed for airports • Noise and visual pollution • Aircraft expensive to build and maintain • No flexibility of routes • Airlines expensive to run • Can carry only small loads compared with VLCC • Aircraft emissions are blamed for a major contribution to greenhouse gases.

Table 13.1 Advantages and disadvantages of sea and air transport

While aircraft have become faster, ocean tankers have become larger. There have also been increases in the size of planes. For example, the Airbus A380, the world's largest passenger plane, can carry about 555 people – more than the Boeing 747 jumbo. On the other hand, Concorde, the world's only commercial supersonic plane, has been taken out of circulation. It was the noise (breaking the sound barrier) that made Concorde uneconomical. It was not permitted to fly supersonically over land, and that limited its flight paths.

Figure 13.4 Brunei Airlines jet

The jet engine

The jet engine is perhaps the most significant recent innovation in long-distance transport ever. The jet is safer, easier to maintain, better suited for longer distances, and more fuel efficient than the propeller. Jet aircraft have a much higher power to- weight ratio, which enables longer range, faster travel, and bigger payloads. Every year an estimated 320 million people meet at professional and corporate events after travelling by air. Of the world's $12 trillion of merchandise trade, 35% by value was shipped by air in 2006.

Most global trade is by maritime shipping, but air transport fills an important niche in just-in-time production systems. In Brazil, known for its primary goods exports, air cargo in 2000 accounted for 0.2% of total export volume by weight, but almost 19% by value. Prime examples of sectors benefiting from air transport are semiconductors and fashion.

Inexpensive and frequent air services have allowed countries like Chile, Colombia, and Kenya to sell agricultural and horticultural products to markets in Europe, the Middle East and North America. For example, Kenya today has a third of the global market for cut flowers. By contrast, Bangladesh's lack of cold storage facilities and refrigerated air cargo capacity has blunted its opportunities to export high-value fruits and vegetables to the Middle East.

In 2005 tourism receipts in low- and middle-income countries were about $200 billion, thanks mostly to inexpensive air travel. Between 1990 and 2005, tourist arrivals in sub-Saharan Africa increased by 10%, and in China grew almost 10% annually. Vietnam now has about 4 million visitors annually –16 times as many as in 1990.

Containers

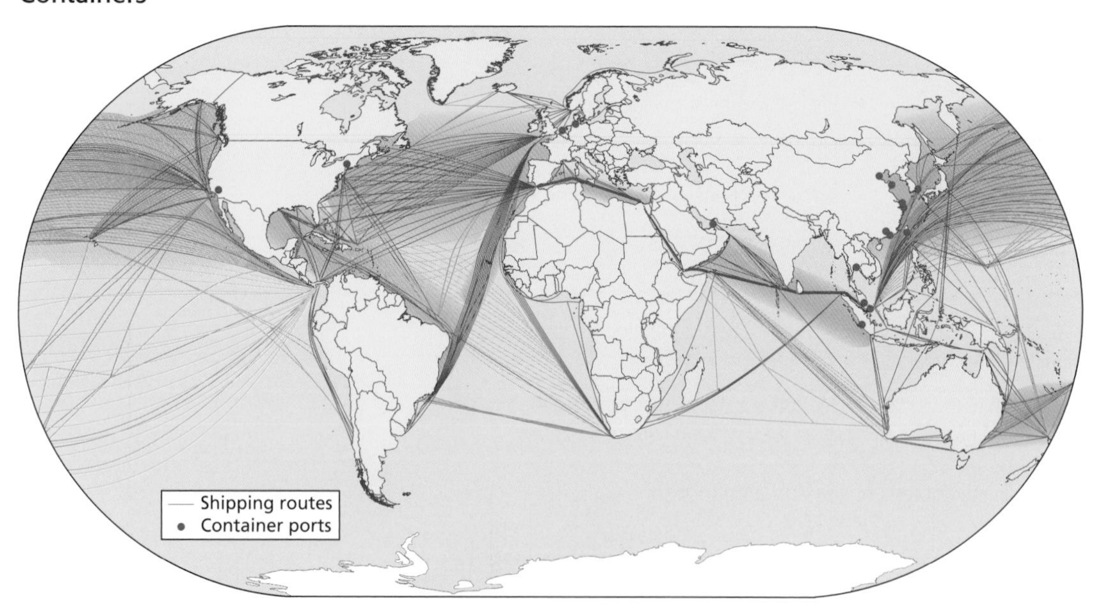

Figure 13.5 Shipping routes and container ports

About 90% of non-bulk cargo worldwide is transported in containers stacked on trucks, rail wagons, and freight ships. In 2007 more than 18 million containers made more than 200 million trips. Cargo shipped is now measured in TEU or 40-foot equivalent units (FEU). A TEU is the measure of a box 6 metres long and 2.5 metres wide, with a maximum gross mass of 24 tonnes.

The Japanese government was the first to support the expansion of containerization. In 1967 it built the world's first container terminals in the Tokyo-Yokohama and Osaka-Kobe areas. Once the infrastructure facilities were in place, container traffic took off. By the end of 1968 the Japan–US route was crowded with container ships.

Container transport has increased at enormous rates. The boxes keep getting larger, with the standard FEU size giving way to 15- and 18-metre boxes that allow trucks to haul more freight on each trip. The world's fleet is expanding steadily, with the capacity of pure container ships rising by 10% annually between 2001 and 2005. The size of the vessels has been increasing, too. Dozens of vessels able to carry 4,000 FEU joined the fleet in 2006.

Figure 13.6 Containers at Tokyo port

Case study: Indian Railways

Almost all rail operations in India are handled by a state-owned organization, Indian Railways. The rail network covers a total length of 63,140 kilometres and transports over 6 billion passengers and over 350 million tonnes of freight annually.

The railway industry exhibits increasing returns to scale in two ways. First, network economies and economies of density lead to size advantages at the firm level. Second, rail transport operations are almost universally combined with the supply of infrastructure services, granting rail firms a natural monopoly, at least locally.

Given the importance of the railways for economic development and the enormous market power of rail firms, it is not surprising that many rail companies are state owned. The biggest of these is Indian Railways. It is the world's largest commercial or utility employer, with more than 1.6 million employees. In 2002 it ran 14 444 trains daily, 8,702 of them for passengers, and owned 216 717 wagons, 39 263 coaches, and 7,739 locomotives.

Founded in 1853 as a system of 42 lines, it was nationalized as one unit in 1951. Vertical integration of Indian Railways is not confined to the bundling with infrastructure services. It owns and runs factories for locomotives, coaches, and even their parts. Long transport distances on the Indian subcontinent should give the railways a stronger competitive edge over roads. Indeed, Indian Railways makes 70% of its revenues and most of its profits from freight, subsidizing the loss-making passenger sector.

The overpricing of freight services is one reason it has lost business to roads in recent years.

Curtailing the potential to provide low-cost freight transport over long distances are extensive social obligations. A major part of the passenger transport deficits covered by freight are urban and suburban losses in Chennai, Kolkata, and Mumbai.

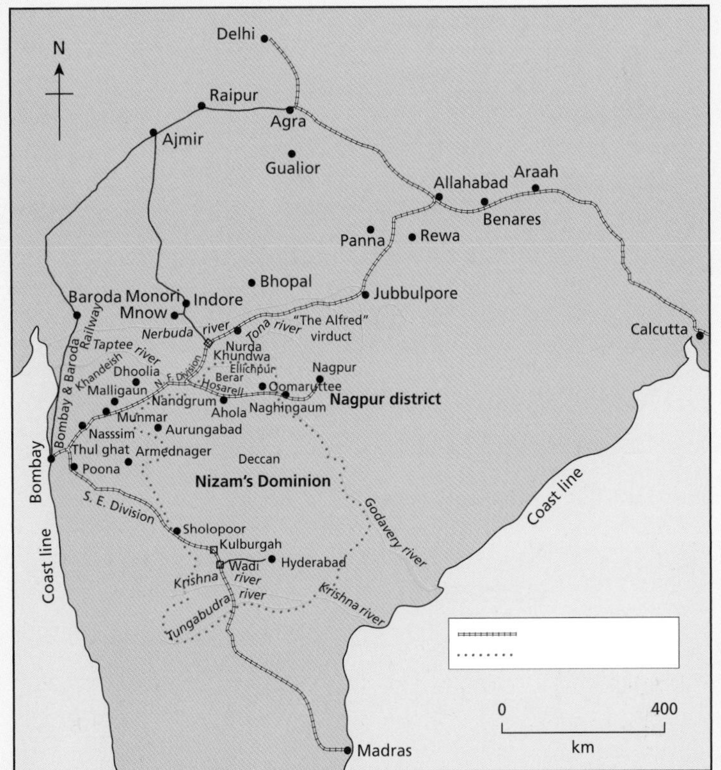

Figure 13.7 The Indian railway network in 1870

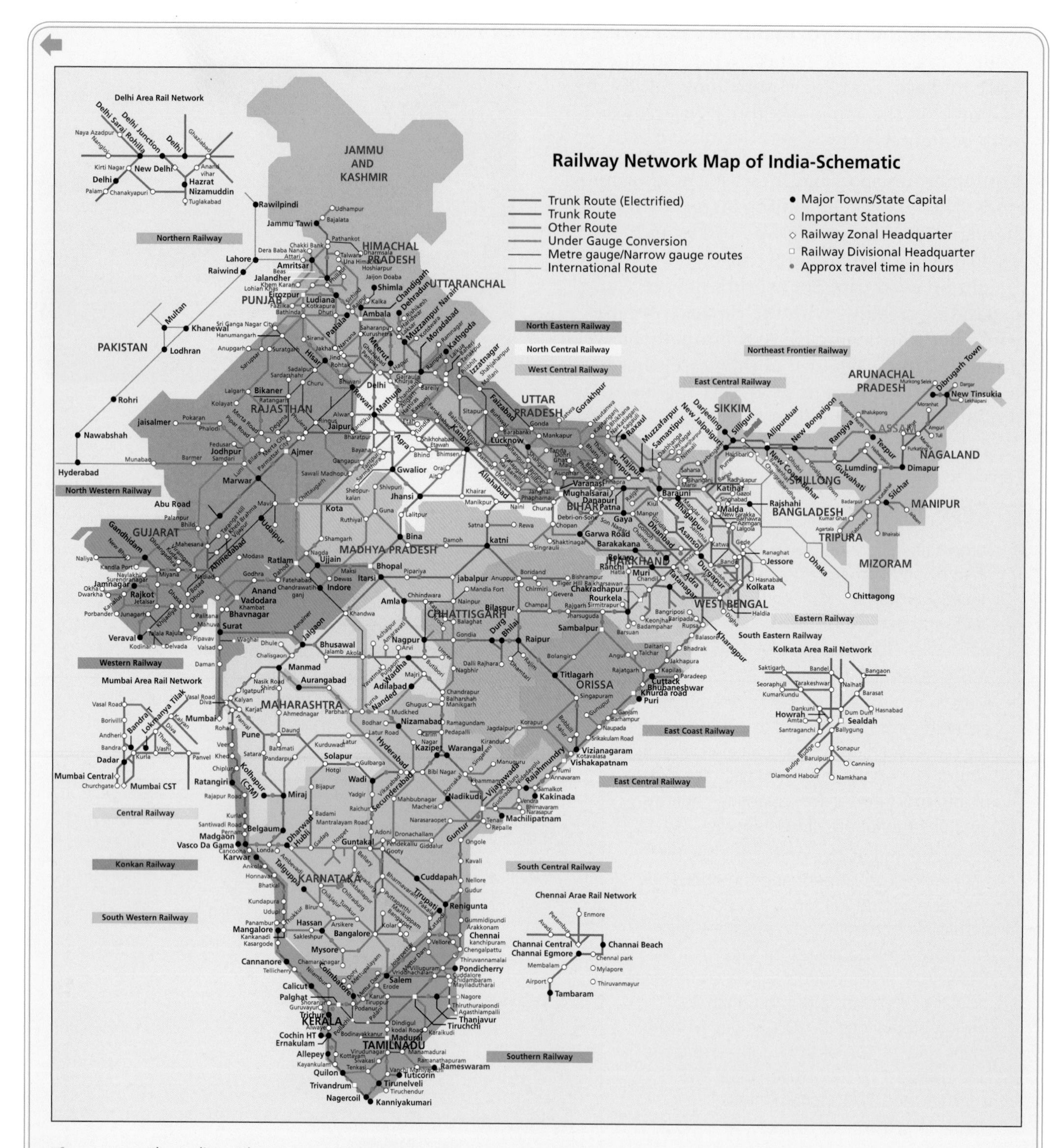

Figure 13.8 The Indian railway network in 2010

To research

Go to the 1909 map of the Indian railway system, at http://en.wikipedia.org/wiki/File:India_railways1909a.jpg

Compare it with the other two maps, Figures 13.7 And 13.8.

Geography and topography limit the ever-increasing size of ships: The Panama is being expanded to allow ships up to 12,000 TEUs to pass. Container ships have an absolute size, limited by the depth of the Straits of Malacca, linking the Indian Ocean to the Pacific Ocean. This limits a ship to dimensions of 470 metres long and 60 metres wide.

Case study: *Railways in Africa*

Transport is a vital element of development and socio-economic growth, and transport infrastructure remains a pillar of development for accelerating growth and reducing poverty. Africa is lagging significantly behind in the growth of regional trade, particularly because of the lack of reliable and adequate transport. Indeed, existing transport facilities for trade are completely outward looking.

To improve transport within Africa, the United Nations proclaimed two Transport and Communication Decades in Africa (1978–88 and 1991–2000) with a view to focusing the efforts of African states and their development partners on the specific issues of transport and communications in Africa.

The African rail transport network is estimated at 89 380 kilometres for a land surface area of 30.19 million square kilometres, or at a density of 2.96 kilometres per 1,000 square kilometres. The network shows very little interconnectivity,

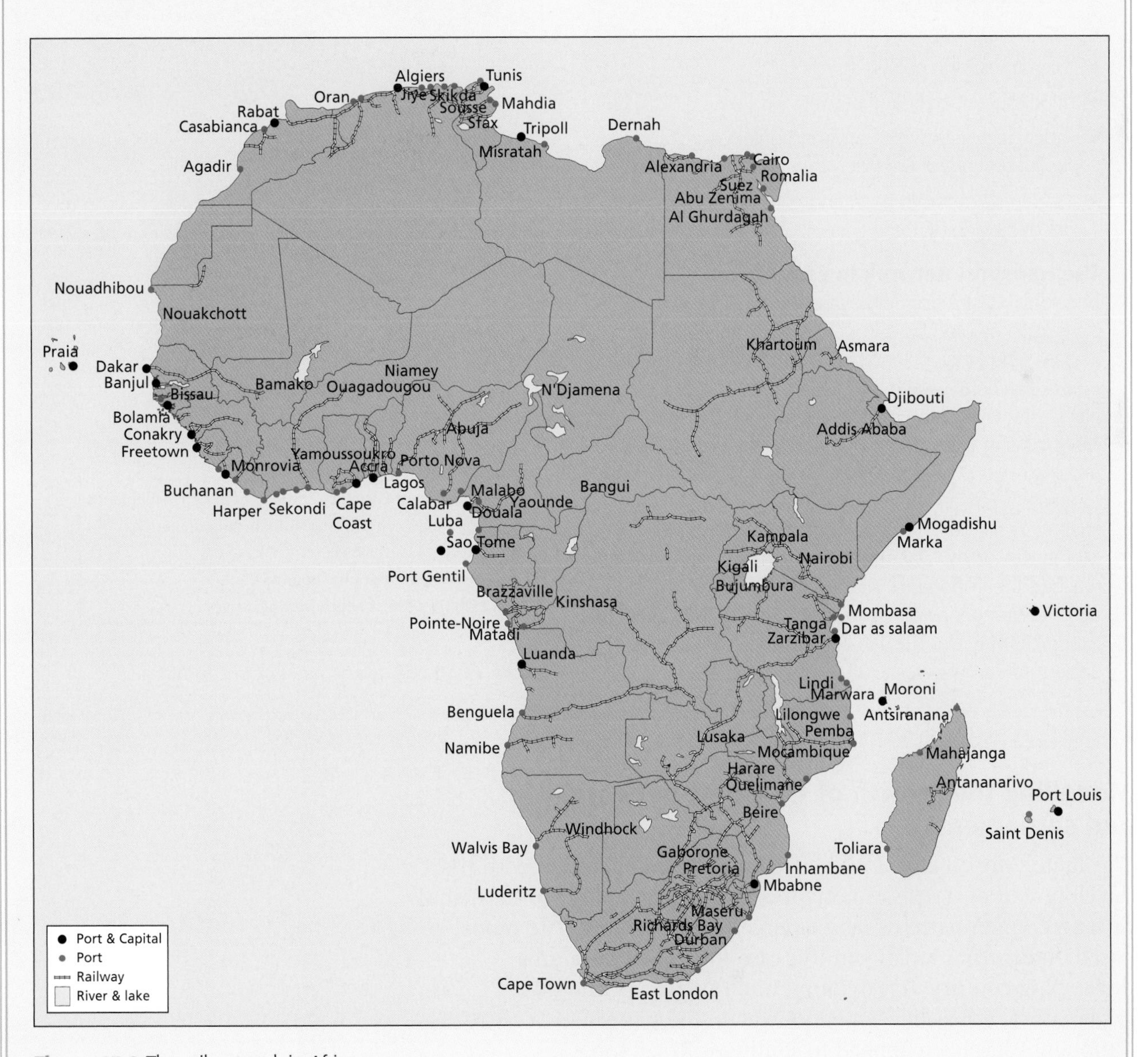

Figure 13.9 The rail network in Africa

particularly in West and Central Africa. Seventeen of the African countries have no railway. Added to this is the heterogeneous nature of rail track gauges, which differ within the same sub-region.

In spite of the major investments made over the 1970s and 1980s in infrastructure and rolling stock, the role of railways has continued to decline nationally and sub-regionally. Railways are facing competition from road transport over the long haulage distances in which they enjoyed a comparative advantage. Moreover, railway companies were (and some still continue to be) characterized by bureaucracy, overstaffing and low productivity.

There are a number of transport development challenges. These include:

- inadequate infrastructural network
- inadequate financing
- lack of appropriate human and institutional capacities
- high transport costs
- inadequate safety and security
- poor degree to which environmental issues and pollution control are taken into account
- underuse of information and communication technologies (ICT).

The transport network in East Africa

In colonial East Africa, railways were built inland from ports such as Mombasa in Kenya, and Tanga, Dar-es-Salaam, and Lindi in Tanzania. These linked the ports to areas of mineral wealth and export potential. Branch lines were extended to the copper deposits of the Ruwenzori Mountains in western Uganda and to the commercial farming regions of Kenya. Lines were not generally built to areas of African settlement. Eventually the separate lines were linked by transverse routes, for example Mombasa–Moshi–Dar-es-Salaam, although these never reached the importance of the original axes. Even in post-independence Africa, railways were built to allow the export of minerals. The Tan-Zam railway from Dar-es-Salaam to Zambia was to allow the export of copper.

Over the last 50 years or so, rail has been giving way to road transport for long-distance transport. The railways remain important, but major highways have been built to run parallel to them. These have tended to confirm the networks and hierarchies established in earlier periods, although they may extend them somewhat.

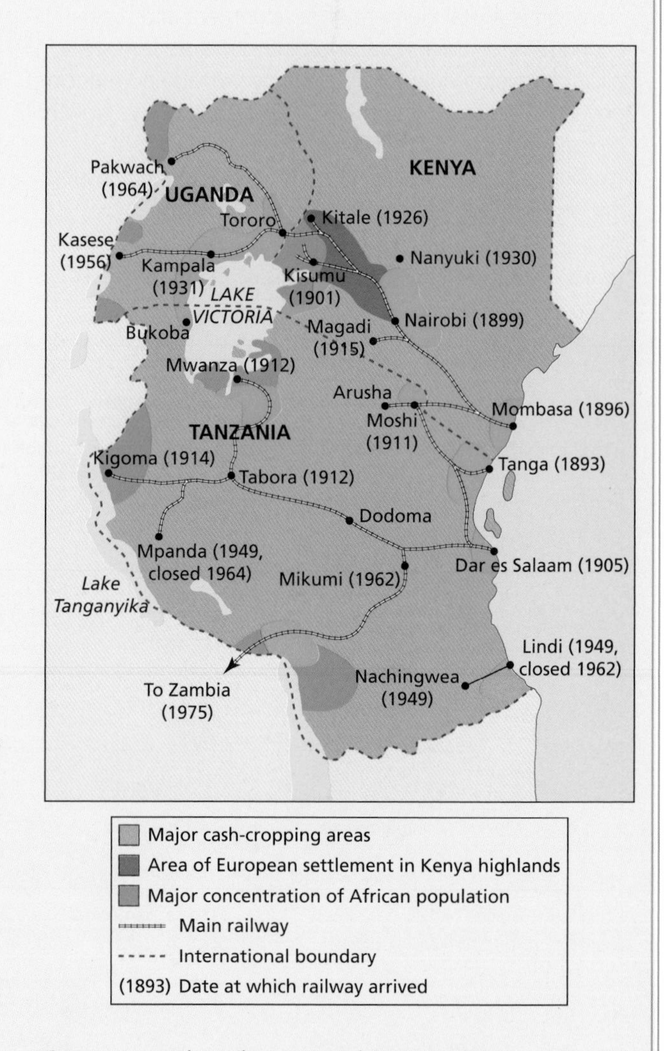

Figure 13.10 The railway network in East Africa

Modelling the growth of transport networks and settlements

The Taafe, Morrill and Gould (1970) model is concerned with the development of transport networks in the developing world. It shows how settlement patterns may emerge (Figure 13.11). The model is based on external forces – in this case the exploitation of a developing country. It is dynamic and functional, and it is explanatory. It was developed to explain the pattern in West Africa, although it has been applied to southeast Brazil too.

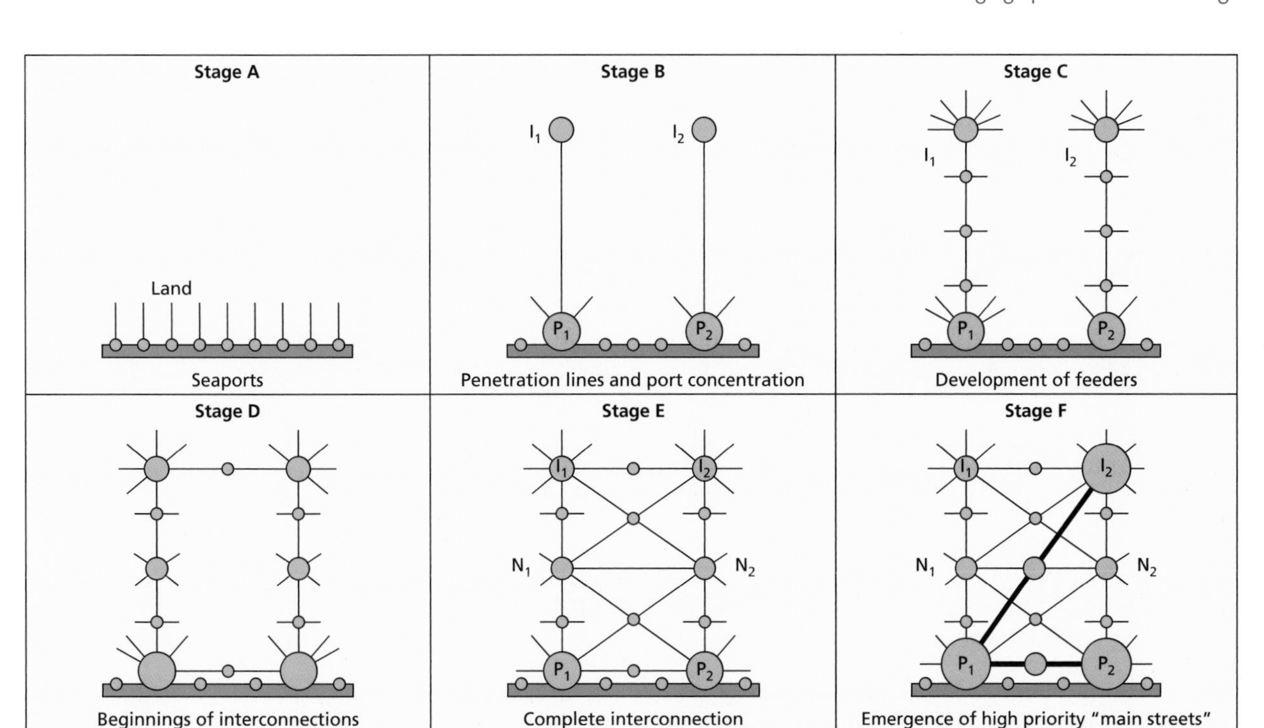

Stage A shows a scatter of small ports and trading posts along the sea coast. Only small indigenous fishing craft and irregularly scheduled trading vessels provide lateral intercommunications. Each port has an extremely limited hinterland.

Stage B shows that market areas have expanded for two ports (P_1 and P_2). Transports costs into the hinterland are reduced for these two ports. Port concentration begins (P_1 and P_2). Feeder routes begin to focus on the major ports and interior centres I_1 and I_2).

Stage C The major ports begin to enlarge their hinterlands at the expense of the smaller ports. Feeder development continues and small nodes develop along the main lines of penetration.

Stage D Certain of the nodes (N_1 and N_2) capture the hinterlands of the smaller nodes on each side. Feeder lines continue to develop and some of the larger feeders begin to link up.

Stage E Lateral links occur until all the ports, interior centres, and main nodes are linked. There are the beginnings of the development of national trunk routes or "main streets", again increasing the connectivity of the network.

Stage F The main streets have reached their full development; an urban hierarchy has emerged.

Figure 13.11 The Taafe, Morrill and Gould model of transport development

To do:

a Briefly describe the evolution of the rail network in either India or East Africa. How does it compare with the model developed by Taafe et al (Figure 13.11)?

b Visit http://www.worldmapper.org/posters/worldmapper_map30_ver5.pdf

  i dentify the two countries with the highest overall amount of rail passenger kilometres per year.

  ii Comment on the countries that do not have a rail network.

  iii Which country has (i) the highest and (ii) the lowest annual railway journeys in kilometres per person per year?

Extension and density of networks

Phone networks

According to the International Telecommunication Union (ITU), at the end of 2009 the world had two mobile-phone subscriptions for every three people. The rapid spread of mobile phones contrasts sharply with the decline of fixed-line phones. The number of landlines per 100 people peaked at 19.5 in 2006, and had fallen to

17.8 by the end of 2009 (Figure 13.12). As recently as 2002, less than a 10th of the world's population used the Internet. According to the ITU, more than a quarter now do.

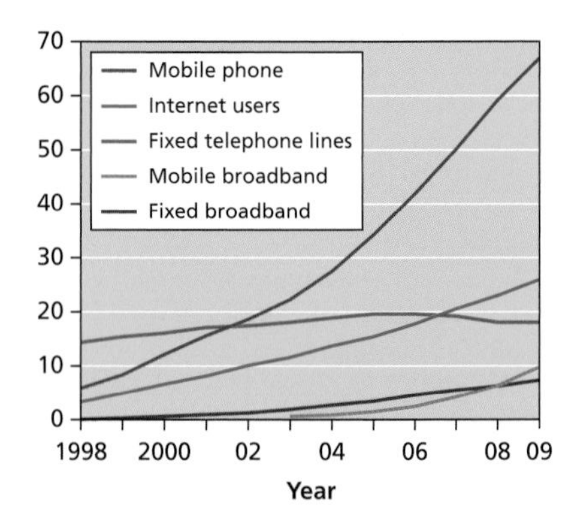

Figure 13.12 Telecommunications: global subscriptions per 100 people

There is a huge inequality in access to telephone lines, although it is decreasing. In 1990 there were over 500 million landlines, 63% of them in North America, western Europe and Japan. By 2002 there were over 1 billion landlines, almost double the 1990 figure. North America, western Europe and Japan still contained most landlines, but there were very large increases in China, India and Brazil. In contrast, much of Africa had relatively little access to landline telephones.

The changes in mobile (cellular) phone ownership are important. In 1990 only 12 million people subscribed to a mobile phone network. Most of these were in the USA (some 5 million), the UK (1 million), Japan, Canada and China (all under 1 million). Large parts of the world had no mobile phone network: the Middle East, central Asia, south Asia and central Africa.

By 2002 there had been a huge growth in mobile networks and mobile phone ownership. The number of mobile phones had increased 100-fold. The largest users included the USA, western Europe, China and Japan, and there has been a huge increase in the use of mobile phones in India and some growth in North Africa and South America. Although there has been some growth in Africa overall, spatial coverage remains limited, as shown in Figure 13.15.

The number of subscribers surged nearly 25% annually between 2000 and 2008. Mobile penetration stood at only 12% in 2000, but was over 60% by the end of 2008.The impressive growth in the number of mobile subscribers is mainly due to developments in some of the world's largest markets. The BRIC economies of

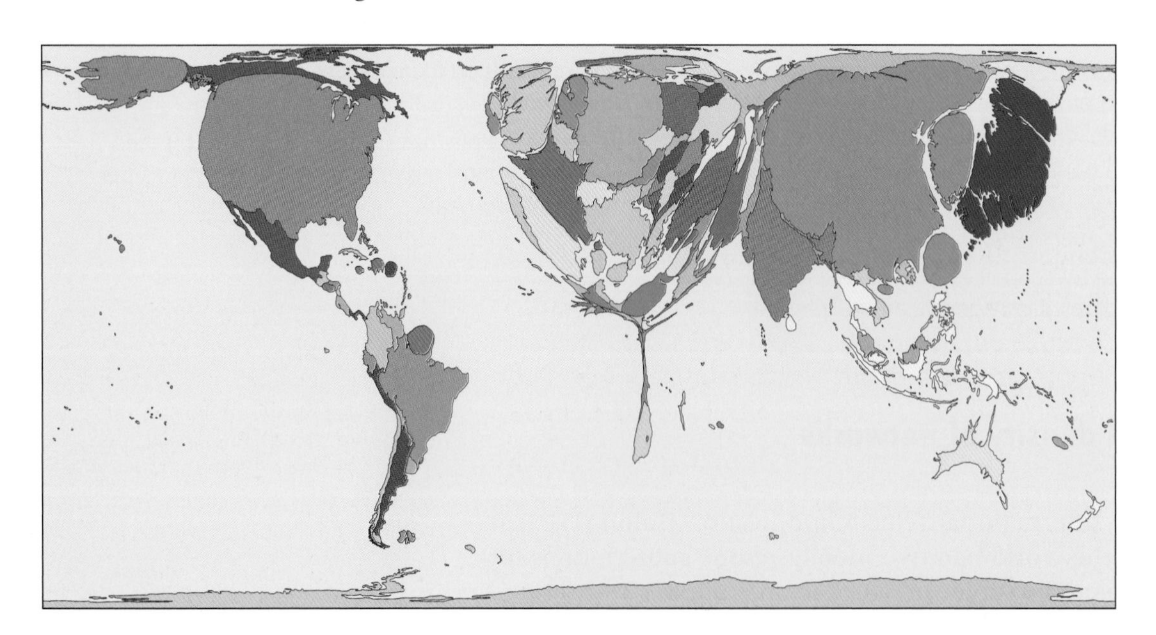

Figure 13.13 Landline phone access in 2002

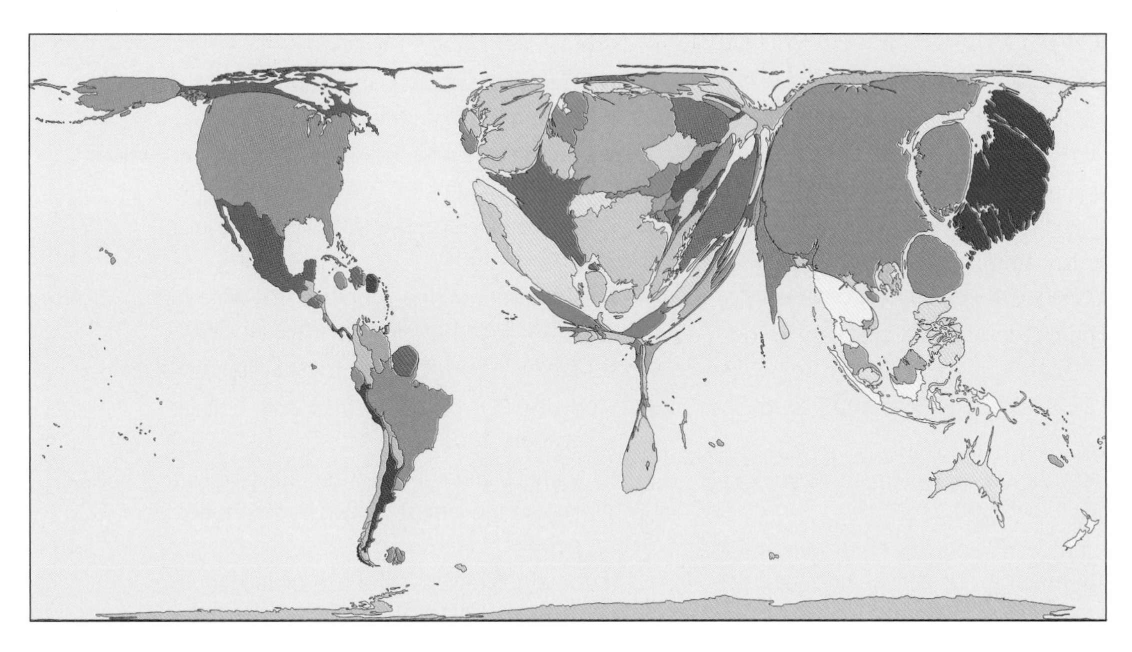

Figure 13.14 Mobile phone access in 2002

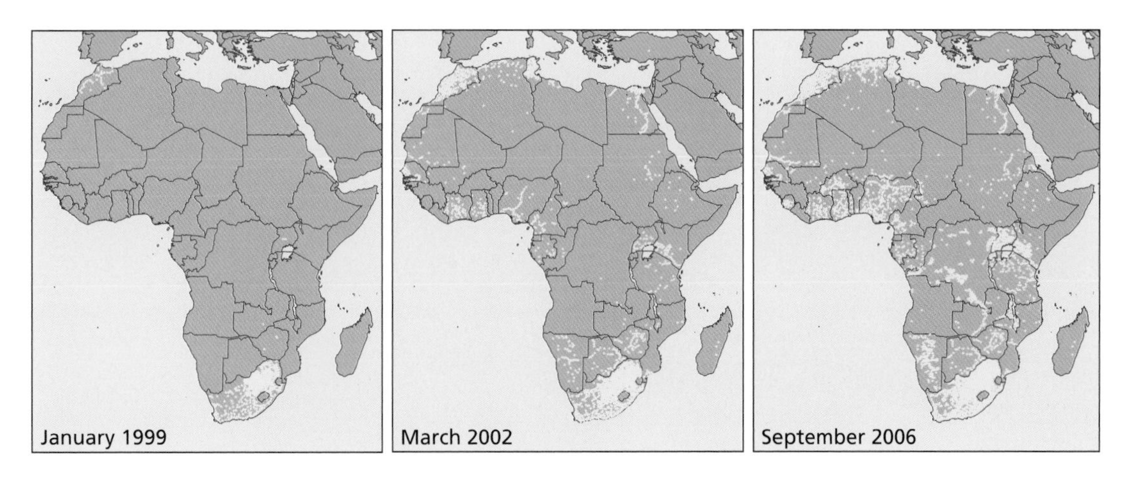

Figure 13.15 The diffusion of mobile phones in Africa

Brazil, Russia, India and China are expected to have an increasingly important impact in terms of population, resources and global GDP share. These economies alone accounted for over 1.3 billion mobile subscribers by the end of 2008.

China surpassed the 600 million mark by mid-2008, representing by far the world's largest mobile market. India had some 296 million mobile subscribers in 2008, but with a relatively low penetration rate of about 20%, the country offers great potential for growth.

Mobile telephony is changing people's lives. Apart from providing communication services to previously unconnected areas, mobile applications have opened the doors to innovations such as m-commerce, to access pricing information for rural farmers, and to pay for goods and services.

TOK Link

How would the map of access to phones vary if it were mobile phones rather than landline phones?

Suggest ways in which access to phones can improve a person's quality of life.

To do:

Visit www.worldmapper.org

a Choose map 331, Telephone lines 1990, and click on "Open PDF poster".

- **i** Identify the countries that were the most connected in terms of landlines in 1990.
- **ii** State how many landlines per 1,000 people there were in (a) North America and (b) central Africa.
- **iii** Describe the global variations in access to landlines.

b Choose map 332, Telephone lines 2002. Study the PDF poster.

- **i** Describe how the overall pattern changes between 1990 and 2002.
- **ii** State the number of landlines there were in 2002 in (a) North America, (b) central Africa and (c) south and east Africa.
- **iii** Identify the regions that showed the greatest increase in landlines between 1990 and 2002.
- **iv** Identify the regions that showed least change between 1990 and 2002.
- **v** Suggest reasons to explain the changes (or lack of changes) in the regions outlined.

c Open the PDF poster for map 333, Cellular subscribers 1990.

- **i** Comment on the countries that were high subscribers to cellular phones in 1990.
- **ii** Which regions had the most cellphone users?

d Study the PDF poster for map 334, Cellular subscribers 2002.

- **i** Identify, and comment on, the regions that have shown the largest growth in cellphones per 1,000 people.
- **ii** In which region is the use of cellphones per 1,000 people least?

e Describe the growth and spread of mobile phones in Africa, as shown in Figure 13.14.

Telephone calls

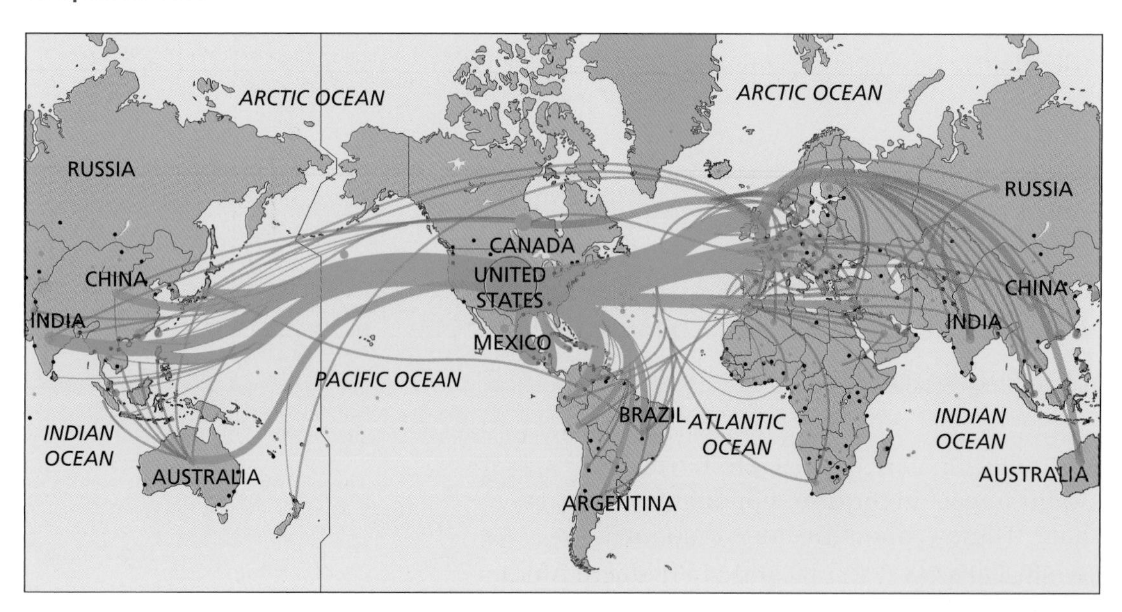

Figure 13.16 Global landline telephone traffic, 2007

The map (Figure 13.16) shows the annual flow of intercontinental calls by fixed landline telephones (not mobile phones) in 2007. Clearly the greatest volume of traffic is between North America and Europe, followed by North America and Southeast Asia. There also large flows between North America and the Caribbean and Latin America. There are relatively few flows between Africa and the other continents.

A number of factors can help explain these patterns:

- **Population size** – countries with small populations, such as Greenland, are unlikely to generate a large number of calls.

To research

The transmission and flow of images and finance

Visit www.asksource.info/pdf/framework2.pdf for Improving health, connecting people: the role of ICTs in the health sector of developing countries and for Key developments in the ICT sector http:www.ictregulationtoolkit.org/en/section.2196.html

- **Population density** – within the USA, for example, there is a small flow to and from Alaska but a very large flow to and from northeastern USA.
- **Wealth** – wealthy countries, such as Japan and the USA, can afford more phones than poorer countries in Africa.
- **Lack of landline infrastructure** – in Africa this has led to mobile phones being far more popular; Africa is one of the fastest growing markets for mobile phones, and landlines may therefore never achieve the same prominence here as in other regions.
- **Trading partners** – countries within a trading bloc, such as the EU, are likely to generate large volumes of calls.
- **TNC or MNC activities** – companies with offices and factories in several different countries are likely to create large volumes of calls between those countries.
- **Migration** – there is likely to be a high volume of calls between the area a migrant moves to and their home country. However, the country of origin may be relatively poor and have relatively few phones.
- **Colonial history** – it is likely that there will be political and historic ties between a former colonial power and its former colonies. The UK and the former British Empire is a good example.
- **Language** –the volume of calls is likely to be greater between countries that share the same language.

Cheapest fixed line broadband as proportion of monthly income			
Rank	**Country**	**Price % of monthly income**	**B'band subscriptions per 1000 inhabitants**
1	Macao	0.30	23.42
2	Israel	0.33	25.8
3	Hong Kong	0.49	29.34
4	USA	0.5	27.1
5	Singapore	0.58	23.71
Most Expensive			
1	Cr.African Rep	3891	No data
2	Ethopia	2085	No data
3	Malan	2038	0.02
4	Guinea	1546	No data
5	Niger	967	0.01

- The Central African Republic is the most expensive place to get a fixed broadband connection – costing nearly 40 times the average monthly wage.
- Macao in China is cheapest at 0.3% of the average monthly wage.
- Niger becomes the most expensive when landlines and mobiles are taken into account.

The digital divide

The Internet is the fastest growing tool of communications ever. Radio took 38 years to reach its first 50 million users; television took 13 years, and the Internet just 4 years. This is partly explained by the fact that the world population was less than 3 billion when radio was introduced, and disposable incomes were much lower. The Internet was developed in 1983. According to Worldmapper, by 1990 just 3 million people had access to it worldwide, 73% of them in the United States, 15% in western Europe and most of the rest in Canada, Australia, Japan, Korea and Israel. In 1990 there was practically no access elsewhere. However, by 2002 there were 631 million Internet users and their worldwide distribution had changed dramatically. There was large growth in Asia–Pacific, southern Asia, South America, China and eastern Europe. Smaller growth was recorded in northern Africa, southeastern Africa and the Middle East.

The **digital divide** among households appears to mainly depend on two factors, namely income and education. People with higher incomes and education levels are more likely to have access to ICT.

The digital divide – inequality in the ICT network infrastructure and distribution of the IT knowledge, skills and resources necessary to access online services and information among different sections of a modern society.

The digital divide refers to the inequalities in opportunities between individuals, households, businesses and nations to access ICT. The digital divide also occurs between urban and rural areas, and between different regions of a country. For example:

- Over 75% of Internet users come from rich countries, which account for just 14% of the world's population.
- In Thailand 90% of Internet users live in urban areas.
- In Chile 74% of Internet users are under 35 years old.
- In Ethiopia 86% of Internet users are male.

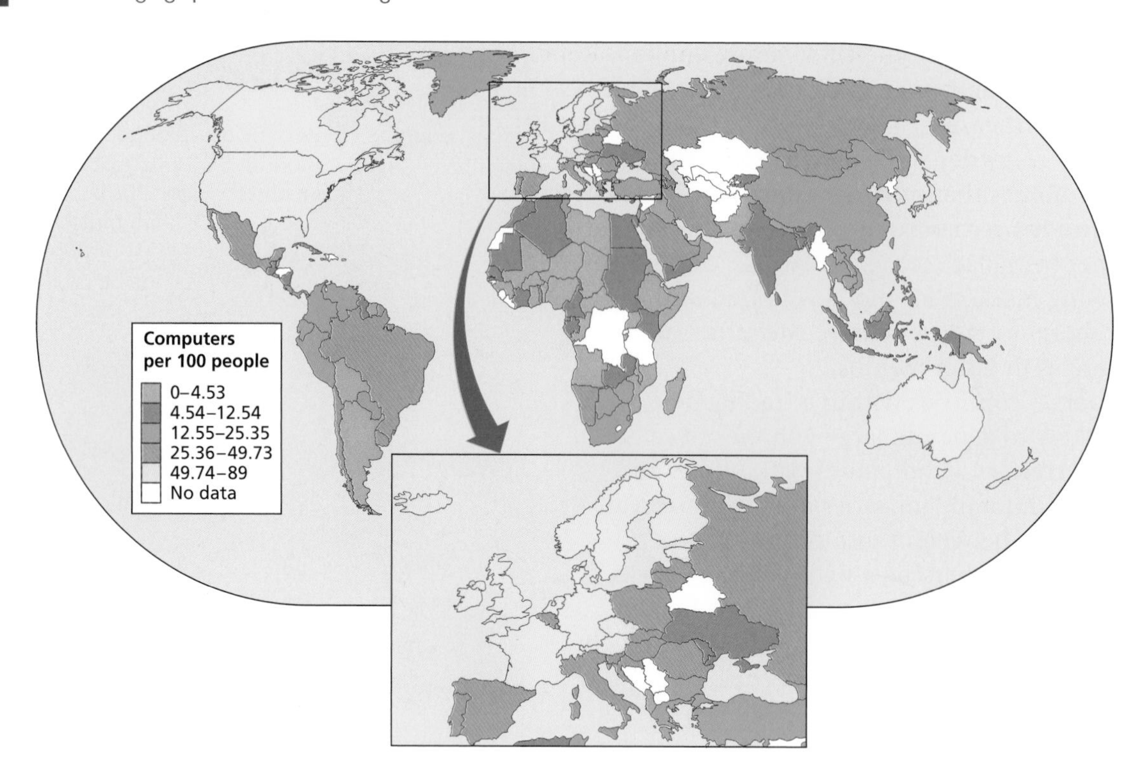

Figure 13.17 The world's digital divide

Instead of reducing inequalities between people, the digital divide may well have reinforced them. There is a widening gap between rich and poor countries, and even within rich countries such as the USA, Internet users are more likely to be white, middle class and male. Many people do not have access to ICT and therefore cannot benefit from the knowledge-based economy. To date there has been little action from rich countries to ensure that the benefits of ICT are extended to people in poorer countries, regions and areas.

The digital access index (DAI)

The digital access index (DAI) measures the overall ability of individuals in a country to access and use ICT. It consists of eight variables grouped into five categories:

- infrastructure – combined fixed and mobile teledensity
- affordability – Internet access price as % of per capita GNI
- knowledge – adult literacy and combined enrolment up to tertiary level
- quality – international Internet bandwidth in bits per capita and % broadband customers
- usage – Internet users per 100 population.

In 2003, leaders at the United Nations World Summit on the Information Society (WSIS) agreed that by 2015,more than half the world should be able to access the Internet. The global net population grew from 381 million to 872 million between 2000 and 2004. Developing nations' share of that number grew from 21% to 39%. However, the annual growth in the number of net users has been dropping since 1998. This means the WSIS goal of 50% of the globe accessing the net by 2015 is in danger. Access to the Internet is not is not just about computers. Many people in developing nations are

To research

Mapping the growth of the internet

Visit http://news.bbc.co.uk/1/hi/technology/8552410.stm for Mapping the growth of the internet.

Choose two countries and compare their experience of internet growth.

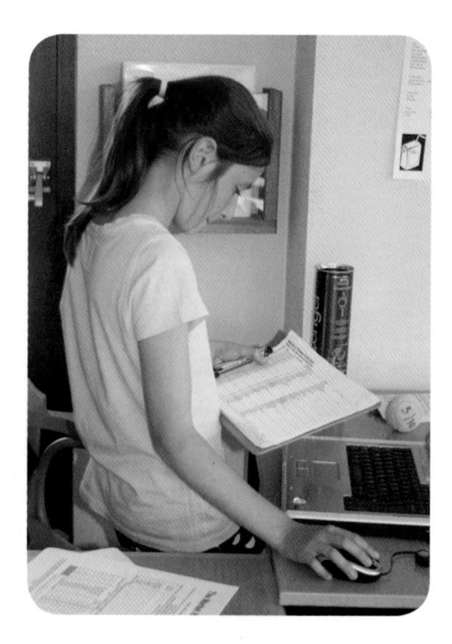

Figure 13.18 The digital divide: in MEDCs young children routinely use computers for data logging

enjoying the benefits of access to information and communication via mobiles. Mobile net bypasses costly physical infrastructure. Just 38% of developing nation schools are online; only 1% in Africa. Almost all schools are online in developed countries. In addition, just 20% of the 153 million global broadband connections are in developing nations. Richer countries have greater access to bigger bandwidth, too.

To research

Visit http://www.itu.int/wsis/tunis/newsroom/stats/ for some interesting statistics about the digital divide.

How do the graphs of telephone traffic compare with the graph of world Internet use?

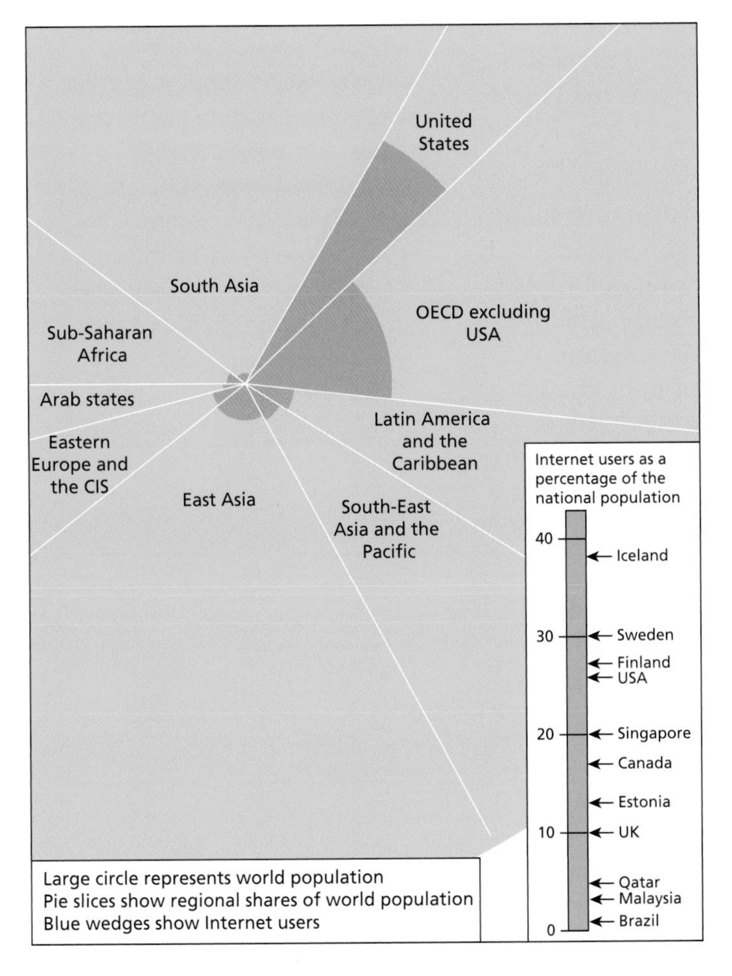

Figure 13.19 Global variations in Internet use

To do:

a Visit www.worldmapper.org. Choose maps 335 and 336, Internet users in 1990 and 2002, and click on Open PDF poster.

i Identify, and comment on, the countries with the most Internet users.

ii Outline the changes that had occurred by 2002.

b Figure 13.19 shows an early representation of Internet users and proportion of population.

i Identify (i) the region in which most people live. (ii) the region with the smallest proportion of the world's population, (iii) the region with the highest proportion of Internet users, and (iv) the region with the lowest proportion of Internet users.

ii Suggest reasons for the patterns you have described in relation to population size and proportion of Internet users.

The role of ICT in Civil Societies

ICT can be used by civil societies for many purposes (see Avaaz.org for its use its in raising popular awareness – page 386). Civil societies include mass organisations, trade unions, faith-based organisations, NGOs, academe, social movements and campaign networks. Through ICT they have been all to reach people all over the world and mobilise support for their activities. ICT can help people to gain information about campaign issues, to mobilise community networks, and to foster dialogue between individuals and organisations.

Loans

Many countries provide loans for developing countries. The main pattern is a transfer from richer countries to poorer countries. However, the definition of a loan is that it is a transfer of money or skills that requires repayment over a set time. Every year the World Bank lends over US$24 billion to developing country governments to fund projects for economic development and poverty reduction.

In many poor countries, economic and social infrastructure such as electricity, gas, transportation and communications services is underdeveloped. In addition there may be issues related to population growth, environmental degradation, disease and conflict. To address these issues, the Millennium Development Goals (MDGs) have been set as common goals, and individual countries have launched a range of measures to achieve them (see also chapter 2). Most of these goals, such as poverty reduction, universal primary education and environmental sustainability, are planned to be reached by 2015.

In theory, official development assistance (ODA) loans promote efficient use of the borrowed funds, since they require repayment and the donor is unlikely to provide loans for projects likely to fail. Moreover, they place a relatively small financial burden on the donor government as they are paid back over time.

Gross domestic product (GDP) – the value of all final goods and services produced within a nation in a given year. The measure is relatively easy to compute and use compared with the GNI.
Gross national income (GNI) – the value of goods and services produced within a country, together with the balance of income and payments from or to other countries; now used in preference to gross national product (GNP).

Case study: Japan and loans

Japan's loan policy can be divided into bilateral aid, in which assistance is given directly to a developing country, and multilateral aid, which is provided through international organizations. Japanese loans are given with an interest rate that varies from 0.01% for the least developed countries to 1.7% for upper-middle-income countries, with a repayment period of between 15 and 40 years.

Types of loan provided by Japan

1. **Project loans** – to finance projects such as roads, power plants, irrigation, water supply and sewerage facilities.
2. **Engineering services (E/S) loans** – to finance the survey and planning stages of a project (a feasibility study).
3. **Financial intermediary loans (two-step loans)** – to promote small- and medium-scale enterprises in manufacturing and agriculture; known as two-step loans because there are two or more steps before the end-beneficiaries receive the funds.
4. **Structural adjustment loans (SALs)** – to improve economic policies and implement structural adjustment for overall economies.
5. **Commodity loans** – to support the balance of payments and economic stability of recipient countries; often used to import commodities such as industrial or agricultural machinery and raw materials, fertilizer and pesticides.
6. **Sector program loans (SPLs)** – to support development policies in prioritized sectors of developing countries. Local currency (counterpart) funds are utilized for public investments for sector-specific improvements.

Most Japanese aid is offered to countries in Asia, although some is offered elsewhere. Good examples of projects include the following.

- Small-scale irrigation management in eastern Indonesia. Eastern Indonesia has a long dry season and low annual rainfall. As a result, agricultural productivity is low and the region is the poorest in

Indonesia. Loans have been provided to develop irrigation, extend farming technology and introduce agricultural water management, thereby increasing productivity and reducing poverty.

- Mass rapid transport system in Delhi. This project aims to improve the infrastructure for economic growth by helping construct a subway line, and elevated and surface railroads. The system is expected to carry 2.26 million passengers a day, and will reduce traffic congestion and improve air quality.
- Integrated reforestation in Tunisia. Much of Tunisia is semi-arid. Forest cover has been reduced by natural disasters and overlogging. A loan has been given to support forest maintenance, replanting, and constructing water and soil conservation facilities including reservoirs.

Debt repayment

(See also chapter 2, page 47, The impact of aid and debt relief.)

External debt refers to the part of a country's total debt that is owed to creditors outside the country. The creditors can include national governments, international organizations such as the World Bank and the IMF, and multinational companies. Those in debt can include national governments, companies and individuals. External (or foreign) debt is different from public debt, which is the money or credit owed by any government, organization or individual to others in the same country.

Debt can be expressed in terms of (a) debt to GDP ratio or (b) foreign debt to exports ratio.

Rank	Country	External debt (US $ million)	External debt per capita (US$)
1	USA	13 450 000	42 343
2	UK	9 088 000	150 673
3	Germany	5 208 000	63 350
4	France	5 021 000	78 453
5	Netherlands	2 452 000	146 826
22	China	347 100	259
24	Turkey	253 200	3288
25	India	232 500	201
26	Brazil	216 100	1091
28	Mexico	177 000	1594
80	Jamaica	11 550	4125
88	Kenya	7,729	198
99	Zimbabwe	5,821	529
107	Ethiopia	4,229	49.7
161	Haiti	428	47.5

Table 14.1 Selected levels of debt (CIA World Factbook, 2010)

There is concern that many poor countries have borrowed more money than they can pay back. The figures in the table suggest that external debt might not just be related to poor countries. There are a number of reasons why debts have accumulated. These include:

To research

Visit http://www.odakorea.go.kr and click on "Korea's ODA at a glance" to see how Korea has changed from a recipient of loans to a donor, and to investigate its plans for the future.

To do:

Study Table 14.1.

a Identify the country with the largest debt (i) in absolute terms and (ii) per person.

b Identify the countries with the lowest debt (i) in absolute terms and (ii) per person.

c Approximately how many times larger is the USA's total debt compared with (a) China's (b) Kenya's and (c) Haiti's?

d Comment on the figures shown in the table.

- the legacy of colonialism – the transfer of debts to the newly independent country, e.g. Indonesia in 1949 (from the Netherlands) and Haiti in 1804 (from France)
- inappropriate lending and spending in the 1960s and 1970s
- excessive interest charges imposed by creditors
- the oil crisis of 1973, which caused poor countries to borrow heavily
- high levels of military spending
- political corruption.

Should we be more concerned with rich countries' debt or with poor countries' debt? Give reasons for your opinion.

Following independence, many countries were loaned large sums of money in order to develop their infrastructure and import substitution industries. Large-scale lending to Mexico resulted in a debt crisis, with Mexico threatening to default on its loans. Lending was replaced with structural adjustment programmes (SAPs), which focused attention on developing the economy and reducing spending, often in areas such as health and education.

Since 2000 there have been a number of situations in which the poorest countries have had some of their debts cancelled by the IMF and the World Bank. However, not all organizations have cancelled debts. A World Bank/IMF study found that, in Tanzania and Uganda, debt relief was linked to improvements in education. Elsewhere, debt relief was linked to improvements in access to water and to healthcare, for example in Burkina Faso.

In 2005 debt cancellation for 18 Heavily Indebted Poor Countries (HIPCs) reduced their debts by about two-thirds. Following the 2004 South Asian tsunami, debt relief was announced for 12 of the countries affected. However, others, such as Sri Lanka and Indonesia, continued to be in debt and had annual debt service bills of $493 million and $1.9 billion respectively. In 2006 the $40 billion owed by the 18 HIPCs to the World Bank, IMF and African Development Bank was written off. Compared with the debt that some rich countries have, this sum does not seem very large. Countries that owe money to the Asian Development Bank and the Inter-American Development Bank did not have their debts cancelled.

Critics of debt cancellation argue that structural adjustment policies should continue. However, this undermines attempts to provide healthcare and education, and may cause a country to remain impoverished.

Development aid

Figure 14.1 shows that the main donors of international aid are the rich countries in North America, Europe, Australia, New Zealand and Japan. In contrast, the main recipients of aid are in the poor countries. Highest levels would appear to be in much of sub-Saharan Africa, eastern Europe and Russia and Southeast Asia.

The largest donors are the USA and Japan, although as a percentage of their GNP each donates less than 0.25% of GNI. France and the UK are the next largest donors, giving less than 0.5% of their GNI. The largest donors (in relation to GNI) are the Scandinavian countries of Norway, Denmark and Sweden.

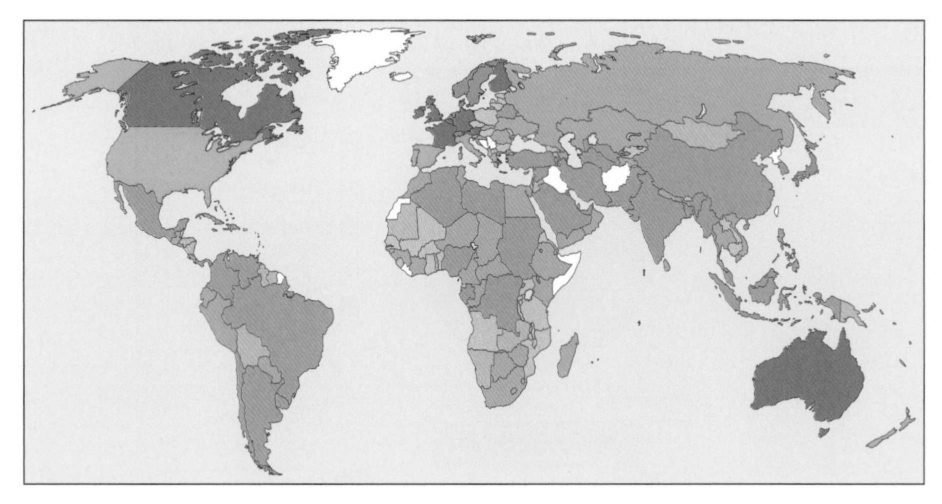

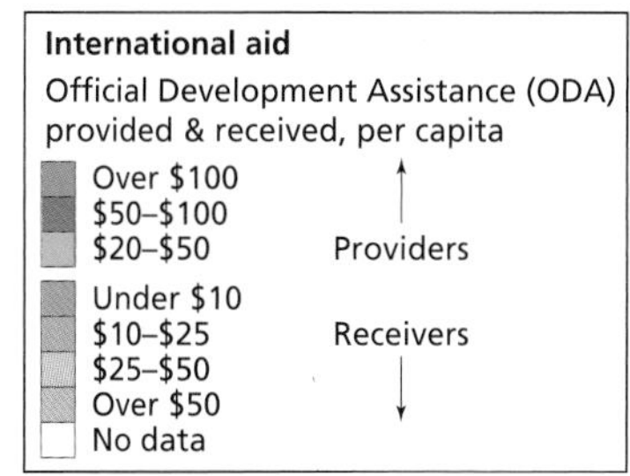

Figure 14.1 International aid, 2002

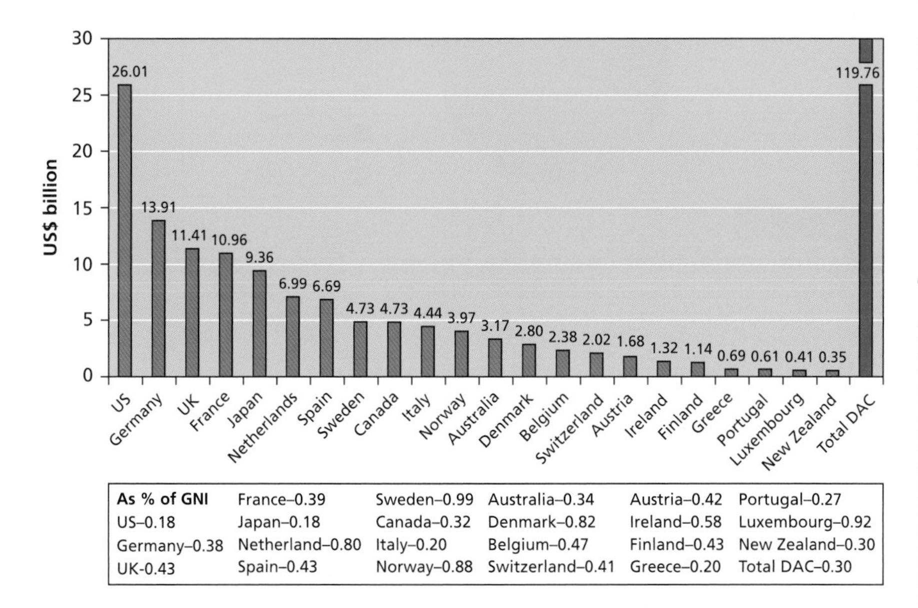

As % of GNI	France–0.39	Sweden–0.99	Australia–0.34	Austria–0.42	Portugal–0.27
US–0.18	Japan–0.18	Canada–0.32	Denmark–0.82	Ireland–0.58	Luxembourg–0.92
Germany–0.38	Netherland–0.80	Italy–0.20	Belgium–0.47	Finland–0.43	New Zealand–0.30
UK-0.43	Spain–0.43	Norway–0.88	Switzerland–0.41	Greece–0.20	Total DAC–0.30

Figure 14.2 Record development aid, 2008

Donor nations' wealth increased through the 1990s until 2008. Levels of aid should have increased too. However, during the 1990s levels of aid actually fell, although they rose again after 2000. Some of the increase in aid in recent years has been due to debt write-off. Aid for the poorest countries remained relatively steady during this period. Thus, in relative terms aid to the poorest countries has declined. In 2009, despite the global financial crisis, levels of aid increased slightly.

Remittances

Migrants help their homelands by remitting cash on a vast scale. The map of global remittances (Figure 14.6) shows that the region that receives the most in remittances is south Asia, in particular India, Pakistan and Bangladesh. In these countries the value of remittances (see also page 46) is greater than the amount of international aid that they receive. In contrast, most

To do:

Study Figure 14.2.

Comment on the donation of aid by countries (a) in absolute terms and (b) as a proportion of their GNI.

To do:

Study Figure 14.3, which shows Net ODA, Breakdown of bilateral ODA by income group and by region, Top ten recipients of bilateral ODA, and Bilateral ODA by sector.

a How did the amount of aid change between 2007 and 2008?

b Comment on the main recipients of aid.

c How does receipt of aid vary with (i) income group and (ii) geographic variations?

d What are the main uses to which this aid has been put?

To research

Visit the Development Assistance Committee (DAC, www.oecd.org/dac), which is the principal body through which the OECD deals with issues related to cooperation with developing countries. Find out about aid statistics and aid effectiveness.

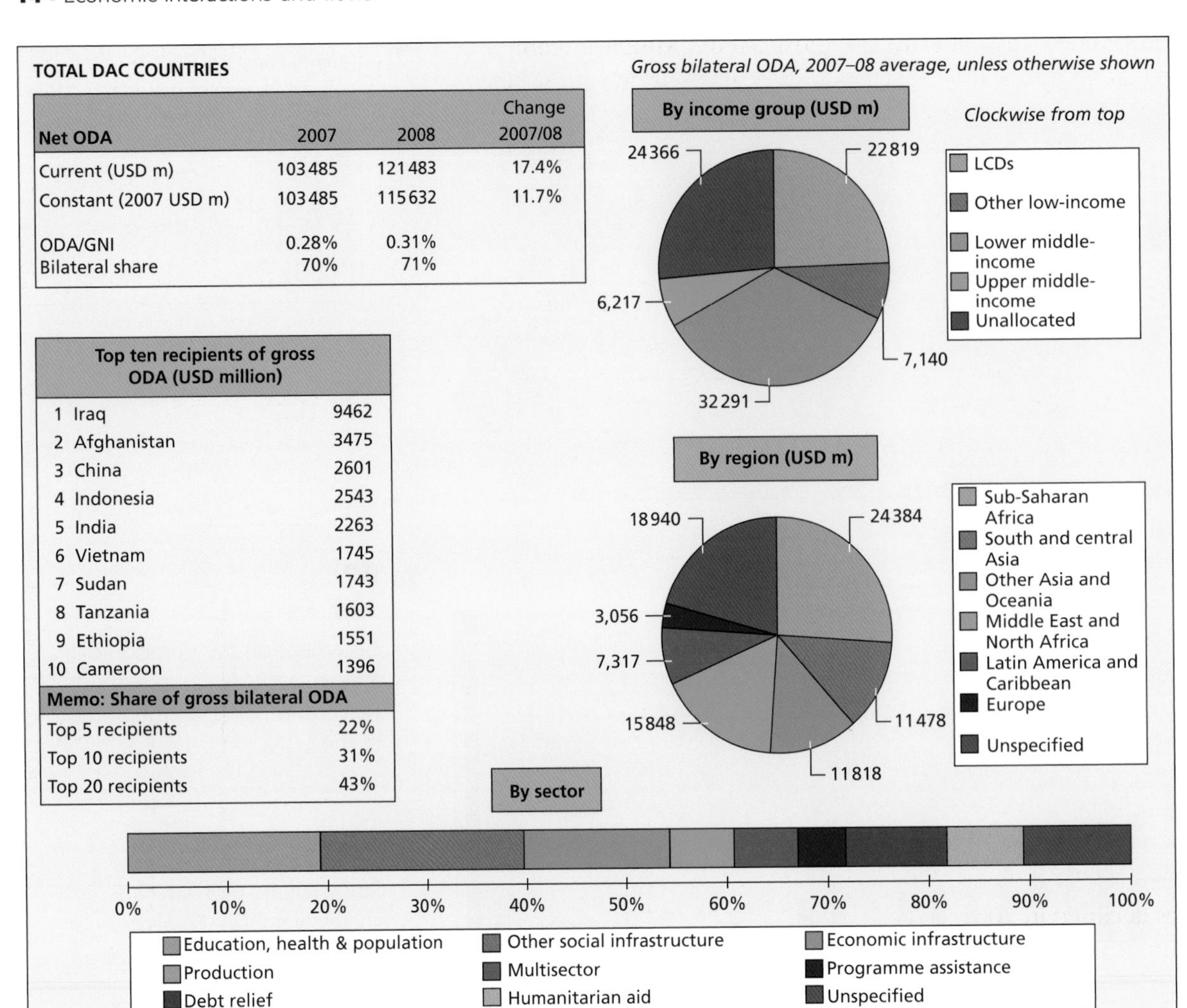

TOTAL DAC COUNTRIES

Net ODA	2007	2008	Change 2007/08
Current (USD m)	103485	121483	17.4%
Constant (2007 USD m)	103485	115632	11.7%
ODA/GNI	0.28%	0.31%	
Bilateral share	70%	71%	

Top ten recipients of gross ODA (USD million)	
1 Iraq	9462
2 Afghanistan	3475
3 China	2601
4 Indonesia	2543
5 India	2263
6 Vietnam	1745
7 Sudan	1743
8 Tanzania	1603
9 Ethiopia	1551
10 Cameroon	1396
Memo: Share of gross bilateral ODA	
Top 5 recipients	22%
Top 10 recipients	31%
Top 20 recipients	43%

Figure 14.3 Overseas development aid statistics

of Africa and the Caribbean receive a relatively small amount in remittances. Sub-Saharan Africa appears to be worst off. The pattern is different from the usual rich–poor divide in a number of ways. For example, the low value of remittances received in eastern Europe and in an arc of countries through Turkey to Kazakhstan makes this pattern unusual.

The importance of remittances

After the attacks of 9/11 in 2001, officials in America and elsewhere started tracking cross-border flows of money from migrants. A great deal of new data emerged on the economics of migration. Some 200 million people working abroad affect the lives of their compatriots at home. The impact, it turns out, is huge and benign. Migrants move more capital to poorer countries than do western aid efforts. The World Bank says that foreign workers sent $328 billion from richer to poorer countries in 2008, more than double the $120 billion in official aid flows from OECD members. India got $52 billion from its diaspora, more than it took in foreign direct investment.

To research

Visit http://www.oecd.org/countrylist/0,3349,en_2649_34447_25602317_1_1_1_1,00.html which shows recipient aid charts.

Click on, for example, Afghanistan.

- **a** Describe how Afghanistan's aid receipts changed between 2005 and 2007.
- **b** Identify the main donors to Afghanistan. Using a pie chart or bar chart, show the main donors (and their relative donations) to Afghanistan.
- **c** Outline the main uses of aid to Afghanistan (in terms of sectors the money was used for).
- **d** Investigate the receipt aid charts for contrasting countries of your choice.

Updates about aid are available at http://www.oecd.org/department/0,2688,en_2649_33721_1_1_1_1_1,00.html

However, not all the cash goes to the most needy. Middle-income places like Poland and Mexico are big recipients. Nor do remittances to poor countries always reach the poorest. Most migrants need some funds and education to get away, so their families are often slightly richer than average.

Even if some remittance money is "wasted" (in development terms) on consumer goods, the flow of cash boosts demand in recipient economies and supplies precious hard currency. In emergencies, relatives abroad can respond with material aid and cash. And some remittances are spent on developmentally useful things like education and health.

The UN's 2009 Human Development Report calls for the freer movement of labour. By crossing a border, most migrants find a richer, longer, healthier and better-educated life than they would otherwise have had. The report also makes the case that migrants send home useful values as well as cash.

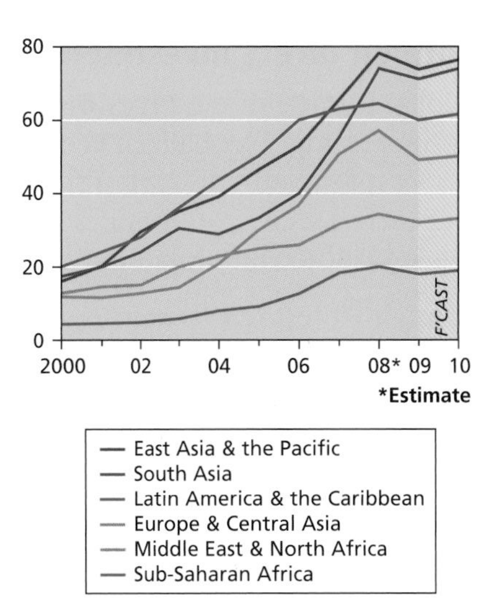

Figure 14.4 Remittances, inflows, $bn

About 3% of the world's population lives in a country other than the one in which they were born, and the World Bank says cash remittances may have peaked in 2008. In some regions, notably Latin America, remittances declined in 2009, although funds from migrants in the Gulf remained steady. An OECD study in 2009 found that migrants have suffered from the downturn in prosperous economies.

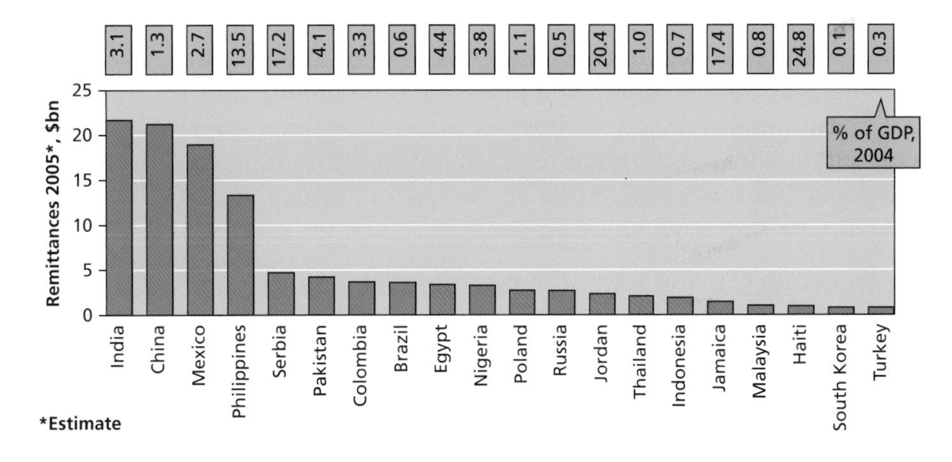

Figure 14.5 Remittances by country, 2005

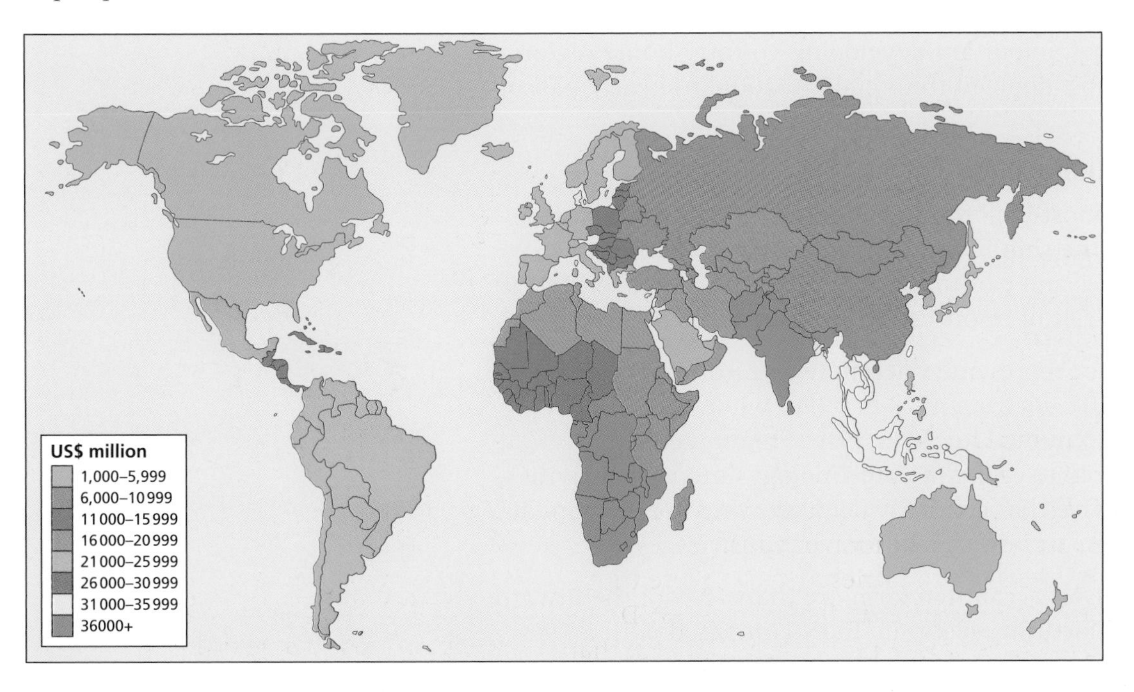

Figure 14.6 Worldwide remittance flow to developing countries, 2006

Foreign direct investment (FDI)

Global foreign direct investment (FDI) is an important feature of an increasingly globalized economic system. FDI flows have increased steadily overall since 1980, despite some declines in the early 1980s, 1990s and 2000s. In 2006, global FDI inflows were $1.306 trillion, close to the 2000 record of $1.411 trillion.

While FDI inflows to developed countries rose by 45% to reach $857 billion, flows to developing countries and the transition economies of southeastern Europe and the Commonwealth of Independent States (CIS) reached record levels of $379 billion and $69 billion respectively. The USA regained its position as the largest single host country, followed by the UK and France.

Among developing economies, apart from the traditionally largest recipients of Hong Kong (China) and Singapore, Turkey ranked fourth after a large FDI increase in 2006, while inflows doubled to $29 billion in the transition economy of the Russian Federation. The European Union (EU) was the largest host region, with $531 billion accounting for 41% of total FDI inflows in 2006, followed by North America with $244 billion. South, east and southeast Asia were the third-largest recipient region with $200 billion, accounting for 15% of total FDI inflows. Developed countries still remained the leading source of FDI, investing 84% of global outflows.

In past decades, world inward foreign direct investment grew more than 10-fold, from $1.2 trillion to $12 trillion in 2006. Developed countries hosted about three-fourths of world inward FDI, although the share of developing countries has increased. However, the least developed countries (LDCs) still remain marginal.

Outward FDI from developing countries accounted for 13% of the global total in 2006. South, east and southeast Asia constitute the most important developing-country home region, whose stock almost doubled from 2000 to 2006, to nearly half of the USA.

The industrial pattern of FDI

Since 1980, the most important change in the industrial pattern of FDI has been the shift towards services and away from primary industries and manufacturing. Recently, however, FDI in extractive industries of resource-rich countries has increased.

The services sector represented nearly two-thirds of the global FDI (61%) in 2005, up from 49% in 1990, while the share of manufacturing accounted for 30%. FDI in manufacturing is increasingly geared to capital and technology-intensive activities, while FDI in services has generally been growing in both capital-intensive and human resource-intensive industries.

Developed countries accounted for more than 70% of the inward FDI in the manufacturing sector in 2005, compared with a 79% share in the services sector. Inward FDI flows in the primary

To do:

- **a** State the value of remittances to India and China, as shown in Figure 14.5.
- **b** Identify the countries in which remittances account for (i) more than 20% of GDP and (ii) 15–20% of GDP.
- **c** Describe the trend and regional differences in remittances between 2000 and 2010, as shown in Figure 14.4.
- **d** Describe the pattern of remittances as shown in Figure 14.6.

Foreign direct investment – investment by a company into the structures, equipment or organizations of a foreign country. It does not include investment in shares of companies of other countries.

sector more than tripled between 1990 and 2005, but the larger share continues to originate from developed countries.

FDI driven by rising demand for various commodities, particularly oil and metal minerals, started to grow significantly in some regions in 2004, especially in mining and oil-related industries in Africa and Latin America.

Highest (% of world total)	Lowest (% of world total)
USA 16.75	Kenya 0.01
UK 7.54	Sri Lanka 0.02
China 5.79	Iran 0.02
France 5.22	Kuwait 0.03
Belgium 4.78	Cuba 0.04

Table 14.2 FDI inflows, 2007–11

To do:

a Study Figure 14.7, which shows FDI inflows and outflows.

i Comment on the FDI inflows by region.

ii Compare the inflows with the outflows (origins).

b Study Figure 14.8, which shows uses of FDI.

i Compare and contrast the main uses of FDI in (i) developing countries, (ii) economies in transition and (iii) developed countries.

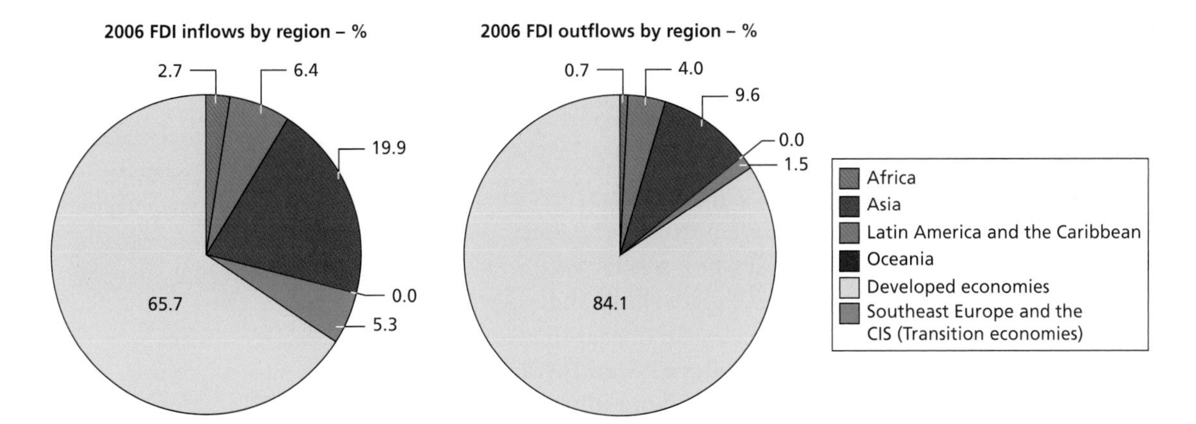

Figure 14.7 FDI inflows and outflows

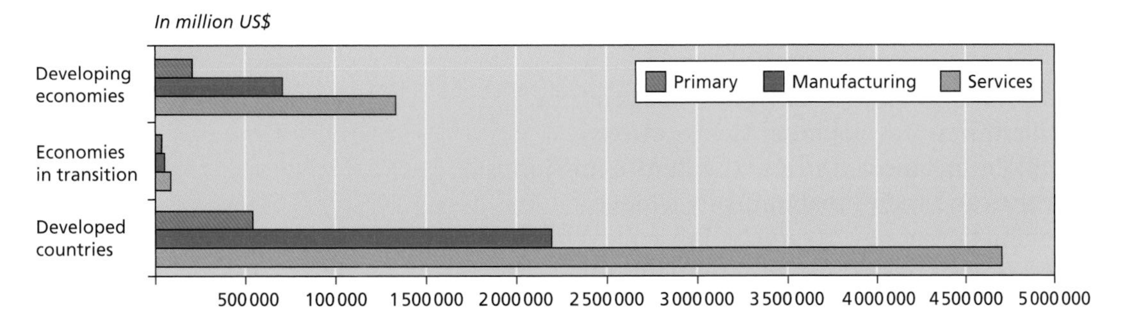

Figure 14.8 Uses of FDI, 2005

Repatriation of profits

Repatriation of profits is the movement of profits made in a business or investment in a foreign country back to the country of origin. Profits are normally repatriated to protect against expropriation, or to take advantage of currency fluctuation. Profit repatriation is an important factor that determines whether FDI in another country is actually profitable for the parent firm.

Profit repatriation – to return foreign-earned profits or financial assets back to the company's home country.

Profit repatriation legalities vary from country to country. For example, when the Volkswagen Group earns huge profits anywhere in the world, it takes a share back home to Germany, after converting it into euros. China is the global business hub today and is ranked very high on the FDI targets list. This is because China has slowly and successfully liberalized its economy and laws to suit FDI needs. It is possible legally to repatriate up to 90% of annual profits from China, provided certain norms are fulfilled, such as setting up local offices in China and creating a reserve account of at least 10% of total net profit.

Although India is attractive, it is lagging behind with economic liberalization and reforms. Nevertheless, it allows free repatriation of profits once all local and central tax liabilities are met.

To research

Explore the UNCTAD document Development and globalization facts and figures, 2008 at http://www.unctad.org/en/docs/gdscsir20071_en.pdf

UNCTAD's *World Investment Report* analyses current FDI trends and the activities of TNCs, and provides policy recommendations. The report is available at www.unctad.org/wir

Use it to find out about:

- trends in FDI
- regional trends
- the largest TNCs
- country reports for contrasting countires that interest you.

The role of governments, world trading organizations and financial institutions

The World Bank

The World Bank is not a bank in the usual sense, but comprises two institutions owned by 187 member countries, the International Bank for Reconstruction and Development (IBRD) and the International Development Association (IDA).

The World Bank – a source of financial and technical assistance to developing countries around the world. Its mission is to fight poverty by providing resources, sharing knowledge and building capacity.

The World Bank was established in 1944, and has its headquarters in Washington DC. It has more than 10 000 employees and more than 100 offices worldwide. Until 1967 the bank undertook a relatively low level of lending. From 1968 to 1980 the bank focused on the needs of people in the developing world, and greatly increased the size and number of its loans to borrowers. Bank policy shifted toward measures such as building schools and hospitals, improving literacy and agricultural reform.

The period 1980–89 was dominated by lending to service third-world debt. Structural adjustment policies (SAPs) aimed at streamlining the economies of developing nations were a large part of World Bank policy during this period.

Since 1989 its policy has changed, and its current focus is on the achievement of the Millennium Development Goals (MDGs), lending primarily to middle-income countries. The Bank's mission is to aid developing countries and their inhabitants to achieve development and the reduction of poverty, including achievement of the MDGs, by helping countries develop an environment for investment, jobs and sustainable growth, thus promoting economic growth through investment and enabling the poor to share the fruits of economic growth.

Criticisms of the World Bank

Many non-governmental organizations and academics say that the World Bank's free-market reform policies are harmful to economic development. In *Masters of Illusion: The World Bank and the Poverty of Nations* (1996), Catherine Caufield argued that western practices are adopted and traditional economic structures and values abandoned. A second assumption is that poor countries cannot modernize without money and advice from abroad.

To research

Visit www.worldbank.org and click on the Financial crisis button to find out more about the debt crisis and the bank's response.

Find out what have been the other main criticisms of the World Bank and the IMF.

Another criticism of the World Bank is the way in which it is governed. While the World Bank represents 187 countries, it is run by a small number of rich countries. In addition, the bank has dual roles that are contradictory: that of a political organization and that of a practical organization. As a political organization the bank must meet the demands of donor and borrowing governments. As a practical organization, it must be neutral, specializing in development aid, technical assistance and loans. Moreover, critics say that it focuses too much on the growth of GDP and not enough on living standards.

The International Monetary Fund

The IMF's stated objectives are to stabilize international exchange rates and facilitate development. Created in July 1944, the Fund started with 45 members but it now has 186. Member countries contribute to a pool from which other member countries with a payment imbalance may borrow on a temporary basis. The IMF was important when it was first created because it helped the world stabilize the economic system. The IMF's influence in the global economy steadily increased as it accumulated more members.

The International Monetary Fund (IMF) – the organization that oversees the global financial system by following the economic policies of its member countries, in particular those with an impact on the exchange rate and the balance of payments.

To deal with the 2008 global financial crisis, the IMF agreed to sell some of its gold reserves. In addition, in 2009 at the G20 London Summit it was decided that the IMF would require additional financial resources to meet the prospective needs of its member countries during the crisis. The G20 leaders pledged to increase the IMF's supplemental cash 10-fold to $500 billion, and to allocate to member countries another $250 billion via special drawing rights.

To research

Visit www.imf.org for more information about the IMF.

Member states with balance of payment problems may request loans to help fill gaps between what they earn and/or are able to borrow from other lenders and what they need to spend in order to operate. In return, these countries usually launch reforms such as structural adjustment programmes.

Criticisms of the IMF

Criticisms of the IMF include the following:

- IMF policymakers have supported military dictators friendly to American and European corporations, and appear unconcerned about democracy, human rights and labour rights.
- One of the main SAP conditions placed on troubled countries is that the governments sell up as much of their national assets as they can, normally to western corporations at a big discount.
- The IMF sometimes advocates "austerity programmes", increasing taxes even when an economy is weak, in order to generate government revenue.
- The IMF's response to a crisis is often delayed, and it tends only to react to them (or even create them) rather than prevent them.
- Historically the IMF's managing director has been European (and the president of the World Bank has been from the USA). The IMF is for the most part controlled by the major western nations.

The World Trade Organization

The World Trade Organization (WTO) deals with the rules of trade between nations at a global or near-global level. Based in Geneva, Switzerland, the WTO has 153 members, representing more than 97% of total world trade. The WTO is:

- an organization for liberalizing trade
- a forum for governments to negotiate trade agreements
- a place for settling trade disputes
- a system of trade rules.

The WTO began life on 1 January 1995, but its trading system is over 50 years older. Since 1948 the General Agreement on Tariffs and Trade (GATT) provided the rules for the system. The last and largest GATT round was the Uruguay Round (1986–94), which led to the WTO's creation.

Whereas GATT had mainly dealt with trade in goods, the WTO and its agreements now cover trade in services, and in traded inventions, creations and designs (intellectual property). The agreements provide the legal ground rules for international commerce, which aim to help trade flow as freely as possible. Nevertheless, trade relations often involve conflicting interests.

The WTO is currently hosting new negotiations under the Doha Development Agenda (or Doha Round), which was launched in 2001 to enhance equitable participation of poorer countries which represent a majority of the world's population. During the Doha Round, the US government blamed Brazil for being inflexible, and the EU for impeding agricultural imports. Brazil's president argued that progress would only be achieved if the richest countries (especially the US and countries in the EU) made deeper cuts in their agricultural subsidies, and opened their markets for agricultural products.

Figure 14.9 The Doha Round logo

The Doha Round was to be an ambitious effort to make globalization more inclusive and help the world's poor, particularly by slashing barriers and subsidies in farming. The negotiations have been highly contentious and by late 2010 agreement had not been reached.

Labour flows

Although capital is mobile and will move quickly, labour tends to be less mobile for cultural and linguistic reasons. Relative to capital, labour is subject to more political restrictions and barriers. Unlike unskilled labour, skilled labour earns higher economic returns where it is abundant. This explains the clustering of talented people in cities in wealthy countries.

Although international migration still captures the media's greatest attention, by far the largest flows of people are between places in the same country, from economically lagging to leading rural areas, for example from western Kenya to the Central Highlands and from Bihar in India to the Punjab. A large share of this migration is temporary.

When people move across national borders, they often do not travel far: most international migration takes place within world regional "neighbourhoods", particularly between developing countries. Movements of capital and labour are driven by the benefits of agglomeration. Economic migration can be a positive and selective process. Labour mobility, when driven by economic forces, leads to greater concentration of people and talent in places of choice, and adds more to agglomeration benefits in these places than to congestion costs.

International labour mobility, particularly unskilled labour, declined after the mass movements of the 19th century and only recently began to rise. Following a peak in the late 19th century, the mobility of labour across borders declined with the rise of economic barriers at the onset of the Great Depression and the Second World War. The history of international migration can be divided into four distinct periods: mercantile, industrial, **autarkic** and post-industrial.

During the mercantile period, from 1500 to 1800, most people moving around the world were Europeans. During the industrial period that followed, sometimes called the first period of economic globalization, an estimated 48 million emigrants, between 10 and 20% of the population, left Europe. A long period of autarky and economic nationalism began in 1910. Unprecedented restrictions were placed on trade, investment and immigration, stifling the international movement of capital and labour.

Post-industrial migration patterns

The post-industrial period of migration began in the 1960s, characterized by new forms, no longer dominated by flows out of Europe. Today, about 200 million people in all countries are foreign born, roughly 3% of the world population. Poor and middle-income countries now send the most emigrants, led by Bangladesh, China, Egypt, India and Mexico. Nevertheless, Italy, Germany and the UK still rank near the top, each accounting for 3–4 million emigrants.

The pattern of international migration is changing, from south–north to south–south. Although the top three receiving countries are the USA, Germany and France, and migration of labour from the low- and middle-income countries of the south to the wealthy countries of the north is still large, the Côte d'Ivoire, India, Iran, Jordan and Pakistan are now among the top 15 destinations. Whereas only 30% of immigrants to the United States, 20% to France and 10% to Germany come from countries with which they share a border, 81% of immigrants to Côte d'Ivoire, 99% to Iran and 93% to India are from neighbouring countries.

Cross-border migration within sub-regional neighbourhoods flows to countries that act as economic engines of growth in developing regions, e.g. Côte d'Ivoire in West Africa, South Africa in southern Africa, and Thailand in the Greater Mekong region of south Asia.

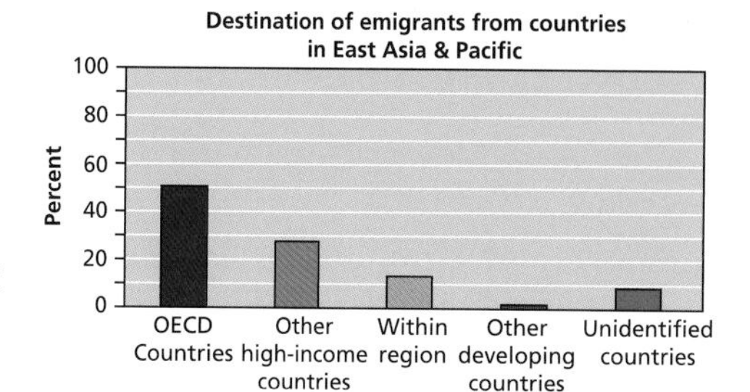

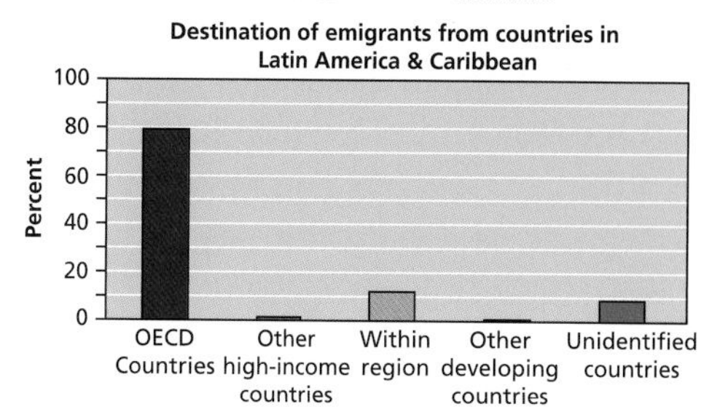

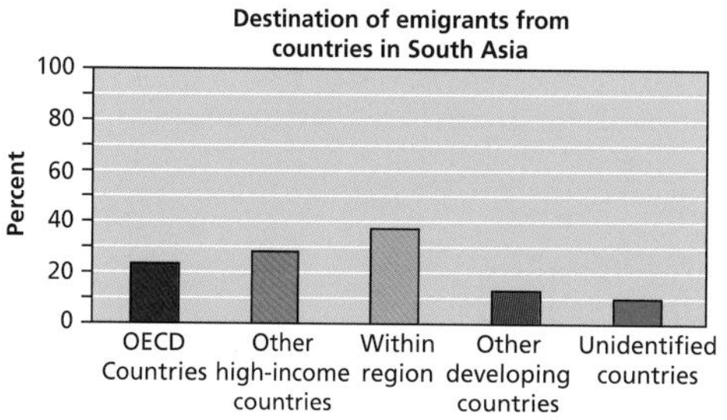

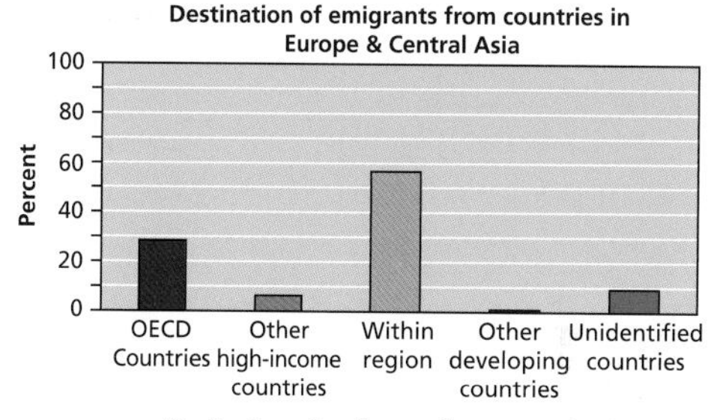

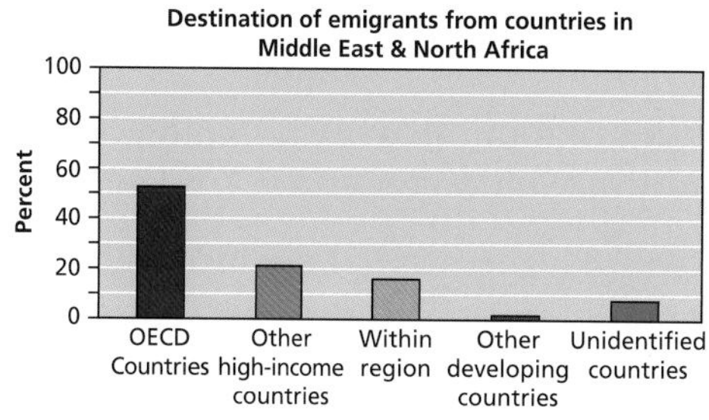

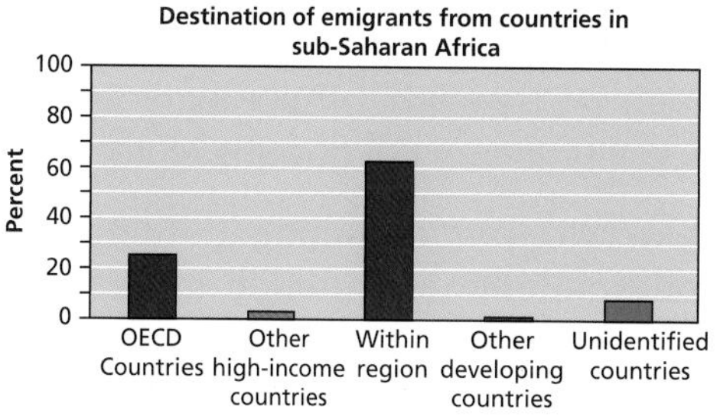

Figure 14.10 Destinations of migrants

Autarky – self-sufficiency.

Case study: Cross-border migration in the Greater Mekong region

The Greater Mekong region – with 315 million people – comprises Cambodia, Laos, Myanmar, Thailand, Vietnam and the Guangxi and Yunnan provinces of China. Thailand's higher wages, faster growth and more favourable social and political climate attract people trying to escape poverty. For Thailand, the migrants are a reservoir of cheap and flexible labour and a boost to its economic competitiveness. Thailand is estimated to have 1.5 million to 2 million migrants. Removing them could reduce Thailand's GDP by about 0.5%.

Migrants are disproportionately young, of working age, and male. Those from Myanmar are, on average, less educated and less literate than the average for the population of origin. In contrast, migrants from Cambodia are slightly more highly educated than the population back home.

Remittances from Thailand to Cambodia, Laos and Myanmar are estimated at $177–315 million a year. In Cambodia they are important for 91% of the households in some regions. Much of this migration is, however, irregular and unregulated, increasing the vulnerability of the migrants, most of whom do not use social services because they fear deportation. One of the biggest problems is ensuring access to schooling for children, who also suffer from a lack of healthcare. Similarly, migrant adults rarely receive health treatment, and migrant children rarely receive vaccinations.

Despite the benefits of labour mobility, facilitating legal flows of people has been slow. The absence of an adequate legal and policy framework, typical of regional neighbourhoods in developing countries, increases the costs (and risks) of migration and reduces its benefits.

To research

Visit http://blogs.worldbank.org/files/growth/image/Trends%20is%20female.JPG for a graph showing the female share of migration.

1 Identify the region in which the female share of international migration is (a) highest and (b) lowest.

2 Describe the trend of female migration in (a) Africa, (b) North America and (c) Asia.

3 Comment on the areas that have seen the largest changes in female migration between (a) 1995 and 2000 and (b) 2000 and 2005.

The role of ICT in outsourcing

According to the OECD's working party on the information economy (WPIE), rapid technological developments in ICT (information and communication technology) are having a profound impact on the way economic activity is organized in general, and in particular on how the ICT sector and ICT-related activities are organized.

Rapid advances in ICT are increasing the tradability of many business services and are also creating new tradable services. Continuing efforts to liberalize trade and investment in services are further enhancing the tradability of services. India in particular has received much attention in the context of the ICT-enabled outsourcing and "offshoring" of services. IT and ICT-enabled offshoring of services activities to China is a recent development in the ongoing globalization of services.

As services are becoming more tradable and increasingly independent of location, firms are starting to offshore certain business functions, such as administrative support units and research and consultancy services, to countries with relatively lower labour costs and a talented workforce, in order to focus on their core activities and increase their competitive advantage. Due to the wage-cost advantage and the large pool of English-speaking skilled labour, India has become a prime location for IT and ICT-enabled services offshoring in recent years.

To do:

Study Figure 14.10, which shows the destination of migrants from different parts of the world.

a State the percentage of migrants from (a) East Asia and the Pacific and (b) Latin America and the Caribbean that go to (i) OECD countries and (ii) stay within the region.

b Describe the pattern of destinations for migrants from (a) the Middle East and North Africa and (b) sub-Saharan Africa.

c Suggest reasons to explain the destinations of migrants from (a) Europe & Central Asia and (b) South Asia.

Most exports and imports (around 80%) of business services and computer and information services originate in OECD countries, and OECD countries account for the largest shares of exports and imports of these services. But other countries, especially China and India, are also accounting for a significant and increasing share.

Offshoring – relocating some part of a firm's activity to another country.
Value chain – a chain of activities used to create a product whereby each step in the chain gives the product an increased value.
Outsourcing – the process of subcontracting part of a firm's business to another company, in order to save money.

Share of employment potentially affected by outsourcing

Close to 20% of total employment could potentially be affected by ICT-enabled offshoring of services. This includes activities coming into a country as well as those leaving a country, and those generated domestically. Incoming offshored services activities would bring about an increase in the share of employment potentially affected by offshoring, whereas services activities that leave the country would bring about a relative decline in the share.

In 2001 offshoring potentially affected about 11% of total employment in the USA. The share of occupations potentially affected by offshoring in the EU15 increased from 17.1% in 1995 to 19.2% in 2003. For Canada it was around 19.5% until 2001, after which it declined to 18.6% in 2003. Canada has served as an offshoring location for the USA, but has become less important as other locations such as India have started to emerge.

Services now account for around two-thirds of output and foreign direct investment in most developed countries, and for up to 20–25% of total international trade. The importance of services in international trade remains comparatively modest because many services have only recently become tradable, and many others remain non-tradable.

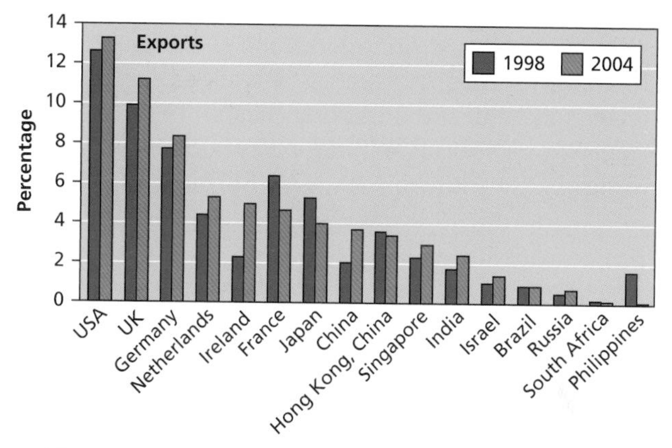

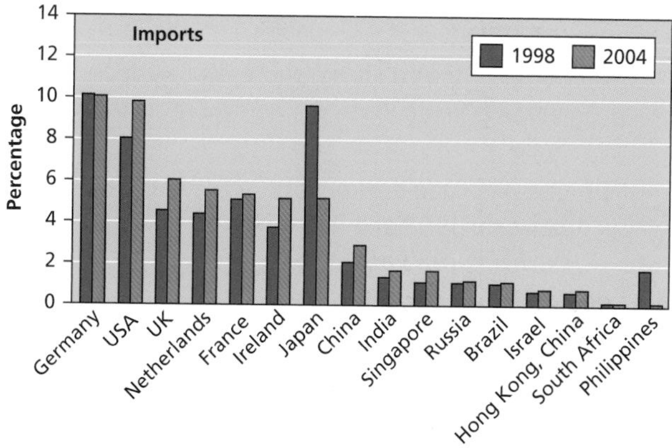

Figure 14.11 Share of total exports and imports of ICT-enabled services, selected countries, 1998 and 2004

China and outsourcing

For China, the main question is whether it will be able to evolve from a manufacturing powerhouse and the world's largest exporter of ICT goods, to a global services exporter. It is argued that this is unlikely to happen unless it improves the skills and quality of its graduates: despite a large labour pool, there may be a shortage of graduates suitable to work in globally engaged activities in IT and ICT-enabled services as they lack the relevant language, cultural and corporate culture skills.

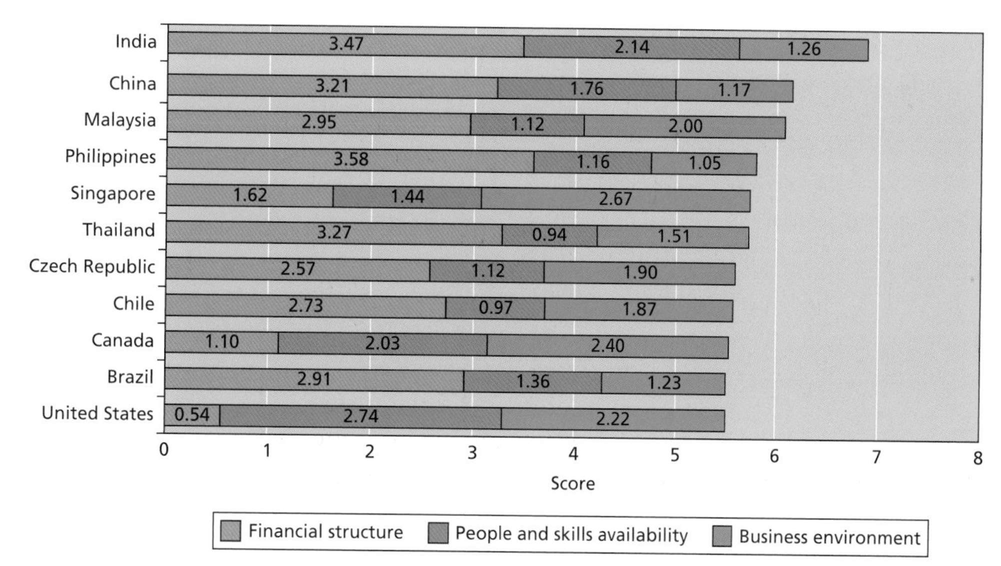

Figure 14.12 Kearney 2005 ranking for top services offshore locations

While outsourcing is increasing in China, Chinese manufacturing multinationals have started to offshore some of their activities abroad, to enter foreign markets and be closer to their customers. It is likely that these activities of Chinese firms abroad will continue to grow in the future.

Case study: *Neusoft*

Neusoft Group is China's largest IT software service provider. Founded in 1991 and based in Shenyang, Neusoft has 8,000 employees and provides software and services, medical systems and IT education and training. Its sales and service network covers over 40 cities in China, and it has branches in the USA and Japan. It has set up software parks in Shenyang, Dalian, Chengdu and Nanhai for R&D and HR development. Neusoft offers products and services to over 8,000 customers, including telecommunications, government (including social security and tax management systems), enterprise and e-commerce, communications, education and finance. It is estimated that Neusoft's international software outsourcing sales exceeded US$ 60 million in 2005.

(adapted from http://www.neusoft.com)

To do:

a Study Figure 14.11, showing export and import shares of ICT-enabled services for selected countries.

- **i** Identify the countries that are the largest (a) exporters and (b) importers of ICT-enabled services in (i) 1998 and (ii) 2004.
- **ii** Compare the countries whose exports decreased between 1998 and 2004 with those whose exports increased between 1998 and 2004.
- **iii** Compare, and comment on, the countries whose imports decreased between 1998 and 2004 with those whose imports increased between 1998 and 2004.

b Study Figure 14.12, showing the Kearney ranking for top services offshore locations.

- **i** Identify the country with:
  - the best service location
  - the best financial structure
  - the best people and skills availability.
- **ii** Compare the composition of the structure for the Philippines and the USA.
- **iii** Comment on the data shown in Figure 14.12.

15 Environmental change

The effects of agro-industrialization on the environment

The food market is truly global. Farming has become increasingly intensive, large scale and globalized in the drive for cheaper food. Advances in technology and communications have combined with falls in the costs of transport to transform the way in which food is sourced. The concentration of power in retailing and food processing has affected those farming on a smaller scale. Increasingly, modern farming methods are having a negative impact on the environment. The term "agro-industrialization" refers to the large-scale, intensive, high-input, high-output, commercial nature of much modern farming.

Since the 1950s, a revolution has taken place in the food industry. Every step in the process – how food is grown, harvested, processed, distributed, retailed and cooked – has changed. Until the Second World War, farmers were the major players in the food industry. After the war they were given grants and subsidies but these were merely to stop them going out of business. Many therefore intensified, increased efficiency and adopted labour-saving technologies such as agro-chemicals, machinery and high-yielding varieties (HYVs) of plants.

Agro-industrialization has increased food production but is a major consumer of energy and contributor to greenhouse gas emissions, air pollution, water pollution, land erosion and loss of biodiversity.

Improved yields and environmental impacts

Food processors usually want large quantities of uniform-quality produce or animals at specific times. This is ideally suited to intensive farming methods. In the last 50 years, wheat yields have increased from 2.6 to 8 tonnes per hectare, barley from 2.6 to 5.8 tonnes. In just two decades, new production methods have increased a dairy cow's average yield from 4,000 litres to 5,800 litres a year.

Unfortunately, intensive farming requires the heavy use of synthetic chemicals and methods that lead to land degradation and animal welfare problems. Air pollution and greenhouse gas emissions from farming cost more than £1.1 billion annually. About 10% of the UK's greenhouse gas emissions come from the methane from livestock digestion and manure and nitrous oxide from fertilized land. Animals are reared on production lines. The spread of disease is a problem. Cox's apples receive an average of 16 pesticide sprays. Lettuces imported to the UK from Spain, Turkey, Zimbabwe and Mexico are sprayed on average 11.7 times.

Figure 15.1 Farming in Barbados – the original rainforest is now found only in steep ravines

Cleaning up the chemical pollution and repairing the habitats caused by industrial farming costs up to £2.3 billion a year in the UK alone. It now costs water companies £135–200 million to remove pesticides and nitrates from drinking water. In the USA it is estimated that the costs of agriculture (pesticides, nutrient runoff, soil loss, etc.) could be as high as $16 billion

($96 per hectare) for arable farming and $714 million for livestock. Soil erosion alone (not all caused by farming) costs up to $6 billion in terms of sedimentation of dams, harbours and fish farms.

Gas	Contribution to climate change (%)	Livestock emissions (billion tonnes carbon dioxide equivalent)	Livestock emissions as % of total anthropogenic
Carbon dioxide	70	2.70	9
Methane	18	2.17	37
Nitrous oxide	9	2.19	64

Table 15.1 The role of livestock in GHG emissions

The global food industry has a massive impact on transport. Food distribution now accounts for between a third and 40% of all UK road freight. The food system has become almost completely dependent on crude oil. This means that food supplies are vulnerable, inefficient and unsustainable. It is estimated that a kilogram of blueberries exported to the UK by plane from New Zealand produces the same emissions as boiling a kettle 268 times.

Between 1978 and 1998, the distance food was transported increased by 50%. Transporting animals long distances to slaughter has made it almost impossible to contain outbreaks of serious diseases such as foot and mouth. Journeys of 300–600 kilometres to slaughter are not unusual for animals today, and the average journey to abattoir has been estimated at 160 kilometres. In order to be transported long distances food must be heavily processed, packaged, or chemically preserved.

To research

Visit http://www.wspa-usa.org and go to the "What we do" section and Resources, for farm animal welfare reports. Download the 2008 report "Eating our future – the environmental impact of industrialized animal agriculture".

Find out about the environmental impact of industrial animal agriculture.

Food miles – A measure of the distance food travels from its source to the consumer. This can be given either in units of actual distance or of energy consumed during transport.

Case study: *Water problems and global farming in Kenya*

The shores of Lake Naivasha in the "Happy Valley" area of Kenya are now blighted. Environmentalists blame the water problems on pollution from pesticides, excessive use of water on the farms, and deforestation caused by migrant workers in the growing shanty towns foraging for fuel.

British and European-owned flower companies grow vast quantities of flowers and vegetables for export, but the official Kenyan water authority, regional bodies, human rights and development groups as well as small-scale farmers have accused flower companies near Mount Kenya of "stealing" water which would normally fill the river. Kenya's second largest river, the Ngiro, is a life-sustaining resource for nomadic farmers, but it also sustains big business for flower farms supplying UK supermarkets. According to the head of the water authority, the 12 largest flower firms may be taking as much as 25% of water normally available to more than 100 000 small farmers.

The flower companies are thereby exporting Kenyan water – this is known as "virtual water". A flower is 90% water. Kenya is one of the driest countries in the world and is exporting water to some of the wettest. The flower companies, which employ 55 000 labourers, are in direct competition with the peasant farmers for water, and the biggest companies pay the same as the smallest peasant for water.

The greatest impact is being felt on the nomadic pastoralists in the semi-arid areas to the north and east of Mt Kenya. The flower farms have taken over land that the pastoralists used and there is now less water.

In 2010, as a result of the eruption of the volcano under the Eyjafjallsjokull in Iceland, flights from Kenya to Europe were shut down. Around a million kilograms of fresh produce is normally shipped out of Kenya every night, three-quarters of it to Europe and more than a third of this to Britain. With most European airports closed, an estimated $8 million-worth of flowers had to be destroyed. On average, it is estimated that the Kenyan horticultural industry lost an average of $3 million a day. Kenya Airways also lost about $1 million a day.

Meat	Processed food	Fruit	Vegetables	Common goods
Beef 1,857	Sausages 1,382	Figs 379	Avocados 154	Pair of jeans 2,900
Pork 756	Processed cheese 589	Plums 193	Corn 109	Hamburger 766
Chicken 469	Eggs 400	Bananas 103	Beans 43	Glass of milk 53
	Fresh cheese 371	Apples 84	Potatoes 31	Cup of coffee 37

Table 15.2 Gallons of water needed to produce 500 grams of various foods and selected items

Land-use change and loss of biodiversity

Farmland makes up the greatest use of land on earth. It is also increasing in cover. In Brazil, for example, the Amazon Basin, the *cerrado* (savanna woodland) and the Atlantic forest (with a high number of endemic species) are at risk of conversion to farmland. In 2004 soya production in areas of the Amazon and the Atlantic forest amounted to over 21 million tonnes. The expansion of cocoa in the 1970s, soya in the 1990s and biofuels in the 2000s has reduced the Atlantic forest to less than 10% of its original size. The *cerrado* is threatened by the expansion of cattle ranching, which is also closely related to soya production.

Figure 15.2 Food display – global food in a supermarket

Food miles and a Christmas dinner

The concept of food miles describes how far food has travelled before it appears on a plate. However, it is also important to consider how the food has been transported and even packaged – frozen, for example has higher energy costs. Critics also argue that transport cost is only part of the environmental impact of food production – there could be other costs such as reduction in biodiversity, eutrophication, decreased water quality and increased risk of flooding.

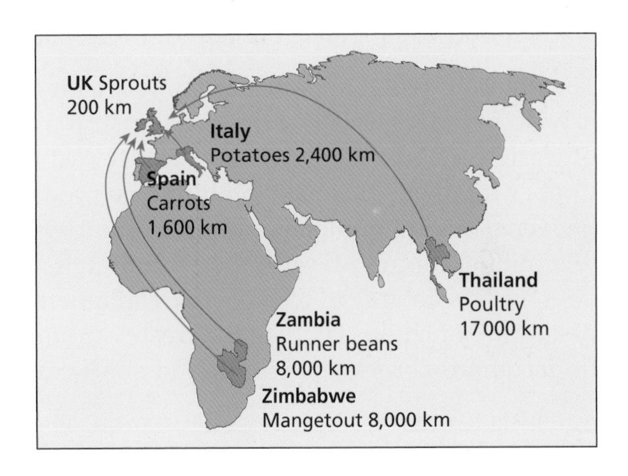

Figure 15.3 The wastefulness of a Christmas dinner

The ingredients of a traditional Christmas meal bought from a supermarket may have cumulatively travelled more than 24 000 miles or 37 000 kilometres, according to a report, Eating Oil. Buying the ingredients in a London supermarket, the report found that poultry could have been imported from Thailand and travelled nearly 17 000 kilometres, runner beans came from Zambia (nearly 8,000 kilometres), carrots from Spain (1,600 kilometres), mangetout from Zimbabwe (over 8,000 kilometres), potatoes from Italy (2,400 kilometres), and sprouts from Britain, where they were transported around the country before reaching the shop (200 kilometres) (Figure 15.4). By the time trucking to and from warehouses to stores was added, the total distance the food had moved was over 38 000 kilometres, or the equivalent of travelling around the world once. Transporting ingredients such great distances makes food supplies vulnerable.

Figure 15.4 Flowers grown for UK supermarkets

Be a critical thinker

Which is best for the environment – organic food that has been flown a long distance or locally produced food treated with fertilizers and pesticides?

It is usually thought that it is best to source food locally, because the amount of greenhouse gas (GHG) used to transport it is reduced. But the food may have other hidden sources of GHG, for instance if fertilizers and chemicals (pesticides, herbicides) have been used in its production. One might have one "good" environmental impact and one "negative" one. Weighing up their overall relative impacts might not be easy.

To do:

a Explain the meaning of the term "agro-industrialization".

b Outline the effects of agro-industrialization on the physical environment.

c Describe the movement of food for the Christmas dinner as shown in Figure 15.3.

d Explain the meaning of the term "food miles".

e How does agro-industrialization affect food miles?

Environmental degradation

The impact of mining

Raw materials that are mined are normally classified into four groups:

- metals, such as iron ore and copper
- industrial minerals, such as lime and soda ash
- construction materials, such as sand and gravel
- energy minerals, such as coal, oil and natural gas.

Construction minerals are the largest product of the mining industry, being found and extracted in almost every country.

The environmental impacts of mining are diverse. Habitat destruction is widespread, especially if opencast or strip mining is used. Disposal of waste rock and "tailings" may destroy vast expanses of ecosystems. Copper mining is especially polluting: to produce 9 million tonnes of copper (world production levels in the 1990s) creates about 990 million tonnes of waste rock. The Bingham copper mine in Utah, USA, covers an area of over 7 square kilometres. Even the production of a tonne of china clay (kaolin) creates a tonne of mica, 2 tonnes of undecomposed rock and 6 tonnes of quartz sand.

To research

Visit http://www.fallsbrookcentre.ca/cgi-bin/calculate.pl for a food miles calculator. You could work out the cost of getting food to you from (a) the USA, (b) India and (c) New Zealand.

Visit www.agriculturalproductsindia.com and click on the "Agro Scenario" tab for information on the agro-industry in India, and www.fao.org/DOCREP/005/Y4383E/y4383e0d.htm for a detailed account of the effects of agro-industrialization on the valleys of Chincha and Mantaro, Peru.

Smelting causes widespread deforestation. The Grande Carajas Project in Brazil removes up to 50 000 hectares of tropical forest each year. The Serra Pelada gold mine was briefly the world's largest alluvial gold deposit. Between 1980 and 1986 up to 100 000 miners worked at the mine, which became famous as a result of pictures by Sebastiao Salgado. The mine is now abandoned and filled with polluted water. High levels of mercury have been recorded in local watercourses and in the fish that live there.

There is widespread pollution from many forms of mining. The pollution results from the extraction, transport and processing of the raw material, and affects air, soil and water. Water is affected by heavy metal pollution, acid mine drainage, eutrophication and deoxygenation. Moreover, dust can be an important local problem. The use of mercury to separate fine gold particles from other minerals in riverbed sediments leads to contamination in many

Figure 15.5 Mining has a significant impact on the environment

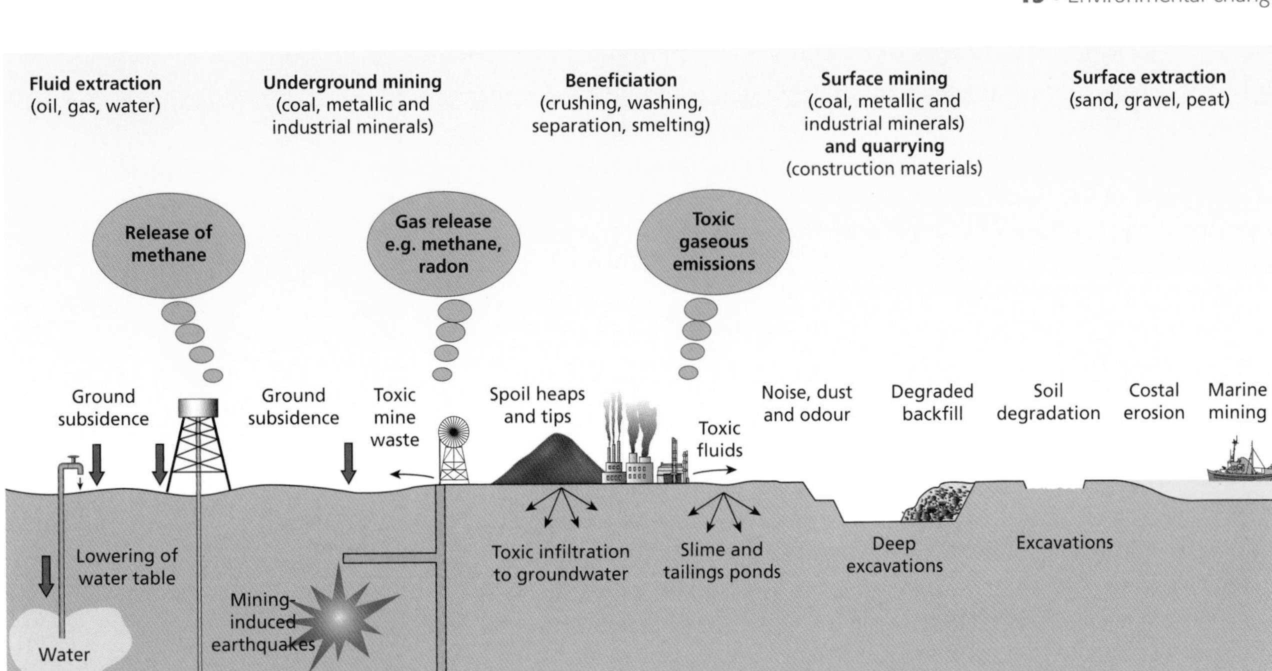

Figure 15.6 Methods of extracting geological resources and their potential impact on the environment

rivers. In Brazil, up to 100 tonnes of mercury have been introduced into rivers by gold prospectors. Mercury is highly toxic and accumulates in the higher levels of the food chain, and can enter the human food chain.

Derelict land that results from extraction produces landforms of various sizes, shapes and origin. A major subdivision is between excavations and heaps. The latter can be visually intrusive and have a large environmental impact. Heaps may be composed of blast furnace slag, fly-ash from power stations, or spoils of natural materials (overmatter), such as the white cones associated with china clay workings, oil shale wastes, and colliery spoil heaps.

Figure 15.7 Mining at Callow Rock Quarry

The impact of air travel

Transport as a whole produces about 25% of the world's CO_2 discharges. Within transport, aviation accounts for about 13%. Surface transport, by contrast, produces 22%. Carbon dioxide emissions from shipping are double those of aviation and increasing at an alarming rate.

Airlines' emissions are especially damaging because the nitrogen oxides from jet-engine exhausts help create ozone, a potent greenhouse gas, and partly because the trails that aircraft leave behind them help make the clouds that can intensify the greenhouse effect. The impact of airplane emissions (condensation trails or contrails) is enhanced because they are emitted directly into the upper atmosphere.

To do:

a Study Figure 15.6. Outline the potential impacts of the extraction of geological resources.

b Outline the impact of the increasing volume of air freight.

Contrails are believed to be responsible for about half of the aviation industry's impact on climate. A ban on night flights would significantly reduce the aviation industry's impact on the climate. The warming effect of aircraft is much greater when they fly in the dark, because of the effects of the condensation trails they leave. Aircraft contrails enhance the greenhouse effect because they trap heat in the same way as clouds.

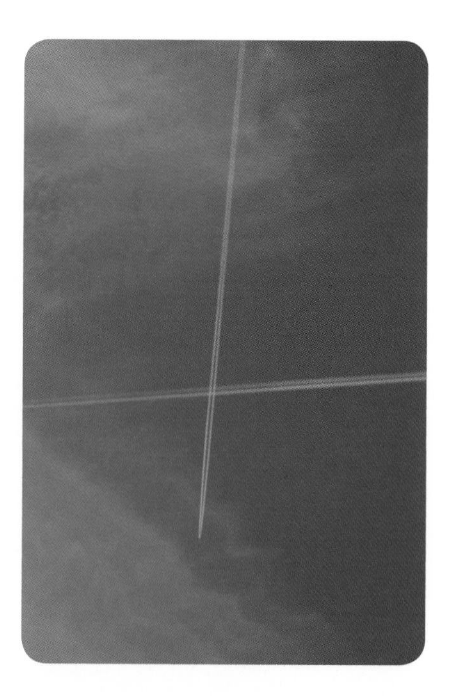

Figure 15.8 Contrails

Case study: Maquiladora *development in Mexico*

Mexico has attracted many US-owned companies to build low-cost assembly plants in places such as Cuidad Juarez, Nuevo Laredo and Tijuana. These factories, called *maquiladora* operations, are foreign owned but employ local labour. Since 1989, over 2,000 US firms have set up in Mexico's border cities. The main attractions are low labour costs, relaxed environmental legislation and good access to US markets.

Although firms are required by Mexican law to transport hazardous substances back to the USA, illegal dumping in Mexico is common. Air and water pollution are increasing as a result. Despite the environmental problems, many Mexicans are in favour of the *maquiladoras*, as they bring investment, money and jobs to northern Mexico. Over 500 000 people are employed in these factories.

Be a critical thinker

Is the relocation of polluting industries the main cause of pollution in Mexico?

A study that investigated the relationship between *maquiladoras*, air pollution and human health in Paso del Norte found that particulate emissions from *maquiladoras* undoubtedly have significant impacts on human health, in particular respiratory disease. However, it found that particulate emissions generate health damage of similar magnitudes regardless of the source, and *maquiladoras* are clearly not the region's leading sources of particulates. Unpaved roads, vehicles and brick kilns were the main sources of particulate emissions. Given that vehicles and brick kilns emit far more combustion-related fine particulates than *maquiladoras*, they inflict more health damage. The study found no evidence that health damage attributable to *maquiladoras* disproportionately affected the poor. However, brick kilns were far more likely in poor areas.

To research

Use Google Images to find images of the Serra Pelada gold mine. Comment on the images you can find.

Jared Diamond's book *Collapse: How Societies Choose to Fail or Succeed* (2005. Viking) has some excellent case studies on the environmental impacts of civilizations.

Polluting industries and relocation to LEDCs

Western nations have long been siting their polluting industries in developing countries, often with disastrous consequences. In 1984 the American-owned Union Carbide company released toxic gas from its pesticide plant in Bhopal, India, killing thousands. Developed countries have more robust green laws, greater social supervision and more effective governments; pollution emissions are higher in developing countries, where environmental regulations and their enforcement are weaker. These less regulated environments give richer nations a chance to export their waste and pollution. The environmental vulnerability of developing countries to pollution is as a result of their underdeveloped systems as well as their need for the economic benefits of the polluting industries.

The West exports high quantities of waste to LEDCs, and has done so since the 1960s. High fees associated with waste processing and pollution emissions make it uneconomical to process the waste locally. Exporting waste allows firms to earn money from governments in the developed world, cutting costs and avoiding local regulations, while the exporters earn income from selling the rubbish. At the same time, developing countries get a source of raw materials. China is the world's second largest consumer of plastic; one tonne of synthetic resin costs around US$1,420, but a tonne of imported plastic, discarded in the west, costs only US$515. The work of sorting the waste is hard and dirty, but for many it is more lucrative than alternatives such as farm labouring.

When profits are to be made, there will always be someone willing to risk others' health by importing trash, and many more who will endanger their own to sort it: it is simple economics. The low cost of waste processing and the large profits to be made in China make it a lucrative industry. Meanwhile, government oversight is weak and punishment is mainly in the form of fines that go directly to government rather than to the victims of pollution. As a result, companies and individuals involved can keep on polluting.

To do:

Suggest reasons why rich countries export waste materials and/or polluting industries to poorer countries.

Using an atlas, find out whereabouts in Mexico Tijuana, Cuidad Juarez and Nuevo Laredo are located. What can you conclude about the location of most *maquiladora* industries?

Case study: *The case of Trafigura*

Another example of the export of waste is that of the oil trader, Trafigura, who were accused of dumping hundreds of tonnes of poisonous waste in the Ivory Coast. In August 2006 the ship *Probo Koala* was due to be cleaned in Amsterdam, but workers refused on the grounds of health and safety. The vessel then sailed to the Ivory Coast where its load was unloaded at night on to 12 lorries. As much as 400 tonnes of waste was transported to 18 sites around Abidjan. The waste included highly poisonous hydrogen sulphide. Over a matter of weeks, thousands of residents in Abidjan became ill, and 10 are believed to have died. Up to 30 000 African residents took legal action against the Swiss-based firm Trafigura, and in 2009 the company paid £32 million compensation in an out-of-court settlement, and £100 million to help clean up the waste, without admitting liability. This dumping would have been illegal in Europe, and many people thought that it demonstrated the double standards of western countries with regard to the environment. The company has now been fined £840 000 by a Netherlands court.

To research

Visit www.trafigura.com/

Can you find any information on the *Probo Koala* incident?

Look up *Probo Koala* using a search engine such as Google. What does this tell you about the reliability of using the Internet as a source of information?

Transboundary pollution: acid rain

Acid rain – or, more precisely, acid deposition – is the increased acidity of rainfall and dry deposition, as a result of human activity. The term "acid rain" was introduced as long ago as the 1850s for rain with a pH of less than 5.65. Rainfall is naturally acid because it absorbs carbon dioxide in the atmosphere and becomes a weak carbonic acid, with a pH between 5 and 6. The pH of "acid rain" can be as low as 3.0.

Rain becomes more than usually acid because of air pollution. The major causes of this pollution are the sulphur dioxide and nitrogen oxides produced when fossil fuels such as coal, oil and gas are burned. Sulphur dioxide and nitrogen oxides are released into the atmosphere, where they can be absorbed by the moisture and become weak sulphuric and nitric acids. Most natural gas contains little or no sulphur and causes less pollution.

Coal-fired power stations are the major producers of **sulphur dioxide**, although all processes that burn coal and oil contribute. Vehicles, especially cars, are responsible for most of the nitrogen oxides in the atmosphere. Some come from the vehicle exhaust itself, but others form when the exhaust gases react with the air. Exhaust gases also react with strong sunlight to produce poisonous ozone gas, which damages plant growth and, in some cases, human health.

There is a saying that "one man's muck (waste) is another man's brass (source of wealth)". Is this a fair summary of the export of waste/polluting industries to poor countries?

pH scale – a measure of a substance's acidity or alkalinity. A pH of 7 is neutral, less than 7 is acidic and more than 7 is alkaline. The pH scale is **logarithmic**, so a decrease of one pH unit represents a 10-fold increase in acidity. Thus pH 4 is 10 times more acidic than pH 5.

Dry and wet acid deposition

Dry deposition typically occurs close to the source of emission and causes damage to buildings and structures. **Wet deposition**, by contrast, occurs when the acids are dissolved in precipitation, and may fall at great distances from the sources. Wet deposition has been called a 'trans-frontier' pollution, as it crosses international boundaries with disregard.

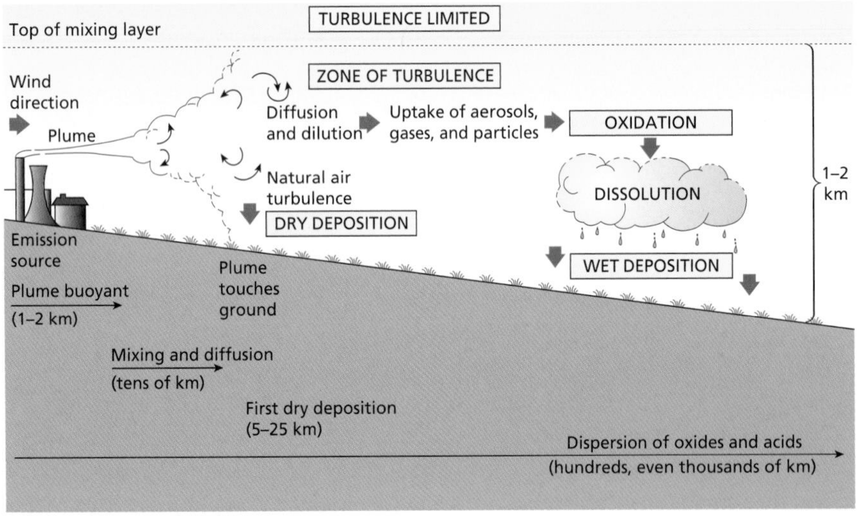

Figure 15.9 Dry and wet deposition

Snow and rain in the northeast USA have been known to have pH values as low as 2.1. In the eastern USA as a whole the average annual acidity values of precipitation tend to be around pH 4 (Figure 15.10). As a general rule, sulphur oxides have the greatest effect, and are responsible for about two-thirds of the problem. Nitric oxides account for most of the rest. However, in some regions, such as Japan and the west coast of the USA, the nitric acid contribution may be of relatively greater

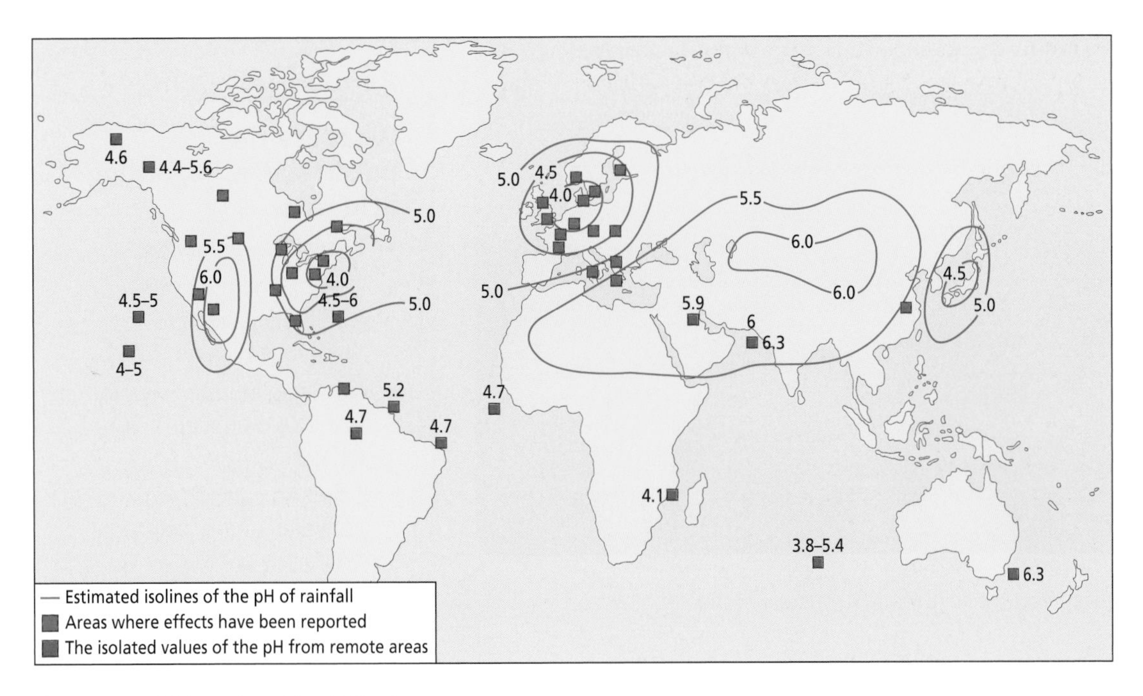

Figure 15.10 Map of world acidification

importance. Worldwide emissions of SO_2 are declining while those of NOx are increasing, partly as a result of increased car ownership.

Acidification has a number of effects:

- Buildings are weathered (Figure 15.11).
- Metals, especially iron and aluminium, are mobilized by acidic water, and flushed into streams and lakes.
- Aluminium damages fish gills.
- Forest growth is severely affected.
- Soil acidity increases.
- Lakes become acidic and aquatic life suffers.
- There are links (as yet unproven) with the rise of senile dementia.

Figure 15.11 The impact of acid rain on stonework

The effects of acid deposition are greatest in areas with high levels of precipitation (causing more acidity to be transferred to the ground) and those with base-poor (acidic) rocks which cannot neutralize the deposited acidity.

Reducing the impacts of acid deposition

Various methods are used to try to reduce the damaging effects of acid deposition. One of these is to add powdered limestone to lakes to increase their pH values. However, the only really effective and practical long-term treatment is to curb the emissions of the offending gases. This can be achieved in a variety of ways:

- by reducing the amount of fossil fuel combustion
- by using less sulphur-rich fossil fuel
- by using alternative energy sources that do not produce nitrate or sulphate gases (e.g. hydropower or nuclear power)
- by removing the pollutants before they reach the atmosphere.

Figure 15.12 A natural source of acidification: Soufrière volcano, Montserrat

Increasing environmental awareness

Civil society is composed of all the civic and social organizations or movements that form the basis of a functioning society. Many of these organizations are active in the field of environmental awareness and conservation. They include:

- local groups fronted by individuals (such as Tim Smit of the Eden Project, Cornwall, UK, and Gerald Durrell who founded the Durrell Trust for Conservation)
- international groups (such as Greenpeace and the World Wide Fund for Nature)
- public servants such as politicians and scientists (such as Al Gore, the former US vice-president and author of *An Inconvenient Truth*, and Diane Fossey, made famous by the film *Gorillas in the Mist*).

In most cases there is a conflict between the need for economic development and the need for environmental conservation or management. Diane Fossey argued for the protection of the mountain habitats in Rwanda and Burundi that were home to the great silverback gorillas. Population growth, civil conflict, and the illegal trade in forest products had led to a decline in forest cover and a reduction in the gorillas' habitat.

There is an urgent need for strategic thinking and planning, especially in some of the world's most valuable biomes, such as coral reefs. This needs to be done in a sustainable way, with the cooperation of the indigenous people.

Greenpeace

Greenpeace is an international environmental organization founded in Vancouver, Canada, in 1971. Its confrontational approach has secured it a high public profile, and helped develop strong support for the organization. It has tackled many issues such as waste disposal,

Be a critical thinker

What are the uncertainties regarding acid rain?

- Rainfall is naturally acidic.
- There are natural sources of acidification, such as volcanoes (Figure 15.12).
- No single industry/country is the sole emitter of SO_2/NOx.
- Cars with catalytic convertors have reduced emissions of NOx.
- Different types of coal have variable sulphur content.

To do:

a Identify the main gases responsible for acid deposition.

b Outline the difference between wet deposition and dry deposition.

c Outline the natural causes of acid deposition.

d Describe the main impacts of acidification.

e Suggest how it is possible to manage acidification.

Civic society – any organization or movement that works in the area between the household, the private sector and the state to negotiate matters of public concern. Civic societies include non-government organizations (NGOs), community groups, trade unions, academic institutions and faith-based organizations.

GREENPEACE

Figure 15.13 The Greenpeace logo

deforestation, nuclear power, harvesting of seal cubs, and industrial pollution. Greenpeace's goal is "to ensure the continuing ability of the earth to nurture life in all its diversity". It has a presence in over 40 countries.

Greenpeace's main interests at present include:

- stopping climate change (global warming)
- preserving the oceans (including stopping whaling and seabed trawling)
- saving ancient forests
- peace and nuclear disarmament
- promoting sustainable farming (and opposing genetic engineering)
- eliminating toxic chemicals, including from electronic (E-) waste.

Greenpeace has been variously criticized, by governments, industrial and political lobbyists and other environmental groups, for being too radical, too mainstream (or not radical enough), for allegedly using methods bordering on eco-terrorism, for causing environmental damage, and for valuing non-human causes over human causes.

The Worldwide Fund for Nature (WWF)

Outside English-speaking countries, WWF is known as the Worldwide Fund for Nature (WFN). Formerly the World Wildlife Fund, the WWF was initially concerned with the protection of endangered species, but now includes all aspects of nature conservation, including landscapes (the environments in which species live). It has over 5 million supporters globally, and is increasingly concerned with the fight against environmental destruction.

The WWF is interested in climate change and global warming; forests; freshwater ecosystems; marine ecosystems; species and biodiversity; sustainability; agriculture; toxins; macro-economic policies, and trade and investment. The WWF works in recognized geographic areas, such as continents and countries, but also in ecoregions, large-scale geographic regions under threat from development. Examples include the Alps ecoregion and the Mekong ecoregion.

The WWF works with governments, NGOs, local peoples and businesses to find ways to protect the earth.

To research

Visit http://www.greenpeace.org.uk/ to find out more about the work of Greenpeace in any selected country. Choose from the drop-down box to select the region of interest to you.

Figure 15.14 The WWF logo

To research

Visit http://www.panda.org/about_wwf/index.cfm to find out more about the work of WWF. Outline their current concerns.

To do:

For a civil society organization that you have studied, outline the way it has improved environmental management.

The homogenization of urban landscapes

The evolution of uniform urban landscapes is the result of a variety of factors:

- improvements in communications technology (television, Internet, etc.), so that people in cities around the world are aware of opportunities and trends in other cities
- the increase in international migration and the spread of ideas and cultures
- time–space convergence, which allows improved interactions between places in a decreasing amount of time
- the desire of global brands (TNCs) such as McDonalds, Coca Cola and Starbucks to reach new markets

- improvements in standards of living and aspirations to be part of a global network of urban centres
- globalization of economic activity, culture (art, media, sport and leisure activities) and political activity.

Many urban landscapes look very similar. Tall towers are a feature of many cities, such as Toronto, Kuala Lumpur, Beijing and New York. Industrial estates and science parks are increasingly globalized, as TNCs outsource their activities to access cheap labour, vital raw materials and potential markets. Much appears to have changed about the city since the mid-1970s, with cities having undergone dramatic transformations in their physical appearance, economy, social composition, governance, shape and size.

So are urban areas around the world converging in form? Are we seeing a globalized urban pattern, or do local and national characteristics still prevail? In Los Angeles, for example. there is a wide array of sites in compartmentalized parts of the inner city: Vietnamese shops and Hong Kong housing in Chinatown; a pseudo-Soho of artist's lofts and galleries; wholesale markets; urban homelessness in the Skid Row district; an enormous muraled barrio (shanty town) stretching eastwards toward east Los Angeles; and the intentionally gentrifying South Park redevelopment zone. Many large cities have their Chinatowns and other ethnic or racial areas. Individual cities are anything but homogeneous. The point is that cities are increasingly globalized, increasingly heterogeneous, and thus more similar now because they are all more diverse.

Seoul – homogenized city or independent trader?

Seoul is a good example of the debate on the homogenization of urban landscapes. On the one hand, it fits the theory of a homogenized landscape – there are global firms (such as McDonald's) in Seoul, just as there are Korean firms like Hyundai and Samsung located in other countries. The CBD is characterized by skyscrapers and international firms such as Barclays and Tesco.

Figure 15.15 Seoul

Figure 15.16 McDonald's sign from Tokyo

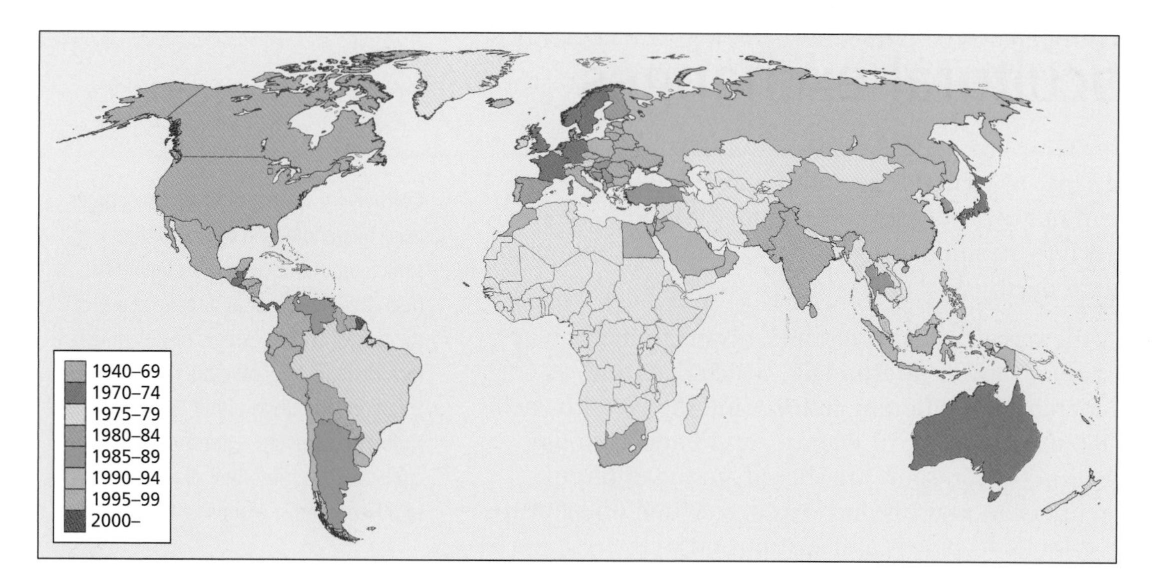

Figure 15.17 Date of opening of McDonald's worldwide

There are high-rise apartments and edge-of-town developments, and decentralization, such as at Gyeonggi-do and Pangyo on the south side of Seoul.

On the other hand, there has been a massive urban redevelopment centred on the restoration of the Cheong Gye Cheon River in downtown Seoul. This restoration has not just been of a river but also has historic, cultural and touristic-economic value. Murals along the side of the river recount some of the most important events to occur in Seoul over the last 600 years, and the river has become an important focus for Seoul residents and visitors – rather like Trafalgar Square in London – partly because it is stressing the individuality and uniqueness of Seoul, and of Korea.

To do:

Explain the evolution of uniform urban landscapes.

To what extent are urban landscapes uniform?

To research

Visit http://news.bbc.co.uk/1/hi/magazine/4602953.stm for an article on clone towns – the extent to which high streets in the UK are all the same – and http://www.docstoc.com/docs/27528924/Clone-Town-Britain for guidance on how to do fieldwork on clone towns.

To what extent are the towns and cities near where you live clone towns?

Be a critical thinker

What type of urban area is unique?

All urban areas have something in common. All have something unique. Urban areas that are less westernized or less globalized might be expected to be more different from one another than those that are more globalized. Perhaps Arab cities are different in structure and environment from western cities? Perhaps most large western cities are becoming more Arabic? Western and Arabic cultures may be merging in some urban areas, or co-existing peacefully. For example, the urban landscape of Bandar Seri Begawan (Brunei) is dominated by a mosque. Equally unusual is that a high proportion of people live in Bandar's water village, yet there is evidence of western culture even there.

16 Sociocultural exchanges

Culture gives us a sense of "who we are" and where we belong. It gives us a sense of our own identity and identity with others. Culture is a process rather than a thing, and is constantly shifting and changing rather than fixed.

Culture – a system of shared meanings used by people who belong to the same community, group or nation to help them interpret and make sense of the world. These systems of meanings include language, religion, custom and tradition, and ideas about "place".
Cultural diffusion – the spread of cultural ideas from their place of origin to other regions, groups or nations.

Cultural diffusion, the spread of cultural traits, occurs in many ways. It may occur when two cultures intermingle, which occurred historically when members of different cultures interacted with each other through trade, intermarriage or warfare, and happens today when, for example, interests in sport are shared among different countries. Cultural diffusion may also be forced, as when one culture defeats another and forces its beliefs and customs on to the conquered people. An example of this is cultural imperialism (see page 374), when the culture of one nation is promoted in another. It reflects an assumption that the culture being imposed is somehow superior to the one being supplanted.

Cultural diffusion today is taking hold around the world as cultural ideas spread through communications technology and the mass media. Globalization is regarded as a key process in driving culture towards a global model. Media TNCs and the movement of workers and tourists aid this process.

An emerging global culture

It is commonly accepted that the world is changing fast, and the rate of this change is probably greater than ever before. New technologies such as the Internet and satellite communications mean that the world is becoming more global and more interconnected. The increased speed of transport and communications, the increasing intersections between economies and cultures, the growth of international migration and the power of global financial markets are among the factors that have changed everyday lives in recent decades.

Proponents of the idea of an emerging global culture suggest that different places and cultural practices around the world are converging and becoming ever more similar. A global culture might be the product of two very different processes:

- the export of supposedly "superior" cultural traits and products from advanced countries, and their worldwide adoption ("westernization", "Americanization", "modernization")
- the mixing, or hybridization, of cultures through greater interconnections and time–space compression (the shrinking of the world through transport links and technological innovation), leading to a new universal cultural practice.

Figure 16.1 Cultural diversity in London

Language

There are a number of languages that have over 100 million native speakers. These include English, Mandarin, Spanish, Portuguese, Hindi, Arabic, Russian and Bengali. English has become one of the dominant world languages, but it has major variations in vocabulary and accents from country to country and also from region to region. The United Nations has six official languages: Arabic, Chinese, English, French, Russian and Spanish. Language evolves over time, and new words come into use just as others disappear. Some languages such as Inuktan, the language of the Inughuit people of NW Greenland, is likely to disappear by about 2020–25, while others, such as Irish, have undergone a major revival since the 1970s when interest in Irish culture and identity increased. New variations, such as SMS language, have also developed.

Religion

According to a 2005 survey in the *Encyclopaedia Britannica*, there are five major global religions: Christianity, Islam, Hinduism, Chinese folk religion and Buddhism. While Christianity and Islam can claim to be truly global, the remaining three are more regional in their distribution.

Christianity and Islam have both used political networks and military strength to expand their spheres of influence. The spread of Christianity across the globe can be linked to the expansion of European colonies throughout Africa, Asia and the America. Christianity gave a shared culture to areas linked by colonial powers, so that economic and political integration (domination) coincided with cultural integration (domination). In the case of the British Empire, improvements in communications, such as the telegraph with its submarine cables that connected the colonies to the Mother Country, allowed more effective links than had been previously possible, and enabled the transmission of cultural information. An education system based on British curricula and textbooks further helped the British culture into its colonies.

As some religions expand, others fade over time. Zoroastrianism – the philosophy and religion based on the teachings of the prophet Zoroaster – was once the dominant religion in Persia. It may have been the world's first monotheistic (one God) religion, and many Jewish, Christian and Muslim beliefs are said to derive from Zoaoastrianism. There are now probably fewer than 200 000 followers of Zoroastrianism, mostly in parts of India.

In other areas, notably China, there are many folk religions that draw on aspects of mythology. There are believed to be nearly 400 million followers of Chinese folk religion, which is a combination of religious practices, worship of ancestors, Taoism and Buddhism. There are also remnants of Neolithic religions such as the worship of the sun, moon, earth and stars.

Music

Music lends itself to globalization because it is one of the few popular modes of cultural expression that is not dependent on written or

> ***Be a critical thinker***
>
> **What is 'culture'?**
>
> Culture is an increasingly important concept within geography. It is a complicated concept with a range of meanings, and it is important to all human populations. Culture varies from region to region, with some areas being relatively similar, and others offering greater diversity. Cities are often culturally diverse, which is reflected in the population, services and built environment. In spite of cultural diffusion, localized cultures survive and new cultures can still be generated.
>
> Culture is the way of life of a particular society or group of people. Among other factors, it includes beliefs, behaviours, customs, traditions, rituals, dress, language, art, music, sport and literature.

spoken language for its primary impact. The production, distribution and consumption of music have a particular geography. The global music industry is dominated by transnational corporations, with the USA and the UK dominating domestically generated popular music. "World music" is now a significant component of the marketing strategies of these corporations, and exposes global audiences to local musical traditions from around the world. Migrations of people have also had cultural impacts on music, evidenced in increasingly "hybridized" forms.

Television

Until recently, television programmes tended to be produced primarily for domestic audiences within national boundaries, and could be subjected to rigorous governmental control. However, with the advent of cable, satellite and digital technologies, in addition to political and legal deregulation in many states, several television channels are now globally disseminated, and to some extent circumvent national restrictions. The USA, France, Germany and the UK are the major exporters of television programmes, but Brazil, Mexico, Egypt, Hong Kong, and Spain are increasing their output.

Sport

Sports are forms of cultural expression that are becoming increasingly globalized, as well as increasingly commodified. Football/soccer is the most obvious example, but similar trends can be observed in US major league baseball. The New York Yankees are a global icon; many major league players hail from countries such as Cuba, the Dominican Republic, Puerto Rico and Costa Rica; the sport is becoming increasingly globalized through television coverage and its inclusion as an Olympic sport.

Tourism

Tourism is one of the most obvious forms of globalization. Once again, the geography of tourism is skewed, since it is dominated by people of all classes from rich countries. It can also be exploitative, particularly through the growth of international sex tourism and the dependency of some poor countries on the exploitation of women. However, it is a form of international cultural exchange that allows large numbers of people to experience other cultures and places. It also locks specific destinations into wider international cultural patterns.

Consumer culture

Consumer culture is a culture permeated by consumerism, or the buying and spending of consumers. It is a social and economic order that is driven by money and therefore closely tied to capitalism, and it focuses on the systematic creation and fostering of a desire to purchase goods or services in ever greater amounts. Consumerism has been a feature of the richer countries since mass production made goods available in huge quantities and at reasonable prices, but has been more fully realized through the process of globalization.

To do:

a Explain the meaning of the term "culture".

b Examine the ways in which the international movement of workers, migrants and commodities can lead to cultural diffusion.

To research

Visit http://geography.about.com/od/culturalgeography/a/culturehearths.htm, for a discussion of culture hearths and cultural diffusion (and some useful links to other sites). Identify the seven original cultural hearths. How do they compare with modern cultural healths?

Consumerism – the behaviours, attitudes, and values associated with the consumption of material goods.

Case study: *IKEA – the spatial and temporal spread of a TNC*

Figure 16.2 An IKEA store in Reykjavik

IKEA International is one of the world's biggest retailers of furniture, home furnishings and housewares. The company designs its own items, and sells them in more than 313 IKEA stores in 37 countries. The company also sells its merchandise through mail order and the Internet. IKEA aims to offer high-quality items at low prices. To save money for itself and its customers, the company buys items in bulk, ships and stores items unassembled using flat packaging, and has customers assemble many items on their own at home.

Figure 16.3 The IKEA logo

IKEA was founded in 1943 in Sweden by Ingvar Kamprad. In 1947, it issued its first mail-order catalogue. In 1953 Kamprad bought a small furniture factory and opened a furniture and home-furnishing showroom in Älmhult. In 1963 the first IKEA store outside Sweden opened, near Oslo in Norway. In 1965 Kamprad opened a store on a greenfield site just outside Stockholm. It is still the largest of all the stores, with some 33 000 square metres of total space and 15 000 square metres of selling space. It sold most of the furniture in flat-pack form, and most of the components of each piece of furniture could be put together by the customers. Another innovative tool was a self-service method of selling, whereby customers walked around the whole store and selected items by themselves.

The IKEA formula was a success, and more IKEA stores were opened in Sweden. In 1969 a store was opened in Denmark. The first IKEA outside Scandinavia was opened in Switzerland in 1973, and in Germany in 1974. Soon there were 10 IKEA stores in five European countries. In 1975 the first IKEA in Australia opened; in 1976 a store opened in Canada; and in 1978 a store was placed in Singapore. Worldwide growth occurred in the 1980s and 1990s. IKEA opened stores in the Canary Islands in 1980, in France and in Iceland in 1981, in Saudi Arabia in 1983. The first store in the USA was in Philadelphia in 1985.

The rapid expansion in a decade and a half changed the pattern of sales. Whereas in 1975, the Scandinavian markets represented around 85% of the company's total sales, by 1990 this proportion had dropped to just over 26%. Sales in Germany had risen to more than 27% of the company's total; the rest of Europe contributed 34%, and stores in other regions accounted for just over 12% of the total.

In 1997, IKEA introduced its website. By the end of the 20th century there were few causes for concern about the company's future, although some feared that saturation point in the number of stores had already been reached in some countries – for example Sweden, Germany, Belgium, and the Netherlands.

To research

1 Visit http://franchisor.ikea.com/worldmap/interactive.html for an interactive world map (using Google Earth) of all of IKEA's stores.

- **a** Describe the distribution of IKEA stores worldwide.

2 Read the IKEA timeline at http://franchisor.ikea.com/showContent.asp?swfId=concept4

- **a** Make a summary timeline describing key moments in the globalization of IKEA.

3 Study the data on sales and number of stores for IKEA, at http://franchisor.ikea.com/showContent.asp?swfId=facts1

- **a** Describe the changes in (i) turnover and (ii) number of visits to IKEA between 1954 and 2009.

4 The data table overleaf shows the date when the first IKEA store was opened in each country where IKEA operates.

- **a** Draw a map to show the decade in which stores were opened (i.e. the 1950s, 1960s, 1970s, 1980s, 1990s and 2000s).
- **b** Describe the spatial (areal) and temporal

(time) changes in the distribution of IKEA stores.

c Suggest reasons why there are no IKEA stores in (i) Africa and (ii) South America.

d Suggest reasons for the high density of IKEA stores in Europe.

e Comment on the type of urban location that these first IKEA stores are located in, in most of the countries listed below. Suggest reasons for their choice of city.

1958 Sweden – Älmhult
1963 Norway – Oslo (Nesbru)
1969 Denmark – Copenhagen (Ballerup)
1973 Switzerland – Zürich (Spreitenbach)
1974 Germany – Münich (Eching)
1975 Australia – Artamon
1975 Hong Kong – Hong Kong (Tsim Sha Tsui)
1976 Canada – Vancouver (Richmond)
1977 Austria – Vienna (Vösendorf)
1978 Netherlands – Rotterdam (Sliedrecht)
1978 Singapore – Singapore
1980 Spain – Gran Canaria (Las Palmas)
1981 Iceland – Reykjavik
1981 France – Paris (Bobigny)
1983 Saudi Arabia – Jeddah
1984 Belgium – Brussels (Zaventem and Ternat)
1984 Kuwait – Kuwait City
1985 United States – Philadelphia
1987 United Kingdom – Manchester (Warrington)
1989 Italy – Milan (Cinisello Balsamo)
1990 Hungary – Budapest
1991 Poland – Platan
1991 Czech Republic – Prague (Zlicin)
1991 United Arab Emirates – Dubai
1992 Slovakia – Bratislava
1994 Taiwan – Taipei
1996 Finland – Esbo
1996 Malaysia – Kuala Lumpur
1998 China – Shanghai
2000 Russia – Moscow (Chimki)
2001 Israel – Netanya
2001 Greece – Thessaloniki
2004 Portugal – Lisbon
2005 Turkey – Istanbul
2006 Japan – Tokyo (Funabashi)
2007 Romania – Bucharest
2007 Cyprus – Nicosia
2009 Ireland – Dublin

Visit IKEA at www.ikea.com for a list of all stores and information about new stores in 2009.

Sociocultural integration

The scattering of a population, or diaspora, originally referred to the dispersal of the Jewish population from Palestine in AD 70. It is now used to refer to any dispersal of a populations formerly concentrated in one place. Examples include:

- the forced resettlement of Africans in the slave trade
- imperial diasporas during colonial expansion, e.g. Indian labourers who migrated to the sugar plantations of South Africa
- professional and business diasporas, such as the movement of Indians and Japanese overseas today
- cultural diasporas such as the movement of migrants of African descent from the Caribbean.

Diaspora – the forced or voluntary dispersal of any population sharing common racial, ethinic or cultural identity, after leaving their settled territory and migrating to new areas.

The Chinese diaspora

Approximately 40 million people of Chinese origin live in sizeable numbers in at least 20 countries. Large concentrations are in

Singapore (2.6 million), Indonesia (7.6 million), Malaysia (6.2 million), Thailand (7 million) and the USA (3.4 million). Historically, Chinese migration began in the 10th century with the expansion of maritime trade. During periods of colonialism large numbers of Chinese moved into Singapore and Mauritius – the latter encouraged by the French. With globalization, Chinese migration for professional and business reasons has increased.

The Chinese diaspora has significant economic power, through remittances, as well as financial power invested in the host countries. The growth and development of Chinatowns throughout the world illustrate the assimilation of the Chinese into the societies in which they exist. Chinatowns are an important identification of culture, place and identity.

The Irish diaspora

The Irish diaspora consists of Irish migrants and their descendants in countries such as the USA, the UK, Australia, New Zealand, Canada and those of continental Europe. The diaspora contains over 80 million people, more than 14 times the population of Ireland.

Emigration has been a constant theme in the development of the Irish nation, and has touched the lives of people in every part of Ireland. The economic and social prosperity of the country has been affected positively, through monies sent home from abroad, and negatively, through the loss of so many talented young Irish people. Irish emigrants have also had an enormous impact on the development of the countries in which they settled.

The USA was the most popular destination in the 19th century, and Irish migration there reached a peak of 1.8 million in 1891. By 1951 the number of Irish in Britain overtook the US figure and by 1981 there were four times as many there as in the USA. Since 1981 there have been major fluctuations in the figures. There was an increase in Irish migrations to the US during the 1980s, a drop in the numbers going to Britain and a rise in numbers going to other EU countries. There were also high rates of return and an overall fall in absolute numbers of emigrants.

With improvements in Ireland's economic situation and a fall in Irish birth rates (since the 1980s), the period of high emigration is fast becoming part of Irish history. Nowadays, fewer than 18 000 Irish people leave each year, and many of these will eventually return to Ireland.

The Irish in Britain

- Of all Irish-born people living abroad, 75% are in Britain.
- About 1.7 million people in Britain were born to Irish parents.
- The third-generation Irish community in Britain could number as many as 6 million people.

The Irish in the USA

- Of the total US population, 10.8% claim Irish ancestry – the equivalent of seven times the population of Ireland itself.
- In 2000, Irish-born people in the US numbered over 270 000.

To research

Visit www.newyorkirishcenter.org for the New York Irish centre. The newsletter is a good source of information.

What aspects of Irish culture does the New York Irish Centre offer and support?

- States with the largest Irish-American populations are California, New York, Pennsylvania, Florida and Illinois.
- In 2009, Irish-Americans were the largest ancestral group in Maryland, Delaware, Massachusetts and New Hampshire.
- Irish Americans number over 44 million, making them the second largest ethnic group in the country, after German Americans.

To do:

With the use of examples, explain what is meant by the term "diaspora".

Cultural diffusion and indigenous groups

Cultural change in Tibet

Tibet's population believes that the Chinese invasion of their country has imposed colonial rule and eroded part of their culture. They desire the right to political self-determination and an end to over 50 years of Chinese rule.

Tibet is one of the most remote and isolated parts of the world. It is often described as "the roof of the world", being located high up in the Himalayas. For much of the early 20th century (1911–49) it functioned like an independent country. However, following the 1949 Chinese Revolution and the creation of the People's Republic of China, Tibet's independence largely vanished. China attacked Tibet in 1950, and in 1951 Tibet signed an agreement giving China control over Tibet's external relations, the establishment of the Chinese military in Tibet, in return for guaranteeing Tibet's political system.

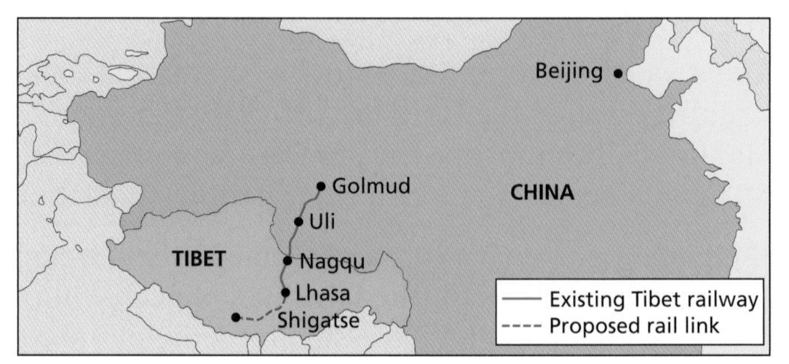

Figure 16.4 The route of the Chinese railway into Tibet

During an uprising in 1959, Tibet's Buddhist spiritual leader, the Dalai Lama, fled and sought refuge in northern India. About 80 000 Tibetans followed. In 1965 the Chinese government created the Tibetan Autonomous Region (TAR).

Tibetan culture is changing. The Chinese government has encouraged migration of the ethnic Han Chinese population into the region. The building of the China–Tibet railway, finished in 2005, has speeded up the migration of Han people. The railway was almost exclusively built using Han labour, and many see it as a political tool, not only allowing for increased migration of the Han into Tibet, but also allowing China to increase its military presence in the area. The railway lets China exploit Tibet's resources and move them to China's urban-industrial complex. Many Tibetans also claim that the ethnic Han are given more of the better-paid jobs in the area.

To do:

Briefly explain how Chinese activities are affecting Tibetan culture.

Cultural change in the Andaman Islands

The Andaman Islands, in the Bay of Bengal, are governed by India. The indigenous population has steadily declined from about 5,000 in the 18th century, at the time of first outsider contact. The island chain was colonized by British settlers in 1858 and used for most of the next century as a penal colony.

There are around 500 islands, 38 of which are inhabited. Indian settlers have poured in since independence; only about 1,000 of its 356 000 population are from the original tribes. One of the tribes, the Sentinelese, still live uncontacted on a remote island. They were filmed firing arrows at a helicopter which went to check whether they had survived after the 2004 tsunami.

To research

Visit http://news.bbc.co.uk/1/hi/world/asia-pacific/6940182.stm

How have conditions changed since the building of the railway?

Tribes on some islands have retained their distinct culture by dwelling deep in the forests and resisting – with arrows – colonizers, missionaries and documentary makers. However, the construction of trunk roads from the 1970s has opened up the region to new forms of development.

The Jarawa tribe is under serious threat as a result of tourist developments. Since 2000, the number of tourist trips into their jungle reserve has grown rapidly. Although notices at the entrance to the forest instruct visitors not to stop or allow the Jarawa into their vehicles, take photographs, feed the tribesmen or give them clothing, there is evidence that these practices are happening.

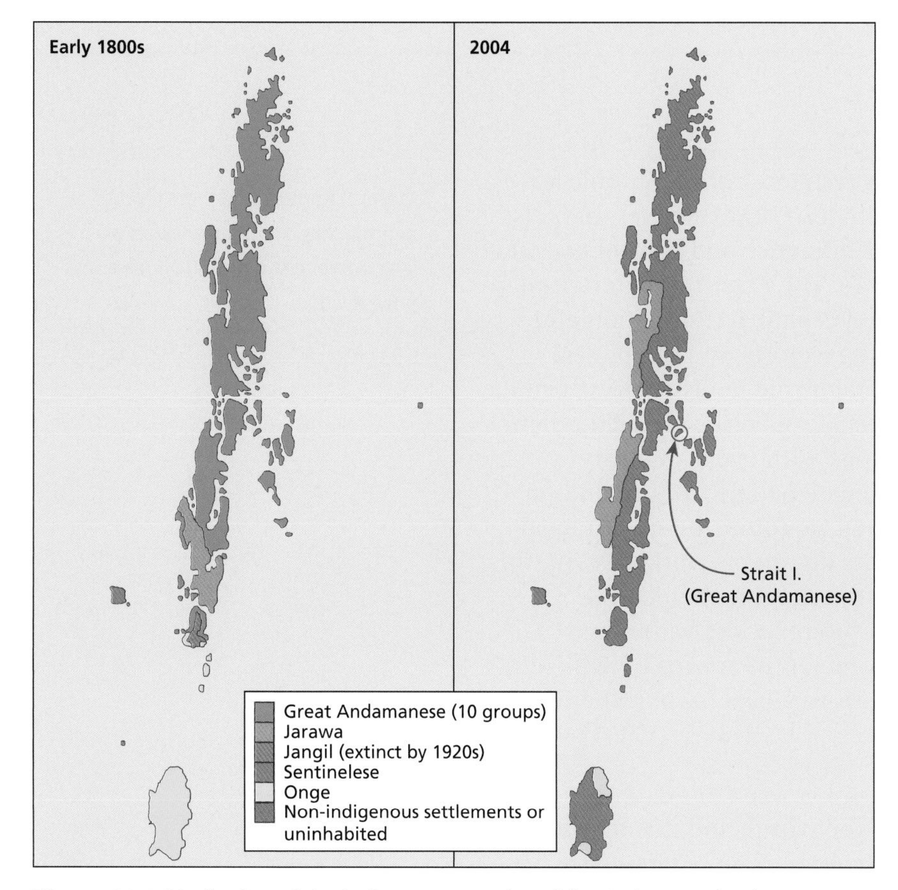

Figure 16.5 Distribution of the indigenous peoples of the Andaman Islands, 1800s and 2004

To do:

Describe and explain how Jarawa culture could be affected by tourism and other developments.

To research

Visit www.survivalinternational.org, the website of Survival International, the movement for tribal peoples. Click on Tribes and campaigns, and look up the Jarawa. Research the threats to the Jarawa.

Visit http://www.guardian.co.uk/world/2010/feb/04/ancient-language-extinct-speaker-dies to read an article about the death of the last person to speak the Andaman language, Bo.

Plans to encourage more tourists will bring an influx of workers and settlers to the area, increasing the pressure on the Jarawa and their land, exposing them to diseases – such as swine flu – to which they have no immunity, and to alcohol, which has ravaged other tribes. Trips to see the Jarawa – officially on the pretext of visiting mud volcanoes that lie within the reserve – have become big business on the islands. The number of visitors from mainland India has increased in the last few years, as cheap flights to the Andamans have replaced the four-day boat journey that was once the only cost-effective way to get there.

Figure 16.6 The Jarawa of the Andaman Islands

Be a critical thinker

What is cultural imperialism, and how has it changed over time?

Proponents of the cultural imperialism thesis date its inception to the industrial colonialism phase. It was during this phase that colonialism reached its zenith, peaking just prior to the First World War, when the British Empire reached its maximum territorial extent. It is usually the case that the cultural imperialist is a large, economically or militarily powerful nation (like Britain during this period) and the victim country is a smaller or less affluent one (like India). However, the end of formal colonialism in the second half of the 20th century did not spell the end of cultural imperialism.

The world is becoming more uniform and standardized, through a technological, commercial and cultural synchronization emanating from the West, and globalization is tied up with modernity. Cultural imperialism has become an economic process as well as a political one. It is forged by TNCs that represent the interests of the elite, especially those of the USA.

Cultural imperialism

Cultural imperialism – the practice of promoting the culture, values or language of one nation in another, less powerful one.

Global cultural imperialism today has resulted from economic forces, as when the dominant culture (usually the USA) captures markets for its commodities and thereby gains influence and control over the popular consciousness of other cultures. The export of entertainment is one of the most important sources of capital accumulation and global profits, displacing manufacturing exports. In the political sphere, cultural imperialism plays a major role in dissociating people from their cultural roots and traditions of solidarity, replacing them with media-created needs which change with every publicity campaign. The political effect is to alienate people from traditional community bonds and from each other. As countries are attracted by and brought under the influence of the dominant world system, they are pressured and sometimes bribed into shaping their social institutions to correspond to, or even promote, the values and structures of the dominant system. Some of the means by which this happens occur through language, tourism, global brands, the media and democracy.

Language

There are around 6,000 languages in the world, and this figure may drop to 3,000 by 2100. Approximately 60% of these languages have fewer than 10,000 speakers; a quarter have under 1,000. English is becoming the world language. Although Mandarin is more widely spoken as a first language, if second-language speakers are taken into account the total number of English speakers is close to a billion. English is the medium of communication in many important fields including air travel, finance and the Internet. Two-thirds of all scientists write in English; 80% of the information stored in electronic retrieval systems is in English; 120 countries receive radio programmes in English; and at any given time over 200 million students are studying English as an additional language. It is an official language in much of Africa, the Pacific, and South and Southeast Asia.

Tourism

Tourism is now the world's largest industry. The journey of many British people to the Costa del Sol, Spain, where they practise cultural traits such as drinking beer and eating fish and chips while lying on crowded beaches surrounded by tall buildings, is a stereotype which captures the essence of this type of standardization. Another example of stereotyping is the actions of German tourists, securing the sunbeds by swimming pools early in the morning.

Global brands

Behind the growth in the influence of TNCs is the rise of global consumer culture built around world brands. McDonald's, for example, operates over 26 000 outlets in 119 countries. In 1997 it opened one outlet every four hours. Coca-Cola is sold in nearly every country. It is a transcultural item, yet it is very much linked with US culture.

Media

National media systems are being superseded by global media complexes. Around 20 to 30 large TNCs dominate the global entertainment and media industry, all of which are from the West, and most of which are from the USA. These include giants such as Time-Warner, Disney (Figure 16.7), News Corporation, Universal and the BBC.

Figure 16.7 The USA's cultural imperialism via the Disney brand

Democracy

The spread of liberal democracy has been profound and is now practised in the vast majority of nation states across the planet. Underlying this diffusion is the Western enlightenment belief that it is the most desirable form of governance.

Criticisms of cultural imperialism

It has been argued that the concept of cultural imperialism ascribes globalization with too much determining power. The power of locality, and of local culture, is thus overlooked. Moreover, a variation on the cultural imperialism argument sees the creation of a universalized hybrid culture. This type of culture is homogeneous but not entirely Western in nature. The impact of contact is not one-way. For example, the British drink tea because of the British imperial connection with India, and a number of words in the English language, such as *bungalow*, *shampoo*, *thug* and *pyjamas*, are borrowed from languages of the subcontinent. The influence of Black American and Hispanic dialects on rap, the most popular music globally at present, and the fact that football (soccer), which diffused through the British Empire, is thought to have been invented in China, are further examples of universalized hybrids.

To do:

Outline ways in which culture may become diluted and/or homogenized.

With the use of examples, explain what is meant by the term "cultural imperialism".

To research

Visit www.culturalpolitics.net/popular_culture. This site raises questions about American culture in a global context. Click on Cultural imperialism.

17 Political outcomes

Some analysts believe that nations are far less important than they once were. They argue that the increasing flow of people, capital, goods and ideas across international boundaries leads to loss of sovereignty and the demise of the nation state. At the same time, the growth of trading blocs and transnational corporations heralds a new world order in which individual countries are less important than before.

Trading blocs

A trading bloc is an arrangement among a group of nations to allow free trade between member countries but to impose tariffs (charges) on other countries that may wish to trade with them. Examples of trading blocs include the European Union (EU), the Association of Southeast Asian Nations (ASEAN), the North American Free Trade Agreement (NAFTA) and MERCOSUR, the common market of South America.

Many trading blocs were established after the Second World War, as countries used political ties to further their economic development. Within a trading bloc, member countries have free access to each others' markets. Thus in the EU, the UK has access to Spanish markets, German markets, and so on. In return, Spain, Germany and the other countries of the EU have access to Britain's market. Being a member of a trading bloc is beneficial as it allows greater market access – in the case of the EU this amounts to over 470 million wealthy consumers.

Some critics believe that trading blocs are unfair, as they deny non-members access to certain markets. Thus, for example, countries from the developing world have more limited access to the rich markets of Europe. This makes it harder for them to trade, and to develop. In order to limit the amount of protectionism, the World Trade Organization tries to promote free trade, which would allow all producers in the world equal access to all markets.

Case study: *The North American Free Trade Agreement (NAFTA)*

The North American Free Trade Agreement (NAFTA) between USA, Canada and Mexico was signed in 1994, creating one of the largest free-trade zones in the world. The zone was the first to join countries from MEDCs and LEDCs. It was an agreement to phase out restrictions on the movement of goods, services and capital between the three countries by 2010. Its aim is to:

- eliminate trade barriers
- promote economic competition
- increase investment opportunities
- improve cooperation between the USA, Canada and Mexico.

The population of the three NAFTA countries is approximately 456 million, with an average per capita GNP of $35,490.

Until 1982 Mexico followed a policy of government-sponsored industrialization, based on import substitution industries (ISIs). However, financial crises through overspending in the 1970s and 1980s forced the Mexican government to seek aid from the USA, the World Bank and the International Monetary Fund. Aid was provided, at a price: Mexico was forced to rearrange its economy along free market lines.

The government was keen to agree, partly in order to receive the aid and partly for fear of being ignored by

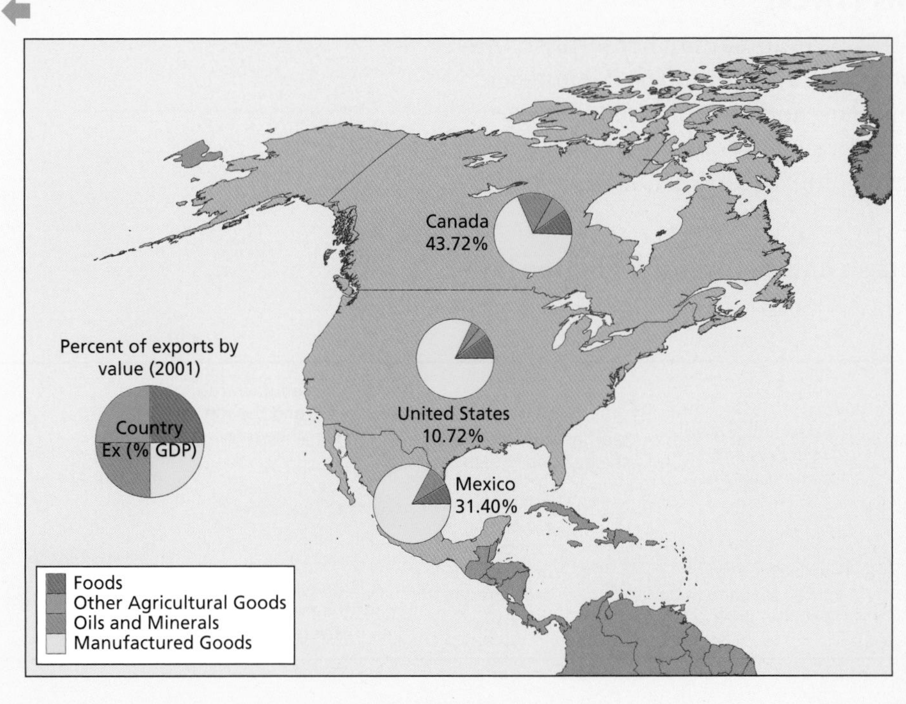

Figure 17.1 NAFTA, composition of exports as percentage of GDP

the USA. Mexico hoped that by joining NAFTA economic growth would follow, employment would increase and it would take off as an NIC. The USA managed to include clauses to protect American workers, and agricultural treaties were arranged bilaterally between the countries.

However, there has been opposition to NAFTA. US critics warned that assembly jobs would be lost to low-cost locations in Mexico. Others argue that it will not necessarily bring economic growth. For example, experience in Canada has shown that:

- many small firms have closed, due to competition with lower-cost US firms
- many firms left Canada for lower-cost areas in the USA
- mergers and takeovers have led to increased unemployment.

With respect to Mexico, US industries move to Mexico to take advantage of its ultra-cheap labour. This has the effect of:

- creating unemployment in the USA
- reinforcing a low-wage mentality in Mexico (incomes in the *maquiladora* sector have risen by only 15% since 1994).

In addition, the removal of border restrictions on trade affects up to 15 million farmers, which means that US and Canadian grain producers can dump their surpluses in Mexico. The removal of subsidies and decline in communal ownership of the land has forced uncompetitive Mexican peasants out of agriculture. On the other hand, Mexico has become an important meat market for the USA. This is partly due to the rise in GDP per capita.

According to NAFTA, Mexico's rural areas will become export orientated: industrial and service growth has replaced agriculture. Along the US–Mexican border there are about 2,000 US-owned, labour-intensive, export-orientated assembly plants, employing about 500 000 Mexican labourers. Poverty rates in Mexico have fallen and there has been a rise in real income. However, where there has been growth its value has been questioned. Many workers are children, wages are low and working conditions unsafe. Critics argue that Mexico could be exploited even more as a huge ultra-cheap labour supply.

Environmentalists point to Mexico's poor record of enforcing environmental laws. They fear that Mexico may become a dumping ground for hazardous material and point out that Mexico's rivers and air are already heavily polluted.

To do:

Study Figure 17.1, which shows the composition of exports from the three NAFTA countries, as a percentage of GDP.

a Compare and contrast the composition of exports of the three countries.

b Comment on the relative contribution of exports by value.

To research

Visit http://news.bbc.co.uk/1/hi/business/4510792.stm for a guide to world trade blocs, and www.ftaa-alca.org/alca_e.asp

What are APEC and FTAA?

What is the Cairns group? What commodity do they trade in? What is unusual about the Cairns group?

Transnational corporations (TNCs)

A transnational corporation (TNC) or multinational enterprise (MNE) is an organization with operations in a large number of countries. Generally, research and development, and decision-making, are concentrated in the core areas of developed countries, while assembly and production are based in developing countries and depressed, peripheral regions.

TNCs provide a range of advantages and disadvantages for the host country (see Figure 17.2).

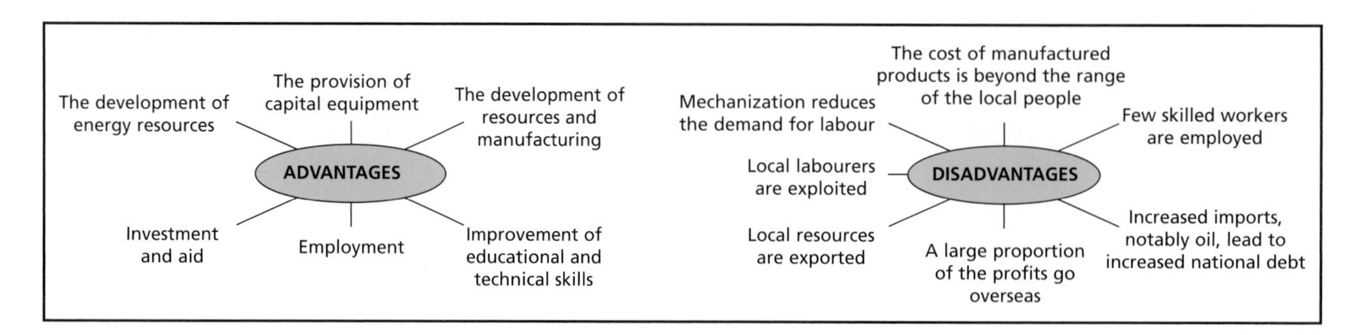

Figure 17.2 TNCs – the balance sheet

TNC power

The sheer scale of the economic transactions that TNCs make around the world and the effect they have on urban, regional, and national economies gives them tremendous power. Thus TNCs have become planned economies with vast **internal markets**.

- Up to one-third of all trade is made up of internal transfers of TNCs. These transfers produce money for governments via taxes and levies.
- Economic power comes from the **ownership of assets**.
- Over 50 million people are employed by TNCs.
- Although many governments in developing countries own their own resources, TNCs still control the marketing and transport of goods.

TNCs and the world's economic crises

Reduced demand and increased competition create unfavourable economic conditions. In order to survive and prosper, TNCs have used three main strategies:

- **rationalization** – a slimming down of the workforce, which involves replacing people with machines
- **reorganization** – including improvements in production, administration and marketing, such as an increase in the subcontracting of production
- **diversification** – developing new products.

Part of the reason for the decline of sovereignty in some countries is the sheer economic size and dominance of some TNCs, as shown in the following table.

To research

Visit www.unctad.org.wir for UNCTAD's world investment report. Click on the tab "The largest TNCs". Comment on the information you find.

Rank	Company/country	Annual sales/GDP ($US billions) (2007)	Rank	Company/country	Annual sales/GDP ($US billions) (2007)
1	USA	13 811	44	Singapore	161
2	Japan	4,376	59	Bangladesh	67
3	Germany	3,297	79	Kenya	29
4	China	3,280	109	Uganda	11
5	UK	2,727	146	Zimbabwe	3

Rank	Company/country	Annual sales/GDP ($US billions) (2007)
1	Wal-Mart, USA	379
2	Exxon-Mobil, USA	358
3	Royal Dutch Shell, Netherlands	355
4	Toyota, Japan	204
5	Chevron, USA	204

Table 17.1 The size of TNCs in annual sales/GDP

Case study: *Imperial Chemical Industries (ICI)*

ICI was formed in 1926 and had its headquarters in the UK. It employs about 29 000 people worldwide and had sales of about £4.8 billion in 2006. Its main products are chemicals, fertilizers, explosives, insecticides, paints and non-ferrous metals. ICI was seen as one of the flagships of British industry and its fortunes were seen as a barometer of the nation's fortunes.

In the 1920s and 1930s ICI helped develop Perspex and Dulux paints, and then in the 1940s it moved into pharmaceuticals. By the late 1970s it was moving away from bulk chemicals into the high-value chemicals industries. In the early 1990s ICI moved its pharmaceuticals, agrochemicals and biological products into a new company (the Zeneca Group). Its sales and profits now depended on four main markets: the UK; the rest of western Europe; North America; and Australia and the Far East. In 2008 it became a subsidiary of the multinational Dutch chemical group Akzo Nobel, the world's largest global paints and coatings company and a major producer of speciality chemicals. The company has 54 000 employees in 80 countries.

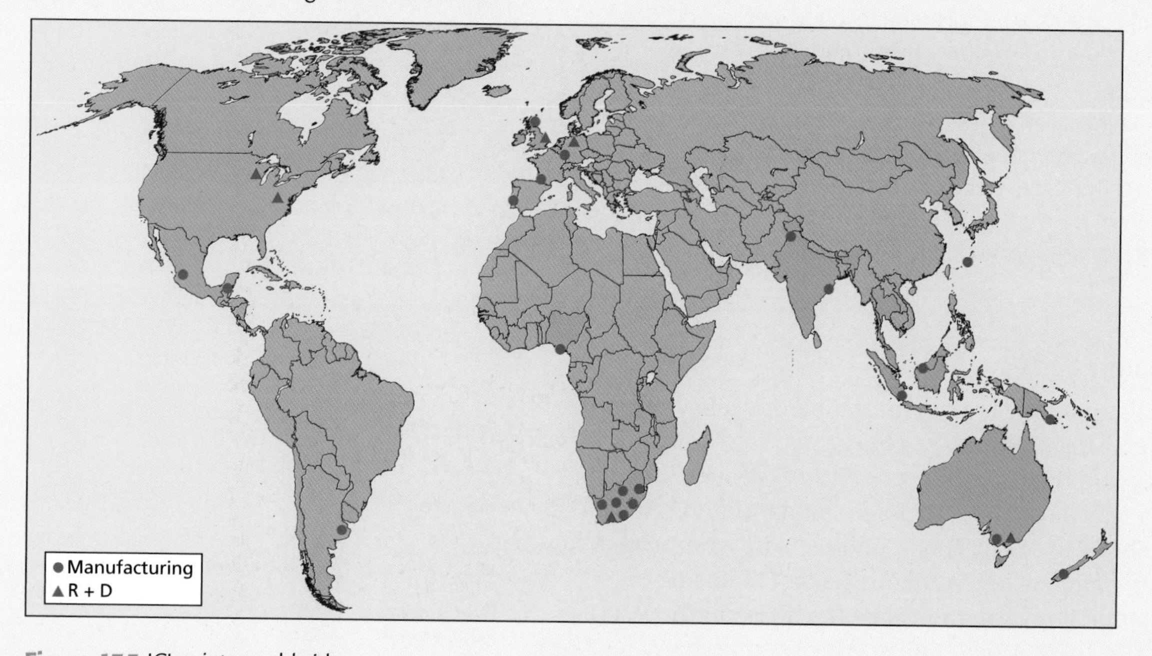

Figure 17.3 ICI paints worldwide

Case study: *Tata Steel*

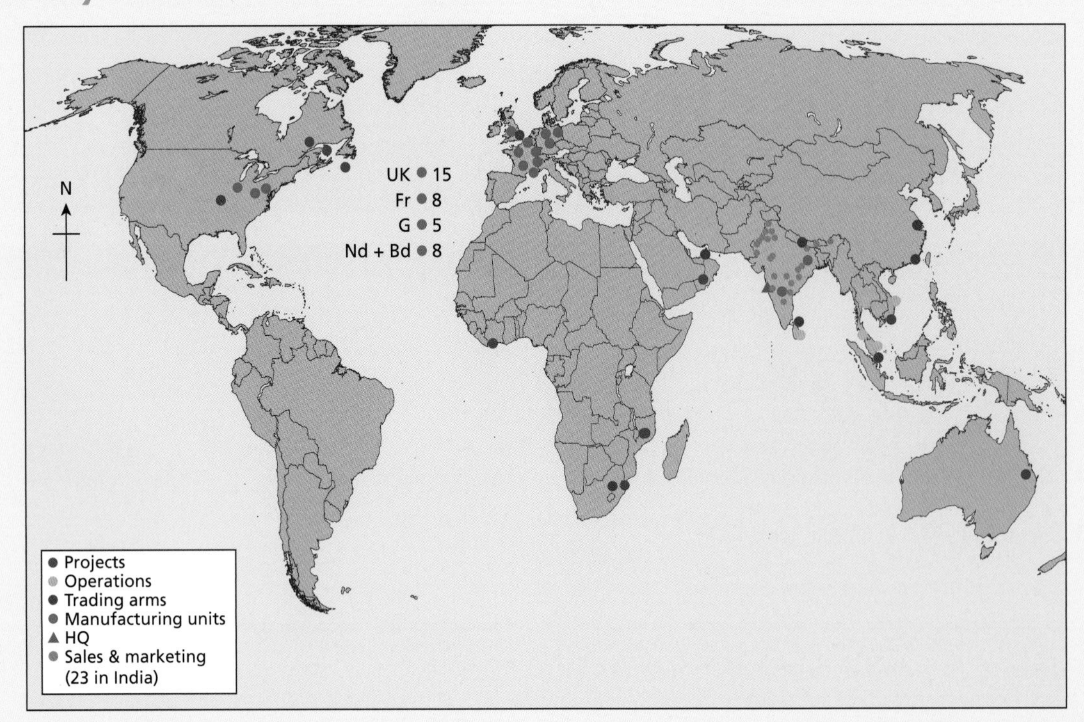

Figure 17.4 The Tata Group

The Tata Steel Group is one of the world's largest steel producers. With a presence in nearly 50 countries, the group, which includes Corus, Tata Steel, Tata Steel Thailand and NatSteel Asia, has over 86 000 employees across five continents with a crude steel production capability of 30 million tonnes.

In 1907 the businessman Jamesetji Nusserwanji Tata established Tat Steel in Jamshedpur, India. Today Tata Steel operates a 5-million-tonne crude steel production plant and a variety of finishing plants there.

With the acquisition of Corus in 2007, the Group now has steel production facilities in the UK and the Netherlands, as well as in India and Southeast Asia. It operates Millennium Steel (renamed as Tata Steel Thailand) and NatSteel Holdings in Singapore. It also has joint venture operations in Australia, South Africa and Mozambique.

Tata steel has an ambitious target to achieve a capacity of 100 million tonnes by 2015. About half of this would be acquired through mergers and half through new developments. Nevertheless, the company is not without its critics, who claim that the drive for growth and profits is at the expense of its once ahead-of-its-time philanthropy. For example, in 1912 the company introduced an eight-hour working day (the norm in the UK at the time was 12 hours). Moreover, it introduced paid leave in 1920 (which only became legally binding in India in 1945), and started a provident fund for its workers in 1920, which became law for all employers in India in 1952. Tat's Steel furnaces have never been affected by industrial action.

In 2009 Corus announced that it was cutting 3,500 jobs worldwide, including about 2,500 in the UK, due to a substantial fall in demand.

Regulatory bodies

A regulatory body is an organization or agency responsible for exercising authority or control over some area of human activity in a regulatory or supervisory capacity. Regulatory bodies may be set up to enforce standards and safety, to oversee use of public goods, and regulate commerce. They include national governments, trading blocs and international organizations like the IMF and the WTO. However, there is widespread criticism that many regulatory

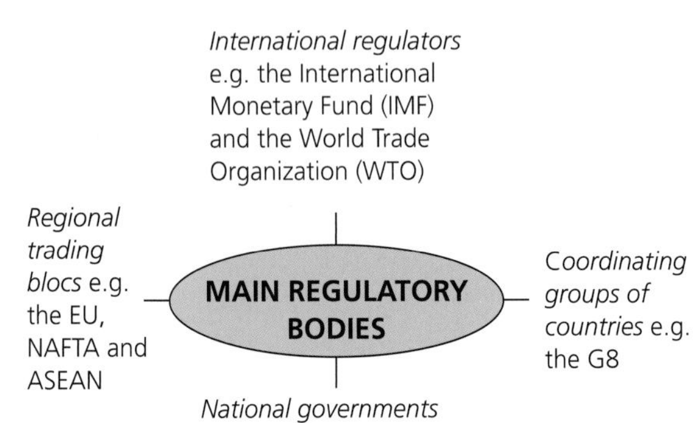

Figure 17.5 The main regulatory bodies

bodies have limited power, and that when faced with a powerful MEDC or TNC they capitulate to their wishes.

Much of the world's trade and money exchange is run by stock exchanges and the world's main banks. For example, Barclays Capital is the investment-banking sector of Barclays Bank. It deals with over £360 billion of investment through its 33 offices located worldwide. Its regional headquarters are located mostly in MEDCs such as London, Paris, Frankfurt, New York and Tokyo. Hong Kong is the exception, although it is an important financial centre, like most of the other cities on the list.

Responses to loss of sovereignty

The concept of the nation state was a western one that has spread throughout the world as a result of colonialism. Crucial to the development of the nation state was the creation of a national identity that cut across class. Nation building involved a variety of factors, such as a common language, an education system, national communications networks, national symbols, and promotion of national culture. Nationalism may be a response to threats or perceived threats from external forces or influences that are seen to be undermining a nation's sovereignty.

A **nation state** is an ideal form consisting of a homogeneous group of people governed by their own state. However, nation states are rarely unified, culturally. Most of them are the result of invasions, settlements and migrations, and contain within their borders people of different cultural and ethnic origins. Moreover, the impact of international migration and the process of globalization challenge even further the idea of the nation state, calling into question whether nations can ever achieve sovereignty in a homeland only for people who belong to their imagined community.

Many nations, such as the Palestinians, Kurds, and Basques, are currently engaged in political movements to win sovereignty over a cultural homeland. Nationalism tends to focus on local cultural histories and is strongly linked to the language, ethnicity, and religion of a particular place.

Nationalism in Ireland

Nationalism in Ireland shows many features. At one extreme were the political freedom fighters, or terrorists, and at the other were the members of the *gaelteacht*, the Irish-speaking regions of Ireland.

Between these two extremes were a variety of programmes to develop the Irish sense of nationality. In schools, the curriculum delivered the Irish sagas in English, History and Irish lessons. The two main sports, Gaelic football and hurling, were unique to Ireland and the Irish diaspora.

To research

Visit http://news.bbc.co.uk/1/hi/business/7850596.stm for the full story of the Corus job closures, and those of other large TNCs worldwide.

Visit http://globalpolicy.org and go to Social and economic policy/International trade and development/Transnational corporations for tables and charts on TNCs.

To do:

- **a** Describe the distribution of Tata's manufacturing units, as shown in Figure 17.4.
- **b** Suggest varied reasons for the limited presence of Tata in South America, Africa and Australia.
- **c** To what extent do you agree that the advantages of TNCs outweigh the disadvantages for host countries?

Nationalism – a political movement or belief that holds that a nation has the right to an independent political development based on a shared history and common destiny.

Nation – a community of people whose members are bound together by a sense of solidarity rooted in an historic attachment to a homeland and a common culture, and by a consciousness of being different from other nations.

State – an independent political unit with territorial boundaries that are recognized by other states.

Case study: *Nationalism in Indonesia*

The island of Timor was first colonized by the Portuguese in 1520. In 1613 the Dutch took control of the western portion of the island, and Portugal and the Netherlands then fought over the island until 1860 when a treaty divided it, granting Portugal the eastern half. In the 20th century Australia and Japan fought each other on Timor during the Second World War; nearly 50 000 East Timorese died during the subsequent Japanese occupation.

In 1949, the Netherlands gave up its colonies in the Dutch West Indies, including West Timor, creating the nation of Indonesia. East Timor remained under Portuguese control until 1975, when the Portuguese abruptly pulled out after over 450 years of colonization. The sudden Portuguese withdrawal left the island vulnerable. Nine days after the Democratic Republic of East Timor was declared an independent nation, it was invaded by Indonesia and annexed on 16 July 1976. Only Australia officially recognized the annexation. Nevertheless, Indonesia's invasion was sanctioned by the USA and other western countries, which had cultivated Indonesia as a trading partner and cold-war ally. In contrast, the leader of the East Timor pro-independence party was Marxist.

Indonesia's invasion and its brutal occupation of East Timor largely escaped international attention. This was because East Timor was small, isolated and poor. East Timor's resistance movement was suppressed by Indonesian military forces, and more than 200 000 Timorese were reported to have died from famine, disease and fighting since the annexation. In 1996 two East Timorese activists received the Nobel Peace Prize for their efforts to gain freedom peacefully for East Timor.

After Indonesia's hard-line president Suharto left office in 1998, his successor, B. J. Habibie, unexpectedly announced his willingness to hold a referendum on East Timorese independence. Following intense fighting, on 30 August 1999, 78.5% of the population voted to secede from Indonesia. After enormous international pressure, Indonesia finally agreed to allow UN forces into East Timor.

The UN Transitional Authority in East Timor (UNTAET) governed the territory for nearly three years. A parliament was elected in 2001 and a constitution assembled, and on 20 May 2002, nationhood was declared. Charismatic rebel leader José Alexandre Gusmão, who was imprisoned by Indonesia from 1992 to 1999, was overwhelmingly elected the nation's first president on 14 April 2002.

The first new country of the millennium, East Timor is also one of the world's poorest. The Indonesian militias had destroyed its meagre infrastructure in 1999, and the economy, primarily made up of subsistence farming and fishing, is in a shambles. East Timor's offshore gas and oil reserves promised the only real hope for lifting it out of poverty, but a dispute with Australia over the rights to the oil reserves in the East Timor Sea have stalled efforts to develop the fields. The oil and gas fields lie much closer to East Timor than to Australia, but a 1989 deal between Indonesia and Australia set the maritime boundary along Australia's continental shelf, which gives it control of 85% of the sea and most of the oil. East Timor wants the border redrawn halfway between the two countries, and estimates that this would allow it to earn $12 billion over the next 30 years, as opposed to $4.4 billion. Australia, however, has refused to negotiate.

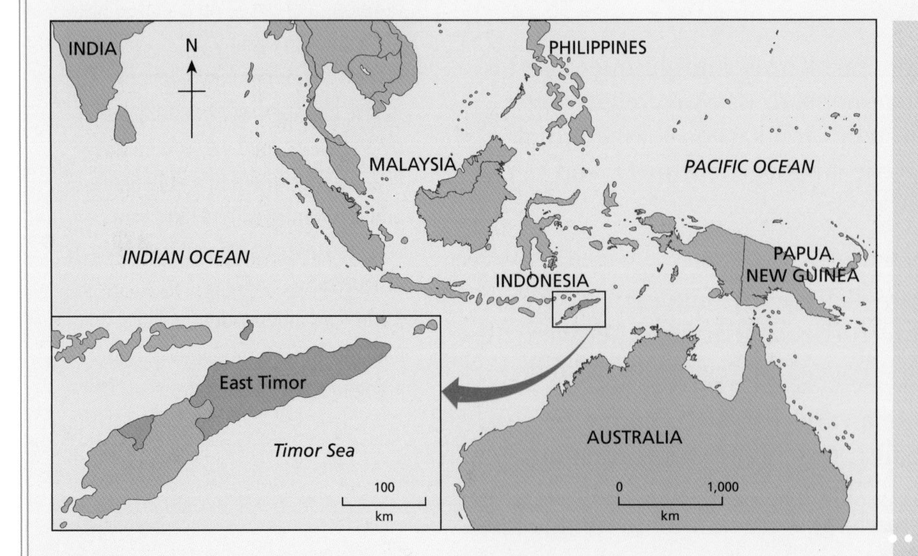

Figure 17.6 The location of East Timor

To research

Visit http://news.bbc.co.uk/1/hi/world/asia-pacific/country_profiles/1508119.stm for a country profile of East Timor.

Visit http://tsjc.asiapacificjustice.org/index.htm, and research the argument between East Timor and Australia regarding oil and gas in the Timor Sea. Use the Background and FAQ icons.

Case study: *Nationalism in Canada*

The Inuit are the indigenous people of the Canadian Arctic, and they have lived in the lands north of the Canadian forests for at least 4,000 years, mainly in small communities near the Arctic coast. Until recently, Inuit depended on caribou, seals, whales and fish for food and clothing.

The Inuit had no political control over their homelands until 1999, when Canada's parliament approved a new territory called Nunavut. Nunavut, which means "our land" in the Inuit language, was carved out of the Canadian Northwest Territories, in the treeless tundra of Canada's far north, and covers an area the size of western Europe. It has a population of about 25 000. Nunavut was created out of a desire to bring government closer to the Inuit. The creation of Nunavut returned to the Inuit control over their own affairs.

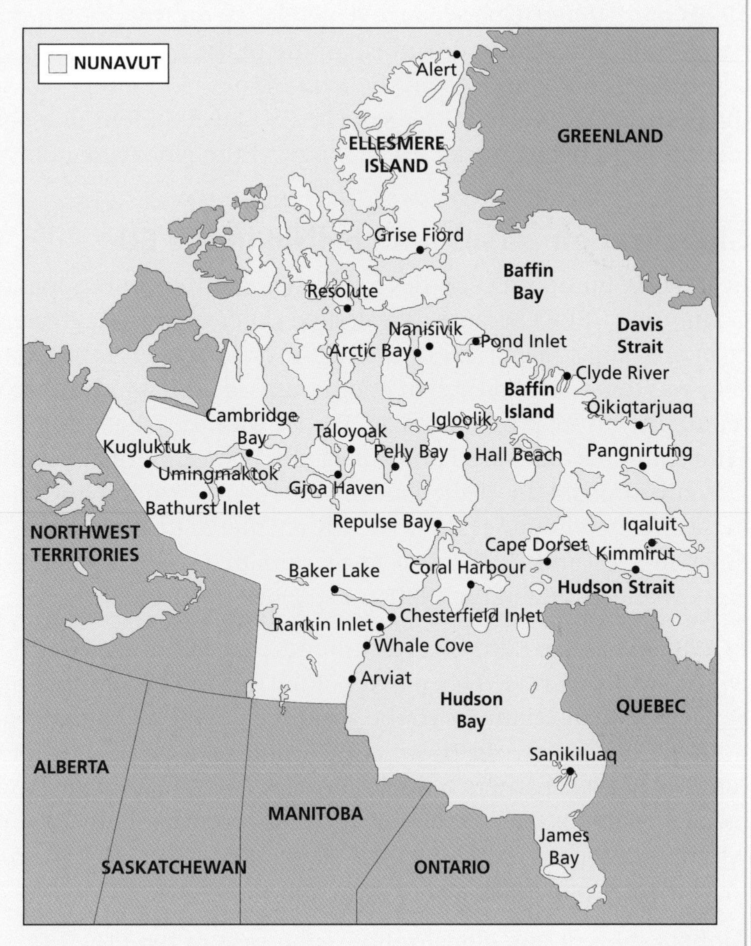

Figure 17.7 The location of Nunavut

To research

Visit www.gov.nu.ca/cley for Nunavut's Department of Culture, Language, Elders and Youth. Visit www.aboriginalcanada.gc.ca and go to the topic "Nunavut's language, heritage and culture" for information.

Case study: *Nationalism in the Former Yugoslav Republic (FYR)*

The Socialist Federal Republic of Yugoslavia was a confederation of six republics created after the Second World War. Ethnic tensions had existed in the region since the First World War, but had been kept under control by the central socialist government. By the early 1980s tensions resurfaced in the midst of economic hardship, and the country was facing rising nationalism among its various ethnic groups.

In 1991 Yugoslavia found that several of its component republics wanted independence and for a while, the central government fought to keep the nation together. When Croatia and Slovenia declared their independence from Yugoslavia in June, war broke out, ending 40 years of peaceful co-existence. The wars ended in 1995 in various stages, mostly resulting in international recognition of new sovereign territories – Croatia, Serbia, Slovenia, Bosnia-Herzegovina and Macedonia.

The wars were among the most brutal civil wars of recent times, and caused huge economic disruption and destruction. They have become infamous for the war crimes they involved, including mass ethnic cleansing. Many key participants were subsequently charged with war crimes. Infighting still continues in the region.

Economically, the campaign to buy "Guaranteed Irish" helped sales of Irish companies. The government's import substitution policies of the 1920s and 1930s helped reduce dependency on Britain. On the other hand, Ireland has been described as one of the most globalized countries in the world, given the amount of direct foreign investment it has attracted. It is hard to reconcile the development of a national identity with becoming an integral part of the global economy.

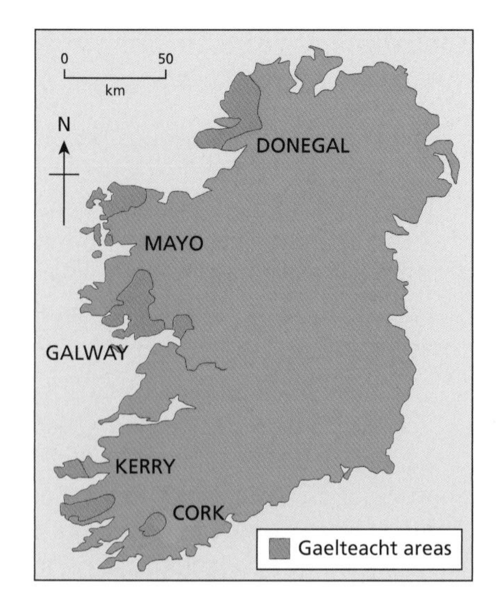

Figure 17.8 *Gaelteacht* areas of Ireland

Globalization versus nationalism in the EU

In 1957 six European countries founded the European Economic Community (EEC) because they desired closer union and greater economic and social progress. One of the main reasons for creating the EEC was that trade had grown enormously since 1945. Another was to reduce the future prospect of war. The six countries that formed the European Coal and Steel Fund – Belgium, France, Germany, Italy, Luxembourg and the Netherlands – went on to form the EEC as defined in the Treaty of Rome (1957). The UK attempted to join but was turned down in 1961 and 1967. It was finally admitted in 1973 when the EEC was expanded, and in the 1980s it expanded further still.

In 1986 the Single European Act introduced a rule of majority decisions, which greatly increased the powers of the Council of Ministers and the Parliament. It also introduced the goal of removing all barriers to trade by 1992. The Maastricht Treaty (1991) confirmed the agenda for the removal of trade barriers, a single currency (the "euro") and a range of social regulations, and established the European Union. In 2002 the euro was introduced into 12 of the then 15 members. The EU was expanded to 25 members in 2004, and to 27 in 2007.

To do:

Explain the difference between a nation and a state.

With the use of an example, examine the role of nationalism as a country tries to gain control of its resources and/or culture.

Globalization or anti-globalization – the case of the EU

The growth of the European Union would appear to be a strong symbol of globalization. Member nations have given up some of their sovereignty and political power to a multinational government. The EU has moved beyond mere economic integration and has achieved some political, social and cultural integration.

European integration since1945 has partly been an attempt to prevent a world war from ever occurring again. (Part of the desire to include Turkey in the European Union is to integrate a large Islamic state into the EU and reduce the possibility of war between Islam and the West.)

However, there has been reaction against the growth of the EU and its imposition of economic, political and social regulations. The UK and Denmark, for example, opted out of the single currency, deciding to retain their own. In Spain, Catalonia and Galicia have achieved significant autonomy, while the Basque Country has not, largely in response to the violence of the independence-seeking party, ETA.

As the EU has expanded it has become more diverse. Economically, socially and culturally it is more varied and divided than ever before. This diversity means that integration is likely to be less complete than when there were fewer member countries. Being large may help economic prospects (a larger market, for example) and political ones (less chance of war), but national identity and regional cooperations are likely to become more important over time.

To research

Visit the homepage of the European Union at http://europa.en and go to Policy areas.
Enlargement to find out which countries opened negotiations in 2010, and which other countries are likely to enter negotiations to join the EU.

Globalization versus regionalism

While globalization of economic activity has certainly occurred, and there is evidence of a new international division of labour, political and cultural values have often created a new feeling of regional identity. Within major trading blocs such as the EU there are very strong nationalist tendencies, such as within Spain and the UK.

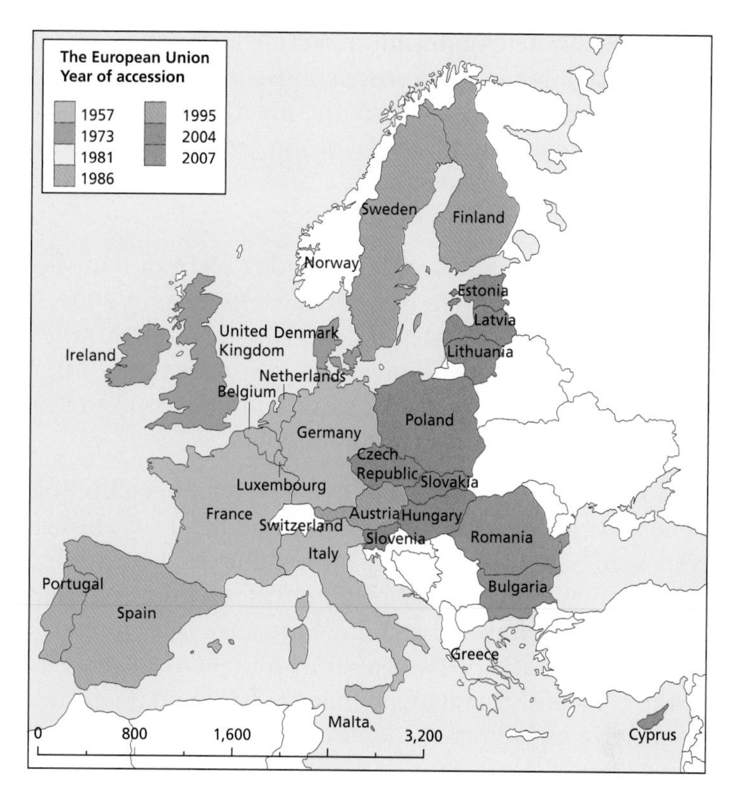

Figure 17.9 Changes in the EU over time

Anti-globalization movements

The anti-globalization movement (AGM) is a general term used to describe a wide variety of protestors, lobbyists and pressure groups. The AGM has attracted attention due to the protests it has mounted during summits in places such as Seattle, Tokyo and Munich. The movement developed during the 1990s, following the actions of the Zapatista National Liberation Army in Chiapas, Mexico. Some 3,000 people, who had lost land in the process of development, took control of the main towns of Chiapas. Moreover, protectionism for farmers had been removed and their livelihoods had suffered.

The AGM reached a global audience at the 1999 WTO trade talks in Seattle, USA, when some 60 000 people arrived in the city, home to Starbucks and Microsoft, and protested. The trade talks were abandoned. In 2001 the World Social Forum was established in Brazil, involving large numbers of diverse groups. The terrorist attacks on the World Trade Centre (the Twin Towers attacks of 9/11) divided people. Some reacted by becoming more nationalistic and prepared to defend their national space, while others saw it as an inevitable result of US global economic and political dominance.

Nevertheless, the AGM lacks focus. Some within the AGM are in favour of globalization, but at a slower pace. Others object to the economic and political power that some rich countries wield. Some see the work of organizations such as the World Bank as pedantic and stifling the needs of poor countries.

To do:

a Describe the changes in the size of the European Union over time.

b To what extent do you think the EU is an example of globalization or regionalism?

People's Global Action

People's Global Action (PGA) is a network for spreading information and coordinating actions between grassroots movements around the world. These diverse groups share an opposition to capitalism and a commitment to direct action and civil disobedience as the most effective form of struggle. The PGA grew out of the international Zapatista gatherings in 1996 and 1997, and was formed as a portal for direct and unmediated contact between autonomous groups.

Its first conference took place in 1998, when movements from all over the world met in Geneva and launched a worldwide coordination of resistance against the global market economy and

To research

Visit http://europa.eu, the official website of the European Union. Use the A-Z index and click on the History of the EU icon to create a timeline for the expansion of the EU.

Access statistical information from Eurostat, the Statistical Department Office within the EU.

the World Trade Organization (WTO). Later that year, hundreds of coordinated demonstrations, actions and street parties took place on all five continents, against the meeting of the G8 and the WTO. From Seattle to Genoa, many of the groups and movements involved with PGA have been a driving force behind the global anti-capitalist mobilizations.

A second international conference took place in Bangalore, India, in 1999 and the third in Cochabamba, Bolivia, in 2001. There have been regional conferences in Latin America, North America, Asia and Europe, and three caravans of movements: the Intercontinental Caravan, the Colombian Black Communities tour and the Peoples' Caravan from Cochabamba to Colombia.

PGA is not an organization and has no members, but it does aim to be an organized network. Contact points for each region are responsible for disseminating information and convening the international and regional conferences; an informal support group helps with fundraising; and there are a website, numerous email lists, and a secretariat. The basis of unity and political analysis is expressed in the constantly evolving manifesto, based on a rejection of destructive globalization, domination and discrimination.

The PGA's detailed manifesto includes sections on each of the following:

- economic globalization, power and the "race to the bottom"
- exploitation, labour and livelihoods
- gender oppression
- indigenous peoples' fight for survival
- oppressed ethnic groups
- onslaught on nature and agriculture
- culture, knowledge and technology
- education and youth
- militarization, migration and discrimination.

Attempts to control migration

International migration has changed much in recent years. Four general trends can be identified:

- Migration is becoming more global, in the sense that more countries are affected at the same time and the diversity of areas of origin is increasing.
- Migration is accelerating, with the number of movements growing in volume in all major regions.
- Migration is becoming more differentiated, with no one type of movement dominating a country's flows, but instead with combinations of permanent settlers, refugees, skilled labour, economic migrants, students, retirees, arranged brides, and so on.
- Migration is being feminized, with women not only moving to join earlier male migrants but now playing a much fuller part in their own right, notably among labour migrants themselves as well as often being dominant in refugee flows.

To research

Visit www.nadir.org/nadir/ and click on Global Action news. What is the PGA currently doing? Investigate one or more of the actions that the PGA has been involved in.

Visit http://www.avaaz.org/en, the website of Avaaz, an international civic organization that promotes activism on issues such as human rights, climate change ansd religious conflict. Its stated mission is to "ensure that the views and values of the world's people inform global decision-making". The organization operates in 13 languages, and claims more than 3 million members worldwide. The word "avaaz" means "voice".

Figure 17.10 The avaaz website

What issues are they currently involved in?

Use the campaigns icon to investigate one or more campaigns that avaaz has been involved in.

Consider signing up as a member and getting involved in issues that are important to you.

These trends have implications for policy makers. There are new challenges for governments for providing for migrants, but there is also increased hostility in receiving countries. Increasing globalization and growing diversity of migrants make it harder for governments to restrict migration.

Migration is important for the growth of an economy. In the USA, economic prosperity is associated with the ability to get a high level of labour, both skilled and unskilled. The USA's population is ageing, and the immigrant population accounted for more than 50% of the growth in the labour market in the 1990s. Nevertheless, many people in the USA want to control migration.

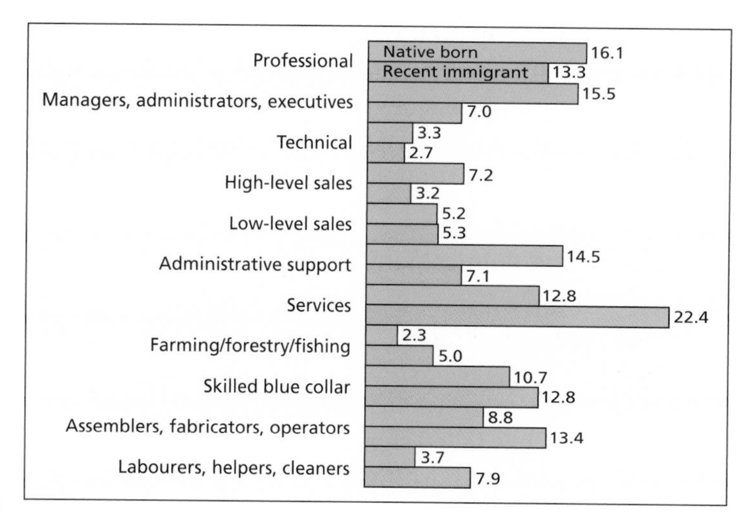

Figure 17.11 Migrants and jobs in the USA, % by occupational group

Illegal immigration to the USA

Illegal immigration to the USA refers to the act of foreign nationals voluntarily residing in the USA in violation of US immigration and nationality law. Illegal immigration carries a civil penalty. Punishment can include fines, imprisonment and deportation. It is estimated that there were between 11.5 and 12 million illegal immigrants in the US in 2006. Their mode of illegal entry into the country is believed to break down as follows:

- **Entered legally with inspection**
  - → Non-immigrant visa overstayers, 4 to 5.5 million
  - → Border-crossing card violators, 250 000 to 500 000
- **Entered illegally without inspection**
  - → Evaded the immigration inspectors and border patrol, 6 to 7 million

Visa overstay

A traveller is considered a "visa overstay" once he or she remains in the USA after the time of admission has expired. Visa overstays tend to be somewhat more educated and better off financially than those who crossed the border illegally.

Fraudulent marriage

People have long used sham marriages as a way to enter the USA.

Border crossing

Each year, an estimated 200 000 to 400 000 illegal immigrants try to make the 24–48-kilometre hike through the wilderness to reach cities in the USA. These people often employ expert criminal assistance – smugglers who promise a safe passage into the USA.

Entry by sea ports

In 1993, 283 Chinese immigrants attempted entry into the USA via a sea vessel. Ten of them arrived dead.

Slavery and prostitution

Indian, Russian, Thai and Chinese women have reportedly been brought to the USA under false pretences to be then used as sex

To research

Visit http://www.opendemocracy.net/people-migrationeurope/article_1274.jsp for the Globalization of migration control. Comment on the views expressed in this article.

Visit www.iom.int for the International Organization for Migration. What are the main concerns of the IOM?

Visit http://www.open2.net/blogs/society/index.php/2009/06/12/paradox_migration_control?blog=10, and outline the paradoxes of migration control.

Visit http://www.open2.net/immigrants/migrantmap.html for an interactive map regarding migrants in many parts of the world. Describe the migration of population in an area that interests you.

slaves. As many as 50 000 people are illicitly trafficked into the USA annually, according to a 1999 CIA study. Trafficking in women plagues the USA as much as it does underdeveloped nations. Organized prostitution networks have migrated from metropolitan areas to small cities and suburbs.

Immigration and enforcement

Illegal migration on the US–Mexico border is concentrated around big border cities such as El Paso and San Diego, which have extensive border fencing and enhanced border patrols. Stricter enforcement of the border in cities has failed significantly to curb illegal immigration, instead pushing the flow into more remote regions and increasing the cost to taxpayers of each arrest from $300 in 1992 to $1,700 in 2002. The cost to illegal immigrants has also increased: they now routinely hire coyotes, or smugglers, to help them get across.

In 2005, the US House of Representatives voted to build a separation barrier along parts of the border not already thus protected. A later vote in 2006 included a plan to blockade 860 miles (1,380 kilometres) of the border with vehicle barriers and triple-layer fencing, along with granting an "earned path to citizenship" to the 12 million illegal aliens in the US and roughly doubling legal immigration (from its 1970s levels). In 2007 Congress approved a plan calling for more fencing along the Mexican border, with funds for approximately 700 miles (1,100 kilometres) of new fencing.

In Australia, migration laws and regulations set the criteria and standards that foreign nationals must meet if they wish to travel to and remain in Australia for a period of time.

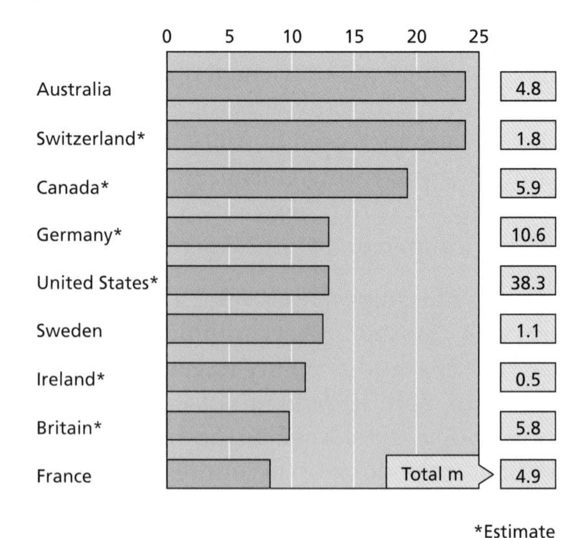

Figure 17.12 Foreign-born populations in selected OECD countries, % of total population

To do:

a Study Figure 17.11 and 17.12.

i Identify the country with (i) the largest absolute number of foreign-born and (ii) the largest relative number of foreign-born population.

ii Identify the most common job sector for (i) native-born workers and (ii) foreign-born workers in the USA.

b Study Figure 17.12.

i Compare and contrast the employment structure of the native-born workers with those of the foreign-born workers.

ii Suggest reasons why a country would want to limit the amount of in-migration to within its borders.

18 Global interactions at the local level

Globalization and glocalization

The increasing presence of McDonald's restaurants worldwide is an example of globalization, while the changes in the menus of the restaurant chain that are designed to appeal to local tastes are an example of glocalization, for example McBurritos in Mexico. Although McDonald's are famed for their uniformity, they have been localized, indigenized and incorporated into traditional cultural forms and practices. The chain has become a local institution in the sense that it has blended into the urban landscape, and its meaning varies from place to place. In Beijing, for example, it has lost its American role as a place of fast and cheap food, and become a place for a special family outing for the middle classes, where customers may linger for hours.

Globalization – "the growing interdependence of countries worldwide through the increasing volume and variety of cross-border transactions in goods and services and of international capital flows, and through the more rapid and widespread diffusion of technology" (IMF)

Glocalization – a term invented in order to emphasize that the globalization of a product is more likely to succeed when the product or service is adapted to the specific locality or culture in which it is marketed.

A model of globalization

There are a number of different views of globalization, the most common of which are those of the hyperglobalists, the sceptics and the transformationalists. Each group has a different view of the significance of globalization and its impact on cultural diversity.

Hyperglobalists

Hyperglobalists believe that this is a new geographical era. They argue that the nation state is no longer important; instead, there is a single global market supported by transnational networks of production, trade and finance. They believe that forms of government above the level of the state – for example trading blocs such as the European Union – are increasingly important.

Hyperglobalists see in this the erosion of the power of the state, and thus the victory of capitalism over socialism. This view was widely supported following the collapse of the former Soviet Union, when socialist hyperglobalists saw globalization as aggressive and regressive.

Hyperglobalists believe that economic forces are dominant in an integrated global economy, and suggest that there is a new world order based on consumerism. They believe that this is leading to a spread of liberal democracy. Hyperglobalists also take the line that globalization leads to homogeneity of culture, the Americanization of the world economy and culture.

However, the financial crisis of 2008 and the involvement of national governments in an attempt to support their economies suggest that the nation state is not yet dead, and that hyperglobalists may have to modify their view.

Sceptics

Sceptics question whether globalization is anything new. They believe that the world was just as integrated in the 19th century. They also claim that if the hyperglobalists are correct, then there would be uninterrupted flows of labour, trade and capital. However, labour is relatively immobile, and protectionist policies limit the amount of free trade. Sceptics believe that national governments are still the most important players. The rise of China, India and Iran as emerging powers is due, in large part, to government policies. In addition, trade blocs promote regionalism rather than globalization.

Sceptics take the stance that the hyperglobalists are only interested in increasing their market share in the new global economy. This leads to the marginalization of the poor. The sceptics' view is that cultural heterogeneity will continue, although homogeneity of culture may occur within a single nation, such as China or Iran.

Transformationalists

The hyperglobalists and the sceptics represent two extremes of the globalization continuum. The transformationalists lie somewhere in between. These academics believe that globalization is real, and is changing society. They consider that such change is an extension of colonial relations. They also believe that the role of national governments is changing rather than being made redundant.

Cultural exchanges lead to hybrids. Politically, nation states have to take into account international and supranational bodies, such as the European Parliament. Economically, production, trade and finance are interlinked in global networks. According to transformationalists, the state is also actively engaged in its own economic and cultural issues, which produces diversity and increased unevenness. There is thus more differentiation in global society, politics and economies.

To do:

Comment on the arguments of the hyperglobalists, sceptics and transformationalists. Which do you think provides the best summary of the significance of globalization on cultural diversity?

	Hyperglobalist	Sceptic	Transformationalist
What is happening?	The global era	Increased regionalism	Unprecedented interconnectedness
Central features	Global civilization based on global capitalism and governance	Core-led regionalism makes globe less interconnected than in late 19th century	"Thick globalization": high intensity, extensity and velocity of globalization
Driving processes	Technology, capitalism and human ingenuity	Nation states and the market	"Modern" forces in unison
Patterns of differentiation	Collapsing of welfare differentials over time as market equalizes	Core-periphery structure reinforced, leading to greater global inequality	New networks of inclusion/ exclusion that are more complex than old patterns
Conceptualization of globalization	Borderless world and perfect markets	Regionalization, internationalization and imperfect markets	Time–space compression and distancing that rescale interaction
Implications for the nation state	Eroded or made irrelevant	Strengthened and made more relevant	Transformed governance patterns and new state imperatives
Historical path	Global civilization based on new transnational elite and cross-class groups	Neo-imperialism and civilizational clashes through actions of regional blocs and neoliberal agenda	Indeterminate: depends on construction and action of nation states and civil society
Core position	Triumph of capitalism and the market over nation states	Powerful states create globalization agenda to perpetuate their dominant position	Transformation of governance at all scales and new networks of power

(Murray, W. 2006. *Geographies of globalization*, London, UK. Routledge)

Table 18.1 Three theses of globalization – a schema

Globalization and glocalization in the manufacturing sector

Increasingly, many commercial activities have become globalized. This is especially the case where transport and communications links are good. The food industry is global, as are tourism, entertainment, energy and the clothing industry, to name just a few. Nevertheless, although some products are global – Coca Cola and McDonald's restaurants, for example – in many cases there are local variations.

	Globalization	Glocalization
Organization	Worldwide	Concentrated in the Triad (EU, North America and Japan – traditionally the world's largest market)
Locational requirements	Comparative advantage and economies of scale	Depressed regions of major international trade blocs, where there is a source of cheap labour
Labour and management	Foreign managers in senior ranks; spatial division of labour	Very difficult for foreign managers to reach senior ranks
Market	Production for world markets	Production for local or regional markets
	Geographically dispersed	Geographically concentrated
	Export-orientated strategy	Export-orientated strategy

Table 18.2 A comparison between globalization and glocalization in the manufacturing sector

Globalization aims at a worldwide intra-firm division of labour. In this strategy, activities are established in many sites spread over the world, based on a country's comparative advantage. A manufacturer striving for globalization aims to secure the supply of its inputs by localizing production of these inputs at the most favourable locations. Thus, labour-intensive production of components will be situated in low-wage areas, while the production of high-tech and high-value-added parts will require a skilled or well-educated workforce. In a European context, this would mean locating research facilities in core areas and assembly plants in peripheral areas. A globalization strategy will promote a division of labour.

Glocalization aims to establish a geographically concentrated inter-firm division of labour in the three main trading blocs: Japan and Southeast Asia, the USA and the EU, collectively known as the Triad. Manufacturers striving for glocalization are building their comparative advantage on close interaction with suppliers and dealers, as well as with other relevant actors such as banks and governments. Two essential elements stand out in a firm's glocalization strategy:

- the decentralization of production to hierarchical networks of local subcontracting
- a high degree of control over supply and distribution.

Figure 18.1 A foreign car showroom in the UK

The strategy for glocalization involves the attempt of a manufacturer to become accepted as a "local citizen" in a different trade bloc, while transferring as little control as possible over its strategic activities. Glocalization is first of all a political, and only in the second place a business location strategy. A manufacturer aiming for glocalization will localize activities in a different trade bloc area only if:

- it otherwise risks being treated as an "outsider" and so subject to trade or investment barriers, and thus stands to lose market share

To do:

a Define the term glocalization.

b Using examples, explain how and why global producers have adapted their products for local markets.

- the inevitable compromise in costs and control will allow it to produce competitively, i.e. there are low labour costs or regional assistance.

Adoption of globalization

Globalization is a very uneven process. For example, the diffusion of telecommunications and ICT has left vast numbers of people without access to either. These are mostly poor, rural people who have missed out on the advantages of globalization. Globalization is not a homogenous process or feature – its outcomes vary markedly across the world. Indeed, the growth of globalization has led to increased inequalities between nations, regions, urban and rural areas, and within cities. Globalization has marginalized and excluded many people from its benefits.

Looking back at economic development between 1945 and 1980 (the modernization era), there are certain groups of people who benefited little economically. Economic growth was accompanied by exclusion. One group of people who have not been able to benefit from globalization are women. Of the 1.3 billion people in poverty, 70% are women. Along with women, landless labourers also failed to make much progress and were also unable to benefit from globalization.

Since 1970 many poor countries have become poorer. There are also clear differences between urban and rural areas, with large urban areas much better off. Nevertheless, within urban areas, those in low-income areas, such as shanty towns, are much less able to benefit from globalization than the wealthy in rich areas.

Globalization has made some places worse off. This is because of what is termed "capital flight". This means that wealthy companies can decide to invest in some places, remove investment from others, close factories, and open new locations on the basis of where they can take advantage of changes in tax regimes, pay negotiations, government incentives, and the availability of grants and loans. To an extent, some countries, especially poor countries, are at the mercy of wealthy TNCs. Who is part of the globalization process and who is out of the globalization process may therefore be determined by the TNC and the nation's government, rather than its people.

Thus the adoption of strategies for globalization may be influenced by many factors, including levels of wealth, landlessness, gender, TNC–government negotiations, and sociocultural factors, such as whether the product is acceptable to a local market.

Globalization and the Middle East

In the Middle East, globalization has often been perceived as largely equivalent to westernization, and is widely regarded as an external threat rather than as an opportunity. Globalization for the Middle East has coincided with wars, intrusive US power, renewed economic dependency and continuing insecurity. It has also strengthened Islamic fundamentalism.

In the Middle East, there are at least three different attitudes toward globalization:

To research

Visit http://mpra.ub.uni-muenchen.de/10025/ for the article "On the Emergence of Glocalization". Read the article. and make notes on the interactions of global and local processes and responses.

Visit www.cosmeticsdesign-europe.com and investigate how the cosmetics industry is responding to glocalization.

- Some see it as a form of imperialism and an invasion of their culture.
- Others see it as a form of scientific and technological modernization, which can exist alongside Arab–Islamic culture.
- Others call for a form of globalization that is compatible with people's national and cultural interests.

For radical Islamists, the West's imperialistic domination of the Muslim world, support for Israel and the invasions of Iraq and Afghanistan have caused Muslims to fear and hate the West. The Islamist view is that globalization threatens to undermine Islam. Human rights, freedom and democracy are viewed as tools of power which serve the interests of Western nations, in particular the USA.

However, it is not quite so simple. For example, following the terrorist attacks on the World Trade Center on 11 September 2001 (9/11), there were scenes of rejoicing in Palestinian refugee camps in Lebanon, while at the same time the Palestinian leader Yasser Arafat announced his country's genuine sorrow and unreserved condemnation of the attack.

While most of the world deplored the attacks, some believed that the USA got what it deserved. They saw the US government as arrogant and hypocritical, pointing out that it had ignored some good international agreements such as a world criminal court, the abolition of land mines, and the Kyoto protocol (the USA is the world's largest producer of greenhouse gases). In addition, the USA had refused to pay its dues to the UN, and had cut its aid to the world's poorest. While it had been eager to prosecute African and Balkan war criminals, it refused to allow its own citizens to submit to an international court. The USA has supported regimes in parts of Africa, Asia and Latin America when it has suited its purpose. For example, it supported Kuwait, attacked Libya and Iraq, and continued to impose economic sanctions against Iraq. The USA has been criticized over its Middle East policy, appearing to have been lenient with Israel and not doing enough to push the Palestinian cause.

Be a critical thinker

McDonalds and globalization

- Does the spread of fast food undermine local food producers?
- Does the spread of fast food undermine local dietary patterns?
- Does the spread of fast food create local environmental problems, such as water shortages?
- Does the spread of fast food create sociocultural problems, such as the change in farming patterns and conflict between people of different generations?

Some researchers believe that while the extremes of globalization may be modified to fit local conditions, they may change local cultural habits. For example, there is evidence to suggest that McDonalds has caused small but significant changes in Asian dietary patterns. The Japanese rarely ate food with their hands, but this is now acceptable and commonplace. Whereas McDonalds is a fast-food restaurant in the USA, it is somewhere to go for a leisurely meal in places such as Seoul, Taipei and Beijing.

McDonald's has opened restaurants in New Delhi and Mumbai. Environmentalists believe that foreign investment in farming and the fast food industry is destroying not only the Indian environment but also its traditional way of life. Environmentalists believe that foreign fast-food chains encourage people to eat a diet based on meat, which the country cannot afford. They say that breeding large numbers of animals will make it difficult for India to feed itself. The animals which give the meat, milk and eggs are fed on grain. The same amount of grain would feed five times as many people if they ate it themselves. Also, 1,700 litres of water are needed to produce 500 grams of chicken. That is 20 times the water that an average Indian family would need each day.

New political unions

The USA lost no time in rewarding its new allies in the war against terrorism, including those in the Middle East. It passed a long-delayed free-trade agreement with Jordan, and cut Egypt's official foreign debt in half in 1991, partly as a reward for its help in the Gulf War. Jordan also received a three-year, $164-million loan programme.

Previously, when Iraq invaded Kuwait in 1990, Egypt gave prompt and full support to an American-led alliance, decisively tipping the Arab world against Saddam Hussein. Egypt's government was not about to let down a superpower that provides $2 billion a year in military and economic aid.

Following 9/11, Saudi Arabia cut off ties with the Taliban, accusing it of recruiting and training impressionable Saudis to take part in international terrorism. Nevertheless, Bin Laden's religious views and his opposition to American forces in Saudi Arabia appealed to some Saudi dissidents.

Hence, it is not quite true to say that the Middle East is united against the perceived threat of globalization or westernization.

Case study: *Wal-Mart and adaptation to globalization*

Wal-Mart is the world's biggest retailer and one of the world's largest companies. It was founded by Sam Walton, with the concept of "low prices always" at the core of the business. The first discount store opened in 1962 in Rodgers, Arkansas, USA. By 1985 there were 874 stores, and now there are over 5,000 stores worldwide.

Wal-Mart became an international company in 1991, and there are now over 7,000 stores (3,000 of them overseas) in more than 10 countries (USA, Canada, Mexico, Puerto Rico, China, Korea, Japan, Argentina, Germany and the UK). However, although Wal-Mart is the model of a globalized company, offering global products to a global market, it has adapted, where necessary, to local conditions. This helps explain how it has been able to survive and thrive overseas.

Figure 18.2 Wal-Mart in China

Wal-Mart in China

In 1996 Wal-Mart decided to enter the Chinese market. As its development progressed, the company realized that Chinese tastes were different from those elsewhere. In particular, Chinese customers required leafy vegetables, which could only be purchased locally. This complicated Wal-Mart's normal strategy of "global centralized" purchasing. The vegetable section in Chinese Wal-Marts is double the size of that of American Wal-Marts, and the products are different. In the Shenzen Wal-Mart it is possible to purchase chicken feet, stewed pork ribs and pickled lettuce, products not normally found in American Wal-Marts.

Many products had to be sourced locally, partly because of the poor transport system in China, and partly because of government regulations; Chinese rules state that products such as alcohol and tobacco have to be purchased locally. The result is that about 85% of products sold in Wal-Mart China come from some 14,000 Chinese suppliers.

Although different from the US model, operations in Wal-Mart China have been very successful, with profits of over $670 million in 2004.

Wal-Mart in Germany

Wal-Mart entered the German market in 1988. At that time, Germany already had discount retailers such as Aldi and Lidl. Customer service was identified as one way that Wal-Mart could differentiate itself from its competitors. Wal-Mart took over approximately 100 hypermarkets in its first year.

In the USA, Wal-Mart customers are met by friendly "greeters", who smile and give them a shopping trolley. It is an attempt to personalize the shopping experience. This did not go down well in Germany, where the greeters were regarded as superficial, and some stores removed the greeters altogether.

One development that took place in Germany was the introduction of "singles shopping nights" on Fridays, in many German Wal-Mart stores. This proved very successful – Friday night profits increased by 25% and at least 30 couples were formed. The idea was transferred to a number of stores in Canada and Korea. However, Wal-Mart closed in Germany in 2006, following dispute over labour costs.

Globalized and localized production

Local commercial production	Globalized production
Benefits	**Benefits**
Producer • Increased market access and sales • Possibly more farm-gate sales	*Producer* • Ability to produce foods cheaply and to a uniform standard
Consumer • Fresh food • Local products "in season" • Reduced air miles • Smaller carbon footprint	*Consumer* • Cheap food available year round • All types of products available year round • Competition between producers keeps main costs down
Local economy • Improved local farming economy • Multiplier effects, e.g. demand for fertilizers, vets, farm equipment	*Local economy* • May provide large amounts of a single product to a major TNC • Specialization allows intensification and increased production
Costs	**Costs**
Producer • Increasing cost of oil makes cost of inputs higher • Greater emphasis on quality may make production less profitable	*Producer* • Increased air miles • Higher costs of inputs, especially fertilizers and oil • Profit margins increasingly squeezed
Consumer • Higher cost of local farm products • Less choice "out of season"	*Consumer* • Increased costs are likely to be passed on to the consumer • Indirect costs such as pollution control, eutrophication of streams, soil erosion, declining water quality
Local economy • Cost of subsidies to maintain farming, e.g. payments to encourage farming in environmentally friendly ways	*Local economy* • Undercuts local farmers who may quit farming • Producers are vulnerable to changes in demand and at the mercy of TNCs

Table 18.3 Globalized and localized food production

Alternatives to globalization

Global civil society is extraordinarily heterogeneous. Groups that comprise it can be liberal, democratic and peaceful, while others are illiberal, anti-democratic and violent. These groups may be large organizations, such as Oxfam, or relatively small, such as Operation Hunger in South Africa. Even global groups that advocate progressive

Civil society – the arena of collective action by independent organizations or groups, based on shared interests, purposes and values.

values – development non-governmental organizations (NGOs) for example – may sometimes act in ways that run counter to those values, or may not be completely independent of the state.

The rise of NGOs

The perception that global institutions, such as the World Bank and the IMF, are undemocratic and do not help all people equally has led to a global civil society movement that is attempting to regulate the global system from below. This has witnessed a massive rise in NGOs representing the needs of many "victims" of globalization. The statistics are impressive:

- A survey of NGOs in 22 nations showed that they employed 19 million workers, recruited 10 million volunteers and generated $1.1 trillion in revenue.
- In 1960, every country had citizens participating in 122 NGOs – by 1990, the number had increased to over 500.
- In western Europe, 66% of NGOs have been formed since 1970.
- There are over 2 million NGOs in the USA, 75% of which have been formed since 1968.
- In eastern Europe, 100 000 non-profit organizations appeared between 1989 and 1995.
- In Kenya, over 250 NGOs appear every year.
- In 1909, there were just 176 international NGOs; by 2000, there were over 29 000, 60% of which had been formed since the 1960s.

However, caution is required. Evidence from South Africa suggests that many small-scale NGOs and local bottom-up development schemes fold after a short time. The figures must be treated with care.

Several broad alliances have emerged within NGOs, such as the global environmental movement, the anti-globalization movement, and the global women's movement. Well-known individual NGOs include Greenpeace, The Fair Trade Network, Stop The War Coalition, Globalize Resistance, Oxfam, CAFOD, Amnesty International and Médicins sans Frontières. Each of these has different aims and methods, but all agree that major globalizing bodies such as the World Bank, the IMF and the G8 countries are pushing an agenda that favours rich western countries at the expense of others.

At an individual level, some people have decided to boycott GM crops. Others, during the recent increase in oil prices, boycotted garages owned by Shell and BP. Others choose to do something positive – buying Fair Trade products is one way of helping producers in poor countries at the expense of large TNCs.

While the role of global civil society institutions should not be overstated (they are generally much less powerful than governments, international organizations and the private sector), there are plenty of recent examples of where global civil society groups have been a force for progressive social change. The International Campaign against Landmines and the Jubilee 2000 campaign for debt relief are two of the best known and most successful. More generally, parts of global civil society have succeeded in putting new issues and ideas on to the international agenda, and in effecting changes in national and international policies. They have helped to improve the transparency and, to some extent, the accountability of global institutions, and to mobilize public awareness and political engagement.

Important areas where global civil society is trying to have an impact include:

- creating a more level playing field for the global South
- supporting free media and access to information
- making global civil society more accountable and transparent
- establishing a new relationship with global institutions.

Case study: *Civil society, Shell and Ogoniland, Nigeria*

In 1979 Nigeria was at the peak of an oil boom. Oil brought in $25 billion that year and external debt was less than $10 billion. Within a few years, however, Nigeria had gone from boom to bust, and has yet to recover. Shell is by far the largest oil company in Nigeria, operating mainly in the Niger Delta, and has long been the focus of many protests. The Ogonis are one of many ethnic groups living in the region, and they are calling for a greater say in how their land is used.

Shell is responsible for nearly half the country's output of 2 million barrels a day, and Nigeria is as dependent on oil as it ever was. It accounts for 80% of export earnings and 90% of government revenue. Additionally, Shell is the leading partner in a proposed liquefied natural gas (LNG) project. This promises to be the most important source of foreign exchange in Nigeria since the development of the oil industry.

The text below is taken from the newspaper advertisements taken out by Greenpeace, The Body Shop International, Friends of the Earth and Chaos Communications. It raises a broad spectrum of issues:

- economic (should the public buy Shell products?)
- environmental (degradation and pollution)
- social (poor people unable to defend themselves)
- cultural (the Ogoni way of life).

DEAR SHELL, THIS IS THE TRUTH. AND IT STINKS.

For over thirty years, the activities of the Nigerian government, Shell and other multinational oil companies have led to the widespread degradation and pollution of the region's lakes, rivers, land and air. The Ogoni are mostly farmers and fishermen, who need their land and water to live. The oil spills and pollution must be cleared up and the lands restored.

Shell must take responsibility for their part in this pollution. We believe that Shell has an obligation to operate to the highest environmental and social standards. We do not believe that Shell has done so in Nigeria.

Please heed the words of Ken Saro-Wiwa himself, writing from his prison cell before his execution on 10 November 1995: "I believe that only a boycott of Shell products and picketing of garages can call Shell to their responsibility to the Niger Delta. I remain hopeful that men and women of goodwill can come to the assistance of the poor deprived in Ogoni and other parts in the Niger Delta who are in no position to defend themselves against a multinational such as Shell."

The case for Shell

Shell also took out an advertisement. These are some of its points.

CLEAR THINKING IN TROUBLED TIMES

There are certainly environmental problems in the area, but as the World Bank's Survey has confirmed, in addition to the oil industry, population growth, deforestation, soil erosion and overfarming are also major environmental problems there.

In fact, Shell and its partners are spending US$100 million this year alone on environment-related projects, and US$20 million on roads, health clinics, schools, scholarships, water schemes and agricultural support projects to help the people of the region. And, recognizing that solutions need to be based on facts, they are sponsoring a $4.5 million independent survey of the Niger Delta.

But another problem is sabotage. In the Ogoni area – where Shell has not operated since January 1993 – over 60% of oil spills were caused by sabotage, usually linked with claims for compensation. And when contractors have tried to

deal with these problems, they have been forcibly denied access.

It has also been suggested that Shell should pull out of Nigeria's Liquefied Natural Gas project. But if we do so now, the project will collapse...A cancellation would certainly hurt the thousands of Nigerians who will be working on the project, and the tens of thousands more benefiting in the local economy. The environment, too, would suffer, with the plant expected to cut greatly the need for gas flaring in the oil industry. It's only the people and the Nigerian Government of that time who will pay the price.

And what would happen if Shell pulled out of Nigeria altogether? The oil would certainly continue flowing. The business would continue operating. The vast majority of employees would remain in place. But the sound and ethical business practices synonymous with Shell, the environmental investment, and tens of millions of dollars spent on community programmes would all be lost. Again, it's the people of Nigeria that you would hurt.

It's easy enough to sit in our comfortable homes in the West, calling for sanctions and boycotts against a developing country. But you have to be sure that knee-jerk reactions won't do more harm than good.

Some campaigning groups say that we should intervene in the political process in Nigeria. But even if we could, we must never do so. Politics is the business of governments and politicians. The world where companies use their economic influence to prop up or bring down governments would be a frightening and bleak one indeed.

We'll keep you in touch with the facts.

Case study: *Quality of life in a non-globalized society*

The Popoluca Indians consist of about 30 000 people who live in a groups of small villages in the state of Vera Cruz, Mexico. The Popoluca generally work as farmers and hunter-gatherers, and their society appears to be somewhat primitive by modern standards. The technology that they use is relatively simple, there is no mechanization, and they use only human labour.

The Popoluca Indians practise a form of agriculture that resembles shifting cultivation, known as the **milpa system** (*milpa* meaning field). They cultivate over 200 species, including maize, beans, cucurbits, papaya, squash, watermelon, tomatoes, oregano, coffee and chilli. The Popolucas have developed this system into a fine art that mimics the natural rainforest. The variety of a natural rainforest is repeated by the variety of shifting cultivation. For example, lemon trees, peppervine and spearmint are **heliophytes,** light seeking, and prefer open conditions, not shade. Coffee, by contrast, is a **sciophyte,** and prefers shade, while the mango tree requires damp conditions.

The close associations found in natural conditions are also seen in the Popolucas' farming system. The Popolucas show a huge amount of ecological knowledge and management. For example, maize and beans go well together, as maize extracts nutrients from the soil while beans return them. Tree trunks and small trees are left because they are useful for returning nutrients to the soil and preventing soil erosion. They are also used as a source of material for housing, hunting spears, and medicines.

Although poor in a monetary sense, the Popoluca Indians have a very good quality of life, as measured by health and diet. Unemployment and crime rates are also low. In all, their farming system uses 244 species of plant, and their food animals include chickens, pigs and turkeys. These are also used for barter and exchange for money, and their waste is used as manure. Rivers and lakes are used for fishing and catching turtles. Thus it is not entirely a subsistence lifestyle, since wood, fruit, turtles and other animals are traded for seeds, mainly maize.

Pressures on the Popoluca

About 90% of Mexico's rainforest has been cut down in recent decades, largely for new forms of agriculture. This is partly a response to Mexico's huge international debt and attempts by the government to increase agricultural exports and reduce imports. The main new forms of farming are:

- cattle ranching for export
- plantations or cash crops, such as tobacco.

However, these new methods are not necessarily suited to the physical and economic environment. Tobacco needs protection from sunlight and excess moisture, and the soil needs to be very fertile. The rainforest is cleared but frequently left bare, and this leads to soil erosion (see chapter 3). Unlike the milpa system, cash crops such as pineapple, sugar cane and tobacco require large inputs of fertilizer, insecticides and pesticides. These inputs are expensive, and costs are rising rapidly.

Ranching prevents the natural succession of vegetation, because of a lack of seed from nearby forests and the grazing effects of cattle. Grasses and a few legumes become dominant. Whereas one hectare of rainforest supports about 200 species of tree and up to 10 000 individual plants, a hectare of rangeland supports just one cow and one or two types of grass. But it is profitable in the short term because land is available, and it is supported by the Mexican government.

Extensive **monoculture** is not only mechanized and costly but it also leads to problems of soil deterioration and microclimatic change. Yet there is little pressure to improve efficiency because it is easy to clear new forest.

The Mexican rainforest can be described as a "**desert covered by trees**". Under natural conditions it is dynamic, but its **resilience** depends on the level of disturbance. Sustainable development of the rainforest requires the management and use of its natural structure and diversity, namely local species, local knowledge and skills rather than the type of farming developed elsewhere and then imported.

To research

Visit Survival International's website at http://www.survivalinternational.org/

Find out about the Penan of Malaysia and the Obo of Ethiopia. Outline the pressures they face. Describe their traditional life style.

Visit http://www.beautyworlds.com/mexicanbutterflies.htm for the memories of an entomologist who stayed with the Popolucas, and www.jstor.org/pss/663215 for a review of *A Primitive Mexican Economy* by George Foster (1942). Contrast his views and comments with your lifestyle, listing the main differences and similarities.

Part 4 Preparing for the exam

You should be aware of the composition of the exams. Higher level students write three papers, which account for 25%, 35%, and 20% of the total mark. Internal assessment accounts for the remaining 25%. For standard level candidates, they write two papers, worth 40% and 35%, and an Internal Assessment worth 25% of the total marks.

The Diploma Programme uses several methods to assess work produced by students. It mainly uses summative assessment, which provides an overview of previous learning, and is concerned with measuring student achievement. The approach to assessment used by the Diploma Programme is criterion related. It looks at students' work by their performance in relation to identified levels of attainment. Markbands are a statement of expected performance against which which responses are judged. Each level descriptor corresponds to a range of marks to differentiate student performance.

19 Essay writing guidelines

The relative importance of essays at HL and SL

IB exams consist of different approaches to assessment, including extended responses. The advice given here is directed towards conventional full-length essays, which are compulsory in Papers 1 and 3. In both cases, one essay carries a relatively heavy mark weighting, as shown below:

- Paper 1, Section B HL 6.25%, SL 10% of total marks
- Paper 3 (HL only) 20% of total marks. In this exam you will have one hour to answer the question, which appears as parts (a) and (b). These may be linked to the same topic, for example Economic Interactions, or may be independent. Either way, you should approach the two parts separately and assume that the examiner will not cross-credit them, i.e. transfer marks from one to the other if information is misplaced.

Interpreting the essay title

1. Underline the keywords in the title.
2. Go through the checklist below to check each aspect against your essay title to see if it is relevant or not. This will ensure that you give the essay title its broadest interpretation. The title may be brief and leave you to think creatively and to comment on specific aspects of the subject which are not actually mentioned in the title but which are relevant to it. For example, if the question asks you to comment on the global variation in fertility rate, you would need to write about variations in time as well as space.

Checklist

Note that not all the items in this checklist will be relevant to your essay.

- **Location** – poor/rich countries, rural/urban areas, tropical/ temperate regions

- **Issues** – positive/negative; human/physical; environmental/social/demographic/political/economic
- **Scale** – global, regional, national, sub-national, local, household
- **Time** – long-term/short-term; past, present, future

Planning

Planning is important. Reasons why you should plan your essay include:

- it allows you to order your thoughts before writing.
- you can return to the essay plan and insert new points as you get inspiration while writing.
- it presents a logical sequence of ideas that the reader can easily follow.
- examiners have little time and will credit a well-structured answer that is easy to follow.
- it allows you to focus on the question and make sure that the content is relevant.

Structure of the essay

Introduction

The introductory paragraph gives an interpretation of the title, defines terms, indicates the slant or the direction of the argument and generally sets the scene.

The main body of the essay

Make sure that each paragraph in this part of your essay presents a distinct point or idea. The opening line of each paragraph should clearly indicate its content. The remainder of the paragraph elaborates on that point.

Examples, case studies and illustrations, such as sketch maps and diagrams, should appear in this section.

Conclusion

Here you should return to the essay title and provide an overview of your response. The conclusion should not contain new ideas; it should round off an argument and summarize the key features of the content.

The language of IB exams

It is recommended that you become familiar with the command words and other terms listed and defined below.

They are all found in IB geography exam questions – misinterpretation costs marks.

Analyse	break down in order to bring out the essential elements or structure
Annotate	add brief notes to a diagram or graph
Classify	arrange or order by class or categories
Compare	give an account of the similarities between two (or more) items or situations, referring to both (or all) of them throughout
Compare and contrast	give an account of similarities and differences between two (or more) items or situations, referring to both (or all) of them throughout
Construct	display information in a diagrammatic or logical form
Contrast	give an account of the differences between two (or more) items or situations, referring to both (or all) of them throughout

Define	give the precise meaning of, for example, a word, phrase, concept or physical quantity
Describe	give a detailed account
Determine	obtain the only possible answer
Discuss	offer a considered and balanced review that includes a range of arguments, factors or hypotheses. Opinions or conclusions should be presented clearly and supported by appropriate evidence
Distinguish	make clear the differences between two or more concepts/items
Draw	represent by means of a labelled, accurate diagram or graph, using a pencil. A ruler (straight edge) should be used for straight lines. Diagrams should be drawn to scale. Graphs should have points correctly plotted (if appropriate) and joined in a straight line or smooth curve
Estimate	obtain an approximate value
Evaluate	make an appraisal by weighing up the strengths and limitations
Examine	consider an argument or concept in a way that uncovers the assumptions and interrelationships of the issue
Explain	give a detailed account, including reasons or causes
Identify	find an answer from a number of possibilities
Justify	give valid reasons or evidence for an answer or conclusion
Label	add labels to a diagram
Outline	give a brief account or summary
State	give a specific name, value or other brief answer without explanation or calculation
Suggest	propose a solution, hypothesis or other possible answer
To what extent	consider the merits or otherwise of an argument or concept. Opinions and conclusions should be presented clearly and supported with empirical evidence and sound argument

Source: *Adapted from the Diploma Programme Geography Guide, IBO*

Exam-speak – common terms that confuse

Verbs

Referring to	mentioning or using
Influence	explain the effect of one thing upon another
Modify	change
Respond to	take action

Nouns

Outcome	consequence/result
Benefits/advantages	positive outcomes
Costs/disadvantages	negative outcomes
Impacts/effects	usually dramatic outcomes
Issues	important and controversial results
Problems	difficulties
Pressures/conflicts	undesirable competition
Challenges	difficulties which may be overcome
Opportunities	potential benefits
Trend	change over time (on a graph)
Pattern	distribution in space
Feature	a distinct part, e.g. a cliff is a coastal feature
Process	the actions or changes that occur between parts
Relationship	a two-way interaction

Adjectives

Global	the whole world
Regional	global regions, e.g. Asia-Pacific
National	belonging to one country
Local	the immediate area or district
Possible	likely to happen
Probable	very likely to happen
Economic	relates to business, finance, employment
Social	relates to human welfare e.g. housing and health
Cultural	relates to language, customs, religion and moral codes
Political	relates to the actions of governments
Demographic	relates to populations e.g. fertility rate
Environmental	relates to the physical environment

External markbands–SL and HL

Paper 1 and 2 markbands

These markbands areto be used for papers 1 and 2 at both standard level and higher level

	AO1	AO2	AO3	AO4	Paper 1 Section B	Paper 2
Level descriptor	**Knowledge/ understanding**	**Application/analysis**	**Synthesis/ evaluation**	**Skills**	**Marks 0–15**	**Marks 0–10**
A	No relevant knowledge; no examples or case studies	No evidence of application; the question has been completely misinterpreted or omitted	No evaluation	None appropriate	0	0
B	Little knowledge and/ or understanding, which is largely superficial or of marginal relevance; no or irrelevant examples and case studies	Very little application; important aspects of the question are ignored	No evaluation	Very low level; little attempt at organization of material; no relevant terminology	1–3	1–2
C	Some relevant knowledge and understanding, but with some omissions; examples and case studies are included, but limited in detail	Little attempt at application answer partially addresses question	No evaluation	Few or no maps or diagrams, little evidence of skills or organization of material; poor terminology	4–6	3–4
D	Relevant knowledge and understanding, but with some omissions; examples and case studies are included, occasionally generalized	Some attempt at application; competent answer although not fully developed, and tends to be descriptive	No evaluation or unsubstantiated evaluation	Basic maps or diagrams, but evidence of some skills; some indication of structure and organization of material; acceptable terminology	7–9	5–6
E	Generally accurate knowledge; and understanding, but with some minor omissions examples and case studies are well chosen, occasionally generalized	Appropriate application; developed answer that covers most aspects of the question	Beginning to show some attempt at evaluation of the issue, which may be unbalanced	Acceptable maps and diagrams; appropriate structure and organization of material; generally appropriate terminology	10–12	7–8
F	Accurate, specific, well-detailed knowledge and understanding; examples and case studies are well chosen and developed	Detailed application; well-developed answer that covers most or all aspects of the question	Good and well-balanced attempt at evaluation	Appropriate and sound maps and diagrams; well structured and organized responses; terminology sound	13–15	9–10

Paper 3 markbands

Part (a)

Level descriptor	Knowledge/ understanding AO1	Application/analysis AO2	Skills AO4	Marks 0–10
A	No relevant knowledge, or inappropriate	The question has been completely misinterpreted or omitted	None appropriate	0
B	Little relevant knowledge and/or understanding	Important aspects of the question are ignored	Little attempt at organization of material	1–3
C	Some relevant knowledge and understanding	Answer partially addresses the question	Some indication of structure or organization	4–6
D	Generally accurate knowledge and understanding	Answer is developed and covers most aspects of the question	Appropriate structure with generally appropriate terminology	7–8
E	Accurate, relevant knowledge and understanding	Well-developed answer that covers most or all aspects of the question	Well-structured response with sound terminology	9–10

Part (b)

Level description	Knowledge/ understanding AO1	Application/ analysis AO2	Synthesis/ evaluation AO3	Skills AO4	Marks 0–10
A	No relevant knowledge, or inappropriate	The question has been completely misinterpreted or omitted	No synthesis/ evaluation	None appropriate	0
B	Little relevant knowledge and/or understanding	Important aspects of the question are ignored	Little attempt at synthesis/evaluation	Little attempt at organization of material	1–4
C	Some relevant knowledge and understanding	Answer partially addresses the question	Basic synthesis/basic or unsubstantiated evaluation	Some indication of structure or organization	5–8
D	Generally accurate know ledge and understanding	Answer is developed and covers most aspects of the question	Synthesis that may be partially undeveloped/ evaluation that may be partially unsubstantiated	Appropriate structure with generally appropriate terminology	9–12
E	Accurate, relevant know ledge and understanding	Well-developed answer that covers most or all aspects of the question	Clear, developed synthesis/clear, substantiated evaluation	Well-structured response with sound terminology	13–15

20 Internal assessment: advice to students and teachers

Why fieldwork matters

Fieldwork is an essential part of learning geography and is compulsory for both HL and SL students. It is referred to as "Internal Assessment" (IA), which means that it will be marked by your teacher and moderated by an external IB examiner.

Your fieldwork investigation is important because it will:

- help you make sense of some of the more difficult aspects of the subject
- improve your overall grade, especially if you don't perform so well in the external exams
- provide useful case study material when answering an external exam question
- provide research skills which will be useful in higher education or employment.

Internal assessment – the essentials

- IA counts for 20% of the total marks at HL and 25% at SL.
- It requires 20 hours of class time (including fieldwork).
- Group work is allowed for data collection.
- Fieldwork reports are written individually.
- Each report must be no more than 2,500 words in length.
- It must be related to a topic on the syllabus.

Fieldwork research methods

Information must come from the student's own observations and measurements collected in the field. This **"primary information"** must form the basis of each investigation.

Fieldwork should provide sufficient information to enable adequate interpretation and analysis.

Common errors

- The report exceeds the 2,500 word limit.
- The chosen topic has no spatial element.
- The chosen topic is not geographical.
- The chosen topic does not relate to the syllabus.
- The fieldwork question is too simplistic.
- The information is collected only from the internet.
- The survey area is too large and covers the whole region.
- The fieldwork information is insufficient to answer the fieldwork question.
- The analysis is purely descriptive.

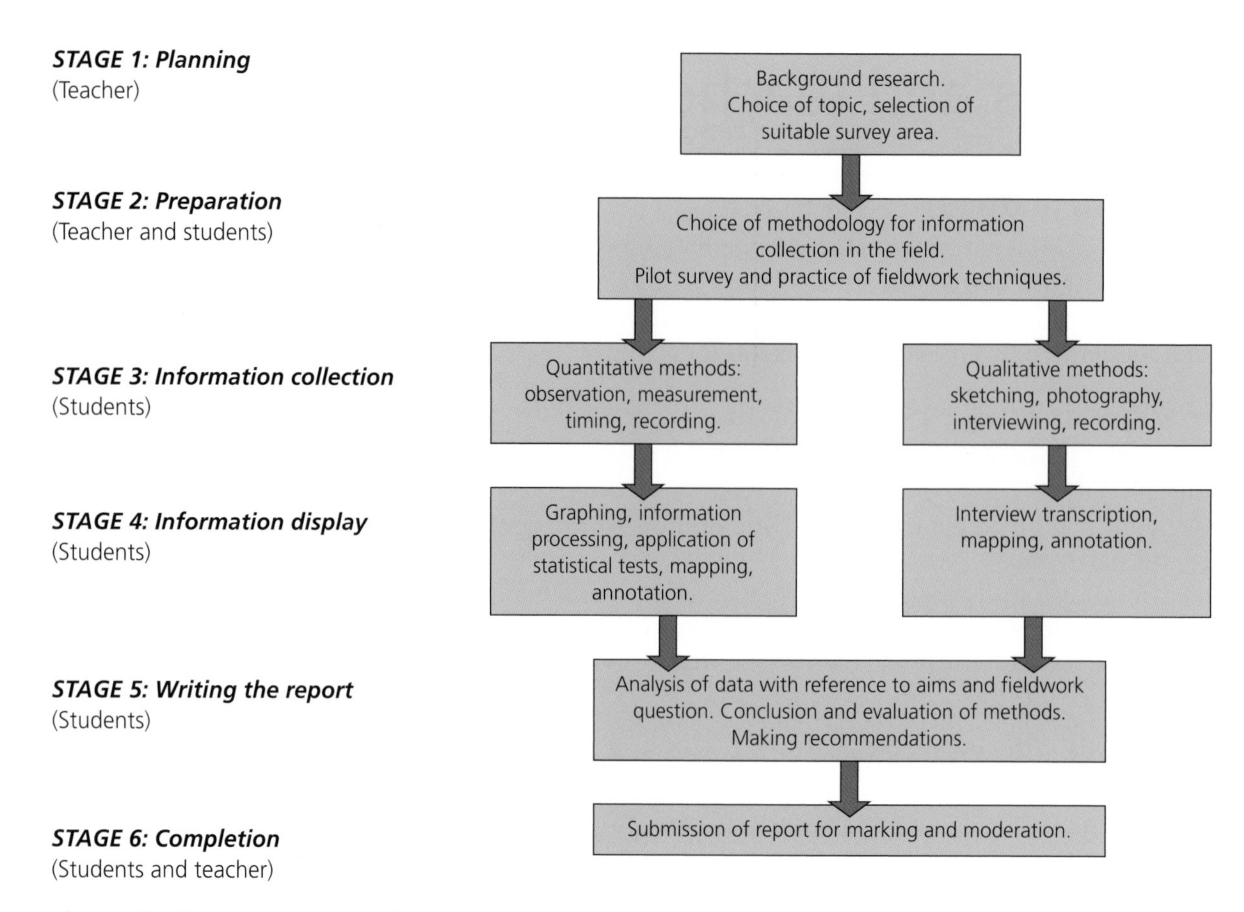

Figure 20.1 Stages in undertaking fieldwork for internal assessment

Stage 1 (teacher only)

Background research

The success of your students' fieldwork will depend on your careful planning and preparation. The following resources are essential reading before you start.

- The Diploma Programme Geography Guide (2009 version with first exam in May 2011) can be found on the subject page of the online curriculum centre (OCC) at **http://occ.ibo.org**, a password-protected IB website designed to support IB teachers. The subject guide can also be purchased from the IB store at **http://store.ibo.org**.
- Additional publications such as teacher support materials, subject reports, internal assessment guidance and grade descriptors can also be found at the OCC.
- The Examiner's reports provide teachers with an overall review of investigations undertaken in a large range of schools and make recommendations for future investigations. The IB Online Curriculum Centre is a discussion forum where geography teachers often exchange ideas on fieldwork.

Choosing the right topic

The fieldwork topic must be related to the syllabus, and the most suitable topics are found within the Optional Themes. The core and

HL extension have very few topics which are suitable owing to their global scale. The investigation must be:

- focused upon a clearly defined fieldwork question
- confined to a small area and on a local scale
- spatial
- based on the collection of primary information in the field
- manageable in terms of the area covered, the time allowed and the 2,500 word limit
- able to fulfil the assessment criteria.

Choosing the right site

The viability and success of the fieldwork is determined by careful planning and preparation. It is essential that you select the survey area in advance of the fieldwork investigation to ensure that it fulfils the following criteria:

- It is on a local scale, but the area covered is large enough for sufficient information to be collected.
- The area can be covered by the students in the time allocated.
- All sites within the area are accessible at all times of day and at all seasons.
- The land is open to the public and research is permitted.

Where fieldwork is restricted to the school site, many successful investigations can be undertaken; for example, surveys of footpath erosion, microclimate, infiltration/ground compaction and waste management.

The role of the teacher, the group and the individual student

It is advisable for you to choose the fieldwork topic and test its viability before embarking on the class exercise.

In general, the most successful undertakings are those involving group work, with the initial planning done by the teacher. The choice of topic, the scale of the investigation, the area covered and the time allowed will be determined by the number of students available to carry out the work. Fieldwork methods used to collect information should be chosen by you, and the techniques and equipment should be practised prior to the investigation by the students. Once the fieldwork is over and the information made available to all members of the class, students should work individually and no further collaboration is allowed.

Stage 2 (teacher and students)

Devising the fieldwork question

The fieldwork question forms a basis to the research, which should allow for an investigative rather than descriptive approach. The question should be clearly focused, unambiguous and answerable. If the question is simplistic and the answer obvious, it is unlikely to be worthy of execution. However, research topics which have uncertain outcome are still perfectly viable.

Collecting the right information!

Fieldwork must involve the collection of primary information. Primary information may be qualitative or quantitative, or a combination of both. In the case of a traffic survey, qualitative data might include photographs, interviews with pedestrians and the subjective assessment of perceived traffic hazard by the student. Quantitative information might include traffic counts, traffic delay times, length of tailback, noise levels in decibels or a survey of suspended particulate matter in the atmosphere. Secondary or published information not collected by the students themselves may be used to supplement primary information. It must not form the basis of the report.

Stage 3 (over to the student)

Once your teacher has done the initial planning and preparation, it is over to you to undertake the task of information collection. Remember that this is a one-off opportunity: the stormy conditions during which you collected your wave data cannot be repeated.

Collecting and justifying your fieldwork methods

You must be aware of all the techniques involved and be able to critically evaluate each of them. Before you start collecting information and before you leave the survey site, make sure you have:

- marked for the sites of information collection
- recorded the date and time of collection
- recorded the weather conditions or any special event occurring on the day that might affect the results
- recorded the technique of handling a particular instrument, where it is placed, the time interval between readings, the advantages and disadvantages of the technique
- justified the choice of survey sites and their number/frequency/location
- justified the choice of method used for information collection
- justified the sampling technique used.

Stage 4 (students)

How to display your fieldwork information

Your fieldwork data should be displayed next to the text to which it refers and should not be confined to the end of the report. Use the table below as a guide.

Method	Do	Don't
Maps	✓ Include a map of the survey sites. ✓ Show your results at specific survey sites on this map. ✓ Annotate your map with brief comments.	× Include a national map: it is irrelevant. × Include scruffy maps drawn with pencil.
Graphs	✓ Wherever possible, place a series of graphs on the same page for comparison. ✓ Use a variety of graphical techniques. ✓ Use transparent overlay maps to show spatial relationships.	× Use a monotonous series of pie charts to represent your data page by page. × Download maps from the internet without first modifying or adapting them for your purpose.

Photos	✓ Make sure that each photograph shows the time it was taken, its location and its orientation.	× Include photos of your friends and teacher unless they are strictly relevant to the investigation.
Sketches	✓ Make sure they are fully labelled/annotated and dated.	× Include these unless relevant.
Method		
Generally	Make sure that all illustrations are properly referenced. Use a range of techniques, but make sure each is suitable. Map information wherever possible.	

Stage 5 (students)

Writing your report

Your report should be structured using the assessment criteria shown below. Note that criterion C can be represented by illustrative material in any part of the report. Assessment of this criterion is not confined to one section. The mark allocation, and the recommended and approximate number of words for each criterion, are both given below.

Criteria			
A	**Fieldwork question**	**3 marks**	**300 words**
	This should be concise and clear to the reader. There should be one question only. You should comment briefly on the geographic context, explaining why that particular area of survey was chosen. It is essential to include a map showing the area under investigation. You should state the syllabus section to which the investigation relates.		
B	**Method(s) of investigation**	**3 marks**	**300 words**
	You should describe the method(s) used to collect information. The methods should be justified, which means explaining sampling techniques, the time chosen, the specific location and any other relevant information such as weather conditions.		
C	**Quality and treatment of information collected**	**5 marks**	***N/A**
	There is a range of possible techniques of data display that you might use in any investigation, but make sure that they are clear and effective. The type of method used will be determined by the nature of the particular investigation.		
D	**Written analysis**	**10 marks**	**C + D: 1,350 words**
	In the written analysis you demonstrate your knowledge and understanding by interpreting and explaining the information collected in relation to the fieldwork question. This involves recognizing spatial patterns and trends found in the information collected. Where appropriate, you should attempt to explain anomalies.		
E	**Conclusion**	**2 marks**	**200 words**
	You should summarize the findings of your fieldwork investigation. There should be a clear, concise statement answering the fieldwork question. It is acceptable for the conclusion to state that the findings do not match any of your preliminary judgments or projections.		

F	**Evaluation**	**3 marks**	**300 words**
	You should review the methods you used to collect the information in the field. You should include any factors which threatened the validity of the data, such as an abnormal weather event. Suggest viable and realistic ways in which the study might be improved in the future.		
G	**Formal requirements**	**4 marks**	**N/A**
	The written report must meet the following five formal requirements of organization and presentation: • The work is within the 2,500 word limit. • The report is neat and well structured. • The pages are numbered. • All sources are correctly referenced. • All illustrations are numbered, fully integrated into the body of the report and not placed in an appendix. General guidance on IB policy to referencing and sourcing can be found in the subject guide.		
	Finished report	**30 marks**	**<2,500 words**

*Criterion C assesses information display and does not include a word count (except for large-sized annotations).

Stage 6 (students and teacher)

Completion of the fieldwork report by the student

Complete this checklist before you submit your fieldwork report.

Tasks	Completed
The candidate name and number is stated on the front cover	
The report is bound or held together securely in a folder	
All plastic pockets have been removed	
There is a contents page	
All the pages are numbered	
All illustrations have figure numbers	
All illustrations are close to the relevant text	
All sources are referenced	
The appendix contains only raw information	
The report has a fieldwork question	
All methods of information collection are fully justified	
All maps have normal conventions of title, scale, north point and key	
The analysis refers to the fieldwork question and the information collected	
There is a conclusion	
The evaluation makes recommendations for improvements	

Examples of different methods of information collection

Investigation using primary (qualitative and quantitative) and secondary methods of information collection

Title	An investigation of gentrification in area A in town X
Aim	To determine the effects of gentrification on area A in town X and to examine local attitude towards it
Fieldwork question	How has gentrification brought social, economic and environmental changes to town X?
Syllabus theme	Urban environments
Conceptual basis	See chapter 11, The consequences of urbanization
Methods of information collection	• Socio-economic patterns and changes in area A – use secondary information from census or housing type and price from estate agents. Survey of local streets to record car type/age • Compare information with averages for town X or adjoining areas • In-migration – questionnaire survey of residents to discover occupation, length of residence and motives for moving • Local attitudes – questionnaire with long-term local residents to determine attitude to affluent newcomers and perception of long-term residents of local changes • Housing survey – evaluate housing condition and record signs of renovation and devise a housing quality index • Environmental quality survey – litter, vandalism, landscape quality, dereliction, noise pollution. In transect across area X and adjoining areas • Economic changes – local facility survey – classify and map new shops and services, new bars, restaurants, good transport services
Methods of information display	• Annotated photos (qualitative) • Maps showing scores for survey sites using overlay (quantitative) • Graphical profiles for landscape quality/dereliction • Classification and mapping of shops and services
Analytical techniques	• Chi-squared test to investigate significance of socio-economic changes in one area over time or in space • Location quotients to identify above-average level of housing renovation scores

21 Skills for fieldwork

Sampling

A sample is a representative body of data. A large number of items (the total population) can be represented by a small subsection (the sample) when it is impractical or impossible to measure the total population. Sampling is therefore an efficient use of time and resources, enabling you to make statements about a total population by using a representative section. Sampling makes fieldwork investigations manageable.

There are different types of sampling, and each has its own strengths and weaknesses. The two main types are **spatial** sampling and **temporal** sampling. Spatial sampling refers to samples taken from a variety of locations. Temporal sampling refers to samples taken over different time periods. Both can be used, for example when monitoring microclimate changes with increasing distance from a city centre between summer and winter.

Both types of sampling can be subdivided into three main subtypes: **random, systematic** and **stratified** (Figure 19.2). Before you select one or more types of sampling, you should consider the following questions:

- What is the population being studied, and in what area/time?
- What is the minimum size of sampling needed to produce reliable information and results?
- What is the most appropriate form of sampling for the enquiry?

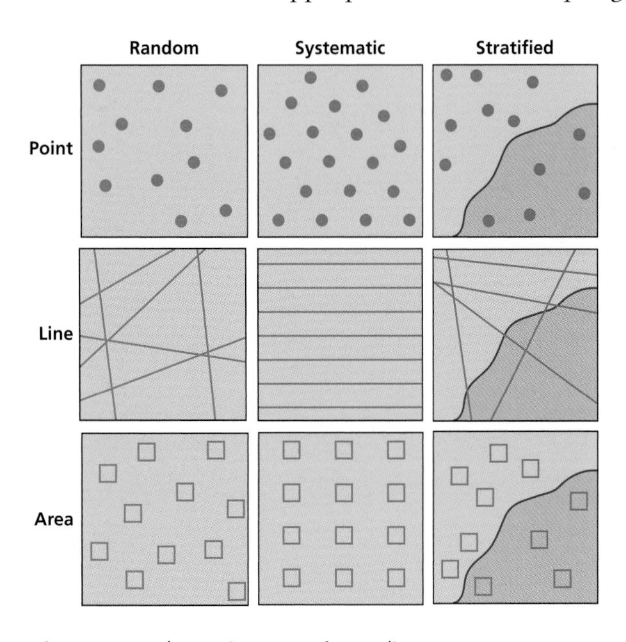

Figure 21.1 The main types of sampling

Random sampling

In a random sample, each item has an equal chance of being picked. Samples are often picked by using a random number table (Figure 19.3). This is a table with no bias in the sequence of numbers.

Once a number is chosen it can be related to a map, a grid reference, an angular direction and distance. Although fair, the random sample may miss important parts of the survey area. It is also very time consuming to do properly.

17	42	28	23	17	59	66	38	61	02	10	78
33	53	70	11	54	48	63	50	90	37	21	46
77	84	87	67	39	95	85	54	97	37	33	41
61	05	92	08	29	94	19	96	50	01	33	85
50	14	30	85	38	97	56	37	08	12	23	07
27	26	08	79	61	03	62	93	23	29	26	04
03	64	59	07	42	95	81	39	06	41	29	81
17	08	72	87	46	75	73	00	26	04	66	91
40	49	27	56	48	79	34	32	81	22	60	53

Figure 21.2 A random number table

Systematic sampling

Systematic sampling is much quicker and easier than random sampling. Items are chosen at regular intervals, e.g. every 5 metres, every 10th person, and so on. However, it is possible that a systematic sample will miss out important features. For examples, in a survey of soil moisture and temperature in a ploughed field, if the surveys are taken on every ridge (or every furrow) and disregard other important microclimates, the results will be biased. The major problem with this type of sampling is that it can easily give a biased result because the sample is too small, and as a result large areas are left out of the sample.

Stratified sampling

If you know that there are important sub-groups in an area, for example different rock types that could influence soil types or farming types, it is possible to make a representative sample that takes account of all the sub-groups in the study area. It is also possible to "weight" the sample so that there is a proportionate number of samples related to the relative size of each sub-group.

Sample size

Determining the appropriate size of a sample is a critical matter. In depends on the nature and aims of the investigation but also on the time available (and other practical considerations such as access, land ownership, safety, and so on). In general, a sample size of 30 or more constitutes a large sample. Samples below 10 are often unreliable, but it may be difficult, if not impossible, to survey 30 sites on a river safely, or to collect data from 30 different settlements.

Descriptive statistics

There are many types of statistics, some of them extremely easy and some very complex. The most basic are simple **descriptive** statistics. These include the **mean** or **average**, the **maximum, minimum, range** (maximum–minimum), the **mode** (the most frequently occurring number, group or class) and the **median** (the middle value when all the numbers are placed in ascending or descending rank order).

There are also four different types of data:

- **Nominal** data refer to named objects , such as rock types, land uses, dates of floods,
- famines, and so on.
- **Ordinal** or **ranked** data are placed in ascending or descending order; for example, settlement hierarchies are often expressed in terms of ranks. Spearman's Rank correlation coefficient (see below) is used to compare two sets of ranked data such as IMR and PPP.
- **Interval** or **ratio** data refer to real numbers. Interval data have no true zero (as in the case of temperature, which can be in °C or °F), whereas ratio data possess a true zero (as in the case of rainfall).

Summarizing data

The **mode** refers to the group or class that occurs most often. In this case every value occurs once, so there is no mode. If, however, two values are 18 936 (for instance) the mode would be 18 936.

Another statistic is the **median**. This is the middle value when all the data are placed in ascending or descending order. In this case, because there are two middle values (the 10th and 11th values) we take the average of these two. In this case they are 20 199 and 18 936, so the median (middle) value is 19 567.5 (thousands), which is not actually a value in the data set.

Country	Tourist arrivals
France	79 083
Spain	58 451
USA	51 063
China	49 600
Italy	41 058
UK	30 654
Germany	23 569
Mexico	21 353
Austria	20 261
Russia	20 199
Ukraine	18 936
Turkey	18 916
Canada	18 265
Malaysia	17 547
Greece	16 039
Hong Kong	15 821
Poland	15 670
Thailand	13 882
Portugal	11 282
Netherlands	10 739
	∑ = 552 388

The **mean** or average is found by totalling (∑) the values (x) for all observations and then dividing by the total number of observations (n), thus ∑x/n. In this case the average number of tourist arrivals/country is 552 388/20 = 27 619.4 (thousands)

(*Economist Pocket World In Figures*, 2009)

Figure 21.3 Countries with the most tourist arrivals

Summarizing groups of data

In some cases, the data we collect is in the form of a group: e.g. daily rainfall, slope angles or ages may be recorded as 0–4, 5–9, 10–14, 15–19, etc.

The data below shows daily rainfall in a tropical area. To make recording simpler, groups of 5 millimetres have been used. Finding an average is slightly more difficult. We use the mid-point of the group and multiply this by the frequency.

Daily rainfall (mm)	Mid-point	Frequency	Mid-point x frequency
0–4	2	20	40
5–9	7	42	294
10–14	12	24	288
15–19	17	12	204
20–24	22	2	44
Total		n = 100	∑x = 870
Mean = ∑x/n	= 870/100	= 8.7	

Figure 21.4 Daily rainfall for a tropical area

The **modal group** is the one that occurs with the most frequency, i.e. 5–9 millimetres. The **median** or middle value will be the average of the 50th and 51st values when ranked: these are both in the 5–9 millimetre group.

Measures of dispersion

The **range** is the difference between the **maximum** (largest) and the **minimum** (smallest) value. In the example of tourist arrivals (Figure 9.4), the maximum is 79 083 and the minimum is 10 739; hence the range is 79 083 − 10 739, namely 68 344. An alternative measure is the **interquartile range (IQR)**. This is similar to the range but only gives the range of the middle half of the results – in this way the extremes are omitted. The **interquartile range** is found by removing the top and bottom **quartiles** (quarters) and stating the range that remains. The top quartile is found by taking the 25% highest values and finding the mid-point between the top 25% and the next point. The lower quartile is found by taking the 25% lowest values and finding the mid-point between these and the next highest value. The first quartile is termed Q1, and the third quartile Q3.

The interquartile range in the case of tourist arrivals is thus midway between the 5th and 6th values (i.e. halfway between 41 058 and 30 654, namely 35 856) and midway between the 15th and 16th values, (i.e. halfway between 16 039 and 15 821, namely 15 930). The result is 35 856 − 15 930, which equals 19 926, showing a much smaller variation than when all values (including extremes) are included, i.e. the range.

Not every case is as easy. For example, there may be 21, 22 or 23 figures, rather than in this case where the number of observations is divisible by 4. In those situations we have to make an informed guess at where the quartile would be. If we take the case of 21 observations, then the quartiles are at 5¼ and 16¾. The figure for Macau is added to the example (10 683) to give a 21st value.

The principle is the same as before. Find the values which represent 25% and 75% of the values. Then, find half the difference between the bottom of the top 25% and the next value below. Then find half the difference between the top of the lowest 25% and the next value above.

The 25% value is found 1/4 of the way between 41 058 and 30 654 (i.e. 5¼ along the scale, as there are now 21 values), while the 75% value (16¾) lies three-quarters of the way between 15 821 and 15 670. Thus the first quartile is found by taking a quarter of the difference of 41 058 and 30 654 from 41 058, i.e.

$$41{,}058 - \frac{(41{,}058 - 30{,}654)}{4} = 41{,}058 - 2601 = 38{,}457$$

Q1 is midway between 38,457 and 30,654, i.e. 34,555.5

The 75% value is found by taking three-quarters of the difference of 15 821 and 15 670 from 15 821, i.e.

$$15\,821 - \frac{3x\,(15\,821 - 15\,670)}{4} = 15\,821 - 113.25 = 15\,707.75$$

Q3 Is located midway between 15 821 and 15 707.25, namely 15 764.125.

Thus, the interquartile range is 34 555.5 − 15 764.125, i.e. 18 791.375.

In the next case there are 22 observations, as the data have been extended to include the 22nd largest number of tourist arrivals, Thailand at 8 659.

The 25% and 75% values now are found at 5½ and 16½ (as each quarter is 5½ in size, i.e. 22/4). Thus the 25% is found halfway between the 5th and 6th figures, 41 058 and 30 654 (i.e. 35 856) and the 75% is found halfway between the 17th and 18th values, 15 670 and 13 882 (i.e. 14 776). Hence Q1 is found halfway between the 25% value and the next value below, i.e. midway between 35 856 and 30 654, namely 33 255. Q3 is found halfway between the 75% value and the next value above, i.e. the midpoint between 14 776 and 15 670, namely 15 223.

Thus the interquartile range in this case is 33 255 − 15 233 = 18 022 (thousands).

Standard deviation

Another way of showing grouping around a central value is by using the standard deviation. This is one of the most important descriptive statistics, because:

- it takes into account all the values in a distribution
- it is necessary for probability and for more complex statistics. It measures dispersal of figures around the mean, and is calculated by first measuring the mean and then comparing the difference of each value from the mean.

It is based on the ideas of probability. If a number of observations are made then we would expect most to be quite close to the average, few much larger or smaller, and with equal proportions above and below the mean. In the case of tourist arrivals, however, there are between two and three times as many countries below the mean as above it. How does this compare with the value for the average HDI?

The formula for the standard deviation = $\sqrt{\Sigma(x-\bar{x})^2/n}$

where x refers to each observation, $\bar{x}$ to the mean, n the number of points, and $(x-\bar{x})^2$ tells us to take the mean from each observation, and then to square the result. The following example shows the working out.

	x	$\bar{x}$	$(x-\bar{x})$	$(x-\bar{x})^2$
Country	Tourist arrivals			
France	79 083	27 619.4	51 463.6	2 648 502 125
Spain	58 451	27 619.4	30 831.6	950 587 558.6
USA	51 063	27 619.4	23 443.6	549 602 381
China	49 600	27 619.4	21 980.6	483 146 776.4
Italy	41 058	27 619.4	13 438.6	180 595 970
UK	30 654	27 619.4	3,034.6	9 208 797.16
Germany	23 569	27 619.4	−4,054.4	16 438 159.36
Mexico	21 353	27 619.4	−6,266.4	39 267 768.96
Austria	20 261	27 619.4	−7,358.4	54 146 050.56
Russia	20 199	27 619.4	−7,420.4	55 062 336.16
Ukraine	18 936	27 619.4	−8,683.4	75 401 435.56
Turkey	18 916	27 619.4	−8,703.4	75 749 171.56
Canada	18 265	27 619.4	−9,354.4	87 504 799.36

Malaysia	17 547	27 619.4	−10 072.4	101 453 241.8
Greece	16 039	27 619.4	−11 580.4	134 105 664.2
Hong Kong	15 821	27 619.4	−11 798.4	139 202 242.6
Poland	15 670	27 619.4	−11 949.4	142 788 160.4
Thailand	13 882	27 619.4	−13 737.4	188 716 158.8
Portugal	11 282	27 619.4	−16 337.4	266 910 638.8
Netherlands	10 739	27 619.4	−16 880.4	284 947 904.2
	$\Sigma = 552\ 388$			$\Sigma = 6\ 483\ 337\ 340$

Figure 21.5 Tourist arrivals by country

Thus, the standard deviation is found by putting the figures into the formula:

$$s(s) = \sqrt{\frac{6\ 483\ 337\ 340}{20}} = \sqrt{324\ 166\ 867} = \text{approximately } 18\ 004$$

Thus the average deviation of all values around the mean (27 619.4) is 18 004. This gives a more accurate figure than the range or the IQR, as it takes into account all values and is not as affected by extreme values. Given normal probability we would expect that *c*.68% of the observations will fall within one standard deviation of the mean, *c*.95% within two standard deviations of the mean, and *c*.99% within three standard deviations (Figure 21.6). Here we can see quite clearly that there are a few countries well above mean (and some are over the mean plus two standard deviations, whereas others are within one standard deviation of the mean).

To do:

The data below are rainfall records for Niamey, Niger (part of the Sahel region):

570, 705, 582, 807, 432, 630, 552, 567, 407, 379, 650, 641, 558, 748, 413, 621, 599, 296, 392, 384.

a Illustrate the data by means of (i) a line graph and (ii) a dispersion diagram.

b State (i) the maximum, (ii) the minimum, (iii) the range and (iv) the interquartile range of the rainfall for the 20-year period.

c Which of these descriptive statistics is the best one, in your opinion? Explain your answer.

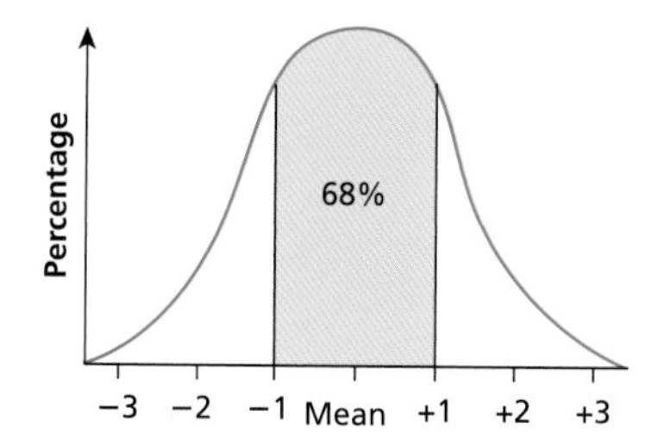

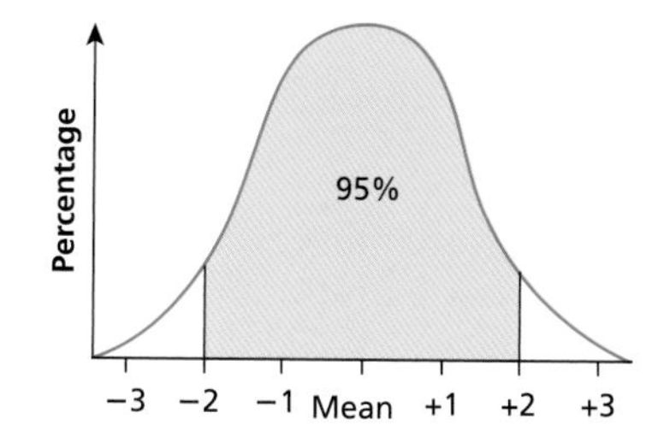

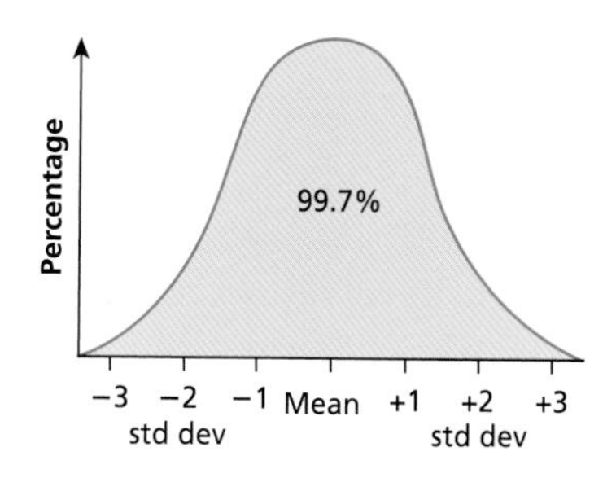

Figure 21.6 Confidence limits and standard deviations

Confidence limits are based on normal probability. This assumes that 50% of the values are above the average (mean) and 50% are below the average. It also assumes that most of the values are within one standard deviation of the mean. Probability states that in a normal distribution:

- 68% of samples lie within +/−1 standard deviation of the mean
- 95% of samples lie within +/−2 standard deviations of the mean
- 99.9% of samples lie within +/−3 standard deviations of the mean.

In other words, there is less than a 1 in a 100 chance that the mean lies outside the sample mean +/−3 standard deviations, and less than a 1 in 20 chance that the true population mean lies outside the sample mean +/−2 standard deviations.

The Gini coefficient

The Gini coefficient is a measure of dispersion, or inequality. It can be used to show inequality in wealth or in land distribution (Figure 21.7). It is usually defined mathematically, based on the Lorenz Curve as shown below for population distribution in Egypt and the UK. It is the ratio of the area that lies between the diagonal line (line of equality) and the curved line over the total area under the diagonal line.

The Gini coefficient ranges between 0 to 1; it is sometimes multiplied by 100 to give values between 0 and 100. A low Gini coefficient indicates a more equal distribution, with 0 corresponding to complete equality, while higher Gini coefficients indicate more unequal distribution, with 1 corresponding to complete inequality.

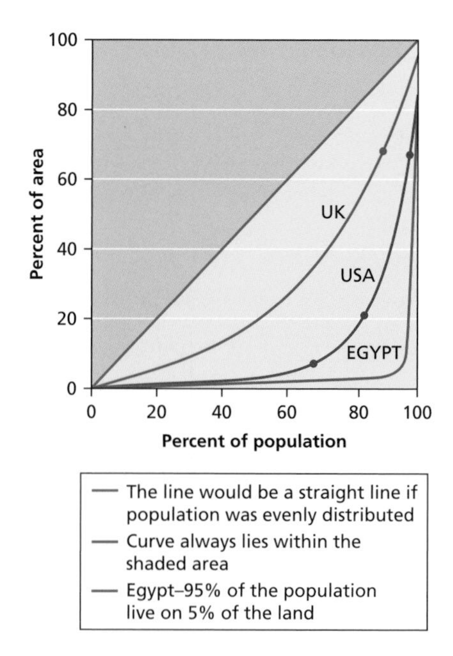

Figure 21.7 The Gini coefficient

Inferential statistics

Inferential statistics use results from surveys to make estimates or predictions, i.e. they make an inference about a total population or about some future situation. To understand inferential statistics it is important to grasp three related concepts, namely probability, significance and sampling.

Probability

One of the main tasks of inferential statistics is to establish the likelihood of a particular event or value occurring – this is known as probability. Probability is measured on a scale from 0 to 1. The value 1 represents absolute certainty (e.g. that everyone will eventually die), whereas the value 0 represents absolute impossibility (e.g. that a non-American citizen will become President of the USA). In statistics, probability (p) is often expressed as a percentage:

- p = 0.05 (a 1 in 20 chance) is a 95% level of probability
- p = 0.01 (a 1 in 100 chance) is a 99% level of probability
- p = 0.001 (a 1 in 1,000 chance) is a 99.9% level of probability.

Significance

Significance relates to the probability that a hypothesis is true. In statistics it is convention to use a **null hypothesis** (a negative statement that we aim to disprove). A null hypothesis might state, for example, that there is no difference in water quality above and below a sewage outlet. The **alternative hypothesis** (also known as the research hypothesis) would state that "there is a difference between the water quality above and below a sewage outlet". The probability at which it is decided to reject the null hypothesis is known as the **significance level**. The significance level indicates the number of times that the observed differences could be caused by chance. The practice is to refer to results as "significant", "highly significant" and "very highly significant" respectively when significant at the 95%, 99% and 99.9% levels of significance. This means there is a 1 in 20, 1 in 100 and 1 in 1,000 chance (probability) of the result occurring by chance.

Spearman's Rank correlation coefficient (Rs)

This is one of the most widely used statistics in geography. It is relatively quick and easy to do. It only requires that data are available on the ordinal (ranked) scale, although other data can be transformed into ranks very simply. It is called a "rank" correlation because only the ranks are correlated, not the actual values. The use of Rs allows us to decide whether or not there is a significant correlation (relationship) between two sets of data. In some cases it is obvious whether a correlation exists or not. However, in most cases it is not so clear-cut, and to avoid subjective comments we use Rs to bring in a degree of statistical accuracy.

Sample	Organic content (OC)	Moisture content (MC)
1	3.8	15
2	4.7	22
3	6.2	30
4	3.9	18
5	5.4	24
6	7.1	29
7	6.2	26
8	4.6	20
9	4.6	25
10	5.1	20

Figure 21.8 Table of significance for Spearman's Rank

Procedure

1. State the **null hypothesis (Ho)** that is there is no **significant relationship** between organic content (OC) and moisture content (MC). The alternative hypothesis (H1) is that there is a significant relationship between the two variables.
2. Rank both sets of data from high to low, i.e. highest value gets rank 1, second highest 2, and so on. In the case of joint ranks, find the average rank, e.g. if two values occupy positions two and three they both take on rank 2.5, if three values occupy positions four, five and six, they all take rank 5.
3. Using the formula $Rs = 1 - \frac{6\ \Sigma d^2}{\text{a. } n^3 - n}$ work out the correlation, where "d" refers to the difference between ranks and "n" the number of observations.
4. Compare the computed Rs with the critical values in the statistical tables.

Once we have the computed value, we compare it to the critical values. For a sample of 10, these values are 0.564 for 95% significance and 0.746 for 99% significance. In this example, it is clear that the relationship is strong, i.e. there is more than a 99% chance that there is a relationship between the data. The next stage would be to offer explanations for the relationship.

Worked example	Rank		Rank		Difference in ranks	
Sample	OC	MC	OC	MC	(d)	d^2
1	3.8	15	10	10	0	0
2	4.7	22	6	6	0	0
3	6.2	30	2.5	1	1.5	2.25
4	3.9	18	9	9	0	0
5	5.4	24	4	5	−1	1
6	7.1	29	1	2	−1	1
7	6.2	26	2.5	3	−0.5	0.25
8	4.6	20	7.5	7.5	0	0
9	4.6	25	7.5	4	3.5	10.25
10	5.1	20	5	7.5	−2.5	6.25
						$\Sigma d^2 = 21$

$$Rs = 1 - \frac{6\ \Sigma d^2}{n^3 - n} = 1 - \frac{6 \times 21}{10_3 - 10} = 1 - \frac{126}{990} = 1 - 0.13 = 0.87$$

Figure 21.9 Spearman's rank worked example

It is important to realize that Spearman's Rank has a number of weaknesses, which must be borne in mind. First, it requires a sample of not less than seven observations. Second, it tests for linear relationships and would give an answer of 0 for data such as river discharge and frequency (shown in the diagrams), which follow a curvilinear pattern, with few very low or very high flows and a large number of medium flows. Third, it is easy to make meaningless correlations, as between the success of English cricket teams and IMR in India. Fourth, the question of scale is always important. As shown on the final diagram, a survey of river sediment rates and discharge for the whole of a drainage system may give a strong correlation, whereas analysis of just the upper catchments gives a much lower result. In Figure 21.10 the total sample produces a weak positive correlation, whereas a subsample of three produces no correlation.

As always, statistics are tools to be used. They are only part of the analysis, and we must be aware of their limits.

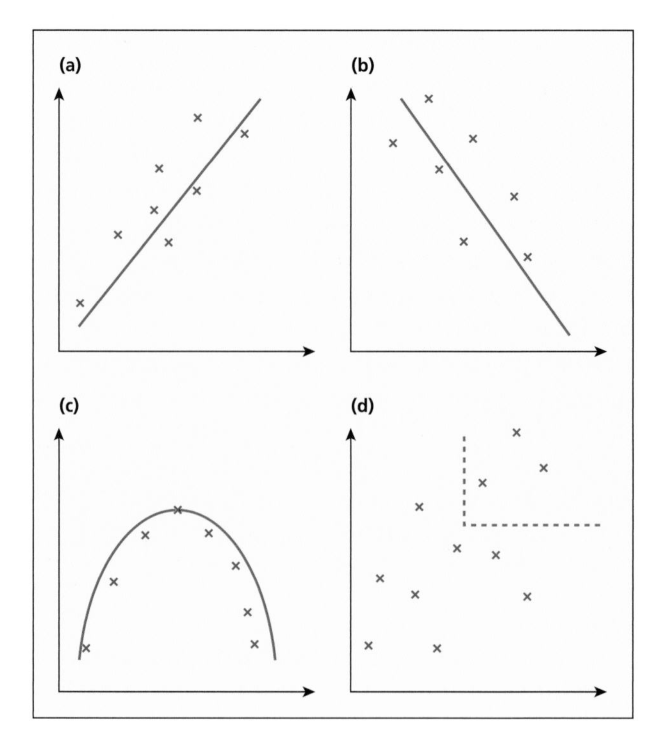

Figure 21.10 (a) Positive correlation, (b) Negative correlation, (c) Curvilinear correlation, (d) The importance of scale

To do:

Using the data below for slope angle and soil depth:

- **a** State the null hypothesis.
- **b** Set out the data in a table, as shown in the worked example.
- **c** Rank both sets of data from high to low (highest = rank 1).
- **d** Work out the difference in ranks.
- **e** Find the square of the differences.
- **f** Add up the figures in the final column to find $\sum d^2$.
- **g** Using this figure, work out the correlation between slope angle and soil depth $Rs = \frac{1 - 6\sum d^2}{n^3 - n}$
- **h** Compare your answer with the critical values in the table. How significant is your result?

Slope	Soil
angle	depth (cm)
5.5	60
11.5	48
12	33
13	18
25	22
22	29
18	24
26	24
20	12
3	60

Figure 21.11 Slope–soil data set

	Significance level	
N	95%	99%
4	1.00	-
5	0.90	1.00
6	0.83	0.94
7	0.71	0.89
8	0.64	0.83
9	0.60	0.78
10	0.56	0.75
12	0.51	0.71
14	0.46	0.65
16	0.43	0.60
18	0.40	0.56
20	0.38	0.53
22	0.36	0.51
24	0.34	0.49
26	0.33	0.47
28	0.32	0.45
30	0.31	0.42

Figure 21.12 Levels of significance

The nearest neighbour index (NNI)

Part of the study of ecosystems (and vegetation) is concerned with distributions in space and over time. We can describe the **spatial distribution** of selected features in an area (such as settlements, factories, trees) by looking at a map or making a survey. This may lead us to conclude that some types of distribution are scattered, dispersed or concentrated. However, the main weakness with the visual method is that it is **subjective**, and individuals differ in their interpretation of the pattern. Some **objective** measure is required, and this is provided by the nearest neighbour index (NNI).

Three main types of pattern can be distinguished: **uniform or regular**; **clustered or aggregated**; and **random**. These are shown on Figure 21.13. The points may represent individual tree, types of shop, or types of services, etc.). If the pattern is regular, the distance between any one point and its nearest neighbour should be approximately the same as from any other point. If the pattern is clustered, many points will be found a short distance from each other and large areas of the map will be without any points. A random distribution normally has a mixture of some clustering and some regularity.

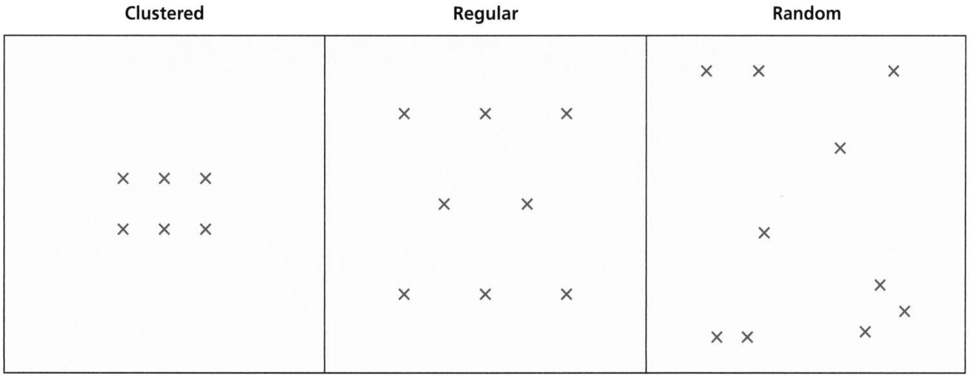

Clustered | Regular | Random

Figure 21.13 Three nearest neighbour patterns

The technique most commonly used to analyse these patterns is the NNI. It measures the spatial distribution of points, and is derived from the average distance between each point and its nearest neighbour. This figure is then compared to computed values, which state whether the pattern is regular (NNI = 2.15), clustered (NNI = 0) or random (NNI = 1.0). Thus, a value below 1.0 shows a tendency towards clustering; a value of above 1.0 shows a tendency towards uniformity.

The formula for the NNI looks somewhat daunting at first, but, like most statistics, is extremely straightforward providing care is taken.

NNI or Rn = $2\bar{D}\sqrt{(N/A)}$

where $\bar{D}$ is the average distance between each point and its nearest neighbour, calculated by finding $\sum d/N$ (d refers to each individual distance), N the number of points under study and A the size of the area under study. It is important that you use the same units for distance and area, e.g. metres or kilometres, but not a mixture.

Strombosia tree	Nearest neighbour	Distance (m)
A	F	17
B	C	16
C	B	16
D	E	11
E	D	11
F	E	11
		$\sum d = 83$

Figure 21.14 Nearest neighbour distribution data

For example, a survey of the distribution of Strombosia trees in rainforest vegetation is shown in Figure 21.16. Other trees are shown by a dot.

Formula

NNI or Rn = $2\bar{D}\sqrt{(N/A)}$

$\bar{D} = \Sigma d/N = 83/6 =$ c. 14

$Rn = 2 \times 14 \times \sqrt{(6/1600)}$

$Rn = 1.71$

The results can vary between 0 and 2.15. There is a continuum of values, and any distribution lies somewhere between the two extremes (Figure 21.15). The answer above suggests a significant degree of regularity.

Nearest neighbour significance ranges

Now complete the same exercise for the Douglas Fir plantation.

There are important points to bear in mind when using it:

- Two or more subpatterns (one clustered, one regular) may suggest a random result.
- What is the definition of, for example, a tree? Do you include all individuals,
 or just those above a certain size?
- Why do we take the nearest neighbour? Why not the third or fourth nearest?
- The choice of the area, and the size of the area studied, can completely alter the result and make a clustered pattern appear regular, and vice-versa.
- Although the NNI may suggest a random pattern, it may be that the controlling factor, e.g. soil type or altitude, is itself randomly distributed, and that the vegetation is in fact located in anything but a random fashion.

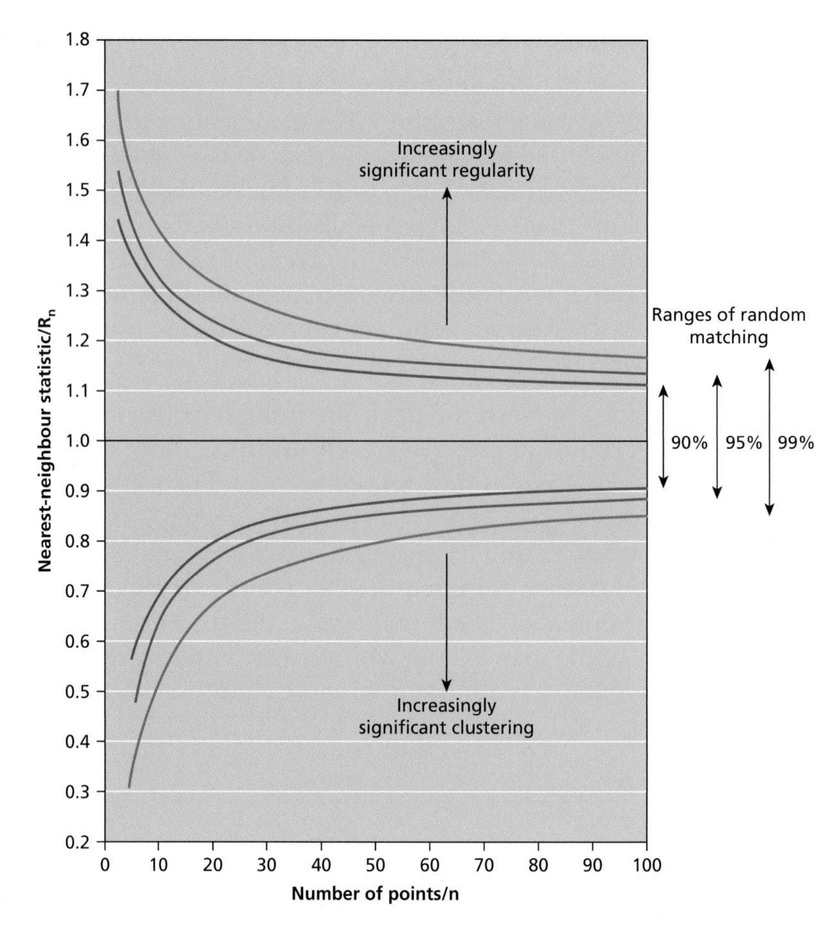

Figure 21.15 Nearest neighbour significance

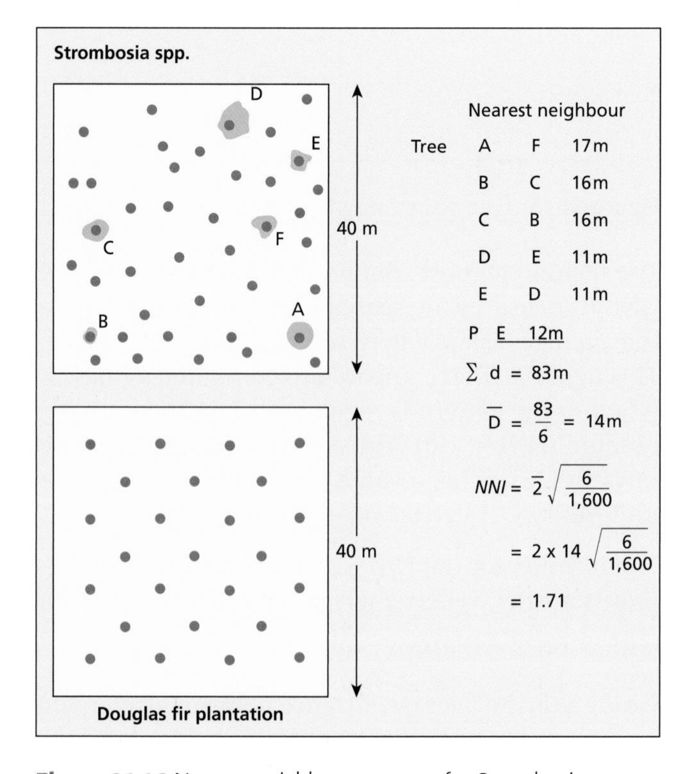

Figure 21.16 Nearest neighbour patterns for Strombosia (rainforest spp.) and Douglas Fir plantation

The Chi-squared test (X2)

The Chi-squared test is used to test whether there is a significant difference between data. For example, we can use it to test whether there is any difference between countries' levels of CO_2 emissions. A common test is to see whether there are significant differences in levels of well-being between areas.

The Chi-squared test can only be used on data with the following characteristics:

- Data must be on the interval or ratio scale (i.e. it has a precise numerical value) and can be grouped into categories.
- The expected frequency in any one category must be greater than 5.

Method

1 State the hypothesis being tested – there is a **significant** difference between an observed pattern and an expected pattern. It is convention to give a **null hypothesis** (a negative test), that is, that there is **no significant difference** between the patterns.
2 Tabulate the data as shown in the example below. The data being tested for significance are known as the "observed" frequency, and the column is headed "O".
3 Calculate the "expected" number of frequencies that you would expect to find. These go in column "E".
4 Calculate the Chi-squared statistic using the formula:
X2 = Σ(O − E)2/E, where:

- X2 is the Chi-squared statistic
- Σ is the sum of
- O refers to the observed frequencies
- E are the expected frequencies.

5 Calculate the degrees of freedom. This is quite simply one less than the total number of observations (N), i.e. N − 1.
6 Compare the calculated figure with the critical values in the significance tables, using the appropriate degrees of freedom. Read off the probability that the data frequencies you are testing could have occurred by chance.

Example

Figure 21.17 provides data on levels of CO_2 emissions among a number of countries.

Country	Million tonnes of carbon dioxide
USA	6,044.0
China	5,005.7
Russia	1,523.6
India	1,341.8
Japan	1,256.8
Germany	808.0
Canada	638.8
UK	586.7
South Korea	465.2
Italy	449.5

Figure 21.17 Levels of CO_2 emissions among a number of countries

1 The null hypothesis Ho states that there is **no significant variation** in the amount of CO_2 emission by country.

The alternative hypothesis (H1) states that there is a **significant difference** in amount of CO_2 emission by country.

2 If there is no difference in the amount of CO_2 emission by country, they should all have roughly the same amount. That means they will all have the about the average. The expected frequency is thus the same as the average frequency, which is (6,044 + 5,005.7 + 1,523.6 + 1,341.8 + 1,256.8 + 808 + 638.8 + 586.7 + 465.2 + 449.5) / 10 = 18 120.1/10 = 1,812

3 O E (O−E) $(O-E)^2$ $(O-E)^2/E$

Country	Observed emissions	Expected emissions			
USA	6,044.0	1,812	4,232	17 909 824	9,884
China	5,005.7	1,812	3,193.7	10 199 719.69	5,628
Russia	1,523.6	1,812	−288.4	83 174.56	45.9
India	1,341.8	1,812	−470.2	221 088.04	122
Japan	1,256.8	1,812	−555.2	308 247.04	170
Germany	808.0	1,812	−1004	1 008 016	556
Canada	638.8	1,812	−1,173.2	1 376 398.24	759.6
UK	586.7	1,812	−1,225.3	1 501 360.09	828.6
South Korea	465.2	1,812	−1,346.8	1 813 870.2	1,001
Italy	449.5	1,812	−1,362.5	1 856 406.25	1,025
					∑ 20,020

Figure 21.18 Observed and expected emissions by country

4 Degrees of freedom (df) = (N−1) = (10−1) = 9

5 The critical values for 9 df are:

0.10	*0.05*	*0.01*	*0.001*
14.68	16.92	21.67	27.88

Clearly, the computed value of 20 020 exceeds the critical values, even at the 0.001 level of significance. This means that there is less than a 1 in a 1,000 (0.001) chance that, given the figures above, there is no difference in the emission of CO_2 by country. Therefore we would reject the null hypothesis and accept the alternative hypothesis. This means that there is a significant difference in emission of CO_2 by country.

NB The next stage is to offer explanations for the results. Remember that the statistic is only used as a means of clarification: it is not an end in itself but a means to help you to explain.

The location quotient

The location quotient is a measure of the relative concentration of an industry in a particular area. It is one of the easiest descriptive statistics available, and is computed in the following way:

$$\frac{\text{Number of people employed in industry A in area X}}{\text{Number of people employed in area X}}$$

$$\frac{\text{Number of people employed in industry A in the whole country}}{\text{Total number of people employed in the country}}$$

In some cases you may see it written as

$$\frac{Ri/Ni}{R/N}$$

where Ri = number employed in industry A in area X

Ni = number employed in area X

R = number employed nationally in industry A

N = total number employed nationally.

A location quotient of:

- > 1.0 reveals that a region has an above-average representation of that particular industry
- 1.0 shows that an area has the same as the national average (e.g. we might expect all areas in a country to have roughly the same proportion of workers in services such as gas, electricity and refuse collection)
- < 1.0 reveals an underrepresentation of a particular form of industry.

For example, if 3,000 people are employed in education in Oxford (UK) out of a total labour force of 60 000, compared with a national total of 300 000 employed in education and a total labour force of 30 million, the location quotient would be as follows:

$$\frac{\frac{3,000}{60\ 000}}{\frac{300\ 000}{30\ 000\ 000}} = \frac{0.05}{0.01} = 5$$

which means that the number of people working in education in Oxford is five times greater than the national average. It would be concluded that education is of great significance to the local economy.

Graphical techniques

Bar charts

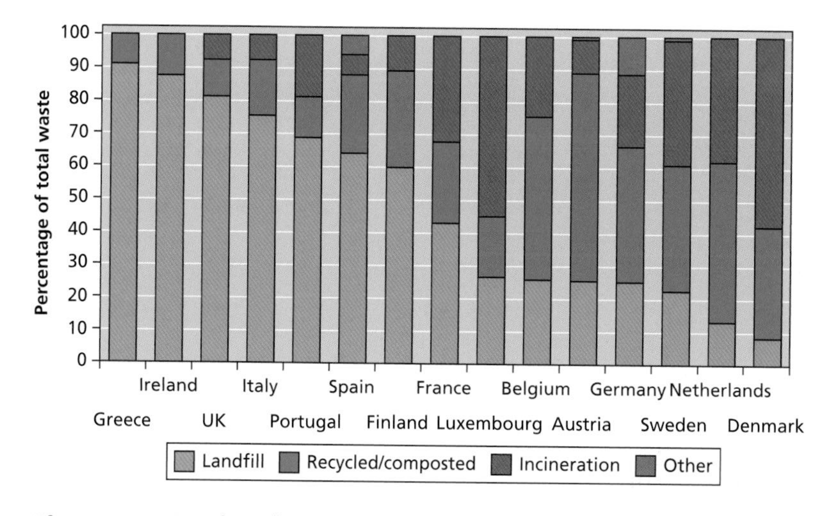

Figure 21.19 Bar chart showing waste management in the European Union

Bar charts are one of the simplest ways of representing data (Figure 21.19). Each bar in a bar chart is of a standard width, but the length or height is proportional to the value being represented.

There is a range of bar chart types. A **simple** bar chart shows a single factor. A **multiple** bar chart can be used to show changing frequency over time, e.g. monthly rainfall figures. A **compound** bar chart involves the subdivision of simple bars. For example, the bar might be proportional to sources of pollution, and may be subdivided on the basis of its composition.

Pie charts

Pie charts are subdivided circles. They are used to show variations in the composition of a feature. Pie charts can also be drawn as proportional in size, to show variations in size of a feature, e.g. the proportion of arable, pastoral and mixed farmland in different areas.

The following steps should be taken when making a pie chart:

1. Convert the data into percentages.
2. Convert the percentages into degrees by multiplying by 3.6 and rounding up or down to the nearest whole number.
3. Draw the appropriately located circles on a map or diagram.
4. Subdivide the circle into sectors using the figures obtained in step 2.
5. Differentiate the sectors by means of different shading.
6. Draw a key explaining the scheme of shading and/or colours.
7. Give the diagram a title.

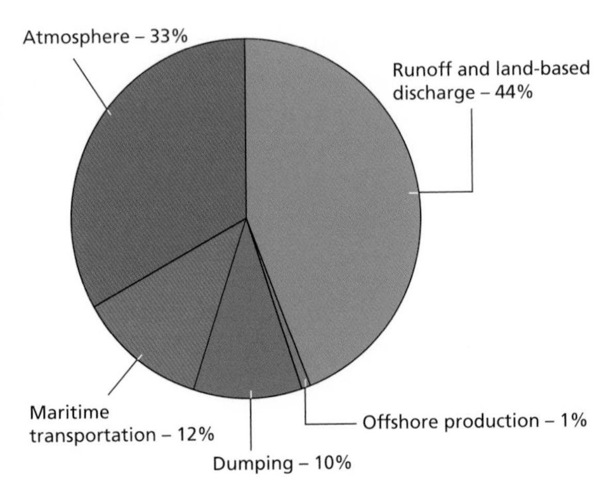

Figure 21.20 Pie chart showing sources of pollutants from human activities entering the sea

Dispersion diagrams

A dispersion diagram is useful for showing the range of a data set, their tendency to group or disperse, and for comparing two sets of data (Figure 21.21). It involves plotting the values of a single variable on a vertical axis. The horizontal axis shows the frequency. The resulting diagram shows the frequency distribution of a data set. They can also be used to determine the median value, modal value and the interquartile range.

A kite diagram is a form of line graph which allows you to view the relative distribution of different species along a transect (Figure 21.22). It is commonly used to show variations in sand dune succession, for example. The value for each reading is shown above and below a horizontal line. First, plot a series of bars representing the relative abundance of each species at each location. Then join all the tops of the bars to form the upper half of the kite, and all the bottom parts of the bars to form the bottom half of the kite. A half-kite diagram is when the readings are shown above the horizontal line only.

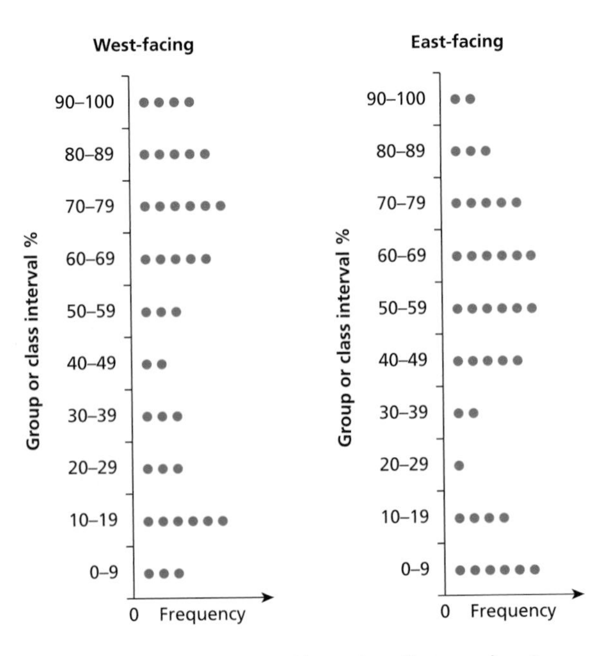

Figure 21.21 Dispersion diagram showing lichen cover on east- and west-facing gravestones

Line graphs

Line graphs can be quite simple graphs that are used to show changes over time (e.g. temperature change related to the enhanced greenhouse effect) or over distance (e.g. type of pollution related to distance from water source).

In all line graphs there is an independent and a dependent variable. The independent variable is plotted on the horizontal or x axis, while the dependant variable is plotted on the vertical or y axis.

Multiple or compound line graphs can show changes in more than one variable, for example changes in energy use over time. Such diagrams can reveal interesting relationships between the variables. On such graphs, data can be plotted in a number of different forms – in absolute terms, relative terms, percentage terms or cumulative terms.

Flow lines

Flow lines show the volume of transfer between different groups or places (see the telephone flow map in chapter 13 of the book (Figure 13.16). Migration rates and direction could be shown using flow lines. In many cases absolute data are used, but it is possible to use relative data.

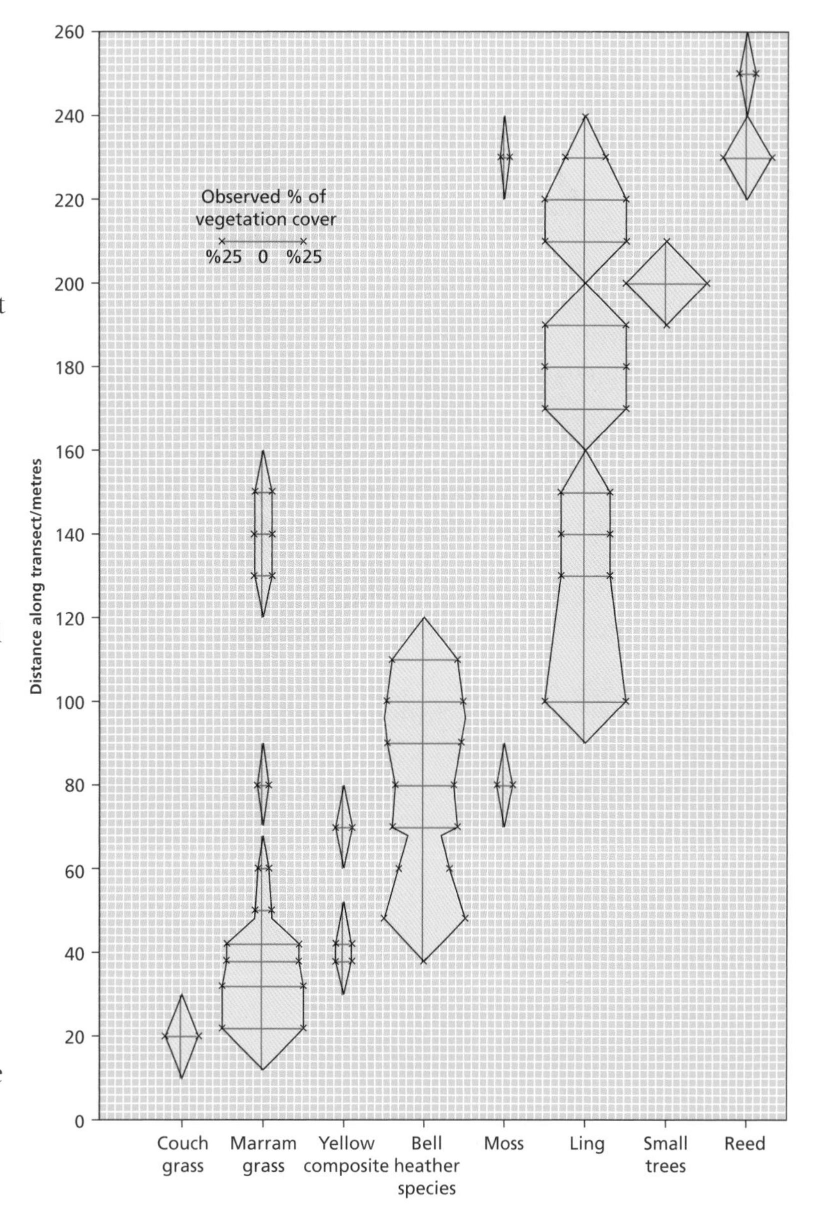

Figure 21.22 Kite diagram showing vegetation succession on a sand dune, Studland Bay, UK

As with all graphical techniques, it is important to:

- keep the background as simple as possible, so as to avoid clutter
- choose an appropriate scale, so that extreme values can be shown without any loss of clarity
- provide a key, and give a title to the diagram.

Triangular graphs

Triangular graphs are used to represent data that can be divided into three parts, e.g. soil (sand, silt and clay) and population (youth, adult and elderly). It requires the data to have been converted into percentages, and the percentages must add up to 100%. On Figure 21.24 point A has 70% silt, 10% sand and 20% clay.

The main advantage of a triangular graph is that it allows a large amount of data to be shown on one diagram. In many cases, once the data have been plotted on to a triangular graph, groupings become evident. In the case of soil texture, there are established soil textural groups. Triangular graphs can be tricky to construct, as it easy to get confused. However, with care they can provide a reliable way of classifying large amounts of data which have three components.

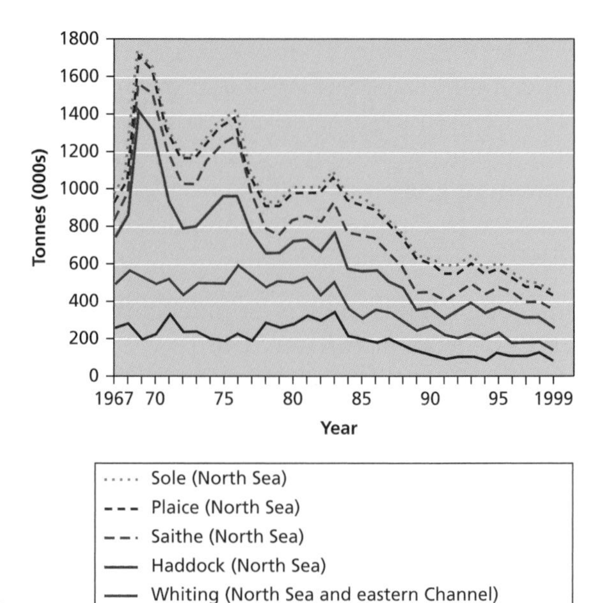

Figure 21.23 Line graph showing the decline of North Sea fisheries

Semi-log and double log graphs

Semi-log and double log graphs can be daunting at first. They allow scientists to compare small-scale features (or populations) with large-scale ones. This would not be as easy on an ordinary line graph (see Figure 21.23).

The logarithmic scale compresses the range of values. It gives more space to smaller values but compresses the space available for the larger values – you can see this clearly when you compare the space available for large and small values on the line graph with the semi-log graphs in Figure 21.25. It can also be used to show relative growth over time.

In a semi-log (also known as log-normal), one scale– normally the vertical – is logarithmic, while the other – normally the horizontal – has a "regular" linear scale. In a double-log (also called log-log), both scales are logarithmic. In the logarithmic parts of the scale, each of the cycles is logarithmic. This means that each cycle on the scale increases by the power of 10. For example, in the first cycle, values may be 1, 2, 3, 4, etc., whereas in the second cycle they would be 10, 20, 30, 40, etc., and in the third cycle 100, 200, 300, 400, etc., and so on.

It is important to realize that the vertical axis does not begin at 0 but some factor of 1, e.g. 0.1, 100, 1,000, etc. In a semi-log graph the

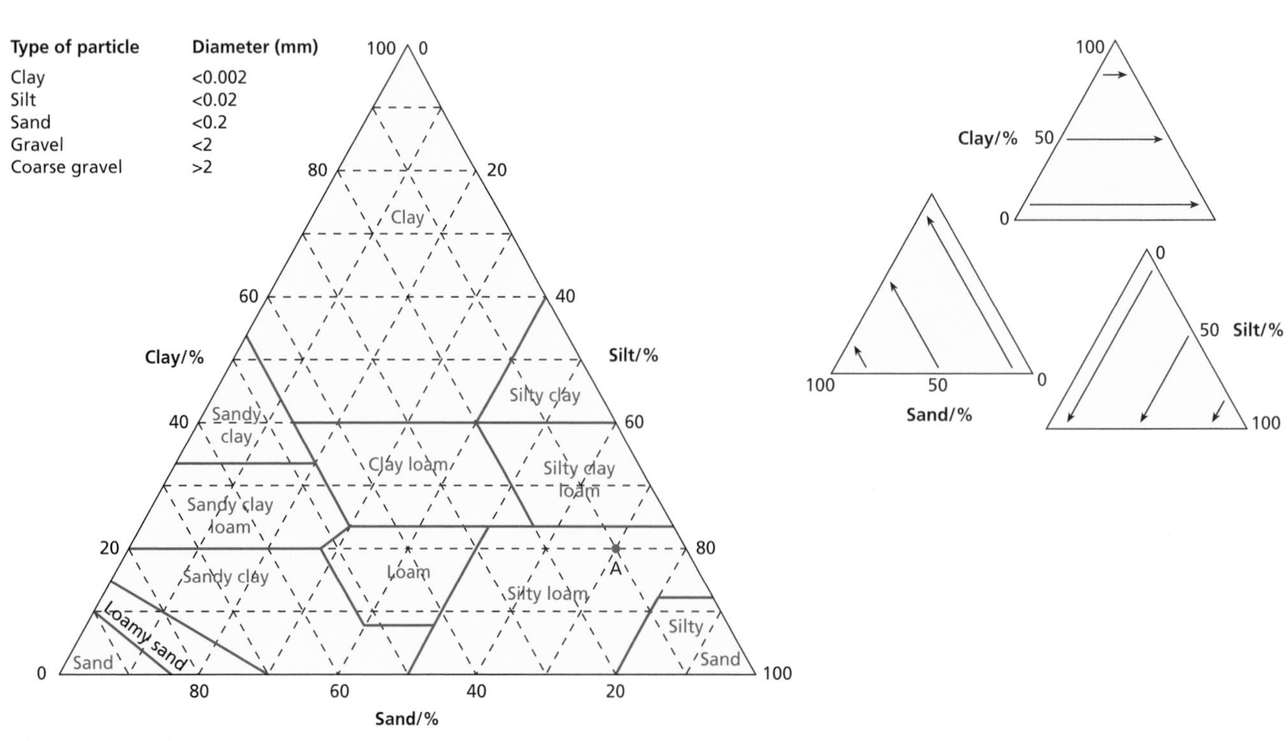

Figure 21.24 Triangular graphs

horizontal axis can begin at any number and could even be nominal data such as the names of the months of the year.

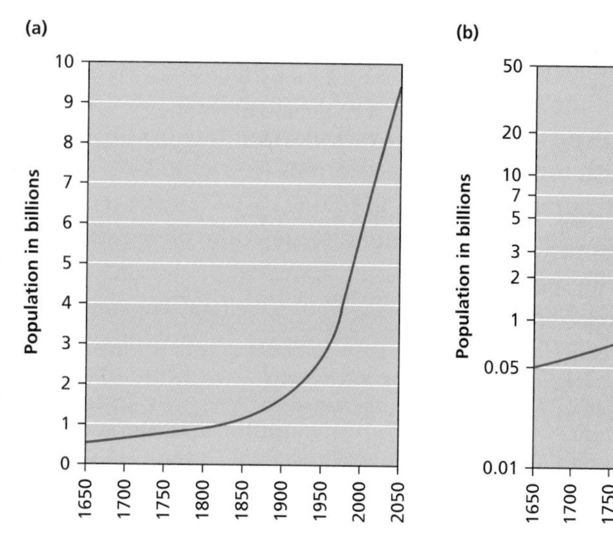

Figure 21.25 Normal and Semi-log graph

Scatter graphs

Scatter graphs show how two sets of data are related to each other, for example population size and number of services, or distance from the source of a river and average pebble size. To plot a scatter graph, decide which variable is independent (population size/ distance from the source) and which is dependent (number of services/average pebble size). The independent variable is plotted on the horizontal or x axis and the dependent on the vertical or y axis. For each set of data, project a line from the corresponding x and y axis (see Figure 21.26), and where the two lines meet mark a dot or an x.

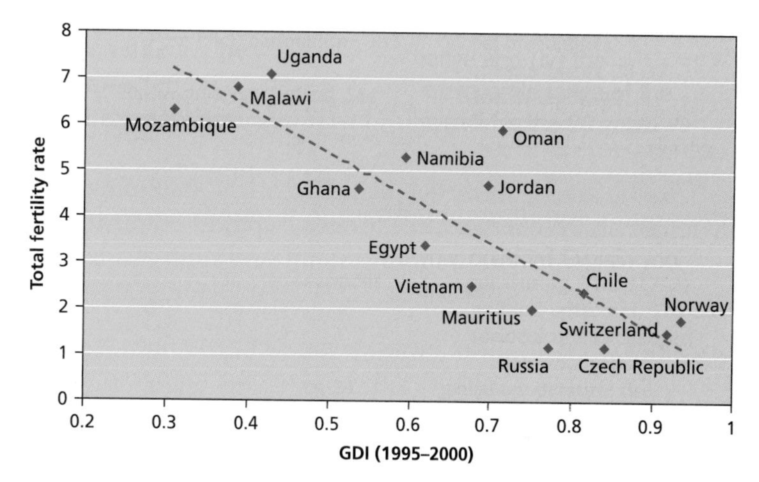

Figure 21.26 Scatter graph

Choropleth maps

Choropleth maps use variations in colour or different densities of black and white shading. To construct a choropleth map, follow these steps:

1 Look at the range of data and divide it into classes. There should be no less than four classes and no more than eight.
2 Allocate a colour to each class. The convention is that shading gets darker as values increase.
3 Now apply each colour to the relevant areas of the map.
4 Provide a key, scale and north point.

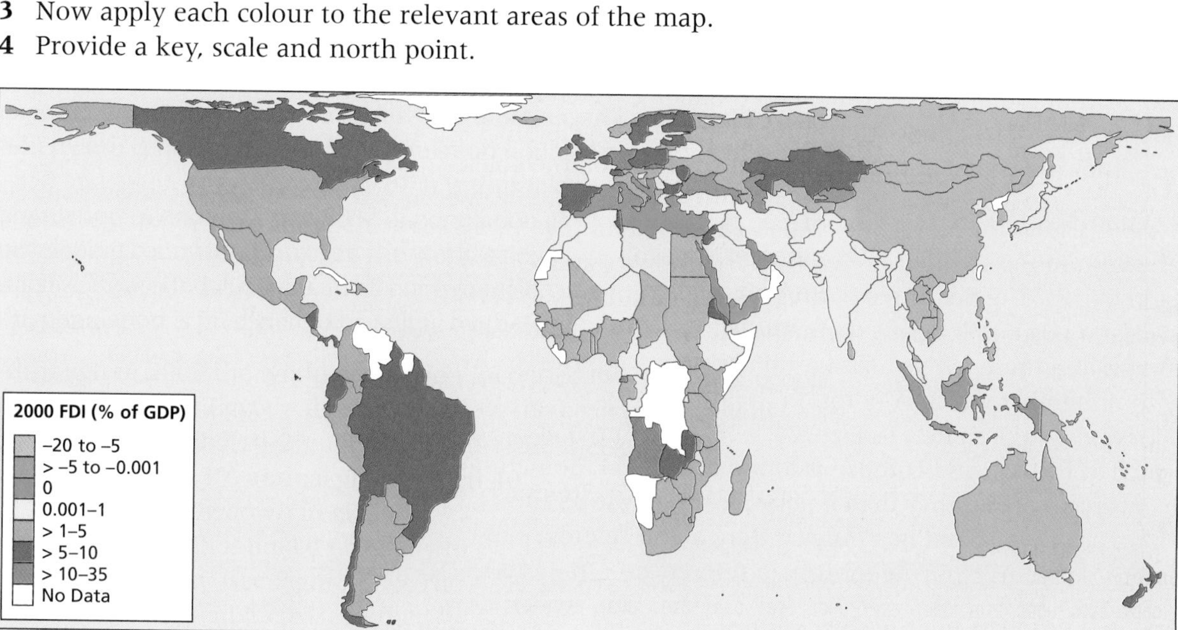

Figure 21.27 Choropleth map

There are important things to consider before making a choropleth map. For example, the map shown (Figure 21.27) suggests uniform conditions throughout the USA or Australia. It exaggerates the role of boundaries, e.g. between France and Spain. Data can only occur in one category. Groupings can be in arithmetic intervals (e.g. 0–4, 5–9, 10–14, etc.), geometric intervals (e.g. 1–2, 3–4, 5–8, 9–16, 17–32, etc.) or at "natural breaks", by dividing the data into roughly equal groupings and using statistical variations, such as mean and standard deviation.

Map reading

Scale

Most of the topographical maps (sometimes called Ordnance Survey or OS maps) that we use are either at a 1:50 000 or a 1:25 000 scale. On a 1:50,000 map, 1 centimetre on the map relates to 50 000 centimetres on the ground. On a 1:25,000 map every 1 centimetre on the map refers to 25 000 centimetres on the ground.

In every kilometre there are 100 000 centimetres (1,000m × 100 cm). Hence:

- on a 1:50 000 map every 2 centimetres corresponds to 1 kilometre
- on a 1:25 000 map every 4 centimetres corresponds to 1 kilometre.

A 1:25 000 map is more detailed than a 1:50 000 map, and is therefore an excellent source for geographical enquiries. Maps on the 1:50 000 scale provide a more general overview of a larger area. You may come across other scales, e.g. 1:10 000 and 1:2,500.

Grid lines on maps make measuring easier. Grid lines are the regular horizontal and vertical lines you can see on a topographical map. The horizontal lines are called **northings** and the vertical lines are called **eastings**. They help to pinpoint the exact location of features on a map.

Latitude and longitude

In an atlas, the lines of **latitude** (the northings) tell you how far north or south of the equator you are. Each line of latitude is divided into 60 minutes, and each minute into 60 seconds. Lines of latitude are parallel to each other. Lines of **longitude** (the eastings) tell you how far east or west a place is from the Greenwich Meridian (London), which is given the value of 0. Lines of longitude stretch from the North Pole to the South Pole – they are widest at the equator but converge on the poles.

Grid and square references

Grid references are the six-figure references that locate precise positions on a map. The first three figures are the eastings, and these tell us how far a position is across the map. The last three figures are the northings, and these tell us how far up the map a position is. An easy way to remember which way round the numbers go is "along the corridor and up the stairs". In Figure 21.28, the church at Rose Hill is located at 691046 and Dundee is found at 760044. Sometimes a feature covers an area rather than a point, for example all the villages or areas of woodland in a district. Here a grid reference is inappropriate, so we use four-figure square references:

- The first two numbers refer to the eastings.
- The last two numbers refer to the northings.

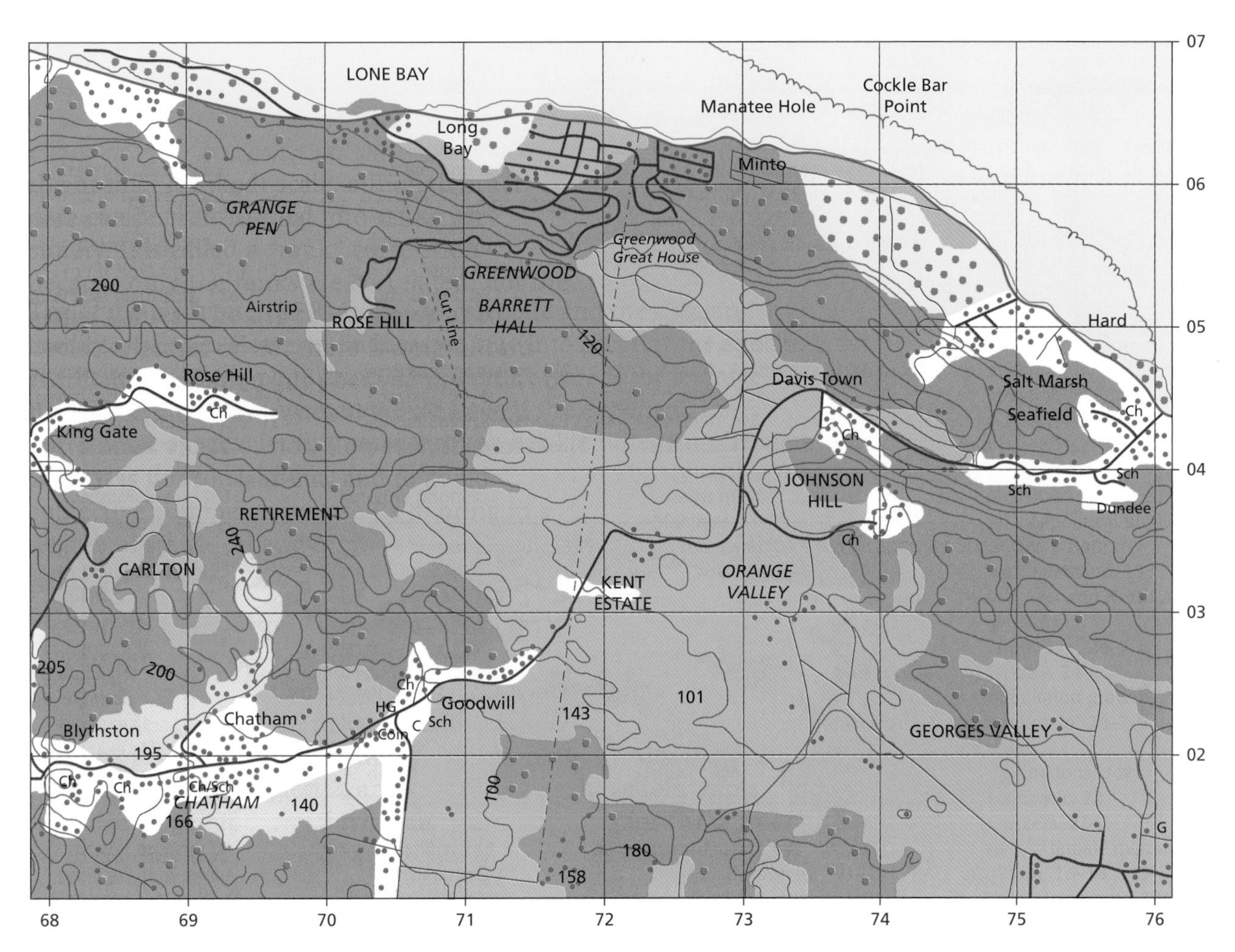

Figure 21.28 Part of a 1:50 000 OS map of Jamaica

- The point where the two grid lines meet is the bottom left-hand corner of the square. Thus in Figure 21.28, most of the village of Seafield is found in 7504. Some features may occur in two or more squares, for example Long Bay is found in squares 7006 and 7106.

Direction

Direction can be expressed in two ways:

- by compass points, e.g. southwest
- by compass bearings or angular directions, e.g. 045°.

Sixteen compass points are commonly used. Some of these are shown in Figure 21.29. Compass bearings are more accurate than compass points but can be quite confusing. Compass bearings show variations from magnetic north. This is slightly different from the grid north shown on a topographical map (which is the way in which the northings go). True north is different again – this is the direction of the North Pole.

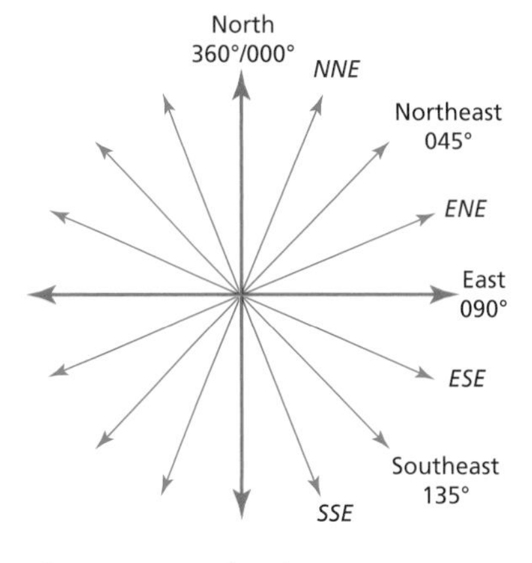

Figure 21.29 Direction

Relief and gradient

Contour lines

A **contour line** is a line that joins places of equal height.

- When the contour lines are spaced far apart the land is quite flat.
- When the contour lines are very close together the land is very steep (when the land is too steep for contour lines a symbol for a cliff is used).

To do:

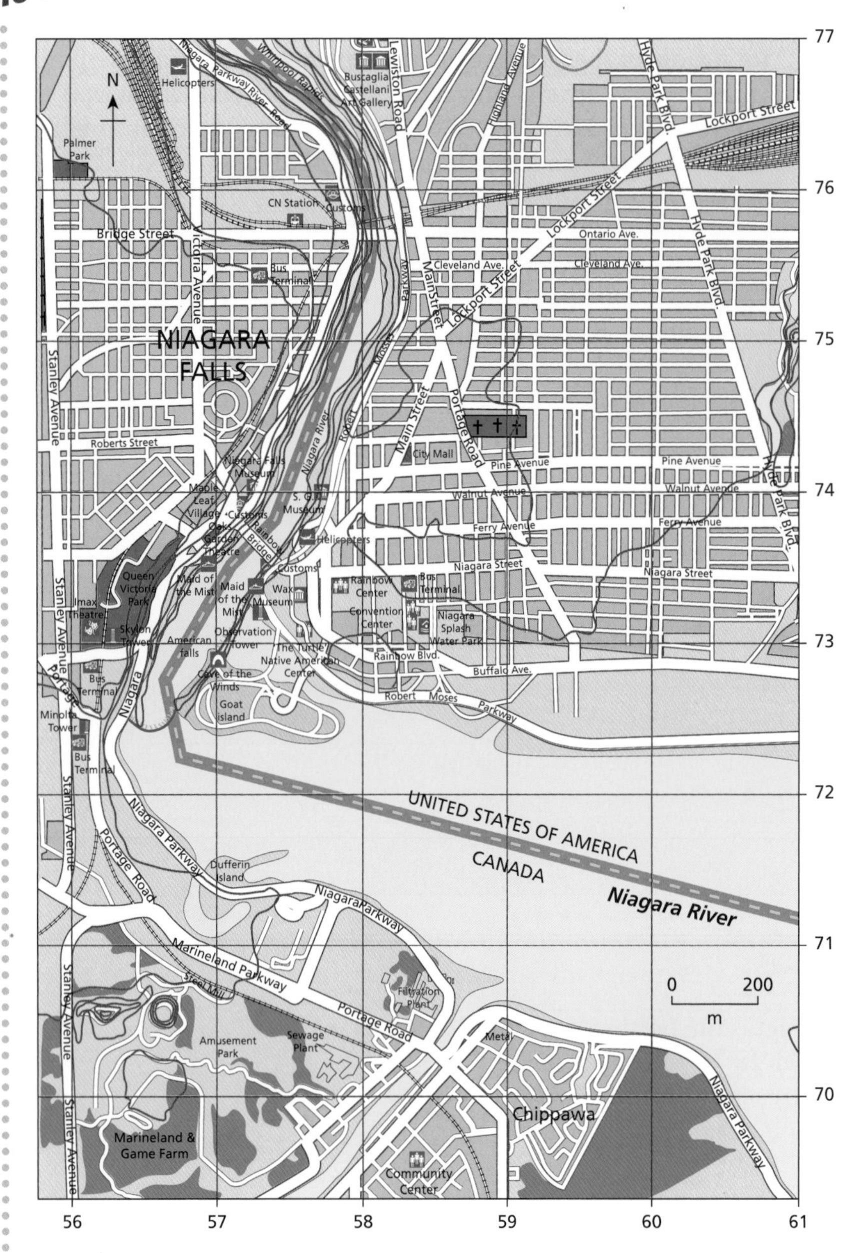

Figure 21.30 Map of Niagara

Study figure 21.30, the map of Niagara.

a Which way is the river flowing?

b Compare and contrast the road network on the USA side of the river with that on the Canadian side.

c Using map evidence, suggest how the river has been used for human activities.

d Describe the nature of the valley below the waterfall.

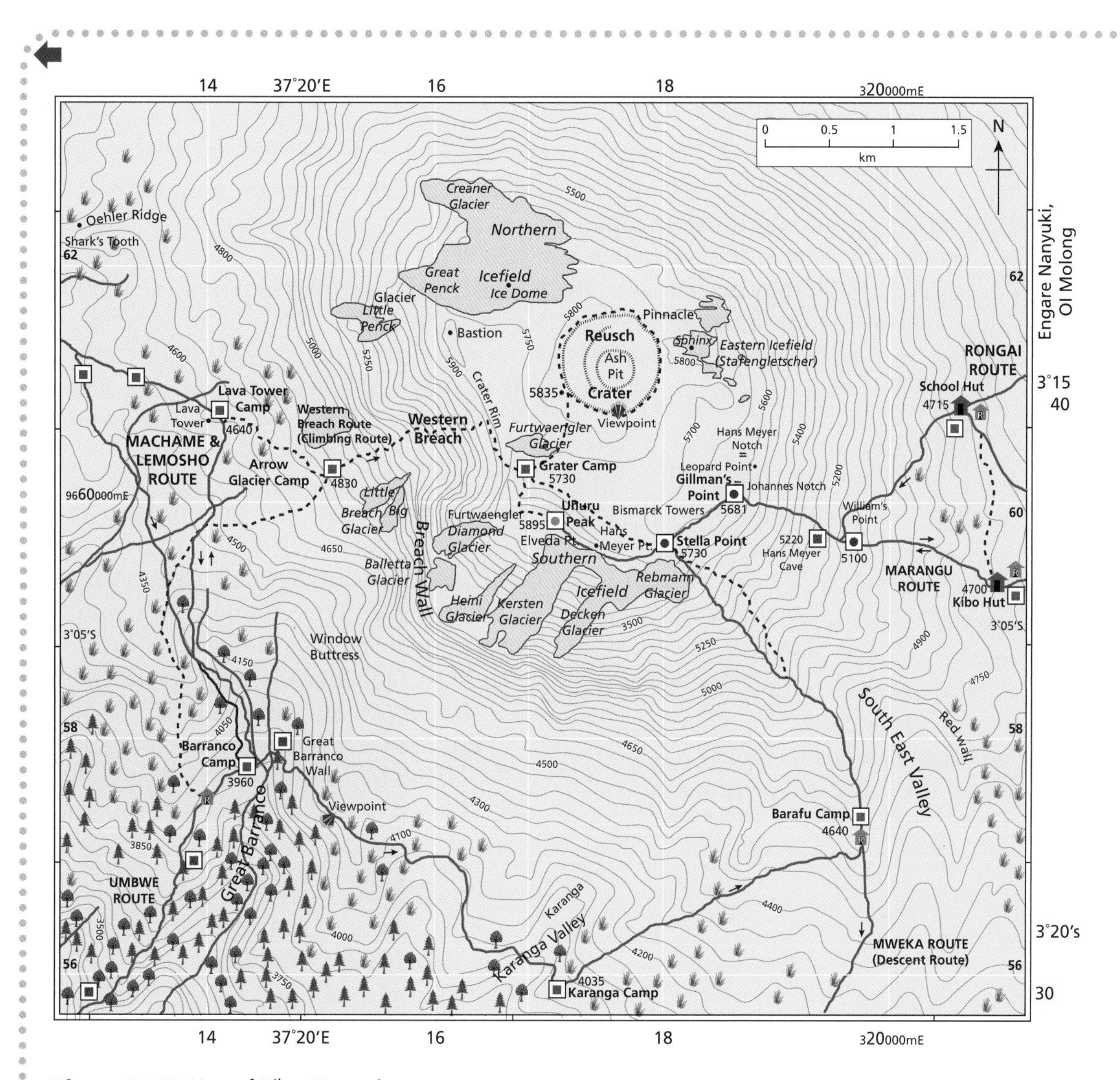

Figure 21.31 Map of Kibo, Tanzania

a What is the highest point shown on the map?

b Describe the distribution of glaciers and ice-fields as shown on this extract. Describe the relief and topography of this extract.

c Draw a sketch section from 140600 to 200600 to show the main features of the relief of the area. A grid for the sketch section has been provided (Figure 21.32).

Figure 21.32 Grid for cross-section

- When contour lines are close together at the top, and then get farther apart, it suggests a concave slope.
- When contour lines are close at the bottom and farther apart at the top, it suggests a convex slope.

Gradients

The gradient of a slope is its steepness. We can get a rough idea of the gradient by looking at the contour pattern. If the contour lines are close together the slope is steep, and if they are far apart the land is quite flat. However, these are not very accurate descriptions. To measure gradient accurately we need two measurements:

- the vertical difference between two points (worked out using the contour lines or spot heights)
- the horizontal distance between two places – this may not be a straight line (for example, a meandering stream would not be straight).

Working out gradient

Make sure that you use the same units for both vertical and horizontal measurements. Divide the difference in horizontal distance (D) by the height (H). If the answer is, for example, "10", express it as "1:10" ("one in ten"); or "5" as "1:5" ("one in five"). This means that for every 10 metres along you rise 1 metre, or for every 5 metres in length the land rises (or drops) 1 metre. Alternatively, divide the height (H) by the difference in horizontal distance (D) and multiply by 100% (H/D × 100%). This expresses the gradient as a percentage.

Sample investigations

Footpath erosion

Soil degradation can be very dramatic. On a local scale, soil erosion or footpath erosion is less dramatic but can yield excellent data for geographical investigations. Some tourist honeypots are heavily scarred by footpath erosion, and there is evidence of compaction and erosion on most sports fields.

Figure 21.33 Footpath erosion

Many simple but excellent projects can be carried out on soils. For example, we could investigate the following:

- How does erosion vary across a footpath?
- How are soil moisture content and organic content related? Do they vary from the centre of the footpath?
- Does the level of footpath erosion vary with the number of users?

A number of techniques can be used for presenting and analysing the data. For example:

- A scatter graph and a line of best fit show variations in moisture and organic content.
- Spearman's Rank correlation coefficient tests whether there is a relationship between variables.
- Triangular graphs show variations in the composition of the soil.
- A cross-section shows local variations in relief.

Soils allow us to obtain a great deal of information from a small area of study. Choose contrasting areas – one used heavily, the other used less heavily. Using a linear sample (transect), take samples from the centre of the footpath and at 0.5-metre intervals up to 3 metres either side of the path.

To record variations in relief, use a tape and metre rule. Place the tape across the footpath and measure down from the tape every 10 centimetres to assess the extent to which the footpath has been eroded.

Two of the most interesting aspects of soils are the moisture content and the organic content. Measuring these is quite easy. Take a sample of soil with a soil auger or a small trowel. Place a handful of soil in an airtight bag. To measure moisture content, weigh a sample of soil (S1). Place it in an oven and burn it at 100°C for 24 hours. Reweigh the sample (S2).

To work out the moisture content of the sample, use this formula:

$$\text{moisture content} = \frac{S1 - S2}{S1} \times 100\%$$

To work out organic content, take the sample S2 and burn it over a bunsen burner at maximum heat for 15 minutes. Reweigh the burned sample (S3). Organic content is found by using this formula:

$$\text{organic content} = \frac{S2 - S3}{S2} \times 100\%$$

Test whether organic content and/or moisture content vary with distance from the centre of the path, and whether there is any correlation between organic content and moisture content.

Strengths of this investigation

- A small area can yield excellent results.
- A wide range of methods and techniques can be used.
- Measurements are accurate and scientifically worked out.
- It links physical and human geography.
- It raises issues about management, and there is a strong "values" section, i.e. why people prefer certain areas to others.

Limitations of this investigation

- Erosion is a long-term process, and this investigation does not focus on process but on form (shape).
- Laboratory work requires supervision.
- It might not be possible to measure both sides of a path, e.g. one alongside a river.

Measuring microclimate around a school or other area

We can learn a great deal about microclimate from simple studies around a school – or almost any building. We could test whether:

- rainfall varies under different types of vegetation and due to buildings
- north-facing places are colder (in the northern hemisphere) than south-facing places (the opposite will be true in the southern hemisphere)
- it is warmer near sources of heat (school kitchens, factories)
- it is colder near sources of water such as streams and canals
- contrasts are more noticeable in winter.

Figure 21.34 Maximum–minimum thermometer

Figure 21.35 Rain gauge

A number of techniques can be used, such as **line graphs** to show differences in north and south-facing walls over the course of a day; **dispersion diagrams** to show rainfall totals under different vegetation types for a number of days; and **simple descriptive statistics**, such as the **mean**, **mode** and **median** to describe data.

The area to be studied should include a variety of land-use types such as an exposed area, areas of different types of vegetation, areas close to buildings with heat sources, areas close to a source of water, and north and south-facing locations. To investigate rainfall totals, place a number of rain gauges in each of the different types of area to be investigated. To study temperature, use maximum–minimum thermometers to compare north- and south-facing walls. To investigate changes throughout the day, the thermometers should be checked hourly.

Figure 21.36 Measuring wind speed

Strengths of this investigation

- It is very local in scale, but could be expanded to investigate part of an urban microclimate.
- It is easy to set up.
- Measurements are very accurate.
- It helps understanding of larger-scale aspects of climate.

Limitations of this investigation

- It is important to remember that large-scale factors, such as weather systems, are the main determinants of weather, not local factors.
- Quite heavy rainfall is needed for the best results.
- A range of land-use types is desirable.
- It is time consuming to measure temperature hourly.
- The best contrasts are in the early morning and during cold winter periods.

Figure 21.37 Building height and accessibility

Examining the characteristics of a CBD

A number of ideas could be tested about CBDs or other retail areas, for example:

- the CBD is the most accessible part of the city
- building heights increase in the CBD
- land use is mostly commercial – there is little residential or manufacturing land use in the CBD
- pedestrian flows are greatest in the CBD
- there are many parking restrictions in the CBD
- there is a concentration of social and leisure facilities in the CBD.

To collect the data, a base map is needed. It is always best if you can compare at least two different areas. Much of the data is collected through observation, e.g. land use, height of buildings, pedestrian flows, a questionnaire to shoppers to find out where they have come from and what they are in the CBD for, and parking restrictions. You may also use some secondary data, such as bus timetables, data on the value of land, and tourist information maps that may provide information of social and leisure facilities.

Figure 21.38 Shinjuku, Tokyo: a concentration of high-rise buildings

A wide range of techniques can be used to present the data. For example, **flow diagrams** can show bus routes and frequency; **line graphs** can show building height; **isolines** and **flow diagrams** can show pedestrian flows; and **land-use maps** and **pie charts** can show type of land use by street, e.g. high- and low-order goods.

Strengths of this investigation

- It allows a wide range of data to be collected.
- The data can be collected quite easily.
- Many techniques can be used.
- CBDs are generally easy to get to.
- It allows a CBD to be compared with a theoretical model.
- Most candidates are familiar with a CBD.

Limitations of this investigation

- Pedestrian flows vary by day and by time – shopping areas are busy by day, social areas in the evening.
- CBDs are also tourist areas.
- It may be difficult to find out the land use on upper floors and in basements.
- There are many forms of transport into the CBD, not just public transport.
- It may be difficult to classify functions, as some companies may offer multi-functions.

Figure 21.39 Vertical zoning in the CBD

Index